AF799467

Stress Challenges and Immunity in Space

Alexander Choukér
Editor

Stress Challenges and Immunity in Space

From Mechanisms to Monitoring and Preventive Strategies

Second Edition

Springer

Editor
Alexander Choukér
Laboratory of Translational Research Stress and Immunity
Department of Anaesthesiology
Hospital of the University of Munich (LMU)
Munich
Germany

ISBN 978-3-030-16995-4 ISBN 978-3-030-16996-1 (eBook)
https://doi.org/10.1007/978-3-030-16996-1

© Springer-Verlag Berlin Heidelberg 2012
© Springer Nature Switzerland AG 2020
Chapters 6, 20 and 28 are Open Access. These chapters are licensed under the terms of the Creative Commons Attribution 4.0 International License (http://creativecommons.org/licenses/by/4.0/). For further details see license information in the chapters.
This work is subject to copyright. All rights are reserved by the Publisher, whether the whole or part of the material is concerned, specifically the rights of translation, reprinting, reuse of illustrations, recitation, broadcasting, reproduction on microfilms or in any other physical way, and transmission or information storage and retrieval, electronic adaptation, computer software, or by similar or dissimilar methodology now known or hereafter developed.
The use of general descriptive names, registered names, trademarks, service marks, etc. in this publication does not imply, even in the absence of a specific statement, that such names are exempt from the relevant protective laws and regulations and therefore free for general use.
The publisher, the authors, and the editors are safe to assume that the advice and information in this book are believed to be true and accurate at the date of publication. Neither the publisher nor the authors or the editors give a warranty, express or implied, with respect to the material contained herein or for any errors or omissions that may have been made. The publisher remains neutral with regard to jurisdictional claims in published maps and institutional affiliations.

This Springer imprint is published by the registered company Springer Nature Switzerland AG
The registered company address is: Gewerbestrasse 11, 6330 Cham, Switzerland

To my wife
Martina

To our children
Marie-Thérèse, Maxime Amédé, Émile Victor
and Alphonse Loïc

for support and understanding,
for making all possible

Acknowledgments 2nd Edition

The compilation of 40 chapters was only possible with the outstanding support, unending patience and extreme flexibility of the authors to provide, to update and to refine the content in the best possible and feasible way to the overall context. This exchange has been a very rewarding task and all authors in this book deserve my sincerest appreciation for their extremely valuable, professional and always kind collaboration throughout the many iterations of the review process.

In science, just as in science related to space and space flight, the emerging topics can only be addressed when a solid "chain of contributors" asks new exciting questions and helps raise funds to enable groundbreaking research or discoveries to generate new insights, knowledge and technologies. Only these interdependent achievements from devoted researchers, students, technicians and administrators all over the world will continue to produce new interventions to maintain or to restore health in humans, in space and on Earth. It is especially important to me to acknowledge all the participating volunteers in space and space analogue environments, and the patients who devoted time and undertook risks by giving their consent to enable research protocols and the progress of science. Without their support, such developments would not be possible neither could a second edition ever be realized.

The continuous support and highly valuable input from all members of the European Space Agency (ESA) Topical Team "Stress and Immunity", their constructive criticism and advice have again been major sources of inspiration for the realization of this new edition. I am very grateful to the ESA for the long-term ongoing support of the Topical Team, which was pivotal in developing this topic further, as reflected in this volume. Special thanks go to Dr. Dominique Moser who was very supportive in the finalization and proof read period.

This project would not have been possible without the institutional support from the Department of Anesthesiology at the Hospital of the Ludwig-Maximilians-Universität and the financial support from the German National Space Program (DLR) as well as all other space and non-space agencies and bodies which have funded the researchers and authors in this book.

Sincerely,

Alexander Choukér

Acknowledgments 1st Edition

The support of many colleagues, partners and international collaborators is much acknowledged, as without their support this book project would have not been completed. All contributing authors to this book deserve my highest appreciation for their work and for the extremely positive and kind collaboration during the preparative and review periods. The continuous input from all members of the ESA Topical Team "Stress and Immunity", their constructive criticism and advice have been major sources of inspiration and the cornerstone for the realization of this book.

Special thanks go to Prof. Dr. Sarah Baatout and to Prof. Dr. Manfred Thiel for helping to generate the necessary momentum during the kick-off phase; and to Dr. Ines Kaufmann, Dr. Alex P. Salam, Dr. Andrew Dobney and to Dr. Chris Choukér who were extremely supportive in the finalization period. This project would also not have been possible without the generous institutional support from the Department of Anesthesiology at the Ludwig-Maximilians-Universität, and I want to especially thank the director Prof. Dr. Bernhard Zwissler for providing full encouragement and support at all times during this process.

My appreciation also goes to the scientists, doctors and operators working in space, in space analogues and extreme environments, as well as in clinical studies, and to the space agencies and funding institutions, who altogether have provided the intellectual input, experimental performance and the financial means to realize the achievements presented in this volume. This acknowledgment extends to all participating volunteers and patients, as well as to the staff and students working in all the laboratories who provide critical and highly important contributions towards the further evolution of the field of stress and immune research in space and on Earth.

Preface

Thank you for your interest and for reading this second edition of the book "Stress Challenges and Immunity in Space". I would like to warmly welcome the new readers and am deeply indebted to the loyal readers who read the first edition already. I am very grateful for their interest, the positive feedback and criticisms that were instrumental in shaping these research topics and in spreading the enthusiasm for this inter- and cross-disciplinary field of research. Encouraged by reader comments and by the expanding body of knowledge in the area of stress and immunity, together with an emerging interest in space and human space exploration beyond the Earth orbit, the publisher, the authors and I were compelled to embark together on another "journey" with this second edition.

The complex nature of stress and stress responses and immunity, especially in light of extreme living conditions such as in space, need to be addressed on the one hand in a very detailed fashion, but on the other hand, in a more holistic perspective. While all organ systems are obviously interconnected in the human body, these interactions can be strongly affected by the space life conditions (the "space exposome"). I hope this second edition of "Stress Challenges and Immunity in Space" is an attempt in approaching this multifaceted issue even more comprehensively. This has not only resulted in significant updates of the existing chapters with new knowledge and new perspectives, but also from the creation of several new chapters, partly stemming from more basic research, which connects the physiological systems and research areas in a meaningful way to reflect progresses in science. This also resulted in a new summary chapter displaying all the interconnected areas and key information and directions at a glance.

This book will hopefully continue to serve as a handy resource and as an inspiration for a more integrative approach in the field of stress and immunity, highlighting the bidirectional scientific, technological and health benefits of research in space and on Earth.

The authors and I look forward to your comments to this new edition.

Munich, Germany
2019

Alexander Choukér

Contents

Part I

Introduction

The Human Space Exploration Endeavour: A Personal View from Inside

1

David Parker

Humanity's road into the solar system is now under construction, and the humans that will ride along that road will be exposed to many physical and psychological challenges. Uniquely among space activities, exploration is driven by the combination of curiosity and opportunity: the curiosity to venture into the unknown and the opportunity to bring back to Earth discoveries and perspectives only possible by exploring with humans and robots. To ensure our human explorers return safely is both a moral responsibility and a scientific puzzle—a puzzle which researchers around the world have been tackling with enthusiasm and increasing success.

The European Space Agency's (ESA) exploration strategy—approved by Ministers back in 2014—focuses on three destinations: Low Earth Orbit, the Moon and Mars. Through the International Space Station (ISS), Low Earth Orbit has proven its value as a place for humans to live and work. In particular, it has allowed sustained physiology research of value back on Earth as well as in preparing for more distant voyages.

The Moon has immense scientific and practical potential, much increased given the discoveries made since the days of Apollo. Space agencies around the world concur that it is the next target for sustained human exploration, being—in comparison with the ISS—a few days instead of a few hours away from the Earth's surface.

Finally, the Red Planet is a key target not least because it has much to reveal about the potential for past or present life elsewhere in the universe. Establishing how to sustain humans on a multi-year mission to Mars is the horizon goal of the current research discussed in this book.

But while exploration without science is merely tourism, exploration is much more than only science. As the ultimate proving ground for important technologies—energy, robotics, life support systems—it is also a vital source of innovation.

D. Parker (✉)
Human and Robotic Exploration, Noordwijk, The Netherlands
e-mail: David.Parker@esa.int

© Springer Nature Switzerland AG 2020

A. Choukér (ed.), *Stress Challenges and Immunity in Space*,
https://doi.org/10.1007/978-3-030-16996-1_1

And this innovation undoubtedly encompasses the practical value to humanity of the knowledge and techniques needed to sustain our human explorers. Put another way, the manifold challenges of space exploration drives the creativity of scientists and engineers to seek understanding and find solutions to problems, which in turn repay us through the increased knowledge of our society.

Everything we do in exploration is in international cooperation. In difficult times, the simple fact that the ISS partnership endures is a hope for a brighter future. Astronauts see no borders from space. A brighter future depends on exciting the citizens of tomorrow with these values as well as the value of science, technology and maths. The members of the European astronaut corps, and indeed astronauts and cosmonauts from around the world, amply demonstrate these ideals.

All of these motivations are embedded within ESA's European Exploration Envelope Programme, approved at the Luzern Council of Ministers in 2016.

It is not about Moon versus Mars or humans versus robots, but rather it is one unified, ambitious but realistic vision for extending humanity's presence into the solar system; and achieving this through strong global partnerships.

First and foremost, Europe continues to be a committed partner to the ISS. Further important research will be conducted by a succession of European astronauts working aboard the European Columbus research laboratory: fundamental science, applied science and feed-forward exploration science. We are also building on the systems that provide power, propulsion, water and oxygen to NASA's Orion space exploration vehicle, destined to undertake the first human voyage into deep space for 50 years.

Scouting ahead of the human explorers, our ExoMars orbiter is seeking understanding of the mysterious trace gases (including methane) first spotted by ESA's Mars Express and which could imply active geology or biology on the Red Planet. Working with our international partners, we will also launch the ExoMars rover mission, the first dedicated life search rover that will join a veritable armada of Mars missions arriving in 2021. Beyond that, bringing pristine geological samples back to Earth with a robotic could be a key scientific and technological step towards human exploration of Mars.

Meanwhile, we also want our human explorers to learn to live and work in the much more distant and hostile environment of cis-lunar space. Many experts think that the next step for human exploration could be a Lunar Orbital Platform; a Gateway orbiting the moon. Operational by the middle of the next decade and located a thousand times further out in space than the ISS, this could be humanity's most distant research station, a stepping stone to more distant voyages and a base camp for humans to return to the Moon.

The Moon. *The eighth continent*, untroubled by humans for more than 50 years, but silently awaiting our return. Back when the Apollo astronauts visited the equatorial regions for just a few days at a time, we knew nothing about the possibility of water in its deeply shadowed polar craters. The Moon is a museum of 4.5 billion years of solar system history: but so far, we have only visited the entrance foyer and the gift shop. We haven't explored its potential for planetary science, astronomy or human research; let alone its inclusion into our future space economy.

Given all this, it is not surprising that the Gateway and the wider exploration of the Moon is generating increasing interest in the science community and beyond. Undoubtedly, it will further advance the research topics—from fundamental understandings of the cell´s function to the integrative and interdisciplinary insights on the effects of stress on the immune and associated organ systems—as discussed in this second edition of Alexander Choukér's volume. There are new discoveries and new history awaiting humanity—it is time for us to take the next steps in life sciences and human exploration!

Space Travel: A Personal View from Above

2

Thomas Reiter

One of the disciplines that is in the focus of scientists is the human physiology. It's not too surprising how much gravity influences the function of our body from a macroscopic- down to the microscopic level of biochemical processes within each cell—the absence of gravity opens completely new insights into physiological and biochemical processes. It is remarkable, how the human body is able to adapt to weightlessness, increased radiation, altered atmospheric constituents and circadian rhythm. ISS—this multidisciplinary laboratory in space is an ideal environment to understand these effects, and consequently might help us to reveal the root-causes of some widespread diseases, e.g. diseases of the cardiovascular system, the demineralization of bones and the deficiencies of the immune system.

Human space flight is also the source of great fascination. The view of our planet from above and the view of the starry sky is just overwhelming, nurtures our curiosity and drives us to continuously expand our boundaries, to find answers to ever new questions about our origins and the physical principles that govern our universe. Curiosity is a deeply human quality, which has always played a central role in our development.

The progression and further evolution of our technical and scientific knowledge is the merit of generations of engineers and researchers, who have been working in the area of spaceflight and who will continue to push the limits of technology and science. I have no doubt that we will again see humans on the surface of the moon. And in two or three decades we will travel to even more distant destinations like our neighbouring planet Mars. Still a number of technological challenges need to be solved, and a range of medical issues, which are linked to long-term exposure to the space-environment, have to be understood.

Spaceflight is an interdisciplinary regime with a direct impact on science, technology and industrial capabilities. In this context we should not lose sight of the

T. Reiter (✉)
Human Spaceflight and Operations (HSO), Darmstadt, Germany
e-mail: Thomas.reiter@esa.int

© Springer Nature Switzerland AG 2020

A. Choukér (ed.), *Stress Challenges and Immunity in Space*,
https://doi.org/10.1007/978-3-030-16996-1_2

cultural aspect of human spaceflight. There are destinations, which can probably never be reached by humans, but by automated or robotic probes. However, humans with their unique combination of cognitive-, sensory- and dexterous capabilities cannot be replaced by robots, not in the near- and maybe not even in the far future. Anyhow, a machine will hardly be able to share any feelings when executing tasks on the surface of a remote celestial body, or share its emotions when looking back to this beautiful planet, with all its colours, textures, oceans, forests and deserts—a multitude of impressions that an astronaut registers, assimilates and never forgets.

3 Entering a New Era of Holistic Research in Establishing Groundwork for Future Human Space Exploration: Perspectives from the ESA-Topical Team "Stress and Immunity"

Alexander Choukér, Sarah Baatout, Patrizia Campolongo, Jean-Pol Frippiat, Jay Gopalakrishnan, Ines Kaufmann, Nicola Montano, Siegfried Praun, Dominique de Quervain, Benno Roozendaal, Gustav Schelling, Manfred Thiel, Detlef Thieme, Antoine Viola, Judith-Irina Buchheim, Alex Salam, and Anne Guo

---

Alexander Choukér, Sarah Baatout, Patrizia Campolongo, Jean-Pol Frippiat, Jay Gopalakrishnan, Ines Kaufmann, Nicola Montano, Siegfried Praun, Dominique de Quervain, Benno Roozendaal, Gustav Schelling, Manfred Thiel, Detlef Thieme, Antoine Viola are the TT-members and Judith-Irina Buchheim, Alex Salam and Anne Guo are co-authors of this chapter.

The European Space Agency (ESA) supports teams of international experts in "Topical Teams." ESA-Topical Teams are open structures led by European researchers to address a scientific field in which gravity and access to space or planetary bodies constitute as cornerstones of their research. The members of the Topical Team "Stress and Immunity" played a crucial role in the realization of this book and its second edition and authored this prelude collectively as a group.

A. Choukér (✉) · J.-I. Buchheim · A. Guo
Laboratory of Translational Research Stress and Immunity, Department of Anaesthesiology, Hospital of the University of Munich (LMU), Munich, Germany
e-mail: Alexander.Chouker@med.uni-muenchen.de

S. Baatout
Radiobiology Unit, Belgian Nuclear Research Centre, SCK•CEN, Mol, Belgium

P. Campolongo
Department of Physiology and Pharmacology, Sapienza University of Rome, Rome, Italy

J.-P. Frippiat
Stress Immunity Pathogens Laboratory, EA 7300, Faculty of Medicine, Lorraine University, Vandœuvre-lès-Nancy, France

J. Gopalakrishnan
Laboratory for Centrosome and Cytoskeleton Biology, Institute of Human Genetics, University Hospital, Heinrich-Heine-University, Düsseldorf, Germany

© Springer Nature Switzerland AG 2020

A. Choukér (ed.), *Stress Challenges and Immunity in Space*,
https://doi.org/10.1007/978-3-030-16996-1_3

I. Kaufmann
Munich Clinic Neuperlach, Munich, Germany

N. Montano
Department of Clinical Sciences and Community Health, Ospedale Maggiore Policlinico, University of Milan, Milan, Italy

S. Praun
VF Services GmbH, Absam, Austria

D. de Quervain
Division of Cognitive Neuroscience, Department of Psychology, Transfaculty Research Platform and University Psychiatric Clinics, University of Basel, Basel, Switzerland

B. Roozendaal
Department of Cognitive Neuroscience, Radboud University Medical Center, Nijmegen, The Netherlands

Donders Institute for Brain, Cognition and Behaviour, Radboud University Nijmegen, Nijmegen, The Netherlands

G. Schelling
Department of Anesthesiology, University Hospital of Munich, LMU, Munich, Germany

M. Thiel
Department of Anesthesiology and Intensive Care Medicine, Medical Faculty of Mannheim, University of Heidelberg, Mannheim, Germany

D. Thieme
Institut für Dopinganalytik und Sportbiochemie (IDAS), Kreischa, Germany

A. Viola
PPRS, Rouffach, France

A. Salam
Nuffield Department of Medicine, University of Oxford, Oxford, UK

3.1 Introduction

We are currently entering a new era of manned space flight, which will expand human presence to the moon and Mars as outlined in the roadmaps of the International Space Exploration Coordination Group (ISECG, www.globalspaceexploration.org). This roadmap builds on the vision for "a coordinated human and robotic exploration of our solar system" within this century, and envisions permanent human colonization on the moon and on our neighboring planet. This goal set by the established and emerging space countries and agencies, as well as by private entrepreneurs, has further fueled the pace and created impact in this new "race to space" to moon again and to Mars, following the first era in exploring the Earth orbit and the landing on the moon in the past century. These new strategies mandate that we also need to enter a new era of holistic research to better understand human beings subjected to extreme environments for longer periods of time.

3.2 Millions of Years on Earth: 60 Years in Space

For centuries, mankind has been struggling to understand the profound complexity governing the principles of life and the universe. This quest has taken him on scientific journeys far and wide: from the delicate structure of our DNA to the hellish and chaotic depths of our sun. Scientific, artistic, and social discoveries are what drive humans, and what distinguish us from all other species. One of the fundamental questions that still troubles us is how life began on this planet and whether it exists elsewhere in the universe. This deep desire to understand and search for life has taken humans on exploratory journeys to the extremes of our planet - from the depths of our oceans to the heights of our mountains- and into to space, escaping the clutches of Earth's gravitational pull.

Yet in 1960, before a human was launched into space, it was not even clear if one could survive in a zero-gravity environment. At no point in our evolution had we been prepared for such extreme and manifold environmental stresses. From the moment life began in the "pre-biotic soup" some three billion years ago, all life on Earth, Eukaryotes, Prokaryotes, and Archaea alike, have been shaped by the universal force of gravity. Within a matter of few minutes into space, one is faced with the most complex forces and must successfully cope with the absence of gravity. Since Gagarin's historic 108-min voyage, others have survived for months not only in weightlessness but also in extreme isolation and confinement. However, adapting to such hostile and unnatural conditions is not without any repercussions and is accompanied by adverse physiological and psychological effects, which over the last decades, have been shown to impact almost all organ systems. While our presence has extended beyond low Earth orbit to the moon, manned exploration beyond Earth's vicinity into the depths of our solar system as stated above requires a much more detailed understanding of the adaptation of human beings to extreme environments. Major questions remain: What are the principles and most important environmental and social threats to physical and mental health during long-duration space flight missions and how can we prevent and mitigate the adverse effects from adaptation to these threats?

3.3 The Biological Definition of Stress

It was Hans Selye who first used the term "stress" in the 1930s to describe how a biological system might adjust to the challenges and demands associated with major environmental changes (Selye 1936). He realized that when a complex organism is challenged by noxious conditions, the resulting symptoms are independent of the quality of the conditions, i.e., the qualitative end result of different stressor types is the same. Rather, it is the quantitative effects that vary. He also recognized that stressful conditions directly affect neural pathways, such as the autonomic nervous system, but also indirectly affect other organ systems, e.g., the immune system. The steps involved in the adaptation process to chronic stress are gradual and the biological system either builds up resistance and maintains a healthy physiological and psychological equilibrium, or succumbs to the stress, resulting in disequilibrium and eventual disease. Stress research has expanded tremendously since then, and Selye probably never imagined that it would transcend Earth's boundaries. Space flight is associated with a very distinct and unique combination of stressors: zero gravity, radiation, altered microbial flora, isolation, confinement, altered circadian rhythm, and closed-loop environments. Such stressors will be experienced in the extreme during interplanetary travel. These combined and multifactorial challenges affect many organ functions with different time slopes of adaptation and can result in health issues critical to mission success. For example, the neurovestibular and cardiovascular systems seem to adapt within days and weeks, respectively (Aubert et al. 2016). However, the significant brain morphological changes (van Ombergen et al. 2018) and altered immune responses that are observed are much less predictable in their degree and "dose-response" in relation to mission duration and exposition to the stressors in space. Moreover, changes in the immune system can influence by secondary effects other physiological systems and even feed-back with neural pathways in a bidirectional manner (Tracey 2009). This expands the current view that the physical and social environment stressors interact in the "*ecological* loop" with the nervous system and further with hormonal and immunological responses, the so-called *macro*organismal loop. The latter is connected by the immune system to the *micro*organismal loop and the intrinsic regulation of microbial changes (Irwin and Cole 2011).

Although astronauts, cosmonauts, and taikonauts are exceptionally well selected, trained and healthy individuals, some are now known to be particularly susceptible to the stressors of space flight. When challenged by complex stressful conditions, individuals react differently and adjustments to the extreme conditions can fail. The"milieu intérieur" (Claude Bernard 1813–1878) is no longer able to maintain "coordinated physiological processes which maintain most of the steady states in the organism," as they "are so complex and so peculiar to living beings – involving, as they may, the brain and nerves, the heart, lungs, kidneys and spleen, all working cooperatively" (Cannon 1932). This concept of "homeostasis" is extended further by the notion of homeodynamics, "the stability of the internal milieu toward perturbation" (Lloyd et al. 2001).

3.4 Researching the Effects of Stress on the Organism

Although studying specific cellular models and simple biological organisms under conditions of simulated weightlessness, increased radiation, or isolation and confinement can help unravel the neurophysiological consequences of standardized emotional and physiological strains, no organ, especially in the case of humans, can be considered as a stand-alone entity. For this reason, new integrative and holistic approaches to the understanding of stress responses and individual predispositions and reactions to stress have started to evolve. With the help of research on the International Space Station and in analogous conditions and environments—bedrest, or group isolation and confinement in chamber studies (i.e., former Mars-500 habitat, now SIRIUS) or field operational conditions (i.e., Antarctica or subaquatic habitats)—the impact of distinct emotional and physical stressors, or a combination thereof, can be investigated. This will eventually help with the understanding of the incremental effects of stress on organ allostasis, from an allostatic load to overload with subsequent exhaustion and failure to re-establishing an appropriate equilibrium.

3.5 Selecting the "Right Stuff"

Given that the reaction to stress can vary between individuals and even within one individual at different times, how can we design strategies to meet the astronauts' individual needs under evolving and unpredictable conditions? How can we provide personalized medicine? This may prove very difficult and would require new technologies and devices. Should we select astronauts based on the presence of genetic characteristics that confer resistance to stress? The new emerging technological tools of molecular biology, such as next-generation sequencing, single-cell genomics, and exosome research, will help to uncover the genetic and epigenetic (e.g., DNA methylation, post-transcriptional regulation) explanations for (mal) adaption and the corresponding therapeutic consequences. Today, organoids are very promising representatives of human models for space research as they represent the miniature of human organs constituting the complexities similar to human tissues. For instance, the human brain is a very susceptible organ when subjected to extreme conditions. Recent years have seen enormous progress in developing human brain organoids reflecting events of early brain development and maturation. Due to the fact that (brain) organoids are amendable for experimental purposes and are economical, clearly they hold promises to uncover aspects that are practically impossible to investigate in real humans. Moreover, autologous organoids are opening up new opportunities for investigating and modeling an individual's organ capacity (Clevers 2016) as affected by external stressors in metabolizing drugs and allowing dose adjustments according to individual responses. Such new bioengineering technologies will also "revolutionize the field of regenerative medicine" (Madl et al. 2018) on Earth and may probably be just as needed when humans aim to expand into outer space. This altogether will help us better predict, prevent, and treat the space voyager and to help him or her coping with external stressors while paving the way to individually and fully controlled human metabolism in enabling a hibernating state, for example, during space travel (Choukèr et al. 2018).

Genetic, metabolic, and bioengineering approaches have the potential to select and *de*select candidates. This would have important psychological, social, and ethical implications. While "reading genes" does not equate to "understanding genes" and the complexity of a human being goes far beyond his genetic heritage, identification of single-gene polymorphisms that appear to correlate with a higher predisposition for physiological and behavioral stresses should not disqualify a potential space flight candidate. Although polymorphisms in genes, such as genes regulating sleep (Goel et al. 2009), traumatic memory encoding, or DNA repair, may confer vulnerability, individuals may have unidentified genetic resistance to other space-related stress factors as well as behavioral coping strategies that may mitigate the genetic risk.

"The right stuff" (Wolfe 2008) seems very likely to be a highly complex mix of gene and environmental interactions. Given ethical implications, the use of genetic analyses or other emerging technologies are *not* permitted in the selection of candidates for space flight, *but* rather, to identify possible risks in order to personalize the frequency and mode of physiological and psychological assessments and countermeasures in space, as well as during rehabilitation upon return to Earth. There is much left to qualify and quantify, but with time we will refine the physiological, psychological, and pharmaceutical factors and interventions that will allow humans to travel interplanetary distances.

3.6 Research in Space for Benefits on Earth

The study of healthy humans experiencing high levels of stress in confinement and isolation or in other space analogous environments allows us to establish clear causal links between stress and physiological disequilibrium and disease. Understanding the interaction between stress and the human body and mind will lead to better healthcare not only for astronauts but also for the vast majority of us who will never escape gravity's pull. Often undermined are sex differences, which have to come more into the focus (Jaillon et al. 2017) as well as the effect of stress on aging of the immune system (Prather et al. 2018) and autoimmunity (Giancheechi et al. 2018). Space and space analogous environments provide an excellent platform for such studies since these processes appear to be accelerated and more profound.

In the future of manned space exploration, the pace of research in understanding these effects will be accelerated through new diagnostic and healthcare tools and regimens that are necessary to enable a personalized, circadian-adjusted precision medicine to balance the individuals' allostatic load. An important question remains though: to which degree will the required human–robotic interactions and artificial intelligence affect the human in space and how they can assist in better adaptation? Along the way, these new developments will not only benefit our space agencies but also the wider society at large. Stress has the ability to alter the function of virtually every single organ system and cell type in the human body, therein lie many new opportunities in understanding the profound complexities governing the principles of life and the adaptation responses of an individual. After all, every single person on this planet experiences stress and no one is completely immune to its effects.

Acknowledgments The authors acknowledge the funding of this Topical Team by the European Space Agency (ESA, contract 4000109114/13/NL/PG)) and express their thanks to Dr. Jennifer Ngo-Anh (ESA) for her support. The funding of the authors by national space agencies or other national or international funding institutions is acknowledged and specified in the authors' individual chapters of this volume.

References[1]

Aubert AE, Larina I, Momken I, Blanc S, White O, Prisk GK, Linnarsson D (2016) Towards human exploration of space: the THESEUS review series on cardiovascular, respiratory, and renal research priorities. NPJ Microgravity 2:16031

[1] The overarching topics addressed in this chapter will be presented also in the parts II–VI of this volume. In addition, the references have been used to compile this chapter.

Cannon WB (1932) "Homeostasis." The wisdom of the body. Norton, New York, NY
Choukèr A, Bereiter-Hahn J, Singer D, Heldmaier G (2018) Hibernating astronauts-science or fiction? Pflugers Arch. https://doi.org/10.1007/s00424-018-2244-7
Clevers H (2016) Modeling development and disease with organoids. Cell 165(7):1586–1597
Gianchecchi E, Delfino DV, Fierabracci A (2018) NK cells in autoimmune diseases: linking innate and adaptive immune responses. Autoimmun Rev 17(2):142–154
Goel N, Banks S, Mignot E, Dinges DF (2009) PER3 polymorphysm predicts cumulative sleep homeostatic but not neurobehavioral changes to chronic partial sleep deprivation. PLoS One 4(6):e5874
Irwin MR, Cole SW (2011) Reciprocal regulation of the neural and innate immune systems. Nat Rev Immunol 11(9):625–632
Jaillon S, Berthenet K, Garlanda C (2017) Sexual dimorphism in innate immunity. Clin Rev Allergy Immunol. https://doi.org/10.1007/s12016-017-8648-x
Lloyd D et al (2001) Why homeodynamics, not homeostasis? Sci World J 1:133–145
Madl CM, Heilshorn SC, Blau HM (2018) Bioengineering strategies to accelerate stem cell therapeutics. Nature 557(7705):335–342
Prather AA, Epel ES, Portela Parra E, Coccia M, Puterman E, Aiello AE, Dhabhar FS (2018) Associations between chronic caregiving stress and T cell markers implicated in immunosenescence. Brain Behav Immun 73:546–549
Selye H (1936) A syndrome produced by diverse noxious agents. Nature 138:30–32
Tracey KJ (2009) Reflex control of immunity. Nat Rev Immunol 9(6):418–428
Van Ombergen A, Jillings S, Jeurissen B, Tomilovskaya E, Rühl RM, Rumshiskaya A, Nosikova I, Litvinova L, Annen J, Pechenkova EV, Kozlovskaya IB, Sunaert S, Parizel PM, Sinitsyn V, Laureys S, Sijbers J, Zu Eulenburg P, Wuyts FL (2018) Brain tissue-volume changes in cosmonauts. N Engl J Med 379(17):1678–1680
Wolfe T (2008) The right stuff, 2nd edn. Picador, London. Revised edition. ISBN-10: 0312427565; ISBN-13: 978-0312427566

Part II

Stress and Immunity Research: A Link Between Space and Earth

What Is Stress?

4

Bruce S. McEwen and Ilia N. Karatsoreos

'I am stressed out' is non-accusatory, apolitical and detached. It is a good way to keep the peace and, at the same time, a low-cost way to complain.
America's Latest Export: A Stressed-Out World. By Richard A. Shweder. Published: January 26, 1997.

4.1 Introduction

Stress is a word that is used throughout the world and it has many meanings. There is "good stress" and "bad stress." Some would prefer to use "stress" to refer only to the experience and consequences of a situation when one is unable to cope physically or psychologically with the challenge (Cohen et al. 2007; Lazarus and Folkman 1984). Physiologically, cortisol and adrenalin are stress hormones, and the "fight or flight" response is usually the focus of discussions of stress. But that is only part of the story. There are multiple biological mediators besides the adrenal stress hormones that are responsible for adaptation in situations that evoke the "fight or flight" response (McEwen and Stellar 1993; Sterling and Eyer 1988) and help us stay alive, not only in our daily lives and in extreme conditions on Earth but also in space. But these same mediators also contribute to pathophysiology when overused and dysregulated, resulting in allostatic load and overload (McEwen 1998; McEwen and Wingfield 2003).

The brain is the central organ of stress and adaptation because it determines not only what is threatening, or at least different and potentially threatening, in a new situation but also determines the physiological and behavioral responses (Fig. 4.1). Alterations in brain structure and function by experiences throughout life determine how each individual will respond to new events. But there are also important contributions from genes; individual life-style habits reflecting items, such as sleep quality and quantity; diet, exercise, and substance abuse; adverse early life experiences

B. S. McEwen · I. N. Karatsoreos (✉)
Harold and Margaret Milliken Hatch Laboratory of Neuroendocrinology, The Rockefeller University, New York, NY, USA

Department of Integrative Physiology and Neuroscience, Sleep and Performance Research Center, Washington State University, Pullman, WA, USA
e-mail: ilia.karatsoreos@wsu.edu

© Springer Nature Switzerland AG 2020

A. Choukér (ed.), *Stress Challenges and Immunity in Space*,
https://doi.org/10.1007/978-3-030-16996-1_4

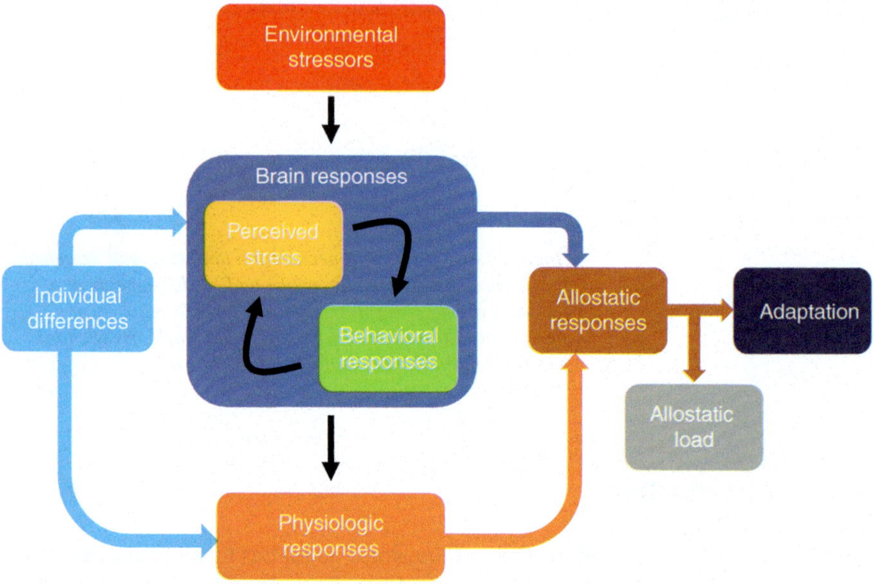

Fig. 4.1 Environmental stressors engage brain systems that perceive the stressful situation, activating both behavioral responses and physiological responses. These responses can be shaped by individual differences that may arise from genetic polymorphisms or from experience (e.g., early life environment, cumulative exposures). The behavioral and physiologic responses are an integral component of allostatic responses which promote adaptation and also contribute to allostatic load. Long-term buildup of allostatic load can lead to allostatic overload and breakdown of this integrated set of responses, promoting pathology

(see also following Chap. 6) that set life-long patterns of behavior and physiological reactivity; and exposure to toxic agents in the environment.

One purpose of this chapter is to describe the concepts of allostasis and allostatic load and overload as a way of making the discussion of the physiology and psychobiology of stress more precise and biologically based and related to life style and health-related behaviors, as well as the stressful experiences themselves. The other purpose of this chapter is to highlight ways in which the brain and its architecture play a key role in how individuals respond to new challenges. The relevance of the adaptation to stressors will be discussed in relation to the challenges of space flight and adaptation to microgravity providing the further basis for the understanding, why, along with effects on the brain, the immune systems and health can be affected under such condition.

4.2 Types of Stress

The term "stress" is perhaps one of the most overused terms in general life. Everyone has a different definition of stress, from the "stressful" day at work to posttraumatic stress disorder (PTSD). For the purposes of this chapter, we will focus on the following definitions of stress: good stress, tolerable stress, and toxic stress (Box 4.1).

Box 4.1 Types of Stress

Positive stress

- A personal challenge that has a satisfying outcome
- Result: Sense of mastery and control
- HEALTHY BRAIN ARCHITECTURE
- Good self-esteem, judgment, and impulse control

Tolerable stress

- Adverse life events buffered by supportive relationships
- Result: Coping and recovery
- HEALTHY BRAIN ARCHITECTURE
- Good self-esteem, judgment, and impulse control

Toxic stress

- Unbuffered adverse events of greater duration and magnitude
- Result: Poor coping and compromised recovery
- Result: Increased life-long risk for physical and mental disorders
- COMPROMISED BRAIN ARCHITECTURE
- Dysregulated physiological systems

Good stress, also known as "eustress," is a term used in popular language to refer to the experience of rising to a challenge, taking a risk and feeling rewarded by an often positive outcome. Good self-esteem and good impulse control and decision-making capability, all functions of a healthy architecture of the brain, are important here. Even adverse outcomes can be "growth experiences" for individuals with such positive, adaptive characteristics.

Perhaps halfway down the spectrum of stress is "tolerable stress," which refers to those situations where bad things happen, but an individual is able to cope, often with the aid of family, friends and other individuals who provide support. Here we can also introduce the term "distress" that refers to the uncomfortable feeling related to the nature of the stressor and the degree to which the individual feels a lack of ability to influence or control the stressor (Lazarus and Folkman 1984).

Finally, "toxic stress," on the extreme end of the stress spectrum, refers to the situation in which bad things happen to an individual who has limited support. In this case, previous life experience may have resulted in neurophysiological changes that can impair the development of good impulse control and judgment and adequate self-esteem, all known to help modulate the negative effects of stress. Here, the degree and/or duration of "distress" may be greater. With "toxic stress," the inability to cope is likely to have adverse effects on behavior and physiology, and this will result in a higher degree of allostatic overload, as will be explained later in this chapter.

4.3 The Concepts of Allostasis, Allostatic Load, and Overload in Health, Disease, and Aging

When discussing the term "stress," it is the negative connotations that seem to be the primary focus. However, the "stress response" is a necessary set of physiological changes that help an organism cope with a threat to homeostasis, and indeed potentially a threat to survival. The body responds to many experiences by releasing chemical mediators, for example, catecholamines that increase heart rate and blood pressure. These mediators promote adaptation to simple acts like getting out of bed in the morning or climbing a flight of stairs or more complex acts, like giving a lecture or a musical performance. However, chronically increased heart rate and blood pressure can cause pathophysiological changes. For example, in the cardiovascular system, over time, these changes can cause pathophysiological conditions like atherosclerosis, that can result in strokes and myocardial infarctions (Cohen et al. 2007).

Because allostaic mediators are paradoxically involved in both protection and damage, and also because the word "stress" has ambiguities and connotations that interfere with its precise use, the term "allostasis" was introduced (Sterling and Eyer 1988) to refer to the active process by which the body responds to daily events and maintains homeostasis—allostasis literally means "achieving stability through change" (Box 4.2). A useful way to disambiguate the terms is that homeostasis employs small changes about a fixed set point. On the other hand, in allostasis the set point is modulated by context, and the brain plays a central role since prior experiences can be used to inform the magnitude and duration of the potential stressor. However, chronically engaged allostatic responses can lead to pathophysiology, we introduced the term "allostatic load or overload" (see distinction in Box 4.2 and below) to refer to the "wear and tear" that results from either too much stress or from inefficient management of allostasis, such as not turning off the response when it is no longer needed (McEwen 1998; McEwen and Stellar 1993; McEwen and Wingfield 2003).

Box 4.2 Definitions

Homeostasis is the stability of physiological systems that maintain life, used here to apply strictly to a limited number of systems such as pH, body temperature, glucose levels, and oxygen tension that are truly essential for life and are therefore maintained within a range optimal for each life history stage.

Allostasis is achieving stability through change, a process that supports homeostasis, i.e., those physiological parameters essential for life defined above, as environments and/or life history stages change. This means that the "setpoints" and other boundaries of control must also change. There are primary mediators of allostasis such as, but not confined to, hormones of the hypothalamo–pituitary–adrenal (HPA) axis, catecholamines, and cytokines. Allostasis also clarifies an inherent ambiguity in the term "homeostasis" and distinguishes between the systems that are essential for life ("homeostasis") and those that maintain these systems in balance ("allostasis") as environment and life history stage change.

Allostatic state: The allostatic state refers to altered and sustained activity levels of the primary mediators, e.g., glucocorticoids, that integrate physiology and associated behaviors in response to changing environments and challenges such as social interactions, weather, disease, predators, and pollution. An allostatic state results in an imbalance of the primary mediators, reflecting excessive production of some and inadequate production of others (Koob and LeMoal 2001). Examples are hypertension, a perturbed cortisol rhythm in major depression or after chronic sleep deprivation, chronic elevation of inflammatory cytokines and low cortisol that increases risk for autoimmune and inflammatory disorders. Allostatic states can be sustained for limited periods if food intake and/or stored energy such as fat can fuel homeostatic mechanisms. For example, bears and other hibernating animals preparing for the winter become hyperphagic as part of the normal life cycle and at a time (summer and early autumn) when food resources can sustain it.

Allostatic load and allostatic overload: The cumulative result of an allostatic state (e.g., a bear putting on fat for the winter) is allostatic load. It can be considered the result of the daily and seasonal routines organisms have to obtain food and survive and extra energy needed to migrate, molt, brood, etc. Within limits, these are adaptive responses to seasonal and other demands. However, if one superimposes additional loads of unpredictable events in the environment such as disease, human disturbance, and social interactions, then allostatic load can increase dramatically. Type 1 allostatic overload occurs when energy demands exceed energy income as well as what can be mobilized from stores. Type 2 allostatic overload occurs if energy demands are not exceeded and the organism continues to take in or store as much or even more energy than it needs. This may be a result of stress-related food consumption, choice of a fat-rich diet, or metabolic imbalances (prediabetic state) that favors fat deposition. There are other cumulative changes in other systems, e.g., neuronal remodeling or loss in hippocampus, atherosclerotic plaques, left ventricular hypertrophy of the heart, glycosylated hemoglobin, and other proteins by advanced glycosylation end products as a measure of sustained hyperglycemia. High cholesterol with low HDL may also occur, and chronic pain and fatigue, e.g., in arthritis or psoriasis, associated with imbalance of immune mediators.

Other forms of allostatic load/overload involve not shutting off the response efficiently, or not engaging an adequate response in the first place (McEwen 1998). Perhaps one of the most significant contributors to allostatic load is repeated engagement of the stress response in an unconstrained or poorly managed manner. Having many stressful events and many stress responses contributes to general "wear and tear" on the body and brain (McEwen 1998). Likewise, not habituating to the recurrence of the same stressor and thus dampening the allostatic response can also lead to overexposure of the brain and body to the mediators of allostasis (McEwen 1998). It is intriguing to posit, and several studies in humans and non-human animals demonstrate, links between accumulation of allostatic load and aging (Piazza et al. 2018). Indeed, it is obvious that as one ages one is more likely to be exposed to more challenges, and thus increased cumulative allostatic load. Many of the mediators and processes discussed below, from circadian rhythms and sleep to immune function and oxidative stress, are involved in organismal and cellular senescence. There is also significant new data from the field of telomere biology that shows that decreased telomere length is associated with stress and allostatic load (Ahrens et al. 2016; Tomiyama et al. 2012; Zalli et al. 2014). Given the recent findings that spaceflight can lengthen telomeres, this is an exciting area for future investigations (Garrett-Bakelman et al. 2019).

4.4 The Complex Interactions of Allostatic Mediators

Protection and damage are the two contrasting sides of the physiology involved in defending the body against the challenges of daily life, whether or not we call them "stressors." That is, while the evolutionary advantage of allostatic responses is adaptation to environmental challenge, repeated or unmanaged allostatic challenges can lead to damage by the very mediators that are supposed to protect the organism. The primary reason for this Janus-faced aspect of allostasis and allostatic load is that there are myriad mediators acting in multiple interacting webs of physiological regulation. Besides adrenalin and noradrenalin, there are many mediators that participate in allostasis, and they are linked together in a network of regulation that is nonlinear, meaning that each mediator has the ability to regulate the activity of the other mediators, sometimes in a biphasic manner (Fig. 4.2). For example, glucocorticoids produced by the adrenal cortex in response to adrenocorticotropic hormone (ACTH) from the pituitary gland are the other major "stress hormones." Yet, pro- and anti-inflammatory cytokines are produced by many cells in the body, and they regulate each other and are, in turn, regulated by glucocorticoids and catecholamines. That is, whereas catecholamines can increase pro-inflammatory cytokine production (Bierhaus et al. 2003), glucocorticoids are known to inhibit this production (Sapolsky et al. 2000). Yet, there are exceptions, e.g., pro-inflammatory effects of glucocorticoids that depend on dose and cell or tissue type (Munhoz et al. 2010). The parasympathetic nervous system also plays an important regulatory role in this nonlinear network of allostasis (see Chap. 8), since it generally opposes the sympathetic nervous system and, for example, slows the heart, and it also has anti-inflammatory effects (Borovikova et al. 2000; Thayer and Lane 2000) .

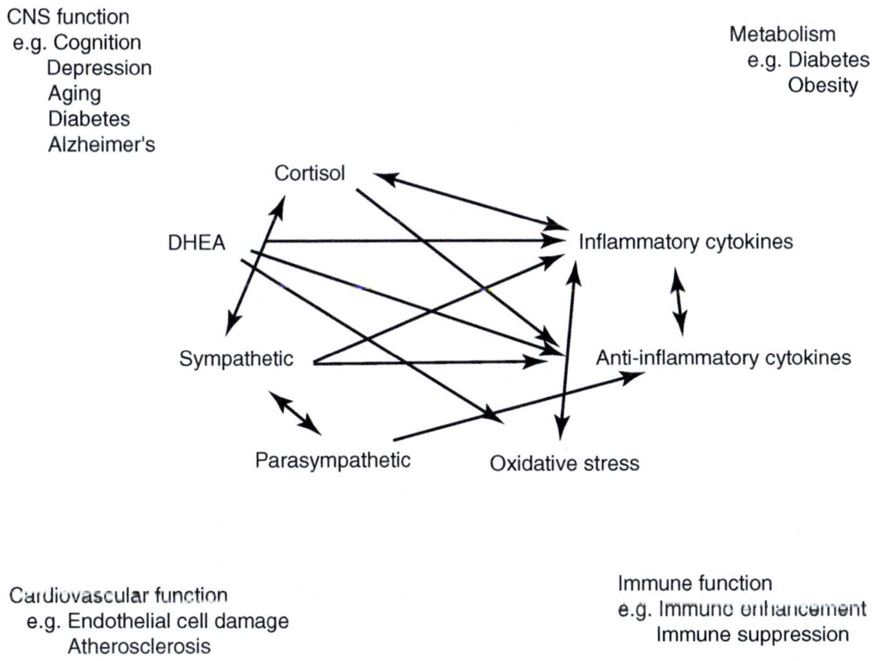

Fig. 4.2 Multiple interacting mediators and nonlinearity of interactions between them. Arrows represent direct and indirect regulatory influences of one mediator system upon the other systems. At the corners of the figure are listed some of the body systems that are concurrently affected by these mediators and their dysregulation

What this nonlinearity means is that when any one mediator is increased or decreased, there are compensatory changes in the other mediators that depend on time course and level of change of each of the mediators (McEwen 2006). Unfortunately, biomedical technology cannot yet measure all components of this system simultaneously and must rely on measurements of only a few of them in any one study, or their secondary consequences (McEwen and Seeman 1999). Yet the nonlinearity must be kept in mind in interpreting the results. One approach is to "tap into" these mediators, and their surrogates, and obtain a broader picture of the network of allostasis is the "allostatic load battery" (McEwen and Seeman 1999; Seeman et al. 2010).

A further important aspect of the mediators of allostasis is the biphasic nature of many of their effects, a concept embodied by the term "hormesis" (Calabrese 2008) and represented very clearly for cortisol (Joels 2006) and for pro- and anti-inflammatory cytokines, e.g., interleukin-6 (Campbell et al. 1993; Moidunny et al. 2010; Patterson 1992).

4.5 Stress: An Ecological Perspective

The operation of allostasis in the natural world provides some insight into how animals use this response to their own benefit or for the benefit of the species. As an example of allostasis, in springtime, a sudden snowstorm causes stress to birds and disrupts mating, as stress hormones are pivotal in directing the birds to suspend reproduction, to find a source of food and to relocate to a better mating site or at least to delay reproduction until the weather improves (Wingfield and Romero 2000). As an example of allostatic load, bears preparing to hibernate for the winter eat large quantities of food and put on body fat to act as an energy source during the winter (Nelson 1980). This accumulation of fat is then used, to survive the winter and provide food for gestation of young. In contrast, the fat accumulation that occurs in bears that are captive in zoos and eating too much (partially out of boredom) while not exercising (McEwen and Wingfield 2003), is an example of "allostatic overload"; a more extreme condition that is associated with pathophysiology and is all-too-common in our own species.

Yet, allostatic overload can also have a useful purpose for the preservation of the species, such as in migrating salmon or the marsupial mouse, that die of excessive stress after mating—the stress and allostatic load are being caused for salmon, in part, by the migration up the rapidly flowing rivers, and also because of physiological changes that represent accelerated aging and include suppression of the immune system (Gotz et al. 2005; Maule et al. 1989). One beneficial result of eliminating the adult salmon is freeing up food and other resources for the next generation. In the case of the marsupial mouse, it is only the males that die after mating, and the hypothesized mechanism is a response to mating that reduces the binding protein for glucocorticoids (corticosteroid binding globulin, CBG) and renders them much more active throughout the body, including likely suppressive actions on the immune defense system (Cockburn and Lee 1988).

4.6 Being "Stressed Out" in the Modern World and How It Affects the Way We Live

As noted by Shweder in the quote at the beginning of this chapter, while "stress" is an ambiguous term, the common experience of being "stressed out" has somewhat more specific meaning in terms of behavior and physiology and has, as its core, the elevation of some of the key systems that lead to allostatic overload—e.g., cortisol, sympathetic activity and pro-inflammatory cytokines, with a decline in parasympathetic activity. When we are "stressed out," we often feel frustrated, anxious and even angry and we are likely to eat comfort foods, drink in excess of normal, smoke, if we are so inclined, and neglect regular moderate physical activity. We are also likely to sleep poorly. Indeed, poor or inadequate sleep is a frequent result of being "stressed out." Sleep deprivation produces an allostatic overload that can have deleterious consequences and a more detailed examination of this and circadian disruption via shift work and jet lag reveals how many aspects of health can be adversely affected. Indeed, it is our health-damaging behavior resulting from being "stressed out" that are among the main contributors to allostatic load and overload. Among those, circadian disruption may be among the most potent.

4.7 Allostatic Consequences of the Modern World's Disruption of Circadian Clock and Sleep

Because the brain is the master regulator of the neuroendocrine, autonomic, and immune systems, as well as behavior (McEwen 1998), alterations in brain function by chronic stress can have direct and indirect effects on the cumulative allostatic overload. One of the key systems in the brain and body that regulate these varied physiological and behavioral variables is the circadian system (see also Chap. 9). Based in the suprachiasmatic nucleus (SCN) of the hypothalamus, the brain's clock controls rhythms in the rest of the brain and body through both neural and diffusible signals. Biological clocks at the molecular level have been detected in almost every body organ and tissue so far examined, and these clocks are synchronized by the SCN directly (by way of neural connections) or indirectly through hormonal signals (e.g. cortisol, melatonin) or behavioral outputs (e.g. feeding). The SCN also regulates the timing of sleep, and sleep and circadian systems interact to regulate rest–activity cycles and keep an organism in synchrony with the external environment. As such, disruption of these key homeostatic systems could clearly contribute to allostatic overload.

Reduced sleep duration has been associated with increased body mass and obesity in the NHANES study (Gangwisch et al. 2005). Sleep restriction to 4 h of sleep per night increases blood pressure, decreases parasympathetic tone, increases evening cortisol and insulin levels, and promotes increased appetite, possibly through the elevation of ghrelin, a pro-appetitive hormone, along with decreased levels of leptin (Spiegel et al. 1999, 2004; Van Cauter et al. 1997). Moreover, pro-inflammatory cytokine levels are increased with sleep deprivation, along with decreased performance in tests of psychomotor vigilance, and this has been reported to result from a modest sleep restriction to 6 h per night (Vgontzas et al. 2004).

Circadian disruption has sometimes been overlooked as a separate yet related phenomenon to sleep deprivation. In modern industrialized societies, circadian disruption can be induced in numerous ways, the most common of which are shift work and jet lag. A longitudinal study in a cohort of nurses in night shift work found that exposure to night work can contribute to weight gain and obesity (Niedhammer et al. 1996). Moreover, alternating shift work is an independent risk factor for the development of obesity in a large longitudinal study of male Japanese shift workers (Niedhammer et al. 1996). Numerous mouse models have also contributed to our understanding of the relationship between circadian disruption and metabolism, with CLOCK mutant mice showing altered basal metabolism and a tendency towards obesity and metabolic dysregulation (Niedhammer et al. 1996), while normal C57Bl/6 mice housed in a 20-h photoperiod (T20) of 10-h light:10-h dark, show accelerated weight gain and disruptions in metabolic hormones (Karatsoreos et al. 2011). While this approach clearly disrupts many body rhythms, it is important to note that in many models of circadian disruption, sleep loss is not observed, while in others sleep loss is a hallmark. In the T20 cycle described above, mice who are disrupted show no sleep loss, but instead reduced sleep quality, more fragmented sleep, and a loss of "niche appropriate" sleep; that is they sleep more in the phase of the light–dark cycle (Phillips et al. 2015). In humans, even short duration misaligned circadian rhythms lead to acute changes in glucose homeostasis and metabolic function that are similar to prediabetic states (Scheer et al. 2009). Thus, understanding how mistimed sleep, not only sleep loss, affects metabolic function will be critical, especially under conditions of long-term space flight.

In addition to the effects on metabolic systems, the circadian system plays a key role in the immune function (see Chap. 9). Since the original work by Halberg et al. (1960), over the past 60 years evidence has accumulated that time of day and the circadian clock impact immune function (Fonken et al. 2015; Fortier et al. 2011; Labrecque and Cermakian 2015; Scheiermann et al. 2018). A recent study by Gagnidze et al. (2016) takes these findings a step further by attempting to modulate potential underlying mechanisms of these circadian effects on immunity. Using vesicular stomatitis virus (VSV) intranasal infection as a model of viral-induced encephalitis, it was demonstrated that time of day has significant effects on survival, with mice infected at the start of their day phase showing significantly increased mortality, while mice infected at the start of their night phase showing significantly enhanced survival. Inactivation of REV-erb-alpha, an important molecular circadian clock factor, in the olfactory bulb completely mimicked the effect of day time infection, significantly reducing survival. While this study is one of many that demonstrates the importance of the circadian clock in regulating immune responses, disrupted circadian timing has also been shown to modulate normal responses. Using circadian disruption as way to increase allostatic load, several groups have demonstrated that the response to lipopolysaccharide (LPS, a component of the bacterial cell wall that mimics a bacterial infection) is sensitive to disrupted circadian timing (Adams et al. 2013; Castanon-Cervantes et al. 2010; Rahman et al. 2015). In the T20 model of disruption, while basal plasma cytokine responses are not significantly altered, the cytokine response to LPS challenge is clearly dysregulated (Phillips et al. 2015). Thus, disrupting the circadian clock may serve as a low-grade allostatic load that alters immune responses, and once again demonstrates that circadian disruption may alter allostatic responses, and lead to distorted or inefficient physiological responses to additional environmental challenges.

While the metabolic and immune consequences of circadian disruption are clear, there are neurobehavioral consequences as well. Behaviorally, circadian disruption can contribute to cognitive impairments. In animal models of disrupted circadian timing, effects have been observed in both emotionality and cognitive functions. For instance, in the T20 cycle model of disruption, mice show an increase in impulsive-like behaviors, altered responses in novel environments (e.g. open field), and impaired cognitive flexibility (Karatsoreos et al. 2011). These changes are accompanied by reductions in medial prefrontal cortex (mPFC) pyramidal neuron dendritic complexity (Karatsoreos et al. 2011). Similarly, in a T7 light–dark cycle, mice show significantly impaired emotional function, coupled with reduced hippocampal learning, and reduced hippocampal long-term potentiation (LTP) (LeGates et al. 2012). Human studies reveal impairments that are very similar both qualitatively and quantitatively. In a study of long recovery vs. short recovery flight crews, it was found that short recovery crews had impaired performance in a psychomotor task, reacting more slowly, and with more errors when compared to a long recovery crew (Cho 2001). Again, understanding the mechanisms involved in the neurobehavioral impairments of circadian disruption will be critical in the isolated conditions that are experienced during space flight. Eventually, when humans colonize Mars, they will be faced with a slightly longer solar day (24 h 37 min). This may present an important consideration in the housing environments of those early explorers.

4.8 The Brain as a Target of Stress and Allostatic Load and Overload

The brain is a target of stress and stress hormones and the processes of allostasis and allostatic load and overload are exemplified by how different brain regions respond to acute and chronic stressors. Because the hippocampus was the first higher brain center that was recognized as a target of stress hormones, it has figured prominently in our understanding of how stress impacts brain structure and behavior. Effects of stress on the amygdala and prefrontal cortex will then be summarized.

4.8.1 The Hippocampus

The hippocampus plays a key role in learning and remembering declarative and spatial information, as well as processing the contextual aspects of emotional events, and regulating visceral functions, including the Hypothalamic–pituitary–adrenal (HPA) axis. The hippocampus, which is interconnected with the amygdala and prefrontal cortex (Petrovich et al. 2001), contains receptors for adrenal steroids, and for major metabolic hormones such as insulin, leptin, ghrelin, and insulin-like growth factor 1 (IGF-1), all which have effects on the hippocampus (McEwen 2007). Specifically, these mediators can enhance cognitive processes, affect mood and motivation, and promote excitability and neuroprotection. Yet, these same mediators can have deleterious effects on the hippocampus under conditions associated with chronic stress and allostatic overload, the most extreme being head trauma, seizures, stroke, and diabetes (Gold et al. 2007; McEwen 2007; Sapolsky 1992). Insulin action and insulin resistance in the hippocampus are very important for cognitive function, depending only partly on modulation of glucose uptake (Biessels and Reagan 2015; Grillo et al. 2015).

A number of animal models demonstrate that chronic stressful experiences (e.g., prolonged immobilization, housing in dominance hierarchies, early maternal separation) can remodel hippocampal neurons and result in changes in the morphology of the hippocampus. Within the hippocampus, input from the entorhinal cortex to the dentate gyrus is ramified by connections between the dentate gyrus and the CA3 pyramidal neurons. Hence, one granule neuron innervates, on average, 12 CA3 neurons, and each CA3 neuron innervates, on average, 50 other CA3 neurons via axon collaterals, as well as 25 inhibitory cells via other axon collaterals. The net result is a 600-fold amplification of excitation, as well as a 300-fold amplification of inhibition, that provides some degree of control of the system (McEwen 1999).

As to why this type of circuitry exists, the dentate gyrus-CA3 system is believed to play a role in the memory of event sequences, although long-term storage of memory occurs in other brain regions. But, because the DG-CA3 system is so delicately balanced in its function and vulnerable to damage, there is also adaptive structural plasticity: that is, CA3 pyramidal cells undergo a reversible remodeling of their dendrites in conditions such as hibernation and chronic stress. The role of this plasticity may be to protect against permanent damage (McEwen 1999, 2007).

Another type of structural plasticity involves replacement of neurons via neurogenesis. The sub-granular layer of the dentate gyrus contains cells that have some properties of astrocytes (e.g., expression of glial fibrillary acidic protein) and which give rise to granule neurons (Seri et al. 2001). After Bromodeoxyuridine (5-bromo-2-deoxyuridine, BrdU) administration to label DNA of dividing cells, these newly born cells appear as clusters in the inner part of the granule cell layer, where a substantial number, i.e. 5000–9000 per day, will subsequently differentiate into granule neurons within just 7 days and contribute in multiple ways to the function and plasticity of the hippocampus (Cameron et al. 1993; Cameron and McKay 2001; Gould and Gross 2002; Cameron and Schoenfeld 2018). There are many hormonal, neurochemical and behavioral modulators of neurogenesis and cell survival in the dentate gyrus, including estradiol, androgens, IGF-1, antidepressants, voluntary exercise, and hippocampal-dependent learning (Duman et al. 2001; Gould et al. 1999; Trejo et al. 2001; van Praag et al. 1999) (Okamoto et al. 2012). With respect to stress, certain types of acute stress and many chronic stressors suppress neurogenesis or cell survival in the dentate gyrus, and the mediators of these inhibitory effects include excitatory amino acids acting via N-methyl-d-aspartic acid (NMDA) receptors and endogenous opioids (for review, see McEwen 2007).

An additional form of neuroplasticity is the remodeling of dendrites in the hippocampus. Chronic restraint stress causes retraction and simplification of dendrites in the CA3 region of the hippocampus (McEwen 1999). Such dendritic reorganization is found in both dominant and subordinate rats undergoing adaptation of psychosocial stress in the visible burrow system, which is independent of adrenal size (McKittrick et al. 2000). What this particular result emphasizes is that it is not adrenal size or presumed amount of physiological stress per se that determines dendritic remodeling, but a complex set of other interacting factors that modulate neuronal structure. Indeed, in species of mammals that hibernate (see also Chap. 5), dendritic remodeling is a reversible process, and it occurs within hours of the onset of hibernation in European hamsters and ground squirrels (Magarinos et al. 2006; Popov et al. 1992). Moreover, it is reversible within hours of wakening of the animals from torpor (Magarinos et al. 2006). This implies that reorganization of the cytoskeleton is taking place rapidly and reversibly and that changes in dendrite length and branching are not "damage" but a form of structural plasticity associated with altered gene expression and mediated by multiple cellular and molecular pathways (McEwen et al. 2015).

The plasticity of the hippocampus in animal models has fostered studies of the human hippocampus that have revealed functional and structural changes that are consistent with the animal model studies even though it is not possible to look at individual neurons in the living human brain (McEwen and Gianaros 2010). Functionally, spatial memories activate the human hippocampus, as in the London cab driver studies (Maguire et al. 1997). Further, intensive learning experiences are associated with increases in hippocampal volume, further suggesting dynamic experience-dependent neuroplasticity in the hippocampus (Draganski et al. 2006). Smaller hippocampal volume has been reported in Cushing's disease, recurrent major depression, Type 2 diabetes, posttraumatic stress disorder, a prolonged history of high perceived stress, chronic systemic inflammation, chronic jet lag, lack of regular exercise and low self-esteem (McEwen and Gianaros 2010). With respect to exercise, there is evidence that physical fitness is associated with larger hippocampal volume (Erickson et al. 2009, 2011).

4.8.2 Amygdala

A critical function of the amygdala in stressor-related processing involves the fear, anxiety and aggression. Sensory input is relayed through thalamic and cortical-thalamic pathways to the basolateral area and then to the central nucleus. As a primary output nucleus, the central nucleus turns on adrenalin and cortisol output and freezing behavior (LeDoux 2003). The central nucleus is also networked with cortical areas involved in stressor-related processing, including the anterior cingulate cortex, ventromedial prefrontal cortex, and orbital prefrontal cortex (McEwen and Gianaros 2010).

Chronic immobilization stress of the type that causes retraction of dendrites in the CA3 region of the hippocampus produces dendritic growth in neurons in basolateral amygdala (Vyas et al. 2002). Moreover, chronic stress of this type not only impairs hippocampal dependent cognitive function but also enhances amygdala-dependent unlearned fear and fear conditioning processes that are consistent with the opposite effects of stress on hippocampal and amygdala structure (Chattarji et al. 2015; Conrad et al. 1999, Wood et al. 2008).

Chronic stress also increases aggression between animals living in the same cage, and this is likely to reflect another aspect of hyperactivity of the amygdala (Wood et al. 2008). Moreover, chronic corticosterone treatment in drinking water produces an anxiogenic effect in mice (Ardayfio and Kim 2006; Karatsoreos et al. 2010), an effect that could be due to the glucocorticoid enhancement of corticotrophin releasing factor (CRF) activity in the amygdala (Makino et al. 1994).

As is the case for the hippocampus, neuropsychological and imaging studies on the human amygdala have revealed biphasic changes in volume in acute versus chronic depression (Frodl et al. 2003; Sheline et al. 1998), as well as increased functional activity (Drevets et al. 1992; Sheline et al. 2001). Moreover, functional amygdala reactivity to facial expressions is increased by sleep deprivation (Yoo et al. 2007) and by an early life history of lower socioeconomic status (Gianaros et al. 2008b) and leads to a predisposition to cardiovascular disease (Gianaros et al. 2008a). Moreover, not only is emotional reactivity affected, but consolidation of fear-related memories is also impacted in humans by sleep deprivation (Menz et al. 2013).

4.8.3 Prefrontal Cortex

The prefrontal cortex is networked with the amygdala and exerts downstream control over amygdala activity, as well as midbrain and brainstem functions, including functions of control, as opposed to learned helplessness (Amat et al. 2005; McEwen and Gianaros 2010), and autonomic balance (Buchanan et al. 2010). Chronic stress also causes functional and structural changes in the medial prefrontal cortex, particularly in areas of anterior cingulate, prelimbic, infralimbic, and orbitofrontal regions (McEwen and Morrison 2013). For example, chronic stress causes dendritic shortening in medial prefrontal cortex (Dias-Ferreira et al. 2009; Liston et al. 2006; Radley et al. 2004; Wellman 2001), and also produce dendritic growth in orbitofrontal cortex (Liston et al. 2006). Taken together with the differential effects of the same stressors on the hippocampus and amygdala, these actions of stress are reminiscent of recent work on experimenter versus self-administered morphine and amphetamine, in which different, and sometimes opposite, effects were seen on dendritic spine density in orbitofrontal cortex, medial prefrontal cortex, and hippocampus CA1 (Robinson and Kolb 1997). For example, amphetamine self-administration increased spine density on pyramidal neurons in the medial prefrontal cortex and decreases spine density on orbitofrontal pyramidal neurons (Crombag et al. 2005).

Behavioral correlates of chronic restraint stress (CRS)-induced remodeling in the prefrontal cortex include impairment in attention set shifting (Liston et al. 2006) and other tests of cognitive flexibility and decision making (Dias-Ferreira et al. 2009), possibly reflecting structural remodeling in the medial prefrontal cortex. In the T20 circadian disruption paradigm which may represent a chronic low-level allostatic load, we have observed remodeling of neurons in the prelimbic region of the medial prefrontal cortex of mice which resemble those observed in chronic stress, including a loss of dendritic branching and reduction in length of the apical dendrites, though no decrease in spine density has been observed (Karatsoreos et al. 2011). Behaviorally, these circadian disrupted animals show normal acquisition of a spatial navigation task but show impaired cognitive flexibility and increased error rates when made to switch their learning to a new location. This suggests that chronic circadian disruption may not impair simple cognitive tasks; more complex tasks requiring flexibility may be impaired.

In the human prefrontal cortex, there are reported changes in functional connectivity in medical students under stress that relate to reduced performance on a test of cognitive flexibility like the attention set shifting study in rodents; these changes are reversible with vacation, supporting the concept of reversible neuroplasticity (Liston et al. 2009).

4.9 Acute Versus Chronic Stress

The essence of the concept of allostasis and allostatic load/overload is that the mediators of stress and adaptation have different effects in situations of acute stress versus chronic stress. We have discussed the effects of various types of stressors for the body as a whole and also for three key brain regions. One way of visualizing the range of effects of various stressful events for the hippocampus is as an inverted U in which acute levels of adrenal steroids acting together with excitatory amino acids and other intracellular mediators do the following things in space and time:

1. Low physiological levels acutely enhance excitability of neurons (Diamond et al. 1992; Pavlides et al. 1996) and enhance contextual fear memory (see Chap. 7);
2. Higher physiological and supraphysiological levels have the opposite effect and acutely suppress excitability (Diamond et al. 1992; Joels 2006; Pavlides et al. 1996);
3. Chronic stress or chronic elevation of glucocorticoids produces "adaptive plasticity" involving the changes in circuitry described in the previous section on the hippocampus;
4. Uncontrolled events such as head trauma, seizures, and stroke activate both adrenal steroids and excitatory amino acids in a way that leads to damage and neuronal loss (Sapolsky 1992).

4.10 Acute and Chronic Stress Effects on Immune Function

Other physiological systems show differences in effects of acute versus chronic stress. Effects of stress on delayed-type hypersensitivity (DTH) in the immune system show that acute restraint stress enhances DTH, whereas chronic restraint stress for 21 days suppresses it (Dhabhar and McEwen 1997, 1999; Dhabhar 2018; Prather et al. 2018). Moreover, the acute stress effect shows a dose-response in that an acute restraint plus shaking of the animal produces a larger enhancement of DTH than acute restraint stress alone (Dhabhar and McEwen 1997, 1999). In the hippocampus, chronic restraint for 21 days produced the remodeling of dendrites of CA3 neurons described, whereas lesser durations of chronic restraint stress did not cause such remodeling (Magarinos and McEwen 1995). Yet a shorter (10 day) period of a more intense stress, namely, chronic immobilization, caused CA3 dendrites to shrink (Dhabhar 2018; Prather et al. 2018; Vyas et al. 2002).

4.11 Stress Controllability: An Important Consideration

The essence of the difference between "tolerable" vs. "toxic" stress (Box 4.1) is the sense of control or lack thereof, as well as the quality and quantity of social support and integration (McEwen and Gianaros 2010). As noted earlier in this chapter, "tolerable stress" refers to those situations where bad things happen but the individual with healthy brain architecture is able to cope, often with the aid of family, friends, and other individuals who provide support. In contrast, "toxic stress" refers to the situation in which bad things happen to an individual who has limited support and who may also have brain architecture that reflects effects of adverse early life events that have led to impaired development of good impulse control and judgment and adequate self-esteem (Shonkoff et al. 2009) (http://developingchild.harvard.edu/initiatives/council/). Thus, with toxic stress, the inability to cope is likely to have adverse effects on behavior and physiology, and this will result in a higher degree of allostatic overload.

The sense of control has several important neurobiological ramifications. Learned helplessness is accompanied by a failure of top-down prefrontal cortical control of midbrain serotonergic activity (Amat et al. 2005). Moreover, lack of control of tail shock stress in a classical eye blink conditioning paradigm revealed a stress effect that inhibited performance in females, but enhanced performance in stressed males (Wood and Shors 1998); giving the rats control of the tail shock stress abolished the stress effect on performance as well as the sex difference (Shors et al. 2007).

Lack of control is often exacerbated by, and related to, low self-esteem and low self-esteem has been linked to a smaller hippocampus and elevated cortisol secretion, with lack of habituation to repeated public speaking challenges (Kirschbaum et al. 1995; Pruessner et al. 1999, 2005). It remains to be seen what other aspects of brain function are altered under these conditions and whether interventions that improve an individual's self-esteem and locus of control can alter brain structure and function in measurable ways.

4.12 Space Flight, Microgravity, and Stress: A Unique Allostatic Load by the "Space Exposome"

Space flight, with reduced gravity and all the behavioral changes that are required of the astronaut, is very likely to change brain structure and function, as well as affecting hormone output, energy expenditure, cardiovascular as well as autonomic and immune system function (Chaps. 5, 8–11) (Alperin et al. 2017; Blanc et al. 2001; Hasan et al. 2018; Roberts et al. 2017; Strollo et al. 1998; Tipton et al. 1996). Both short- and long-duration space flight is associated with shifts in adrenocortical and immune system function (Crucian et al. 2011; Stowe et al. 2011). For instance, motion sickness induced by parabolic flight, is linked with reduced activity of the endocannabinoid system in humans (Chouker et al. 2010), an important regulator of the stress response. The experience of microgravity, which increases glucocorticoid levels (Macho et al. 1993) has been shown to increase dendritic growth by acting through serum- and glucocorticoid-inducible kinase 1 (David et al. 2005). Seven days of microgravity has been shown to result in a loss of proteins in hippocampus (Sarkar et al. 2006). Further studies of brain regions involved in cognitive function and emotional control are warranted. This is especially true for prolonged space flight where neuroendocrine adaptations to shorter space missions may be diminished (Strollo et al. 1998; Tipton et al. 1996). Exposure to microgravity has been experimentally demonstrated to alter the length of the free-running (i.e. endogenous) period in insects (Hoban-Higgins et al. 2003), as well as altering the nycthermal distribution of energy expenditure in rats (Blanc et al. 2001) and changes of the body core temperature of about 1°C that developed gradually over 2.5 months in crew exposed to 6 months space flight on the ISS (Stahn et al. 2017; see also Chap. 26). Importantly, circadian rhythms appear to function up to 90 days in space but weaken thereafter, along with disruptions in sleep (Monk et al. 2001). Consequences of circadian and sleep disruption such as those discussed earlier in this article must be undertaken to fully appreciate the metabolic, neurological, behavioral, and immune system consequences of weightlessness and prolonged space missions. In this respect, space is providing a complex set of allostatic loads as a consequence of being exposed to the "space exposome" the individual has to adapt to. The "space exposome" can be defined—in extension of the definition by Wild (2012)—as the sum of every exposure to which the crew member is subjected in their mission, which become a function of the quality, the intensity, and duration of the impact to affect immunity (Crucian et al. 2018). Some of those effects can also be selectively mirrored in so-called space analog environments to assess some of the allostatic loads in extreme living conditions such as in Antarctica or in controlled isolation or bed-rest studies (Chaps. 36 and 37). Here the effects on stress hormone release, circadian rhythms, and sleep, as well as metabolism and temperature regulations have been conducted (Gander et al. 1991; Harris et al. 2010; Jacubowski et al. 2015; Mendt et al. 2017; Pattyn et al. 2017).

4.13 Conclusion: Toward a Better Understanding of Stress Effects on Brain and Immune, Metabolic, Cardiovascular, and Other Body Systems

Stress is a word that has communication value in our "stressed-out" world because it identifies a state-of-mind that is common among many people and provides an explanatory excuse for being harried and in a rush, as well as a means of generating understanding and sympathy. This chapter has described a physiological and neurobiological foundation that should permit a better understanding of the consequences of acute and chronic stress on the body and brain; it has placed this foundation in the context of life-long influences of the social, as well as physical environments. The brain is the key to what these experiences do to the body because it determines threat and controllability and activates the behavioral and physiological responses, as well as the lifestyle choices and health-related behaviors that result from our experiences. We have seen that the brain is also a target of stress and changes in brain architecture to determine how the brain will respond.

The experiences of space flight are both physical and psychological stressors and likely sources of allostatic load and overload, particularly when they persist over weeks and months, and there is great need to better understand the ability of the body and brain to adapt to both the acute and chronic aspects of these experiences.

The analysis provided in this chapter and in other chapters in this volume will help in this quest by directing attention to the role of the brain, its structural and functional plasticity, and its interaction with the immune system as well as metabolic, cardiovascular, and other body systems in relation to health and disease.

Moreover, the view on this interacting nature of both brain-dependent biological hormonal and neural mediators and peripheral functions with other direct gravitational, metabolic, microbial, or radiation-dependent consequences of the stressful living conditions in space is important. For example, it is known that the circadian clock is very important in tissue responses to radiation-induced damage and cancer (Dakup and Gaddameedhi 2017; Wang et al. 2017). What could be the consequences of the combined allostatic load of disrupted circadian timing, increased radiation exposure, impaired inflammatory regulation, and increased psychological stress on the response of tissues and organs? Understanding that these systems interact, sometimes in unpredictable ways, is important as it provides us an insight into the complex processes of adaptation of human organ-systems. This is of importance in order to find the appropriate means to prevent the occurrence, or repair the negative consequences, of prolonged allostatic states and resulting allostatic overload. Such states can result from chronic disease, and also in space. The relevance of understanding these processes for crew on their way to long-duration exploration class missions is important not only in space but also here on Earth. Thus, knowledge of how organisms adapt to severe environmental challenges has both basic science and practical applied impacts in many different realms of exploration.

References

Adams KL, Castanon-Cervantes O, Evans JA, Davidson AJ (2013) Environmental circadian disruption elevates the IL-6 response to lipopolysaccharide in blood. J Biol Rhythm 28:272–277

Ahrens KA, Rossen LM, Simon AE (2016) Relationship between mean leucocyte telomere length and measures of allostatic load in us reproductive-aged women, NHANES 1999–2002. Paediatr Perinat Epidemiol 30:325–335

Alperin N, Bagci AM, Lee SH (2017) Spaceflight-induced changes in white matter hyperintensity burden in astronauts. Neurology 89:2187–2191

Amat J, Baratta MV, Paul E, Bland ST, Watkins LR, Maier SF (2005) Medial prefrontal cortex determines how stressor controllability affects behavior and dorsal raphe nucleus. Nat Neurosci 8:365–371

Ardayfio P, Kim K-S (2006) Anxiogenic-like effect of chronic corticosterone in the light-dark emergency task in mice. Behav Neurosci 120:249–256

Bierhaus A, Wolf J, Andrassy M, Rohleder N, Humpert PM et al (2003) A mechanism converting psychosocial stress into mononuclear cell activation. Proc Natl Acad Sci U S A 100:1920–1925

Biessels GJ, Reagan LP (2015) Hippocampal insulin resistance and cognitive dysfunction. Nat Rev Neurosci 16:660–671

Blanc S, Geloen A, Normand S, Gharib C, Somody L (2001) Simulated weightlessness alters the nycthemeral distribution of energy expenditure in rats. J Exp Biol 204:4107–4113

Borovikova LV, Ivanova S, Zhang M, Yang H, Botchkina GI et al (2000) Vagus nerve stimulation attenuates the systemic inflammatory response to endotoxin. Nature 405:458–462

Buchanan TW, Driscoll D, Mowrer SM, Sollers JJI, Thayer JF et al (2010) Medial prefrontal cortex damage affects physiological and psychological stress responses differently in men and women. Psychoneuroendocrinology 35:56–66

Calabrese EJ (2008) Hormesis and medicine. Br J Clin Pharm 66:594–617
Cameron HA, McKay RDG (2001) Adult neurogenesis produces a large pool of new granule cells in the dentate gyrus. J Comp Neurol 435:406–417
Cameron HA, Schoenfeld TJ (2018) Behavioral and structural adaptations to stress. Front Neuroendocrinol 49:106–113
Cameron H, Woolley C, McEwen BS, Gould E (1993) Differentiation of newly born neurons and glia in the dentate gyrus of the adult rat. Neuroscience 56:337–344
Campbell IL, Abraham CR, Masliah E, Kemper P, Inglis JD et al (1993) Neurologic disease induced in transgenic mice by cerebral overexpression of interleukin 6. Proc Natl Acad Sci U S A 90:10061–10065
Castanon-Cervantes O, Wu M, Ehlen JC, Paul K, Gamble KL et al (2010) Dysregulation of inflammatory responses by chronic circadian disruption. J Immunol 185:5796–5805
Chattarji S, Tomar A, Suvrathan A, Ghosh S, Rahman MM (2015) Neighborhood matters: divergent patterns of stress-induced plasticity across the brain. Nat Neurosci 18:1364–1375
Cho K (2001) Chronic 'jet lag' produces temporal lobe atrophy and spatial cognitive deficits. Nat Neurosci 4:567–568
Chouker A, Kaufmann I, Kreth S, Hauer D, Feuerecker M et al (2010) Motion sickness, stress and the endocannabinoid system. PLoS One 5:e10752
Cockburn A, Lee AK (1988) Marsupial femmes fatales. Nat Hist 97:40–47
Cohen S, Janicki-Deverts D, Miller GE (2007) Psychological stress and disease. JAMA 298:1685–1688
Conrad CD, Magarinos AM, LeDoux JE, McEwen BS (1999) Repeated restraint stress facilitates fear conditioning independently of causing hippocampal CA3 dendritic atrophy. Behav Neurosci 113:902–913
Crombag HS, Gorny G, Li Y, Kolb B, Robinson TE (2005) Opposite effects of amphetamine self-administration experience on dendritic spines in the medial and orbital prefrontal cortex. Cereb Cortex 15:341–348
Crucian B, Stowe R, Quiriarte H, Pierson D, Sams C (2011) Monocyte phenotype and cytokine production profiles are dysregulated by short-duration spaceflight. Aviat Space Environ Med 82:857–862
Crucian BE, Chouker A, Simpson RJ, Mehta S, Marshall G et al (2018) Immune system dysregulation during spaceflight: potential countermeasures for deep space exploration missions. Front Immunol 9:1437
Dakup P, Gaddameedhi S (2017) Impact of the circadian clock on UV-induced DNA damage response and photocarcinogenesis. Photochem Photobiol 93:296–303
David S, Stegenga SL, Hu P, Xiong G, Kerr E et al (2005) Expression of serum- and glucocorticoid-inducible kinase is regulated in an experience-dependent manner and can cause dendrite growth. J Neurosci 25:7048–7053
Dhabhar FS (2018) The short-term stress response – mother nature's mechanism for enhancing protection and performance under conditions of threat, challenge, and opportunity. Front Neuroendocrinol 49:175–192
Dhabhar FS, McEwen BS (1997) Acute stress enhances while chronic stress suppresses cell-mediated immunity in vivo: a potential role for leukocyte trafficking. Brain Behav Immun 11:286–306
Dhabhar F, McEwen B (1999) Enhancing versus suppressive effects of stress hormones on skin immune function. Proc Natl Acad Sci U S A 96:1059–1064
Diamond DM, Bennett MC, Fleshner M, Rose GM (1992) Inverted-U relationship between the level of peripheral corticosterone and the magnitude of hippocampal primed burst potentiation. Hippocampus 2:421–430
Dias-Ferreira E, Sousa JC, Melo I, Morgado P, Mesquita AR et al (2009) Chronic stress causes frontostriatal reorganization and affects decision-making. Science 325:621–625
Draganski B, Gaser C, Kempermann G, Kuhn HG, Winkler J et al (2006) Temporal and spatial dynamics of brain structure changes during extensive learning. J Neurosci 26: 6314–6317

Drevets WC, Videen TO, Price JL, Preskorn SH, Carmichael ST, Raichle ME (1992) A functional anatomical study of unipolar depression. J Neurosci 12:3628–3641
Duman RS, Nakagawa S, Malberg J (2001) Regulation of adult neurogenesis by antidepressant treatment. Neuropsychopharmacology 25:836–844
Erickson KI, Prakash RS, Voss MW, Chaddock L, Hu L et al (2009) Aerobic fitness is associated with hippocampal volume in elderly humans. Hippocampus 19(10):1030–1039
Erickson KI, Voss MW, Prakash RS, Basak C, Szabo A et al (2011) Exercise training increases size of hippocampus and improves memory. Proc Natl Acad Sci U S A 108:3017–3022
Fonken LK, Frank MG, Kitt MM, Barrientos RM, Watkins LR, Maier SF (2015) Microglia inflammatory responses are controlled by an intrinsic circadian clock. Brain Behav Immun 45:171–179
Fortier EE, Rooney J, Dardente H, Hardy MP, Labrecque N, Cermakian N (2011) Circadian variation of the response of T cells to antigen. J Immunol 187:6291–6300
Frodl T, Meisenzahl EM, Zetzsche T, Born C, Jager M et al (2003) Larger amygdala volumes in first depressive episode as compared to recurrent major depression and healthy control subjects. Biol Psychiatry 53:338–344
Gagnidze K, Hajdarovic KH, Moskalenko M, Karatsoreos IN, McEwen BS, Bulloch K (2016) Nuclear receptor REV-ERBalpha mediates circadian sensitivity to mortality in murine vesicular stomatitis virus-induced encephalitis. Proc Natl Acad Sci U S A 113:5730–5735
Gander PH, Macdonald JA, Montgomery JC, Paulin MG (1991) Adaptation of sleep and circadian rhythms to the Antarctic summer: a question of zeitgeber strength. Aviat Space Environ Med 62:1019–1025
Gangwisch JE, Malaspina D, Boden-Albala B, Heymsfield SB (2005) Inadequate sleep as a risk factor for obesity: analyses of the NHANES I. Sleep 28:1289–1296
Garrett-Bakelman FE, Darshi M, Green SJ, Gur RC, Lin L, Macias BR, McKenna MJ, Meydan C, Mishra T, Nasrini J, Piening BD, Rizzardi LF, Sharma K, Siamwala JH, Taylor L, Vitaterna MH, Afkarian M, Afshinnekoo E, Ahadi S, Ambati A, Arya M, Bezdan D, Callahan CM, Chen S, Choi AMK, Chlipala GE, Contrepois K, Covington M, Crucian BE, De Vivo I, Dinges DF, Ebert DJ, Feinberg JI, Gandara JA, George KA, Goutsias J, Grills GS, Hargens AR, Heer M, Hillary RP, Hoofnagle AN, Hook VYH, Jenkinson G, Jiang P, Keshavarzian A, Laurie SS, Lee-McMullen B, Lumpkins SB, MacKay M, Maienschein-Cline MG, Melnick AM, Moore TM, Nakahira K, Patel HH, Pietrzyk R, Rao V, Saito R, Salins DN, Schilling JM, Sears DD, Sheridan CK, Stenger MB, Tryggvadottir R, Urban AE, Vaisar T, Van Espen B, Zhang J, Ziegler MG, Zwart SR, Charles JB, Kundrot CE, Scott GBI, Bailey SM, Basner M, Feinberg AP, Lee SMC, Mason CE, Mignot E, Rana BK, Smith SM, Snyder MP, Turek FW (2019) The NASA Twins Study: a multidimensional analysis of a year-long human spaceflight. Science 364(6436):eaau8650. https://doi.org/10.1126/science.aau8650
Gianaros PJ, Hariri AR, Sheu LK, Muldoon MF, Sutton-Tyrrell K, Manuck SB (2008a) Preclinical atherosclerosis covaries with individual differences in reactivity and functional connectivity of the amygdala. Biol Psychiatry 65(11):943–950
Gianaros PJ, Horenstein JA, Hariri AR, Sheu LK, Manuck SB et al (2008b) Potential neural embedding of parental social standing. Soc Cogn Affect Neurosci 3:91–96
Gold SM, Dziobek I, Sweat V, Tirsi A, Rogers K et al (2007) Hippocampal damage and memory impairments as possible early brain complications of type 2 diabetes. Diabetologia 50:711–719
Gotz ME, Malz CR, Dirr A, Blum D, Gsell W et al (2005) Brain aging phenomena in migrating sockeye salmon *Oncorhynchus nerka nerka*. J Neural Transm 112:1177–1199
Gould E, Gross CG (2002) Neurogenesis in adult mammals: some progress and problems. J Neurosci 22:619–623
Gould E, Beylin A, Tanapat P, Reeves A, Shors TJ (1999) Learning enhances adult neurogenesis in the hippocampal formation. Nat Neurosci 2:260–265
Grillo CA, Piroli GG, Lawrence RC, Wrighten SA, Green AJ et al (2015) Hippocampal insulin resistance impairs spatial learning and synaptic plasticity. Diabetes 64:3927–3936
Halberg F, Johnson EA, Brown BW, Bittner JJ (1960) Susceptibility rhythm to *E. coli* endotoxin and bioassay. Proc Soc Exp Biol Med 103:142–144

Harris A, Marquis P, Eriksen HR, Grant I, Corbett R et al (2010) Diurnal rhythm in British Antarctic personnel. Rural Remote Health 10:1351

Hasan KM, Mwangi B, Keser Z, Riascos R, Sargsyan AE, Kramer LA (2018) Brain quantitative MRI metrics in astronauts as a unique professional group. J Neuroimaging 28:256–268

Hoban-Higgins TM, Alpatov AM, Wassmer GT, Rietveld WJ, Fuller CA (2003) Gravity and light effects on the circadian clock of a desert beetle, Trigonoscelis gigas. J Insect Physiol 49:671–675

Jacubowski A, Abeln V, Vogt T, Yi B, Chouker A et al (2015) The impact of long-term confinement and exercise on central and peripheral stress markers. Physiol Behav 152:106–111

Joels M (2006) Corticosteroid effects in the brain: U-shape it. Trends Pharmacol Sci 27:244–250

Karatsoreos IN, Bhagat SM, Bowles NP, Weil ZM, Pfaff DW, McEwen BS (2010) Endocrine and physiological changes in response to chronic corticosterone: a potential model of the metabolic syndrome in mouse. Endocrinology 151:2117–2127

Karatsoreos IN, Bhagat S, Bloss EB, Morrison JH, McEwen BS (2011) Disruption of circadian clocks has ramifications for metabolism, brain, and behavior. Proc Natl Acad Sci U S A 108:1657–1662

Kirschbaum C, Prussner JC, Stone AA, Federenko I, Gaab J et al (1995) Persistent high cortisol responses to repeated psychological stress in a subpopulation of healthy men. Psychosom Med 57:468–474

Koob GF, Le Moal M (2001) Drug addiction, dysregulation of reward, and allostasis. Neuropsychopharmacology 24:97–129

Labrecque N, Cermakian N (2015) Circadian clocks in the immune system. J Biol Rhythm 30:277–290

Lazarus RS, Folkman S (eds) (1984) Stress, appraisal and coping. Springer, New York, NY

LeDoux J (2003) The emotional brain, fear, and the amygdala. Cell Mol Neurobiol 23: 727–738

LeGates TA, Altimus CM, Wang H, Lee HK, Yang S et al (2012) Aberrant light directly impairs mood and learning through melanopsin-expressing neurons. Nature 491:594–598

Liston C, Miller MM, Goldwater DS, Radley JJ, Rocher AB et al (2006) Stress-induced alterations in prefrontal cortical dendritic morphology predict selective impairments in perceptual attentional set-shifting. J Neurosci 26:7870–7874

Liston C, McEwen BS, Casey BJ (2009) Psychosocial stress reversibly disrupts prefrontal processing and attentional control. Proc Natl Acad Sci U S A 106:912–917

Macho L, Jezova D, Jurcovicova J, Kvetnansky R, Vigas M, Serova LB (1993) Effect of space flight on the development of endocrine functions in rats. Endocr Regul 27:17–22

Magarinos AM, McEwen BS (1995) Stress-induced atrophy of apical dendrites of hippocampal CA3c neurons: comparison of stressors. Neuroscience 69:83–88

Magarinos AM, McEwen BS, Saboureau M, Pevet P (2006) Rapid and reversible changes in intrahippocampal connectivity during the course of hibernation in European hamsters. Proc Natl Acad Sci U S A 103:18775–18780

Maguire EA, Frackowiak RSJ, Frith CD (1997) Recalling routes around London: activation of the right hippocampus in taxi drivers. J Neurosci 17:7103–7110

Makino S, Gold PW, Schulkin J (1994) Corticosterone effects on corticotropin-releasing hormone mRNA in the central nucleus of the amygdala and the parvocellular region of the paraventricular nucleus of the hypothalamus. Brain Res 640:105–112

Maule AG, Tripp RA, Kaattari SL, Schreck CB (1989) Stress alters immune function and disease resistance in chinook salmon *(Oncorhynchus tshawytscha)*. J Endocrinol 120:135–142

McEwen BS (1998) Protective and damaging effects of stress mediators. N Engl J Med 338: 171–179

McEwen BS (1999) Stress and hippocampal plasticity. Annu Rev Neurosci 22:105–122

McEwen BS (2006) Protective and damaging effects of stress mediators: central role of the brain. Dial Clin Neurosci Stress 8:367–381

McEwen BS (2007) The physiology and neurobiology of stress and adaptation: central role of the brain. Physiol Rev 87(3):873–904

McEwen BS, Gianaros PJ (2010) Central role of the brain in stress and adaptation: links to socioeconomic status, health, and disease. Ann N Y Acad Sci 1186:190–222
McEwen BS, Morrison JH (2013) The brain on stress: vulnerability and plasticity of the prefrontal cortex over the life course. Neuron 79:16–29
McEwen BS, Seeman T (1999) Protective and damaging effects of mediators of stress: elaborating and testing the concepts of allostasis and allostatic load. Ann N Y Acad Sci 896:30–47
McEwen BS, Stellar E (1993) Stress and the Individual: mechanisms leading to disease. Arch Intern Med 153:2093–2101
McEwen BS, Wingfield JC (2003) The concept of allostasis in biology and biomedicine. Horm Behav 43:2–15
McEwen BS, Bowles NP, Gray JD, Hill MN, Hunter RG et al (2015) Mechanisms of stress in the brain. Nat Neurosci 18:1353–1363
McKittrick CR, Magarinos AM, Blanchard DC, Blanchard RJ, McEwen BS, Sakai RR (2000) Chronic social stress reduces dendritic arbors in CA3 of hippocampus and decreases binding to serotonin transporter sites. Synapse 36:85–94
Mendt S, Maggioni MA, Nordine M, Steinach M, Opatz O et al (2017) Circadian rhythms in bed rest: monitoring core body temperature via heat-flux approach is superior to skin surface temperature. Chronobiol Int 34:666–676
Menz MM, Rihm JS, Salari N, Born J, Kalisch R et al (2013) The role of sleep and sleep deprivation in consolidating fear memories. NeuroImage 75:87–96
Moidunny S, Dias RB, Wesseling E, Sekino Y, Boddeke HW et al (2010) Interleukin-6-type cytokines in neuroprotection and neuromodulation: oncostatin M, but not leukemia inhibitory factor, requires neuronal adenosine A1 receptor function. J Neurochem 114:1667–1677
Monk TH, Kennedy KS, Rose LR, Linenger JM (2001) Decreased human circadian pacemaker influence after 100 days in space: a case study. Psychosom Med 63:881–885
Munhoz CD, Sorrells SF, Caso JR, Scavone C, Sapolsky RM (2010) Glucocorticoids exacerbate lipopolysaccharide-induced signaling in the frontal cortex and hippocampus in a dose-dependent manner. J Neurosci 30:13690–13698
Nelson RA (1980) Protein and fat metabolism in hibernating bears. Fed Proc 39:2955–2958
Niedhammer I, Lert F, Marne MJ (1996) Prevalence of overweight and weight gain in relation to night work in a nurses' cohort. Int J Obes Relat Metab Disord 20:625–633
Okamoto M, Hojo Y, Inoue K, Matsui T, Kawato S et al (2012) Mild exercise increases dihydrotestosterone in hippocampus providing evidence for androgenic mediation of neurogenesis. Proc Natl Acad Sci U S A 109:13100–13105
Patterson PH (1992) The emerging neuropoietic cytokine family: first CDF/LIF, CNTF and IL-6; next ONC, MGF, GCSF? Curr Opin Neurobiol 2:94–97
Pattyn N, Mairesse O, Cortoos A, Marcoen N, Neyt X, Meeusen R (2017) Sleep during an Antarctic summer expedition: new light on "polar insomnia". J Appl Physiol (1985) 122:788–794
Pavlides C, Ogawa S, Kimura A, McEwen B (1996) Role of adrenal steroid mineralocorticoid and glucocorticoid receptors in long-term potentiation in the CA1 field of hippocampal slices. Brain Res 738:229–235
Petrovich GD, Canteras NS, Swanson LW (2001) Combinatorial amygdalar inputs to hippocampal domains and hypothalamic behavior systems. Brain Res Rev 38:247–289
Phillips DJ, Savenkova MI, Karatsoreos IN (2015) Environmental disruption of the circadian clock leads to altered sleep and immune responses in mouse. Brain Behav Immun 47:14–23
Piazza JR, Stawski RS, Sheffler JL (2018) Age, daily stress processes, and allostatic load: a longitudinal study. J Aging Health 2018:898264318788493
Popov VI, Bocharova LS, Bragin AG (1992) Repeated changes of dendritic morphology in the hippocampus of ground squirrels in the course of hibernation. Neuroscience 48:45–51
van Praag H, Christie BR, Sejnowski TJ, Gage FH (1999) Running enhances neurogenesis, learning, and long-term potentiation in mice. Proc Natl Acad Sci 96:13427–13431
Prather AA, Epel ES, Portela Parra E, Coccia M, Puterman E et al (2018) Associations between chronic caregiving stress and T cell markers implicated in immunosenescence. Brain Behav Immun 73:546–549

Pruessner JC, Hellhammer DH, Kirschbaum C (1999) Low self-esteem, induced failure and the adrenocortical stress response. Personal Individ Differ 27:477–489
Pruessner JC, Baldwin MW, Dedovic K, Renwick RM, Mahani NK, Lord C et al (2005) Self-esteem, locus of control, hippocampal volume, and cortisol regulation in young and old adulthood. NeuroImage 28:815–826
Radley JJ, Sisti HM, Hao J, Rocher AB, McCall T et al (2004) Chronic behavioral stress induces apical dendritic reorganization in pyramidal neurons of the medial prefrontal cortex. Neuroscience 125:1–6
Rahman SA, Castanon-Cervantes O, Scheer FA, Shea SA, Czeisler CA et al (2015) Endogenous circadian regulation of pro-inflammatory cytokines and chemokines in the presence of bacterial lipopolysaccharide in humans. Brain Behav Immun 47:4–13
Roberts DR, Albrecht MH, Collins HR, Asemani D, Chatterjee AR et al (2017) Effects of spaceflight on astronaut brain structure as indicated on MRI. N Engl J Med 377:1746–1753
Robinson TE, Kolb B (1997) Persistent structural modifications in nucleus accumbens and prefrontal cortex neurons produced by previous experience with amphetamine. J Neurosci 17:8491–8497
Sapolsky R (1992) Stress, the aging brain and the mechanisms of neuron death, vol 1. MIT Press, Cambridge, p 423
Sapolsky RM, Romero LM, Munck AU (2000) How do glucocorticoids influence stress responses? Integrating permissive, suppressive, stimulatory, and preparative actions. Endocr Rev 21:55–89
Sarkar P, Sarkar S, Ramesh V, Hayes BE, Thomas RL et al (2006) Proteomic analysis of mice hippocampus in simulated microgravity environment. J Proteome Res 5:548–553
Scheer FA, Hilton MF, Mantzoros CS, Shea SA (2009) Adverse metabolic and cardiovascular consequences of circadian misalignment. Proc Natl Acad Sci U S A 106:4453–4458
Scheiermann C, Gibbs J, Ince L, Loudon A (2018) Clocking in to immunity. Nat Rev Immunol 18:423–437
Seeman T, Epel E, Gruenewald T, Karlamangla A, McEwen BS (2010) Socio-economic differentials in peripheral biology: cumulative allostatic load. Ann N Y Acad Sci 1186:223–239
Seri B, Garcia-Verdugo JM, McEwen BS, Alvarez-Buylla A (2001) Astrocytes give rise to new neurons in the adult mammalian hippocampus. J Neurosci 21:7153–7160
Sheline YI, Gado MH, Price JL (1998) Amygdala core nuclei volumes are decreased in recurrent major depression. Neuroreport 9:2023–2028
Sheline YI, Barch DM, Donnelly JM, Ollinger JM, Snyder AZ, Mintun MA (2001) Increased amygdala response to masked emotional faces in depressed subjects resolves with antidepressant treatment: an fMRI study. Biol Psychiatry 50:651–658
Shonkoff JP, Boyce WT, McEwen BS (2009) Neuroscience, molecular biology, and the childhood roots of health disparities. JAMA 301:2252–2259
Shors TJ, Mathew J, Sisti HM, Edgecomb C, Beckoff S, Dalla C (2007) Neurogenesis and helplessness are mediated by controllability in males but not in females. Biol Psychiatry 62:487–495
Spiegel K, Leproult R, Van Cauter E (1999) Impact of sleep debt on metabolic and endocrine function. Lancet 354:1435–1439
Spiegel K, Tasali E, Penev P, Van Cauter E (2004) Brief communication: sleep curtailment in healthy young men is associated with decreased leptin levels, elevated ghrelin levels, and increased hunger and appetite. Ann Intern Med 141:846–850
Stahn AC, Werner A, Opatz O, Maggioni MA, Steinach M et al (2017) Increased core body temperature in astronauts during long-duration space missions. Sci Rep 7:16180
Sterling P, Eyer J (1988) Allostasis: a new paradigm to explain arousal pathology. In: Fisher S, Reason J (eds) Handbook of life stress, cognition and health. John Wiley & Sons, New York, NY, pp 629–649
Stowe RP, Sams CF, Pierson DL (2011) Adrenocortical and immune responses following short- and long-duration spaceflight. Aviat Space Environ Med 82:627–634
Strollo F, Strollo G, More M, Bollanti L, Ciarmatori A et al (1998) Hormonal adaptation to real and simulated microgravity. J Gravit Physiol 5:P89–P92

Thayer JF, Lane RD (2000) A model of neurovisceral integration in emotion regulation and dysregulation. J Affect Disord 61:201–216
Tipton CM, Greenleaf JE, Jackson CG (1996) Neuroendocrine and immune system responses with spaceflights. Med Sci Sports Exerc 28:988–998
Tomiyama AJ, O'Donovan A, Lin J, Puterman E, Lazaro A et al (2012) Does cellular aging relate to patterns of allostasis? An examination of basal and stress reactive HPA axis activity and telomere length. Physiol Behav 106:40–45
Trejo JL, Carro E, Torres-Aleman I (2001) Circulating insulin-like growth factor I mediates exercise-induced increases in the number of new neurons in the adult hippocampus. J Neurosci 21:1628–1634
Van Cauter E, Polonsky KS, Scheen AJ (1997) Roles of circadian rhythmicity and sleep in human glucose regulation. Endocr Rev 18:716–738
Vgontzas AN, Zoumakis E, Bixler EO, Lin H-M, Follett H et al (2004) Adverse effects of modest sleep restriction on sleepiness, performance, and inflammatory cytokines. J Clin Endocrinol Metab 89:2119–2126
Vyas A, Mitra R, Rao BSS, Chattarji S (2002) Chronic stress induces contrasting patterns of dendritic remodeling in hippocampal and amygdaloid neurons. J Neurosci 22:6810–6818
Wang H, van Spyk E, Liu Q, Geyfman M, Salmans ML et al (2017) Time-restricted feeding shifts the skin circadian clock and alters UVB-induced DNA damage. Cell Rep 20:1061–1072
Wellman CL (2001) Dendritic reorganization in pyramidal neurons in medial prefrontal cortex after chronic corticosterone administration. J Neurobiol 49:245–253
Wild CP (2012) The exposome: from concept to utility. Int J Epidemiol 41:24–32
Wingfield JC, Romero LM (2000) Adrenocortical responses to stress and their modulation in free-living vertebrates. In: Coping with the environment: neural and endocrine mechanisms. Oxford University Press, New York, NY, pp 211–234
Wood GE, Shors TJ (1998) Stress facilitates classical conditioning in males, but impairs classical conditioning in females through activational effects of ovarian hormones. Proc Natl Acad Sci U S A 95:4066–4071
Wood GE, Norris EH, Waters E, Stoldt JT, McEwen BS (2008) Chronic immobilization stress alters aspects of emotionality and associative learning in the rat. Behav Neurosci 122:282–292
Yoo S-S, Gujar N, Hu P, Jolesz FA, Walker MP (2007) The human emotional brain without sleep – a prefronta amygdala disconnect. Curr Biol 17:R877–RR78
Zalli A, Carvalho LA, Lin J, Hamer M, Erusalimsky JD et al (2014) Shorter telomeres with high telomerase activity are associated with raised allostatic load and impoverished psychosocial resources. Proc Natl Acad Sci U S A 111:4519–4524

Environmental Stress: Mitochondria as Targets and Stressors in Cellular Metabolism

5

Jürgen Bereiter-Hahn and Marina Jendrach

Abbreviations

ATG	Autophagy-related genes
Drp1	Dynamin related protein 1
ER/sER	Endoplasmic reticulum, smooth ER
HIF	Hypoxia inducible factor
Hsp	Heat shock protein
HUVEC	Human umbilical vein endothelial cells
I/R	Ischemia-reperfusion
IPC	Ischemic preconditioning
LACTB	Lactamase B
MAPK	Mitogen activated protein kinase
Mfn	Mitofusin
MOAS	Mitochondria on a string
Opa1	Optic atrophy protein
PASMC	Pulmonary arterial smooth muscle cells
PGC	Peroxisome proliferator-activated receptor γ (coactivator 1α and 1ß)
PHD	Prolylhydroxylase
PINK1	PTEN-induced putative kinase 1
SCN	Suprachiasmatic nuclei

J. Bereiter-Hahn (✉)
Institute for Cell Biology and Neurosciences, Frankfurt am Main, Germany
e-mail: bereiter-hahn@bio.uni-frankfurt.de

M. Jendrach
Institut für Neuropathologie, Charité—Universitätsmedizin Berlin, Berlin, Germany
e-mail: marina.jendrach@charite.de

© Springer Nature Switzerland AG 2020

A. Choukér (ed.), *Stress Challenges and Immunity in Space*,
https://doi.org/10.1007/978-3-030-16996-1_5

5.1 Introduction

On cell level "*stress consists of any physico-chemical alteration of the environment that interferes with cell functioning, potentially or actually producing damage. Stress elicits active cell responses that, according to cell type, and to type and extent of damage, aim at cell survival and/or cell – reparative response*" (Cerella et al. 2010). The quality of stress seems to be of minor significance for the resulting symptoms (Selye 1936); therefore, stresses generated on Earth could well be used as predictors for stresses expected during space missions.

Stresses are unavoidable. Keeping an organism alive requires chemical and physical reactions and these produce entropy, i.e. dysfunctional molecules. Dysfunctional genes or molecules themselves are stressors for the organism, cells, and tissues. Therefore, stresses originate not only from the environment but also from the activity of cellular organelles themselves. A large range of countermeasures against restriction of function and to cope with stresses has been developed: Repair mechanisms, increased resistance against deleterious stresses, elimination of dysfunctional molecules, organelles, and cells. These countermeasures warrant homeostasis of cells and tissues and maintenance of the physiological functions within an organism.

Mitochondria are key players in the production of physiological stresses as well as in fighting the consequences of external and internal attacks to cellular viability. Furthermore, they are also targets for external stressors like radiation damage (see Chap. 20), hypoxia (Chap. 16), and nuclear dysfunction because more than 90% of their proteins are coded within the nucleus. Mitochondria are central in the network of processes resulting in apoptosis, in aging and in neurodegenerative diseases. The relative amount of mitochondria per cell varies depending on the energy requirement of the specific cell types and mitochondria often are in close contact to structures with high demand of ATP like the endoplasmatic reticulum (ER), the basal convolute in proximal kidney tubules, synaptic structures in nerves and they are wrapped around the nucleus as seen in many cells in culture, for review see (Bereiter-Hahn et al. 2008). Although mitochondria mostly exist as single entities, they may fuse and divide continuously and by this mitochondrial material (proteins as well as DNA) underlies a continuous mixing process. Sometimes mitochondria fuse to form a few huge, branched organelles (Fig. 5.1a, b). These observations justify the term *chondriome* because all single mitochondria contribute continuously to the whole mitochondrial system within a cell.

5.2 Free Radicals: Stressors Produced Within Cells

Reactive oxygen species (ROS) are produced by mitochondria and peroxisomes, whereas mitochondria generate the majority of ROS in the respiratory chain complexes I and III. ROS can act as signaling molecules as well as damage molecules, depending on the ROS species and concentration (Scialo et al. 2013). One of the main targets of ROS are the mitochondria themselves. Damage to mitochondria

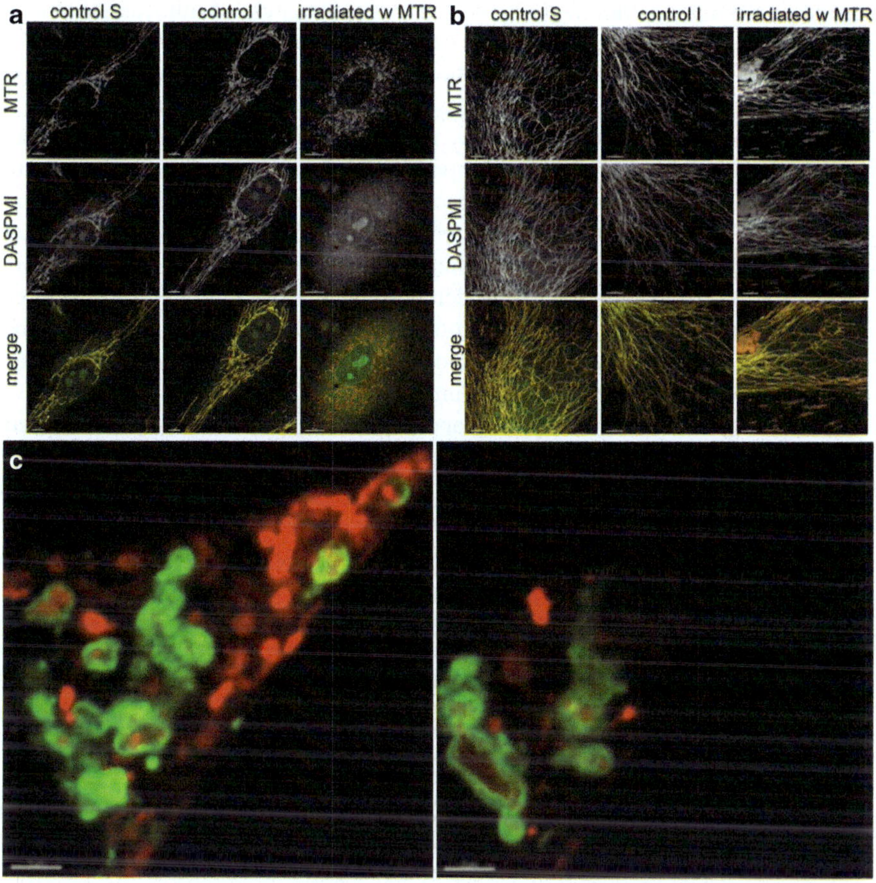

Fig. 5.1 Comparison of the reaction of mitochondria in young (proliferation active) HUVEC and senescent HUVEC (stationary, nonproliferating) to external stress: irradiation by green light which is strongly absorbed from mitochondria selectively stained with Mitotracker Red (MTR, 25 nM). Controls were stained but not irradiated (control S), or irradiated without MTR staining (control I), or MTR stained and irradiated (0.3 J/cm^2). 8 h after irradiation, the mitochondrial membrane potential was analyzed using the dye DASPMI. Both controls in both cell cultures show strong DASPMI fluorescence in the mitochondria, indicating a high membrane potential. (**a**) In proliferation of active ("young") cell cultures mitochondria appear shorter and less interconnected than in senescent cells (**b**) and 8 h after irradiation of MTR stained cells mitochondria are fragmented, appearing as red points in the MTR image and are no longer stainable with DASPMI which was released to the cytoplasm revealing loss of membrane potential. (**b**) Senescent (stationary) HUVEC cultures. Equal relative mitochondrial MTR fluorescence in young and old HUVECs was reached by an increase in MTR staining concentration for old cells from 25 to 40 nM. The very long mitochondria are intensively interconnected. After 8 h, mitochondria in both controls and irradiated cells appeared unaltered and exhibited comparable DASPMI fluorescence, indicating that the membrane potential remained intact (From Mai et al. 2010 with permission). (**c**) Young HUVEC transfected with GFP-LC3 and stained with MTR, 72 h after irradiation: Extensive mitophagy is seen: read mitochondrial remnants are surrounded by green fluorescence from GFP-LC3-II which participated in the formation of autophagic vesicles. Two confocal laser scan images taken at different focus levels from the same cell area (From Mai et al. 2012 Suppl., with permission)

increases ROS production and this creates a vicious circle of mitochondrial ROS production and ROS-induced oxidative damage to mitochondria, which can contribute to aging as was summarized by the mitochondrial free-radical hypothesis of aging (Harman 1956, 1972).

Mitochondria are especially sensitive to ROS due to their proximity to the place of its generation. Furthermore, mitochondrial DNA (mtDNA) is not protected by histones and it possesses poor repair capacities compared to nuclear DNA (Weissman et al. 2007). This correlates with data implying that damaged mtDNA can induce premature aging in animal models (Trifunovic et al. 2004, 2005). Numerous experimental studies in old cells and tissues demonstrate increased ROS levels, mutations of mtDNA, an increase of oxidatively modified cellular and mitochondrial proteins, and increased accumulation of damaged mitochondria, thus supporting the mitochondrial free-radical hypothesis of aging.

As a comparison between young and old, postmitotic human umbilical vein endothelial cells (HUVEC) showed an age-dependent increase of ROS. Consistent with these data, old cells exhibited decreased mitochondrial membrane potential and increased oxidative damage to the mtDNA. Furthermore, cellular proteins of old HUVEC were also characterized by oxidative modifications (carbonylation, AGE modifications) (Unterluggauer et al. 2007; Jendrach et al. 2005). Interestingly, fibroblasts isolated from long- and short-lived birds (pigeon embryonic fibroblast (PEF) and chicken embryonic fibroblasts (CEF), respectively) showed no increase in ROS in old age and in correlation with this data no increase in ROS-mediated damage (carbonylated proteins and damage to mtDNA) was detected in old avian cells (Strecker et al. 2010).

A direct test of the mitochondrial free-radical hypothesis of aging was a short-term treatment of young HUVEC with hydrogen peroxide. The transient increase in the ROS content after hydrogen peroxide addition caused a concentration-dependent induction of deletions and strand breaks of the mtDNA. Consistent with data showing premature aging as a result of mtDNA damage (Trifunovic et al. 2004, 2005), replicative lifespan of the cell population treated with hydrogen peroxide was evidently reduced compared to untreated cells in vitro (Jendrach et al. 2008). Thus, these findings support the mitochondrial free-radical hypothesis of aging for human cells in vitro and imply an important role for ROS in human aging.

5.3 Fission and Fusion of Mitochondria Represent Dual Stress Responses

Following the rule that the extent of stress and not the quality of stress determines the reaction, responses of endothelial cells to a very general stimulation—shear stress—can be compared to a stress regime hitting selectively mitochondria in endothelial cells. Bretón-Romero et al. (2014) exposed endothelial cells of various origins (human umbilical vein, bovine aorta) to fluid shear stress (12 dyn/cm^2). This treatment resulted in mitochondrial fission within 30 min, increased ROS production, decreased respiration, and increase in mitochondrial membrane potential. Fission is preceded by recruitment of dynamin-related protein 1 (Drp1) to the outer mitochondrial membrane, most probably as a consequence of the higher level of Ca^{2+} provoked by the shear stress acting along the axis ROS—mitochondrial Ca-uniporter (Alevriadou et al. 2017). Fragmentation of mitochondria by Drp1-mediated fission is a very general reaction following stress exposure of cells (Youle and Van Der Bliek 2012) and may precede apoptosis. However, apoptosis is not a necessary consequence of mitochondrial fragmentation, and fission is not a one-way process as was shown by application of ROS to the culture medium of human endothelial cells. This induces extensive fragmentation of mitochondria, but within 2 days fragmented mitochondria either became degraded by mitophagy or they fused again and by this return to "normality" when fusion and fission events balance each other. On the transcriptional level, fusion and fission factors become elevated and Peroxisome proliferator-activated receptor gamma coactivator 1 (PGC-1) based mitochondrial biogenesis may be stimulated demonstrating the transient character of mitochondrial impairment under the experimental stimuli used (Jendrach et al. 2008).

Fusion and fission underlie both the exchange of proteins and mtDNA between mitochondria (Muster et al. 2010; Busch et al. 2006; Sato et al. 2006; Ono et al. 2001; Ishihara et al. 2003; Arimura et al. 2004). This immense mixing process runs through the filter of mitochondrial membrane potential, only those parts retain the ability to fuse and to become reintegrated into the chondriome which are able to develop membrane potential (Priault et al. 2005; Elmore et al. 2001; Legros et al. 2002) while dysfunctional mitochondria are prone to degradation via autophagy (Twig et al. 2008; Mai et al. 2012).

After depolarization, Parkin—which plays a critical role in ubiquitination—is recruited and binds to stabilized PTEN-induced putative kinase 1 (PINK1) on the outer mitochondrial membrane (Lazarou et al. 2012; Okatsu et al. 2015), initiating selective elimination of dysfunctional parts of mitochondria (Narendra et al. 2010; Bereiter-Hahn and Jendrach 2010). The mechanism of gathering dysfunctional proteins to one zone which then can be selected and eliminated is unclear. Measurements of protein diffusion within mitochondria (Sukhorukov et al. 2010; Sukhorukov and Bereiter-Hahn 2009) revealed the existence of mobile and immobile fractions, but these data were not achieved in relation to functionality. Monte Carlo simulations using computational randomness algorithms showed that cristae junctions are prone to act as regulators for protein diffusion within the membrane (Sukhorukov and Bereiter-Hahn 2009). But nothing is known whether these junctions act to trap dysfunctional proteins.

Elongation of mitochondria has been found to provide increased resistance against disturbing metabolic conditions, and prevents from autophagic degradation (Gomes et al. 2011). Senescent endothelial cells no longer shuttle their mitochondria between cell center and periphery. Endothelial cells in culture are flat and well spread, covering a large area, which is much larger in old cells than in those still proliferating. Two adaptive mechanisms improve survival with this microstructure: ATP production largely is fueled by glycolysis (Bereiter-Hahn et al. 1995; Bretón-Romero et al. 2014; Tillmann and Bereiter-Hahn 1986) and in senescent cells mitochondria form a stable network of very long and interconnected mitochondria (Fig. 5.1b). Fission is strongly reduced (Jendrach et al. 2005), thus this chondriome must survive with its dysfunctional proteins. Cessation of intracellular mitochondrial trafficking preserves these elongated and interconnected mitochondria from PINK1/Parkin-mediated accumulation in the perinuclear region associated with autophagy (Vives-Bauza et al. 2010). On the other hand the extremely large size achieved by extensive fusion spreads electron transport and ATP production throughout the whole chondriome, i.e. the proton gradient can spread over the whole structure and where ever intact oxidative phosphorylation sites are present, those are fueled to produce ATP. A simple experiment visualizes this situation (Fig. 5.1): When irradiated with green light, mitochondria selectively stained with the fluorochrome "Mitotracker Red" loose membrane potential and fragment in a Drp1-dependent manner in proliferating cells with many medium sized or small mitochondria, and extensive mitophagy is evoked (Fig. 5.1c), whilst the large mitochondrial network in senescent cells remains stable and keeps a high level of membrane potential (Fig. 5.1b). Thus the extensive connection compensates the irradiation damage. The mechanism of this behavior is based on upregulation of the kinase PINK1 following aging-related downregulation of Drp1 (Mai et al. 2010). PINK1 becomes bound to the outer mitochondrial membrane and phosphorylates the mitochondrial chaperon TRAP1 which confers resistance against oxidative stress-induced apoptosis and reduces mitochondrial ROS production (Hua et al. 2007).

Elongation of mitochondria not only results from downregulation of Drp1 but also requires the interaction mitofusin (MFN1) and optic atrophy protein 1 (OPA1) in mammals (Romanello and Sandri 2015). The fact that OPA1 acts as a physical sensitivity sensor in skeletal muscle and its deficiency promotes degeneration of mitochondria, sarcopenia, and premature aging (Tezze et al. 2017) underlines the protective role of mitochondrial elongation increasing stress resistance. OPA1 deficiency only in skeletal muscle is sufficient to induce a systemic catabolic program and early death via ER stress (Tezze et al. 2017).

A molecule which is intensively discussed for its proposed anti-aging activities is resveratrol (see Chaps. 13 and 33). A recent study on the cardiomyocyte cell line H9c2 which can be transferred to senescence in presence of D-galactose reveals—in addition to its multiple effects on mitochondrial mass—the strong influence of resveratrol on the control of mitochondrial fission and fusion (Ren et al. 2017): Extensive mitochondrial elongations are considered to reduce the efficiency of ischemic preconditioning (IPC) in senescent cardiomyocytes. Resveratrol restores IPC activity in D-galactose-induced senescence in this cardiomyocyte cell line by increasing Drp1 expression. In this situation senescent cells also become more sensitive against Carbonylcyanide 3-chlorophenylhydrazone (CCCP)-induced depolarization (Ren et al. 2017).

5.4 Hypometabolism as a Countermeasure to Avoid Mitochondrial Stresses

The differentiation and proliferative status of cells is closely linked to energy metabolism. Mitochondria play a central role in this control process. Aged cells or quiescent cells are characterized by low metabolism but can be activated for entering a proliferative state. Among mitochondrial proteins the LACTB/PSID equilibrium

(Keckesova et al. 2017; Torrano and Carracedo 2017) is a key player for quiescence/proliferation transition. The mitochondrial protein lactamase beta (LACTB) potently inhibits proliferation of tumor cells and prevents formation of tumor cells from normal tissue by altering mitochondrial lipid metabolism via reduction of mitochondrial phosphatidylethanolamine synthesis (Keckesova et al. 2017). LACTB is a protease with functional connection to phosphatidylserin decarboxylase (PSID). Cells over-expressing LACTB show reduced mitochondrial membrane potential and ATP levels and increased ROS production. Evasion from this metabolic state typical for proliferative quiescence can be supported either by downregulating LACTB or disconnecting PISD from the negative control by LACTB (Torrano and Carracedo 2017).

Organisms might be exposed to extreme environmental situations, i.e. heat or cold, lack of nutrients, long lasting diving and apnea or any forms of lack of oxygen. In all these situations reduction of metabolism provides a chance to survive such adverse conditions as long as they are transient (Storey 2015; Gorr 2017). Hypometabolism is a physiological state with reduction of protein synthesis, growth and proliferation activities on the energy consumer side and corresponding reduction of ATP delivery on the side of "energy production" by metabolic pathways producing ATP. Energy demand is very different in different organs; therefore, hypometabolism is not a general property of the same extent in all organs of an organism, even not within all tissues of an organ. Functional reductions of some organs are life threatening (e.g. brain, heart) while others can be tolerated much better, i.e. in skeletal muscle. Mitochondria are the main source of ATP in aerobic organisms. Their activity will be restricted by low oxygen concentration (in case of hypoxia) as well by internal organismic factors reducing electron transport and ATP production. These intrinsic conditions are responsible for hypometabolism in response to fasting or by torpor in estivating and hibernating animals.

5.4.1 Hypoxia

The basic form of forced hypometabolism is hypoxia (see Chap. 16). However, hypoxia cannot be defined by a single limit of oxygen concentration. Isolated mitochondria, regardless of their origin, may respire with almost constant speed down to very low oxygen concentrations (Gnaiger et al. 1998), whilst whole cell respiration may follow kinetics close to Michaelis Menten kinetics as we showed for adhering human proximal kidney tubule cells (hPTC) in culture (Fig. 5.2), but the distal tubule cells were almost insensitive to oxygen concentration variation and reached their 50% respiratory activity at lower oxygen saturation (19% instead of the K_m at 26% O_2 saturation of hPTC).

The difference of respiration of mitochondria and of whole cells on oxygen concentration may be due to the additional membrane barrier and surrounding cytoplasm restricting diffusion of oxygen into mitochondria and to regulation of respiration by energy-consuming activities within cells (Gnaiger et al. 1998). Within living cells many factors control respiration and which are missing after mitochondria have been isolated.

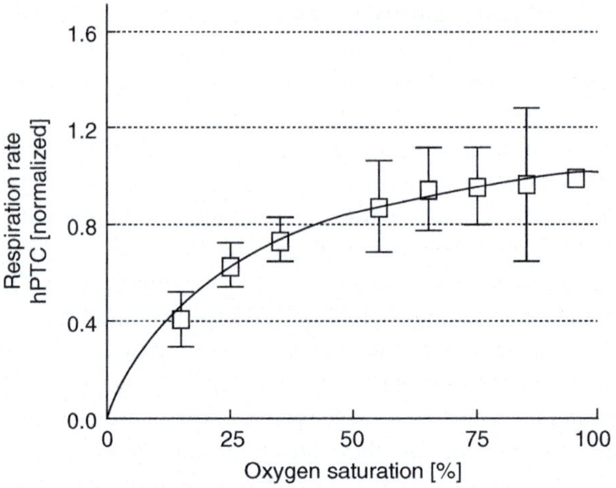

Fig. 5.2 Oxygen consumption of human proximal kidney tubule cells adhering to a solid substrate in relation of oxygen concentration of the culture medium follows Michaelis Menten kinetics. Because of the attachment to a solid surface diffusion ways of oxygen are very small and cytoskeleton organization corresponds better to the in situ situation than in suspended cells used for the Oroboros oxygraph (Luttropp et al. 2011)

Because most of the oxygen is consumed in mitochondria, these organelles are key control elements. Hypoxic responses in metabolism go in parallel with rearrangements of proteins in respiratory complexes to keep mitochondria intact during low oxygen conditions, inhibition of mitochondrial fusion due to MFN2 degradation by the mitochondrial ubiquitin ligase MARCH5 (Kim et al. 2015) and with increased mitophagy (Fuhrmann and Brune 2017). At least three organs act as sensors for whole body hypoxia in mammals, the carotid body, lung, and epidermis. In all these cases mitochondrial ROS trigger the response involving activation of hypoxia-induced factor-1 (HIF-1) (Guzy et al. 2005). Hampering electron transport by reduced oxygen supply increases ROS production. ROS induce contraction of pulmonary arterial smooth muscle cells (PASMC) by causing release of Ca^{2+} from smooth ER (sER) and thus activating myosin kinase which promotes myosin-actin-based contractions (Weir et al. 2005). This sequence is specific for PASMC, most other smooth muscle cells reduce their contractile force in response to hypoxia and by this facilitating oxygen delivery to hypoxic tissues (Weir et al. 2005).

Oxygen supply of epidermis cells in humans is provided exclusively from the surrounding air, and this cell type consumes a very small amount of oxygen (Stucker et al. 2002) but epidermis plays a pivotal role in systemic response to hypoxia (Boutin et al. 2008). Hypoxia activates dimerization of the transcription factor HIF-1α with HIF-1β and by this shifts energy metabolism from oxygen consumption towards glycolysis (Papandreou et al. 2006; Tello et al. 2011) and starts transactivation of hundreds of target genes that are vital for proliferation, glycolysis and metastasis, as are e.g. phosphoglycerate kinase-1 (PGK1), carbonic anhydrase 9 (CAIX), vascular endothelial growth factor (VEGF), glucose transporter 1 (Glut1) and plasminogen activator inhibitor 1 (Pai1) (Semenza 2012; Hamanaka et al. 2016). This process strongly depends on mitochondrial ROS production: In mouse epidermis under normoxic conditions degradation of the transcription factor HIF-1α by prolyl hydroxalases (Kaelin Jr. and Ratcliffe 2008) and ubiquitinilation keeps its level low despite continuous expression as revealed by mRNA levels. Low oxygen, mitochondrial ROS production and tricarboxylic cycle intermediates have been shown to inhibit prolylhydroxylases (PHDs) and thus stabilize HIF-1α which then can dimerize with HIF-1ß and activate HIF-target genes. In addition the dimer promotes cutaneous vasodilation directing blood flow toward the body periphery and by this potentiates the hypoxic responses in the internal organs (Hamanaka et al. 2016). Renal HIF-2α increases renal erythropoietin production and finally supports chronic hypoxia tolerance, enhances erythrocyte production and changes in metabolism which seem to be involved in reduced cancer mortality in high altitude (hypobaric conditions) (Thiersch et al. 2017) and immune sensitization (Feuerecker et al. 2018). On the other hand hypoxic areas in tumors contain elevated HIF-levels activating tumor growth while inhibition or loss of HIF-1α decreased tumor growth (Semenza 2012).

Insufficient oxygen supply is a severe health risk, i.e. for coronary artery diseases and hypertension. Ischemia and reperfusion (I/R) evoke tissue damage e.g. in brain, heart, kidney, intestine and traumatic/hemorrhagic shock. The severity of tissue damage (e.g. infarct volume) depends on the extent and duration of ischemia. Although extravasation and infarct can be identified only after start of reperfusion, the underlying events have to be described in a holistic way, understanding ischemia and reperfusion as one process. I/R represent severe mitochondrial stress on the organ and tissue level. Reduced availability of oxygen increases mitochondrial ROS production, decreases mitochondrial membrane potential and interacts with mitochondrial Ca^{2+} handling. Only high concentrations of ROS may be detrimental by oxidizing proteins, and initiating apoptosis when opening the mitochondrial permeability transition pores, whilst at low concentration ROS are important signaling molecules. Anaerobic proton production during ischemia results in acidification of the cytoplasm which is buffered by Na^+/H^+ exchange between the extracellular and intracellular space where Na^+ accumulates because Na^+/K^+ exchange does not work anymore because of reduced ATP availability (Murphy et al. 1991, 1999). In this situation the Na^+/Ca^{2+} exchange becomes reversed, Ca^{2+} enters the cytoplasm, its reuptake by the ER is hampered. ATP generated by glycolysis might be used to generate mitochondrial membrane potential used for taking up Ca^{2+} into mitochondria. On reperfusion the reestablished mitochondrial and plasmalemmal membrane potentials are used to extrude H^+ and Na^+ from the cytoplasm and Ca^{2+} from mitochondria. Close apposition of sER and mitochondria is the structural basis of regulated Ca^{2+} shuttling between both. Release of Ca^{2+} from the sER and mitochondria cause cytoplasmic overload by Ca^{2+}, inducing opening of the mitochondrial permeability transition pore (MPT), break down of mitochondrial membrane potential and finally release of cytochrome c, inducing apoptosis or other forms of cell death reviewed by (Murphy and Steenbergen 2008; Kalogeris et al. 2012). This pathway is not restricted to myocytes, it strongly affects endothelial cells as well, where it is potentiated by eNOS and by shear stress also promoting fragmentation of mitochondria (Giedt et al. 2012). This reperfusion injury is a multifactorial event comprising rise in ROS, Ca^{2+} overload, opening of the MPT (Ong et al. 2015), endothelial dysfunction and finally inflammatory responses (Kalogeris et al. 2012). However, the sensitivity of cells against I/R differs depending on their differentiation state (Hidalgo et al. 2018).

I/R-related damage can be diminished by preconditioning which may reduce ischemic acidosis, decrease ATP consumption, retard decline of mitochondrial membrane potential and release of agonists binding to G-protein receptors and activation of a series of kinases (e.g. MAPK, ERK, AKT, PKC) (Murphy and Steenbergen 2008). In humans, cardiac preconditioning is lost beyond 1 h until sustained ischemia, but a second window of increased resistance was described to occur approximately 24 h after preconditioning (Yellon and Baxter 1995; Rizvi et al. 1999). Protein kinases participate in I/R injury in a very complex manner. For cardiomyocytes, Piper et al. (2008) distinguish reperfusion salvaging kinases (RISK) and reperfusion injury-causing kinases (RICK). Inhibition of these kinases attenuates I/R injury (Kalogeris et al. 2012; Park et al. 2001). The RISK pathway PI3K-Akt-eNOS-PKG exerts salvaging function even if activated in a post-conditioning manner during reperfusion by targeting sER and mitochondria simultaneously and protecting MTP from opening (Piper et al. 2008). Another principle of preconditioning is exposure to heat which initiates increased expression of heat shock proteins (Hsp) which exert multiple salvaging activities (Hutter et al. 1994; Sammut et al. 2001; Sammut and Harrison 2003). In rats heating to >41°C for 35 min induced Hsp72 providing infarct size reduction on I/R-induced damage (Hutter et al. 1994), such a salvaging effect was also observed on overexpression of Hsp70 (Williamson et al. 2008).

Under ischemic/hypoxic situations mitochondrial functions can be stabilized with those substances reducing ROS and interfering with the deleterious pathways, e.g. by activation of Mn-superoxide dismutase (Jung et al. 2009). A substance with a broad range of activities against mitochondrial damage is melatonin. It reduces ROS production, keeps high mitochondrial membrane potential, reduces cytochrome c release and membrane permeability, and thus apoptosis. These actions have been shown to protect neuronal tissues from injury during the reperfusion phase (Hamada et al. 2010; Andrabi et al. 2004; Revuelta et al. 2016; Wakatsuki et al. 2001; Watanabe et al. 2004), however more recent results indicate that probably only astrocytes and not neurons may be rescued (Berger et al. 2016). The beneficial influence of melatonin also led to reduction of myocardial infarct in rats (Hu et al. 2017; Petrosillo et al. 2006) and reperfusion injury in liver and other (Ma et al. 2017). See in this chapter below at Sect. 5.5.

Hibernating animals regularly undergo I/R cycles. Blood flow of tissues declines dramatically during torpor, but during arousal from torpor reperfusion takes place. Hibernators during the hibernation season appear to be resistant to the deleterious effects (Kurtz et al. 2006) which even might extend to euthermic situations as was shown for the arctic ground squirrel (*Spermophillus parryii*) (Dave et al. 2006).

5.4.2 Calorie Restriction and Torpor

Calorie restriction (reduction by approx. 30% of the ad libitum level and otherwise balanced nutrition; abbrev. CR) is the most successful intervention prolonging lifespan. This effect has been shown for almost all organisms from yeast to fungi, worms, insects to humans. Also longevity of cell cultures can be prolonged by reduced glucose availability. CR decreases electron flux, mitochondrial membrane potential and ROS production and thus the deleterious actions of free radicals (Lopez-Lluch et al. 2006) while retaining ATP production. This goal is reached by activation of mitochondrial biogenesis stimulated by peroxisome proliferation activated receptor coactivator 1 (PPAR 1) and the silent information regulator (SIRT1) stimulating PGC-1α (Masuda et al. 2017). Cells shift from excessive turnover to lowered basic energy metabolism with high efficiency of increased mass of mitochondria, but less ROS production and by this attenuate age-dependent endogenous oxidative damage (Lopez-Lluch et al. 2006). Similar results have been described for humans exposed to 25% CR either with or without physical exercise. Individuals with CR showed decrease in energy expenditure and reduced DNA damage although the activity of key mitochondrial enzymes of the tricarboxylic acid (TCA) cycle and electron transport chain were unchanged, but expression of some genes (e.g. SIRT1, eNOS, PPAR, GC1A) involved in mitochondrial function was enhanced (Civitarese et al. 2007).

In parallel CR activates autophagy (Wojtovich et al. 2012) and mitophagy in particular, the second strategy attenuating oxidative stress by eliminating dysfunctional mitochondria. However, mitochondrial elimination by mitophagy may be counteracted by elongation of mitochondria: starvation increases cAMP levels activating PKA which renders Drp1 inactive by phosphorylation as was shown for mouse embryonic fibroblasts (Gomes et al. 2011).

It could be shown that overexpression of three specific autophagy genes (ATG5, ATG8F and ATG12), but not of the Lysosomal-associated membrane protein 1 (LAMP-1) resulted in an enhanced protection against oxidative stress, an increased mitochondrial fitness and an extended replicative life span of two in vitro aging cell models (HUVEC and CEF) (Mai et al. 2012; Strecker et al. 2010).

Torpor is characterized by an intrinsically controlled reduction of metabolic rate (MR) (measured as oxygen consumption per body weight; heart rate or ventilation frequency may also be used as a measure of MR), linked with a controlled decrease of body temperature, which in small mammals may closely approach low ambient temperature but in larger mammals may become reduced by a few degrees only. These changes certainly involve mitochondria. Whether mitochondria themselves play a key role in the control of torpor or whether they are primarily targets of control mechanisms (e.g. respiratory substrate availability) is still under debate. How should mitochondrial activity changes during torpor and inter-torpor bouts be characterized, what represents the correct baseline? Most studies on mitochondrial changes during torpor addressed these questions only marginally, i.e. mitochondria isolated from organs in small animals act at very low temperature, should their activity be measured at 37°C or at lower temperatures? Furthermore, metabolic activities differ among different organs. Another problem is the influence of isolation of mitochondria on their metabolic activities, which may differ considerably from the situation in situ. Measurements of oxygen consumption of tissue slices did not compare properly with state 3 and state 4 differences revealed from isolated mitochondria (Gallagher and Staples 2013).

In topor MR decrease by 65–98% below basal MR results from a suppression of body functions (demand of metabolic energy) and a reorganization of metabolic pathways i.e. a shift from glycolysis to lipolysis, reduction of mitochondrial activity, and suppression of gene expression and protein synthesis. MR reduction obviously is limited to a basic level of oxygen consumption per unit body weight common to all mammals. Seasonal hibernation and daily torpor are considered to rely on the same principles although MR in daily torpor often does not go as deep as in hibernation.

Denning bears show a moderate reduction in body temperature only, they preserve some reduced reactivity to environmental stimuli but do not interrupt their period of quiescence if not being disturbed (continuous torpor). Some authors still consider strong reduction of body temperature a key element for torpor, neglecting that reduction of body temperature follows metabolic rate decrease, which therefore represents the basal event (Heldmaier et al. 2004; Watts et al. 1981). This type of torpor is considered a useful model for humans subjected to a state of hypometabolism.

On the side of ATP availability the shift from glycolysis to lipolysis (β-oxidation) and reduced mitochondrial activity are limiting the supply. Reduction of glycolytic enzymes is achieved by posttranslational modifications. Reduction of succinate dehydrogenase (SDH) is at least in part due to high levels of oxaloacetate during torpor (Armstrong and Staples 2010). During entrance into torpor of the 13-lined ground squirrel (*Ictidomys tridecemlineatus*) respiration of liver mitochondria is rapidly suppressed by 70% relative to the interbout situation (Mathers et al. 2017). Whilst the total amount of proteins of the electron transport chain remains unchanged, electron flux through complexes I–IV and II–IV was suppressed by 40/60% respectively, but complexes III–IV and IV remained unaltered (Mathers et al. 2017). These observations coincide with reduction of state 3 respiration (state of phosphorylation) of mitochondria from the liver and skeletal muscle of 13-lined ground squirrels with succinate as a substrate (Brown et al. 2012). When measured at 37°C the decline was 70% and 30% respectively, cooling mitochondria down to 10° reduced state 3 respiration by 88% relative to IBE (interbout euthermia). Also state 4 respiration is reduced, although less.

Suppression of metabolism appears to be invoked quickly during entrance into torpor—when body temperature still is high—but reversal during arousal is slow, therefore Staples (Staples 2014) proposes enzyme-mediated posttranslational modification of electron transport complexes e.g. by phosphorylation or acetylation. This still is an open question because no torpor specific protein phosphorylation has been identified in 13-lined ground squirrel liver mitochondria, although F1-ATPase and long chain-specific acyl-CoA dehydrogenase and ornithine transcarbamylase showed seasonal (summer/winter) differences in phosphorylation of threonine and serine residues (Chung et al. 2013).

Liver mitochondria from torpid animals showed increased ROS production with succinate and malate-glutamate at 37°C but this strongly depended on temperature. Thus at low body temperature which is typical for torpid squirrels, most probably ROS production will decline to lower levels than in euthermic conditions (Brown et al. 2012). Together with increased level of antioxidants during torpor and arousal in hibernators, this explains the resistance against I/R injury mentioned above which became even more prominent by comparing preservation of liver grafts from rats and squirrels stored at 4°C up to 72 h and then reperfused with 37°C buffer in vitro. Lactate dehydrogenase (LDH) release within 60 min was the measure to assess integrity of the grafts. After 72 h cold storage, livers from hibernators showed better maintenance of mitochondrial respiration and bile production and released 37-fold less LDH than rat livers and only 1/10 of summer livers (Lindell et al. 2005). This observation is only one example for improved stress resistance of hibernator's organs. Hibernators cannot afford to arouse with dysfunctional organs, thus mechanisms have been developed for resistance against ionizing radiation and organ disuse. The exact mechanisms how this is achieved are not known yet, but some key factors can be identified: the low ROS production and general reduction of metabolism during torpor, strong antioxidant production in early arousal and stimulation of autophagy during arousal. Similarities between intermittent ischemia used for preconditioning against I/R and remodeling related to hibernation have been termed "myocardial hibernation" (Kalogeris et al. 2012), however this terminology is a bit misleading if we consider the profound changes occurring during torpor.

5.4.3 Comparison of Fasting and Torpor

In both situations (Fig. 5.3), fasting (long lasting CR) and in torpor nutrient uptake is interrupted, requiring similar adaptations of metabolism affecting mitochondria which finally support longevity and resistance against external stresses (Raffaghello et al. 2010). Comparison of metabolic rates during euthermic fasting and torpor in dwarf Siberian hamsters (*Phodopus sungorus*) and in euthermic mice showed lowered state 3 respiration in both situations but quantitatively different. The same was true for ROS production with glutamate as substrate but with succinate only mitochondria from torpid animals showed increased ROS production as 37°C (Brown et al. 2012). Shift from glycolysis to lipolysis and inhibition of growth hormone release as well as Insulin-like growth factor (IGF)-1 activity are reactions in common to both metabolic states. These result e.g. in reduction of blood glucose levels and activities of the different MAPKinases. Stabilization of blood glucose levels also under starvation is reached by serotonin release from pancreatic cells (Wyler et al. 2017). Differences are apparent for organ integrity, i.e. muscle mass and strength which will be reduced on fasting but seems to persist even during long periods of torpor in hibernation when tissue integrity is widely preserved. This difference can be explained in part by the differences in autophagy, which becomes strongly activated during fasting whilst it is reduced during torpor and becomes stimulated only during phases of arousal and readaptation to euthermic conditions.

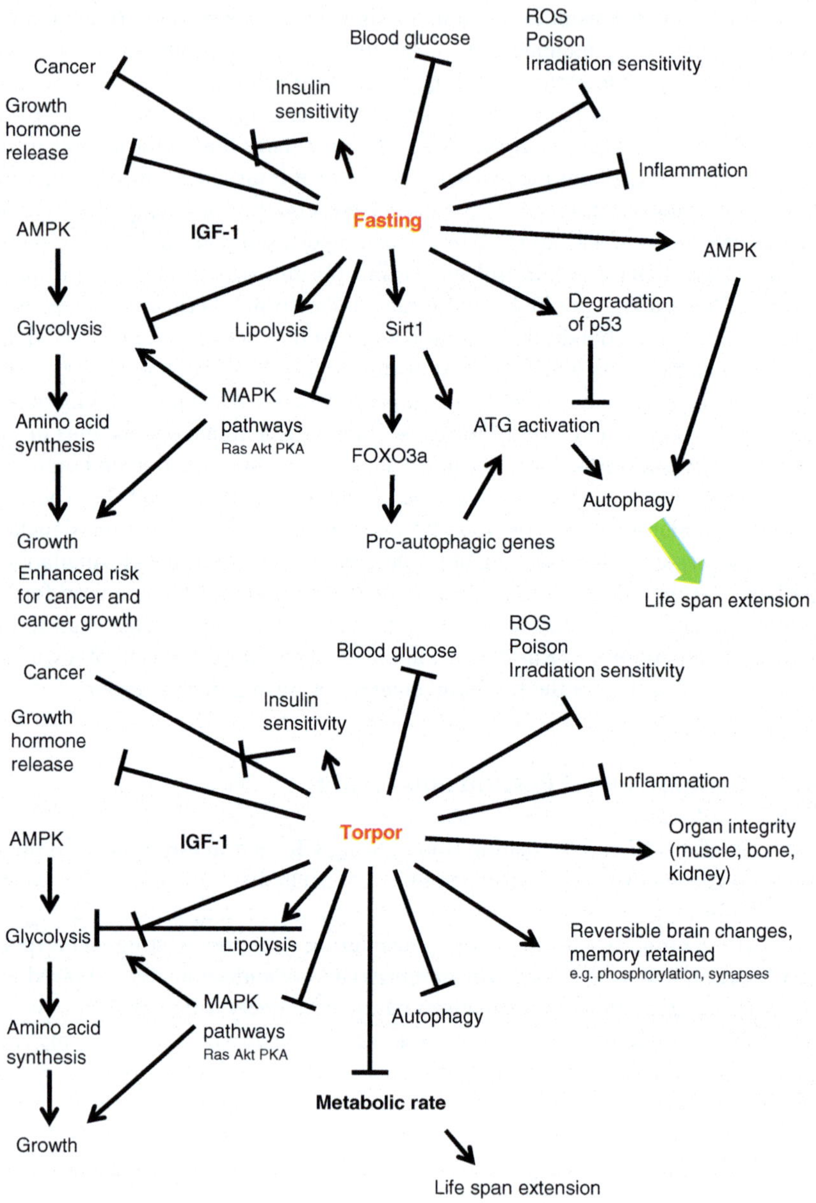

Fig. 5.3 Main metabolic pathways affected by starvation (fasting) and torpor. In both cases food intake is strongly reduced or absent which results in similar inhibitions of metabolism i.e. inhibition of glycolysis, insulin-like growth factor (IGF), and mitogen-activated protein kinase (MAPK) pathways and thus growth. Two main differences are obvious: metabolic rate becomes severely reduced in torpor but not in fasting, and autophagy is activated in fasting conditions by several pathways, the Sirt1-Foxo-way activating pro-autophagic gene, proteasomal degradation of p53 a potent inhibitor of autophagy, or via cAMP-dependent kinase (AMPK) and by the direct Sirt1-based activation of autophagy-related genes (ATG). This is not the case as long as torpor lasts; however, on arousal autophagy also increases. Both situations result in prolonged life span. The fasting way because of degradation of dysfunctional cells and their renewal when feeding situation improved, in torpor reduction of metabolic rate adds to life span

A further factor preserving tissue integrity is the high concentration of antioxidants in hibernators, of H_2S in particular. Autophagy together with apoptosis is a process supporting longevity because disintegration of dysfunctional proteins and cells acts as a rejuvenation process when followed by regeneration phase with nutrition sufficient to fuel all life processes. This is also shown by the close relationship between mitochondrial biogenesis stimulated by PGC-1α and muscle growth (Romanello et al. 2010). cAMP-dependent kinase (AMPK) can stimulate mitochondrial biogenesis via PGC-1α and as well autophagy when phosphorylated as is the case on starvation (of mice) (Romanello et al. 2010). ROS production becomes reduced on CR as well as during torpor adding to tissue integrity and this might also be one factor improving resistance against stresses i.e. improved tolerance of ionizing irradiation (see Chaps. 20 and 28).

5.5 Neurodegeneration: A Consequence of Mitochondrial Stress and Dysfunction?

In the last years mitochondria are also getting more and more in a focus of attention in aging-progressive, neurodegenerative diseases as Alzheimer's Disease (AD), Parkinson's Disease (PD), *Frontotemporal dementia* (FTD), and Huntington's Disease (HD). Studies on cells and tissues from patients with these neurodegenerative diseases, as well as work with the corresponding cell and animal models implicate mitochondrial involvement in the development and/or progression of these diseases with common features as increased ROS, impaired mitochondrial physiology, and disturbed equilibrium of mitochondrial dynamics resulting in increased cell death and neurodegeneration.

In cells derived from an AD mouse model (Swedish and London double mutant), increased levels of ROS, reduced mitochondrial membrane potential, and decreased ATP synthesis were found, with no reduction in the number of mitochondria (Hauptmann et al. 2009). Induction of ROS production by addition of respiratory complex I and III inhibitors resulted in vitro and in vivo in elevated levels of soluble amyloid beta, one of the hallmark proteins of the disease (Leuner et al. 2012). In correlation, blockage of ROS production protected against cognitive decline and oxidative stress (Manczak et al. 2010; Mcmanus et al. 2011). Increased presence of amyloid beta correlates with mitochondrial fragmentation (Calkins et al. 2011), whilst blockage of mitochondrial fission resulted in vitro and in vivo in increased mitochondrial fitness, improved behavioral, and reduced amyloid beta accumulation (Joshi et al. 2018; Baek et al. 2017; Cassidy-Stone et al. 2008).

Increased fission was observed in cell and animal models of AD (Wang et al. 2016), HD (Costa et al. 2010; Shirendeb et al. 2011; Song et al. 2011) and PD. Overexpression of alpha-synuclein, a hallmark protein in PD or the loss of the functional PD-associated proteins PINK1, Parkin, and DJ1 resulted in mitochondrial fragmentation especially under stress. Reduced activity of respiratory chain complex I and increased oxidative stress has been found in the affected brain regions of PD patients such as the substantia nigra as well as in PD-cell and animal models (Hoepken et al. 2008; Gautier et al. 2008). Furthermore, the serine-threonine kinase

PINK1 is located in the mitochondria; Mutations of this gene, e.g. the D309G point mutation, underlies the autosomal recessive PD variant PARK6 (Kessler et al. 2005; Valente et al. 2004). Accordingly, in PINK1 knockout mice a significantly reduced ATP production as well as reduced mitochondrial membrane potential was observed. Furthermore, loss of functional PINK1 conveyed increased sensitivity to oxidative and proteosomal stress, mediating fragmentation and aggregation of mitochondria (Gispert et al. 2009; Mai et al. 2010). In addition, reduction of PINK1 correlated with a downregulation of key autophagic genes, including Beclin1, LC3, and LAMP-2, giving raise to increased apoptosis in response to starvation in an amino acid-free medium (Parganlija et al. 2014).

Also models for frontotemporal dementia with Parkinsonism linked to chromosome 17 (FTDP-17) exhibit impaired mitochondrial fitness and reduced intra- and intermitochondrial dynamics as well as an increased sensitivity against oxidative stress and neurodegeneration (Schulz et al. 2012; Esteras et al. 2017). The disease hallmark protein TDP-43 aggregates in the inner mitochondrial membrane in motor neurons and cortical neurons from patients with frontotemporal dementia (FTD) or amyotrophic lateral sclerosis (ALS). These aggregates impair complex I and mitochondrial fitness, increase of ROS formation, mitochondrial fission and sensitivity against oxidative stress (Wang et al. 2016; Zhang et al. 2017).

5.6 Innate Immune Response

Peripheral macrophages remove different kinds of pathogens and debris by phagocytosis and are the main players in the nonspecific immune defense (innate immunity) by producing and releasing pro- and anti-inflammatory cytokines. In the brain neuroinflammation is observed in many neurodegenerative diseases including AD, PD, FTD and HD (Heneka et al. 2014; Heppner et al. 2015). Increased release of pro-inflammatory cytokines, mainly by microglia the CNS-resident myeloid cells, contributes to the development and the progression of the neurodegenerative pathology. One emerging function of mitochondria in the innate immune response is the regulation of one of the key pro-inflammatory cytokines: interleukin1β (IL1β). Its precursor is processed by a multi-protein complex, the so-called inflammasome where NLRP3 is the sensor component, leading to the mature cytokine (Walsh et al. 2014). Work on myeloid cells (microglia and macrophages) shows that NLRP3 is often colocalizing with mitochondria and that mitochondrial fragmentation and increased ROS production activate the inflammasome (Zhou et al. 2011; Saitoh et al. 2008; Ye et al. 2017; Lee et al. 2016; Lodder et al. 2015). Production of TNF-α, another important pro-inflammatory cytokine, is also increased by ROS via stabilization of HIF-1alpha (Wang et al. 2010). This could be mediated by increased levels of succinate, one of the products of the TCA cycle (Tannahill et al. 2013). Therapeutically in correlation with the molecular data ROS blockage (Park et al. 2015) or application of resveratrol (see Sect. 5.3) (Zhao et al. 2017) resulted in improved wound healing.

Next to ROS also ROS-independent activation of the innate immunity occurs: mitochondrial debris (e.g. mitochondrial DNA, mitochondrial N-formyl peptides)

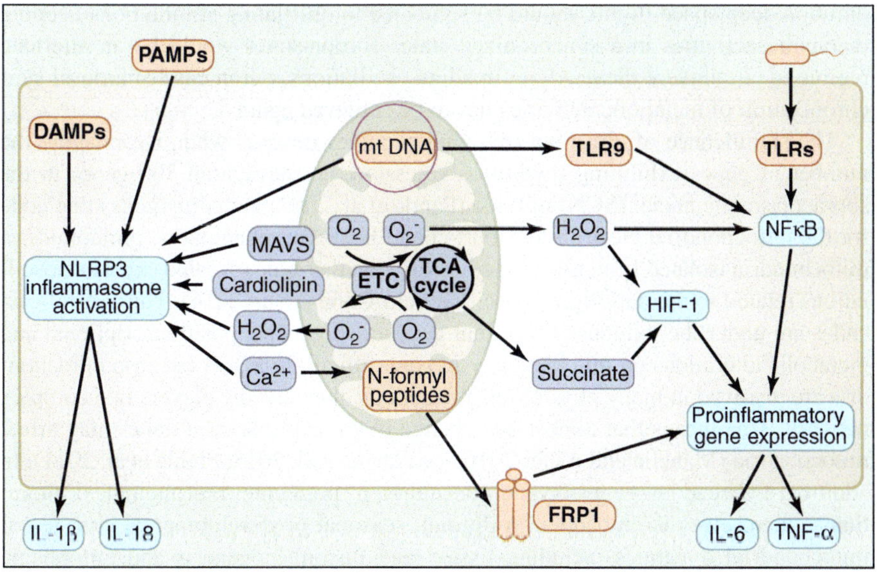

Fig. 5.4 Mitochondria in the innate immune response. Mitochondria contribute and modulate the cytokine production of macrophages and microglia in different ways. The NLRP3-inflammasome is a multiprotein complex that generates the mature forms of IL1ß and IL18 from their precursor proteins. Mitochondria-generated reactive oxygen species (ROS), released mtDNA and externalization of cardiolipin trigger activation of the inflammasome. ROS and succinate are also increasing production of TNF-α via stabilization of HIF-1α (From Weinberg et al. 2015 with permission)

activates the immune response of myeloid cells (Zhang et al. 2010). Furthermore, following loss of mitochondrial membrane potential cardiolipin is transferred from the inner to the outer mitochondrial membrane, associates with NLRP3 and stimulates thus IL1β production (Iyer et al. 2013). A new observation is that mitochondria can be transmitted via exosomes from myeloid cells to T cells, thus potentially altering the immune response within different cell populations (Hough et al. 2018). Since mitochondria play such important and multiple roles in inflammation, their role and impact has to be considered not only as a bio-energetic organelles and in their biosynthetic functions, but also as immune cell signaling elements (Fig. 5.4) (Weinberg et al. 2015).

5.7 Circadian Influences on Mitochondrial Function

Metabolic activities in organisms are modulated by seasonal and by diurnal cycles (see also Chap. 9). These cycles are adjusted to light/dark cycles in the environment. If external synchronizing oscillators are missing the cycles continue, but with changing duration. Shifts in circadian cycle phase are experienced as physiological stress, e.g. in jet lag. Total lack of circadian activity cycles impose a strong impact on physiological regulation, therefore also astronauts on a space station or on long duration deep space flights should be subjected to oscillating stimulations keeping metabolic activities in a synchronized state. Torpor states would be an alternate regime to circumvent the need of circadian oscillations, but in case of arousal synchronization of metabolic activities has to be achieved again.

The significance of circadian rhythmic becomes obvious when we imagine the number of genes exhibiting rhythmic expression, i.e. more than 300 genes in the suprachiasmatic nuclei (SCN) of mice (Panda et al. 2002) including genes that code for the mitochondrial electron transport chain. However, quantitative proteomics of mitochondria isolated from mice killed at different day time revealed extensive oscillations related to the dark/light phase (Neufeld-Cohen et al. 2016). The control cascades are under the influence of diurnal genetic, epigenetic, posttranscriptional and metabolic alterations on mitochondria and are difficult to assess because mitochondria are involved in many physiological reactions, they are key players in a complex network. This means that almost any change in the expression of genes may affect mitochondria (Manella and Asher 2016; De Goede et al. 2018; Stehle et al. 2011). In addition, the readout—e.g. oxygen consumption, proteome, morphological alterations—themselves is embedded in diurnal, seasonal or developmental changes of mitochondrial dynamics including fission and fusion, biogenesis and mitophagic degradation and locomotion from the perinuclear region towards cell periphery and reverse, and environmental factors i.e. availability of oxygen and substrates.

Because of the multiple factors controlling circadian mitochondrial activities, discussion should concentrate on downstream factors affecting mitochondria directly. Cycling of Drp1 seems to be such an immediately acting protein. Schmitt et al. (2018) demonstrated circadian ATP and Drp1 changes during 24 h after synchronisation of nonproliferating cell cultures of human skin fibroblasts and fibroblasts from Drp1-deficient or clock-deficient mice. The results argue for circadian changes mediated by Drp1. Extension of this elegant study to other cell types and to aging would further improve our understanding. Cell type specific differences in the fission and fusion dynamics are shown by maintenance of long mitochondria in dividing HUVEC as compared to reports describing extended fission of mitochondria prior to mitosis in fibroblasts (Bereiter-Hahn et al. 2008).

A further factor involved in systemic circadian rhythm control as well as with immediate influence on mitochondria is melatonin. Cyclic melatonin production in the pineal gland links light/dark cycles to endocrine signalling cascades of the organism. The nocturnal high levels of melatonin in serum result from pineal melatonin production and release. Although the pineal gland is not the only source of melatonin which can be synthesized from tryptophan within mitochondria of many organs, the pineal gland is the organ controlling circadian cycles in concert with clock genes and controlling the pacemakers in the SCN in the anterior hypothalamus. Melatonin is found in all organisms from bacteria and plants to protozoa and mammals. Corresponding to this ubiquitous distribution, a broad range of physiological activities has been ascribed to this small molecule, e.g. regulation of the sleep/wake cycle (Onder and Green 2018), of sexual reproduction cycles, retardation of aging (Paradies et al. 2017) and prevention and improvement of Type-2 diabetes mellitus (T2DM), some forms of cancer (Proietti et al. 2017) and neurodegenerative diseases

(Mayo et al. 2017; Tapias et al. 2009). All these effects have been well reviewed recently (De Goede et al. 2018; Stehle et al. 2011; Manella and Asher 2016; Mayo et al. 2017). Pineal melatonin synthesis becomes strongly reduced with age, impaired synthesis is observed in demented patients or those affected by Alzheimer's disease, Smith-Magenis syndrome, autism spectrum disorder and sleep phase disorders (Stehle et al. 2011). The common denominator of all these diseases is mitochondrial dysfunction.

Uptake of melatonin into cells and mitochondria is mediated via several receptors, including the widely distributed seven transmembrane G protein-coupled MT1 and MT2 receptors in the plasma membrane (Yasuo et al. 2009; Yang et al. 2014). Furthermore, accumulation in phospholipid layers of different cellular membranes modulate their fluidity (Dies et al. 2015; Mayo et al. 2017) and allows for diffusive uptake into the cytoplasm in addition to receptor-mediated signalling. Finally, melatonin might exert its action within the mitochondrial matrix, however, the way shuttling melatonin to the interior of mitochondria still requires clarification. Inhibition of Ca^{2+}-dependent Drp1 translocation to the mitochondria by melatonin rescues these organelles from fission (Xu et al. 2016) and thus may improve their stability and function (see Sect. 5.3).

Melatonin acts as a potent scavenger of free radicals and antioxidant in mitochondria (Tan et al. 2007; Reiter et al. 2003). It stabilizes and improves ATP production, and formation of components of the electron transport chain. The recreation potential of nocturnal sleep phases at least in part is explained by regeneration of the chondriome due to rising melatonin levels (see Chap. 9).

5.8 Summary

Stress is part of life of any organism: stresses may originate from biochemical processes within the cells as well as from impacts exerted by the environment among which disturbance of diurnal and seasonal rhythms, radiation effects, mental and physiological disturbances and organ degeneration by reduced encroachment in confinement are those of particular significance in deep space exploration. The knowledge of mutual stress responses, their mechanisms and countermeasures will be a prerequisite for successful manned space missions. Mitochondria as the key players in the complex network of cellular functions are thus preferentially suited for cellular stress management. Many neuronal diseases and pathologies of the immune system are consequences of stresses affecting mitochondrial function. On the other hand countermeasures exist against increased ROS levels, to get rid of dysfunctional mitochondria or mitochondrial components and to compensate for reduced functionality. These countermeasures are based on mitochondrial dynamics, increased production of ROS scavengers, reduction of metabolic activities (e.g. caloric restriction, torpor, preconditioning for ischemic situations), and elimination of dysfunctional mitochondria and cells via autophagy and apoptosis, respectively. Depending on the type of extra- or intracellular stress initiation respectively enhancement of these countermeasures can contribute to healthy aging as well as long-distance space travel.

Acknowledgments This compilation is based on experiments which have been funded by the following organizations: DFG (grant BE423/23-2), BMBF Grant 0312826 and Gambro Dialysatoren GmbH, Hechingen. The EU (Integrated Project MiMage CT 2004-512020), the BMBF (NGFNplus NeuroNet Parkinson and GerontoMitosys 0315584A), and the Parkinson Initiative of the University Frankfurt Medical School. Vereinigung der Freunde und Förderer der Goethe Universität. During many years of research, material has been exchanged with many colleagues for Austria, Germany, United Kingdom, and United States.

References

Alevriadou BR, Shanmughapriya S, Patel A, Stathopulos PB, Madesh M (2017) Mitochondrial Ca(2+) transport in the endothelium: regulation by ions, redox signalling and mechanical forces. J R Soc Interface 14:pii: 20170672

Andrabi SA, Sayeed I, Siemen D, Wolf G, Horn TF (2004) Direct inhibition of the mitochondrial permeability transition pore: a possible mechanism responsible for anti-apoptotic effects of melatonin. FASEB J 18:869–871

Arimura S, Yamamoto J, Aida GP, Nakazono M, Tsutsumi N (2004) Frequent fusion and fission of plant mitochondria with unequal nucleoid distribution. Proc Natl Acad Sci U S A 101:7805–7808

Armstrong C, Staples JF (2010) The role of succinate dehydrogenase and oxaloacetate in metabolic suppression during hibernation and arousal. J Comp Physiol B 180:775–783

Baek SH, Park SJ, Jeong JI, Kim SH, Han J, Kyung JW, Baik SH, Choi Y, Choi BY, Park JS, Bahn G, Shin JH, Jo DS, Lee JY, Jang CG, Arumugam TV, Kim J, Han JW, Koh JY, Cho DH, Jo DG (2017) Inhibition of Drp1 ameliorates synaptic depression, abeta deposition, and cognitive impairment in an Alzheimer's disease model. J Neurosci 37:5099–5110

Bereiter-Hahn J, Jendrach M (2010) Mitochondrial dynamics. Int Rev Cell Mol Biol 284:1–65

Bereiter-Hahn J, Stubig C, Heymann V (1995) Cell cycle-related changes in F-actin distribution are correlated with glycolytic activity. Exp Cell Res 218:551–560

Bereiter-Hahn J, Voth M, Mai S, Jendrach M (2008) Structural implications of mitochondrial dynamics. Biotechnol J 3:765–780

Berger HR, Morken TS, Vettukattil R, Brubakk AM, Sonnewald U, Wideroe M (2016) No improvement of neuronal metabolism in the reperfusion phase with melatonin treatment after hypoxic-ischemic brain injury in the neonatal rat. J Neurochem 136:339–350

Boutin AT, Weidemann A, Fu Z, Mesropian L, Gradin K, Jamora C, Wiesener M, Eckardt KU, Koch CJ, Ellies LG, Haddad G, Haase VH, Simon MC, Poellinger L, Powell FL, Johnson RS (2008) Epidermal sensing of oxygen is essential for systemic hypoxic response. Cell 133:223–234

Bretón-Romero R, Acín-Perez R, Rodríguez-Pascual F, Martínez-Molledo M, Brandes RP, Rial E, Enríquez JA, Lamas S (2014) Laminar shear stress regulates mitochondrial dynamics, bioenergetics responses and Prx3 activation in endothelial cells. Biochim Biophys Acta 1843:2403–2413

Brown JC, Chung DJ, Belgrave KR, Staples JF (2012) Mitochondrial metabolic suppression and reactive oxygen species production in liver and skeletal muscle of hibernating thirteen-lined ground squirrels. Am J Physiol Regul Integr Comp Physiol 302:R15–R28

Busch KB, Bereiter-Hahn J, Wittig I, Schagger H, Jendrach M (2006) Mitochondrial dynamics generate equal distribution but patchwork localization of respiratory complex I. Mol Membr Biol 23:509–520

Calkins MJ, Manczak M, Mao P, Shirendeb U, Reddy PH (2011) Impaired mitochondrial biogenesis, defective axonal transport of mitochondria, abnormal mitochondrial dynamics and synaptic degeneration in a mouse model of Alzheimer's disease. Hum Mol Genet 20:4515–4529

Cassidy-Stone A, Chipuk JE, Ingerman E, Song C, Yoo C, Kuwana T, Kurth MJ, Shaw JT, Hinshaw JE, Green DR, Nunnari J (2008) Chemical inhibition of the mitochondrial division dynamin

reveals its role in Bax/Bak-dependent mitochondrial outer membrane permeabilization. Dev Cell 14:193–204
Cerella C, Diederich M, Ghibelli L (2010) The dual role of calcium as messenger and stressor in cell damage, death, and survival. Int J Cell Biol 2010:546163
Chung DJ, Szyszka B, Brown JC, Huner NP, Staples JF (2013) Changes in the mitochondrial phosphoproteome during mammalian hibernation. Physiol Genomics 45:389–399
Civitarese AE, Carling S, Heilbronn LK, Hulver MH, Ukropcova B, Deutsch WA, Smith SR, Ravussin E (2007) Calorie restriction increases muscle mitochondrial biogenesis in healthy humans. PLoS Med 4:E76
Costa V, Giacomello M, Hudec R, Lopreiato R, Ermak G, Lim D, Malorni W, Davies KJ, Carafoli E, Scorrano L (2010) Mitochondrial fission and cristae disruption increase the response of cell models of Huntington's disease to apoptotic stimuli. EMBO Mol Med 2:490–503
Dave KR, Prado R, Raval AP, Drew KL, Perez-Pinzon MA (2006) The Arctic ground squirrel brain is resistant to injury from cardiac arrest during euthermia. Stroke 37:1261–1265
De Goede P, Wefers J, Brombacher EC, Schrauwen P, Kalsbeek A (2018) Circadian rhythms in mitochondrial respiration. J Mol Endocrinol 60:R115–R130
Dies H, Cheung B, Tang J, Rheinstadter MC (2015) The organization of melatonin in lipid membranes. Biochim Biophys Acta 1848:1032–1040
Elmore SP, Qian T, Grissom SF, Lemasters JJ (2001) The mitochondrial permeability transition initiates autophagy in rat hepatocytes. FASEB J 15:2286–2287
Esteras N, Rohrer JD, Hardy J, Wray S, Abramov AY (2017) Mitochondrial hyperpolarization in ipsc-derived neurons from patients of FTDP-17 with 10+16 MAPT mutation leads to oxidative stress and neurodegeneration. Redox Biol 12:410–422
Feuerecker M, Crucian BE, Quintens R, Buchheim JI, Salam AP, Rybka A, Moreels M, Strewe C, Stowe R, Mehta S, Schelling G, Thiel M, Baatout S, Sams C, Chouker A (2018) Immune sensitization during one year in the antarctic high altitude concordia environment. Allergy 74(1):64–77
Fuhrmann DC, Brune B (2017) Mitochondrial composition and function under the control of hypoxia. Redox Biol 12:208–215
Gallagher K, Staples JF (2013) Metabolism of brain cortex and cardiac muscle mitochondria in hibernating 13-lined ground squirrels *Ictidomys tridecemlineatus*. Physiol Biochem Zool 86:1–8
Gautier CA, Kitada T, Shen J (2008) Loss of PINK1 causes mitochondrial functional defects and increased sensitivity to oxidative stress. Proc Natl Acad Sci U S A 105:11364–11369
Giedt RJ, Yang C, Zweier JL, Matzavinos A, Alevriadou BR (2012) Mitochondrial fission in endothelial cells after simulated ischemia/reperfusion: role of nitric oxide and reactive oxygen species. Free Radic Biol Med 52:348–356
Gispert S, Ricciardi F, Kurz A, Azizov M, Hoepken HH, Becker D, Voos W, Leuner K, Muller WE, Kudin AP, Kunz WS, Zimmermann A, Roeper J, Wenzel D, Jendrach M, Garcia-Arencibia M, Fernandez-Ruiz J, Huber L, Rohrer H, Barrera M, Reichert AS, Rub U, Chen A, Nussbaum RL, Auburger G (2009) Parkinson phenotype in aged PINK1-deficient mice is accompanied by progressive mitochondrial dysfunction in absence of neurodegeneration. PLoS One 4:E5777
Gnaiger E, Lassnig B, Kuznetsov AV, Margreiter R (1998) Mitochondrial respiration in the low oxygen environment of the cell. Effect of ADP on oxygen kinetics. Biochim Biophys Acta 1365:249–254
Gomes LC, Di Benedetto G, Scorrano L (2011) During autophagy mitochondria elongate, are spared from degradation and sustain cell viability. Nat Cell Biol 13:589–598
Gorr TA (2017) Hypometabolism as the ultimate defence in stress response: how the comparative approach helps understanding of medically relevant questions. Acta Physiol (Oxf) 219:409–440
Guzy RD, Hoyos B, Robin E, Chen H, Liu L, Mansfield KD, Simon MC, Hammerling U, Schumacker PT (2005) Mitochondrial complex III is required for hypoxia-induced ros production and cellular oxygen sensing. Cell Metab 1:401–408

Hamada F, Watanabe K, Wakatsuki A, Nagai R, Shinohara K, Hayashi Y, Imamura R, Fukaya T (2010) Therapeutic effects of maternal melatonin administration on ischemia/reperfusion-induced oxidative cerebral damage in neonatal rats. Neonatology 98:33–40

Hamanaka RB, Weinberg SE, Reczek CR, Chandel NS (2016) The mitochondrial respiratory chain is required for organismal adaptation to hypoxia. Cell Rep 15:451–459

Harman D (1956) Aging: a theory based on free radical and radiation chemistry. J Gerontol 11:298–300

Harman D (1972) The biologic clock: the mitochondria? J Am Geriatr Soc 20:145–147

Hauptmann S, Scherping I, Drose S, Brandt U, Schulz KL, Jendrach M, Leuner K, Eckert A, Muller WE (2009) Mitochondrial dysfunction: an early event in Alzheimer pathology accumulates with age in AD transgenic mice. Neurobiol Aging 30:1574–1586

Heldmaier G, Ortmann S, Elvert R (2004) Natural hypometabolism during hibernation and daily torpor in mammals. Respir Physiol Neurobiol 141:317–329

Heneka MT, Kummer MP, Latz E (2014) Innate immune activation in neurodegenerative disease. Nat Rev Immunol 14:463–477

Heppner FL, Ransohoff RM, Becher B (2015) Immune attack: the role of inflammation in Alzheimer disease. Nat Rev Neurosci 16:358–372

Hidalgo A, Glass N, Ovchinnikov D, Yang S-K, Zhang X, Mazzone S, Chen C, Wolvetang E, Cooper-White J (2018) Modelling ischemia-reperfusion injury (Iri) in vitro using metabolically matured induced pluripotent stem cell-derived cardiomyocytes. Appl Bioeng 2:026102

Hoepken HH, Gispert S, Azizov M, Klinkenberg M, Ricciardi F, Kurz A, Morales-Gordo B, Bonin M, Riess O, Gasser T, Kogel D, Steinmetz H, Auburger G (2008) Parkinson patient fibroblasts show increased alpha-synuclein expression. Exp Neurol 212:307–313

Hough KP, Trevor JL, Strenkowski JG, Wang Y, Chacko BK, Tousif S, Chanda D, Steele C, Antony VB, Dokland T, Ouyang X, Zhang J, Duncan SR, Thannickal VJ, Darley-Usmar VM, Deshane JS (2018) Exosomal transfer of mitochondria from airway myeloid-derived regulatory cells to T cells. Redox Biol 18:54–64

Hu J, Zhang L, Yang Y, Guo Y, Fan Y, Zhang M, Man W, Gao E, Hu W, Reiter RJ, Wang H, Sun D (2017) Melatonin alleviates postinfarction cardiac remodeling and dysfunction by inhibiting Mst1. J Pineal Res 62. https://doi.org/10.1111/jpi.12368

Hua G, Zhang Q, Fan Z (2007) Heat shock protein 75 (Trap1) antagonizes reactive oxygen species generation and protects cells from granzyme M-mediated apoptosis. J Biol Chem 282:20553–20560

Hutter MM, Sievers RE, Barbosa V, Wolfe CL (1994) Heat-shock protein induction in rat hearts. a direct correlation between the amount of heat-shock protein induced and the degree of myocardial protection. Circulation 89:355–360

Ishihara N, Jofuku A, Eura Y, Mihara K (2003) Regulation of mitochondrial morphology by membrane potential, and Drp1-dependent division and Fzo1-dependent fusion reaction in mammalian cells. Biochem Biophys Res Commun 301:891–898

Iyer SS, He Q, Janczy JR, Elliott EI, Zhong Z, Olivier AK, Sadler JJ, Knepper-Adrian V, Han R, Qiao L, Eisenbarth SC, Nauseef WM, Cassel SL, Sutterwala FS (2013) Mitochondrial cardiolipin is required for Nlrp3 inflammasome activation. Immunity 39:311–323

Jendrach M, Pohl S, Voth M, Kowald A, Hammerstein P, Bereiter-Hahn J (2005) Morpho-dynamic changes of mitochondria during ageing of human endothelial cells. Mech Ageing Dev 126:813–821

Jendrach M, Mai S, Pohl S, Voth M, Bereiter-Hahn J (2008) Short- and long-term alterations of mitochondrial morphology, dynamics and mtDNA after transient oxidative stress. Mitochondrion 8:293–304

Joshi AU, Saw NL, Shamloo M, Mochly-Rosen D (2018) Drp1/Fis1 interaction mediates mitochondrial dysfunction, bioenergetic failure and cognitive decline in Alzheimer's disease. Oncotarget 9:6128–6143

Jung JE, Kim GS, Narasimhan P, Song YS, Chan PH (2009) Regulation of Mn-superoxide dismutase activity and neuroprotection by Stat3 in mice after cerebral ischemia. J Neurosci 29:7003–7014

Kaelin WG Jr, Ratcliffe PJ (2008) Oxygen sensing by metazoans: the central role of the hif hydroxylase pathway. Mol Cell 30:393–402

Kalogeris T, Baines CP, Krenz M, Korthuis RJ (2012) Cell biology of ischemia/reperfusion injury. Int Rev Cell Mol Biol 298:229–317

Keckesova Z, Donaher JL, De Cock J, Freinkman E, Lingrell S, Bachovchin DA, Bierie B, Tischler V, Noske A, Okondo MC, Reinhardt F, Thiru P, Golub TR, Vance JE, Weinberg RA (2017) Lactb is a tumour suppressor that modulates lipid metabolism and cell state. Nature 543:681–686

Kessler KR, Hamscho N, Morales B, Menzel C, Barrero F, Vives F, Gispert S, Auburger G (2005) Dopaminergic function in a family with the Park6 form of autosomal recessive Parkinson's syndrome. J Neural Transm (Vienna) 112:1345–1353

Kim HJ, Nagano Y, Choi SJ, Park SY, Kim H, Yao TP, Lee JY (2015) Hdac6 maintains mitochondrial connectivity under hypoxic stress by suppressing March5/Mitol dependent Mfn2 degradation. Biochem Biophys Res Commun 464:1235–1240

Kurtz CC, Lindell SL, Mangino MJ, Carey HV (2006) Hibernation confers resistance to intestinal ischemia-reperfusion injury. Am J Physiol Gastrointest Liver Physiol 291:G895–G901

Lazarou M, Jin SM, Kane LA, Youle RJ (2012) Role of PINK1 binding to the Tom complex and alternate intracellular membranes in recruitment and activation of the E3 ligase Parkin. Dev Cell 22:320–333

Lee HY, Kim J, Quan W, Lee JC, Kim MS, Kim SH, Bae JW, Hur KY, Lee MS (2016) Autophagy deficiency in myeloid cells increases susceptibility to obesity-induced diabetes and experimental colitis. Autophagy 12:1390–1403

Legros F, Lombes A, Frachon P, Rojo M (2002) Mitochondrial fusion in human cells is efficient, requires the inner membrane potential, and is mediated by mitofusins. Mol Biol Cell 13:4343–4354

Leuner K, Schutt T, Kurz C, Eckert SH, Schiller C, Occhipinti A, Mai S, Jendrach M, Eckert GP, Kruse SE, Palmiter RD, Brandt U, Drose S, Wittig I, Willem M, Haass C, Reichert AS, Muller WE (2012) Mitochondrion-derived reactive oxygen species lead to enhanced amyloid beta formation. Antioxid Redox Signal 16:1421–1433

Lindell SL, Klahn SL, Piazza TM, Mangino MJ, Torrealba JR, Southard JH, Carey HV (2005) Natural resistance to liver cold ischemia-reperfusion injury associated with the hibernation phenotype. Am J Physiol Gastrointest Liver Physiol 288:G473–G480

Lodder J, Denaes T, Chobert MN, Wan J, El-Benna J, Pawlotsky JM, Lotersztajn S, Teixeira-Clerc F (2015) Macrophage autophagy protects against liver fibrosis in mice. Autophagy 11:1280–1292

Lopez-Lluch G, Hunt N, Jones B, Zhu M, Jamieson H, Hilmer S, Cascajo MV, Allard J, Ingram DK, Navas P, De Cabo R (2006) Calorie restriction induces mitochondrial biogenesis and bioenergetic efficiency. Proc Natl Acad Sci U S A 103:1768–1773

Luttropp D, Schade M, Baer PC, Bereiter-Hahn J (2011) Respiration rate in human primary renal proximal and early distal tubular cells in vitro: considerations for biohybrid renal devices. Biotechnol Prog 27:262–268

Ma Z, Xin Z, Di W, Yan X, Li X, Reiter RJ, Yang Y (2017) Melatonin and mitochondrial function during ischemia/reperfusion injury. Cell Mol Life Sci 74:3989–3998

Mai S, Klinkenberg M, Auburger G, Bereiter-Hahn J, Jendrach M (2010) Decreased expression of Drp1 and Fis1 mediates mitochondrial elongation in senescent cells and enhances resistance to oxidative stress through PINK1. J Cell Sci 123:917–926

Mai S, Muster B, Bereiter-Hahn J, Jendrach M (2012) Autophagy proteins Lc3b, Atg5 and Atg12 participate in quality control after mitochondrial damage and influence life span. Autophagy 8:47–62

Manczak M, Mao P, Calkins MJ, Cornea A, Reddy AP, Murphy MP, Szeto HH, Park B, Reddy PH (2010) Mitochondria-targeted antioxidants protect against amyloid-beta toxicity in Alzheimer's disease neurons. J Alzheimers Dis 20(Suppl 2):S609–S631

Manella G, Asher G (2016) The circadian nature of mitochondrial biology. Front Endocrinol (Lausanne) 7:162

Masuda K, Jue T, Ray HRD (2017) Mitochondrial biogenesis induced by exercise and nutrients: implication for performance and health benefits. Indo J Sci Technol 2(2):2528-1410

Mathers KE, Mcfarlane SV, Zhao L, Staples JF (2017) Regulation of mitochondrial metabolism during hibernation by reversible suppression of electron transport system enzymes. J Comp Physiol B 187:227–234

Mayo JC, Sainz RM, Gonzalez-Menendez P, Hevia D, Cernuda-Cernuda R (2017) Melatonin transport into mitochondria. Cell Mol Life Sci 74:3927–3940

Mcmanus MJ, Murphy MP, Franklin JL (2011) The mitochondria-targeted antioxidant mitoq prevents loss of spatial memory retention and early neuropathology in a transgenic mouse model of Alzheimer's disease. J Neurosci 31:15703–15715

Murphy E, Steenbergen C (2008) Mechanisms underlying acute protection from cardiac ischemia-reperfusion injury. Physiol Rev 88:581–609

Murphy E, Perlman M, London RE, Steenbergen C (1991) Amiloride delays the ischemia-induced rise in cytosolic free calcium. Circ Res 68:1250–1258

Murphy E, Cross H, Steenbergen C (1999) Sodium regulation during ischemia versus reperfusion and its role in injury. Circ Res 84:1469–1470

Muster B, Kohl W, Wittig I, Strecker V, Joos F, Haase W, Bereiter-Hahn J, Busch K (2010) Respiratory chain complexes in dynamic mitochondria display a patchy distribution in life cells. PLoS One 5:E11910

Narendra DP, Jin SM, Tanaka A, Suen DF, Gautier CA, Shen J, Cookson MR, Youle RJ (2010) PINK1 is selectively stabilized on impaired mitochondria to activate Parkin. PLoS Biol 8:E1000298

Neufeld-Cohen A, Robles MS, Aviram R, Manella G, Adamovich Y, Ladeuix B, Nir D, Rousso-Noori L, Kuperman Y, Golik M, Mann M, Asher G (2016) Circadian control of oscillations in mitochondrial rate-limiting enzymes and nutrient utilization by period proteins. Proc Natl Acad Sci U S A 113:E1673–E1682

Okatsu K, Kimura M, Oka T, Tanaka K, Matsuda N (2015) Unconventional PINK1 localization to the outer membrane of depolarized mitochondria drives Parkin recruitment. J Cell Sci 128:964–978

Onder Y, Green CB (2018) Rhythms of metabolism in adipose tissue and mitochondria. Neurobiol Sleep Circadian Rhythms 4:57–63

Ong SB, Samangouei P, Kalkhoran SB, Hausenloy DJ (2015) The mitochondrial permeability transition pore and its role in myocardial ischemia reperfusion injury. J Mol Cell Cardiol 78:23–34

Ono T, Isobe K, Nakada K, Hayashi JI (2001) Human cells are protected from mitochondrial dysfunction by complementation of DNA products in fused mitochondria. Nat Genet 28:272–275

Panda S, Antoch MP, Miller BH, Su AI, Schook AB, Straume M, Schultz PG, Kay SA, Takahashi JS, Hogenesch JB (2002) Coordinated transcription of key pathways in the mouse by the circadian clock. Cell 109:307–320

Papandreou I, Cairns RA, Fontana L, Lim AL, Denko NC (2006) Hif-1 mediates adaptation to hypoxia by actively downregulating mitochondrial oxygen consumption. Cell Metab 3:187–197

Paradies G, Paradies V, Ruggiero FM, Petrosillo G (2017) Mitochondrial bioenergetics decay in aging: beneficial effect of melatonin. Cell Mol Life Sci 74:3897–3911

Parganlija D, Klinkenberg M, Dominguez-Bautista J, Hetzel M, Gispert S, Chimi MA, Drose S, Mai S, Brandt U, Auburger G, Jendrach M (2014) Loss of PINK1 impairs stress-induced autophagy and cell survival. PLoS One 9:E95288

Park KM, Chen A, Bonventre JV (2001) Prevention of kidney ischemia/reperfusion-induced functional injury and Jnk, P38, and Mapk kinase activation by remote ischemic pretreatment. J Biol Chem 276:11870–11876

Park JH, Kang SS, Kim JY, Tchah H (2015) The antioxidant N-acetylcysteine inhibits inflammatory and apoptotic processes in human conjunctival epithelial cells in a high-glucose environment. Invest Ophthalmol Vis Sci 56:5614–5621

Petrosillo G, Di Venosa N, Pistolese M, Casanova G, Tiravanti E, Colantuono G, Federici A, Paradies G, Ruggiero FM (2006) Protective effect of melatonin against mitochondrial dysfunction associated with cardiac ischemia-reperfusion: role of cardiolipin. FASEB J 20:269–276

Piper HM, Abdallah Y, Kasseckert S, Schluter KD (2008) Sarcoplasmic reticulum-mitochondrial interaction in the mechanism of acute reperfusion injury. viewpoint. Cardiovasc Res 77:234–236

Priault M, Salin B, Schaeffer J, Vallette FM, Di Rago JP, Martinou JC (2005) Impairing the bioenergetic status and the biogenesis of mitochondria triggers mitophagy in yeast. Cell Death Differ 12:1613–1621

Proietti S, Cucina A, Minini M, Bizzarri M (2017) Melatonin, mitochondria, and the cancer cell. Cell Mol Life Sci 74:4015–4025

Raffaghello L, Safdie F, Bianchi G, Dorff T, Fontana L, Longo VD (2010) Fasting and differential chemotherapy protection in patients. Cell Cycle 9:4474–4476

Reiter RJ, Tan DX, Mayo JC, Sainz RM, Leon J, Czarnocki Z (2003) Melatonin as an antioxidant: biochemical mechanisms and pathophysiological implications in humans. Acta Biochim Pol 50:1129–1146

Ren X, Chen L, Xie J, Zhang Z, Dong G, Liang J, Liu L, Zhou H, Luo P (2017) Resveratrol ameliorates mitochondrial elongation via Drp1/Parkin/PINK1 signaling in senescent-like cardiomyocytes. Oxidative Med Cell Longev 2017:4175353

Revuelta M, Arteaga O, Montalvo H, Alvarez A, Hilario E, Martinez-Ibarguen A (2016) Antioxidant treatments recover the alteration of auditory-evoked potentials and reduce morphological damage in the inferior colliculus after perinatal asphyxia in rat. Brain Pathol 26:186–198

Rizvi A, Tang XL, Qiu Y, Xuan YT, Takano H, Jadoon AK, Bolli R (1999) Increased protein synthesis is necessary for the development of late preconditioning against myocardial stunning. Am J Phys 277:H874–H884

Romanello V, Sandri M (2015) Mitochondrial quality control and muscle mass maintenance. Front Physiol 6:422

Romanello V, Guadagnin E, Gomes L, Roder I, Sandri C, Petersen Y, Milan G, Masiero E, Del Piccolo P, Foretz M, Scorrano L, Rudolf R, Sandri M (2010) Mitochondrial fission and remodelling contributes to muscle atrophy. EMBO J 29:1774–1785

Saitoh T, Fujita N, Jang MH, Uematsu S, Yang BG, Satoh T, Omori H, Noda T, Yamamoto N, Komatsu M, Tanaka K, Kawai T, Tsujimura T, Takeuchi O, Yoshimori T, Akira S (2008) Loss of the autophagy protein Atg16l1 enhances endotoxin-induced IL-1beta production. Nature 456:264–268

Sammut IA, Harrison JC (2003) Cardiac mitochondrial complex activity is enhanced by heat shock proteins. Clin Exp Pharmacol Physiol 30:110–115

Sammut IA, Jayakumar J, Latif N, Rothery S, Severs NJ, Smolenski RT, Bates TE, Yacoub MH (2001) Heat stress contributes to the enhancement of cardiac mitochondrial complex activity. Am J Pathol 158:1821–1831

Sato A, Nakada K, Hayashi J (2006) Mitochondrial dynamics and aging: mitochondrial interaction preventing individuals from expression of respiratory deficiency caused by mutant mtDNA. Biochim Biophys Acta 1763:473–481

Schmitt K, Grimm A, Dallmann R, Oettinghaus B, Restelli LM, Witzig M, Ishihara N, Mihara K, Ripperger JA, Albrecht U, Frank S, Brown SA, Eckert A (2018) Circadian control Of Drp1 activity regulates mitochondrial dynamics and bioenergetics. Cell Metab 27:657–666.E5

Schulz KL, Eckert A, Rhein V, Mai S, Haase W, Reichert AS, Jendrach M, Muller WE, Leuner K (2012) A new link to mitochondrial impairment in tauopathies. Mol Neurobiol 46:205–216

Scialo F, Mallikarjun V, Stefanatos R, Sanz A (2013) Regulation of lifespan by the mitochondrial electron transport chain: reactive oxygen species-dependent and reactive oxygen species-independent mechanisms. Antioxid Redox Signal 19:1953–1969

Selye H (1936) A syndrome produced by diverse nocuous agents. Nature 138:32

Semenza GL (2012) Hypoxia-inducible factors in physiology and medicine. Cell 148:399–408

Shirendeb U, Reddy AP, Manczak M, Calkins MJ, Mao P, Tagle DA, Reddy PH (2011) Abnormal mitochondrial dynamics, mitochondrial loss and mutant huntingtin oligomers in Huntington's disease: implications for selective neuronal damage. Hum Mol Genet 20:1438–1455

Song W, Chen J, Petrilli A, Liot G, Klinglmayr E, Zhou Y, Poquiz P, Tjong J, Pouladi MA, Hayden MR, Masliah E, Ellisman M, Rouiller I, Schwarzenbacher R, Bossy B, Perkins G, Bossy-Wetzel

E (2011) Mutant huntingtin binds the mitochondrial fission GTPase dynamin-related protein-1 and increases its enzymatic activity. Nat Med 17:377–382
Staples JF (2014) Metabolic suppression in mammalian hibernation: the role of mitochondria. J Exp Biol 217:2032–2036
Stehle JH, Saade A, Rawashdeh O, Ackermann K, Jilg A, Sebesteny T, Maronde E (2011) A survey of molecular details in the human pineal gland in the light of phylogeny, structure, function and chronobiological diseases. J Pineal Res 51:17–43
Storey KB (2015) Regulation of hypometabolism: insights into epigenetic controls. J Exp Biol 218:150–159
Strecker V, Mai S, Muster B, Beneke S, Burkle A, Bereiter-Hahn J, Jendrach M (2010) Aging of different avian cultured cells: lack of ROS-induced damage and quality control mechanisms. Mech Ageing Dev 131:48–59
Stucker M, Struk A, Altmeyer P, Herde M, Baumgartl H, Lubbers DW (2002) The cutaneous uptake of atmospheric oxygen contributes significantly to the oxygen supply of human dermis and epidermis. J Physiol 538:985–994
Sukhorukov VM, Bereiter-Hahn J (2009) Anomalous diffusion induced by cristae geometry in the inner mitochondrial membrane. PLoS One 4:E4604
Sukhorukov VM, Dikov D, Busch K, Strecker V, Wittig I, Bereiter-Hahn J (2010) Determination of protein mobility in mitochondrial membranes of living cells. Biochim Biophys Acta 1798:2022–2032
Tan DX, Manchester LC, Terron MP, Flores LJ, Reiter RJ (2007) One molecule, many derivatives: a never-ending interaction of melatonin with reactive oxygen and nitrogen species? J Pineal Res 42:28–42
Tannahill GM, Curtis AM, Adamik J, Palsson-Mcdermott EM, Mcgettrick AF, Goel G, Frezza C, Bernard NJ, Kelly B, Foley NH, Zheng L, Gardet A, Tong Z, Jany SS, Corr SC, Haneklaus M, Caffrey BE, Pierce K, Walmsley S, Beasley FC, Cummins E, Nizet V, Whyte M, Taylor CT, Lin H, Masters SL, Gottlieb E, Kelly VP, Clish C, Auron PE, Xavier RJ, O'neill LA (2013) Succinate is an inflammatory signal that induces IL-1beta through HIF-1alpha. Nature 496:238–242
Tapias V, Escames G, Lopez LC, Lopez A, Camacho E, Carrion MD, Entrena A, Gallo MA, Espinosa A, Acuna-Castroviejo D (2009) Melatonin and its brain metabolite N(1)-acetyl-5-methoxykynuramine prevent mitochondrial nitric oxide synthase induction in Parkinsonian mice. J Neurosci Res 87:3002–3010
Tello D, Balsa E, Acosta-Iborra B, Fuertes-Yebra E, Elorza A, Ordonez A, Corral-Escariz M, Soro I, Lopez-Bernardo E, Perales-Clemente E, Martinez-Ruiz A, Enriquez JA, Aragones J, Cadenas S, Landazuri MO (2011) Induction of the mitochondrial Ndufa4l2 protein by HIF-1alpha decreases oxygen consumption by inhibiting complex I activity. Cell Metab 14:768–779
Tezze C, Romanello V, Desbats MA, Fadini GP, Albiero M, Favaro G, Ciciliot S, Soriano ME, Morbidoni V, Cerqua C, Loefler S, Kern H, Franceschi C, Salvioli S, Conte M, Blaauw B, Zampieri S, Salviati L, Scorrano L, Sandri M (2017) Age-associated loss of OPA1 in muscle impacts muscle mass, metabolic homeostasis, systemic inflammation, and epithelial senescence. Cell Metab 25:1374–1389.E6
Thiersch M, Swenson ER, Haider T, Gassmann M (2017) Reduced cancer mortality at high altitude: the role of glucose, lipids, iron and physical activity. Exp Cell Res 356:209–216
Tillmann U, Bereiter-Hahn J (1986) Relation of actin fibrils to energy metabolism of endothelial cells. Cell Tissue Res 243:579–585
Torrano V, Carracedo A (2017) Quiescence-like metabolism to push cancer out of the race. Cell Metab 25:997–999
Trifunovic A, Wredenberg A, Falkenberg M, Spelbrink JN, Rovio AT, Bruder CE, Bohlooly YM, Gidlof S, Oldfors A, Wibom R, Tornell J, Jacobs HT, Larsson NG (2004) Premature ageing in mice expressing defective mitochondrial DNA polymerase. Nature 429:417–423
Trifunovic A, Hansson A, Wredenberg A, Rovio AT, Dufour E, Khvorostov I, Spelbrink JN, Wibom R, Jacobs HT, Larsson NG (2005) Somatic mtDNA mutations cause aging phenotypes without affecting reactive oxygen species production. Proc Natl Acad Sci U S A 102:17993–17998

Twig G, Elorza A, Molina AJ, Mohamed H, Wikstrom JD, Walzer G, Stiles L, Haigh SE, Katz S, Las G, Alroy J, Wu M, Py BF, Yuan J, Deeney JT, Corkey BE, Shirihai OS (2008) Fission and selective fusion govern mitochondrial segregation and elimination by autophagy. EMBO J 27:433–446

Unterluggauer H, Hutter E, Voglauer R, Grillari J, Voth M, Bereiter-Hahn J, Jansen-Durr P, Jendrach M (2007) Identification of cultivation-independent markers of human endothelial cell senescence in vitro. Biogerontology 8:383–397

Valente EM, Abou-Sleiman PM, Caputo V, Muqit MM, Harvey K, Gispert S, Ali Z, Del Turco D, Bentivoglio AR, Healy DG, Albanese A, Nussbaum R, Gonzalez-Maldonado R, Deller T, Salvi S, Cortelli P, Gilks WP, Latchman DS, Harvey RJ, Dallapiccola B, Auburger G, Wood NW (2004) Hereditary early-onset Parkinson's disease caused by mutations in PINK1. Science 304:1158–1160

Vives-Bauza C, Zhou C, Huang Y, Cui M, De Vries RL, Kim J, May J, Tocilescu MA, Liu W, Ko HS, Magrane J, Moore DJ, Dawson VL, Grailhe R, Dawson TM, Li C, Tieu K, Przedborski S (2010) PINK1-dependent recruitment of Parkin to mitochondria in mitophagy. Proc Natl Acad Sci U S A 107:378–383

Wakatsuki A, Okatani Y, Shinohara K, Ikenoue N, Fukaya T (2001) Melatonin protects against ischemia/reperfusion-induced oxidative damage to mitochondria in fetal rat brain. J Pineal Res 31:167–172

Walsh JG, Muruve DA, Power C (2014) Inflammasomes in the CNS. Nat Rev Neurosci 15:84–97

Wang D, Malo D, Hekimi S (2010) Elevated mitochondrial reactive oxygen species generation affects the immune response via hypoxia-inducible factor-1alpha in long-lived Mclk1+/− mouse mutants. J Immunol 184:582–590

Wang W, Wang L, Lu J, Siedlak SL, Fujioka H, Liang J, Jiang S, Ma X, Jiang Z, Da Rocha EL, Sheng M, Choi H, Lerou PH, Li H, Wang X (2016) The inhibition Of Tdp-43 mitochondrial localization blocks its neuronal toxicity. Nat Med 22:869–878

Watanabe K, Wakatsuki A, Shinohara K, Ikenoue N, Yokota K, Fukaya T (2004) Maternally administered melatonin protects against ischemia and reperfusion-induced oxidative mitochondrial damage in premature fetal rat brain. J Pineal Res 37:276–280

Watts PD, Øritsland NA, Jonkel C, Ronald K (1981) Mammalian hibernation and the oxygen consumption of a Denning Black Bear (*Ursus americanas*). Comp Biochem Physiol A Physiol 69:121–123

Weinberg SE, Sena LA, Chandel NS (2015) Mitochondria in the regulation of innate and adaptive immunity. Immunity 42:406–417

Weir EK, Lopez-Barneo J, Buckler KJ, Archer SL (2005) Acute oxygen-sensing mechanisms. N Engl J Med 353:2042–2055

Weissman L, De Souza-Pinto NC, Stevnsner T, Bohr VA (2007) DNA repair, mitochondria, and neurodegeneration. Neuroscience 145:1318–1329

Williamson CL, Dabkowski ER, Dillmann WH, Hollander JM (2008) Mitochondria protection from hypoxia/reoxygenation injury with mitochondria heat shock protein 70 overexpression. Am J Physiol Heart Circ Physiol 294:H249–H256

Wojtovich AP, Nadtochiy SM, Brookes PS, Nehrke K (2012) Ischemic preconditioning: the role of mitochondria and aging. Exp Gerontol 47:1–7

Wyler SC, Lord CC, Lee S, Elmquist JK, Liu C (2017) Serotonergic control of metabolic homeostasis. Front Cell Neurosci 11:277

Xu S, Pi H, Zhang L, Zhang N, Li Y, Zhang H, Tang J, Li H, Feng M, Deng P, Guo P, Tian L, Xie J, He M, Lu Y, Zhong M, Zhang Y, Wang W, Reiter RJ, Yu Z, Zhou Z (2016) Melatonin prevents abnormal mitochondrial dynamics resulting from the neurotoxicity of cadmium by blocking calcium-dependent translocation of Drp1 to the mitochondria. J Pineal Res 60:291–302

Yang WC, Tang KQ, Fu CZ, Riaz H, Zhang Q, Zan LS (2014) Melatonin regulates the development and function of bovine sertoli cells via its receptors MT1 and MT2. Anim Reprod Sci 147:10–16

Yasuo S, Yoshimura T, Ebihara S, Korf HW (2009) Melatonin transmits photoperiodic signals through the MT1 melatonin receptor. J Neurosci 29:2885–2889

Ye J, Jiang Z, Chen X, Liu M, Li J, Liu N (2017) The role of autophagy in pro-inflammatory responses of microglia activation via mitochondrial reactive oxygen species in vitro. J Neurochem 142(2):215–230

Yellon DM, Baxter GF (1995) A "second window of protection" or delayed preconditioning phenomenon: future horizons for myocardial protection? J Mol Cell Cardiol 27:1023–1034

Youle RJ, Van Der Bliek AM (2012) Mitochondrial fission, fusion, and stress. Science 337:1062–1065

Zhang Q, Raoof M, Chen Y, Sumi Y, Sursal T, Junger W, Brohi K, Itagaki K, Hauser CJ (2010) Circulating mitochondrial damps cause inflammatory responses to injury. Nature 464:104–107

Zhang Y, Schmid B, Nikolaisen NK, Rasmussen MA, Aldana BI, Agger M, Calloe K, Stummann TC, Larsen HM, Nielsen TT, Huang J, Xu F, Liu X, Bolund L, Meyer M, Bak LK, Waagepetersen HS, Luo Y, Nielsen JE, Holst B, Clausen C, Hyttel P, Freude KK (2017) Patient IPSC-derived neurons for disease modeling of frontotemporal dementia with mutation in CHMP2B. Stem Cell Rep 8:648–658

Zhao P, Sui BD, Liu N, Lv YJ, Zheng CX, Lu YB, Huang WT, Zhou CH, Chen J, Pang DL, Fei DD, Xuan K, Hu CH, Jin Y (2017) Anti-aging pharmacology in cutaneous wound healing: effects of metformin, resveratrol, and rapamycin by local application. Aging Cell 16:1083–1093

Zhou R, Yazdi AS, Menu P, Tschopp J (2011) A role for mitochondria in NLRP3 inflammasome activation. Nature 469:221–225

The Impact of Everyday Stressors on the Immune System and Health

6

Annina Seiler, Christopher P. Fagundes, and Lisa M. Christian

6.1 Stress, Immunity, and Health

The central nervous system (CNS), endocrine system, and immune system are complex systems that interact with each other. Stressful life events and the negative emotions they generate can dysregulate the immune response by disturbing the sensitive interplay among these systems (Glaser and Kiecolt-Glaser 2005). Psychoneuroimmunology (PNI) is a field of investigation concerned with the interactions of psychological factors with the neuroendocrine and immune system and consequences for higher brain function and human behavior (Dantzer 2010).

A stressor can be defined as an event that exceeds an individual's perceived ability to cope (Lazarus and Folkman 1984) and can result in an allostatic load and overload (see Chap. 4). Individual differences exist in the extent to which people mount a physiological stress response. Individual differences in stress physiology are, among other things, related to the brain, which plays a critical role in appraising stressors, as well as in modulating immune system reactivity to physical and

A. Seiler (✉)
Department of Consultation-Liaison-Psychiatry and Psychosomatic Medicine, University Hospital Zurich, Zurich, Switzerland
e-mail: annina.seiler@usz.ch

C. P. Fagundes
Department of Psychology, Rice University, Houston, TX, USA

Department of Behavioral Science, The University of Texas MD Anderson Cancer Center, Houston, TX, USA

L. M. Christian
Department of Psychiatry and Behavioral Health, The Ohio State University Wexner Medical Center, Columbus, OH, USA

The Institute for Behavioral Medicine Research, The Ohio State University Wexner Medical Center, Columbus, OH, USA

© The Author(s) 2020

A. Choukér (ed.), *Stress Challenges and Immunity in Space*,
https://doi.org/10.1007/978-3-030-16996-1_6

social threats (Slavich and Irwin 2014). Additionally, certain characteristics of a situation are associated with greater stress responses, including the intensity, severity, controllability, and predictability of the stressor. Physiological reactivity to stressors are commonly observed even after repeated exposure to the same stressor (Dhabhar 2014).

The autonomic nervous system (ANS) and the hypothalamic-pituitary-adrenal (HPA) axis are two major stress-signaling pathways that contribute to immune dysregulation (Glaser and Kiecolt-Glaser 2005). Experiencing a stressful situation, as perceived by the brain, activates the HPA axis and the sympathetic-adrenal medullary axis (SAM), which provokes the release of hormones which modulate immune function including adrenocorticotropic hormone (ACTH), cortisol, growth hormone, prolactin, epinephrine, and norepinephrine (Glaser and Kiecolt-Glaser 2005) (see Chap. 7).

Immunity is the natural or acquired resistance of an organism to bacterial or viral invaders, diseases, or infections, while having adequate tolerance to avoid allergy, and autoimmune diseases. Lymphocytes, including T and B cells are the main type of cells of the immune system. T cells orchestrate the immune response via the production of cytokines and stimulate B cells to produce antibodies and signal killer cells to destroy the antigen-displaying cell (Sompayrac 2016). Helper T cells (Th) can be separated into Th1 cells, which primarily produce IL-2, IFN-γ and TNF, and Th2 cells, which produce IL-4, IL-5, IL-6, IL-10, and IL-13. Typically, type 1 cytokines favor the development of a strong cellular immune response, whereas type 2 cytokines favor a strong humoral immune response (Spellberg and Edwards Jr. 2001). Chronic stress can suppress or dysregulate innate and adaptive immune responses by altering the type 1/type 2 cytokine balance, thereby inducing low-grade inflammation and suppressing the function of immuno-protective cells (Dhabhar 2014).

A primary focus of the field of psychoneuroimmunology has been to understanding the link between stress and inflammatory responses. Although acute inflammation is an adaptive response to physical injury or infection, exaggerated and/or prolonged inflammatory responses are detrimental to health (Dhabhar 2014). Chronic inflammation secondary to long-term stress has been causally linked with risk for numerous diseases, including infectious illnesses, cardiovascular disease, diabetes, certain cancers, and autoimmune disease, as well as general frailty and mortality (Glaser and Kiecolt-Glaser 2005; Dhabhar 2014; Padro and Sanders 2014; Webster Marketon and Glaser 2008). One potential explanation for the mechanism linking chronic stress and inflammation in the onset of a wide range of diseases is that prolonged stressors result in glucocorticoid receptor resistance, which, in turn, causes dysregulated HPA axis function and interferes with the appropriate regulation of inflammation (Cohen et al. 2012).

Animal models have provided compelling evidence that biobehavioral stress mechanisms and their molecular and cellular pathways can cause illness behavior and illness itself. These experimental studies have conclusively demonstrated that exposure to restraint stress triggers exaggerated inflammatory responses (Korte et al. 1992; Ahlers et al. 1980; Bartolomucci et al. 2003). In addition, pharmacological experiments have amply demonstrated that mice injected with proinflammatory cytokines, including IL-1β or TNF, have decreased motor activity, social withdrawal, reduced food and water intake, increased slow-wave sleep, altered cognition, and increased pain sensitivity (Bluthe et al. 2000; Dantzer 2009). These experiments highlight how conditions of chronic inflammation can induce sickness and depressive-like behaviors in response to chronic stress (Dantzer et al. 2008).

6.2 Stress and Wound Healing

Wound healing is a vitally important process during recovery from either injury or surgery. Poor healing is associated with increased risks for wound infections and other complications, patient discomfort, prolonged hospital stays, and delays in one's return to normal activities (Tevis and Kennedy 2013). Converging evidence from observational, experimental, and interventional studies implies that stress and other behavioral factors can impede wound healing processes and compromise immunity via multiple physiological pathways (Kiecolt-Glaser et al. 1998; Gouin et al. 2008; Ebrecht et al. 2004; Pinto et al. 2016; Walburn et al. 2009).

Wound healing progresses through several sequential and overlapping phases, including inflammation, proliferation, and regeneration. Cellular immunity plays an important role in the regulation of wound healing through the production of proinflammatory cytokines and chemokines (e.g., platelet derived growth factor [PDG]; transforming growth factor [TGF]; vascular endothelial growth factor [VEGF]; TNF; IFN-γ; IL-1β; IL-8), which mediate many of the complex interactions involved in wound healing. These factors act as chemo-attractants for the migration of phagocytes and other cells to the wound site, starting the proliferative phase which involves the recruitment and replication of cells necessary for tissue regeneration and capillary regrowth (Gethin 2012). Inflammation is a prerequisite to healing. Proinflammatory cytokines help to protect against infection and prepare injured tissue for repair by enhancing the recruitment and activation of phagocytes. Unfortunately, stress disrupts the production of proinflammatory cytokines that are essential for wound healing and, when dysregulated, impose a considerable delay in wound repair (Gouin and Kiecolt-Glaser 2011).

The clinical relevance of the relationship between stress and impaired wound healing has been demonstrated in several studies. In one experiment, individuals with a "slow healing" speed had higher stress and higher cortisol levels at awakening, implicating a key role of elevated cortisol levels in the process of cutaneous wound healing (Ebrecht et al. 2004). A meta-analysis (Walburn et al. 2009) corroborated these findings, synthesizing 17 articles that documented how stress is significantly associated with impaired healing and dysregulation of biomarkers crucial to wound healing. In addition to this meta-analysis, a statistically significant and moderately strong inverse correlation of $r = -0.42$ (95% CI = −0.51 to −0.32; $p < 0.01$) was calculated between the level of stress and speed of wound healing. These results confirm earlier findings by Kiecolt-Glaser et al. (1995), who observed that women experiencing the long-term stress of caring for a relative with Alzheimer's disease, took 24% longer than controls to heal a small, standardized dermal wound. In addition, the peripheral blood leukocytes of caregivers produced less IL-1β in response to lipopolysaccharide (LPS) stimulation.

Surgical complications (e.g., postoperative pain; permanent disfigurement) pose significant challenges to surgical patients and may contribute to psychological distress, anxiety, and depression (Pinto et al. 2016). In an observational study involving patients undergoing coronary artery bypass graft (CABG) surgery, individuals with more depressive symptoms at discharge had more infections and poorer wound healing over the first six weeks following surgery than individuals who reported less distress (Doering et al. 2005). In addition, the pain associated with surgery can itself generate psychological distress, which has been shown to further influence wound healing. A prospective study involving 17 women who underwent elective gastric bypass surgery revealed that greater acute pain immediately after surgery and persistent pain in the four weeks following surgery both were associated with slower healing (McGuire et al. 2006).

In summary, acute and chronic stressors can negatively impact the wound healing process, by interrupting the inflammatory cascade that is fundamental for wound repair. These findings highlight the importance of addressing patients' psychological needs in a timely manner, if possible both before and immediately after surgery, so as to prevent stress-related immune disruption.

6.3 Stress and Infectious Agents

Stress can also dysregulate humoral and cellular immune responses to pathogens, increasing risk for infectious illnesses including influenza and the common cold (Glaser and Kiecolt-Glaser 2005). The association between psychological stress and susceptibility to the common cold has long been recognized; stress suppresses the host resistance to infection and increases rates of infection (Cohen et al. 1991). Loneliness is another well-established risk factor for poor physical health. In a study of our own, we were able to demonstrate that loneliness predicts self-reported cold symptoms after a viral challenge, suggesting that cold symptoms are more severe among those who feel lonely (LeRoy et al. 2017).

Vaccination against influenza virus reduces both risk and severity of infection, thus decreasing risk for hospitalization and death. Vaccine effectiveness is of particular importance among high-risk groups, including pregnant women and older adults. However, the protective efficacy of antiviral vaccines depends upon their ability to induce both humoral and cell-mediated immune responses (Lambert et al. 2012).

A meta-analysis of 13 studies concluded that the effect of stress on antibody responses to influenza virus vaccination corresponded to adequate antibody responses among 41% of stressed individuals versus 59% of less-stressed individuals with similar effects among older and younger adults (Pedersen et al. 2009). Furthermore, psychological distress and biobehavioral vulnerabilities, which arise from being older or sedentary, have independently been found to alter immune responses to influenza vaccination (Segerstrom et al. 2012). In addition, studies in adults and adolescence have confirmed that negative emotions, including anxiety and depression, can modulate the antibody and T-cell responses to antiviral vaccinations, resulting in suppressed immune responses (O'Connor et al. 2014; Coughlin 2012). Interestingly, a 4-week massage intervention in students embarking on academic examinations was associated with reduced distress and enhanced antibody responses after a hepatitis B vaccine (Loft et al. 2012). Positive effects of other mind-body therapies, including Tai Chi, Qi Gong, meditation, and Yoga, on the immune system and virus-specific antibody responses to vaccines have also been documented in a meta-analysis of 34 studies (Morgan et al. 2014).

Herpes viruses, including herpes simplex virus (HSV) I and II, varicella-zoster virus (VZV), Epstein-Barr virus (EBV), and cytomegalovirus (CMV), assume a latent state after the initial infection (Grinde 2013). After primary infection, the herpes virus continues to reside in B lymphocytes and white blood cells for the life of the individual. Under normal health conditions, reactivation and replication of the EBV virus is prevented by the cellular immune system, largely orchestrated through specific-memory cytotoxic T cells and natural killer (NK); thus, individuals with herpesvirus infections generally remain asymptomatic (Glaser et al. 1993). However, under stressful conditions, suppressive immune activity may be reduced, permitting reactivation of the virus.

The relationships between neuroendocrine activity, immune function, and latent HSV type 1 reactivation were initially documented in animal studies. Among mice infected with HSV type 1, those exposed to a stressor exhibited reactivation of the latent virus, whereas nonstressed mice did not (Padgett et al. 1998). Today, a body of literature in humans confirms that psychosocial stressors predict reactivation of latent viruses (see Chap. 19). For instance, higher self-reported health was associated with lower reactivation of latent herpesviruses and inflammation (Murdock et al. 2016). Meanwhile, increased antibody titers against EBV viral capsid antigen (VCA) have been observed in the context of depression (Bennett et al. 2012), perceived stress (Brook et al. 2017), childhood adversity (Fagundes et al. 2013a), bereavement or divorce (Derry et al. 2012), exam stress (Matalka et al. 2000), attachment anxiety (Fagundes et al. 2014), and perceived discrimination (Christian et al. 2012). Together, these human and animal studies show that stress can modulate the steady-state expression of latent herpesviruses, downregulating specific T-cell responses to the virus to an extent that is sufficient to result in viral activation.

Human immunodeficiency virus (HIV) is similar to herpes viruses, in that the virus remains in a latent state in the body after primary infection. As individuals infected with HIV may have lowered levels of T cells, cells that are important to fight infections, much interest exists in whether chronic stress and depression—that also are known to suppress the human immune system—may affect HIV disease progression. Indeed, there is a substantial body of evidence pointing at a relationship between chronic stress and the rate of HIV disease progression. In particular, stressful life events are considered to exert important impacts on certain biological markers of the disease: viral load and CD4 cell count (Kołodziej 2016). For instance, HIV-infected persons with posttraumatic stress disorders (PTSD) after Hurricane

Katrina were more likely than those without PTSD to have detectable plasma viral loads and CD4 cell counts <200 mm^{-3} at 12 and 14 months, as well as two years post disaster (Reilly et al. 2009).

Major depression is highly prevalent among HIV-positive patients. Depression is associated with, among other factors, increased inflammatory markers (e.g., CRP; IL-1β; IL-6, TNF) (Slavich and Irwin 2014), which may alter the function of lymphocytes and decrease NK activity, contributing to HIV disease progression and mortality in these patients (Arseniou et al. 2014). These findings are corroborated by a study that investigated norepinephrine, cortisol, depression, hopelessness, coping, and life event stress as predictors of HIV progression in a diverse subject sample every 6 months over a period of 4 years. The authors found that norepinephrine, depression, hopelessness, and avoidant coping significantly predicted a greater rate of decrease in CD4 and increase in viral load, demonstrating a robust effect of chronic stress on HIV disease progression (Ironson et al. 2015).

In summary, stress can not only increase susceptibility to illness after exposure to infectious agents but also can inhibit antibody and virus-specific T cell responses to vaccines, permit reactivation of latent herpesviruses, and influence the progression of HIV-related disease.

6.4 Stress and Cardiovascular Disease

Cardiovascular disease (CVD) is a major cause of morbidity and mortality. Chronic low-grade inflammation is implicated in the link between stress and CVD via contributions to the early emergence, progression, and thrombotic complications of atherosclerosis (Liu et al. 2017). IL-6 and CRP, two important biomarkers of inflammation, are thought to be indicative and potentially predictive of atherosclerosis (Nadrowski et al. 2016). Of clinical importance, the biological effects of stress do not exist in isolation, and are often aggravated by unhealthy behaviors including poor diet, inadequate physical activity, tobacco use, and poor adherence to medication (Lagraauw et al. 2015).

Epidemiological research over the last half-century has conclusively linked chronic stress and other psychosocial factors to the increased incidence of coronary artery disease (von Kanel 2012). For instance, individuals exposed to work-related stressors including shiftwork, workplace conflict, and positions typified by high demands combined with low control, exhibit risk for elevations in serum CRP and IL-6 (von Kanel et al. 2008), as well as CVD (Kivimaki and Kawachi 2015). Furthermore, evidence suggests that childhood adversity, particularly severe physical and sexual abuse, confers risk for cardiovascular events, particularly among women (Garad et al. 2017). Similarly, among adults with greater childhood adversity/trauma, elevated risk for depressive symptoms, higher serum CRP, reduced methylation of the IL-6 promoter, and higher serum IL-6 have been observed (Janusek et al. 2017). These results shed light on potential epigenetic mechanisms that could link childhood adversity to disproportionally elevated risks of inflammatory disease in adulthood.

6.5 Stress and Metabolic Disease

Type-2 diabetes mellitus (T2DM) is a chronic metabolic disorder that results from defects in insulin secretion and insulin action (Hackett and Steptoe 2017). Though limited, an emerging body of literature suggests that stress plays a role in the etiology of T2DM, both as a predictor of new-onset T2DM and as a prognostic factor in individuals with existing T2DM (Hackett and Steptoe 2017). Stress-related biological pathways, including chronic activation of the HPA axis, which can lead to dysregulated cortisol output and neuroendocrine dysfunction, have been conjectured to contribute to the pathogenesis of T2DM (Hackett and Steptoe 2017). For instance, insulin resistance frequently develops during acute or chronic stress (Tsuneki et al. 2013). Moreover, obesity commonly co-occurs in patients with T2DM, and visceral adipose tissue (e.g., adipokines) is a major source of inflammation, including CRP, IL-1β and IL-6 (Donath and Shoelson 2011), supporting a link between T2DM and inflammation.

Results from meta-analyses suggest that depression further contributes to an increased risk of diabetes mellitus (Bădescu et al. 2016; Yu et al. 2015). Stress exposure during childhood has also been found to constitute a risk factor for obesity and diabetes. Experiencing an adverse childhood experience increases a child's risk of type 1 diabetes during childhood (Nygren et al. 2015). Likewise, a review of literature revealed a significant association between exposure to childhood adversity and an increased risk of T2DM in adulthood (Huffhines et al. 2016; Hughes et al. 2017), with the effects of neglect and sexual abuse most prominent (Huang et al. 2015). Of particular note, stress can perinatally impair metabolic health in later life. Fetal exposure to high concentrations of maternal glucocorticoids, as well as obesity, have been associated with low birth weight, which in turn is associated with increased risk for hypertension, diabetes, and cardiometabolic diseases during adulthood (Zöller et al. 2015; Capra et al. 2013). However, the mechanisms for this effect are not yet fully understood. One novel potential pathway linking maternal and child weight is the transmission of obesogenic microbes from mother to child (Galley et al. 2014).

6.6 Stress and Cancer

Research over the past 30 years in the field of psychoneuroimmunology has contributed to considerable understanding of the effect of stress on cancer biology, and has identified psychosocial factors including stress, depression, and the lack of social support as risk factors for tumor progression (Moreno-Smith et al. 2010). Stress hormones (e.g., glucocorticoids, norepinephrine, epinephrine) have multiple effects on human tumor biology. Thus, via adrenergic- and glucocorticoid-mediated mechanisms, sympathetic nervous system (SNS)-activation may alter immune defenses mechanisms and anti-tumor immune capabilities with implications for tumor progression (Antoni et al. 2006; Lutgendorf and Andersen 2015; Armaiz-Pena et al. 2013). For instance, exposure to chronic stress (Lamkin et al. 2012) as well as the pharmacological stimulation of SNS pathways with a β-adrenergic agonist (e.g., isoproterenol) (Sloan et al. 2010) in tumor-bearing animals, significantly enhances tumor progression and metastasis, implying a fundamental role of stress hormones and β-adrenergic receptor signaling in both processes. Furthermore, both animal and human studies have consistently revealed that the negative effects of stress on tumor cell dissemination can be abrogated using a β2-adrenergic receptor antagonist (e.g., propranolol), supporting the use of β-blockers to modulate cancer metastasis (Sloan et al. 2010; Shaashua et al. 2017).

Chronic stress can increase inflammation and alter protective immune responses, and thereby may increase susceptibility to certain types of cancer by suppressing type 1 cytokines and protective T cells, and increasing regulatory/suppressor T-cell function (Dhabhar 2014). Correspondingly, increased catecholamine levels have been linked to T lymphocyte apoptosis (Radojevic et al. 2014), altered distribution of NK (see also Chap. 13) and granulocytes, and suppressed NK activity (Elenkov and Chrousos 2002), all important defense mechanisms against tumors and their metastasis (see also Chap. 13). It has become clear that cancer-related systemic inflammation is associated with poor outcomes, independent of tumor stage (Dolan et al. 2017). Several inflammatory mediators including IL-6, IL-12, IFN-γ, and TNF are implicated in tumor growth and progression (Cash et al. 2015; Landskron et al. 2014).

The immune system plays a critical role in the occurrence and progression of immunogenic tumors, including skin cancer (Song et al. 2016). For instance, an increased immune response reflected by enhanced expression of intercellular adhesion molecule (ICAM) 1 and infiltration of CD68+ cells (macrophages) surrounding, or within the tumor, have been observed during basal cell carcinoma (BCC) tumor regression following treatment (De Giorgi et al. 2009; Urosevic et al. 2003). Furthermore, immunosuppression, such as in solid-organ transplant recipients and patients with human immunodeficiency virus (HIV) or hematologic malignant neoplasms, has been clearly linked with increased incidence of non-melanoma skin cancer, including BCC, and squamous cell carcinoma (Song et al. 2016; Jensen et al. 2009).

Importantly, chronic stress can alter the anti-tumor-specific immune response to immunogenic tumors. Our own data demonstrate that emotional maltreatment in childhood and occurrence of a major life event in adulthood, showed poorer immune responses to BCC as indexed by suppressed expression of messenger RNA (mRNA) immune markers (CD25, CD3ε, ICAM-1, and CD68) to BCC (Fagundes et al. 2012). Animal models support these findings; mice under restraint stress developed ultraviolet-light (Illi et al. 2012) -induced squamous cell carcinoma more rapidly and showed a poorer immune response [as assessed by messenger RNA (mRNA) in their tumors] relative to nonstressed control mice (Saul et al. 2005). Taken together, these preclinical and clinical studies provide evidence that behavioral stressors can influence the tumor microenvironment.

While much initial work focused on direct effects of catecholamines and other stress mediators on cancer progression, subsequent work identified that the tumor microenvironment is a critical regulator of cancer progression and metastasis (Landskron et al. 2014; Wang et al. 2017; Berghoff and Preusser 2015). The tumor microenvironment has a pivotal role in regulating tumor cell growth, invasion, and metastasis, specifically through reciprocal cross-talk with infiltrating immune cells (lymphocytes, neutrophils, and macrophages), endothelial cells, mesenchymal stromal cells (fibroblasts and myofibroblasts), and their secretory products, all of which can modulate gene expression and alter the behavior of tumor cells (Mostofa et al. 2017).

6.7 Contextual Factors and Immune-Dysregulation

6.7.1 Stressful Life Events

Many investigators have studied pathways between major life events and inflammation. Caring for a loved one with a chronic medical condition, such as a spouse with dementia, is commonly characterized by significant life changes and social isolation (Holmes and Rahe 1967). The chronic stress of caregiving has been linked with exacerbation of typical age-related increases in serum levels of IL-6 and CRP (Gouin et al. 2012), providing a plausible physiological pathway via which chronic stress may lead to poor health. Analogously, the loss of a spouse is considered one of the most stressful life events one may encounter (Holmes and Rahe 1967). Indeed, bereavement has been associated with increased inflammation (Buckley et al. 2012; Cohen et al. 2015) as well as elevated rates of chronic inflammatory conditions, including type 2 diabetes, cardiovascular disease, and cancer within the first three years following the death (Stahl et al. 2016).

Particularly strong evidence indicates that trauma exposure during adulthood increases risks for psychiatric morbidity and poor health outcomes, and there is emerging evidence that inflammation contributes to this link (Flory and Yehuda 2015). Trauma exposure and posttraumatic stress disorder (PTSD, see Chap. 7) have been linked to increased risks of both depression (Dunn et al. 2017) and cardiovascular disorders (Edmondson and von Kanel 2017). The prevalence of trauma-related inflammation was addressed in a review paper, providing evidence for elevated systemic inflammation in individuals with PTSD, with this effect especially strong among those with comorbid PTSD and depression (Baker et al. 2012). In another study involving survivors of the World Trade Center attacks on September 11, 2001, altered salivary cortisol responses to trauma activation (induced by trauma recollections through a standardized interview) were observed, with stronger effects documented in those with comorbid PTSD and depression (Dekel et al. 2017).

6.7.2 Adverse Childhood Experiences

Early adversity confers risk for physical and mental illness (e.g., depression, cardiovascular disease, type 2 diabetes, cancer) in adulthood (Ziol-Guest et al. 2012; Ehrlich et al. 2016) with more robust effects amongst those experiencing multiple adversities (Hughes et al. 2017). Inflammatory pathways are implicated in these links; meta-analyses of 25 studies concluded that early life adversity contributes to significantly elevated peripheral CRP, IL-6, and TNF in adulthood (Baumeister et al. 2016). Most interestingly, different types of trauma exposure impacted inflammatory markers differentially: physical and sexual abuse were associated with significantly increased TNF and IL-6, but not CRP (Baumeister et al. 2016). Similar results were reported by Lin et al. (2016), who found that adults who had experienced childhood adversity had elevated levels of CRP and were almost three times as likely to have experienced trauma as an adult, relative to those without adverse childhood experiences.

Health behaviors (e.g., smoking and obesity) appear to partially mediate this relationship. For instance, in one study it was shown that early adversity predicted increased smoking and BMI through ongoing chronic stress in young adulthood (Raposa et al. 2014). In the same study, higher BMI predicted higher levels of soluble TNF receptor type II (sTNF-RII) and CRP, suggesting that early adversity contributes to inflammation, in part through ongoing stress and maladaptive health behaviors. In accordance with this, several previous studies provide evidence that specific early adversity, including low socioeconomic status in childhood (Brummett et al. 2013; Hagger-Johnson et al. 2012) and childhood abuse (Matthews et al. 2014), affects CRP through unhealthy behaviors and increased BMI.

Evidence also supports a role for heightened emotional and physiological reactivity to stress, which in turn drives the expression of an increasingly proinflammatory phenotype (Slavich and Irwin 2014). In accordance with this assumption, Shapero et al. (2014) reported that individuals with more severe childhood emotional abuse experienced greater increases in depressive symptoms when confronted with a stressor, implying the importance of emotional abuse as an indicator of reactivity to stressful life events. In addition, individuals who have experienced childhood adversities may have fewer social and psychological resources available to them for coping with stress (Fagundes et al. 2013b).

Taken together, intense and chronic stress experienced during one's developmental years appears to have long-lasting neurobiological effects and increases one's risk of later morbidity (e.g., anxiety, depression, and physical disorders) and mortality (Raposa et al. 2014; Fagundes et al. 2013b). Another important effect is that stress exposure during childhood might alter behavioral and physiological responses to acute and chronic stress in adulthood, which may determine one's later risk of disease.

6.7.3 Pregnancy

The prenatal period is a critical time for neurodevelopment and, as such, represents a period of vulnerability during which a wide range of exposures has been found to exert long-term effects on brain development and behaviors (Christian 2012). Maternal psychosocial stress during pregnancy is associated with risks to maternal health and birth outcomes, as well as to various adverse health and behavioral outcomes in the offspring (Christian 2015). During pregnancy, the immune system undergoes substantial adaptations. Under normal circumstances, pregnancy is characterized by elevations in circulating inflammatory mediators relative to nonpregnancy (Christian and Porter 2014). However, excessive inflammation or deviations in inflammatory trajectories of change across pregnancy have been associated with gestational hypertension, miscarriages, preterm births, and adverse influences on fetal development (Christian 2012).

Stress, anxiety, and depression in pregnancy are considerable risk factors for adverse outcomes for both mothers and babies, and are associated with shorter gestation and impaired fetal neurodevelopment and child outcomes (Christian 2012). Inflammation is a likely mechanism by which stress may promote these negative health outcomes (Christian et al. 2009).

Inflammatory responses to influenza virus vaccine have been shown to be mild, transient, and generally similar in pregnant and nonpregnant women (Christian et al. 2013a). Since it is considered safe and recommended for pregnant women, seasonal influenza vaccination provides a useful model with which to study individual differences in inflammatory responses during pregnancy. In one influenza-virus vaccine study in pregnant women, Christian et al. (2013a) demonstrated that women in the highest percentile of depressive symptoms had markedly higher inflammatory responses, as indicated by elevated serum levels of macrophage migration inhibitory factor (MIF) one week post-vaccination, indicating that women with depressive symptoms may be more vulnerable to negative sequelae of infectious illness during pregnancy. Similarly, in pregnant women, greater EBV reactivation has been reported in association with maternal depression, perceived distress, and perceived racial discrimination. Notably, this effect was significantly stronger among African American women who reported greater racial discrimination (Christian et al. 2012).

Other risk factors can mediate the association between chronic stress and inflammation in pregnant women. Obesity, conceptualized as a physiological stress, has been linked to considerable increases in circulatory inflammatory markers, particularly IL-6, throughout pregnancy and postpartum. Moreover, psychological stress and obesity may interact synergistically, resulting in more pronounced effects among women with both risk factors (Mitchell and Christian 2018). In addition, obesity has been observed to increase the risks of gestational hypertension and gestational diabetes via inflammatory pathways. Most importantly, obesity-induced inflammation is transmitted to the child, and can potentially affect their immune function, metabolism, and cognitive development (Christian 2015). In addition, it is well established that poor sleep triggers inflammation. Accordingly, sleep-induced immune dysregulation has been found to be predictive of preterm birth. This effect, again, was especially pronounced in African American women (Blair et al. 2015). Indeed, the relationship between stress-induced inflammatory responses has been found to be more robust in racial minorities, placing these women at greater risk of delivering their infants preterm (Christian et al. 2013b).

Self-rated health is a reliable predictor of health outcomes including morbidity and mortality (Idler and Benyamini 1997; Nielsen et al. 2008). Indeed, poorer self-rated health has been shown to be associated with significantly higher serum IL-1β and MIF in pregnant women during the second trimester, suggesting an influential role of inflammation on self-rated health prior to the emergence of objective and quantifiable signs of disease (Christian et al. 2013c).

These studies suggest that pregnant women with psychosocial risk factors may experience higher daily exposure to inflammatory mediators. It is critical to identify biological markers, symptoms, and diagnostic thresholds that warrant prenatal intervention, and to develop efficient and valid screening and intervention strategies to prevent stress-related adverse health outcomes in mothers and their offspring.

6.7.4 Aging

Chronic stress has been shown to suppress and dysregulate immune function by affecting immunosenescence (Mathur et al. 2016). The term *immunosenescence* refers to a loss of immune function that typically occurs in elderly individuals. Declining T-cell function is a very well-characterized feature of immunosenescence, which contributes to chronic low-grade inflammation (Wu and Meydani 2008). Typically, elderly individuals (aged 65 years and older) compared with other age groups have two- to fourfold elevations in circulating levels of proinflammatory cytokines, such as IL-6, TNF, CRP, and serum amyloid A (SAA) (Michaud et al. 2013), which in turn suppresses the function of immune-protective cells and disrupts the body's ability to defend itself against bacteria, viruses, and parasites. As a result, age-associated deterioration in immune function contributes to many illnesses and renders older individuals more vulnerable to further assaults on their immune system (e.g., stress, immunocompromising medications, infectious diseases) (Burleson et al. 2002). Dysregulation of the inflammatory pathway may also affect the central nervous system and the pathophysiological mechanisms of neurodegenerative disorders including Alzheimer's disease (McCaulley and Grush 2015).

Epel et al. (2004) demonstrated that chronic stress in healthy premenopausal women was significantly associated with higher oxidative stress, lower telomerase activity, and shorter telomere length. Telomere length shortens with age (Rizvi et al. 2014). Extensive research has revealed that progressive shortening of telomeres leads to senescence, apoptosis, and carcinogenesis, which has been associated with the increased incidence of various diseases and poor survival (Shammas 2011). Moreover, adverse life experiences and lifestyle factors appear to affect the rate of telomere shortening over one's lifespan (Rizvi et al. 2014). In accordance with these findings, in one of their studies Kiecolt-Glaser et al. (2011) demonstrated that childhood adversities have considerable consequences for cell aging in later life, and that the presence of multiple childhood adversities is linked to shorter telomeres, which underlines how adverse childhood experiences can generate continued vulnerability through to older adulthood. These findings have implications for understanding, at a cellular level, how stress gets "under the skin" and may promote the precocious onset of age-related diseases.

6.8 From Daily Life to Space Travel

Spaceflight conditions reflect an extreme and complex environmental challenge, with the potential for multiple aversive consequences for human health. Spaceflight, even when short in duration, can induce a wide range of adverse effects by reason of adaptations to the physical stressors of gravitational changes, radiation, malnutrition, disrupted sleep, and psychological stress (Choukèr 2012). As such, space travel presents an exceptional and intense combination of physical and psychological challenges that also provides a unique opportunity for investigators to examine the susceptibility of the human body to stress and explore interventions to promote psychological and physiological resilience.

In a study investigating spaceflight effects on the immune system in 30 cosmonauts, striking alterations in immune responses during and after space flight were observed, including a reduced percentage of NK, as well as suppressed NK activity by up to 85% relative to pre-flight (Rykova et al. 2008). Similar findings were reported in a study on long-duration spaceflight by 12 Russian cosmonauts, which included significantly suppressed T-cell immunity and exaggerated cytokine production after landing relative to before launch (Morukov et al. 2011). Moreover, alterations in the endocannabinoid system (ECS), which is known to play an important role in the regulation of various physiological functions, including stress regulation, behavior, mood, memory, vegetative control, and immunity, were observed following time on board of the International Space Station (ISS), resulting in an increase in circulating endocannabinoids (Strewe et al. 2012).

Despite the many improvements that have been made to living conditions aboard the ISS and during space travel, the clinical health risk of time in space remains high, as demonstrated by a remarkably compromised immune system and disease-fighting capabilities following exploration missions, placing cosmonauts at a greater risk for disease development, including bacterial and viral infections. In light of these results, full characterization of the shifts in the innate and adaptive immune system after space travel (see Chaps. 11–15) is critical to understand the relationship between microgravity and the stress effects of space flight in human space explorers (Morukov et al. 2011). In turn, studies of stressors encountered in space travel might also advance our understanding of stress in our daily lives.

6.9 Interventions

Given the clear negative impact of stress on immune function and health, interventions addressing stress from a psychosocial, physical, nutritional/dietary, and pharmacological perspective are of clinical importance. To appropriately manage stress in both healthy and ill individuals, comprehensive and multidisciplinary approaches that include psychopharmacological treatment, education, cognitive behavioral therapy, mindfulness-based approaches, and relaxation techniques should be provided at an early stage, particularly in physically ill patients.

A variety of stress-reduction techniques have demonstrated beneficial effects for reducing stress and improving mental health and quality of life, including cognitive behavioral therapy (Antoni et al. 2009), mindfulness-based stress reduction interventions (Gallegos et al. 2015), meditation (Rosenkranz et al. 2016), and yoga (Kiecolt-Glaser et al. 2010). Moreover, psychological interventions including cognitive behavioral stress management (Antoni et al. 2009; Gallegos et al. 2015), meditation (Rosenkranz et al. 2016), and yoga (Kiecolt-Glaser et al. 2010) have been demonstrated to improve immune function in diverse populations, including healthy individuals, women exposed to trauma, and cancer patients. These stress-reduction interventions seem to result in a healthy balance between sympathetic and parasympathetic arousal (Chaoul et al. 2014).

Exercise presents a promising intervention to counteract the deleterious effects of chronic stress (see also Chap. 32). A body of research has already examined the ability of physical/aerobic exercise to enhance immune responses when performed regularly and in moderation (Simpson et al. 2015). Beneficial effects of exercise and lifestyle interventions on stress reduction, inflammation, and overall well-being have been found for healthy working adults group (Kettunen et al. 2015), elderly individuals (Emery et al. 2005), patients with T2DM (Chen et al. 2015), and cancer patients (Zhu et al. 2016).

Undoubtedly, exercise is a powerful behavioral intervention with the potential to improve immune function and health outcomes in the healthy, the obese, and the elderly, as well as in patients specifically having CVD, diabetes, or cancer. Improvements in immunity, resulting from regular exercise of moderate intensity, may be due to reduced inflammation, maintained thymic mass, enhanced immunosurveillance, reduced psychological distress, and improved overall well-being (Simpson et al. 2015).

In summary, a variety of interventions show promise for counteracting the negative effects of psychological stress. The particular intervention (i.e., stress management, physical activity, meditation), which is most beneficial, likely depends upon the outcome of interest and the type of stressors experienced, as well as on the individual's personality characteristics and preexisting primary illness and comorbidities.

6.10 Conclusions

The findings synthesized above highlight the complex interactions that underlie the relationships among stress, neuroendocrine activity, immunity, and health outcomes. Chronic stress and its correlates affect a variety of clinically meaningful immune parameters, including wound healing, antibody responses to vaccines, susceptibility to infectious illnesses, the ability of the immune system to suppress latent viruses, and various inflammatory processes. These effects, in turn, can increase one's risk a variety of physical and mental disorders, including cardiovascular disease, diabetes, certain cancers, and autoimmune disease, as well as general frailty and mortality. Together, these findings provide a robust pathway through which chronic stress and immune dysregulation may contribute to serious adverse health outcomes. Past research provides support for several promising avenues for interventions to prevent stress-induced immune dysfunction. However, further research is warranted to provide individualized intervention strategies.

Acknowledgement *Funding sources*: Annina Seiler has received funding from the Swiss National Science Foundation (SNSF) (P2FRP1-168479). Work on this chapter was also supported by grant R01 NR013661 awarded to Lisa M. Christian by the National Institutes of Health (NIH). The content of this chapter is solely the responsibility of the authors and does not necessarily represent the official views of the National Institutes of Health.

References

Ahlers I, Zahumenska L, Toropila M, Smajda B, Ahlersova E (1980) The effect of season on circadian rhythm of serum and adrenal corticosterone in rats. Act Nerv Super 22:60–61

Antoni MH, Lutgendorf SK, Cole SW, Dhabhar FS, Sephton SE, McDonald PG, Stefanek M, Sood AK (2006) The influence of bio-behavioural factors on tumour biology: pathways and mechanisms. Nat Rev Cancer 6:240–248

Antoni MH, Lechner S, Diaz A, Vargas S, Holley H, Phillips K, McGregor B, Carver CS, Blomberg B (2009) Cognitive behavioral stress management effects on psychosocial and physiological adaptation in women undergoing treatment for breast cancer. Brain Behav Immun 23:580–591

Armaiz-Pena GN, Cole SW, Lutgendorf SK, Sood AK (2013) Neuroendocrine influences on cancer progression. Brain Behav Immun 30(Suppl):S19–S25

Arseniou S, Arvaniti A, Samakouri M (2014) HIV infection and depression. Psychiatry Clin Neurosci 68:96–109

Bădescu SV, Tătaru C, Kobylinska L, Georgescu EL, Zahiu DM, Zăgrean AM, Zăgrean L (2016) The association between diabetes mellitus and depression. J Med Life 9:120–125

Baker DG, Nievergelt CM, O'Connor DT (2012) Biomarkers of PTSD: neuropeptides and immune signaling. Neuropharmacology 62:663–673

Bartolomucci A, Palanza P, Parmigiani S, Pederzani T, Merlot E, Neveu PJ, Dantzer R (2003) Chronic psychosocial stress down regulates central cytokines mRNA. Brain Res Bull 62:173–178

Baumeister D, Akhtar R, Ciufolini S, Pariante CM, Mondelli V (2016) Childhood trauma and adulthood inflammation: a meta-analysis of peripheral C-reactive protein, interleukin-6 and tumour necrosis factor-[alpha]. Mol Psychiatry 21:642–649

Bennett JM, Glaser R, Malarkey WB, Beversdorf DQ, Peng J, Kiecolt-Glaser JK (2012) Inflammation and reactivation of latent herpesviruses in older adults. Brain Behav Immun 26:739–746

Berghoff AS, Preusser M (2015) The inflammatory microenvironment in brain metastases: potential treatment target? Chin Clin Oncol 4:21

Blair LM, Porter K, Leblebicioglu B, Christian LM (2015) Poor sleep quality and associated inflammation predict preterm birth: heightened risk among African Americans. Sleep 38:1259–1267

Bluthe RM, Laye S, Michaud B, Combe C, Dantzer R, Parnet P (2000) Role of interleukin-1beta and tumour necrosis factor-alpha in lipopolysaccharide-induced sickness behaviour: a study with interleukin-1 type I receptor-deficient mice. Eur J Neurosci 12:4447–4456

Brook MJ, Christian LM, Hade EM, Ruffin M (2017) The effect of perceived stress on Epstein-Barr virus antibody titers in Appalachian women. Neuroimmunomodulation 24(2):67–73

Brummett BH, Babyak MA, Singh A, Jiang R, Williams RB, Harris KM, Siegler IC (2013) Socioeconomic indices as independent correlates of C-reactive protein in the National Longitudinal Study of Adolescent Health. Psychosom Med 75:882–893

Buckley T, Morel-Kopp MC, Ward C, Bartrop R, McKinley S, Mihailidou AS, Spinaze M, Chen W, Tofler G (2012) Inflammatory and thrombotic changes in early bereavement: a prospective evaluation. Eur J Prev Cardiol 19:1145–1152

Burleson MH, Poehlmann KM, Hawkley LC, Ernst JM, Berntson GG, Malarkey WB, Kiecolt-Glaser JK, Glaser R, Cacioppo JT (2002) Stress-related immune changes in middle-aged and older women: 1-year consistency of individual differences. Health Psychol 21:321–331

Capra L, Tezza G, Mazzei F, Boner AL (2013) The origins of health and disease: the influence of maternal diseases and lifestyle during gestation. Ital J Pediatr 39:7

Cash E, Sephton SE, Chagpar AB, Spiegel D, Rebholz WN, Zimmaro LA, Tillie JM, Dhabhar FS (2015) Circadian disruption and biomarkers of tumor progression in breast cancer patients awaiting surgery. Brain Behav Immun 48:102–114

Chaoul A, Milbury K, Sood AK, Prinsloo S, Cohen L (2014) Mind-body practices in cancer care. Curr Oncol Rep 16:417

Chen L, Pei JH, Kuang J, Chen HM, Chen Z, Li ZW, Yang HZ (2015) Effect of lifestyle intervention in patients with type 2 diabetes: a meta-analysis. Metab Clin Exp 64:338–347

Choukèr AE (2012) Stress challenges and immunity in space. Springer, Berlin

Christian LM (2012) Psychoneuroimmunology in pregnancy: immune pathways linking stress with maternal health, adverse birth outcomes, and fetal development. Neurosci Biobehav Rev 36:350–361

Christian LM (2015) Stress and immune function during pregnancy: an emerging focus in mind-body medicine. Curr Dir Psychol Sci 24:3–9

Christian LM, Porter K (2014) Longitudinal changes in serum proinflammatory markers across pregnancy and postpartum: effects of maternal body mass index. Cytokine 70:134–140

Christian LM, Franco A, Glaser R, Iams JD (2009) Depressive symptoms are associated with elevated serum proinflammatory cytokines among pregnant women. Brain Behav Immun 23:750–754

Christian LM, Iams JD, Porter K, Glaser R (2012) Epstein-Barr virus reactivation during pregnancy and postpartum: effects of race and racial discrimination. Brain Behav Immun 26:1280–1287

Christian LM, Porter K, Karlsson E, Schultz-Cherry S, Iams JD (2013a) Serum proinflammatory cytokine responses to influenza virus vaccine among women during pregnancy versus non-pregnancy. Am J Reprod Immunol (New York, NY 1989) 70:45–53

Christian LM, Glaser R, Porter K, Iams JD (2013b) Stress-induced inflammatory responses in women: effects of race and pregnancy. Psychosom Med 75:658–669

Christian LM, Iams J, Porter K, Leblebicioglu B (2013c) Self-rated health among pregnant women: associations with objective health indicators, psychological functioning, and serum inflammatory markers. Ann Behav Med 46:295–309

Cohen S, Tyrrell DA, Smith AP (1991) Psychological stress and susceptibility to the common cold. N Engl J Med 325:606–612

Cohen S, Janicki-Deverts D, Doyle WJ, Miller GE, Frank E, Rabin BS, Turner RB (2012) Chronic stress, glucocorticoid receptor resistance, inflammation, and disease risk. Proc Natl Acad Sci U S A 109:5995–5999

Cohen M, Granger S, Fuller-Thomson E (2015) The association between bereavement and biomarkers of inflammation. Behav Med (Washington, DC) 41:49–59

Coughlin SS (2012) Anxiety and depression: linkages with viral diseases. Public Health Rev 34:92

Dantzer R (2009) Cytokine, sickness behavior, and depression. Immunol Allergy Clin N Am 29:247–264

Dantzer R (2010) Psychoneuroendocrinology of stress. In: George FK, Richards FT (eds) Encyclopedia of behavioral neuroscience. Academic, Oxford, pp 126–131

Dantzer R, O'Connor JC, Freund GG, Johnson RW, Kelley KW (2008) From inflammation to sickness and depression: when the immune system subjugates the brain. Nat Rev Neurosci 9:46–56

De Giorgi V, Salvini C, Chiarugi A, Paglierani M, Maio V, Nicoletti P, Santucci M, Carli P, Massi D (2009) In vivo characterization of the inflammatory infiltrate and apoptotic status in imiquimod-treated basal cell carcinoma. Int J Dermatol 48:312–321

Dekel S, Ein-Dor T, Rosen JB, Bonanno GA (2017) Differences in cortisol response to trauma activation in individuals with and without comorbid PTSD and depression. Front Psychol 8:797

Derry HM, Glaser R, Kiecolt-Glaser JK (2012) Marital status is related to Epstein-Barr virus latency in individuals undergoing cancer diagnostic procedures. Brain Behav Immun 26(Supplement 1):S30–SS1

Dhabhar FS (2014) Effects of stress on immune function: the good, the bad, and the beautiful. Immunol Res 58:193–210

Doering LV, Moser DK, Lemankiewicz W, Luper C, Khan S (2005) Depression, healing, and recovery from coronary artery bypass surgery. Am J Crit Care 14:316–324

Dolan RD, McSorley ST, Horgan PG, Laird B, McMillan DC (2017) The role of the systemic inflammatory response in predicting outcomes in patients with advanced inoperable cancer: systematic review and meta-analysis. Crit Rev Oncol Hematol 116:134–146

Donath MY, Shoelson SE (2011) Type 2 diabetes as an inflammatory disease. Nat Rev Immunol 11:98–107

Dunn EC, Nishimi K, Powers A, Bradley B (2017) Is developmental timing of trauma exposure associated with depressive and post-traumatic stress disorder symptoms in adulthood? J Psychiatr Res 84:119–127

Ebrecht M, Hextall J, Kirtley LG, Taylor A, Dyson M, Weinman J (2004) Perceived stress and cortisol levels predict speed of wound healing in healthy male adults. Psychoneuroendocrinology 29:798–809

Edmondson D, von Kanel R (2017) Post-traumatic stress disorder and cardiovascular disease. Lancet Psychiatry 4:320–329

Ehrlich KB, Miller GE, Chen E (2016) Childhood adversity and adult physical health. Developmental psychopathology. John Wiley & Sons, Inc., Hoboken, NJ

Elenkov IJ, Chrousos GP (2002) Stress hormones, proinflammatory and antiinflammatory cytokines, and autoimmunity. Ann N Y Acad Sci 966:290–303

Emery CF, Kiecolt-Glaser JK, Glaser R, Malarkey WB, Frid DJ (2005) Exercise accelerates wound healing among healthy older adults: a preliminary investigation. J Gerontol A Biol Sci Med Sci 60:1432–1436

Epel ES, Blackburn EH, Lin J, Dhabhar FS, Adler NE, Morrow JD, Cawthon RM (2004) Accelerated telomere shortening in response to life stress. Proc Natl Acad Sci U S A 101:17312–17315

Fagundes CP, Glaser R, Johnson SL, Andridge RR, Yang EV, Di Gregorio MP, Chen M, Lambert DR, Jewell SD, Bechtel MA, Hearne DW, Herron JB, Kiecolt-Glaser JK (2012) Basal cell carcinoma: stressful life events and the tumor environment. Arch Gen Psychiatry 69:618–626

Fagundes CP, Glaser R, Malarkey WB, Kiecolt-Glaser JK (2013a) Childhood adversity and herpesvirus latency in breast cancer survivors. Health Psychol 32:337–344

Fagundes CP, Glaser R, Kiecolt-Glaser JK (2013b) Stressful early life experiences and immune dysregulation across the lifespan. Brain Behav Immun 27:8–12

Fagundes CP, Jaremka LM, Glaser R, Alfano CM, Povoski SP, Lipari AM, Agnese DM, Yee LD, Carson WE 3rd, Farrar WB, Malarkey WB, Chen M, Kiecolt-Glaser JK (2014) Attachment anxiety is related to Epstein-Barr virus latency. Brain Behav Immun 41:232–238

Flory JD, Yehuda R (2015) Comorbidity between post-traumatic stress disorder and major depressive disorder: alternative explanations and treatment considerations. Dialogues Clin Neurosci 17:141–150

Gallegos AM, Lytle MC, Moynihan JA, Talbot NL (2015) Mindfulness-based stress reduction to enhance psychological functioning and improve inflammatory biomarkers in trauma-exposed women: a pilot study. Psychol Trauma 7:525–532

Galley JD, Bailey M, Kamp Dush C, Schoppe-Sullivan S, Christian LM (2014) Maternal obesity is associated with alterations in the gut microbiome in toddlers. PLoS One 9:e113026

Garad Y, Maximova K, MacKinnon N, McGrath JJ, Kozyrskyj AL, Colman I (2017) Sex-specific differences in the association between childhood adversity and cardiovascular disease in adulthood: evidence from a national cohort study. Can J Cardiol 33:1013–1019

Gethin G (2012) Understanding the inflammatory process in wound healing. Br J Community Nurs Suppl:S17–S18, S20, S2

Glaser R, Kiecolt-Glaser JK (2005) Stress-induced immune dysfunction: implications for health. Nat Rev Immunol 5:243–251
Glaser R, Pearson GR, Bonneau RH, Esterling BA, Atkinson C, Kiecolt-Glaser JK (1993) Stress and the memory T-cell response to the Epstein-Barr virus in healthy medical students. Health Psychol 12:435–442
Gouin J-P, Kiecolt-Glaser JK (2011) The impact of psychological stress on wound healing: methods and mechanisms. Immunol Allergy Clin N Am 31:81–93
Gouin JP, Hantsoo L, Kiecolt-Glaser JK (2008) Immune dysregulation and chronic stress among older adults: a review. Neuroimmunomodulation 15:251–259
Gouin JP, Glaser R, Malarkey WB, Beversdorf D, Kiecolt-Glaser J (2012) Chronic stress, daily stressors, and circulating inflammatory markers. Health Psychol 31:264–268
Grinde B (2013) Herpesviruses: latency and reactivation – viral strategies and host response. J Oral Microbiol 5. https://doi.org/10.3402/jom.v5i0.22766
Hackett RA, Steptoe A (2017) Type 2 diabetes mellitus and psychological stress [mdash] a modifiable risk factor. Nat Rev Endocrinol 13:547–560
Hagger-Johnson G, Mottus R, Craig LC, Starr JM, Deary IJ (2012) Pathways from childhood intelligence and socioeconomic status to late-life cardiovascular disease risk. Health Psychol 31:403–412
Holmes TH, Rahe RH (1967) The social readjustment rating scale. J Psychosom Res 11:213–218
Huang H, Yan P, Shan Z, Chen S, Li M, Luo C, Gao H, Hao L, Liu L (2015) Adverse childhood experiences and risk of type 2 diabetes: a systematic review and meta-analysis. Metab Clin Exp 64:1408–1418
Huffhines L, Noser A, Patton SR (2016) The link between adverse childhood experiences and diabetes. Curr Diab Rep 16:54
Hughes K, Bellis MA, Hardcastle KA, Sethi D, Butchart A, Mikton C, Jones L, Dunne MP (2017) The effect of multiple adverse childhood experiences on health: a systematic review and meta-analysis. Lancet Public Health 2:e356–ee66
Idler EL, Benyamini Y (1997) Self-rated health and mortality: a review of twenty-seven community studies. J Health Soc Behav 38:21–37
Illi J, Miaskowski C, Cooper B, Levine JD, Dunn L, West C, Dodd M, Dhruva A, Paul SM, Baggott C, Cataldo J, Langford D, Schmidt B, Aouizerat BE (2012) Association between pro- and anti-inflammatory cytokine genes and a symptom cluster of pain, fatigue, sleep disturbance, and depression. Cytokine 58:437–447
Ironson G, O'Cleirigh C, Kumar M, Kaplan L, Balbin E, Kelsch CB, Fletcher MA, Schneiderman N (2015) Psychosocial and neurohormonal predictors of HIV disease progression (CD4 cells and viral load): a 4 year prospective study. AIDS Behav 19:1388–1397
Janusek LW, Tell D, Gaylord-Harden N, Mathews HL (2017) Relationship of childhood adversity and neighborhood violence to a proinflammatory phenotype in emerging adult African American men: an epigenetic link. Brain Behav Immun 60:126–135
Jensen AO, Thomsen HF, Engebjerg MC, Olesen AB, Friis S, Karagas MR, Sorensen HT (2009) Use of oral glucocorticoids and risk of skin cancer and non-Hodgkin's lymphoma: a population-based case-control study. Br J Cancer 100:200–205
von Kanel R (2012) Psychosocial stress and cardiovascular risk : current opinion. Swiss Med Wkly 142:w13502
von Kanel R, Bellingrath S, Kudielka BM (2008) Association between burnout and circulating levels of pro- and anti-inflammatory cytokines in schoolteachers. J Psychosom Res 65:51–59
Kettunen O, Vuorimaa T, Vasankari T (2015) A 12-month exercise intervention decreased stress symptoms and increased mental resources among working adults – results perceived after a 12-month follow-up. Int J Occup Med Environ Health 28:157–168
Kiecolt-Glaser JK, Marucha PT, Malarkey WB, Mercado AM, Glaser R (1995) Slowing of wound healing by psychological stress. Lancet 346:1194–1196
Kiecolt-Glaser JK, Page GG, Marucha PT, MacCallum RC, Glaser R (1998) Psychological influences on surgical recovery. Perspectives from psychoneuroimmunology. Am Psychol 53:1209–1218

Kiecolt-Glaser JK, Christian L, Preston H, Houts CR, Malarkey WB, Emery CF, Glaser R (2010) Stress, inflammation, and yoga practice. Psychosom Med 72:113–121

Kiecolt-Glaser JK, Gouin JP, Weng NP, Malarkey WB, Beversdorf DQ, Glaser R (2011) Childhood adversity heightens the impact of later-life caregiving stress on telomere length and inflammation. Psychosom Med 73:16–22

Kivimaki M, Kawachi I (2015) Work stress as a risk factor for cardiovascular disease. Curr Cardiol Rep 17:630

Kołodziej J (2016) Effects of stress on HIV infection progression. HIV AIDS Rev 15:13–16

Korte SM, Bouws GA, Bohus B (1992) Adrenal hormones in rats before and after stress-experience: effects of ipsapirone. Physiol Behav 51:1129–1133

Lagraauw HM, Kuiper J, Bot I (2015) Acute and chronic psychological stress as risk factors for cardiovascular disease: insights gained from epidemiological, clinical and experimental studies. Brain Behav Immun 50:18–30

Lambert ND, Ovsyannikova IG, Pankratz VS, Jacobson RM, Poland GA (2012) Understanding the immune response to seasonal influenza vaccination in older adults: a systems biology approach. Expert Rev Vaccines 11:985–994

Lamkin DM, Sloan EK, Patel AJ, Chiang BS, Pimentel MA, Ma JC, Arevalo JM, Morizono K, Cole SW (2012) Chronic stress enhances progression of acute lymphoblastic leukemia via beta-adrenergic signaling. Brain Behav Immun 26:635–641

Landskron G, De la Fuente M, Thuwajit P, Thuwajit C, Hermoso MA (2014) Chronic inflammation and cytokines in the tumor microenvironment. J Immunol Res 2014:149185

Lazarus RS, Folkman S (1984) Stress, appraisal, and coping. Springer, New York, NY

LeRoy AS, Murdock KW, Jaremka LM, Loya A, Fagundes CP (2017) Loneliness predicts self-reported cold symptoms after a viral challenge. Health Psychol 36:512–520

Lin JE, Neylan TC, Epel E, O'Donovan A (2016) Associations of childhood adversity and adulthood trauma with C-reactive protein: a cross-sectional population-based study. Brain Behav Immun 53:105–112

Liu Y-Z, Wang Y-X, Jiang C-L (2017) Inflammation: the common pathway of stress-related diseases. Front Hum Neurosci 11:316

Loft P, Petrie KJ, Booth RJ, Thomas MG, Robinson E, Vedhara K (2012) Effects of massage on antibody responses after hepatitis B vaccination. Psychosom Med 74:982–987

Lutgendorf SK, Andersen BL (2015) Biobehavioral approaches to cancer progression and survival: mechanisms and interventions. Am Psychol 70:186–197

Matalka KZ, Sidki A, Abdul-Malik SM, Thewaini AJ (2000) Academic stress – influence on epstein-barr virus and cytomegalovirus reactivation, cortisol, and prolactin. Lab Med 31:163–168

Mathur MB, Epel E, Kind S, Desai M, Parks CG, Sandler DP, Khazeni N (2016) Perceived stress and telomere length: a systematic review, meta-analysis, and methodologic considerations for advancing the field. Brain Behav Immun 54:158–169

Matthews KA, Chang YF, Thurston RC, Bromberger JT (2014) Child abuse is related to inflammation in mid-life women: role of obesity. Brain Behav Immun 36:29–34

McCaulley ME, Grush KA (2015) Alzheimer's disease: exploring the role of inflammation and implications for treatment. Int J Alzheimers Dis 2015:515248

McGuire L, Heffner K, Glaser R, Needleman B, Malarkey W, Dickinson S, Lemeshow S, Cook C, Muscarella P, Melvin WS, Ellison EC, Kiecolt Glaser JK (2006) Pain and wound healing in surgical patients. Ann Behav Med 31:165–172

Michaud M, Balardy L, Moulis G, Gaudin C, Peyrot C, Vellas B, Cesari M, Nourhashemi F (2013) Proinflammatory cytokines, aging, and age-related diseases. J Am Med Dir Assoc 14:877–882

Mitchell AM, Christian LM (2018) Examination of the role of obesity in the association between childhood trauma and inflammation during pregnancy. Health Psychol 37(2):114–124

Moreno-Smith M, Lutgendorf SK, Sood AK (2010) Impact of stress on cancer metastasis. Fut Oncol (London) 6:1863–1881

Morgan N, Irwin MR, Chung M, Wang C (2014) The effects of mind-body therapies on the immune system: meta-analysis. PLoS One 9:e100903

Morukov B, Rykova M, Antropova E, Berendeeva T, Ponomaryov S, Larina I (2011) T-cell immunity and cytokine production in cosmonauts after long-duration space flights. Acta Astronaut 68:739–746

Mostofa AG, Punganuru SR, Madala HR, Al-Obaide M, Srivenugopal KS (2017) The process and regulatory components of inflammation in brain oncogenesis. Biomol Ther 7:pii:E34

Murdock KW, Fagundes CP, Peek MK, Vohra V, Stowe RP (2016) The effect of self-reported health on latent herpesvirus reactivation and inflammation in an ethnically diverse sample. Psychoneuroendocrinology 72:113–118

Nadrowski P, Chudek J, Skrzypek M, Puzianowska-Kuznicka M, Mossakowska M, Wiecek A, Zdrojewski T, Grodzicki T, Kozakiewicz K (2016) Associations between cardiovascular disease risk factors and IL-6 and hsCRP levels in the elderly. Exp Gerontol 85:112–117

Nielsen AB, Siersma V, Hiort LC, Drivsholm T, Kreiner S, Hollnagel H (2008) Self-rated general health among 40-year-old Danes and its association with all-cause mortality at 10-, 20-, and 29 years' follow-up. Scand J Public Health 36:3–11

Nygren M, Carstensen J, Koch F, Ludvigsson J, Frostell A (2015) Experience of a serious life event increases the risk for childhood type 1 diabetes: the ABIS population-based prospective cohort study. Diabetologia 58:1188–1197

O'Connor TG, Moynihan JA, Wyman PA, Carnahan J, Lofthus G, Quataert SA, Bowman M, Caserta MT (2014) Depressive symptoms and immune response to meningococcal conjugate vaccine in early adolescence. Dev Psychopathol 26:1567–1576

Padgett DA, Sheridan JF, Dorne J, Berntson GG, Candelora J, Glaser R (1998) Social stress and the reactivation of latent herpes simplex virus type 1. Proc Natl Acad Sci U S A 95:7231–7235

Padro CJ, Sanders VM (2014) Neuroendocrine regulation of inflammation. Semin Immunol 26:357–368

Pedersen AF, Zachariae R, Bovbjerg DH (2009) Psychological stress and antibody response to influenza vaccination: a meta-analysis. Brain Behav Immun 23:427–433

Pinto A, Faiz O, Davis R, Almoudaris A, Vincent C (2016) Surgical complications and their impact on patients' psychosocial well-being: a systematic review and meta-analysis. BMJ Open 6:e007224

Radojevic K, Rakin A, Pilipovic I, Kosec D, Djikic J, Bufan B, Vujnovic I, Leposavic G (2014) Effects of catecholamines on thymocyte apoptosis and proliferation depend on thymocyte microenvironment. J Neuroimmunol 272:16–28

Raposa EB, Bower JE, Hammen CL, Najman JM, Brennan PA (2014) A developmental pathway from early life stress to inflammation: the role of negative health behaviors. Psychol Sci 25:1268–1274

Reilly KH, Clark RA, Schmidt N, Benight CC, Kissinger P (2009) The effect of post-traumatic stress disorder on HIV disease progression following hurricane Katrina. AIDS Care 21:1298–1305

Rizvi S, Raza ST, Mahdi F (2014) Telomere length variations in aging and age-related diseases. Curr Aging Sci 7:161–167

Rosenkranz MA, Lutz A, Perlman DM, Bachhuber DR, Schuyler BS, MacCoon DG, Davidson RJ (2016) Reduced stress and inflammatory responsiveness in experienced meditators compared to a matched healthy control group. Psychoneuroendocrinology 68:117–125

Rykova MP, Antropova EN, Larina IM, Morukov BV (2008) Humoral and cellular immunity in cosmonauts after the ISS missions. Acta Astronaut 63:697–705

Saul AN, Oberyszyn TM, Daugherty C, Kusewitt D, Jones S, Jewell S, Malarkey WB, Lehman A, Lemeshow S, Dhabhar FS (2005) Chronic stress and susceptibility to skin cancer. J Natl Cancer Inst 97:1760–1767

Segerstrom SC, Hardy JK, Evans DR, Greenberg RN (2012) Vulnerability, distress, and immune response to vaccination in older adults. Brain Behav Immun 26:747–753

Shaashua L, Shabat-Simon M, Haldar R, Matzner P, Zmora O, Shabtai M, Sharon E, Allweis T, Barshack I, Hayman L, Arevalo JMG, Ma J, Horowitz M, Cole SW, Ben-Eliyahu S (2017) Perioperative COX-2 and β-adrenergic blockade improves metastatic biomarkers in breast cancer patients in a phase-II randomized trial. Clin Cancer Res 23(16):4651–4661

Shammas MA (2011) Telomeres, lifestyle, cancer, and aging. Curr Opin Clin Nutr Metab Care 14:28–34

Shapero BG, Black SK, Liu RT, Klugman J, Bender RE, Abramson LY, Alloy LB (2014) Stressful life events and depression symptoms: the effect of childhood emotional abuse on stress reactivity. J Clin Psychol 70:209–223

Simpson RJ, Kunz H, Agha N, Graff R (2015) Exercise and the regulation of immune functions. Prog Mol Biol Transl Sci 135:355–380

Slavich GM, Irwin MR (2014) From stress to inflammation and major depressive disorder: a social signal transduction theory of depression. Psychol Bull 140:774–815

Sloan EK, Priceman SJ, Cox BF, Yu S, Pimentel MA, Tangkanangnukul V, Arevalo JM, Morizono K, Karanikolas BD, Wu L, Sood AK, Cole SW (2010) The sympathetic nervous system induces a metastatic switch in primary breast cancer. Cancer Res 70:7042–7052

Sompayrac L (2016) How the immune system works, 5th edn. Singapore, Wiley Blackwell

Song SS, Goldenberg A, Ortiz A, Eimpunth S, Oganesyan G, Jiang SI (2016) Nonmelanoma skin cancer with aggressive subclinical extension in immunosuppressed patients. JAMA Dermatol 152:683–690

Spellberg B, Edwards JE Jr (2001) Type 1/type 2 immunity in infectious diseases. Clin Infect Dis 32:76–102

Stahl ST, Arnold AM, Chen JY, Anderson S, Schulz R (2016) Mortality after bereavement: the role of cardiovascular disease and depression. Psychosom Med 78:697–703

Strewe C, Feuerecker M, Nichiporuk I, Kaufmann I, Hauer D, Morukov B, Schelling G, Choukèr A (2012) Effects of parabolic flight and spaceflight on the endocannabinoid system in humans. Rev Neurosci 23:673

Tevis SE, Kennedy GD (2013) Postoperative complications and implications on patient-centered outcomes. J Surg Res 181:106–113

Tsuneki H, Tokai E, Sugawara C, Wada T, Sakurai T, Sasaoka T (2013) Hypothalamic orexin prevents hepatic insulin resistance induced by social defeat stress in mice. Neuropeptides 47:213–219

Urosevic M, Maier T, Benninghoff B, Slade H, Burg G, Dummer R (2003) Mechanisms underlying imiquimod-induced regression of basal cell carcinoma in vivo. Arch Dermatol 139:1325–1332

Walburn J, Vedhara K, Hankins M, Rixon L, Weinman J (2009) Psychological stress and wound healing in humans: a systematic review and meta-analysis. J Psychosom Res 67:253–271

Wang M, Zhao J, Zhang L, Wei F, Lian Y, Wu Y, Gong Z, Zhang S, Zhou J, Cao K, Li X, Xiong W, Li G, Zeng Z, Guo C (2017) Role of tumor microenvironment in tumorigenesis. J Cancer 8:761–773

Webster Marketon JI, Glaser R (2008) Stress hormones and immune function. Cell Immunol 252:16–26

Wu D, Meydani SN (2008) Age-associated changes in immune and inflammatory responses: impact of vitamin E intervention. J Leukoc Biol 84:900–914

Yu M, Zhang X, Lu F, Fang L (2015) Depression and risk for diabetes: a meta-analysis. Can J Diabetes 39:266–272

Zhu G, Zhang X, Wang Y, Xiong H, Zhao Y, Sun F (2016) Effects of exercise intervention in breast cancer survivors: a meta-analysis of 33 randomized controlled trails. OncoTargets Ther 9:2153–2168

Ziol-Guest KM, Duncan GJ, Kalil A, Boyce WT (2012) Early childhood poverty, immune-mediated disease processes, and adult productivity. Proc Natl Acad Sci U S A 109(Suppl 2):17289–17293

Zöller B, Sundquist J, Sundquist K, Crump C (2015) Perinatal risk factors for premature ischaemic heart disease in a Swedish national cohort. BMJ Open 5(6):e007308

Open Access This chapter is licensed under the terms of the Creative Commons Attribution 4.0 International License (http://creativecommons.org/licenses/by/4.0/), which permits use, sharing, adaptation, distribution and reproduction in any medium or format, as long as you give appropriate credit to the original author(s) and the source, provide a link to the Creative Commons license and indicate if changes were made.

The images or other third party material in this chapter are included in the chapter's Creative Commons license, unless indicated otherwise in a credit line to the material. If material is not included in the chapter's Creative Commons license and your intended use is not permitted by statutory regulation or exceeds the permitted use, you will need to obtain permission directly from the copyright holder.

Part III

Stress and Immune Allostasis in Space, from Brain to Immune Responses

Neurobiological Mechanisms of Stress and Glucocorticoid Effects on Learning and Memory: Implications for Stress Disorders on Earth and in Space

7

Raquel V. Fornari, Amanda Aerni,
Dominique J.-F. de Quervain, and Benno Roozendaal

7.1 Introduction

Space flight conditions affect human health due to complex environmental challenges ("stressors"). As discussed elsewhere in this book, adequate function of human organisms in space is challenged by biological (e.g. sleep deprivation, pain, see Chap. 9), physical (e.g. microgravity, variable oxygenation status, radiation, Chaps. 16, 17, 20), and psychological (e.g. confinement, work load, Chaps. 22 and 31) stress factors. It has been long recognized that stress leads to an activation of the sympathetic nervous system and hypothalamus–pituitary adrenal (HPA) axis, culminating in the release of catecholamines and glucocorticoids (i.e., cortisol) from the adrenal medulla and cortex, respectively (De Boer et al. 1990; McCarty and

R. V. Fornari
Centro de Matemática, Computação e Cognição, Universidade Federal do ABC, São Bernardo do Campo, Brazil

A. Aerni
Division of Cognitive Neuroscience, Department of Psychology, University of Basel, Basel, Switzerland

D. J.-F. de Quervain
Division of Cognitive Neuroscience, Department of Psychology, University of Basel, Basel, Switzerland

Transfaculty Research Platform, University of Basel, Basel, Switzerland

University Psychiatric Clinics, University of Basel, Basel, Switzerland

B. Roozendaal (✉)
Department of Cognitive Neuroscience, Radboud University Medical Center, Nijmegen, The Netherlands

Donders Institute for Brain, Cognition and Behaviour, Radboud University, Nijmegen, The Netherlands
e-mail: Benno.Roozendaal@radboudumc.nl

© Springer Nature Switzerland AG 2020
A. Choukér (ed.), *Stress Challenges and Immunity in Space*,
https://doi.org/10.1007/978-3-030-16996-1_7

Gold 1981; Roozendaal et al. 1996a). These stress hormones, in concert with numerous other stress mediators, neurotransmitters, and neuropeptides, are known to influence the organism's ability to cope with stress, influencing target systems in the periphery (see also Chaps. 6, 8 and 9). However, they also induce a myriad of effects on the brain. Accordingly, high stress exposure is known to affect emotional regulation, cognitive function, and mood. Glucocorticoid hormones play a crucial role in regulating stress effects on aversive memory and mood. Although such stress and glucocorticoid effects on memory are typically considered to be adaptive responses, they are also critically involved in influencing the development and symptomatology of stress-related (anxiety) disorders (Aerni et al. 2004; Schelling et al. 2004b; Soravia et al. 2006; de Quervain et al. 2017).

A few studies investigated the occurrence of psychological problems during space flight conditions. Noted problems included anxiety, boredom, crew interactions, problems associated with isolation and confinement, and others (Cooper Jr. 1996). Likewise, simulation studies of manned space flight on Earth reported the incidence of anxiety, depression, psychosis, psychosomatic symptoms, emotional reactions related to mission stage, asthenia, and post-flight personality and marital problems (Kanas 1997). It is obvious that an altered emotional, cognitive, or mood status during highly demanding tasks, especially during prolonged periods, could add to the subjective feeling of perceived stress and further impact human health via neuroendocrine and nerval-immune modulatory effects as well as might induce long-lasting impairment of psychological well-being. Several studies have also investigated human cognitive performance in space flight or analog environment (reviewed in Strangman et al. 2014). Although the available evidence fails to strongly support or refute the existence of specific cognitive deficits in low Earth orbit during long-duration space flight, the studies consistently suggest that novel, and thus potentially stressful or arousing, environments (space flight or other) induce variable alterations in cognitive performance across individuals, consistent with known astronaut experiences. This highlights the need to better examine the impact of interindividual variability, and understand its underlying factors, when predicting in-flight cognitive functioning. Particularly, investigations into individual variability in stress experience and responsiveness, such as cortisol secretion, and in-flight cognitive performance or during simulation studies on Earth and how such changes predict or relate to psychological and other health problems should be implemented.

In this chapter, we will describe the current status of our understanding of the neurobiological mechanisms that are involved in regulating stress and glucocorticoid effects on cognitive performance and why these stress hormones might specifically modulate memory processes of emotionally arousing experiences or during emotionally arousing situations. Furthermore, because emotional memory plays a crucial role in the pathogenesis and symptomatology of anxiety disorders, such as posttraumatic stress disorder (PTSD), which is linked to a variety of immune changes as well, we will discuss to what extent the basic findings on glucocorticoid effects on cognitive performance and emotional memory might have clinical implications on Earth and in space.

7.2 Glucocorticoid Effects on Learning and Memory

Early reports on both enhancing and impairing properties of glucocorticoids on memory (Arbel et al. 1994; Beckwith et al. 1986; Bohus and Lissak 1968; Flood et al. 1978; Luine et al. 1993) have indicated that these hormones have complex effects on cognitive functions. More recent studies investigating glucocorticoid effects on distinct memory phases and studies discerning acute from chronic effects helped to disentangle the multifaceted actions of these stress hormones. For example, acute elevations of glucocorticoids are known to enhance the consolidation of memory of new information, but to impair the retrieval of previously stored information (de Quervain et al. 1998, 2017; Roozendaal and McGaugh 1996, 2011). Conditions with chronically elevated glucocorticoid levels are usually associated with impaired cognitive performance and these deficits are thought to result from a cumulative and long-lasting burden on prefrontal and hippocampal function and morphology (McEwen 1998; Roozendaal et al. 2009). Memory deficits observed under such chronic conditions can, however, also result, at least in part, from acute and reversible actions of elevated glucocorticoid levels on memory retrieval processes (Coluccia et al. 2008). Some of the effects of glucocorticoids on cognitive performance might be mediated via influences on immune regulators. It is well established that cytokines as well as other immune factors affect learning and memory functions (Cibelli et al. 2010; Derecki et al. 2010). These findings indicate that an integrated prospective is necessary to fully understand the impact of stress and glucocorticoids on cognition and health. As long-duration space flight is often associated with both high stress and an impacted immune function, such interactions might be especially relevant in these conditions.

7.2.1 Glucocorticoid Effects on Memory Consolidation

Memory consolidation is the process by which a fragile short-term memory trace is transferred into stable long-term memory. However, not all information is equally well transferred into long-term storage. In fact, it is well recognized that especially emotionally arousing (pleasant or unpleasant) experiences are well remembered, even after decades (McGaugh 2003). There is extensive evidence from studies in animals and humans that glucocorticoids, along with other components of the stress response, are critically involved in regulating memory consolidation of emotionally arousing experiences (de Quervain et al. 2017; Het et al. 2005; McGaugh and Roozendaal 2002; Schwabe et al. 2012). Acute systemic administration of glucocorticoids enhances long-term memory consolidation when given either before (Abercrombie et al. 2003; Buchanan and Lovallo 2001) or immediately after a training experience (Cordero et al. 2002; Roozendaal and McGaugh 1996; Wilhelm et al. 2011). In contrast, a blockade of glucocorticoid production with the synthesis inhibitor metyrapone impairs memory consolidation (Maheu et al. 2004; Roozendaal et al. 1996a) and prevents stress- and epinephrine-induced memory enhancement

(Liu et al. 1999; Roozendaal et al. 1996b). Such glucocorticoid effects on memory consolidation follow an inverted U-shaped dose–response relationship: Moderate doses enhance memory, whereas higher doses are typically less effective or may even impair memory consolidation (Roozendaal et al. 1999b).

7.2.1.1 Role of Emotional Arousal in Enabling Glucocorticoid Effects on Memory Consolidation

Findings from animal and human studies suggest that glucocorticoids enhance memory consolidation of emotionally arousing experiences but do not affect memory consolidation of emotionally neutral information. We investigated the importance of emotional arousal in mediating glucocorticoid effects on memory consolidation in rats trained on an object recognition task, a task based on the innate preference of rodents to explore a novel object (Okuda et al. 2004). Although no rewarding or aversive stimulation is used during object recognition training, placing an animal into the training context induces modest novelty-induced stress or arousal (De Boer et al. 1990). However, extensive habituation of rats to the training context prior to the training reduces the arousal level induced by object recognition training. We found that corticosterone, the main glucocorticoid in rodents, administered systemically immediately after training enhanced 24-h retention performance of rats that were not previously habituated to the experimental context (i.e., that were emotionally aroused). In contrast, posttraining corticosterone did not affect 24-h retention of rats that had received extensive prior habituation to the experimental context and, thus, had decreased novelty-induced emotional arousal during training (Okuda et al. 2004) (Fig. 7.1).

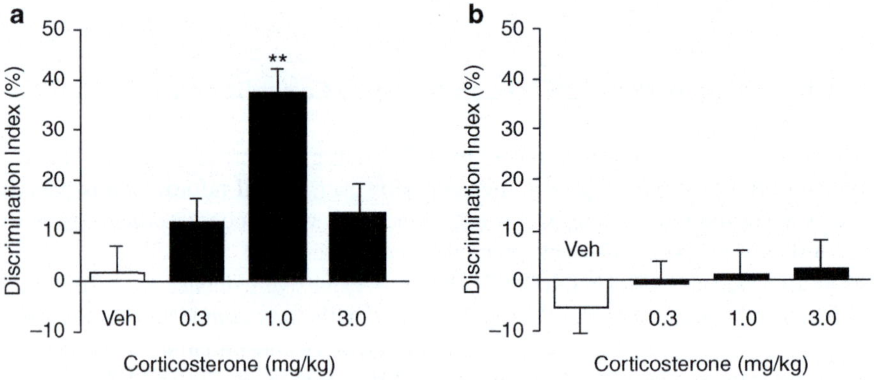

Fig. 7.1 The enhancing effect of posttraining administration of corticosterone on 24-h object recognition performance depends on emotional arousal. Rats received a single injection of corticosterone or vehicle immediately after the 3-min training trial. Corticosterone administered in a dose of 1.0 mg/kg significantly enhanced 24-h object recognition memory of aroused rats in the WITHOUT-habituation condition (**a**) but failed to affect memory of nonaroused rats in the WITH-habituation condition (**b**). $^{**}P < 0.0001$ compared with the corresponding vehicle control group. Data are presented as mean ± SEM. Reprinted from Okuda et al., Proc. Natl. Acad. Sci. U.S.A., Okuda et al. (2004)

A link between the level of emotional arousal at encoding and the efficacy of glucocorticoids in influencing memory consolidation has also been demonstrated in humans. Cortisol administered shortly before or after learning selectively enhances long-term memory of emotionally arousing, but not of emotionally neutral, material (Buchanan and Lovallo 2001; Kuhlmann and Wolf 2006a). Moreover, a cold pressor stress in humans (i.e. placing the arm in ice water for up to 3 min), a procedure that significantly elevates endogenous cortisol levels, enhanced memory of emotionally arousing slides, but did not affect memory of emotionally neutral slides (Cahill et al. 2003). Consistent with these findings, it has been reported that levels of endogenous cortisol at the time of learning correlated with enhanced memory consolidation only in individuals who were emotionally aroused (Abercrombie et al. 2006). Thus, these findings from animal and human studies indicate that endogenous emotional arousal is essential for enabling glucocorticoid effects on memory consolidation.

7.2.1.2 Role of Arousal-Induced Noradrenergic Activation in the Amygdala

Why do stress hormones selectively enhance memory consolidation of emotionally arousing experiences? Our findings suggest that interactions between stress hormones and amygdala activity may be key in determining this selectivity. It is well established that emotional experiences that induce the release of adrenal stress hormones also activate the amygdala (Pelletier et al. 2005). Evidence from a large number of studies in both animals and humans has indicated that the amygdala plays a critical role in the induction of the stress response via influences on hypothalamic and brainstem control centers as well as in the processing of emotionally arousing information, including emotional influences on attention, perception, learning, and memory (LeDoux 2003; Phelps and LeDoux 2005). Extensive evidence indicates that the enhancing effects of stress hormone administration on the consolidation of memory of emotionally arousing experiences involve the amygdala. Experimentally induced damage to the amygdala, particularly basolateral amygdala (BLA), blocks the memory-modulatory effects of glucocorticoids (Roozendaal and McGaugh 1996). Moreover, and in support of this view, a glucocorticoid receptor agonist administered directly into the BLA enhances memory consolidation, whereas that of a glucocorticoid receptor antagonist impairs memory consolidation (Roozendaal and McGaugh 1997b).

Further findings indicate that glucocorticoids require training-associated noradrenergic activation within the BLA to influence memory of emotionally arousing training. In vivo microdialysis, a technique that is used to measure neurotransmitter levels in the brain, indicated that aversive stimulation of electrical footshock induces the release of norepinephrine in the amygdala of rats (McIntyre et al. 2002). Glucocorticoid administration rapidly augments such training-induced norepinephrine levels within the amygdala (McReynolds et al. 2010). On the other hand, blockade of norepinephrine signaling in the BLA with a β-adrenoceptor antagonist prevented the memory-enhancing effect of systemically administered glucocorticoids (Quirarte et al. 1997; Roozendaal et al. 2002, 2006a, b). These findings strongly suggest that because glucocorticoid effects on memory consolidation require noradrenergic activation within the BLA, they only modulate memory under emotionally arousing conditions that induce the release of norepinephrine (Roozendaal et al. 2006b).

Human studies have also provided evidence that stress and glucocorticoid effects on memory enhancement for emotionally arousing experiences require amygdala and noradrenergic activity (Adolphs et al. 1997; Cahill et al. 1995, 1996; Canli et al. 2000; Hamann et al. 1999). An interaction between glucocorticoids and noradrenergic activity within the amygdala in emotionally influenced memory has also been investigated in studies using functional MRI. Van Stegeren et al. (2007) reported that the relationship between amygdala activity during encoding and subsequent long-term memory was greatest for the most emotionally arousing stimuli and for participants with higher endogenous levels of cortisol. Importantly, β-adrenoceptor antagonists blocked both the increase in amygdala activity and the enhanced retention induced by emotional stimuli. Kukolja et al. (2008) furthermore showed increased amygdala activity in response to negatively arousing images after combined administration of reboxetine, a norepinephrine-reuptake inhibitor, and cortisol, but not with either drug alone.

Several experimental findings further suggested that glucocorticoid effects on increasing noradrenergic signaling might have an onset that is too fast to be mediated via transcriptional regulation in the nucleus and involve a rapid, nongenomic mode of action (Dallman 2005; de Kloet 2000; Popoli et al. 2012). Nongenomic glucocorticoid actions likely involve the activation of a membrane-associated variant(s) of the steroid receptor (Dallman 2005; Johnson et al. 2005; Riedemann et al. 2010). Recent findings indicate that such rapid actions of glucocorticoid hormones on memory consolidation involve the endocannabinoid system. The endocannabinoid system is a fast lipid system in the brain and recently emerged as an important stress-response system (Atsak et al. 2012b; Campolongo et al. 2009; Evanson et al. 2010; Hill and McEwen 2009; Hill and Tasker 2012; Morena et al. 2016). Hill et al. (2010) showed that a single glucocorticoid administration rapidly (within 10 min) elevated levels of the endocannabinoid anandamide in the brain. Endocannabinoid signaling appears to be essentially involved in mediating glucocorticoid effects on memory consolidation as the cannabinoid type 1 (CB1) receptor antagonist AM-251 administered into the BLA blocked the ability of posttraining systemic corticosterone to facilitate memory consolidation (Campolongo et al. 2009). As shown in Fig. 7.2a, emotionally arousing training also rapidly increases anandamide levels in the BLA (Morena et al. 2014). To investigate whether glucocorticoid-endocannabinoid interactions on memory consolidation are dependent upon an adrenal steroid receptor on the cell surface, we performed an additional experiment. As shown in Fig. 7.2b, c, AM-251 infused into the BLA blocked the memory-enhancing effects induced by concurrent administration of either a specific glucocorticoid receptor agonist or the membrane-impermeable ligand cort:BSA (Atsak et al. 2015). In contrast, a glucocorticoid receptor antagonist infused into the BLA did not alter the memory-enhancing effects of a CB1 receptor agonist (Fig. 7.2d). Therefore, these findings indicate that glucocorticoid effects on downstream endocannabinoid transmission are required, presumably

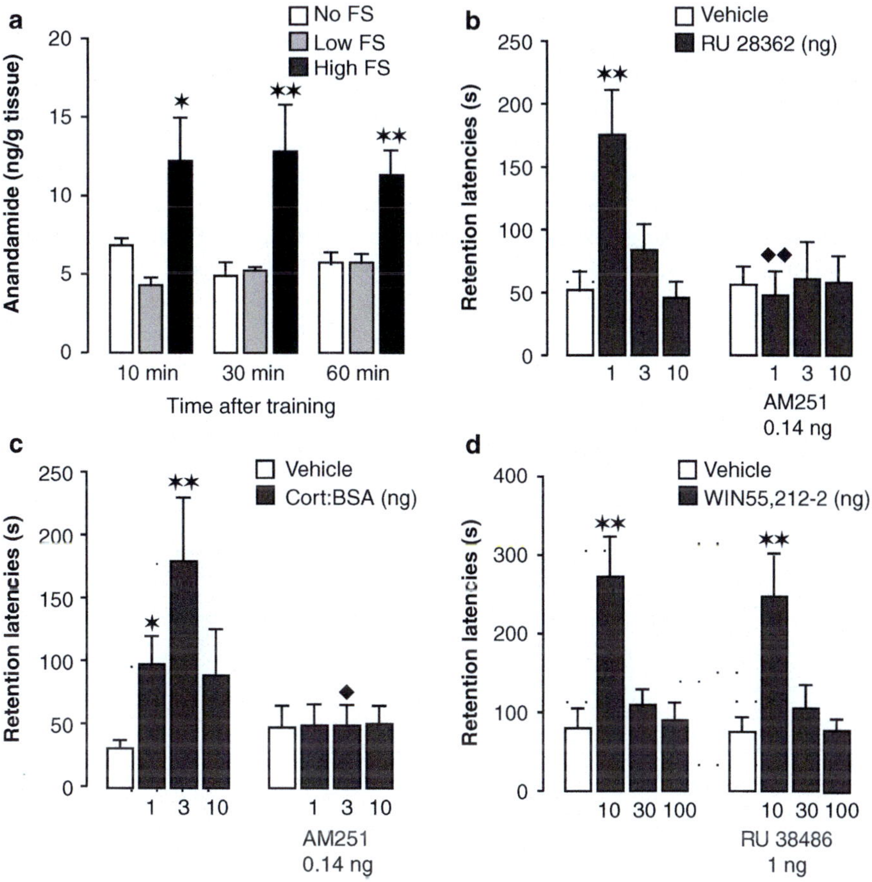

Fig. 7.2 Glucocorticoids interact with the endocannabinoid system of the BLA in enhancing memory consolidation of inhibitory avoidance training. (**a**) Training rats on an inhibitory avoidance task with a high footshock (High FS; 0.45 mA, 1 s), but not low footshock (Low FS; 0.35 mA, 1 s), intensity increased anandamide levels in the amygdala 10, 30, and 60 min after the training. $^*P < 0.05$, $^{**}P < 0.01$ vs. the no footshock (No FS) group. Results represent mean ± SEM. (**b**) Forty-eight-hour retention latencies (in s, mean ± SEM) of rats given immediate posttraining intra BLA infusions of the glucocorticoid receptor agonist RU 28362 (1, 3 or 10 ng in 0.2 μl) alone or together with the CB1 receptor antagonist AM-251 (0.14 ng). $^{**}P < 0.01$ vs. vehicle; ♦♦$P < 0.01$ vs. RU 28362 alone. (**c**) Forty-eight-hour retention latencies of rats given immediate posttraining intra-BLA infusions of the membrane-impermeable glucocorticoid cort:BSA (1, 3 or 10 ng in 0.2 μl) alone or together with AM-251 (0.14 ng). $^*P < 0.05$; $^{**}P < 0.01$ vs. vehicle; ♦$P < 0.05$ vs. cort:BSA alone. (**d**) Forty-eight-hour retention latencies of rats given immediate posttraining intra-BLA infusions of the cannabinoid agonist WIN55,212-2 (10, 30 or 100 ng in 0.2 μl) alone or together with the glucocorticoid antagonist RU 38486 (1 ng). $^{**}P < 0.01$ vs. vehicle. Adapted from Morena et al., Proc. Natl. Acad. Sci. U.S.A., 2014 and Atsak et al., Neuropsychopharmacology (2015)

involving the activation of a glucocorticoid receptor on the cell surface. Subsequent experiments indicated that such glucocorticoid interactions with the endocannabinoid system also mediate the rapid effects of glucocorticoids onto the noradrenergic system (Atsak et al. 2015).

As illustrated in Fig. 7.3, these findings thus suggest that glucocorticoids might bind to a glucocorticoid receptor on the cell surface and rapidly induce the release of endocannabinoids. The released endocannabinoids might then bind to CB1 receptors on GABAergic interneurons and inhibit the release of GABA that can then result in an increased noradrenergic transmission in BLA neurons.

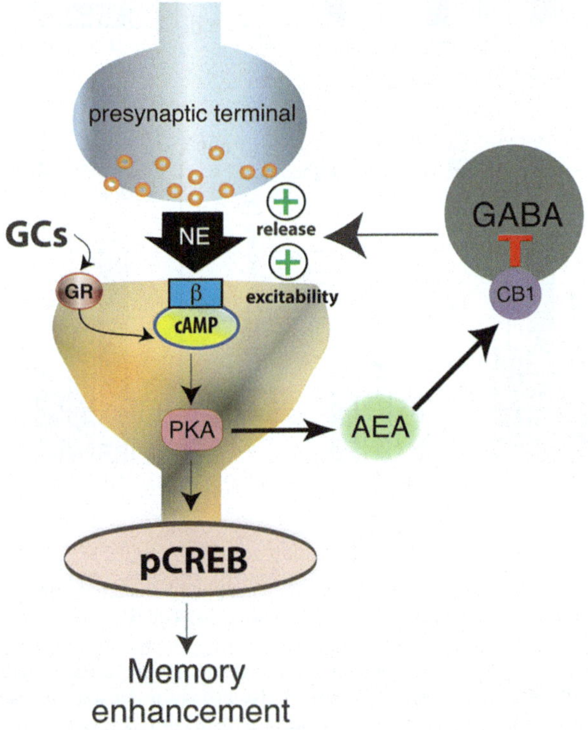

Fig. 7.3 A model illustrating the role of the endocannabinoid system in integrating the effects of glucocorticoids and norepinephrine within the BLA on memory consolidation. Glucocorticoids, released during emotionally arousing situations, bind to a membrane-bound glucocorticoid receptor (GR), and activate the intracellular cAMP/protein kinase A (PKA) signaling cascade. This triggers the release of endocannabinoids, particularly anandamide (AEA). Anandamide then activates CB1 receptors on GABAergic interneurons and thereby inhibit GABA release. This subsequently disinhibits norepinephrine (NE) release, and increases the excitability of pyramidal neurons within the BLA. This overall increases the sensitivity of BLA neurons to the effects of norepinephrine and results in an increased activation of the cAMP/PKA pathway and phosphorylation of cAMP response element-binding (pCREB) protein. These stress hormone effects in the BLA are required for the optimal enhancement of memory for emotionally arousing experiences by influencing information storage processes in other brain regions. Adapted from Atsak et al., Neuropsychopharmacology (2015)

7.2.1.3 Interaction of the Amygdala with Other Brain Regions

Although the BLA might be a critical site involved in encoding, storage, and expression of fear-based memories (LeDoux 2003), an involvement of the BLA in modulating long-term memory formation has been obtained in experiments using many different kinds of training tasks. As these different training experiences are known to engage different brain systems (Izquierdo et al. 1997; McGaugh 2002; Packard et al. 1994; Packard and Knowlton 2002), the BLA-induced modulation undoubtedly involves influences on processing, occurring in these brain regions (McGaugh and Roozendaal 2002; Roozendaal and McGaugh 2011). For example, glucocorticoids administered into the rat hippocampus influence memory of inhibitory avoidance or water-maze spatial training (Roozendaal and McGaugh 1997a; Roozendaal et al. 1999a), which is consistent with a role of the hippocampus in processing spatial/contextual information (Maren and Fanselow 1997; Lisman et al. 2017; Morris et al. 1982; Sacchetti et al. 1999). Importantly, damage to the BLA or the administration of a β-adrenoceptor antagonist into the BLA blocks the enhancing effect of posttraining intra-hippocampal infusions of a glucocorticoid receptor agonist on inhibitory avoidance memory (Roozendaal and McGaugh 1997a; Roozendaal et al. 1999a). Thus, these findings indicate that an intact and functional BLA is required for enabling memory modulation of spatial/contextual information induced by manipulation of glucocorticoid receptor activity in the hippocampus. Other studies indicated similar interactions of the BLA with a wide variety of other brain regions, including the striatum, medial prefrontal cortex, anterior cingulate cortex, and insular cortex, in regulating stress hormone effects on synaptic plasticity and memory consolidation underlying different kinds of emotionally arousing experiences (Packard et al. 1994; McIntyre et al. 2005; Malin and McGaugh 2006; Chavez et al. 2013; Bass et al. 2014; Bermudez-Rattoni 2014; Lovitz and Thompson 2015; Atucha et al. 2017).

Guided by animal research, human neuroimaging studies also provide evidence indicating that amygdala activation influences memory processing in other brain regions. For example, activity of the amygdala and hippocampal/parahippocampal regions are correlated during emotional experiences (Cahill et al. 1996; Hamann et al. 1999; Fastenrath et al. 2014) and such activation is correlated with subsequent retention (Dolcos et al. 2004). Functional connectivity network modeling in humans has further demonstrated that emotionally arousing tasks that activate the amygdala also lead to a consistent and robust coactivation of a specific set of other brain regions, such as the dorsal anterior cingulate cortex, medial prefrontal cortex, anterior insula, caudate nucleus, hypothalamus as well as brainstem and midbrain regions (Hermans et al. 2011). Interestingly, functional connectivity, in particular within the cingulo-opercular "salience" network (Menon 2011; Hermans et al. 2014) is regulated by stress hormones (Hermans et al. 2011). Thus, a critical question is whether salience network activity is involved in the formation of an interoceptive representation of the physiological state evoked by prior salient events, and contributes to long-term memory (Damasio and Carvalho 2013).

7.2.2 Glucocorticoid Effects on Memory Retrieval

Many studies indicate that stress and glucocorticoids not only modulate the strength of newly formed memories but also influence the remembrance of previously acquired information. In contrast to the enhancing effects of glucocorticoids on memory consolidation, these hormones typically impair memory retrieval. In the first study investigating the effects of stress and glucocorticoids on retrieval processes (de Quervain et al. 1998), we reported that 30 min after exposure to footshock stress, rats displayed impaired retention of spatial memory of a water-maze task they had acquired 24 h earlier. The water-maze task is a commonly used learning task in which rats make use of spatial cues in their environment to learn navigating to an invisible escape platform in a tank filled with water. Interestingly, memory performance was not impaired when rats were tested either 2 min or 4 h after the stress exposure. These time-dependent effects on retrieval processes corresponded to the circulating glucocorticoid levels at the time of testing, which suggested that the retrieval impairment was directly related to increased adrenocortical function. In a next step, we have translated these findings to healthy humans and found that a single administration of cortisone impaired the recall of words learned 24 h earlier (de Quervain et al. 2000). Several further studies from different laboratories have indicated that glucocorticoids impair memory retrieval of spatial or contextual memory in rats and declarative (mostly episodic) memory in humans (Buss et al. 2004; Coluccia et al. 2008; de Quervain et al. 2003; Het et al. 2005; Kuhlmann et al. 2005a; Rashidy-Pour et al. 2004; Roozendaal et al. 2004b; Sajadi et al. 2007; Wolf 2008; Wolf et al. 2001). Moreover, increased cortisol levels after psychological stress exposure have also been shown to impair declarative memory retrieval (Buchanan et al. 2006; Domes et al. 2004; Kuhlmann et al. 2005b). Although the vast majority of studies have investigated the effects of stress and glucocorticoids on the retrieval of hippocampus-dependent forms of memory, other studies have shown that stress exposure or glucocorticoid administration also impairs the retrieval of cortex-dependent recognition memory (Barsegyan et al. 2015) and striatal-dependent stimulus-response associations, i.e., habit-like memory (Guenzel et al. 2013; Atsak et al. 2016).

Glucocorticoid effects on memory retrieval are highly comparable to those seen in studies investigating memory consolidation in that the effects depend on emotional arousal. Specifically, it has been shown in studies in humans that emotionally arousing information is especially sensitive to the retrieval-impairing effects of glucocorticoids (Buchanan et al. 2006; Buss et al. 2004; de Quervain et al. 2000, 2007a; Het et al. 2005; Kuhlmann et al. 2005a, b; Kuhlmann and Wolf 2006b; Schwabe et al. 2009; Smeets et al. 2008, 2009). Studies of rats investigating the neural mechanisms underlying this selectivity have indicated that glucocorticoid effects on memory retrieval also depend critically on noradrenergic activity within the brain (Roozendaal et al. 2004a, b). In line with this idea, and illustrated in Fig. 7.4, the β-blocker propranolol administered to healthy humans abolished the impairing effect of cortisone administration on the retrieval of emotionally arousing material (de Quervain et al. 2007a). Another study indicated that propranolol also blocks the

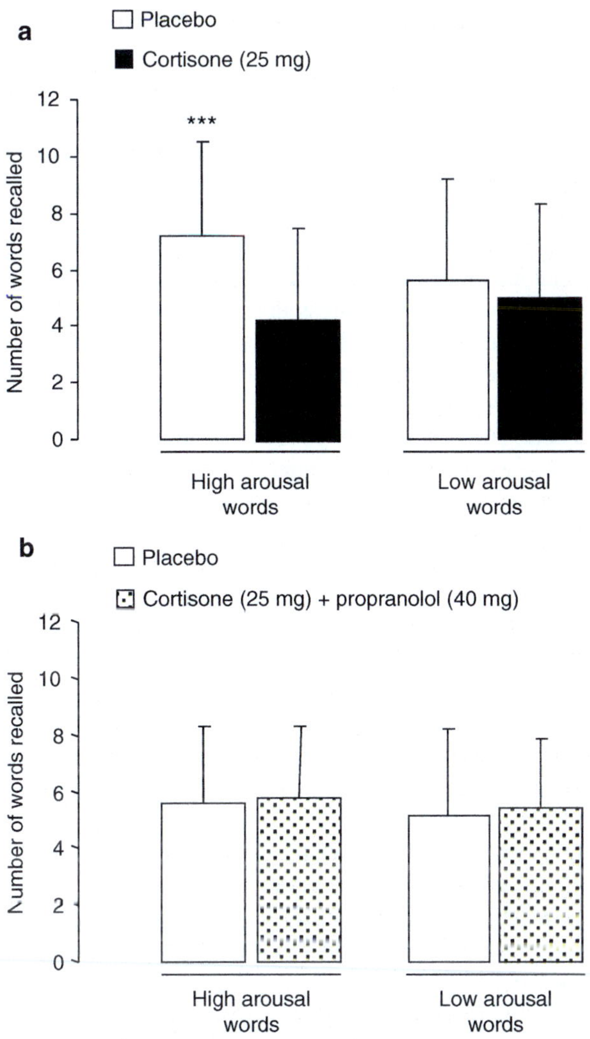

Fig. 7.4 Glucocorticoid effects on memory retrieval depend on emotional arousal. (**a**) Cortisone administered 1 h before recall testing impaired retrieval of high-arousal words, but not of low-arousal words. (**b**) Concurrent administration of the β-blocker propranolol prevented the impairing effect of cortisone on retrieving high-arousal words. Data are presented as mean ± SEM. ***$P < 0.0001$ compared with the corresponding cortisone condition. Adapted from de Quervain et al., Am. J. Psychiatry (2007)

impairing effect of stress induction on memory retrieval (Schwabe et al. 2009). The temporary impairment of glucocorticoids after stress induction or pharmacological administration on memory retrieval is fast and does not seem to depend on gene transcription (Sajadi et al. 2007). Recent findings indicate that endocannabinoid signaling is also involved in mediating glucocorticoid-induced impairment of memory retrieval (Atsak et al. 2012a; Morena et al. 2015).

7.2.3 Glucocorticoid Effects on Working Memory

Stress exposure is also known to impair working memory (Arnsten and Goldman-Rakic 1998; Schoofs et al. 2008). Working memory is a dynamic process whereby information is updated continuously, providing a temporary storage of information (Baddeley 1992; Jones 2002). Evidence from lesion, pharmacological, imaging, and clinical studies indicates that working memory depends on the integrity of the prefrontal cortex (Brito et al. 1982; Fuster 1991; Owen et al. 2005). Although basal levels of endogenous glucocorticoids are required to maintain prefrontal cortical function (Mizoguchi et al. 2004), stress doses of glucocorticoids impair working memory in rats (Roozendaal et al. 2004c). Stress or stress-level cortisol treatment is also known to impair working memory performance in human subjects during demanding tasks that require a high level of arousal (Baddeley 1992; Lupien et al. 1999; Schoofs et al. 2008; Wolf et al. 2001; Young et al. 1999). Importantly, glucocorticoid effects on working memory impairment also depend on interactions with arousal-associated noradrenergic mechanisms (Roozendaal et al. 2004c; Barsegyan et al. 2010). In humans, negative effects of stress induction on working memory (Elzinga and Roelofs 2005; Luethi et al. 2008; Oei et al. 2006; Taverniers et al. 2010; Qin et al. 2012; Schoofs et al. 2008, 2009) and prefrontal cortex activity (Ossewaarde et al. 2011; Qin et al. 2009) are usually observed within the time window in which both glucocorticoid and noradrenergic activity are elevated.

7.3 Modulatory Effects of Glucocorticoids on Emotional Memory: Implications for Anxiety Disorders

From the findings reviewed above we have learned that glucocorticoids enhance the consolidation, but impair the retrieval, of memory of emotionally arousing experiences. Glucocorticoids also impair working memory (Fig. 7.5). Enhanced memory for emotional events is a well-recognized phenomenon, which helps us to remember important information. Although glucocorticoid-induced temporary impairments of memory retrieval and working memory may directly negatively influence task performance during highly stressful conditions, these effects should not *a priori* be regarded in a biological sense as being maladaptive. In fact, these effects may actually aid to an accurate storage of emotionally arousing information by blocking, for example, retroactive interference (Roozendaal 2002). However, whereas the enhanced memory for emotionally arousing events in most cases has a clear adaptive value, in certain circumstances extremely aversive experiences can also lead to highly emotional, traumatic or fearful memories, which contribute to the development and symptoms of psychological problems and anxiety disorders. Therefore, understanding the basic modulatory actions of glucocorticoids on different aspects of cognition may have important implications for understanding and, possibly, treating stress-related (anxiety) disorders. Although we focus in this chapter on the implications of memory-modulatory glucocorticoid effects for anxiety disorders, Earth-bound research on the role of glucocorticoids is likely to have

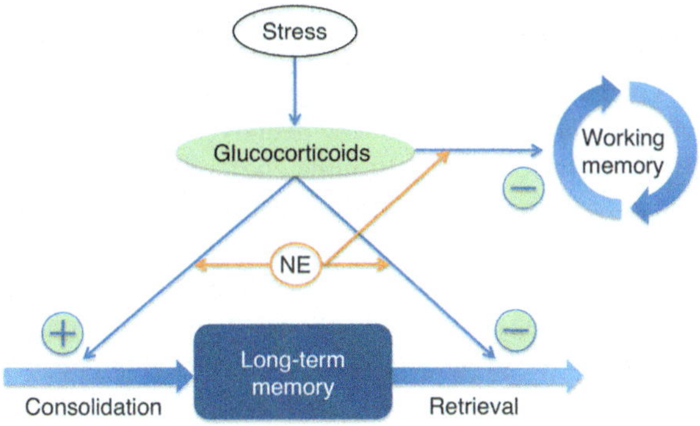

Fig. 7.5 Effects of stress and glucocorticoids on memory functions. Whereas glucocorticoids enhance memory consolidation, they impair memory retrieval and working memory. All of these hormone effects depend on emotional arousal-induced activation of noradrenergic transmission. *NE* norepinephrine. Reprinted from de Quervain et al., Front. Neuroendocrinol. (2009), with permission

important implications for other psychiatric disorders, such as depression or schizophrenia, as well (Belanoff et al. 2001; Newcomer et al. 1998; Rubinow et al. 1984; Starkman et al. 1981).

7.3.1 The Role of Aversive Memories in Anxiety Disorders

Several lines of evidence indicate that after an aversive experience the formation of an aversive memory trace is an important pathogenic mechanism for the development of anxiety disorders, such as PTSD or phobias (LeDoux 2003; Mineka and Oehlberg 2008; Phelps and LeDoux 2005; Pitman 1989, Yehuda and LeDoux 2007). Neuroimaging studies have shown that while the prefrontal cortex seems to be hyporesponsive, amygdala activity in response to viewing aversive information is exaggerated in patients with PTSD as compared to healthy controls and, importantly, correlates positively with later recall of the aversive information, and with PTSD symptom severity (Armony et al. 2005; Dickie et al. 2008; Francati et al. 2007; Heim and Nemeroff 2009; Rauch et al. 2000; Shin et al. 2006, 2005). These findings are in line with the well-known role of the amygdala in the formation of emotional memory, as reviewed above. Furthermore, some evidence indicates that the administration of a β-blocker, which is known to impair the consolidation of memory of emotionally arousing experiences, might be preventive with regard to the development of subsequent PTSD (Pitman et al. 2002). These findings underscore the important pathogenic role of aversive memory formation in the development of PTSD. However, the formation of a strong aversive memory trace is, of course, not sufficient to develop an anxiety disorder. In fact, building strong memories of an aversive event is a primarily adaptive mechanism and even intrusive thoughts (intrusive memory retrieval) and related symptoms are normal reactions in the initial period after an aversive experience. In individuals who do not develop an anxiety disorder, which fortunately is mostly the case, intrusive memory retrieval declines with the passing of time, although a strong aversive memory can still be recalled voluntarily even after a long time. In contrast, in individuals who do develop a chronic anxiety disorder, the aversive memory trace remains easily reactivatable by an aversive cue (e.g. trauma cue), or even spontaneously, leading to uncontrollable aversive memory retrieval and related clinical symptoms (reexperiencing in PTSD).

PTSD is a chronic response to a traumatic event and characterized by the following features: Reexperiencing of the traumatic event, avoidance of stimuli associated with the trauma, and hyperarousal. Reexperiencing symptoms include intrusive daytime recollections, traumatic nightmares, and flashbacks in which components of the event are relived (American Psychiatric Association 1994; Yehuda 2002b). These reexperiencing symptoms result from excessive retrieval of traumatic memories which often retain their vividness and power to evoke distress for decades or even a lifetime. Importantly, traumatic reexperiencing phenomena are again consolidated (reconsolidated) into memory, which cements the traumatic memory trace (see the discussion of the concept of intrusions in the PTSD research literature (e.g. Brewin 2001; Michael et al. 2005a, b). Persistent retrieval, reexperiencing and reconsolidation of traumatic memories is a process that keeps these memories vivid and thereby the disorder alive (see Sect. 7.3.3).

7.3.2 Glucocorticoids and PTSD

7.3.2.1 Preventive Effects of Glucocorticoids with Regard to the Development of PTSD

From the basic studies discussed above, we have learned that acutely elevated glucocorticoids enhance the consolidation of emotional memories. Based on these findings, it can be assumed that elevated glucocorticoid levels at the time of an aversive experience may contribute to the formation of traumatic and fearful memories. This idea is supported by a study on traumatic memories in critically ill patients. These patients often report traumatic memories from intensive care treatment and have a relatively high incidence of chronic stress symptoms and PTSD during follow-up (Schelling et al. 2004b). It was found that the number of traumatic memories from the intensive care unit correlated positively with the amount of cortisol acutely administered to patients undergoing cardiac surgery (Schelling et al. 2003). Theoretically, it might therefore be useful to therapeutically block glucocorticoid signaling immediately after a traumatic incident, as has been proposed for adrenergic signaling (Pitman and Delahanty 2005). However, to block initial consolidation of aversive memories, the anti-glucocorticoid treatment should be given shortly after the aversive event. This is usually not possible and therefore this approach seems difficult to implement. Moreover, there is evidence suggesting that reduced cortisol excretion in response to a traumatic event is actually associated with an increased risk of developing subsequent PTSD (Delahanty et al. 2000; McFarlane et al. 1997; Yehuda et al. 1998). These findings suggest that high glucocorticoid levels after an aversive event might be preventive with regard to the development of PTSD. This idea is strongly supported by studies showing that prolonged (several days) administration of stress doses of cortisol during intensive care treatment reduces the risk for later PTSD (Schelling et al. 2001, 2004a, 2006; Weis et al. 2006). But how do such findings of preventive effects fit with the idea that glucocorticoids enhance the formation of traumatic memories? After initial consolidation of traumatic experiences, which is likely to be enhanced by glucocorticoids, cortisol levels later on may play a crucial role in controlling the amount of retrieved traumatic memories. Specifically, by the known reducing effects of glucocorticoids on memory retrieval, high levels of these hormones may partly interrupt the vicious cycle of retrieving, reexperiencing, and reconsolidating aversive memories, thereby preventing a further cementation of the aversive memory trace. Studies showing that the preventive effects of glucocorticoid administration are also observed when the treatment started already at the time of the traumatic event (Schelling et al. 2004a; Weis et al. 2006) indicate that such an inhibitory effect of glucocorticoids on memory retrieval prevails the potentially enhancing effect on initial consolidation. Taken together, these findings suggest that elevated glucocorticoid levels (endogenously or pharmacologically) act preventively with regard to the development of PTSD. Some evidence suggests that interactions between altered glucocorticoid signaling and inflammatory responses might play a crucial role in the development of and vulnerability to PTSD (Gill et al. 2009). Two recent systematic reviews indicate that prolonged administration of glucocorticoids after a traumatic event is the most effective pharmacological intervention that is currently available for the prevention of PTSD (Amos et al. 2014; Sijbrandij et al. 2015). One review included seven randomized controlled trials of different pharmacological treatments (four with cortisol, three with the β-blocker propranolol, one with the selective serotonin-reuptake inhibitor escitalopram and one with the benzodiazepine temazepam) and reported that cortisol only showed efficacy in preventing PTSD development in adults (Amos et al. 2014). The other review included five placebo-controlled studies with cortisol and showed a large effect of cortisol in reducing the risk of PTSD (Sijbrandij et al. 2015).

7.3.2.2 Glucocorticoids Reduce the Retrieval of Traumatic Memories in Chronic PTSD

In addition to individuals at risk for PTSD, also patients with an established PTSD can show low endogenous cortisol levels (Bierer et al. 2006; Mason et al. 1986; Yehuda 2002a; Yehuda and Bierer 2008; Yehuda et al. 1995, 2007; but see: Pitman and Orr 1990; Young and Breslau 2004). A meta-analysis has shown that low cortisol levels depend on several factors, including gender and trauma type (Meewisse et al. 2007). Low cortisol levels may contribute to a hyper-retrieval of aversive memories. Based on the finding that glucocorticoids impair the retrieval of emotional information, we hypothesized that patients with chronic PTSD might benefit from glucocorticoid treatment. In an initial study, we tested this hypothesis in a small number of patients with chronic PTSD (Aerni et al. 2004). During a 3-month observation period, low-dose cortisol (10 mg per day) was administered orally for

1 month using a double-blind, placebo-controlled, crossover design. The administration of this low dose of cortisol for 1 month does not cause major side effects and does not suppress endogenous cortisol production (Cleare et al. 1999). To assess possible treatment effects on the retrieval of traumatic memories, the patients rated daily the intensity and frequency of the feeling of reliving the traumatic event and the physiological distress felt in response to traumatic memories and nightmares (self-administered rating scales from the Clinician Administered PTSD Scale questions). The results of this study indicated that low-dose cortisol treatment had beneficial effects with regard to reexperiencing symptoms and nightmares and we found evidence for cortisol effects that outlasted the treatment period (Aerni et al. 2004). However, a study, which used a similar design but in a larger group of patients receiving various psychotropic medications (including serotonin- or norepinephrine-reuptake inhibitors), did not find beneficial effects of cortisol (10 mg or 30 mg per day) treatment on PTSD symptoms (Ludascher et al. 2015). A recent randomized, double-blind, placebo-controlled trial in 24 veterans with PTSD showed that cortisol (30 mg) combined with exposure treatment improved treatment retention and outcome (Yehuda et al. 2015).

7.3.3 Possible Mode of Action of Glucocorticoids in the Reduction of Traumatic Memory

The findings of our clinical study suggest that glucocorticoid administration has acute beneficial effects on clinical symptoms by reducing the retrieval of traumatic memory (Aerni et al. 2004). Additionally, we found evidence that symptoms were reduced even after cessation of the treatment period. What might be the underlying mechanism? In PTSD, excessive retrieval of traumatic memory, which may be spontaneous or triggered by a trauma cue, leads to reexperiencing of the traumatic event (Michael and Ehlers 2007). The (re)consolidation of such aversive experiences further cements the aversive memory trace and thereby contributes to the persistence of the disorder (Fig. 7.6a). By inhibiting memory retrieval, cortisol may partly interrupt the vicious cycle of spontaneous retrieving, reexperiencing and reconsolidating traumatic memories in PTSD and, thereby, promote forgetting, a spontaneous process that occurs when memory is not reactivated (Fig. 7.6b).

Fig. 7.6 Model on the role of glucocorticoids in the reduction of aversive (traumatic) memory. (**a**) Excessive retrieval of aversive memories causes reexperiencing symptoms in PTSD. Reconsolidation of such aversive experiences further cements the aversive memory trace. (**b**) Glucocorticoid-induced reduction of the aversive memory trace. By inhibiting memory retrieval, glucocorticoids partly interrupt this vicious cycle of retrieving (1), reexperiencing (2), and reconsolidating (3) aversive memories, which leads to a weakening of the aversive memory trace (4). Furthermore, because the aversive cue is no longer followed by the usual aversive memory retrieval and related clinical symptoms, the cue becomes associated with a nonaversive experience, which is stored as extinction memory (5). Based on the findings of animal studies, glucocorticoids are likely to enhance long-term consolidation of extinction memory (see text for details). Reprinted from de Quervain et al., Front. Neuroendocrinol. (2009), with permission

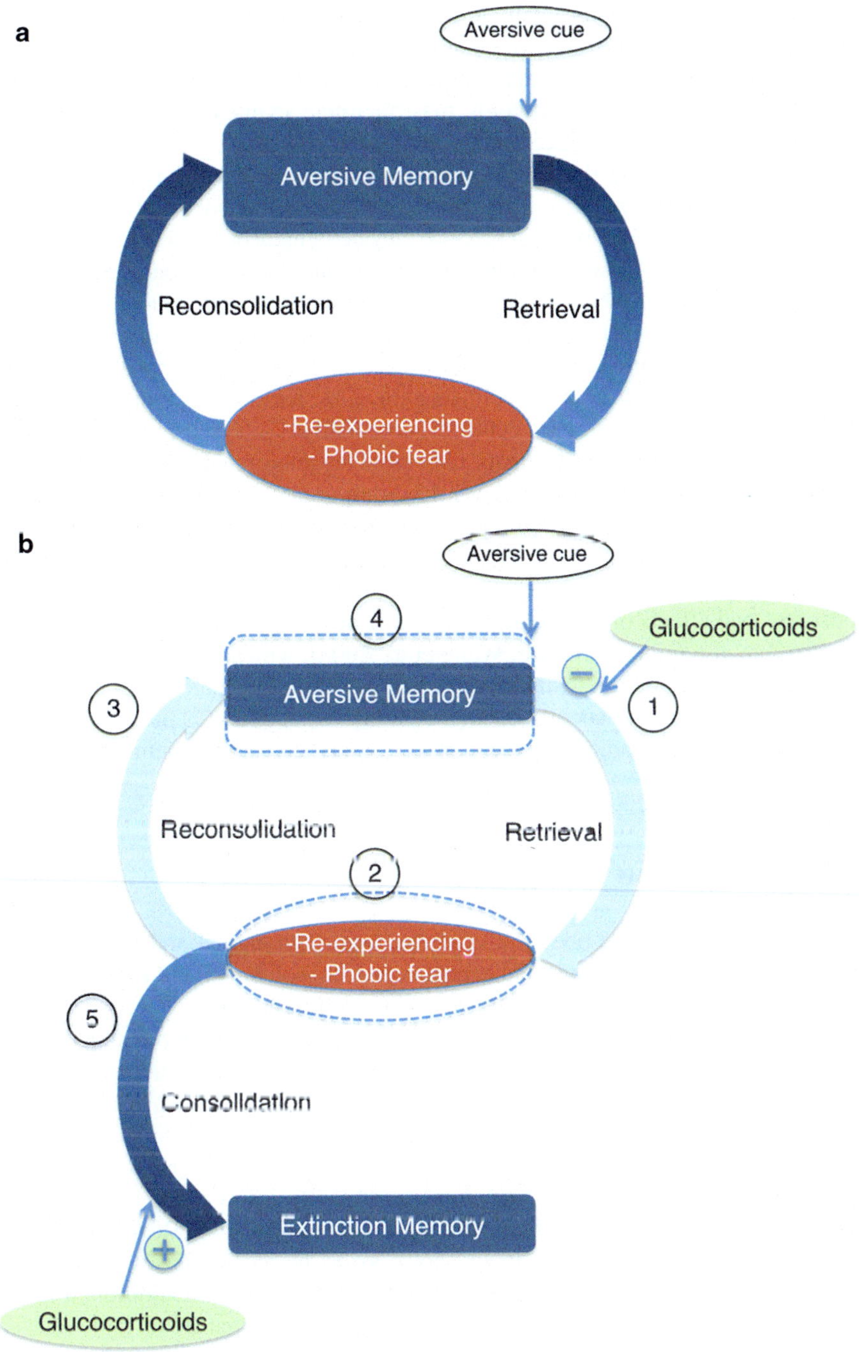
a
Aversive cue
Aversive Memory
Reconsolidation
Retrieval
-Re-experiencing
- Phobic fear
b
Aversive cue
4
Glucocorticoids
3
Aversive Memory
1
Reconsolidation
Retrieval
2
-Re-experiencing
- Phobic fear
5
Consolidation
Extinction Memory
Glucocorticoids

Furthermore, cortisol may facilitate the extinction of aversive memories, as evidenced by animal studies showing that glucocorticoid signaling promotes memory extinction processes (Barrett and Gonzalez-Lima 2004; Bohus and Lissak 1968; Yang et al. 2006). Glucocorticoids may facilitate extinction in two ways: (1) because of the cortisol-induced reduction of memory retrieval, an aversive cue is no longer followed by the usual aversive memory retrieval and related clinical symptoms but, instead, becomes associated with a nonaversive experience which is stored as extinction memory; (2) because elevated glucocorticoid levels are known to enhance the long-term consolidation of memories (Buchanan and Lovallo 2001; Flood et al. 1978; Kovacs et al. 1977; Kuhlmann and Wolf 2006a; Roozendaal 2000), it is possible that glucocorticoids facilitate the storage of corrective experiences. This is supported by animal studies showing that post-retrieval administration of glucocorticoids is able to enhance the consolidation of extinction memory (Abrari et al. 2008; Cai et al. 2006). Theoretically, such post-retrieval (or post-reactivation) glucocorticoid effects may also be interpreted as an inhibition of reconsolidation (Tronson and Taylor 2007; Wang et al. 2008). However, findings in animals suggest that reconsolidation of aversive memory is disrupted by blocking rather than by activating glucocorticoid signaling (Tronel and Alberini 2007). Furthermore, in favor of the memory extinction hypothesis, it has been shown that post-retrieval effects of glucocorticoids on memory are of transient nature and are reversed by a reminder (but see: Wang et al. 2008), which should not occur after inhibited reconsolidation. Although the data currently available rather speak for a facilitating effect of glucocorticoids on memory extinction, it is possible that, perhaps under certain conditions, glucocorticoids may also inhibit memory reconsolidation processes.

7.4 Conclusion

In this chapter, we have reviewed evidence from both animal and human studies indicating that glucocorticoids enhance memory consolidation, but impair memory retrieval and working memory (Fig. 7.5). These stress hormone effects depend on emotional arousal-induced activation of noradrenergic transmission within the BLA and on interactions of the BLA with other brain regions, such as the hippocampus and neocortical structures. Therefore, glucocorticoids, via BLA activation, can modulate memory processes of many different kinds of emotionally arousing experiences.

Enhanced consolidation for emotionally arousing information is an adaptive mechanism, which helps us to retain important information. Reduced memory retrieval and working memory should not *a prior*i be regarded as maladaptive as they support this process of retaining important information. In addition, a reduction of memory retrieval may aid to suppressing behaviors that are no more relevant or even maladaptive. This mechanism is especially important in more chronic situations when the organism is forced to adapt to a changed environment (e.g. during space flight). Under such conditions, also the facilitating effects of glucocorticoids on extinction processes represent an adaptive response, which helps the organism to deal with stressful events (De Kloet et al. 1999; McEwen 1998).

Because emotionally aversive memories play an important role in the development and symptomatology of stress-related (anxiety) disorders, we aimed to translate the basic findings on the effects of glucocorticoids on emotional memory in animals and healthy humans to clinical conditions. Specifically, the findings, which indicated that glucocorticoids reduce memory retrieval and enhance extinction of emotional memories, led us to hypothesize that these stress hormones might be useful in the treatment of anxiety disorders. Clinical studies and studies in animal models of acquired fear indicate that glucocorticoid treatment indeed reduces the retrieval of traumatic memories and enhances extinction processes. These dual actions of glucocorticoids seem to be especially suited for the treatment of acquired fear. By inhibiting memory retrieval, glucocorticoids may partly interrupt the vicious cycle of spontaneous retrieving, reexperiencing and (re)consolidating aversive memories, in patients on Earth, as in space crew members who have undergone an unpredicted, traumatic emergency situation.

More research is needed to better understand the molecular underpinnings of glucocorticoid actions on different memory processes as well as the role of (epi) genetic differences across individuals (de Quervain et al. 2007b; Vukojevic et al. 2014). Such research might further promote the understanding of why some individuals become vulnerable to the development of psychological (mood) problems and anxiety disorders, whereas others are resilient or even gain strength from stressful experiences. This knowledge might be also of considerable importance when selecting crew members for a long-duration mission for space exploration. Moreover, because cognitive dysfunctions, stress disorders, and the effects of glucocorticoids are linked to downstream immune modulation, a more integrative view is needed, also with regard to possible countermeasures. Finally, as stress and stress hormones have a large impact on specific memory functions and space flight conditions are associated with many stressors, such as microgravity, sleep deprivation, pain, variable oxygenation status, radiation and confinement, research has been initiated to investigate these specific cognitive functions in space flight conditions together with the psycho-neuroendocrine and immune status.

Acknowledgments The authors and publisher gratefully acknowledge the permission to reproduce the copyright material (i.e. text extracts and figures from (de Quervain et al. 2009), Frontiers in Neuroendocrinology, Elsevier Inc.) in this book.

References

Abercrombie HC, Kalin NH, Thurow ME, Rosenkranz MA, Davidson RJ (2003) Cortisol variation in humans affects memory for emotionally laden and neutral information. Behav Neurosci 117(3):505–516

Abercrombie HC, Speck NS, Monticelli RM (2006) Endogenous cortisol elevations are related to memory facilitation only in individuals who are emotionally aroused. Psychoneuroendocrinology 31(2):187–196

Abrari K, Rashidy-Pour A, Semnanian S, Fathollahi Y (2008) Administration of corticosterone after memory reactivation disrupts subsequent retrieval of a contextual conditioned fear memory: dependence upon training intensity. Neurobiol Learn Mem 89(2):178–184
Adolphs R, Cahill L, Schul R, Babinsky R (1997) Impaired declarative memory for emotional material following bilateral amygdala damage in humans. Learn Mem 4(3):291–300
Aerni A, Traber R, Hock C, Roozendaal B, Schelling G, Papassotiropoulos A, Nitsch RM, Schnyder U, de Quervain DJ (2004) Low-dose cortisol for symptoms of posttraumatic stress disorder. Am J Psychiatry 161(8):1488–1490
American Psychiatric Association (1994) Diagnostic and statistical manual of mental disorders, fourth edition (DSM-IV). American Psychiatric Association, Washington, DC
Amos T, Stein DJ, Ipser JC (2014) Pharmacological interventions for preventing post-traumatic stress disorder (PTSD). Cochrane Database Syst Rev 7:CD006239
Arbel I, Kadar T, Silbermann M, Levy A (1994) The effects of long-term corticosterone administration on hippocampal morphology and cognitive performance of middle-aged rats. Brain Res 657(1–2):227–235
Armony JL, Corbo V, Clement MH, Brunet A (2005) Amygdala response in patients with acute PTSD to masked and unmasked emotional facial expressions. Am J Psychiatry 162(10):1961–1963
Arnsten AF, Goldman-Rakic PS (1998) Noise stress impairs prefrontal cortical cognitive function in monkeys: evidence for a hyperdopaminergic mechanism. Arch Gen Psychiatry 55(4):362–368
Atsak P, Hauer D, Campolongo P, Schelling G, McGaugh JL, Roozendaal B (2012a) Glucocorticoids interact with the hippocampal endocannabinoid system in impairing retrieval of contextual fear memory. Proc Natl Acad Sci U S A 109(9):3504–3509
Atsak P, Roozendaal B, Campolongo P (2012b) Role of the endocannabinoid system in regulating glucocorticoid effects on memory for emotional experiences. Neuroscience 204:104–116
Atsak P, Hauer D, Campolongo P, Schelling G, Fornari RV, Roozendaal B (2015) Endocannabinoid signaling within the basolateral amygdala integrates multiple stress hormone effects on memory consolidation. Neuropsychopharmacology 40(6):1485–1494
Atsak P, Guenzel FM, Kantar-Gok D, Zalachoras I, Yargicoglu P, Meijer OC, Quirarte GL, Wolf OT, Schwabe L, Roozendaal B (2016) Glucocorticoids mediate stress-induced impairment of retrieval of stimulus – response memory. Psychoneuroendocrinology 67:207–215
Atucha E, Vukojevic V, Fornari RV, Ronzoni G, Demougin P, Peter F, Atsak P, Coolen MW, Papassotiropoulos A, McGaugh JL, de Quervain DJ, Roozendaal B (2017) Noradrenergic activation of the basolateral amygdala maintains hippocampus-dependent accuracy of remote memory. Proc Natl Acad Sci U S A 114(34):9176–9181
Baddeley A (1992) Working memory. Science 255(5044):556–559
Barrett D, Gonzalez-Lima F (2004) Behavioral effects of metyrapone on Pavlovian extinction. Neurosci Lett 371(2–3):91–96
Barsegyan A, Mackenzie SM, Kurose BD, McGaugh JL, Roozendaal B (2010) Glucocorticoids in the prefrontal cortex enhance memory consolidation and impair working memory by a common neural mechanism. Proc Natl Acad Sci U S A 107(38):16655–16660
Barsegyan A, Atsak P, Hornberger WB, Jacobson PB, van Gaalen MM, Roozendaal B (2015) The vasopressin 1b receptor antagonist A-988315 blocks stress effects on the retrieval of object-recognition memory. Neuropsychopharmacology 40(8):1979–1989
Bass DI, Nizam ZG, Partain KN, Wang A, Manns JR (2014) Amygdala-mediated enhancement of memory for specific events depends on the hippocampus. Neurobiol Learn Mem 107:37–41
Beckwith BE, Petros TV, Scaglione C, Nelson J (1986) Dose-dependent effects of hydrocortisone on memory in human males. Physiol Behav 36(2):283–286
Belanoff JK, Kalehzan M, Sund B, Fleming Ficek SK, Schatzberg AF (2001) Cortisol activity and cognitive changes in psychotic major depression. Am J Psychiatry 158(10):1612–1616
Bermudez-Rattoni F (2014) The forgotten insular cortex: its role on recognition memory formation. Neurobiol Learn Mem 109:207–216

Bierer LM, Tischler L, Labinsky E, Cahill S, Foa E, Yehuda R (2006) Clinical correlates of 24-h cortisol and norepinephrine excretion among subjects seeking treatment following the world trade center attacks on 9/11. Ann N Y Acad Sci 1071:514–520

Bohus B, Lissak K (1968) Adrenocortical hormones and avoidance behaviour of rats. Int J Neuropharmacol 7(4):301–306

Brewin CR (2001) A cognitive neuroscience account of posttraumatic stress disorder and its treatment. Behav Res Ther 39(4):373–393

Brito GN, Thomas GJ, Davis BJ, Gingold SI (1982) Prelimbic cortex, mediodorsal thalamus, septum, and delayed alternation in rats. Exp Brain Res 46(1):52–58

Buchanan TW, Lovallo WR (2001) Enhanced memory for emotional material following stress-level cortisol treatment in humans. Psychoneuroendocrinology 26(3):307–317

Buchanan TW, Tranel D, Adolphs R (2006) Impaired memory retrieval correlates with individual differences in cortisol response but not autonomic response. Learn Mem 13(3):382–387

Buss C, Wolf OT, Witt J, Hellhammer DH (2004) Autobiographic memory impairment following acute cortisol administration. Psychoneuroendocrinology 29(8):1093–1096

Cahill L, Babinsky R, Markowitsch HJ, McGaugh JL (1995) The amygdala and emotional memory. Nature 377(6547):295–296

Cahill L, Haier RJ, Fallon J, Alkire MT, Tang C, Keator D, Wu J, McGaugh JL (1996) Amygdala activity at encoding correlated with long-term, free recall of emotional information. Proc Natl Acad Sci U S A 93(15):8016–8021

Cahill L, Gorski L, Le K (2003) Enhanced human memory consolidation with post-learning stress: interaction with the degree of arousal at encoding. Learn Mem 10(4):270–274

Cai WH, Blundell J, Han J, Greene RW, Powell CM (2006) Postreactivation glucocorticoids impair recall of established fear memory. J Neurosci 26(37):9560–9566

Campolongo P, Roozendaal B, Trezza V, Hauer D, Schelling G, McGaugh JL, Cuomo V (2009) Endocannabinoids in the rat basolateral amygdala enhance memory consolidation and enable glucocorticoid modulation of memory. Proc Natl Acad Sci U S A 106(12):4888–4893

Canli T, Zhao Z, Brewer J, Gabrieli JD, Cahill L (2000) Event-related activation in the human amygdala associates with later memory for individual emotional experience. J Neurosci 20(19):RC99

Chavez CM, McGaugh JL, Weinberger NM (2013) Activation of the basolateral amygdala induces long-term enhancement of specific memory representations in the cerebral cortex. Neurobiol Learn Mem 101:8–18

Cibelli M, Fidalgo AR, Terrando N, Ma D, Monaco C, Feldmann M, Takata M, Lever IJ, Nanchahal J, Fanselow MS, Maze M (2010) Role of interleukin-1beta in postoperative cognitive dysfunction. Ann Neurol 68(3):360–368

Cleare AJ, Heap E, Malhi GS, Wessely S, O'Keane V, Miell J (1999) Low-dose hydrocortisone in chronic fatigue syndrome: a randomised crossover trial. Lancet 353(9151):455–458

Coluccia D, Wolf OT, Kollias S, Roozendaal B, Forster A, de Quervain DJ (2008) Glucocorticoid therapy-induced memory deficits: acute versus chronic effects. J Neurosci 28(13):3474–3478

Cooper HS Jr (1996) The loneliness of the long duration astronaut. Air Space 11(2):37–45

Cordero MI, Kruyt ND, Merino JJ, Sandi C (2002) Glucocorticoid involvement in memory formation in a rat model for traumatic memory. Stress 5(1):73–79

Dallman MF (2005) Fast glucocorticoid actions on brain: back to the future. Front Neuroendocrinol 26(3–4):103–108

Damasio A, Carvalho GB (2013) The nature of feelings: evolutionary and neurobiological origins. Nat Rev Neurosci 14(2):143–152

De Boer SF, Koopmans SJ, Slangen JL, Van der Gugten J (1990) Plasma catecholamine, corticosterone and glucose responses to repeated stress in rats: effect of interstressor interval length. Physiol Behav 47(6):1117–1124

De Kloet ER, Oitzl MS, Joels M (1999) Stress and cognition: are corticosteroids good or bad guys? Trends Neurosci 22(10):422–426

Delahanty DL, Raimonde AJ, Spoonster E (2000) Initial posttraumatic urinary cortisol levels predict subsequent PTSD symptoms in motor vehicle accident victims. Biol Psychiatry 48(9):940–947

Derecki NC, Cardani AN, Yang CH, Quinnies KM, Crihfield A, Lynch KR, Kipnis J (2010) Regulation of learning and memory by meningeal immunity: a key role for IL-4. J Exp Med 207(5):1067–1080

Dickie EW, Brunet A, Akerib V, Armony JL (2008) An fMRI investigation of memory encoding in PTSD: influence of symptom severity. Neuropsychologia 46(5):1522–1531

Dolcos F, Labar KS, Cabeza R (2004) Interaction between the amygdala and the medial temporal lobe memory system predicts better memory for emotional events. Neuron 42(5):855–863

Domes G, Heinrichs M, Rimmele U, Reichwald U, Hautzinger M (2004) Acute stress impairs recognition for positive words—association with stress-induced cortisol secretion. Stress 7(3):173–181

Elzinga BM, Roelofs K (2005) Cortisol-induced impairments of working memory require acute sympathetic activation. Behav Neurosci 119(1):98–103

Evanson NK, Tasker JG, Hill MN, Hillard CJ, Herman JP (2010) Fast feedback inhibition of the HPA axis by glucocorticoids is mediated by endocannabinoid signaling. Endocrinology 151(10):4811–4814

Fastenrath M, Coynel D, Spalek K, Milnik A, Gschwind L, Roozendaal B, Papassotiropoulos A, de Quervain DJ (2014) Dynamic modulation of amygdala-hippocampal connectivity by emotional arousal. J Neurosci 34(42):13935–13947

Flood JF, Vidal D, Bennett EL, Orme AE, Vasquez S, Jarvik ME (1978) Memory facilitating and anti-amnesic effects of corticosteroids. Pharmacol Biochem Behav 8(1):81–87

Francati V, Vermetten E, Bremner JD (2007) Functional neuroimaging studies in posttraumatic stress disorder: review of current methods and findings. Depress Anxiety 24(3):202–218

Fuster JM (1991) The prefrontal cortex and its relation to behavior. Prog Brain Res 87:201–211

Gill JM, Saligan L, Woods S, Page G (2009) PTSD is associated with an excess of inflammatory immune activities. Perspect Psychiatr Care 45(4):262–277

Guenzel FM, Wolf OT, Schwabe L (2013) Stress disrupts response memory retrieval. Psychoneuroendocrinology 38(8):1460–1465

Hamann SB, Ely TD, Grafton ST, Kilts CD (1999) Amygdala activity related to enhanced memory for pleasant and aversive stimuli. Nat Neurosci 2(3):289–293

Heim C, Nemeroff CB (2009) Neurobiology of posttraumatic stress disorder. CNS Spectr 14(1 Suppl 1):13–24

Hermans EJ, van Marle HJF, Ossewaarde L, Henckens MJAG, Qin S, van Kesteren MTR, Schoots VC, Cousijn H, Rijpkema M, Oostenveld R, Fernández G (2011) Stress-related noradrenergic activity prompts large-scale neural network reconfiguration. Science 334(6059):1151–1153

Hermans EJ, Henckens MJAG, Joëls M, Fernández G (2014) Dynamic adaptation of large-scale brain networks in response to acute stressors. Trends Neurosci 37(6):304–314

Het S, Ramlow G, Wolf OT (2005) A meta-analytic review of the effects of acute cortisol administration on human memory. Psychoneuroendocrinology 30(8):771–784

Hill MN, McEwen BS (2009) Endocannabinoids: the silent partner of glucocorticoids in the synapse. Proc Natl Acad Sci U S A 106(12):4579–4580

Hill MN, Tasker JG (2012) Endocannabinoid signaling, glucocorticoid-mediated negative feedback, and regulation of the hypothalamic-pituitary-adrenal axis. Neuroscience 204:5–16

Hill MN, Karatsoreos IN, Hillard CJ, McEwen BS (2010) Rapid elevations in limbic endocannabinoid content by glucocorticoid hormones in vivo. Psychoneuroendocrinology 35(9):1333–1338

Izquierdo I, Quillfeldt JA, Zanatta MS, Quevedo J, Schaeffer E, Schmitz PK, Medina JH (1997) Sequential role of hippocampus and amygdala, entorhinal cortex and parietal cortex in formation and retrieval of memory for inhibitory avoidance in rats. Eur J Neurosci 9(4):786–793

Johnson LR, Farb C, Morrison JH, McEwen BS, LeDoux JE (2005) Localization of glucocorticoid receptors at postsynaptic membranes in the lateral amygdala. Neuroscience 136(1):289–299

Jones MW (2002) A comparative review of rodent prefrontal cortex and working memory. Curr Mol Med 2(7):639–647

Kanas N (1997) Psychosocial value of space simulation for extended spaceflight. Adv Space Biol Med 6:81–91

de Kloet ER (2000) Stress in the brain. Eur J Pharmacol 405(1–3):187–198

Kovacs GL, Telegdy G, Lissak K (1977) Dose-dependent action of corticosteroids on brain serotonin content and passive avoidance behavior. Horm Behav 8(2):155–165

Kuhlmann S, Wolf OT (2006a) Arousal and cortisol interact in modulating memory consolidation in healthy young men. Behav Neurosci 120(1):217–223

Kuhlmann S, Wolf OT (2006b) A non-arousing test situation abolishes the impairing effects of cortisol on delayed memory retrieval in healthy women. Neurosci Lett 399(3):268–272

Kuhlmann S, Kirschbaum C, Wolf OT (2005a) Effects of oral cortisol treatment in healthy young women on memory retrieval of negative and neutral words. Neurobiol Learn Mem 83(2):158–162

Kuhlmann S, Piel M, Wolf OT (2005b) Impaired memory retrieval after psychosocial stress in healthy young men. J Neurosci 25(11):2977–2982

Kukolja J, Schläpfer TE, Keysers C, Klingmüller D, Maier W, Fink GR, Hurlemann R (2008) Modeling a negative response bias in the human amygdala by noradrenergic-glucocorticoid interactions. J Neurosci 28(48):12868–12876

LeDoux J (2003) The emotional brain, fear, and the amygdala. Cell Mol Neurobiol 23(4–5):727–738

Lisman J, Buzsáki G, Eichenbaum H, Nadel L, Ranganath C, Redish AD (2017) Viewpoints: how the hippocampus contributes to memory, navigation and cognition. Nat Neurosci 20(11):1434–1447

Liu L, Tsuji M, Takeda H, Takada K, Matsumiya T (1999) Adrenocortical suppression blocks the enhancement of memory storage produced by exposure to psychological stress in rats. Brain Res 821(1):134–140

Lovitz ES, Thompson LT (2015) Memory-enhancing intra-basolateral amygdala clenbuterol infusion reduces post-burst afterhyperpolarizations in hippocampal CA1 pyramidal neurons following inhibitory avoidance learning. Neurobiol Learn Mem 119:34–41

Ludascher P et al (2015) No evidence for differential dose effects of hydrocortisone on intrusive memories in female patients with complex post-traumatic stress disorder—a randomized, double-blind, placebo-controlled, crossover study. J Psychopharmacol 29:1077–1084

Luethi M, Meier B, Sandi C (2008) Stress effects on working memory, explicit memory, and implicit memory for neutral and emotional stimuli in healthy men. Front Behav Neurosci 2:5

Luine VN, Spencer RL, McEwen BS (1993) Effects of chronic corticosterone ingestion on spatial memory performance and hippocampal serotonergic function. Brain Res 616(1–2):65–70

Lupien SJ, Gillin CJ, Hauger RL (1999) Working memory is more sensitive than declarative memory to the acute effects of corticosteroids: a dose-response study in humans. Behav Neurosci 113(3):420–430

Maheu FS, Joober R, Beaulieu S, Lupien SJ (2004) Differential effects of adrenergic and corticosteroid hormonal systems on human short- and long-term declarative memory for emotionally arousing material. Behav Neurosci 118(2):420–428

Malin EL, McGaugh JL (2006) Differential involvement of the hippocampus, anterior cingulate cortex and basolateral amygdala in memory for context and footshock. Proc Natl Acad Sci U S A 103(6):1959–1963

Maren S, Fanselow MS (1997) Electrolytic lesions of the fimbria/fornix, dorsal hippocampus, or entorhinal cortex produce anterograde deficits in contextual fear conditioning in rats. Neurobiol Learn Mem 67(2):142–149

Mason JW, Giller EL, Kosten TR, Ostroff RB, Podd L (1986) Urinary free-cortisol levels in post-traumatic stress disorder patients. J Nerv Ment Dis 174(3):145–149

McCarty R, Gold PE (1981) Plasma catecholamines: effects of footshock level and hormonal modulators of memory storage. Horm Behav 15(2):168–182

McEwen BS (1998) Protective and damaging effects of stress mediators. N Engl J Med 338(3):171–179

McFarlane AC, Atchison M, Yehuda R (1997) The acute stress response following motor vehicle accidents and its relation to PTSD. Ann N Y Acad Sci 821:437–441

McGaugh JL (2002) Memory consolidation and the amygdala: a systems perspective. Trends Neurosci 25(9):456

McGaugh JL (2003) Memory and emotion: the making of lasting memory. Columbia University Press, New York, NY
McGaugh JL, Roozendaal B (2002) Role of adrenal stress hormones in forming lasting memories in the brain. Curr Opin Neurobiol 12(2):205–210
McIntyre CK, Hatfield T, McGaugh JL (2002) Amygdala norepinephrine levels after training predict inhibitory avoidance retention performance in rats. Eur J Neurosci 16(7):1223–1226
McIntyre CK, Miyashita T, Setlow B, Marjon KD, Steward O, Guzowski JF, McGaugh JL (2005) Memory-influencing intra-basolateral amygdala drug infusions modulate expression of Arc protein in the hippocampus. Proc Natl Acad Sci U S A 102(30):10718–10723
McReynolds JR, Donowho K, Abdi A, McGaugh JL, Roozendaal B, McIntyre CK (2010) Memory-enhancing corticosterone treatment increases amygdala norepinephrine and Arc protein expression in hippocampal synaptic fractions. Neurobiol Learn Mem 93(3):312–321
Meewisse ML, Reitsma JB, de Vries GJ, Gersons BP, Olff M (2007) Cortisol and post-traumatic stress disorder in adults: systematic review and meta-analysis. Br J Psychiatry 191:387–392
Menon V (2011) Large-scale brain networks and psychopathology: a unifying triple network model. Trends Cogn Sci 15(10):483–506
Michael T, Ehlers A (2007) Enhanced perceptual priming for neutral stimuli occurring in a traumatic context: two experimental investigations. Behav Res Ther 45(2):341–358
Michael T, Ehlers A, Halligan SL (2005a) Enhanced priming for trauma-related material in posttraumatic stress disorder. Emotion 5(1):103–112
Michael T, Ehlers A, Halligan SL, Clark DM (2005b) Unwanted memories of assault: what intrusion characteristics are associated with PTSD? Behav Res Ther 43(5):613–628
Mineka S, Oehlberg K (2008) The relevance of recent developments in classical conditioning to understanding the etiology and maintenance of anxiety disorders. Acta Psychol (Amst) 127(3):567–580
Mizoguchi K, Ishige A, Takeda S, Aburada M, Tabira T (2004) Endogenous glucocorticoids are essential for maintaining prefrontal cortical cognitive function. J Neurosci 24(24):5492–5499
Morena M, Roozendaal B, Trezza V, Ratano P, Peloso A, Hauer D, Atsak P, Trabace L, Cuomo V, McGaugh JL, Schelling G, Campolongo P (2014) Endogenous cannabinoid release within prefrontal-limbic pathways affects memory consolidation of emotional training. Proc Natl Acad Sci U S A 111(51):18333–18338
Morena M, De Castro V, Gray JM, Palmery M, Trezza V, Roozendaal B, Hill MN, Campolongo P (2015) Training-associated emotional arousal shapes endocannabinoid modulation of spatial memory retrieval in rats. J Neurosci 35(41):13962–13974
Morena M, Patel S, Bains JS, Hill MN (2016) Neurobiological interactions between stress and the endocannabinoid system. Neuropsychopharmacology 41(1):80–102
Morris RG, Garrud P, Rawlins JN, O'Keefe J (1982) Place navigation impaired in rats with hippocampal lesions. Nature 297(5868):681–683
Newcomer JW, Craft S, Askins K, Hershey T, Bardgett ME, Csernansky JG, Gagliardi AE, Vogler G (1998) Glucocorticoid interactions with memory function in schizophrenia. Psychoneuroendocrinology 23(1):65–72
Oei NYL, Everaerd WTAM, Elzinga BM, van Well S, Bermond B (2006) Psychosocial stress impairs working memory at high loads: an association with cortisol levels and memory retrieval. Stress (Amsterdam) 9(3):133–141
Okuda S, Roozendaal B, McGaugh JL (2004) Glucocorticoid effects on object recognition memory require training-associated emotional arousal. Proc Natl Acad Sci U S A 101(3):853–858
Ossewaarde L, Qin S, van Marle HJF, van Wingen GA, Fernández G, Hermans EJ (2011) Stress-induced reduction in reward-related prefrontal cortex function. NeuroImage 55(1):345–352
Owen AM, McMillan KM, Laird AR, Bullmore E (2005) N-Back working memory paradigm: a meta-analysis of normative functional neuroimaging studies. Hum Brain Mapp 25(1):46–59
Packard MG, Knowlton BJ (2002) Learning and memory functions of the Basal Ganglia. Annu Rev Neurosci 25:563–593
Packard MG, Cahill L, McGaugh JL (1994) Amygdala modulation of hippocampal-dependent and caudate nucleus-dependent memory processes. Proc Natl Acad Sci U S A 91(18):8477–8481

Pelletier JG, Likhtik E, Filali M, Pare D (2005) Lasting increases in basolateral amygdala activity after emotional arousal: implications for facilitated consolidation of emotional memories. Learn Mem 12(2):96–102
Phelps EA, LeDoux JE (2005) Contributions of the amygdala to emotion processing: from animal models to human behavior. Neuron 48(2):175–187
Pitman RK (1989) Post-traumatic stress disorder, hormones, and memory. Biol Psychiatry 26(3):221–223
Pitman RK, Delahanty DL (2005) Conceptually driven pharmacologic approaches to acute trauma. CNS Spectr 10(2):99–106
Pitman RK, Orr SP (1990) Twenty-four hour urinary cortisol and catecholamine excretion in combat-related posttraumatic stress disorder. Biol Psychiatry 27(2):245–247
Pitman RK, Sanders KM, Zusman RM, Healy AR, Cheema F, Lasko NB, Cahill L, Orr SP (2002) Pilot study of secondary prevention of posttraumatic stress disorder with propranolol. Biol Psychiatry 51(2):189–192
Popoli M, Yan Z, McEwen BS, Sanacora G (2012) The stressed synapse: the impact of stress and glucocorticoids on glutamate transmission. Nat Rev Neurosci 13(1):22–37
Qin S, Hermans EJ, van Marle HJF, Luo J, Fernández G (2009) Acute psychological stress reduces working memory-related activity in the dorsolateral prefrontal cortex. Biol Psychiatry 66(1):25–32
Qin S, Cousijn H, Rijpkema M, Luo J, Franke B, Hermans EJ, Fernández G (2012) The effect of moderate acute psychological stress on working memory-related neural activity is modulated by a genetic variation in catecholaminergic function in humans. Front Integrative Neurosci 6:16
de Quervain DJ, Roozendaal B, McGaugh JL (1998) Stress and glucocorticoids impair retrieval of long-term spatial memory. Nature 394(6695):787–790
de Quervain DJF, Roozendaal B, Nitsch RM, McGaugh JL, Hock C (2000) Acute cortisone administration impairs retrieval of long-term declarative memory in humans. Nat Neurosci 3(4):313–314
de Quervain DJ, Henke K, Aerni A, Treyer V, McGaugh JL, Berthold T, Nitsch RM, Buck A, Roozendaal B, Hock C (2003) Glucocorticoid-induced impairment of declarative memory retrieval is associated with reduced blood flow in the medial temporal lobe. Eur J Neurosci 17(6):1296–1302
de Quervain DJ, Aerni A, Roozendaal B (2007a) Preventive effect of {beta}-adrenoceptor blockade on glucocorticoid-induced memory retrieval deficits. Am J Psychiatry 164(6):967–969
de Quervain DJ, Kolassa IT, Ertl V, Onyut PL, Neuner F, Elbert T, Papassotiropoulos A (2007b) A deletion variant of the alpha2b-adrenoceptor is related to emotional memory in Europeans and Africans. Nat Neurosci 10(9):1137–1139
de Quervain DJ, Aerni A, Schelling G, Roozendaal B (2009) Glucocorticoids and the regulation of memory in health and disease. Front Neuroendocrinol 30(3):358–370
de Quervain D, Schwabe L, Roozendaal B (2017) Stress, glucocorticoids and memory: implications for treating fear-related disorders. Nat Rev Neurosci 18(1):7–19
Quirarte GL, Roozendaal B, McGaugh JL (1997) Glucocorticoid enhancement of memory storage involves noradrenergic activation in the basolateral amygdala. Proc Natl Acad Sci U S A 94(25):14048–14053
Rashidy-Pour A, Sadeghi H, Taherain AA, Vafaei AA, Fathollahi Y (2004) The effects of acute restraint stress and dexamethasone on retrieval of long-term memory in rats: an interaction with opiate system. Behav Brain Res 154(1):193–198
Rauch SL, Whalen PJ, Shin LM, McInerney SC, Macklin ML, Lasko NB, Orr SP, Pitman RK (2000) Exaggerated amygdala response to masked facial stimuli in posttraumatic stress disorder: a functional MRI study. Biol Psychiatry 47(9):769–776
Riedemann T, Patchev AV, Cho K, Almeida OFX (2010) Corticosteroids: way upstream. Mol Brain 3:2
Roozendaal B (2000) 1999 Curt P. Richter award. Glucocorticoids and the regulation of memory consolidation. Psychoneuroendocrinology 25(3):213–238

Roozendaal B (2002) Stress and memory: opposing effects of glucocorticoids on memory consolidation and memory retrieval. Neurobiol Learn Mem 78(3):578–595
Roozendaal B, McGaugh JL (1996) Amygdaloid nuclei lesions differentially affect glucocorticoid-induced memory enhancement in an inhibitory avoidance task. Neurobiol Learn Mem 65(1):1–8
Roozendaal B, McGaugh JL (1997a) Basolateral amygdala lesions block the memory-enhancing effect of glucocorticoid administration in the dorsal hippocampus of rats. Eur J Neurosci 9(1):76–83
Roozendaal B, McGaugh JL (1997b) Glucocorticoid receptor agonist and antagonist administration into the basolateral but not central amygdala modulates memory storage. Neurobiol Learn Mem 67(2):176–179
Roozendaal B, McGaugh JL (2011) Memory modulation. Behav Neurosci 125(6):797–824
Roozendaal B, Bohus B, McGaugh JL (1996a) Dose-dependent suppression of adrenocortical activity with metyrapone: effects on emotion and memory. Psychoneuroendocrinology 21(8):681–693
Roozendaal B, Carmi O, McGaugh JL (1996b) Adrenocortical suppression blocks the memory-enhancing effects of amphetamine and epinephrine. Proc Natl Acad Sci U S A 93(4):1429–1433
Roozendaal B, Nguyen BT, Power AE, McGaugh JL (1999a) Basolateral amygdala noradrenergic influence enables enhancement of memory consolidation induced by hippocampal glucocorticoid receptor activation. Proc Natl Acad Sci U S A 96(20):11642–11647
Roozendaal B, Williams CL, McGaugh JL (1999b) Glucocorticoid receptor activation in the rat nucleus of the solitary tract facilitates memory consolidation: involvement of the basolateral amygdala. Eur J Neurosci 11(4):1317–1323
Roozendaal B, Quirarte GL, McGaugh JL (2002) Glucocorticoids interact with the basolateral amygdala beta-adrenoceptor—cAMP/PKA system in influencing memory consolidation. Eur J Neurosci 15(3):553–560
Roozendaal B, de Quervain DJ, Schelling G, McGaugh JL (2004a) A systemically administered beta-adrenoceptor antagonist blocks corticosterone-induced impairment of contextual memory retrieval in rats. Neurobiol Learn Mem 81(2):150–154
Roozendaal B, Hahn EL, Nathan SV, de Quervain DJ, McGaugh JL (2004b) Glucocorticoid effects on memory retrieval require concurrent noradrenergic activity in the hippocampus and basolateral amygdala. J Neurosci 24(37):8161–8169
Roozendaal B, McReynolds JR, McGaugh JL (2004c) The basolateral amygdala interacts with the medial prefrontal cortex in regulating glucocorticoid effects on working memory impairment. J Neurosci 24(6):1385–1392
Roozendaal B, Okuda S, de Quervain DJ, McGaugh JL (2006a) Glucocorticoids interact with emotion-induced noradrenergic activation in influencing different memory functions. Neuroscience 138(3):901–910
Roozendaal B, Okuda S, van der Zee EA, McGaugh JL (2006b) Glucocorticoid enhancement of memory requires arousal-induced noradrenergic activation in the basolateral amygdala. Proc Natl Acad Sci U S A 103(17):6741–6746
Roozendaal B, McEwen BS, Chattarji S (2009) Stress, memory and the amygdala. Nat Rev Neurosci 10(6):423–433
Rubinow DR, Post RM, Savard R, Gold PW (1984) Cortisol hypersecretion and cognitive impairment in depression. Arch Gen Psychiatry 41(3):279–283
Sacchetti B, Lorenzini CA, Baldi E, Tassoni G, Bucherelli C (1999) Auditory thalamus, dorsal hippocampus, basolateral amygdala, and perirhinal cortex role in the consolidation of conditioned freezing to context and to acoustic conditioned stimulus in the rat. J Neurosci 19(21):9570–9578
Sajadi AA, Samaei SA, Rashidy-Pour A (2007) Blocking effects of intra-hippocampal naltrexone microinjections on glucocorticoid-induced impairment of spatial memory retrieval in rats. Neuropharmacology 52(2):347–354
Schelling G, Briegel J, Roozendaal B, Stoll C, Rothenhausler HB, Kapfhammer HP (2001) The effect of stress doses of hydrocortisone during septic shock on posttraumatic stress disorder in survivors. Biol Psychiatry 50(12):978–985

Schelling G, Richter M, Roozendaal B, Rothenhausler HB, Krauseneck T, Stoll C, Nollert G, Schmidt M, Kapfhammer HP (2003) Exposure to high stress in the intensive care unit may have negative effects on health-related quality-of-life outcomes after cardiac surgery. Crit Care Med 31(7):1971–1980

Schelling G, Kilger E, Roozendaal B, de Quervain DJ, Briegel J, Dagge A, Rothenhausler HB, Krauseneck T, Nollert G, Kapfhammer HP (2004a) Stress doses of hydrocortisone, traumatic memories, and symptoms of posttraumatic stress disorder in patients after cardiac surgery: a randomized study. Biol Psychiatry 55(6):627–633

Schelling G, Roozendaal B, de Quervain DJ (2004b) Can posttraumatic stress disorder be prevented with glucocorticoids? Ann N Y Acad Sci 1032:158–166

Schelling G, Roozendaal B, Krauseneck T, Schmoelz M, De QD, Briegel J (2006) Efficacy of hydrocortisone in preventing posttraumatic stress disorder following critical illness and major surgery. Ann N Y Acad Sci 1071:46–53

Schoofs D, Preuss D, Wolf OT (2008) Psychosocial stress induces working memory impairments in an n-back paradigm. Psychoneuroendocrinology 33(5):643–653

Schoofs D, Wolf OT, Smeets T (2009) Cold pressor stress impairs performance on working memory tasks requiring executive functions in healthy young men. Behav Neurosci 123(5):1066–1075

Schwabe L, Römer S, Richter S, Dockendorf S, Bilak B, Schachinger H (2009) Stress effects on declarative memory retrieval are blocked by a beta-adrenoceptor antagonist in humans. Psychoneuroendocrinology 34(3):446–454

Schwabe L, Joëls M, Roozendaal B, Wolf OT, Oitzl MS (2012) Stress effects on memory: an update and integration. Neurosci Biobehav Rev 36(7):1740–1749

Shin LM, Wright CI, Cannistraro PA, Wedig MM, McMullin K, Martis B, Macklin ML, Lasko NB, Cavanagh SR, Krangel TS, Orr SP, Pitman RK, Whalen PJ, Rauch SL (2005) A functional magnetic resonance imaging study of amygdala and medial prefrontal cortex responses to overtly presented fearful faces in posttraumatic stress disorder. Arch Gen Psychiatry 62(3):273–281

Shin LM, Rauch SL, Pitman RK (2006) Amygdala, medial prefrontal cortex, and hippocampal function in PTSD. Ann N Y Acad Sci 1071:67–79

Sijbrandij M, Kleiboer A, Bisson JI, Barbui C, Cuijpers P (2015) Pharmacological prevention of post-traumatic stress disorder and acute stress disorder: a systematic review and meta-analysis. Lancet Psychiatry 2:413–421

Smeets T, Otgaar H, Candel I, Wolf OT (2008) True or false? Memory is differentially affected by stress-induced cortisol elevations and sympathetic activity at consolidation and retrieval. Psychoneuroendocrinology 33(10):1378–1386

Smeets T, Wolf OT, Giesbrecht T, Sijstermans K, Telgen S, Joëls M (2009) Stress selectively and lastingly promotes learning of context-related high arousing information. Psychoneuroendocrinology 34(8):1152–1161

Soravia LM, Heinrichs M, Aerni A, Maroni C, Schelling G, Ehlert U, Roozendaal B, de Quervain DJ (2006) Glucocorticoids reduce phobic fear in humans. Proc Natl Acad Sci U S A 103(14):5585–5590

Starkman MN, Schteingart DE, Schork MA (1981) Depressed mood and other psychiatric manifestations of Cushing's syndrome: relationship to hormone levels. Psychosom Med 43(1):3–18

van Stegeren AH, Wolf OT, Everaerd W, Scheltens P, Barkhof F, Rombouts SARB (2007) Endogenous cortisol level interacts with noradrenergic activation in the human amygdala. Neurobiol Learn Mem 87(1):57–66

Strangman GE, Sipes W, Beven G (2014) Human cognitive performance in spaceflight and analogue environments. Aviat Space Environ Med 85(10):1033–1048

Taverniers J, Van Ruysseveldt J, Smeets T, von Grumbkow J (2010) High-intensity stress elicits robust cortisol increases, and impairs working memory and visuo-spatial declarative memory in Special Forces candidates: a field experiment. Stress 13(4):323–333

Tronel S, Alberini CM (2007) Persistent disruption of a traumatic memory by postretrieval inactivation of glucocorticoid receptors in the amygdala. Biol Psychiatry 62(1):33–39

Tronson NC, Taylor JR (2007) Molecular mechanisms of memory reconsolidation. Nat Rev Neurosci 8(4):262–275

Vukojevic V, Kolassa IT, Fastenrath M, Gschwind L, Spalek K, Milnik A, Heck A, Vogler C, Wilker S, Demougin P, Peter F, Atucha E, Stetak A, Roozendaal B, Elbert T, Papassotiropoulos A, de Quervain DJ (2014) Epigenetic modification of the glucocorticoid receptor gene is linked to traumatic memory and post-traumatic stress disorder risk in genocide survivors. J Neurosci 34(31):10274–10284

Wang XY, Zhao M, Ghitza UE, Li YQ, Lu L (2008) Stress impairs reconsolidation of drug memory via glucocorticoid receptors in the basolateral amygdala. J Neurosci 28(21):5602–5610

Weis F, Kilger E, Roozendaal B, de Quervain DJ, Lamm P, Schmidt M, Schmolz M, Briegel J, Schelling G (2006) Stress doses of hydrocortisone reduce chronic stress symptoms and improve health-related quality of life in high-risk patients after cardiac surgery: a randomized study. J Thorac Cardiovasc Surg 131(2):277–282

Wilhelm I, Wagner U, Born J (2011) Opposite effects of cortisol on consolidation of temporal sequence memory during waking and sleep. J Cogn Neurosci 23(12):3703–3712

Wolf OT (2008) The influence of stress hormones on emotional memory: relevance for psychopathology. Acta Psychol (Amst) 127(3):513–531

Wolf OT, Convit A, McHugh PF, Kandil E, Thorn EL, De Santi S, McEwen BS, de Leon MJ (2001) Cortisol differentially affects memory in young and elderly men. Behav Neurosci 115(5):1002–1011

Yang YL, Chao PK, Lu KT (2006) Systemic and intra-amygdala administration of glucocorticoid agonist and antagonist modulate extinction of conditioned fear. Neuropsychopharmacology 31(5):912–924

Yehuda R (2002a) Current status of cortisol findings in post-traumatic stress disorder. Psychiatr Clin North Am 25(2):341–368. vii

Yehuda R (2002b) Post-traumatic stress disorder. N Engl J Med 346(2):108–114

Yehuda R, Bierer LM (2008) Transgenerational transmission of cortisol and PTSD risk. Prog Brain Res 167:121–135

Yehuda R, LeDoux J (2007) Response variation following trauma: a translational neuroscience approach to understanding PTSD. Neuron 56(1):19–32

Yehuda R, Kahana B, Binder-Brynes K, Southwick SM, Mason JW, Giller EL (1995) Low urinary cortisol excretion in Holocaust survivors with posttraumatic stress disorder. Am J Psychiatry 152(7):982–986

Yehuda R, McFarlane AC, Shalev AY (1998) Predicting the development of posttraumatic stress disorder from the acute response to a traumatic event. Biol Psychiatry 44(12):1305–1313

Yehuda R, Teicher MH, Seckl JR, Grossman RA, Morris A, Bierer LM (2007) Parental posttraumatic stress disorder as a vulnerability factor for low cortisol trait in offspring of holocaust survivors. Arch Gen Psychiatry 64(9):1040–1048

Yehuda R et al (2015) Cortisol augmentation of a psychological treatment for warfighters with posttraumatic stress disorder: Randomized trial showing improved treatment retention and outcome. Psychoneuroendocrinology 51:589–597

Young EA, Breslau N (2004) Cortisol and catecholamines in posttraumatic stress disorder: an epidemiologic community study. Arch Gen Psychiatry 61(4):394–401

Young AH, Sahakian BJ, Robbins TW, Cowen PJ (1999) The effects of chronic administration of hydrocortisone on cognitive function in normal male volunteers. Psychopharmacology 145(3):260–266

8 The Autonomic Nervous System

Eleonora Tobaldini, Giorgio Colombo, Alberto Porta, and Nicola Montano

8.1 Introduction

The autonomic nervous system (ANS) is the branch of the nervous system that controls the visceral functions in order to maintain the "homeostasis/homeodynamic" state of the body, being an interface between body, central nervous system (CNS), and external stimuli (Malliani 2000). ANS can be divided into two subsystems, the sympathetic (SNS) and the parasympathetic (PNS) branches. ANS plays a key role in the regulation of several body functions like control of cardiovascular function, gastrointestinal, pulmonary, skin, and genitourinary and immune control, just to cite the most important, that are all essential in order to maintain the physiological body functions. Moreover, the ANS is very sensitive to react upon environmental challenges/stressors to make the body adaptable to the environment (Fig. 8.1). Despite the widely used definition of "autonomic," a lot of evidences strongly support the hypothesis of very complex and integrated control mechanisms operating at different central and peripheral levels coordinating the function of ANS (Montano et al. 2009). Just considering this latter point, it has also been widely demonstrated that several pathological conditions (e.g., cardiovascular diseases such as hypertension, heart failure, and myocardial infarction) are characterized by an impaired autonomic control, consisting of a chronic sympathetic overactivity, due to both afferent

E. Tobaldini · G. Colombo · N. Montano (✉)
Department of Clinical Sciences and Community Health, University of Milan, Milan, Italy

Division of Internal Medicine and Immunology, IRCCS Fondazione Ca' Granda, Ospedale Maggiore Policlinico, Milan, Italy
e-mail: nicola.montano@unimi.it

A. Porta
Department of Biomedical Sciences for Health, University of Milan, Milan, Italy

© Springer Nature Switzerland AG 2020
A. Choukér (ed.), *Stress Challenges and Immunity in Space*,
https://doi.org/10.1007/978-3-030-16996-1_8

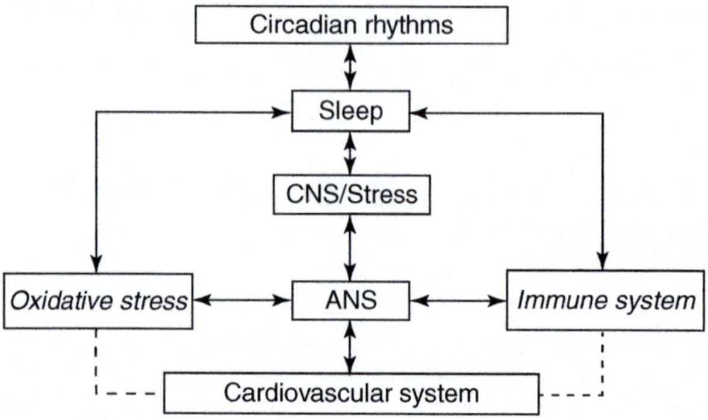

Fig. 8.1 Schematic representation of the central role played by autonomic nervous system (ANS) in being modulated/modulating by stress system in the central nervous system (CNS). This is in turn affected by sleep, which is under the major influence of circadian rhythms. Changes in ANS induced by stress system and/or sleep may affect cardiovascular system not only directly but also by increasing oxidative stress and modifying immune responses. As this relationship is bidirectional, primary changes in oxidative stress and immune responses also may affect ANS and, moving upstream, stress responses and sleep

and efferent pathways activation, characterized by altered responses to stressors stimuli (Lombardi 1986; Guzzetti et al. 1986; Pagani and Lucini 2001).

Therefore assessment of ANS regulation is fundamental in order to identify the pathological basis of such diseases (cardiovascular and noncardiovascular, see also Chaps. 6 and 7) and can also be an important target for pharmacological therapy (Malliani 2000; Montano et al. 2009). Recently, it has been demonstrated that the ANS plays an important role not only in the regulation of vegetative state but also in modulating immune system responses (Tracey 2009), metabolism, and inflammation (Sternberg 2006; Grassi et al. 2007), thus suggesting a complex integrative role of ANS at different control levels.

In conclusion, it is clear how important the assessment of ANS is in physiological and pathological situations, as modifications of the control mechanisms regulate autonomic functions under various stress challenges.

8.2 Methodologies

Since many years, several invasive and noninvasive different techniques have been developed for the analysis of the ANS; however, the attempt to find out a noninvasive, reliable, and reproducible technique is still challenging. Considering all the methods developed for the ANS evaluation, it is possible to identify three main groups of techniques: (a) humoral measures of sympathetic activity based on the quantification of catecholamines, (b) direct assessment of muscle sympathetic nerve activity (MSNA) by means of microneurographic techniques, and (c) indirect measures of sympathetic and parasympathetic modulations derived from the analysis of heart rate variability analysis (HRV).

8.2.1 Catecholamines and Muscle Sympathetic Nerve Activity

For many years, the levels of plasma and urinary catecholamines (epinephrine and norepinephrine) have been considered as a reliable index of sympathetic activity. In 1983, Goldstein et al. reported that hypertensive subjects had higher plasma catecholamines levels compared to healthy subjects (Goldstein et al. 1983). Several factors may influence catecholamine levels and kinetics and have been identified as possible confound elements contributing to the variability among subjects. Moreover, because the plasma levels of catecholamines suggest general and non-organ-specific changes in sympathetic control, more specific regional measures of catecholamines have been developed. In fact, measures of the local rates of noradrenaline release could allow the assessment of organ-specific sympathetic nervous system activity, as the cardiac catecholamines spill over (Esler 1993).

Sympathetic activity could also be directly recorded in humans, using the microneurography techniques (Valbo et al. 1979). The direct recording of sympathetic nerve activity directed to skeletal muscles (MSNA) and skin (SSNA) has provided important information regarding cardiovascular control reflexes in several physiological and pathological conditions (exercise, response to orthostatic stress, cardiovascular and metabolic diseases such as hypertension, heart failure, obstructive sleep apnea) (Narkiewicz and Somers 2003; Wallin and Charkoudian 2007; Kato et al. 2009). Although these techniques are more reliable than the catecholamine measurements in assessing the sympathetic function, the limitation that they are minimally invasive makes this approach unsuitable for studies on large populations (Montano et al. 2009).

Another key observation is the fact that all the techniques mentioned above are capable of providing information on sympathetic branch; they cannot give any information regarding the parasympathetic activity. These two limitations, the invasiveness and the lack of information on parasympathetic nervous system, lead to the development of noninvasive techniques capable of providing information on the sympathovagal balance and on the rhythmical oscillations on the heart period and arterial pressure variability in the time and frequency domain.

8.2.2 Heart Rate Variability Analysis

More than 40 years ago, Lee and Hon (1965) described, in fetal distress, alterations in interbeat intervals preceding evident changes in heart rate itself. Some years later, based on the hypothesis that heart rate and blood pressure exhibit beat-to-beat oscillations around the main values, it was possible to identify these rhythmical components in order to provide reliable information on sympathetic and parasympathetic modulation of heart rate and blood pressure variability (Pagani et al. 1986). Sayers and coworkers (Hyndman et al. 1971; Sayers 1973) introduced the idea of a computational analysis of the variability of heart rate (heart rate variability, HRV) as modulations. Kleiger found further evidence of a strong clinical relevance of HRV in cardiac patients in a pioneering study, where HRV was found to be an independent predictor of mortality after acute myocardial infarction (Kleiger 1987; Malik 1989) provided the evidence of the clinical relevance of this technique. Differently from the time domain measures, HRV analysis is based on the idea that each time series, including the RR (Riva-Rocci) interval time series (tachograms) and the systolic and diastolic arterial pressure time series (systograms and diastograms), can be considered as the sum of different rhythmical oscillations, characterized by specific frequency and amplitude. It has been clearly shown that the assessment of these rhythmical components on cardiovascular signals provides reliable information on the ANS function in healthy people and the diseased (Akselrod et al. 1981; Malliani et al. 1991). In the last few decades, intensive research in this field allowed to better understand the physiological interaction between sympathetic and parasympathetic efferent pathways, the sympatho-sympathetic excitatory reflexes (Malliani and Montano 2002), and the alteration of the sympathovagal balance in pathological conditions that form basis for cardiovascular diseases.

8.2.3 Frequency Domain Analysis

Several methods have been validated for the spectral analysis of HRV and those can be divided into parametric and nonparametric methods. The nonparametric methods use a more simple algorithm (usually a fast Fourier transform, FFT) and the process speed is higher, while the parametric methods allow the identification of spectral components independent of preselected frequency bands, automatic calculation of low- and high-frequency components, and the possibility of precise estimation of spectrum density on small samples (Task Force of the European Society of Cardiology the North American Society of Pacing Electrophysiology 1996). From the time series, the algorithm is capable of distinguishing three main spectral components: very low frequency component (VLF), frequency band below 0.04 Hz, low-frequency component (LF), centered around 0.1 Hz (0.04–0.15 Hz), marker of sympathetic modulation, and a high-frequency component (HF), synchronous with respiration, marker of vagal modulation. Each oscillation can be described by a specific frequency band and amplitude that is expressed in absolute values of power (milliseconds squared) and in normalized units (nu), which represent the relative value of each power component in proportion to the total power minus the VLF component (see Chap. 23 for more technical details and explanations).

8.2.4 Nonlinear Analysis of HRV

In the past few years, a growing interest has been focused on the investigation of nonlinear dynamics of HRV and on the evaluation of the complexity of the cardiovascular control of HRV and blood pressure variability. The hypothesis that ANS control is based on nonlinear dynamics and on the interaction of different subsystems that act at several timescales leads to the growing interest in providing nonlinear analyses and complexity measures of cardiovascular ANS control in physiological and pathological states (Goldberger and West 1987; Goldberger et al. 1988; Kaplan et al. 1991; Porta et al. 2007a, b, c, d; Maestri et al. 2007; Voss et al. 2009; Huikuri et al. 2009). In a recent review, Voss et al. (2009) suggested that the nonlinear dynamics and complexity of a biological system can be due to three main components: (1) different subsystem acting with feedback interactions to adapt to internal and external stimuli, (2) the adaptation of a subsystem to changed conditions (as in pathological conditions or during aging process), (3) when a subsystem failed, the others operate to compensate the lacking control mechanisms. Three main families of nonlinear and complexity analyses can be identified: (1) fractal measures, such as Power-law correlation and detrended fluctuation analysis (DFA); (2) symbolic dynamics, and (3) complexity measures, such as Sample Entropy, Approximate Entropy, Multiscale Entropy, Shannon Entropy, Conditional Entropy, and Corrected Conditional Entropy.

8.2.4.1 Detrend Fluctuation Analysis

Detrended fluctuations analysis (DFA) was introduced by Peng et al. (1995) and it is based on a modified random walk analysis applied to physiological time series (Peng et al. 1995). It quantifies the fractal correlation properties in nonstationary time series and it usually estimates a short-term fractal scaling exponent a1 and a long-term scaling exponent a2, to identify the fluctuations on multilength timescales. a1 and a2 describe, respectively, the short- and the long-term fractal-like scaling properties of the heart period fluctuations. Technically, the time series is divided into windows of the same size and the variability is analyzed in relation to local trend in each window and the process is repeated for all different window sizes. A scaling exponent value of approximately 1 corresponds to 1/f fluctuations.

In healthy subjects, the scaling exponent is approximately 1, thus suggesting a fractal like behavior, while under pathological conditions, this index is reduced, suggesting a loss of fractal-like heart period dynamics (Mäkikallio et al. 2002; Huikuri et al. 2009; Voss et al. 1995, 2009). This technique has been applied for the identification of high-risk cardiac patients: It has been shown by Mäkikallio et al. (Mäkikallio et al. 1997; Huikuri et al. 2000) that a reduction of a1 is a powerful predictor of mortality and that this reduction is associated with vulnerability to ventricular arrhythmias.

8.2.4.2 Symbolic Analysis

In the last few years, a new method called symbolic analysis (SA) has been implemented by several authors (Porta et al. 2001, 2007a, b, c, d; Voss et al. 2009) as a nonlinear tool capable of providing information on autonomic cardiovascular control. SA is based on the conversion of a series into a sequence of symbols suitable for the description of sympathetic and parasympathetic modulation of HR and BPV. Briefly, it is based on the transformation of heart period variability series into a sequence of symbols and the construction of patterns (i.e., words). The complexity of the data time series is determined from the distribution of the words using different mathematical methods (Voss et al. 1995; Wessel et al. 1998, 2000; Porta et al. 2007a, b, c, d).

In the method proposed by Porta et al., the full range of sequences is spread over six levels and patterns of three beats length are constructed. All the patterns are then grouped, without any loss, into four families: the 0 V family (patterns with no variations), 1 V family (patterns with one variation), 2LV family (patterns with two like variations), and 2UV family (patterns with two unlike variations) (Porta et al. 2007a, b, c, d). The rate of occurrence, expressed as a percentage, is then evaluated. The application of this method to evaluate cardiac autonomic control of HRV revealed that the family of 0 V pattern is marker of sympathetic modulation, while the 2UV family is a marker of vagal modulation.

Compared to spectral analysis, this method has some advantages. SA is a nonlinear method, it is independent of the definition of the a priori frequency bands of the heart period oscillations and it is more reliable than spectral analysis in evaluating ANS control during coactivation of the two branches.

It has been shown that SA is capable of tracking the progressive changes of sympathetic modulation occurring during graded head-up tilt test better than more classic spectral analysis (Porta et al. 2007a, b, c, d). The clinical application of this technique allows the identification of high-risk patients after acute myocardial infarction (Voss et al. 1995) and a better predictor of high risk for arrhythmias in post-myocardial infarction patients.

Interestingly, the application of SA for the evaluation of cardiac autonomic dynamics in different physiological situations such as sleep revealed interesting results. In several papers, SA was more reliable than classical linear spectral analysis in detecting changes of cardiac autonomic control during the different sleep stages (NREM vs. REM sleep). This was confirmed both in experimental settings, such as sleep deprivation (Tobaldini et al. 2017) and clinical settings, such as patients with cardiovascular and neurological disorders (Tobaldini et al. 2013, 2015). We can speculate that this phenomenon is related to the presence of nonlinear dynamics during sleep and, possibly, to the coactivation of the sympathetic and vagal branches during specific sleep periods. Keeping in mind these results, we can hypothesize the use of SA as a fundamental tool for the investigation of cardiac autonomic control during wake–sleep cycles in space medicine.

8.2.4.3 Entropy Measures and Cardiac Autonomic Control

In order to assess the regularity (or, its opposite, the irregularity) of the heart period and blood pressure oscillations, in the last few years, several entropy measures have been developed. Complexity analysis can provide the quantification of different aspects of the cardiovascular control mainly related to the organization of different subsystems that cooperate in order to regulate the physiological and pathological responses of cardiovascular ANS. Complexity is measured by evaluating the amount of information carried by a series, i.e., larger the information, greater the complexity (Porta et al. 2001). For the evaluation of complexity of heart period variability, several indices have been developed and applied in physiological and pathological conditions. In physiological conditions, a maneuver known to induce a gradual increase of sympathetic modulation, i.e. the head-up tilt test, caused a reduction of complexity of heart period variability. Moreover, several studies have shown that aging and pathological processes are associated with a decrease in complexity of autonomic cardiovascular control.

Different tools have been validated to quantify the complexity of cardiac autonomic control in health and diseases, such as Sample Entropy, Approximate Entropy, Multiscale Entropy, Shannon Entropy, Conditional Entropy, and Corrected Conditional Entropy. Although a detailed description of these techniques is beyond the scope of this chapter, we will briefly summarize the main application of complexity measures on autonomic analysis.

Approximate Entropy (ApE) and Sample Entropy (SE)

ApE can be considered an index of the overall complexity of a time series (Pincus 1991). It measures the unpredictability of fluctuations in a time series, quantifying the probability that patterns will remain similar for successive incremental comparison (Pincus and Goldberger 1994). Higher the ApE, higher the complexity of the time series, lower the ApE, lower the complexity of the time series (i.e., the time series is more predictable because it contains many repetitive patterns). SE is an improvement of ApE and it quantifies the conditional probability that two sequences of data points similar to each other will remain similar when one consecutive point is included (Voss et al. 2009). Recently, the application of nonlinear tools such as ApE in chronic patients revealed that nonlinear tools are important predictor of events. For instance, in congestive heart failure patients, a decreased ApE was an independent predictor of total mortality and sudden cardiac death during atrial fibrillation (Cygankiewicz et al. 2015). Interestingly, the prognostic role of these nonlinear methods has been demonstrated also in neurological conditions such as acute ischemic stroke (Graff et al. 2013).

Shannon Entropy (ShE), Conditional Entropy (CE), and Corrected Conditional Entropy (CCE)

ShE assesses the complexity of the distribution of the patterns of length L. ShE is an index that describes the shape of the distribution of heart periods: ShE is large if the distribution is flat (all the patterns are identically distributed and the pattern distribution carries the maximum amount of information). On the contrary, ShE is small if there is a subset of patterns that is more likely, while others are missing or infrequent (as it happens in a Gaussian distribution) (Porta et al. 2007a, b, c, d).

CE is a measure of the amount of information carried by the current RR sample when the previous values are known. In other words, it represents the difficulty in predicting future values of RR based on past values of the same series. The CE is 0 when the knowledge of the previous values is helpful in predicting the future values while it is equal to ShE when the knowledge of past values of RR is not helpful to reduce the uncertainty of future RR values. The CCE, proposed by Porta et al. (2001), decreased to 0 only when RR was completely predictable, showed the maximum value when RR is completely unpredictable and showed a minimum when the knowledge of past values is helpful in reducing the uncertainty associated with future RR values.

8.3 Mirroring the Interaction of Stress and the Autonomic Nervous System Through Cardiovascular Read-Out Parameters

8.3.1 Autonomic Oscillators

The pivotal role of the ANS in regulating instantaneous cardiovascular responses to everyday stressors justifies the major interest of investigating variations of cardiovascular autonomic regulatory mechanisms on Earth and in space. Neural control of cardiac function and circulation is mainly regulated by the interaction between sympathetic and parasympathetic efferent pathways. Usually, physiological regulation of cardiovascular control and responses to external stimuli is based on the concept of sympathovagal balance. The system is based on a reciprocal response of the two pathways with an activation of the sympathetic drive accompanied by an inhibition of the vagal outflow and vice versa (Malliani 2000). This balance allows the autonomic cardiovascular system to maintain a homeostatic condition in response to internal and external stimuli (exercise, response to orthostatic challenge, stressors, etc.) and oscillates from a states of quiet, when negative feedback reflexes are predominant, to states of excitation (for physical or mental stress) characterized by central excitatory mechanisms reinforced by peripheral positive feedback reflexes (Malliani et al. 1991; Zucker 1996; Legramante et al. 1999; Malliani 2000; Montano et al. 2009).

However, particular physiological situations, such as exercise, cold face immersion, and apneas (Paton et al. 2005), and also pathophysiological situations, such as the period immediately preceding ventricular major arrhythmias (Guzzetti et al. 2005), are characterized by a coactivation of the two subsystems, thus leading from one side to visceral adaptation to stressful conditions but from the other pathological conditions. While acute activation of the sympathetic nervous system subserves fundamental regulatory functions and it is necessary to human phylogenesis (exemplified by the "fight or flight" concept), chronic sympathetic activation as occurring during stressful conditions (physical and or psychological) may result in profound alterations of neural visceral regulatory functions, leading from functional disturbances up to the development of cardiovascular and noncardiovascular diseases. Short- and long-term spaceflight may challenge cardiovascular autonomic control and, finally, cardiovascular health. Thus, it has become fundamental to obtain relevant information about the mechanisms regulating cardiac autonomic control in space medicine, in order to evaluate the possible implications for cardiovascular health and to tailor ad hoc countermeasures after space flights.

8.3.2 Circadian Oscillators

Finally, all organisms have behavioral and biochemical oscillations over a 24 h period and in mammals the daily rhythm is regulated by two main clocks, a central one and a peripheral one (see also Chap. 9). It has been hypothesized that a loss of synchronization between the central and peripheral clocks may underlie the onset and the progression of cardiovascular disease (Takeda and Maemura 2015). This central clock, maintained at the molecular levels by clock genes, is responsible for the circadian variations that characterize cardiovascular and autonomic functions. In fact, a decrease of heart rate and arterial blood pressure characterizes night and sleep phases compared to wakefulness. These changes reflect the variations of ANS in regulating heart rate and blood pressure during sleep–wake cycle. Several studies showed that ANS oscillates between wake and sleep and it varies across sleep stages. Briefly, during NREM sleep, a vagal predominance is observed, while during REM sleep there is a shift of the sympathovagal balance towards a clear sympathetic predominance (Trinder et al. 2001). The evaluation of circadian rhythmicity of ANS and also taking into account the differences between wakefulness and sleep stages in the space environment remain to be elucidated.

8.3.3 Cardiovascular Autonomic Control During Immobilization Stress (Head-Down Bed Rest)

It is well known that Head-Down Bed Rest (HDBR) is a stressful condition consisting of maintaining subjects strictly in bed, with −6° of head down. HDBR is capable of inducing orthostatic intolerance (OI). HDBR has been widely used in experimental physiology as a simulation of microgravity environment characteristic of space flights on various organ-systems (see also Chap. 36). Experimental animal and human models of physical inactivity, either mild or extreme, showed that sedentary lifestyle that characterized bed rest protocols or spaceflight, alters the balance between sympathetic and vagal control of cardiovascular functions (Hughson and Shoemaker 2014). The short term and long-term HDBR allowed researchers to investigate the effect of microgravity on autonomic cardiovascular functions in terms of heart rate and blood pressure variability, baroreflex responses, plasma and urinary catecholamines levels mimicking short- and long-term space flights.

The major clinical interest in studying HDBR-related changes in autonomic control is the possibility to mimic the weightlessness and microgravity that are known to cause OI and the so-called deconditioning phenomenon, experienced by astronauts after landing, characterized by the reduced orthostatic tolerance, pre-syncope, and syncope. It is worth noting that the physiopathological mechanisms of OI after HDBR still need to be elucidated.

Several studies investigated the autonomic cardiovascular responses of cardiovascular parameters, HRV indices, and baroreflex sensitivity during short-term and long-term HDBR and the recovery period. The main effects of short-term HDBR exposure are an increase of HR and a reduction of the total variability of heart period and of parasympathetic modulation and an impairment of baroreflex function (Hirayanagi et al. 2004; Traon et al. 1998; Crandall et al. 1994). In fact, a reduction of the high-frequency component, HF, marker of vagal modulation, has been observed during short-term HDBR in the late phase of the test and during the early recovery period (Hirayanagi et al. 2004), as well as a reduction in sympathetic modulation more evident in the late phase of the test (Traon et al. 1998). Interestingly, it has been shown that during a 24-h HDBR the heart period variability indices and the baroreflex sensitivity expressed a relevant circadian variation, characterized by an increase of total power and high-frequency component during the evening (Hartikainen et al. 1993).

Interestingly, authors that studied ventricular heterogeneity using the analysis of T-wave alternans, an index of temporal and spatial repolarization heterogeneity, showed that subjects with lower orthostatic tolerance after 5 days of HDBR had higher T-wave alternans during recovery after orthostatic testing (Martín-Yebra et al. 2015).

Animal studies also reported that in an animal model, sympathetic vasoconstrictor responsiveness and NO-mediated inhibition of sympathetic vasoconstriction were not altered during physical inactivity induced by hindlimb unweighting, thus suggesting that short-term physical inactivity does not impair vascular sympathetic control in muscle vascular bed (Just et al. 2015).

The data on long-term HDBR (60–120 days) are more debated; for example, the long-term exposure to HDBR has been shown to both increase (Kamiya et al. 2003) and decrease the muscle sympathetic nerve activity (Ferretti et al. 2009). As to cardiovascular parameters, heart rate was significantly higher after 60 and 120 days of HDBR while blood pressure remained unchanged during the entire period (Kamiya et al. 1999); considering HRV and the baroreflex function, a significant decrease in HRV and baroreflex sensitivity both in the late phase of HDBR and the early phase of recovery period has been observed; however, the HDBR did not alter the rhythmical oscillatory components of heart period and blood pressure variability (Ferretti et al. 2009).

As stated above, the physiopathological mechanisms underlying the OI after HDBR is still unclear. A recent paper by Liu et al. (2015a, b) aimed to determine whether the OI after a 60-day HDBR is related to the autonomic adaptation. Interestingly, authors showed that none of the circulatory patterns (i.e. HR, SBP, HRV parameters) was significantly different between fainters and nonfainters during the entire experimental protocol, thus suggesting that the OI after HDBR does not exhibit specific hemodynamic and autonomic characteristics.

As to the issue of the mechanisms responsible for OI after HDBR, the role of other autonomic reflexes has been proposed. The vestibulosympathetic reflex is important for the postural regulation of blood pressure and it has been hypothesized that HDBR could reduce MSNA through otholit stimulation. In an elegant experimental protocol, healthy subjects underwent head-down rotation with otolith stimulation, before HDBR, after 24-hours HDBR and after prolonged HDBR (36 days). Results showed that prolonged, but not acute, HDBR is able to attenuate the vestibulosympathetic reflex. This result seems to highlights the role of this mechanism in OI (Dyckman et al. 2012).

In addition to the standard frequency domain approach to HRV, the nonlinear analysis of HRV after HDBR revealed a reduction in the complexity of heart period variability, underlying the important effects of the microgravity and weightlessness on autonomic cardiovascular control and organization (Goldberger et al. 1994).

All these results support the hypothesis that a decrease in HRV, thus an increase in cardiovascular sympathetic modulation, associated with other factors, such as hypovolemia and impairment of endothelium-dependent functions at the microcirculation level, could be responsible for the deconditioning effect experienced after weightlessness predisposing subjects to postural hypotension and reduced ability to react to orthostatic stimuli (Kamiya et al. 2003; Coupé et al. 2009).

8.3.4 Cardiovascular Autonomic Control During Acute Gravitational Stress (Parabolic Flight)

Parabolic flights are used to create and reproduce short periods of changing gravity, within a range of 0–1.8 g, with 1 g equal to 9.81 m/s^2. A standard parabolic flight can be divided into five different phases: 1 g (before and after each parabola), hypergravity during the ascending leg of the parabola, microgravity at the apex of parabola, and hypergravity during the descending leg of parabola.

Several studies have investigated the hemodynamic and cardiovascular effects of parabolic flights. These experiments aimed to simulate hypergravity and microgravity characteristics of space missions. During microgravity, there is a redistribution of blood toward the upper body that is capable of stimulating baroreceptors (Lipnicki 2009) and left atrial diameter well as intrathoracic blood volume and systemic blood pressure (Iwase et al. 1999). However, in supine position, microgravity induced a decrease of mean arterial pressure and an increase of left atrial diameter, while during 1 g conditions left atrial diameter increased (Pump et al. 1999).

As to autonomic control, results seem to be more complex. Experiments that evaluated direct sympathetic activity using MSNA recordings revealed that during parabolic flights, in seated position, MSNA was increased under hypergravity before microgravity entry and then suppressed at the onset of microgravity; this phenomenon has been attributed to the loading of the cardiopulmonary volume receptors (Iwase et al. 1998, 1999).

As to cardiovascular autonomic modulation, several studies reported no significant differences in terms of spectral parameters in supine position between the different flight phases; however, in standing position, a predominant vagal modulation was observed in microgravity compared to hypergravity (Seps et al. 2002; Beckers et al. 2003).

The model of parabolic flight has been used to investigate the effects of partial gravity levels (zero, lunar and Martian gravity) on cardiovascular and autonomic control, having in mind the future missions on Mars. It has been recently shown that heart rate and blood pressure increase as gravity increases; on the contrary, the variability of cardiovascular modulation, together with vagal control, decreases as gravity increases. These results suggest a correlation between gravity level and cardiovascular autonomic modulation during parabolic flights (Widjaja et al. 2015).

The hemodynamic and cardiovascular responses in the gravitational changes during parabolic flights may be responsible for the syndromes of orthostatic intolerance that mimic the symptoms after space flight, even after short parabolic flights (Schlegel et al. 2001).

8.3.5 Cardiovascular Autonomic Control During Space Flights

Several studies described that autonomic control of cardiovascular functions is impaired during space flights, representing a challenge for the astronauts and for space missions. The combination of microgravity and cosmic rays exposure represents a great health challenge for cardiovascular medicine during deep-space missions. In fact, all these factors do have important effects on cardiovascular physiology, either during spaceflights and post-flights periods. The synergistic negative consequences of weightlessness and space radiations result in an impaired autonomic control of heart and vasculature functions and, thus, on cardiovascular system (Hughson and Helm 2018).

It has been shown that a short exposure to microgravity, as during the short-term space flights, is capable of inducing a reduction of orthostatic tolerance and increased plasma norepinephrine and epinephrine levels (Fritsch-Yelle et al. 1994), an increase of baroreflex sensitivity in the early phase of microgravity, reducing the respiratory modulation of heart rate and decreasing cardiac baroreflex gain with no effects on arterial blood pressure during post-flights period (Verheyden et al. 2007). The early exposure to microgravity induced a decrease of heart rate both in supine and standing position; diastolic pressure and premature ventricular contractions all were significantly reduced in flight (Fritsch-Yelle et al. 1996) while overall variability, LF/HF and respiratory sinus arrhythmia amplitude and phase were similar to preflight values (Migeotte et al. 2003).

A more recent study on the topic showed that early stages of space flight are characterized by an augmented baroreflex sensitivity with enhanced vagal modulation and an higher blood pressure, compared to pre-flight condition. On the contrary, in late stages, baroreflex sensitivity, vagal modulation measured by HRV and blood pressure were similar to preflight values in supine position. This study also demonstrated that the effects of microgravity on baroreflex sensitivity are dampened by mild physical exercise, possibly because of central reset of baroreflex during physical exercise (Di Rienzo et al. 2008).

Compared to Earth, systolic arterial pressure and heart rate tended to be higher in space with an impairment of vagal baroreflex control, suggesting the idea that microgravity exposure induces a prevalence of sympathetic and a decrease of parasympathetic cardiovascular control (Eckberg et al. 2010). It is possible that the central volume redistribution during the initial phase of space flight is responsible for augmented blood pressure and for the autonomic changes observed. Body fluid loss and the consequent reset of cardiac baroreflex could explain why in mission's late stages blood pressure and autonomic balance were similar to preflight values (Di Rienzo et al. 2008).

The analysis of HRV during long-term exposure to microgravity revealed an alteration of the rhythmic oscillatory components with the evidence of a super-slow wave oscillation, marker of ultradian rhythms, supporting the hypothesis of a new setting point of cardiac autonomic regulation (Baevsky et al. 1998). Similarly, MSNA tended to be higher when recorded in astronauts during spaceflight than on Earth (Eckberg and Neurolab Autonomic Nervous System Team 2003).

A recent study by Eckberg et al. showed an augmented sympathetic activity during space flight at early and late stages. This sympathetic hyperactivity was characterized by higher frequency of MSNA bursts. This hyperactivity was associated with higher vagal control and lower blood pressure during the first day of flight; in two weeks, vagal activity was reduced and blood pressure raised in comparison to preflight values. MSNA activity remained higher even after landing day, suggesting that the autonomic changes considered can be at least in part due to mechanisms of autonomic neuroplasticity (Eckberg et al. 2016).

In the last few years, an increasing interest has also been focused on the autonomic changes in space flights during sleep. The daily mean systolic blood pressure and heart rate during space flight were similar to the preflight values, but during sleep, while no differences from pre-flights values have been observed in heart rate, an increase to over preflight values of systolic arterial pressure was described (Shiraishi et al. 2004). Comparing the different sleep stages, a more pronounced decrease of heart rate was observed during NREM than for REM sleep (Gundel et al. 1999), thus suggesting a possible increase of parasympathetic dominance of cardiac rhythms in space.

While in flight heart rate during sleep is similar to preflight heart rate values, this is not true for in-flight ANS dynamics. HRV and especially its vagal component is reduced during sleep in long duration space flights; there is a positive correlation between the duration of the exposure to microgravity and these modifications. This findings could be easily explained considering two factors: the reduction in total power of HRV could be due to a lower basal stimulation of the autonomic nervous system, whether a higher respiratory rate, as it happens during microgravity exposure, could led to a reduced vagal tone (Xu et al. 2013).

One of the major problems of astronauts who return to Earth after space flights is the possibility to experience orthostatic intolerance and, rarely, syncope, for a deconditioning effect of microgravity on autonomic cardiac control. This is the reason why several studies investigated the autonomic control and the response to orthostatic challenge after space missions (Eckberg et al. 2010). It has been observed that long-term space flights (9 months) have been able to reduce the total power of HRV and to increase the systolic pressure variability, suggesting a reduction of the cardiac vagal control and the baroreflex gain (Cooke et al. 2000). Studying the autonomic response to orthostatism, it has been demonstrated that astronauts who were not able to finish the orthostatic stress (tilt test) had a lower systolic arterial pressure at rest in pre-flight condition while, at the end of the tilt test, heart rate and systolic arterial pressure were lower in nonfinisher than finishers (Sigaudo-Roussel et al. 2002). These subjects also exhibited lower vascular resistance and smaller response to phenylephrine before and after flights and low norepinephrine release during orthostatic challenge after landing. Similarly, the MSNA response to orthostatic challenge was preserved in the finishers (without orthostatic intolerance) but it was reduced in those who experienced orthostatic intolerance, supporting the hypothesis of an impairment of baroreflex control (Mano 2005). These data suggested that the orthostatic intolerance after space flights could be due to a decreased number of alpha-1 adrenergic receptor responsiveness before the flight and a remodeling of the central nervous system that occurs during space flight (Meck et al. 2004).

As we stated above, inter-individual differences can influence microgravity related autonomic changes. A recent study compared ANS and responses to microgravity in ten Chinese and European astronauts. Results revealed that European cosmonauts had a higher pre-flight vagal modulation and a higher HR in the first day of the space mission compared to Chinese astronauts. After landing back on Earth, European cosmonauts experienced orthostatic intolerance and tachycardia, whereas Chinese astronauts did not. This interesting finding can be explained by several factors: a shorter mean duration of Chinese space missions, possible differences in nutrition, a different physical preparation to space missions in the two groups and finally to differences in the ANS per se mainly related to genetic differences (Liu et al. 2015a, b).

Exposure to microgravity can lead to profound changing in cardiovascular function and in autonomic cardiovascular control; every component of cardiovascular system can be influenced by this challenge. A part from the already mentioned modification in cardiovascular autonomic control, it has been described that exposure to microgravity can bring to reduced cardiac mass, worsening of vascular function with stiffer arteries and endothelial dysfunction. All these modifications force the cardiovascular system to work in sub-optimal conditions, leading to reduced cardiovascular performance and maybe to accelerated cardiovascular senescence.

Human cardiovascular system usually works in a 1-g environment, protected from radiation coming from the outer space by the magnetosphere and the atmosphere. Spaceflight missions are characterized by profoundly different condition if compared to living on Earth surface: not only as far as microgravity is concerned, but also regarding to radiation exposure. High-energy radiation is a spaceflight-related risk factor for cardiovascular disease synergistic to microgravity: even low doses of radiation elevate the cardiovascular mortality-associated risk. It has been demonstrated that accelerated atherogenesis and endothelial dysfunction related to radiation exposure could lead to increased cardiovascular mortality. Appropriate shielding is necessary to guarantee a safer environment during spaceflights (Hughson and Helm 2018).

8.4 Stress, Autonomic Nervous Systems and Immune Regulation

Nervous and immune system have unique and similar characteristics: they respond to a huge number of different stimuli and they are able to memorize information in complex network of cells and intercellular signaling. It has been postulated that these two complex systems can talk to each other. In fact, the detection of pathogen fragments, cytokines or other immune molecules by sensory neurons generates immunoregulatory responses through efferent autonomic neuron signaling. A new concept, the inflammatory reflex, is the mainstem of the whole system. In brief, immune stimulation of vagal afferent components by immune-molecules stimulates a reflex response in vagal efferent component that activate anti-inflammatory and immune-regulating responses. In particular, the activation of the vagus lead to augmented levels of acetylcholine in the spleen that suppress the release of TNF from splenic macrophages, thus suppressing serum TNF level. This is only one part of the system and many other components of the ANS have an important role in immune-regulation and consequences on health (see also Chap. 6). For example, it seems that β2-adrenergic receptor stimulation elicits an increased production of anti-inflammatory molecules, including IL-10 and TGF-β, whether α-adrenergic signaling on monocytes and macrophages can result in increased production of TNF or other proinflammatory cytokines (Pavlov et al. 2018).

The pivotal role of the autonomic nervous system (ANS) in regulating instantaneous bodily responses to everyday stressors and especially in space has been described exemplarily on the interaction between the ANS and the cardiovascular systems as above, showing the cardiovascular changes as a critical indicator of the role of the ANS in the regulation of the human stress-homeostasis. Mounting evidence suggests that an imbalance of the ANS and its respective hormones (e.g., catecholamines) is affecting immune responses. For instance, with some evidence that the sympathetic increasing and the vagus decreasing the release of pro-inflammatory cytokines with relevance to diseases (Bellinger et al. 2008). Moreover, it has been described that space flight–associated stressful situations can result in a sympathetic and/or glucocorticoid-mediated immune down regulation (Stowe et al. 2003) which was also associated with reactivation of herpes virus (Stowe et al. 2001). In brief, stressful conditions of psychological or physical nature do modulate key functions of the host's immune responses by neurohumoral, catecholaminergic, and glucocorticoid system together with neural pathways to control the host's immune homeostasis (Tracey 2002, 2009). It appears likely that ANS alterations in space may impinge also upon homeostatic immune mechanisms, thereby modulating key function of innate and adaptive immune responses in space (please see Fig. 8.1 and Chaps. 11–16 of this volume).

8.5 Summary

ANS plays a crucial role as an interface between visceral function and physical/emotional stress challenges. Once regarded only as a source of pure efferent motor output, it is now well-established that it is a complex system based on the integration of several reflexes having both negative and positive feedback characteristics, continuously balancing each other. This reflex activity is directed both upstream, impinging upon CNS structures related to sleep–wake cycle and activating the arousal system, as well as downstream, regulating visceral (cardiovascular, gastrointestinal and genitourinary) functions. Thus, changes in ANS activity may affect sleep and cortical responses to stress and vice versa. Interestingly, it has recently observed that ANS is also able not only to be modulated, but also to modulate immune response, thus strengthening the importance of ANS as a possible target system to elaborate countermeasures to better cope with stressful conditions, such as space environment.

References

Akselrod S, Gordon D, Ubel FA, Shannon DC, Berger AC, Cohen RJ (1981) Power spectrum analysis of heart rate fluctuation: a quantitative probe of beat-to-beat cardiovascular control. Science 213(4504):220–222

Baevsky RM, Moser M, Nikulina GA, Polyakov VV, Funtova II, Chernikova AG (1998) Autonomic regulation of circulation and cardiac contractility during a 14-month space flight. Acta Astronaut 42(1–8):159–173

Beckers F, Seps B, Ramaekers D, Verheyden B, Aubert AE (2003) Parasympathetic heart rate modulation during parabolic flights. Eur J Appl Physiol 90(1–2):83–91

Bellinger DL, Millar BA, Perez S et al (2008) Sympathetic modulation of immunity: relevance to disease. Cell Immunol 252(1–2):27–56

Cooke WH, Ames JE IV, Crossman AA, Cox JF, Kuusela TA, Tahvanainen KU, Moon LB, Drescher J, Baisch FJ, Mano T, Levine BD, Blomqvist CG, Eckberg DL (2000) Nine months in space: effects on human autonomic cardiovascular regulation. J Appl Physiol 89(3):1039–1045

Coupé M, Fortrat JO, Larina I, Gauquelin-Koch G, Gharib C, Custaud MA (2009) Cardiovascular deconditioning: from autonomic nervous system to microvascular dysfunctions. Respir Physiol Neurobiol 169(Suppl 1):S10–S12

Crandall CG, Engelke KA, Pawelczyk JA, Raven PB, Convertino VA (1994) Power spectral and time based analysis of heart rate variability following 15 days head-down bed rest. Aviat Space Environ Med 65(12):1105–1109

Cygankiewicz I, Corino V, Vazquez R, Bayes-Genis A, Mainardi L, Zareba W, de Luna AB, Platonov PG, Trial Investigators MUSIC (2015 Oct 1) Reduced irregularity of ventricular response during atrial fibrillation and long-term outcome in patients with heart failure. Am J Cardiol 116(7):1071–1075. https://doi.org/10.1016/j.amjcard.2015.06.043

Di Rienzo M, Castiglioni P, Iellamo F, Volterrani M, Pagani M, Mancia G (2008) Dynamic adaptation of cardiac baroreflex sensitivity to prolonged exposure to microgravity: data from a 16-day spaceflight. J Appl Physiol 105(5):1569–1575

Dyckman DJ, Sauder CL, Ray CA (2012) Effects of short-term and prolonged bed rest on the vestibulosympathetic reflex. Am J Physiol Heart Circ Physiol 302(1):H368–H374. https://doi.org/10.1152/ajpheart.00193.2011

Eckberg DL, Neurolab Autonomic Nervous System Team (2003) Bursting into space: alterations of sympathetic control by space travel. Acta Physiol Scand 177(3):299–311

Eckberg DL, Halliwill JR, Beightol LA, Brown TE, Taylor JA, Goble R (2010) Human vagal baroreflex mechanisms in space. J Physiol 588(Pt 7):1129–1138

Eckberg DL, Diedrich A, Cooke WH, Biaggioni I, Buckey JC Jr, Pawelczyk JA, Ertl AC (2016) Respiratory modulation of human autonomic function: long term neuroplasticity in space. J Physiol 594(19):5629–5646

Esler M (1993) Clinical application of noradrenaline spillover methodology: delineation of regional human sympathetic nervous responses. Pharmacol Toxicol 73:243–253

Ferretti G, Iellamo F, Pizzinelli P, Kenfack MA, Lador F, Lucini D, Porta A, Narkiewicz K, Pagani M (2009) Prolonged head down bed rest-induced inactivity impairs tonic autonomic regulation while sparing oscillatory cardiovascular rhythms in healthy humans. J Hypertens 27(3):551–561

Fritsch-Yelle JM, Charles JB, Jones MM, Beightol LA, Eckberg DL (1994) Spaceflight alters autonomic regulation of arterial pressure in humans. J Appl Physiol 77(4):1776–1783

Fritsch-Yelle JM, Whitson PA, Bondar RL, Brown TE (1996) Subnormal norepinephrine release relates to presyncope in astronauts after spaceflight. J Appl Physiol 81(5):2134–2141

Goldberger AL, West BJ (1987) Fractals in physiology and medicine. Yale J Biol Med 60(5):421–435

Goldberger AL, Rigney DR, Mietus J, Antman EM, Greenwald S (1988) Nonlinear dynamics in sudden cardiac death syndrome: heart rate oscillations and bifurcations. Experientia 44(11–12):983–987

Goldberger AL, Mietus JE, Rigney DR, Wood ML, Fortney SM (1994) Effects of head-down bed rest on complex heart rate variability: response to LBNP testing. J Appl Physiol 77(6):2863–2869

Goldstein DS, McCarty R, Polinsky RJ, Kopin IJ (1983) Relationship between plasma norepinephrine and sympathetic neural activity. Hypertension 5(4):552–559

Graff B, Gąsecki D, Rojek A, Boutouyrie P, Nyka W, Laurent S, Narkiewicz K (2013 Aug) Heart rate variability and functional outcome in ischemic stroke: a multiparameter approach. J Hypertens 31(8):1629–1636. https://doi.org/10.1097/HJH.0b013e328361e48b

Grassi G, Quarti-Trevano F, Seravalle G, Dell'Oro R (2007) Cardiovascular risk and adrenergic overdrive in the metabolic syndrome. Nutr Metab Cardiovasc Dis 17:473–481

Gundel A, Drescher J, Spatenko YA, Polyakov VV (1999) Heart period and heart period variability during sleep on the MIR space station. J Sleep Res 8(1):37–43

Guzzetti S, Piccaluga E, Casati R, Cerutti S, Lombardi F, Pagani M, Malliani A (1986) Sympathetic predominance in essential hypertension: a study employing spectral analysis of heart rate variability. J Hypertens 6(9):711–717

Guzzetti S, Borroni E, Garbelli PE et al (2005) Symbolic dynamics of heart rate variability a probe to investigate cardiac autonomic modulation. Circulation 112:465–470

Hartikainen J, Tarkiainen I, Tahvanainen K, Mäntysaari M, Länsimies E, Pyörälä K (1993) Circadian variation of cardiac autonomic regulation during 24-h bed rest. Clin Physiol 13(2):185–196

Hirayanagi K, Iwase S, Kamiya A, Sasaki T, Mano T, Yajima K (2004) Functional changes in autonomic nervous system and baroreceptor reflex induced by 14 days of 6 degrees head-down bed rest. Eur J Appl Physiol 92(1–2):160–167

Hughson RL, Shoemaker JK (2014) Autonomic responses to exercise: deconditioning/inactivity. Auton Neurosci 188:32–35

Hughson RL, Helm A, Durante M (2018) Heart in space: effect of the extraterrestrial environment on the cardiovascular system. Nat Rev Cardiol 15(3):167–180

Huikuri HV, Mäkikallio TH, Peng CK, Goldberger AL, Hintze U, Møller M (2000) Fractal correlation properties of R-R interval dynamics and mortality in patients with depressed left ventricular function after an acute myocardial infarction. Circulation 101(1):47–53

Huikuri HV, Perkiömäki JS, Maestri R, Pinna GD (2009) Clinical impact of evaluation of cardiovascular control by novel methods of heart rate dynamics. Philos Transact A Math Phys Eng Sci 367(1892):1223–1238. Review

Hyndman BW, Kitney RI, Sayers BM (1971) Spontaneous rhythms in physiological control systems. Nature 233(5318):339–341

Iwase S, Mano T, Cui J, Kitazawa H, Kamiya A, Miyazaki S, Sugiyama Y, Mukai C, Kohno M, Nagaoka S (1998) Changes in muscle sympathetic nerve activity and effect of breathing maneuvers during microgravity induced by parabolic flight in humans. Environ Med 42(2):152–155

Iwase S, Mano T, Cui J, Kitazawa H, Kamiya A, Miyazaki S, Sugiyama Y, Mukai C, Nagaoka S (1999) Sympathetic outflow to muscle in humans during short periods of microgravity produced by parabolic flight. Am J Phys 277(2Pt 2):R419–R426

Just TP, Jendzjowsky NG, DeLorey DS (2015) Hindlimb unweighting does not alter vasoconstrictor responsiveness and nitric oxide-mediated inhibition of sympathetic vasoconstriction. J Physiol 593(9):2213–2224

Kamiya A, Iwase S, Kitazawa H, Mano T (1999) Muscle sympathetic nerve activity (MSNA) after 120 days of 6 degrees head-down bed rest (HDBR). Environ Med 43(2):150–152
Kamiya A, Michikami D, Fu Q, Iwase S, Hayano J, Kawada T, Mano T, Sunagawa K (2003) Pathophysiology of orthostatic hypotension after bed rest: paradoxical sympathetic withdrawal. Am J Physiol Heart Circ Physiol 285(3):H1158–H1167
Kaplan DT, Furman MI, Pincus SM, Ryan SM, Lipsitz LA, Goldberger AL (1991) Aging and the complexity of cardiovascular dynamics. Biophys J 59(4):945–949
Kato M, Adachi T, Koshino Y, Somers VK (2009) Obstructive sleep apnea and cardiovascular disease. Circ J 73(8):1363–1370. Review
Kleiger RE (1987) Decreased heart rate variability and its association with increased mortality after acute myocardial infarction. Am J Cardiol 59:256–262
Lee ST, Hon EH (1965) The fetal electrocardiogram. IV. Unusual variations in the qrs complex during labor. Am J Obstet Gynecol 92:1140–1148
Legramante JM, Raimondi G, Massaro M, Cassarino S, Peruzzi G, Iellamo F (1999) Investigating feed-forward neural regulation of circulation from analysis of spontaneous arterial pressure and heart rate fluctuations. Circulation 99:1760–1766
Lipnicki DM (2009) Baroreceptor activity potentially facilitates cortical inhibition in zero gravity. NeuroImage 46(1):10–11
Liu J, Li Y, Verheyden B, Chen S, Chen Z, Gay Y (2015a) Is autonomic modulation different between European and Chinese astronauts? PLoS One 10(3):e0120920
Liu J, Li Y, Verheyden B, Chen Z, Wang J, Li Y, Aubert AE, Yuan M (2015b) Orthostatic intolerance is independent of the degree of autonomic cardiovascular adaptation after 60 days of head-down bed rest. Biomed Res Int 2015:896372. https://doi.org/10.1155/2015/896372
Lombardi F (1986) Acute myocardial ischaemia, neural reflexes and ventricular arrhythmias. Eur Heart J 7(Suppl A):91–97
Maestri R, Pinna GD, Accardo A, Allegrini P, Balocchi R, D'Addio G, Ferrario M, Menicucci D, Porta A, Sassi R, Signorini MG, La Rovere MT, Cerutti S (2007) Nonlinear indices of heart rate variability in chronic heart failure patients: redundancy and comparative clinical value. J Cardiovasc Electrophysiol 18(4):425–433
Mäkikallio TH, Seppänen T, Airaksinen KE, Koistinen J, Tulppo MP, Peng CK, Goldberger AL, Huikuri HV (1997) Dynamic analysis of heart rate may predict subsequent ventricular tachycardia after myocardial infarction. Am J Cardiol 80(6):779–783
Mäkikallio TH, Tapanainen JM, Tulppo MP, Huikuri HV (2002) Clinical applicability of heart rate variability analysis by methods based on nonlinear dynamics. Card Electrophysiol Rev 6(3):250–255. Review
Malik M (1989) Heart rate variability in relation to prognosis after myocardial infarction: selection of optimal processing techniques. Eur Heart J 10:1060–1074
Malliani A (2000) Principles of cardiovascular neural regulation in health and disease. Kluwer Academic Publishers, London
Malliani A, Montano N (2002) Emerging excitatory role of cardiovascular sympathetic afferents in pathophysiological conditions. Hypertension 39(1):63–68
Malliani A, Pagani M, Lombardi F, Cerutti S (1991) Cardiovascular neural regulation explored in the frequency domain. Circulation 84(2):482–492. Review
Mano T (2005) Autonomic neural functions in space. Curr Pharm Biotechnol 6(4):319–324
Martín-Yebra A, Caiani EG, Monasterio V, Pellegrini A, Laguna P, Martínez JP (2015) Evaluation of T-wave alternans activity under stress conditions after 5 d and 21 d of sedentary head-down bed rest. Physiol Meas 36(10):2041–2055
Meck JV, Waters WW, Ziegler MG, deBlock HF, Mills PJ, Robertson D, Huang PL (2004) mechanisms of postspaceflight orthostatic hypotension: low alpha1-adrenergic receptor responses before flight and central autonomic dysregulation postflight. Am J Physiol Heart Circ Physiol 286(4):H1486–H1495
Migeotte PF, Prisk GK, Paiva M (2003) Microgravity alters respiratory sinus arrhythmia and short-term heart rate variability in humans. Am J Physiol Heart Circ Physiol 284(6): H1995–H2006

Montano N, Porta A, Cogliati C, Costantino G, Tobaldini E, Casali KR, Iellamo F (2009) Heart rate variability explored in the frequency domain: a tool to investigate the link between heart and behavior. Neurosci Biobehav Rev 33(2):71–80
Narkiewicz K, Somers VK (2003) Sympathetic nerve activity in obstructive sleep apnoea. Acta Physiol Scand 177(3):385–390
Pagani M, Lucini D (2001) Autonomic dysregulation in essential hypertension: insight from heart rate and arterial pressure variability. Auton Neurosci 90(1–2):76–82
Pagani M, Lombardi F, Guzzetti S, Rimoldi O, Furlan R, Pizzinelli P, Sandrone G, Malfatto G, Dell'Orto S, Piccaluga E et al (1986) Power spectral analysis of heart rate and arterial pressure variabilities as a marker of sympatho-vagal interaction in man and conscious dog. Circ Res 59(2):178–193
Paton JFR, Boscan P, Pickering AE, Nalivaiko E (2005) The yin and yang of cardiac autonomic control: vago-sympathetic interactions revisited. Brain Res Rev 49:555–565
Pavlov VA, Chavan SS, Tracey KJ (2018) Molecular and functional neuroscience in immunity. Annu Rev Immunol 36:783–812
Peng CK, Havlin S, Stanley HE, Goldberger AL (1995) Quantification of scaling exponents and crossover phenomena in nonstationary heartbeat time series. Chaos 5(1):82–87
Pincus SM (1991) Approximate entropy as a measure of system complexity. Proc Natl Acad Sci U S A 88(6):2297–2301
Pincus SM, Goldberger AL (1994) Physiological time-series analysis: what does regularity quantify? Am J Phys 266:H1643–H1656
Porta A, Guzzetti S, Montano N, Furlan R, Pagani M, Malliani A, Cerutti S (2001) Entropy, entropy rate, and pattern classification as tools to typify complexity in short heart period variability series. IEEE Trans Biomed Eng 48(11):1282–1291
Porta A, Guzzetti S, Furlan R, Gnecchi-Ruscone T, Montano N, Malliani A (2007a) Complexity and nonlinearity in short-term heart period variability: comparison of methods based on local nonlinear prediction. IEEE Trans Biomed Eng 54(1):94–106
Porta A, Faes L, Masé M, D'Addio G, Pinna GD, Maestri R, Montano N, Furlan R, Guzzetti S, Nollo G, Malliani A (2007b) An integrated approach based on uniform quantization for the evaluation of complexity of short-term heart period variability: application to 24 h holter recordings in healthy and heart failure humans. Chaos 17(1):015117
Porta A, Tobaldini E, Guzzetti S, Furlan R, Montano N, Gnecchi-Ruscone T (2007c) Assessment of cardiac autonomic modulation during graded head-up tilt by symbolic analysis of heart rate variability. Am J Physiol Heart Circ Physiol 293(1):H702–H708
Porta A, Gnecchi-Ruscone T, Tobaldini E, Guzzetti S, Furlan R, Montano N (2007d) Progressive decrease of heart period variability entropy-based complexity during graded head-up tilt. J Appl Physiol 103(4):1143–1149
Pump B, Videbaek R, Gabrielsen A, Norsk P (1999) Arterial pressure in humans during weightlessness induced by parabolic flights. J Appl Physiol 87(3):928–932
Sayers BM (1973) Analysis of heart rate variability. Ergonomics 16(1):17–32
Schlegel TT, Brown TE, Wood SJ, Benavides EW, Bondar RL, Stein F, Moradshahi P, Harm DL, Fritsch-Yelle JM, Low PA (2001) Orthostatic intolerance and motion sickness after parabolic flight. J Appl Physiol 90(1):67–82
Seps B, Beckers F, Aubert AE (2002) Heart rate variability during gravity transitions. Comput Cardiol 29:433–436
Shiraishi M, Kamo T, Kamegai M, Baevsky RM, Funtova II, Chernikova A, Nemoto S, Hotta M, Nomura Y, Suzuki T (2004) Periodic structures and diurnal variation in blood pressure and heart rate in relation to microgravity on space station MIR. Biomed Pharmacother 58(Suppl 1):S31–S34
Sigaudo-Roussel D, Custaud MA, Maillet A, Güell A, Kaspranski R, Hughson RL, Gharib C, Fortrat JO (2002) Heart rate variability after prolonged spaceflights. Eur J Appl Physiol 86(3):258–265
Sternberg EM (2006) Neural regulation of innate immunity: a coordinated nonspecific host response to pathogens. Nat Rev Immunol 6:318–328

Stowe RP, Mehta SK, Ferrando AA, Feeback DL, Pierson DL (2001) Immune responses and latent herpesvirus reactivation in spaceflight. Aviat Space Environ Med 72:884–889

Stowe RP, Sams CF, Pierson DL (2003) Effects of mission duration on neuroimmune responses in astronauts. Aviat Space Environ Med 74:1281–1284

Takeda N, Maemura K (2015 Sep) The role of clock genes and circadian rhythm in the development of cardiovascular diseases. Cell Mol Life Sci 72(17):3225–3234. https://doi.org/10.1007/s00018-015-1923-1

Task Force of the European Society of Cardiology the North American Society of Pacing Electrophysiology (1996) Heart rate variability standards of measurement, physiological interpretation, and clinical use. Circulation 93:1043–1065

Tobaldini E, Brugada J, Benito B, Molina I, Montserrat J, Kara T, Leinveber P, Porta A, Macedo PG, Montano N, Somers VK (2013 Oct 9) Cardiac autonomic control in Brugada syndrome patients during sleep: the effects of sleep disordered breathing. Int J Cardiol 168(4):3267–3272. https://doi.org/10.1016/j.ijcard.2013.04.137

Tobaldini E, Proserpio P, Sambusida K, Lanza A, Redaelli T, Frigerio P, Fratticci L, Rosa S, Casali KR, Somers VK, Nobili L, Montano N (2015 Jun) Preserved cardiac autonomic dynamics during sleep in subjects with spinal cord injuries. Sleep Med 16(6):779–784. https://doi.org/10.1016/j.sleep.2014.12.023

Tobaldini E, Covassin N, Calvin A, Singh P, Bukartyk J, Wang S, Montano N, Somers VK (2017 Apr) Cardiac autonomic control and complexity during sleep are preserved after chronic sleep restriction in healthy subjects. Physiol Rep 5(7):pii:e13197. https://doi.org/10.14814/phy2.13197

Tracey KJ (2002) The inflammatory reflex. Nature 420:853–859

Tracey KJ (2009) Reflex control of immunity. Nat Rev Immunol 9(6):418–428. Review

Traon AP, Sigaudo D, Vasseur P, Maillet A, Fortrat JO, Hughson RL, Gauquelin-Koch G, Gharib C (1998) Cardiovascular responses to orthostatic tests after a 42-day head-down bed-rest. Eur J Appl Physiol Occup Physiol 77(1–2):50–59

Trinder J, Kleiman J, Carrington M, Smith S, Breen S, Tan N, Kim Y (2001 Dec) Autonomic activity during human sleep as a function of time and sleep stage. J Sleep Res 10(4):253–264

Valbo AB, Hagbarth KE, Torebjörk HE, Wallin BG (1979) Somatosensory, proprioceptive, and sympathetic activity in human peripheral nerves. Physiol Rev 59:919–957

Verheyden B, Beckers F, Couckuyt K, Liu J, Aubert AE (2007) Respiratory modulation of cardiovascular rhythms before and after short-duration human spaceflight. Acta Physiol (Oxf) 191(4):297–308

Voss A, Kurths J, Kleiner HJ, Witt A, Wessel N (1995) Improved analysis of heart rate variability by methods of nonlinear dynamics. J Electrocardiol 28(Suppl):81–88

Voss A, Schulz S, Schroeder R, Baumert M, Caminal P (2009) Methods derived from nonlinear dynamics for analysing heart rate variability. Philos Transact A Math Phys Eng Sci 367(1887):277–296. Review

Wallin BG, Charkoudian N (2007) Sympathetic neural control of integrated cardiovascular function: insights from measurement of human sympathetic nerve activity. Muscle Nerve 36:595–614

Wessel N, Schirdewan A, Malik M, Voss A (1998) Symbolic dynamics—an independent method for detecting nonlinear phenomena of heart rate regulation. Biomed Tech 43(Suppl):510–511

Wessel N, Voss A, Kurths J, Schirdewan A, Hnatkova K, Malik M (2000) Evaluation of renormalised entropy for risk stratification using heart rate variability data. Med Biol Eng Comput 38(6):680–685

Widjaja D, Vandeput S, Van Huffel S, Aubert AE (2015 Jun) Cardiovascular autonomic adaptation in lunar and martian gravity during parabolic flight. Eur J Appl Physiol 115(6):1205–1218. https://doi.org/10.1007/s00421-015-3118-8

Xu D, Shoemaker JK, Blaber AP, Arbeille P, Fraser K, Hughson RL (2013) Reduced herat rate variability during sleep in long-duration spaceflight. Am J Physiol Regul Integr Comp Physiol 305(2):R164–R170

Zucker IH (1996) Neural control of the circulation in heart failure and coronary ischaemia: introduction. Clin Exp Pharmacol Physiol 23:685–687

9 Circadian Rhythm and Stress

Mathias Steinach and Hanns-Christian Gunga

9.1 Circadian Rhythm

9.1.1 Definition and Regulation

When the Nobel Prize in Physiology/Medicine was awareded in 2017 to Jeffrey C. Hall, Michael Rosbash, and Michael W. Young, the new era of chronobiology and medicine has deserved the world's attention, as every living organism on Earth seems to be controlled. Their laureat discoveries of molecular mechanisms controlling the circadian rhythm describe and decipher the rhythmicity of physiological variables which is an ubiquitous phenomenon in living systems (Aschoff and Wever 1962a). Even organisms with temporal removal of the external day–night cycle (such as polar animals) or those living in complete darkness (deep-sea-animals, animals living in caves or subterranean habitats) often retain some form of circadian cycle thus suggesting an evolutionary advantage of rhythmic physiology (Beale et al. 2016). Several different rhythms appear to exist, from long-wavelength rhythms like the monthly menstrual cycle to circadian rhythms of many parameters. Several early works of research have described diurnal changes and rhythmicities in biological phenomena like the pulse rate, respiratory gas exchange, or body temperature, exhibiting the peak temperature at the early evening (6 p.m.) and a nadir during the late night and early morning (4 a.m.) (Gierse 1842). This rhythmicity seemed to be innate, as it was unaffected by fasting or physical activity. Also it was shown that isolation from external time cues (zeitgeber) like most notably sun light, but also social interaction, left the circadian rhythm intact, even though its phase period appeared to become elongated (Aschoff and Wever 1962b). It was concluded that there is an endogenous origin of the human circadian rhythm

M. Steinach · H.-C. Gunga (✉)
ZWMB, Berlin, Germany
e-mail: mathias.steinach@charite.de; hanns-christian.gunga@charite.de

© Springer Nature Switzerland AG 2020

A. Choukér (ed.), *Stress Challenges and Immunity in Space*,
https://doi.org/10.1007/978-3-030-16996-1_9

Since then, several similar experiments have revealed an endogenous rhythmicity of approximately 24.5 h, when no external time cues are present (Halberg et al. 1965).

The circadian rhythm is created by a pacemaker system located in the suprachiasmatic nucleus of the anterior hypothalamus (SCN) (Mistlberger and Rusak 1989), which coordinates molecular clocks in each body cell through the control of body temperature, hormone concentrations, and behavior as well as through a transcription/translation feedback loop of clock-related genes (e.g., Per1–2, Cry1–2) influenced by a protein dimer (BMAL1-CLOCK—brain and muscle ARNT-like protein 1 and circadian locomotor output cycles kaput) in all peripheral cells that are similarly regulated by hormones and cytokines (Levi et al. 2008) (Fig. 9.1). Mendelian genes and their respective proteins have been identified for advancing sleep (PER2) and shortening sleep duration (DEC2) (He et al. 2009). A model of this system consists of an SCN oscillator generating a rhythm of approximately 24 h, with input pathways responsible for synchronizing the internal system with external stimuli like the natural light–dark cycle, and output rhythms that are regulated by the pacemaker; the synchronization process that adjusts the internal rhythm to the external zeitgeber is called entrainment (Moore-Ede et al. 1982). As light reaches the retina, this information is transmitted to the circadian system via a direct pathway (Fig. 9.2), the retinohypothalamic tract (RHT), and via an indirect pathway through the intergeniculate leaflet (IGL) (Harrington 1997). It has been suggested that melanopsin-containing retinal ganglion cells, which appear to be neither rods or cones, are the primary circadian photoreceptors in the retina (Ruby et al. 2002), exhibit the most sensitivity to wavelengths of around 460 nm (Warman et al. 2003) and it as has been shown that light of this wavelength induces the greatest changes in circadian light-induced phase shifts and melatonin suppression. Melatonin, a tryptophan metabolite, is secreted by the pineal gland which has afferent and efferent connections to other parts of the circadian system. The pineal gland is innervated through the superior cervical ganglia of the sympathicus, which receives input from the SCN. Melatonin from the pineal gland in turn affects the SCN as it can alter the timing of the circadian rhythms and helps to promote sleep as it inhibits the neurons located in the SCN, thus creating a sleep-permissive condition (Liu et al. 1997).

Previous theories have endorsed a "master–slave model" in a strict hierarchy, where the SCN acts as pacemaker that drives peripheral clock rhythms, which in turn are less or not at all affected by external stimuli. Newer research, however, provides a different understanding, which has been coined an "orchestra model." In this model the SCN acts as the conductor and the peripheral clocks as the various musicians. All peripheral clocks together generate rhythms in physiology. In this sense, not only is the time of eating a potent zeitgeber with regard to social interaction, but there is also growing evidence that the resulting circulating metabolites such as amino acids, carbohydrates, insulin, or nicotinamide adenine dinucleotide (NAD) influence circadian rhythms in both peripheral and central regions (Delezie et al. 2012).

The SCN clock induces other oscillators in other CNS regions and peripheral tissues to produce circadian rhythms, as rest–activity cycle, periodic daily variations in metabolism and rhythmic secretion of hormones (Dibner et al. 2010).

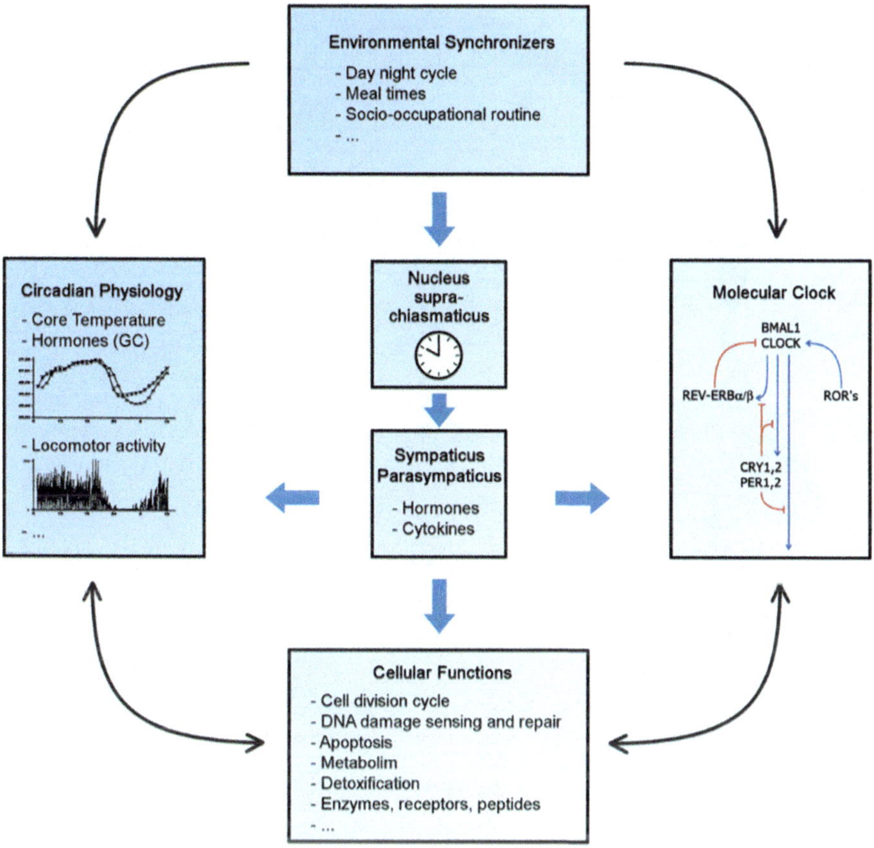

Fig. 9.1 Schematic view of the circadian system. The periods of the suprachiasmatic nuclei are calibrated by the day–night cycle and other environmental synchronizers. The suprachiasmatic nuclei control cellular rhythmicity through their influence on hormonal alterations, locomotor activity, and the body core temperature. Other pathways involve the sympathetic and parasympathetic systems as well as cytokines like TGF and EGF. Molecular clocks in all peripheral cells are controlled by these influences involving transcription/translation feedback loops, in which the BMAL1:CLOCK protein dimer plays a central role. The molecular clocks in turn influence cellular functions and eventually affect circadian physiology (after Filipski et al. 2009, Labrecque and Cermakian 2015 and Son et al. 2018). *BMAL1* brain and muscle Aryl hydrocarbon receptor nuclear translocator (ARNT)-like, *CLOCK* circadian locomotor output cycles kaput, *GC* glucocorticoids, *REV-ERBa/b* BMAL1- and CLOCK regulating transcriptions factors, also known as: nuclear receptor subfamily 1, group D, member 1 (NR1D1), *ROR* RAR-related orphan receptor also known as nuclear receptor subfamily 1, group F, Member 1 (NR1F1), *CRY* cryptochrome, *PER* period

Glucocorticoids, secreted by the adrenocortical steroidogenic cells constitute an important pillar of the stress responsive neuroendocrine system and exhibit a clear circadian rhythm with a daily peak around the time of the sleep–wake transition and minimal levels in the evening and early part of the night and the circadian fluctuation of circulating glucocorticoid levels has been recognized as a synchronizer of

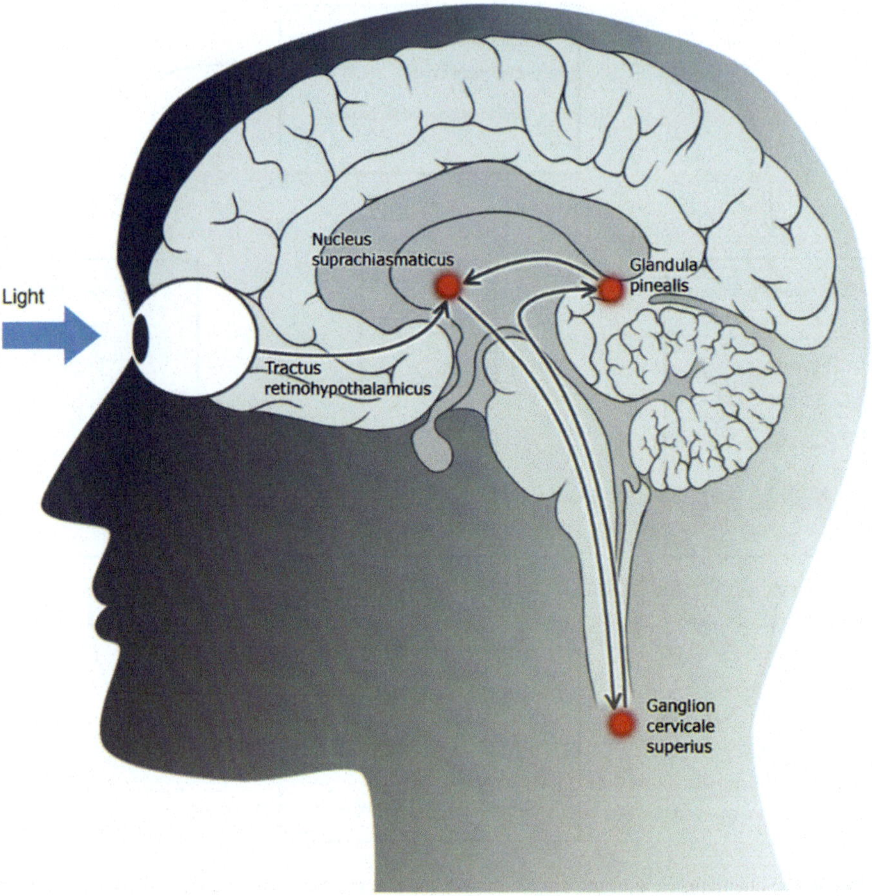

Fig. 9.2 Schematic view of the neural pathways on the effects of light influence on the circadian system. Light information is routed to the suprachiasmatic nuclei (SCN). From there, the inhibition signals reach the Glandula pinealis via the Ganglion cervicale superius. This inhibition disappears in the absence of light (after Reid and Zee 2009)

the circadian system between various tissues and physiological functions such as stress response, metabolism, and energy expenditure (Oster et al. 2017).

The circadian interconnection between various tissues can also be shown through the use of "circadian metabolomics," which allow the simultaneous measurement of multiple cellular events ranging from gene expression to metabolite creation, which have consistently revealed 24-h rhythms in almost every tissue and metabolic pathway (Dyar and Eckel-Mahan 2017).

This "orchestra model" allows external stimuli to fine-tune peripheral rhythms directly, but still relies on entrainment with the SCN (Pilorz et al. 2018). Furthermore, it shows the high interconnection between various physiological functions both central and peripheral, e.g., central and peripheral circadian clocks, metabolism, cardiovascular system, vegetative state, psycosocial behavior, immune function, and functions on a celluar level. Inspite of these new theories, there is still need of further research to clarify the mechanisms of clock-systems and the relationship of central and peripheral oscillators under different conditions.

A model of the circadian clock consists of an endogenous circadian process (process C) and a homeostatic process (process S) and environmental factors. The endogenous process facilitates wakefulness during the day and the consolidation of sleep during the night (Zulley et al. 1981). In addition, the homeostatic process accumulates as a function of prior wakefulness, so that the drive for sleep continues to rise proportionally until it is reduced through sleep, momentarily the only known means to reduce the sleep propensity of the homeostatic process (Dijk et al. 1990). With regard to sleep regulation, it was furthermore suggested that circadian rhythm and immune function, which both rely on a cellular molecular-clock-system, may be codependent on each other, in the sense that circadian rhythm not only influcences immune function (see that subsection below) but also that immune function (or lack thereof) may influence circadian rhythm and provide additional information for the need to sleep (Kurien et al. 2013).

The regulation process is also associated with thermoregulation, as was shown that sleep is usually initiated when the body temperature begins to fall and the awakening is triggered after the body temperature begins to rise again. A hypothermic effect of melatonin that affects sleep propensity has been proved (Krauchi et al. 2000). The induction of sleep through peripheral heat loss via vasodilatation followed by a decline of body core temperature further supports the suggested relationship between the sleep–wake cycle and thermoregulation (see also Chap. 26).

The highly regulated system of the circadian rhythm can be disturbed due to a misalignment between the endogenous circadian phase and the external 24-h-based social and physical environment. Also stressful events or conditions that cause alterations in the regulation of associated mediators like stress hormones can influence the sleep–wake cycle, as described below.

To the end of the aforementioned alteration during stress—especially with regard to hormonal changes like in the HPA axis and circulating glucocorticoids with their influence on alertness and vigilance (see Chaps. 4, 7, and 8)—it becomes obvious that stressful events can have a direct impact on the sleeping pattern and ultimately the circadian rhythm of humans exposed to stress. The following paragraphs illustrate these relationships and different forms of circadian misalignment in stressfull conditions more closely.

9.2 Circadian Misalignment

A misalignment within the circadian rhythm can occur due to changes of the external environment relative to the internal rhythmicity, as it appears in jet lag disorder or shift workers. Such changes can lead to a shortened sleep length that has been shown to be associated with an increased risk of other diseases like obesity, diabetes, hypertension, and a reduced immune function (Cohen et al. 2009). On the other hand, changes of the endogenous circadian system itself can lead to changes in the sleep–wake cycle, as it appears in patients with delayed sleep phase disorder (DSPD) or advanced sleep phase disorder (ASPD) (Sack et al. 2007). In addition, disruptions in sleep can be more subtle yet may have great impact as has the use of artificial illumination until late in the night (i.e., the use and exposure to electrical light, TVs, laptops, mobile devices etc.) for billions of people, leading to delayed bedtimes, later wake-up times, shorter sleep durations, higher incidence of awakenings and increased daytime sleepiness and also disturbing wildlife animals (Aulsebrook et al. 2018). It should be noted in this context that over one-third of adults in the United States of America regularly receive insufficient amounts of sleep (Liu et al. 2016). Circadian disruption can lead to deregulated cell division and eventually to cancer. Jet lag, shift work, and other manifestations of life in modern societies are are associated with higher incidences of malignancies, which are the leading cause of mortality worldwide (Shostak 2017), see Sect. 9.4.4.

9.2.1 Delayed Sleep Phase Disorder

Delayed sleep phase disorder (DPSD) patients have trouble falling asleep and complain about sleepiness in the morning. They will seek advice to readjust their rhythm because of social and occupational demands (Regestein and Monk 1995). These patients may worsen their situation as they may administer self-medication with hypnotics or alcohol. The status may lead to a conditioned insomnia. DSPD is estimated to account for up to 10% of chronic insomnia patients. Reports on epidemiologic data are limited, one study suggests a prevalence of 0.17% in the general population (Schrader et al. 1993), others report higher prevalence in young adults and adolescents (7–14%) (Regestein and Monk 1995). More recent reviews indicate DSPD to be the most common among the circadian rhythm sleep disorders (Nesbitt and Dijk 2014). Stress and occupational requirements that continue into the late evening may perpetuate the delayed sleep phase. Research has shown that patients with DSPD have a hypersensitive suppression of melatonin by light (Aoki et al. 2001); in addition, lack of appropriate light in the morning may exacerbate the condition, as patients tend to sleep later due to the phase shift (Ozaki et al. 1996). Bright light therapy is one of the most commonly used treatments for DSPD (Morgenthaler et al. 2007) as exposure to bright light in the morning for 1–2 h resets the human circadian clock by advancing the phase of the circadian rhythms (Czeisler et al. 1989). Studies have shown that bright light exposure was able to sufficiently suppress melatonin secretion and reset phase effects on the human circadian system (Lewy et al. 1987). While high-intensity bright light (3000–10,000 lux) induces the greatest impact on the circadian system, even modest intensities (50–600 lux) can produce notable phase shifts, if presented to individuals, who have been living in dim-lit environments. Three cycles of exposure to as little as 12 lux for 6.5 h produced phase shifts (Duffy and Wright Jr 2005). And finally, intermittent exposure to bright and dim light has been reported to produce almost as much phase shifting as a continuous exposure (Gronfier et al. 2004)—which in turn means that light of lower intensity and even intermittently presented could alter the circadian phase and could therefore be regarded as a stressor to circadian rhythm if exposed to at an undesired time. The chronic sleep deprivation and circadian disruption of DSPD are associated with molecular dysregulation such as a loss of circadian rhythmicity of the transcripts of clock genes (Nesbitt and Dijk 2014).

The regime of bright light therapy is especially effective if patients avoid bright light in the evening (Rosenthal et al. 1990). The administration of the sleep-inducing hormone melatonin in the evenings has demonstrated to effectively reset the circadian clock (Kamei et al. 2000), but due to the limited number of clinical studies a standardized approach has not yet been developed. Phase estimation through measurement of dim light melatonin onset (DLMO) is important to pursue to guide accurate therapy (Nesbitt and Dijk 2014); so it was shown that timely advancements of DLMO for about 1–2 h provided information of successful intervention with DSPD-patients (Saxvig et al. 2014).

9.2.2 Advanced Sleep Phase Disorder

Patients with advanced sleep phase disorder (ASPD) wake up earlier relative to the external environment and experience feelings of sleepiness in the evening (Reid and Zee 2009). Phase advancement in ASPD is relatively rare in younger adults (Jones et al. 1999) as it seems to be associated with aging (Paine et al. 2014); a prevalence of 1% has been estimated in middle-aged adults. Hereditary correlation has been shown in several family cases (Jones et al. 1999) but the general pathophysiology is yet unclear, although a shortened endogenous period in these patients of less than 24 h has been suggested (Jones et al. 1999). Treatment of ASPD has shown to be effective with bright light therapy in the evening as it improved sleep efficiency and delayed the phase of the endogenous circadian cycle (Lack 1993), some patients, however, had difficulties complying with the regimen (Campbell 1999). It was more recently summarized, that light source as low as 265 lux might be sufficient treatment, especially with regard to patient compliance, although higher intensity light sources at about 4000 lux would be preferable as they lead to a significantly delayed circadian phase of 141 min post-treatment (Auger et al. 2015). Little data exist about the effectiveness of melatonin as a treatment of ASPD and it is unclear if possible sedative effects of melatonin might interfere with daytime function (Reid and Zee 2009) and still in more recent reviews, early-morning administration of melatonin is considered only as option to treat ASPD (Auger et al. 2015).

9.2.3 Free Running Disorder

Free running disorder (FRD) is a condition in which patients exhibit a steady drift of 1–2 h of their sleep–wake period each day, it is also known as non-24-h sleep–wake disorder (N24SWD). As the circadian rhythm continues to drift the patients will be synchronized with external zeitgeber until the misalignment ensues again—the symptoms of insomnia and daytime sleepiness are thus cyclical. If these patients try to adhere to the conventional sleep–wake times, they often complain about insomnia, early morning awakening, and daytime sleepiness (ICSD-2 2005). FRD is commonly reported in blind people who lack entrainment through sunlight zeitgeber (Eliott et al. 1971); an estimated 50% of blind individuals have FRD. FRD seems to be rare in sighted people (Weber et al. 1980), some cases reported FRD in patients following a head injury or psychiatric patients (Wulff et al. 2012). Genetic factors may also play a role. One study showed a significant association between the PER3 polymorphism rs228697 and FRD in sighted individuals (Hida et al. 2014).

FRD patients show positive results to nonphotic entrainment such as social and work schedules; also some blind individuals seem to respond to bright light therapy despite their lack of visual perception. It seems conceivable that these patients suffer from a blindness that does not affect the ganglionic photoreceptors recently associated with photic input as a zeitgeber. In addition, melatonin, administered 1 h before bedtime, seems to be effective in treating FRD in blind individuals (Sack et al. 2000), especially when FRD-patients are relatively in phase with external rhythms (Salva et al. 2017).

9.2.4 Shift Work Disorder

Shift work at night can have a significant impact on the circadian rhythm as it misaligns the individuals' activity from social and physical zeitgeber and hinders the individual to participate in a normal schedule of daily activities. Shift work–sleep disorder (SWSD) typically presents itself as insomnia and daytime sleepiness, chronic fatigue, although the degree of the symptoms may vary, as some shift workers have less difficulty than others and as factors like age, work schedule, access to leisure activities, and other sleep-affecting illnesses, like sleep apnea or narcolepsy, can influence the ability to cope with SWSD (Folkard et al. 1978).

The prevalence of SWSD is potentially high as about 20% of the workforce in industrialized countries is required to attend to night shift work (Presser 1999). More recent summaries report that between 15% and 30% of adult workers are engaged in some type of shift work in western countries (Boivin and Boudreau 2014). Reports suggest that 1–5% of the general population in western countries and approximately 30% of shift workers suffer from SWSD (Drake et al. 2004). Many shift workers report sleep difficulties after night shifts, after awakening from a nonrefreshing sleep and daytime sleepiness that even can persist on the days off after the night shift (Knauth 1981). According to a Canadian survey, 34% of shift workers report difficulties in sleep onset and maintenance compared to 25% of the daytime workers (Hurst 2008). Working night shifts, whether on a fixed or rotating schedule, is associated with the most sleep disruption compared to daytime and evening shift-work (Pilcher and Coplen 2000). A reason for this is a rapid adaptation of the circadian system to the night-shift activity, as it was found in oil-rig-workers (Gibbs et al. 2002) and during simulated night-shifts (Roach et al. 2005). It was shown that melatonin and cortisol adapted more quickly and that clock-gene-expression of PER1 and PER2 reached adjustement after about eight days (James et al. 2007).

During and after shift work, fatigue may ensue and lead to diminished cognitive performance and reduced psychomotor vigilance (Zhou et al. 2012). Thus, it is not surprising that during night shifts, the risk of accidents is significantly greater than during daytime, as a study found an 2.77-fold increase (Swanson et al. 2011).

Shift work may pose a problem for the individuals' activities even though the shift work period has passed. The return to a normal sleep–wake pattern is dependent on the speed and direction of the shift rotation (Reid and Zee 2009).

Gastrointestinal complaints are found more often in shift workers than in day time workers, which are also attributed to changed eating habits mostly dictated by work schedules (Waterhouse et al. 2003). Such complaints include alterations of appetite, constipation, dyspepsia, and abdominal pains. Twenty to seventy-five percentage of shift workers compared to 10–25% of day time workers report such complaints, gastric ulcers occur more than twice as often (Segawa et al. 1987). It has been shown that people working in night shifts exhibit a significantly higher risk of developing cancer (Viswanathan and Schernhammer 2008). Also, disruptions of menstrual cycle and dysmenorrhea have been reported (Labyak et al. 2002), as have been pregnancy problems (Zhu et al. 2004). Furthermore, shift work can also lead to cardiovascular diseases (Thosar et al. 2018). The impact of shift work on the vegetative state and cardiac autonomic control can be evaluated via heart rate variability (HRV) (Skornyakov et al. 2018) as HRV is being widely used as a novel tool to reflect vegetative state as well as physiological and psychological resilience and adaptability in various settings (Rundfeldt et al. 2018) (see also Chaps. 8 and 23).

With regard to metabolic changes, higher plasma concentrations of LDL (>3.3 mmol/L), triglycerides (>1.7 mmol/L) and a lower HDL plasma concentration (<1.0 mmol/L) in a shift working population (Karlsson et al. 2001) have been reported. Also it was found that night shifts are associated with decreased glucose tolerance and increased insulin resistance and lipid intolerance (Lund et al. 2001). A report found that 2 h postprandial breakfast plasma glucose levels in eight shift workers increased from 99.9 ± 4.5 mg/dL when aligned to the circadian cycle to 132 ± 13 mg/dL when misaligned. The glucose level increase occurred despite a concomitant increase of insulin (23.3 ± 5.6–49.9 ± 14.0 μIU/mL), which suggests that the insulin sensitivity had dropped while misaligned. Also an abnormally high level of plasma cortisol, previously mentioned as playing an important role during stress and regulation of the circadian rhythm, was found in misaligned shift workers at the end of the waking and beginning of the sleeping period, which could further contribute to insulin resistance and hyperglycemia (Dinneen et al. 1993). All in all, there is a higher relative risk of 1.84 (95% CI = 1.45–2.34) for the metabolic syndrome associated with shift work compared to day workers with a positive dose-response relationship (Wang et al. 2014).

As mentioned, the possibility of leisure and physical activity of shift workers is limited, as these activities are generally scheduled to a diurnal rhythm and there are little arrangements to meet the demands of the shift worker (Baker et al. 2003). Many shift workers wish to but cannot engage in desired leisure activities as dayworkers; therefore, shift workers have problems maintaining their physical fitness compared to day time workers. Shift-working women can be considered to be at higher risk in this regard because of still existing double pressure of work and domestic responsibilities (Portela et al. 2004). These circumstances of less recreational activities and regular exercise adds up to the already mentioned changes of increased plasma lipids and decreased glucose intolerance.

It also has been shown that physical activity increases duration and quality of nocturnal sleep and that physical activity may be beneficial to the shift worker in improving sleep quality as well as influences their fatigue levels and mediates favorable changes in physiologic functions (Youngstedt 2005). It was found that the amount of slow wave sleep (SWS) is important for brain restoration and that SWS is increased by physical activity (Youngstedt 2005). Thus, the combination of a declined sleeping pattern through shift work in combination with the lack of an otherwise brain restaurative SWS due to less physical activity even worsens the situation.

Recently, abnormalities of neuropeptides such as ghrelin, leptin, and orexin have been associated with sleep restriction, which may in turn affect appetite, the risk of overeating, and eventually obesity (Atkinson and Davenne 2007). Leptin, for example, was found to be 17% lower in misaligned shift worker over the entire misaligned period compared to when aligned normally ($p < 0.001$) (Scheer et al. 2009). These substances may also affect non-exercise thermogenesis as a variable component of energy expenditure which was shown to account for different amounts of weight gain in rats. However, little is known about the influences and relationships between sleep, metabolism, body mass changes, and thermoregulation—and therefore, to close the circle, the circadian rhythm.

Treatments for SWSD consist of improving sleep quality through an optimal sleep environment (quiet and dark) as well as administration of melatonin and bright light therapy in order to realign the circadian rhythm (Morgenthaler et al. 2007) which has to be shown successful to match sleep propensity with the desired sleep time (Sharkey et al. 2001). In bright light therapy, various intensities, ranging from 1200 to 10,000 lux (Smith and Eastman 2008), and different exposure times, both continuous and intermittent (Eastman et al. 1995), have shown to be successful. Wearing shaded eyepieces to avoid sun light exposure at an undesired time can help improve phase adjustment even without bright light therapy. Other countermeasures have been reported to improve alertness during the night shift such as 300 mg caffeine plus a 1–2 h nap planned 3–4 h before night shifts (Schweitzer et al. 2006). Treatment should also encompass the social environment of the individual and therapy plans need to be individualized (Reid and Zee 2009). A recent review indicated the meaningfulness to consider individual chronotype with regard to shift work (Hittle and Gillespie 2018). It should also be mentioned, that idividual personality traits (hardness or resistance against the impact of shift-work) also plays a role and can lead to greater adaptability, less sleep problems and lower depression and anxiety among shift-workers (Natvik et al. 2011).

9.2.5 Jet Lag Disorder

Jet lag disorder (JLD) is a misalignment of the external zeitgeber and the internal circadian rhythm due to long distance travelling across time zones. It should be differentiated from travel fatigue that does not depend on number of travelled time-zones and occures after north- or soundbound travel, although travel (Waterhouse et al. 2007). The arising symptoms may vary and are usually temporary and depend on the number of time zones travelled and the direction of travelling. Eastward travel usually results in trouble falling asleep while westward travel brings difficulty staying awake (Boulos et al. 1995). Travelling eastward appears more difficult because the intrinsic period of the circadian clock in humans is on average longer than 24 h (Czeisler et al. 1999). Other accompanying symptoms are general malaise, gastrointestinal illness, and impaired performance (ICSD-2 2005). JLD, due to the long journeys that often take place in relatively small compartments (economy class), is often accompanied by a nonspecific travel fatigue.

Jet Lag Disorder is a clinical condition with the following criteria for its diagnosis: jet travel across at least two time zones, daytime sleepiness and/or distortion of the sleep cycle along with decreased total sleeptime, generalized fatigue and other symptoms of malaise within 2 days of travel onset, no other condition that could be attributed to the sleep disturbance (American Academy of Sleep Medicine 2014).

Several reports suggest that older individuals may be less prone to JLD and its symptoms (Tresguerres et al. 2001). In contrast, another report showed that middle-aged groups (ages 37–52) had more fragmented sleep and felt less alert when compared to younger individuals (ages 18–25) during a 6 h time advance (Moline et al. 1992).

Treatment usually aims at adjusting the internal circadian system to the new external setting. Therefore behavioral modifications like seeking social and occupational activity according to the present time zone, as well as likewise exposition and avoidance of light can help to diminish the jet lag symptoms (Waterhouse et al. 1997). A general recommendation for travel up to eight time zones would be to seek bright light exposure in the morning after eastward travel and in the evening after westward travel. As in SWSD, the wearing of shaded goggles may help to avoid light exposure at undesired times (Smith et al. 2008). If the time zones crossed exceed 8–10 during eastward travel, it might be more sensible to treat developing jet lag symptoms like a westward travel, since advancing the internal clock is commonly more difficult than delaying it (Takahashi et al. 2002). Entrainment to the new local time can be accelerated using bright light therapy and/or the administration of melatonin (Boulos et al. 1995) at doses of 3–5 mg, taken approximately 2–3 h before bedtime, are considered to be effective (Waterhouse et al. 2007). If possible, travelers might consider to gradually adjust their sleep–wake cycle to the time zone of the destination prior to departure. Individual adjustments might be necessary if the traveler had prior jet lag symptoms from previous journeys or had other atypical circadian orientations like from shift work. As a general rule, the natural circadian rhythm adjusts to the time of destination by approximately 1–2 h per day (Waterhouse et al. 2007). Calculators exist on various websites in order to individually reduce jet lag symptoms, such as www.jetlagrooster.com, as well as applications for mobile devices, but none of these have been evaluated scientifically (Simmons et al. 2015).

9.3 Sleep and Circadian Rhythm Under Extreme Environments Related to Human Space Flight and Exploration Missions

Understanding the conditions of how the circadian clock is paced and also shifted which has been investigated under normal conditions of working life (see above, e.g., shift work disorders and jet lag) provides the basis to investigate and to understand the conditions observed in humans subjected to unusual, extreme and somehow life-threatening extreme environmental stressors. Extreme conditions there can lead to disturbances within the circadian rhythm, and its most obvious reflection, the sleep–wake pattern. Sleep deprivation with its detrimental effects on the organism might occur (Rechtschaffen et al. 2002) that can have significant impact on physical health (An et al. 2018), physiological function (McMahon et al. 2018) and thus professional performance (Munafo et al. 2018). Human curiosity and the search for new territories will lead explorers, scientists, and entrepreneurs into inhospitable regions on Earth and in space. To mimic especially the full scale of mission-related stressor of long-duration exploration missions, investigations on Earth are indispensable. This ranges from extreme deltas of outer temperatures to variable pressurization of spacecrafts and habitats and degrees of confinement.

9.3.1 Hot and Cold Environments

It is commonly known that sleep can be disturbed during a hot summer night. The number of reports about sleep disturbances is proportional to heat waves during summer periods, as reported when a heat wave hit Europe in August 2003 (Ledrans et al. 2004). Researchers have tried to standardize the settings in which sleep under increased temperature conditions could be studied.

Heat exposure prior to sleep, such as from sauna visits or from a hot bath, has shown to enhance slow wave sleep (SWS). The change in body temperature has been attributed to cause the change in SWS. Heat exposure during sleep, with an ambient dry bulb temperature between 31 and 38°C, has been shown to reduce the duration of both SWS and rapid eye movement (REM) sleep (Kendel and Schmidt-Kessen 1973). Also such an increase of temperature often led to sleep disruptions. It seems that heat exposure can lead to different outcomes depending on whether it is administered before or during sleep. Also moderate changes in ambient temperatures can lead to sleep-disturbance and less perceived sleep quality. An increase in ambient temperature from 26 to 30°C led to a longer duration of sleep onset latency, lower duration of SWS, and lower subjective sleep quality among young Chinese students (n = 18, 9 male, 9 female, 23.3 ± 1.8 years) (Lan et al. 2014). Research to scrutinize differences in acclimatization has shown that tolerated heat load during the day leads to an elevation of SWS, while a nontolerated heat strain leads to a diachronic SWS impairment. Sleep stability and REM sleep have shown to be more susceptible to synchronic changes, as these parameters react to nighttime heat strain (Buguet et al. 1998). Daytime exposure to heat-climate led to diachronic changes in sleep, with an increase in SWS which appeared proportional to the heat load in heat acclimatized European people living in dry tropical Africa (Montmayeur and Buguet 1992). Sleeping under warm and humid conditions, however, did not lead to the changes observed in a hot and dry climate. While sleep was also unstable, there was no increase in SWS and REM sleep compared to temperate conditions (Buguet et al. 2002).It is interesting to note, that local cooling of parts of the body (neck and back) was sufficient to alleviate the negative effects of a hot environment (ambient temperature 32°C) and thereby to significantly increase sleep efficiency compared to noncooled subjects (Lan et al. 2018).

Research in cold exposure and sleep has led to controversial results. Some subjects (male) had a very disturbed sleep (Haskell et al. 1981) while the temperature was at 21°C, whereas women showed to have only little alteration in their SWS (Sewitch et al. 1986). Another test revealed no alteration in sleeping pattern, but leads to a longer wake time before subjects fell asleep (Palca et al. 1986). Overall, these results suggest that cold exposure during sleep has detrimental effects on sleep, as it increases restlessness and alters the REM phase. This is also substantiated by results found in nonacclimatized overwinterers in the Arctic, who slept in sleeping bags (insulation factor: 9 clo) and from whom polysomnographic measurements were recorded. Sleeping in the cold caused a cooling until the rectal temperature reached 34.9°C from where it began to rise again due to shivering and body movements. This in turn caused several sleep disruptions and awakenings. REM sleep phases were shorter than in euthermic environments, with the shortening being proportional to the severity of the cold stress (Eichenberger et al. 1993). Interestingly, there was no alteration in sleep pattern and no deprivation of REM phases in acclimatized subjects, who were exposed to nine daily cold baths of 10°C, each lasting 1 h, prior to their departure for the Arctic (Radomski and Boutelier 1982). Acclimatization to cold seems to be protective of sleep disturbances found in unacclimatized subjects.

Data on long-term overwinterers showed a decrease in SWS and REM sleep phases with the oncoming of the polar night and increased again until the summer. These findings are being attributed to the changes in daylight variation; inter-individual variances can also be caused by different levels of physical activity as it was seen in one study where the SWS increased, while the winter was mild and the subjects engaged in daily excursions to a nearby penguin rookery (Buguet et al. 1987).

More recently, efforts have been undertaken to consider all relevant variables that are determinant of adequate thermoregulation—such as air temperature, mean radiant temperature, humidity, surface area, clothing surface temperature, convective heat transfer rate, and the subjects metabolic rate, sufficient for restorative sleep (Joshi et al. 2016). According to these calculations, the air should not exceed sultriness limits, wind speeds should be closely limited, layering of air temperatures between head and ankle should be less than 2 K and there should be a minimum difference between air and radiant temperature as well as the perceived temperatures in different parts of the room, which should not change by more than 0.8 K. Therefore, practical recommendations aim to maintain a microclimate temperature of about 28–30°C, to improve night-time ventilation and cooling also with regard to air quality (CO_2- and humidity levels) and that this can be often achieved by airing of the room through windows pre and during sleep instead of air-conditioners in order to control the energy consumption of households (Lan et al. 2017). This approach will become ever more relevant in the light of increasing ambient temperatures due to climate change with consequences for billions of people (Kim et al. 2014).

9.3.2 Hypobaric/Hyperbaric Pressure

High altitude is associated with a diminished barometric pressure which results in a lowered arterial partial pressure of oxygen. Without additional oxygen, acute mountain sickness (AMS) can occur, depending on the rate of ascent, acclimatization of the subject, and his or her physical exertion. AMS is associated with drowsiness, headaches, a decrease in cognitive and psychomotic functions, and even pulmonary and cerebral edema as the severity intensifies. During sleep, arterial oxygen saturation decreases even further, thus paving the way for AMS to develop. Also central sleep apnea at high altitudes leads to sleep interruption and it was shown that the sleep interruptions occurring in men due to moderate obstructive sleep apnea were replaced and aggravated by central sleep apnea (Burgess et al. 2006). The severity of central sleep apnea can be reduced by descent, supplemental oxygen therapy, and the application of acetazolamide, while sleep quality could be improved by the use of benzodiazepine sedation (Burgess and Ainslie 2016). Insomnia and sleep interruptions seem to be proportional to the altitude with the sleep alteration usually appearing at altitudes above 2000 m and especially during the first 3 weeks of acclimatization (American Academy of Sleep Medicine AASM 2005), so even at moderately low altitudes, such as 2581 m a.s.l., central sleep apneas were frequent findings in healthy mountaineers (Ortiz-Naretto et al. 2018). The sleep disturbances varied from decreases in sleep time, awakenings to decrease of SWS (Nicholson et al. 1988); however, a great range of susceptibility exists (Mizuno et al. 2005) and some subjects seem to be rather impervious to sleep disturbances at high altitudes, while subjects who develop symptoms of AMS show significantly lower sleep efficiency, lower central apnea index, and longer latencies to sleep and rapid eye movement during the nights after a rapid ascent to 3150 m a.s.l. (Tseng et al. 2015). Also ethnic differences seem to be a factor as Tibetian natives exhibited a longer undisturbed sleep time at 5000 m simulated altitude than Han newcomers living in Tibet (Plywaczewski et al. 2003).

Sleep in a hyperbaric environment—such as hyperbaric chamber used in diving medicine—seems to be only mildly affected at shallow depths (30 m), where sleep is slightly disturbed and shortened with a prolonged sleep latency and feelings of fatigue and these effects remain unaltered until depths around 300 m where SWS decreases (Matsuoka et al. 1986). Total sleep time, slow wave sleep, and REM sleep may even increase (Heslegrave et al. 1982). Below 400 m using heliox (mix of helium and oxygen) or trimix (mix of oxygen, nitrogen and helium), the sleep disturbances aggravate further as SWS and REM continue to decrease and awakenings occur (Rostain et al. 1997). Sleep at hyperbaric pressure is highly disturbed, unaffected by the used gas mixture; sleeping patterns return to normal during recompression.

9.3.3 Confinement

Preparation of space missions requires that crewmembers live close to each other for prolonged periods of time in space analogue environments on Earth (see also Chap. 36). Studies regarding the influences of isolation and confined space were conducted with different crew sizes and composition (POLAREMSI, ANTEMSI, ISEMSI, HYDREMSI, EXEMSI, with "EMSI" standing for "European Manned Space Infrastructure," as well as the studies HUBES (HUman BEhavior Study) and SFINCSS (Simulation of a Flight of International Crew on Space Station)). Living in such confined spaces as they exist in current space ships or the International Space Station (ISS) could give rise to problems that emerge from cultural or linguistic disparities, disharmonies in crew composition and cohesion, as well as aggravation of mood and thought disorders. The lack of privacy can play an important role as a social stressor (please see definition of stress above) which may lead to interpersonal conflicts and thus further tenses the situation. Several studies on isolation and confinement have shown that in the isolated condition, experiences of interpersonal conflict like arguing, fighting, and withdrawal increased (Kanas and Manzey 2010). During overwinterings at the German Antarctic Station Georg-von-Neumayer II and III, it was found that there were significant changes in sleep patterns in $n = 54$ investigated participants (37 male, 17 female), with dependencies from overwintering time (decreasing sleep times and sleep efficiencies and increasing number of arousals per night) and local sunshine radiation (38 min longer times in bed and delayed sleep onset with 32 min and offset with 54 min during the darkness phase) while women exhibited a greater deterioration in sleep quality over time than men, possibly due to a greater gender-based susceptibility to the overwintering-stress than men (Steinach et al. 2016)

With regard to the circadian rhythm and sleep, it was found during an isolation study of 7 days that stress under the conditions of isolation and confinement can lead to increased levels of sympathetic activity and increased sleep motor activity, which in turn may lead to disrupted sleep (Kraft et al. 2002). On the other hand, there are also studies on isolation in which the participants experienced no subjective changes in their sleep quality and where no objective changes in their sleep could be found (Tobler and Borbély 1993). It may well be that in such cases, the crew composition was a better match leading to a less stressful environment for the participants and thus less negative effects on their sleep.

9.3.4 Space Flight

Space exploration is still the final frontier of science where researchers encounter conditions very different from the normal environment on Earth. Weightlessness, also called microgravity, is the most significant alteration and as mission time grew longer from the very first journeys in space, the issue of sleep in space had to be addressed. Aside from microgravity, lack of the familiar day–night cycle, narrow space, noises from fans and servomotors, uncomfortable postures and missing proprioceptive feedback, as well as excitement, space motion sickness, and uncomfortable ambient temperatures can lead to a shorter and less qualitative sleep. Some astronauts also reported inefficient workload-arrangements as cause for their poorer sleep (Whitmire et al. 2013). It has been shown that sleep in space is shortened significantly to an average of 6 h, with occasional shortenings as low as 4 h (Santy et al. 1988). In another study, astronauts exhibited in average 1 h less sleep-time during low-orbit space-flight on the ISS than pre-flight (5.4 ± 1.4 h vs. 6.4 ± 1.2 h, $p < 0.01$) and reported about 10% less in sleep-quality, respectively (60.2 ± 21.0 vs. 66.8 ± 17.7, $p < 0.01$) (Flynn-Evans et al. 2016). It is therefore not surprising that hypnotics are the second most used medication onboard a space flight, surpassed only by drugs treating space motion sickness. Over 70% of crew members used drugs for sleep aids while on space shuttle missions and on the ISS with multiple doses applied for 17% of the observed nights (Barger et al. 2014; Wotring 2015).

More often mission times are extended to the other end of the spectrum: Recreational and bedtime are delayed and eventually reduced due to operational demands by mission control (Dijk et al. 2001). This effect ceases during long-term missions, as operational demands and overtime occurs most likely during the first few days and weeks of a mission, so that sleep times return to normal during the course of a long-term mission (Gundel et al. 2001). Nevertheless, even during analogue-studies like the "Mars-500"-isolation experiment, it was found that sleep quality and vigilance degraded over the course of a 520-day simulation (Basner et al. 2013). In addition, also in simulated microgravity on Earth during bed-rest studies, sleep achitecture was dirupted with increased apnea-hypopnea index and decreased breathing stability (Morrison et al. 2017).

The lack of natural zeitgeber and the other changes during a mission in space can lead to a misalignment of the internal circadian rhythm and the work–rest schedule directed by mission control. This misalignment, as in shift workers or as seen in studies on jet lag (see section above), can lead to sleep disturbances, wake time sleepiness and a decrease in well-being and operational performance (Dijk et al. 2001). Increased sleepiness usually coincides with the circadian minimum of the body core temperature—being asleep at that time seems to be important for the restorative function of sleep (Zulley et al. 1981). Chronically restricted sleep time and/or sleep in poor quality can give rise to possible risk in the operation during a mission in space—as it does in all other areas of occupation. It was suggested that two nights of sleep restricted to 6 h or less result in increased response times, more errors in reaction tests, slower performance in arithmetic tests, and impaired working memory functions (Van Dongen et al. 2003). The impairments accumulate if the sleep restriction persists. If the sleep is restricted for a total of 14 nights to less than 6 h, the result on the accumulated decrement in performance is equal to two nights of complete sleep loss (Van Dongen et al. 2003). It has been suggested that the total duration of sleep during a 24 h period is the most important factor to secure the restorative function of sleep, whether it be taken uninterrupted or split into an anchor period and additional smaller periods of naps (Mollicone et al. 2007).

Altered environmental zeitgeber in space will probably meet a challenged thermoregulation, as well. Therefore, in a recent study we investigated whether due to changes in the thermal environment, i.e. space, the role of the core body temperature might be changed. Under terrestrial thermoneutral and resting conditions, heat exchange between an organism and its environment occurs by way of conduction, convection, radiation, and evaporation. If the ambient temperature in a room, for example, is around 27°C the heat exchange happens mainly by radiation and convection (Clark and Edholm 1985). In space, it has been assumed that these mechanisms of heat exchange are challenged due to a lack of convective heat transfer, diminished production, and/or efficiency of evaporation (Fortney et al. 1998; Polyakov et al. 2001). Several authors have touched on the issue of thermoregulation in weightlessness, either in sleep or even in circadian rhythm-related studies (Gundel et al. 1993, 1997; Crandall et al. 1994; Greenleaf 1997; Koscheyev et al. 2000; Dijk et al. 2001). Since vigorous exercise is a special kind of additional stress on the thermoregulatory system, because under such conditions >80% of energy expenditure is converted to heat, we addressed this issue in two joint international research projects called Thermolab and Circadian Rhythms. These studies were conducted in close cooperation with the German Aerospace Agency (DLR), the European Space Agency (ESA), and the National Aeronautics and Space Administration (NASA). The overarching aim of the first study, Thermolab, was to first continuously monitor the changes of core body temperature (CBT) during exercise before, during, and after long-term stays on the ISS. It was found (Stahn et al. 2017) that the astronauts on the ISS increased gradually in about two-and-a-half months their resting body core temperature by about 1°C compared to body core temperatures determined under resting conditions on Earth. This increase in CBT was associated with augmented concentrations of interleukin-1 receptor antagonist (IL 1ra), an important anti-inflammatory protein in human autoimmune and chronic inflammatory diseases (Cartmell et al. 2001). In addition, the thermoregulatory system in humans seems even further challenged in space, because we observed that astronauts' CBT rises higher and faster during vigorous short-term exercise in space than on ground, in some instances exceeding 40°C. Besides the fact that even minor increases in CBT can impair physical and cognitive performance, and that these findings have an impact on astronauts' health and well-being during future long-term spaceflights, it remains unclear how these changes might affect the circadian undulations of core body temperatures of humans in space. Currently, the reasons for the increase in core body temperature and the increase in interleukin-1 receptor antagonist remain unclear as well. Potential candidates—among others—might by the daily physical stress by the vigorous exercise of the astronauts (2 h) to avoid muscular/cardiovascular deconditioning during their stay in weightlessness (Choukèr 2012) and/or the increased radiation exposure on the ISS which is a 100 times higher than on ground (Reitz et al. 2009). The second study on the ISS, Circadian Rhythms, will be presumably finished in 2019. A preliminary data trend analysis showed that (1) a similar increase of resting core body temperature of about 1 °C in the first two-and-a-half months could be observed in these new long-term astronauts on the ISS as seen in the Thermolab Study, and (2) some astronauts showed surprisingly well—circadian preserved circadian rhythms in core body temperature over 36 h, although they were circulating around the Earth and were thereby exposed to 16 days/nights per day. Definitely, further analysis of the Circadian Rhythms data are necessary to draw final conclusions. In this regard, new methods to determine circadian characteristics of astronauts, such as the BodyTime assay using circadian transcriptomics to determine the chronotype, may be beneficial to help maintain health and performance in such challenging environments (Wittenbrink et al. 2018). It is reasonable to assume that these findings will provide further insights into the circadian rhythm regulations of astronauts during a long-term exposures in space.

After all, to ensure fully functional astronauts and their health, the aforementioned sleep disturbing factors need to be removed or at least relieved, like loud noises, uncomfortable quarters etc. Further approaches include adjustment of lighting conditions as countermeasure to disrupted circadian rhythm, such as the use of a solid-state lighting system on the ISS to improve sleep, circadian entrainment, and daytime alertness (Brainard et al. 2016). Also for the ground crew, adjustments might be necessary, for instance to adapt to the longer day on Mars (1 day 37 min). This can be achieved by the application of blue-enriched illumination (Barger et al. 2012).

To secure optimal vigilance during critical tasks (e.g., extravehicular activities), such tasks should be planned in accordance with the astronauts' circadian phase and sleep history to determine the optimal time (Van Dongen 2004). ESA and NASA programs are being adjusted to further study this area and to develop countermeasures to meet the needs of the astronauts (Mallis and DeRoshia 2005). Currently, the ISS has improved the habitability in more comfortable ambient temperature and wind circulation, less disturbing noises, as well as more natural carbon dioxide levels in addition to more comfortable sleeping bags and private sleep quarters to minimize sleep interruptions (Wu et al. 2018).

9.4 The "Circadian Homeostasis": Examples of Physiological Functions, Malfunctions and Countermeasures

Not only do stressful events influence the physiological human circadian rhythm (see Fig. 9.3) and as previously discussed but also do changes within the circadian cycle alter the ability to respond to stress and physical demands on human subjects in unvafourable circadiane time points of such rhythm.This can have strong implications on health, well-being and performance.

As has been mentioned, circadian disruption can have negative impact on vital physiological functions such as the cardiovascular system, vegetative regulation, metabolic functions and influence disease progression and outcome (Morris et al.

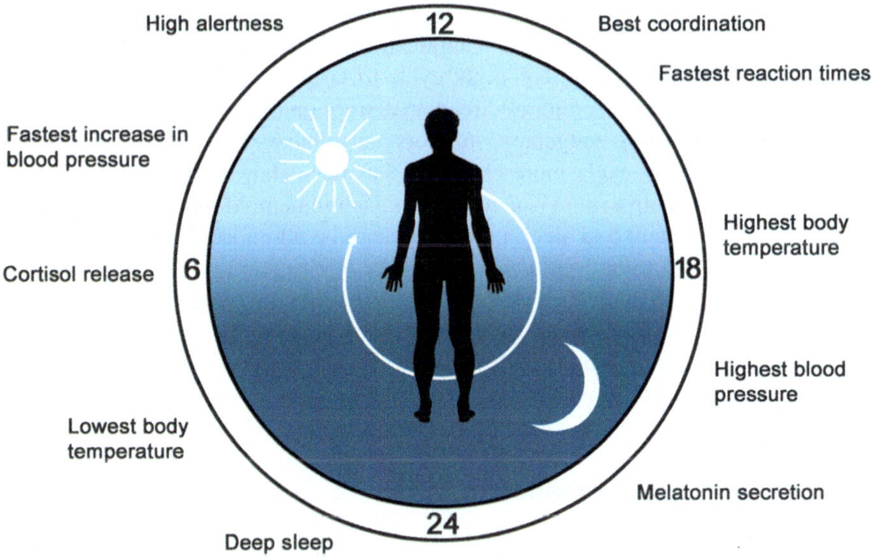

Fig. 9.3 The circadian clock anticipates and adapts our physiology to the different phases of the day. Our biological clock helps to regulate sleep patterns, feeding behavior, hormone release, blood pressure, and body temperature (modified after 2017 Nobel Price for Physiology or Medicine, press release by Nobelförsamlingen, Nobel Assembly at Karolinska Institutet 2017)

2016; Grimaldi et al. 2016; Leproult et al. 2014; Litinski et al. 2009). It has therefore been suggested, to transform the current knowledge, such as the influence of light on mood and alertness or the measurement of circadian biomarkers, into strategies and guidelines to apply appropriate lighting and to improve circadian-dependent health. Specific goals include the reduction of night time light exposure (intensity and wavelength), to target vulnearble populations (children, elderly, night-shift-workers, patient-populations), to implement circadian-appropriate lighting and to develop educational policies and public awareness (Mason et al. 2018). Especially the clinical setting has received special attention as patients are considered to be more susceptible to circadian stress and sleep disturbances than healthy persons. Simple methods can be applied in order to reduce the irregular sound and light exposure on ICU-patients, leading to less disturbed sleep and better outcome (Bion et al. 2018). More specifically, countermeasures should aim to more closely reproduce the natural day–night cycle with high illumination over the day (1000 lux on an overcast day and more than 100,000 lux on a bright sunshine day) and very low illumination during the night (as low as 0.0001 lux on a moonless night) through the use of large windows and adjusted artificial illumination. At night, light disturbances should be minimized, yet allow staff to perform their tasks. Eye masks might help to minimize the impact of disturbing light sources. Noises should be minimized as well through training of staff and removal of unecessary sound-sources to less disturbing places. Even ear-plugs for patients should be considered as well as appropriate timing of various tasks on the patient e. g., examination, washing, turning etc. Furthermore, timing of feeding the patient, mobilization, even through the use of different postures and the application of sedatives should be applied in accordance with the natural day–night cycle to keep zeitgebers at their expected times and thus avoid worsening of circadian disruption (McKenna et al. 2018).

Aside from adjusting zeitgebers, stressors and behavior in order to better follow the natural day–night cycle more closely, it has been shown that pharmacoligical interventions may help to alleviate circadian misalignment. Melatonin as a mediator of circadian alignment and sleep propensity, can be administered enterally. It has shown the potential to both improve the quality of sleep and performance the next day (Cheikh et al. 2018) as well as to treat jet-lag-symptoms (Tortorolo et al. 2015). Also, critically ill patients may benefit from its application to reduce sleep disruption and the occurance of delir (Bily et al. 2015), although more studies seem to be required in this field to provide conclusive recommendations (Lewis et al. 2018).

9.4.1 Physical Performance

As shown above, there is a circadian rhythm in functions that are of importance to athletes, such as alertness, reaction times, and strength. All in all, these abilities seem to peak during early evening, which not coincidentally, is when most world records in running and swimming are set (Atkinson and Reilly 1996); however, different individual chronotype may greatly influence individual physical performance (Winter et al. 2011). A possibility to determine the impact of sleep deprivation on the ability of physical performance is to measure the time to exhaustion (TTE) of human test subjects.

Research has shown that individuals' TTE after sleep deprivation ranged from a 5% improvement to a 4% decrement compared to TTE of non-sleep-deprived individuals (Pickett and Morris 1975), though unknown data about dietary intake and caffeine consumption may limit these findings; in addition, other factors such as motivation or the adrenaline release during physical performance can blunt the effects of sleep-deprivation (Blumert et al. 2007). Sleep deprivation has been shown to reduce the evaporative cooling function as well as dry heat loss in warm environments (Sawka et al. 1984), thus reducing the body's thermoregulatory ability. Another report revealed less running distance covered by eleven male test subjects after one night of sleep deprivation compared to the same group in a control phase with normal sleep. The test subjects, despite running less distance, exhibited similar perceptions of effort (RPE) during both trials, thus indicating that an altered perception of effort (similar perception of lower running speed) accounts for the decrease in performance after sleep deprivation. Results like this might be linked to impaired energy metabolism following sleep deprivation. It was shown that preexercise muscle glycogen stores were decreased after sleep deprivation, suggesting an alteration in substrate availability leading to impaired endurance performance (Skein et al. 2011). In addition, sleep deprivation or restriction can also impair maximal muscle strength in resistive exercise tests, although facilitated motiviation, such as exercising in groups may alleviate the impact (Knowles et al. 2018).

Other studies showed less accuracy in tennis players who were sleep deprived (Reyner and Horne 2013). Not surprisingly, sleep-deprived athletes also tended to more likely attain injuries compared non-sleep-deprived ones (Luke et al. 2011). Although larger studies are still lacking, smaller investigations have shown that recovery is impaired by sleep deprivation in professional athletes (Skein et al. 2013; Chase et al. 2017).

Finally, it is important to note that competitive athletes are at higher risk to suffer from sleep problems, because of arousal and anxiety before or after competition. One study reported that over 60% of athletes had problems with insomnia the night before a competition (Juliff et al. 2015); in addition, athletes often face a combination of stressors such as training and competition schedules, travel-stress, academic and occupational demands, and overtraining, which can increase the risk of sleep disturbances (Watson 2017).

9.4.2 Cognitive Function

Sufficient sleep is necessary for memorizing and recalling, in both motoric and cognitive learning. PET scans revealed that during sleep, the same cerebral areas were active as they were while practicing a motoric function (Marquet et al. 2003). Subjects were able to perform a certain motoric function previously practiced about 11% more sufficiently after sleep compared to subjects, who had to stay awake. Mice exhibited a memory decline after a sleep deficit that was accompanied by oxidative stress in the hippocampus (Silva et al. 2004). Healthy humans exhibited impaired vigilance after 27 and 37 h of sleep deprivation, respectively (McMahon et al. 2018). Chronic sleep deprivation and REM-sleep restriction can result in disrupted hippocampal signaling pathways, impaired hippocampal synaptic plasticity and less consolidation of hippocampal-dependent memories (Prince and Abel 2013). One night of sleep loss has been shown to negatively impact attention and working memory function through the impairment of coupling between the hippocampus and cortical regions (Krause et al. 2017). A possible way by which impaired sleep negatively influences hippocampal function might be through epigenetic modification. Sleep deprivation appears to produce changes in epigenetic markers and patterns of gene transcription with evidence of altered methylation profile in healthy human subjects (Nilsson et al. 2016). Another model approach is that of synaptic homeostasis where sleep serves to reduce the ongoing synaptic potentiation during wakefulness, which would otherwise not be sustainable in terms of energy requirements and which then allows new information to be processed (Tononi and Cirelli 2014).

It also could be shown that sleep inertia—the impaired cognitive performance immediately upon awakening—is also influenced by the circadian rhythm, as the reduction of cognitive function immediately upon awakening is more than threefold more pronounced when subjects are woken at the biological night than during daytime. This suggests an influence of the circadian rhythm to the demands on humans to cope with psychological stress such as in case of on-call emergency workers who need to be alert immediately after awakening. The results of the two previously mentioned relationships between circadian rhythm and physical and cognitive performance, respectively, suggest that night work in a misaligned individual can lead to a reduced performance both mentally and physically. Finally, there is evidence that suggests that long-term sleep deprivation and restriction might increase the risk for neurodegenerative diseases like Alzheimer's disease and dementia (Xie et al. 2013).

9.4.3 Immune Function and Fibrinolysis

Furthermore, studies have supported the reception that sleep loss makes humans more susceptible to infections.

The majority of immune cells express circadian clock genes and physiological functions based on a 24 h rhythm. This impacts cellular functions, such as synthesis and release of cytokines and cytolytic factors, expression of recognition receptors, phagocytosis, migration to inflamed tissue, cytolytic activity, and proliferative response to antigens. Therefore, alterations of circadian rhythms (e.g., environmental disruption similar to shift work) can lead to dysfunctional immune responses (Labrecque and Cermakian 2015).

Accordingly it could be shown that sleep improves the immune responses in human vaccination studies such as influenza or hepatitis A (Spiegel et al. 2002). The protective role of sleep was demonstrated when the morbidity and mortality of experimentally infected rabbits were substantially reduced by a longer sleep duration (Toth et al. 1993). It is of interest that not only pacemaker cells of the circadian regulation system in the hypothalamus, but also rat NK, mouse macrophages, and human leukocytes exhibit rhythmic expression of clock genes that are associated with the sleep–wake cycle (Hayashi et al. 2007). It has been suggested that substances released in relationship with the circadian regulation like hormones such as cortisol, melatonin, or growth hormone not only play a role in regulation of the immune function of immune cells, but also that such hormones—as well as influences from the sympathetic nervous system and the body core temperature-synchronize the peripheral circadian clocks of immune cells and thereby harmonize the functional rhythm of immunity at the cellular level. It is further speculated that sleep deprivation will disrupt the synchrony between immune cells, thus leading to a desynchronization of immune functions and eventually a deregulated immune response. Furthermore, not only do most immune cells, like T-cells, granulocytes, and macrophages, exhibit a daily variation in their function, but also many genes responsible for pathogen recognition and cytokine secretion exhibit circadian rhythms (Keller et al. 2009).

In turn, immune responses can also give feedback to the regulation of the circadian rhythm, most likely through pro-inflammatory cytokines—it could be shown that infections and low-dose lipopolysaccharide (LPS) administration increases sleep in humans (Bryant et al. 2004). In another study, inhibition of TNF by a soluble TNF receptor improved fatigue in patients with rheumatoid arthritis (Pollard et al. 2006). Also, neutralization of TNF reduced daytime sleepiness in patients with sleep apnea syndrome. It could be shown that TNF suppresses the expression of PAR bZip clock-controlled genes Dbp, Tef, and Hlf as well as the period genes Per 1–3 in fibroblasts in the mice liver; TNF also interferes with the clock-controlled gene Dbp in the suprachiasmatic nucleus of mice causing prolonged rest periods (Cavadini et al. 2007). These findings that suggest a high integration of the circadian cycle and immune function are further supported by another report in which a systemic immune response through the in vivo administration of LPS resulted in an upregulation of the core clock genes Per 2 and Bmal1 in equine blood cells, while a treatment utilizing nonsteroidal anti-inflammatory drugs (NSAID) not only inhibited the inflammatory responses but also the upregulation of clock gene expression. While these responses could only be found exclusively in the in vivo model, it suggests a yet unknown pathway between immune modulators and the regulation of circadian clocks in the periphery, which also proposes the high significance of the circadian rhythm to innate immune reactions (Murphy et al. 2007). Furthermore, it was shown that pattern recognition receptors (PRR) of macrophages, which are activated by pathogen-associated molecular patterns (PAMPs), do follow a circadian rhythm, such as the response to bacterial LPS (Gibbs et al. 2012). Likewise, NK exhibited highly disturbed rhythms of IFNγ, granzyme B, and perforin mRNAs an thus of their phsysiologcal function, under repeated jet-lag conditions (Logan et al. 2012).

These research culminated in the understanding, that the circadian clock of the immune system is a very critical element with regard to immune function. The circadian clockwork influences both innate and adaptive immunity, such as responses to pathogen exposure, immune cell movement, lymphocyte development and trafficking, adaptive immune responses, and even the symbiosis between body cells and microbiota such as in the gut, where microbiotic products appear to influence circadian immunity, thus providing a link between disturbed circadian rhythm, changed immune function, and microbiota. Interaction and disruption of the circadian rhythm and immune system may also give rise to diseases like psoriasis or irritable bowel disease that have been associated with shift work, or the circadian aspects of asthma with exacerbated symptoms in the early morning. The increased knowledge of circadian interactions of the immune system provides therapeutic opportunities (chronotherapy) to provide more effective chemotherapy, pain-relief at the time of greatest need, more effective vaccination results and even less rejection response with transplant patients (Scheiermann et al. 2018).

It was shown that there also exists a circadian rhythmicity in some components of the hemostatic system, tissue-type plasminogen activator (tPA) and its inhibitor plasminogen activator inhibitor-1 (PAI-1) show a marked circadian variation in plasma (Andreotti and Kluft 1991). This variation seems to be even more pronounced in combination with stress. It was shown that exhaustion and chronic exposure to life stressors such as chronic overtime work are associated with decreased early morning fibrinolysis and increased fibrinogen levels throughout the day, which in turn would promote thrombus formation and increase the risk of coronary infarction (Van Diest et al. 2002).

The clinical importance of clock gene regulated immune challenges have become more and more also in the focus of clinicians. Critically ill patients are especially susceptible to circadian dysrhythmia due to mistiming or complete loss of sensory input disturbing circadian synchronization with the environment and pathology affecting the clock mechanism at a cellular level, which may relate to inflammatory responses, with septic patients to have abnormal patterns of melatonin secretion compared to nonseptic patients (Mundigler et al. 2002). Possible outcomes of circadian disruption on the intensive care unit (ICU) are the interconnected phenomena of sleep disturbance (sleep fragmentation) and delirium affecting up to 80% of ICU patients and which is know to be associated with increased morbidity and mortality (Ely et al. 2004).

In this context, the terms "Chronopathology" and "Chronofitness" have been coined (McKenna et al. 2018), to emphasize the impact of circadian disruption as a distinct source of pathological processes of the immune system and other funtions and the need to address the internal synchronization of circadian clocks in various tissues and their synchronization with the environment, in order to provide a personalized treatment for critically ill patients as well as general approaches as countermeasures to avoid circadian disruption in the clinical setting (see Sect. 9.4 above).

9.4.4 Cancer

With regard to the relationships discussed, diseases closely connected to the immunological function such as cancer could also be affected by alterations of the circadian rhythm. It was shown that the circadian regulation system also controls functions responsible for drug metabolism and detoxification as well as cellular proliferation such as the cell cycle, DNA repair mechanisms, and apoptosis (Okyar and Lévi 2008). Conventional knowledge attributed tumor suppressor properties to the circadian clock system. However, in turn, the clock can also promote tumorigenesis (Shostak 2017). Nuclear receptors, such as thyroid hormone receptors, steroid receptors, and retinoic acid receptor-related orphan receptors (ROR), regulate numerous physiological processes, such as metabolism, inflammation, and are also involved in the regulation of circadian rhythm (Weikum et al. 2018; Fan et al. 2018), while in turn circadian misalignment may promote and influence the progression of various malignancies (Doan et al. 2017; Fan et al. 2018). In experimental models, chronic jet lag accelerated growth of two transplantable tumors in mice. In that model, the expression rhythms of clock genes in liver and tumor cells were significantly altered (Filipski et al. 2009). It was shown that disruptions of the circadian regulation could interfere with the regulation of oncogenes like c-Myc and p53, leading to a upregulation of c-Myc and a downregulation of p53, thus causing genomic instability, increasing proliferation, and eventually the possible accumulation of mutations favoring carcinogenesis (Fu et al. 2002). Recently, cell cycle and apoptosis regulator 2 (CCAR2) was identified as an important mediator in DNA damage response and regulates transcription and by repressing BMAL1 and CLOCK expression, it was reported to modulate the circadian rhythm (Magni et al. 2018). Notably, CCAR2 expression is deregulated in several tumors showing a link between carcinogenesis and disrupted circadian cycle. Because of this recent research and further studies that have demonstrated the links between circadian rhythm and cell cycle (and the diruption thereof) (Jiang et al. 2016; Huber et al. 2016; Gotoh et al. 2014) has lead to the conclusion that the circadian cycle is an important tumor suppressor and that disrupted circadian rhythms promote tumor development (Fu and Kettner 2013). Accordingly it could be shown that persons working in night shifts have a significantly greater risk of developing cancer like breast, colon, and prostate cancer and non-Hodgkin lymphoma (Viswanathan and Schernhammer 2008).

Likewise, it is notable that certain carcinogenic toxins might not only induce cancer but could also itself alter the circadian rhythm through cytokine responses. Thus, carcinogenic effects of such toxins and changes in the circadian cycle could potentiate themselves as it appears in chronic alcohol abuse or hepatotropic viruses, as they both produce preneoplastic liver lesions and disrupt the circadian cycle in experimental model and in humans (Spanagel et al. 2005). In line with these findings it could be advised that treatment strategies should aim at the prevention or correction of circadian disruption to prevent cancer development and control its progress respectively (Filipski et al. 2009).

Recent studies also showed dynamic circadian oscillations in oxidation-reduction (redox) status with a direct stabilizing feedback of redox cofactors (such as FAD) on the circadian clock protein cryptochrome (Pritchett and Reddy 2017), which in turn also influences immune functions through an interplay between circadian fluctuations and redox status (Wende et al. 2016). In line with these findings, stress, because of its influence on the immune system via hormones like cortisol and through changes of the circadian rhythm on the systemic and peripheral cellular level, should be prevented in order to maintain health and well-being, especially under conditions of space flight, when stressor may act in an additive way.

9.5 Summary

As it was shown in this chapter, the circadian rhythm and its regulation are on the one hand closely connected to and controlled by external influences, most notably the day–night cycle as well as other zeitgeber and on the other hand the circadian rhythm itself has close connections to and influences various physiological functions of the human body. A disturbance of the regulation of the circadian rhythm (e.g., through jet lag or in the shift worker) therefore not only results in obvious effects like sleeping disorders such as difficulties falling asleep, sleeping through and daytime sleepiness which reduce physical and mental capabilities—but it can also lead to pathological changes in the metabolism of carbohydrates and lipids or have effects on vegetative regulation and immune function, thus increasing the susceptibility to diseases like the metabolic syndrome, cardiovascular disorders, or various types of cancer. "*The 2017 Nobel Laureates have uncovered a mechanism controlling a truly fundamental process in physiology, how our cells and bodies keep time. Such time-keeping is essential for our adaptation, and has important implications for human health; not just jetlag, but also the incidence of chronic syndromes, such as cancer, metabolic and sleep disorders, and several neurological conditions*" (presentation speech by Professor Carlos Ibáñez on behalf of the Nobel Assembly at the Karolinska Institute, December 10, 2017). These discoveries and research fueled by these discoveries will help to understand the influence of extreme environments on circadian rhythms. Investigations in man living in extreme climates or being exposed to isolated and confined conditions will comprehend related changes observed in space crews. This knowledge will pave appropriate countermeasures to alleviate the impact of these changes—especially during long-term space missions—on health and performance of astronauts.

References

American Academy of Sleep Medicine (AASM) (2005) International classification of sleep disorders. Diagnostic and coding manual, 2nd edn. American Academy of Sleep Medicine, Westchester

American Academy of Sleep Medicine (AASM) (2014) International classification of sleep disorders, 3rd edn. Amaerican Academy of Sleep Medicine, Darien, IL

An R, Wang J, Ashrafi SA, Yang Y, Guan C (2018) Chronic noise exposure and adiposity: a systematic review and meta-analysis. Am J Prev Med 55(3):403–411

Andreotti F, Kluft C (1991) Circadian variation of fibrinolytic activity in blood. Chronobiol Int 8(5):336–351

Aoki H, Ozeki Y, Yamada N (2001) Hypersensitivity of melatonin suppression in response to light in patients with delayed sleep phase syndrome. Chronobiol Int 18(2):263–271

Aschoff J, Wever R (1962a) Biologische rhythmen und regelung. Bad Oeyenhaus Gespräch 5:1–15

Aschoff J, Wever R (1962b) Spontanperiodik des Menschen bei Ausschluss aller Zeitgeber. Naturwissenschaften 49:337–342

Atkinson G, Davenne D (2007) Relationships between sleep, physical activity and human health. Physiol Behav 90(2–3):229–235

Atkinson G, Reilly T (1996) Circadian variation in sports performance. Sports Med 21(4):292–312

Auger RR, Burgess HJ, Emens JS, Deriy LV, Thomas SM, Sharkey KM (2015) Clinical practice guideline for the treatment of intrinsic circadian rhythm sleep-wake disorders: advanced sleep-wake phase disorder (ASWPD), delayed sleep-wake phase disorder (DSWPD), non-24-hour sleepwake rhythm disorder (N24SWD), and irregular sleep-wake rhythm disorder (ISWRD). An update for 2015. J Clin Sleep Med 11(10):1199–1236

Aulsebrook AE, Jones TM, Mulder RA, Lesku JA (2018) Impacts of artificial light at night on sleep: a review and prospectus. J Exp Zool A Ecol Integr Physiol 329:409–418

Baker A, Ferguson S, Dawson D (2003) The perceived value of time – controls versus shift workers. Time Soc 12:27–39

Barger LK, Sullivan JP, Vincent AS, Fiedler ER, McKenna LM (2012) Learning to live on a Mars day: fatigue countermeasures during the Phoenix Mars Lander mission. Sleep 35(10):1423–1435

Barger LK, Flynn-Evans EE, Kubey A, Walsh L, Ronda JM, Wang W, Wright KP, Czeisler CA (2014) Prevalence of sleep deficiency and use of hypnotic drugs in astronauts before, during, and after spaceflight: an observational study. Lancet 13:904–912

Basner M, Dinges DF, Mollicone D, Ecker A, Jones CW (2013) Mars 520-d mission simulation reveals protracted crew hypokinesis and alterations of sleep duration and timing. Proc Natl Acad Sci U S A 110(7):2635–2640. Erratum in Proc Natl Acad Sci U S A 110(7):2676

Beale AD, Whitmore D, Moran D (2016) Life in a dark biosphere: a review of circadian physiology in "arrhythmic" environments. J Comp Physiol B 186(8):947–968

Bily B, Sabol F, Török P, Artemiou P, Bilecova-Rabajdova M, Kolarcik P (2015) Influence of prophylactic melatonin administration on the incidence of early postoperative delirium in cardiac surgery patients. Anesteziol Intenzivni Med 26:319–327

Bion V, Lowe AS, Puthucheary Z, Montgomery H (2018) Reducing sound and light exposure to improve sleep on the adult intensive care unit: an inclusive narrative review. J Intensive Care Soc 19(2):138–146

Blumert PA, Crum AJ, Ernsting M (2007) The acute effects of twenty-four hours of sleep loss on the performance of national-caliber male collegiate weightlifters. J Strength Cond Res 21(4):1146–1154

Boivin DB, Boudreau P (2014) Impacts of shift work on sleep and circadian rhythms. Pathol Biol (Paris) 62(5):292–301

Boulos Z, Campbell SS, Lewy AJ, Terman M, Dijk DJ, Eastman CI (1995) Light treatment for sleep disorders: consensus report. VII. Jet lag. J Biol Rhythms 10(2):167–176

Brainard GC, Barger LK, Soler RR, Hanifin JP (2016) The development of lighting countermeasures for sleep disruption and circadian misalignment during spaceflight. Curr Opin Pulm Med 22(6):535–544

Bryant PA, Trinder J, Curtis N (2004) Sick and tired: does sleep have a vital role in the immune system? Nat Rev Immunol 4:457–467

Buguet A, Rivolier J, Jouvet M (1987) Human sleep patterns in Antarctica. Sleep 10:374–382

Buguet A, Cespuglio R, Radomski MW (1998) Sleep and stress in man: an approach through exercise and exposure to extreme environments. Can J Physiol Pharmacol 76:553–561

Buguet A, Tapie P, Bisser S, Chapotot F, Banzet S, Bogui P (2002) Sleep in tropical Africa: at the laboratory and in villages without electrical power. J Sleep Res 11(Suppl 1):30

Burgess KR, Ainslie PN (2016) Central sleep apnea at high altitude. Adv Exp Med Biol 903:275–283

Burgess KR, Cooper J, Rice A, Wong K, Kinsman T, Hahn A (2006) Effect of simulated altitude during sleep on moderate-severity OSA. Respirology 11:62–69

Campbell SS (1999) Intrinsic disruption of normal sleep and circadian patterns. In: Turek FW, Zee PC (eds) Regulation of sleep and circadian rhythms. Marcel Dekker, New York, NY, pp 465–486

Cartmell T, Luheshi GN, Hopkins SJ, Rothwell NJ, Poole S (2001) Role of endogenous interleukin-1 receptor antagonist in regulating fever induced by localised inflammation in the rat. J Physiol 531:171–180

Cavadini G, Petrzilka S, Kohler P, Jud C, Tobler I, Birchler T, Fontana A (2007) TNF-alpha suppresses the expression of clock genes by interfering with E-box-mediated transcription. Proc Natl Acad Sci U S A 104(31):12843–12848

Chase JD, Roberson PA, Saunders MJ (2017) One night of sleep restriction following heavy exercise impairs 3-km cycling time-trial performance in the morning. Appl Physiol Nutr Metab 42(9):909–915

Cheikh M, Hammouda O, Gaamouri N, Driss T, Chamari K, Cheikh RB, Dogui M, Souissi N (2018) Melatonin ingestion after exhaustive late-evening exercise improves sleep quality and quantity, and short-term performances in teenage athletes. Chronobiol Int 30:1–13

Chouker A (2012) Stress challenges and immunity in space: from mechanisms to monitoring, and preventive strategies. Springer, Heidelberg

Clark RP, Edholm OG (1985) Man and his thermal environment. E. Arnold, London

Cohen S, Doyle WJ, Alper CM, Janicki-Deverts D, Turner RB (2009) Sleep habits and susceptibility to the common cold. Arch Intern Med 169:62–67

Crandall CG, Johnson JM, Convertino VA, Raven PB, Engelke KA (1994) Altered thermoregulatory responses after 15 days of head-down tilt. J Appl Physiol 77:1863–1867

Czeisler CA, Kronauer RE, Allan JS et al (1989) Bright light induction of strong (type 0) resetting of the human circadian pacemaker. Science 244(4910):1328–1333

Czeisler CA, Duffy JF, Shanahan TL (1999) Stability, precision, and near-24-hour period of the human circadian pacemaker. Science 284(5423):2177–2181

Delezie J, Dumont S, Dardente H, Oudart H, Grechez-Cassiau A, Klosen P, Teboul M, Delaunay F, Pevet P, Challet E (2012) The nuclear receptor REV-ERBalpha is required for the daily balance of carbohydrate and lipid metabolism. FASEB J 26:3321–3335

Dibner C, Schibler U, Albrecht U (2010) The mammalian circadian timing system: organization and coordination of central and peripheral clocks. Annu Rev Physiol 1:517–549

Dijk DJ, Brunner DP, Beersma DG, Borbe'ly AA (1990) Electroencephalogram power density and slow wave sleep as a function of prior waking and circadian phase. Sleep 13(5):430–440

Dijk DJ, Neri DF, Wyatt JK, Ronda JM, Riel E, Ritz-De Cecco A, Hughes RJ, Elliott AR, Prisk GK, West JB, Czeisler CA (2001) Sleep, performance, circadian rhythms, and light-dark cycles during two space shuttle flights. Am J Physiol Regul Integr Comp Physiol 281(5):R1647–R1664

Dinneen S, Alzaid A, Miles J, Rizza R (1993) Metabolic effects of the nocturnal rise in cortisol on carbohydrate metabolism in normal humans. J Clin Invest 92:2283–2290

Doan TB, Graham JD, Clarke CL (2017) Emerging functional roles of nuclear receptors in breast cancer. J Mol Endocrinol 58(3):R169–R190

Drake CL, Roehrs T, Richardson G, Walsh JK, Roth T (2004) Shift work sleep disorder: prevalence and consequences beyond that of symptomatic day workers. Sleep 27(8):1453–1462. 1778–1779

Duffy JF, Wright KP Jr (2005) Entrainment of the human circadian system by light. J Biol Rhythms 20:326–338

Dyar KA, Eckel-Mahan KL (2017) Circadian metabolomics in time and space. Front Neurosci 11:369

Eastman CI, Liu L, Fogg LF (1995) Circadian rhythm adaptation to simulated night shift work: effect of nocturnal brightlight duration. Sleep 18(6):399–407

Eichenberger U, Waber U, Maggiorini M, Oelz O, Bartsch P (1993) Acute high altitude illnesses are not related to periodic breathing and apneas during sleep. In: Sutton JR, Coates J, Houston CS (eds) Hypoxia and mountain medicine. Pergamon Press, Oxford, p 302

Eliott AL, Mills JN, Waterhouse JM (1971) A man with too long a day. J Physiol 212(2):30P–31P

Ely EW, Shintani A, Truman B (2004) Delirium as a predictor of mortality in mechanically ventilated patients in the intensive care unit. JAMA 291:1753–1762

Fan J, Lv Z, Yang G, Liao Tt XJ, Wu F, Huang Q, Guo M, Hu G, Zhou M, Duan L, Liu S, Jin Y (2018) Retinoic acid receptor-related orphan receptors: critical roles in tumorigenesis. Front Immunol 9:1187

Filipski E, Subramanian P, Carrière J, Guettier C, Barbason H, Lévi F (2009) Circadian disruption accelerates liver carcinogenesis in mice. Mutat Res 680(1–2):95–105

Flynn-Evans EE, Barger LK, Kubey AA, Sullivan JP, Czeisler CA (2016) Circadian misalignment affects sleep and medication use before and during spaceflight. NPJ Microgravity 2:15019

Folkard S, Monk TH, Lobban MC (1978) Short and long-term adjustment of circadian rhythms in 'permanent' night nurses. Ergonomics 21(10):785–799

Fortney SM, Mikhaylov V, Lee SM, Kobzev Y, Gonzalez RR, Greenleaf JE (1998) Body temperature and thermoregulation during submaximal exercise after 115-day spaceflight. Aviat Space Environ Med 69:137–141

Fu L, Kettner NM (2013) The circadian clock in cancer development and therapy. Prog Mol Biol Transl Sci 119:221–282

Fu L, Pelicano H, Liu J, Huang P, Lee C (2002) The circadian gene Period2 plays an important role in tumor suppression and DNA damage response in vivo. Cell 111:41–50

Gibbs M, Hampton S, Morgan L, Arendt J (2002) Adaptation of the circadian rhythm of 6-sulphatoxymelatonin to a shift schedule of seven nights followed by seven days in offshore oil installation workers. Neurosci Lett 325:91–94

Gibbs JE, Blaikley J, Beesley S, Matthews L, Simpson KD, Boyce SH, Farrow SN, Else KJ, Singh D, Ray DW (2012) The nuclear receptor REV-ERBalpha mediates circadian regulation of innate immunity through selective regulation of inflammatory cytokines. Proc Natl Acad Sci U S A 109:582–587

Gierse A (1842) Quaemiam sit ratio caloris organic. Dissertation Halle

Gotoh T, Vila-Caballer M, Santos CS, Liu J, Yang J, Finkielstein CV (2014) The circadian factor period 2 modulates p53 stability and transcriptional activity in unstressed cells. Mol Biol Cell 25:3081–3093

Greenleaf JE (1997) Exercise thermoregulation with bed rest, confinement, and immersion deconditioning. Ann N Y Acad Sci 813:741–750

Grimaldi D, Carter JR, Van Cauter E, Leproult R (2016) Adverse impact of sleep restriction and circadian misalignment on autonomic function in healthy young adults. Hypertension 68:243–250

Gronfier C, Wright KP Jr, Kronauer RE, Jewett ME, Czeisler CA (2004) Efficacy of a single sequence of intermittent bright light pulses for delaying circadian phase in humans. Am J Physiol Endocrinol Metab 287:E174–E181

Gundel A, Nalishiti V, Reucher E, Vejvoda M, Zulley J (1993) Sleep and circadian rhythm during a short space mission. Clin Investig 71:718–724

Gundel A, Polyakov V, Zulley J (1997) The alteration of human sleep and circadian rhythms during spaceflight. J Sleep Res 6:1–8

Gundel A, Drescher J, Polyakov VV (2001) Quantity and quality of sleep during the record manned space flight of 438 days. Hum Factors Aerospace Saf 1:87–98

Halberg F, Siffre M, Engeli M, Hillmann D, Reinberg A (1965) Etude en librecours des rythmes circadien du pouls, de l'alternance veillesommeil et de l'estimation du temps pendant les deux mois de séjour souterrain d'un homme adulte jeune. C R Acad Sci Paris 260:1259–1262

Harrington ME (1997) The ventral lateral geniculate nucleus and the intergeniculate leaflet: interrelated structures in the visual and circadian systems. Neurosci Biobehav Rev 21(5):705–727

Haskell EH, Palca JW, Walker JM, Berger RJ, Heller HC (1981) Metabolism and thermoregulation during stages of sleep in humans exposed to heat and cold. J Appl Physiol Respir Environ Exerc Physiol 51:948–954

Hayashi M, Shimba S, Tezuka M (2007) Characterization of the molecular clock in mouse peritoneal macrophages. Biol Pharm Bull 30:621–626

He Y, Jones CR, Fujiki N, Xu Y, Guo B, Holder JL, Rossner MJ, Nishino S, Fu YH (2009) The transcriptional repressor DEC2 regulates sleep length in mammals. Science 325:866–870

Heslegrave RJ, Angus RG, Buguet A (1982) Changes in sleep patterns as a function of hyperbaric exposure and confinement: a preliminary report. Meeting of the Society for Psychophysiological Research, Minneapolis, MN

Hida A, Kitamura S, Katayose Y, Kato M, Ono H, Kadotani H, Uchiyama M, Ebisawa T, Inoue Y, Kamei Y, Okawa M, Takahashi K, Mishima K (2014) Screening of clock gene polymorphisms demonstrates association of a PER3 polymorphism with morningness-eveningness preference and circadian rhythm sleep disorder. Sci Rep 4:6309. https://doi.org/10.1038/srep06309

Hittle BM, Gillespie GL (2018) Identifying shift worker chronotype: implications for health. Ind Health 56(6):512–523

Huber AL, Papp SJ, Chan AB, Henriksson E, Jordan SD, Kriebs A, Nguyen M, Wallace M, Li Z, Metallo CM (2016) CRY2 and FBXL3 cooperatively degrade c-MYC. Mol Cell 64:774–789

Hurst M (2008) Qui dort la nuit de nos jour? Les habitudes de sommeil des canadiens. Stat Can 11–008:42–48

ICSD-2 (2005) The international classification of sleep disorders: diagnostic and coding manual, 2nd edn. American Academy of Sleep Medicine, Westchester

James FO, Cermakian N, Boivin DB (2007) Circadian rhythms of melatonin, cortisol, and clock gene expression during simulated night shift work. Sleep 30:1427–1436

Jiang W, Zhao S, Jiang X, Zhang E, Hu G, Hu B, Zheng P, Xiao J, Lu Z, Lu Y (2016) The circadian clock gene bmal1 acts as a potential anti-oncogene in pancreatic cancer by activating the p53 tumor suppressor pathway. Cancer Lett 371:314–325

Jones CR, Campbell SS, Zone SE et al (1999) Familial advanced sleep-phase syndrome: a short-period circadian rhythm variant in humans. Nat Med 5(9):1062–1065

Joshi SS, Lesser TJ, Olsen JW, O'Hara BF (2016) The importance of temperature and thermoregulation for optimal human sleep. Energ Buildings 131:153–157

Juliff LE, Halson SL, Peiffer JJ (2015) Understanding sleep disturbance in athletes prior to important competitions. J Sci Med Sport 18(1):13–18

Kamei Y, Hayakawa T, Urata J et al (2000) Melatonin treatment for circadian rhythm sleep disorders. Psychiatry Clin Neurosci 54(3):381–382

Kanas N, Manzey D (2010) Space psychology and psychiatry, 2nd edn. Springer, New York, NY, p 96f

Karlsson B, Knutsson A, Lindahl B (2001) Is there an association between shift work and having a metabolic syndrome? Results from a population based study of 27,485 people. Occup Environ Med 58:747–752

Keller M, Mazuch J, Abraham U, Eom GD, Herzog ED, Volk HD, Kramer A, Maier B (2009) A circadian clock in macrophages controls inflammatory immune responses. Proc Natl Acad Sci U S A 106:21407–21412

Kendel K, Schmidt-Kessen W (1973) The influence of room temperature on night sleep in man (polygraphic night-sleep recordings in the climatic chamber). In: Kendel K, Schmidt-Kessen W (eds) Sleep. S. Karger, Basel, pp 423–425

Kim KH, Kabir E, Ara Jahan S (2014) A review of the consequences of global climate change on human health. J Environ Sci Health C Environ Carcinog Ecotoxicol Rev 32(3):299–318

Knauth PR (1981) Duration of sleep related to the type of shiftwork. In: Reinberg AV, Andlauer N (eds) Advances in the biosciences: night and shiftwork biological and social aspects. Pergamon Press, Oxford, pp 161–168

Knowles OE, Drinkwater EJ, Urwin CS, Lamon S, Aisbett B (2018) Inadequate sleep and muscle strength: Implications for resistance training. J Sci Med Sport 21(9):959–968

Koscheyev VS, Leon GR, Hubel A, Nelson ED, Tranchida D (2000) Thermoregulation and heat exchange in a nonuniform thermal environment during simulated extended EVA. Extravehicular activities. Aviat Space Environ Med 71:579–585

Kraft NO, Inoue N, Mizuno K, Ohshima H, Murai T, Sekiguchi C (2002) Physiological changes, sleep, and morning mood in an isolated environment. Aviat Space Environ Med 73(11):1089–1093

Krauchi K, Cajochen C, Werth E, Wirz-Justice A (2000) Functional link between distal vasodilation and sleep-onset latency? Am J Physiol Regul Integr Comp Physiol 278:741–748

Krause AJ, Simon EB, Mander BA (2017) The sleep-deprived human brain. Nat Rev Neurosci 18:404–418

Kurien PA, Chong SY, Ptáček LJ, Fu YH (2013) Sick and tired: how molecular regulators of human sleep schedules and duration impact immune function. Curr Opin Neurobiol 23(5):873–879

Labrecque N, Cermakian N (2015) Circadian clocks in the immune system. J Biol Rhythms 30(4):277–290

Labyak S, Lava S, Turek F, Zee P (2002) Effects of shiftwork on sleep and menstrual function in nurses. Health Care Women Int 23:703–714

Lack LS (1993) Evening light treatment of early morning insomnia. Sleep Res 22:225

Lan L, Pan L, Lian Z, Huang H, Lin Y (2014) Experimental study on thermal comfort of sleeping people at different air temperatures. Build Environ 73:24–31

Lan L, Tsuzuki L, Liu YF, Lian ZW (2017) Thermal environment and sleep quality: a review. Energy Build 149(15):101–113

Lan L, Qian XL, Lian ZW, Lin YB (2018) Local body cooling to improve sleep quality and thermal comfort in a hot environment. Indoor Air 28(1):135–145

Ledrans M, Pirard P, Tillaut H, Vandentorren S, Suzan F, Salines G et al (2004) La canicule d'août 2003: que s'est-il passé ? Rev Prat 54:1289–1297

Leproult R, Holmback U, Van Cauter E (2014) Circadian misalignment augments markers of insulin resistance and inflammation, independently of sleep loss. Diabetes 63:1860–1869

Levi F, Altinok A, Clairambault J, Goldbeter A (2008) Implications of circadian clocks for the rhythmic delivery of cancer therapeutics. Philos Transact A Math Phys Eng Sci 366:3575–3598

Lewis SR, Pritchard MW, Schofield-Robinson OJ, Alderson P, Smith AF (2018) Melatonin for the promotion of sleep in adults in the intensive care unit. Cochrane Database Syst Rev 5:CD012455

Lewy AJ, Sack RL, Miller LS, Hoban TM (1987) Antidepressant and circadian phase-shifting effects of light. Science 235:352–354

Litinski M, Scheer FA, Shea SA (2009) Influence of the circadian system on disease severity. Sleep Med Clin 4:143–163

Liu C, Weaver DR, Jin X et al (1997) Molecular dissection of two distinct actions of melatonin on the suprachiasmatic circadian clock. Neuron 19(1):91–102

Liu Y, Wheaton AG, Chapman DP (2016) Prevalence of healthy sleep duration among adults—United States, 2014. MMWR Morb Mortal Wkly Rep 65:137–141

Logan RW, Zhang C, Murugan S, O'Connell S, Levitt D, Rosenwasser AM, Sarkar DK (2012) Chronic shiftlag alters the circadian clock of NK cells and promotes lung cancer growth in rats. J Immunol 188:2583–2591

Luke A, Lazaro RM, Bergeron MF (2011) Sports-related injuries in youth athletes: is overscheduling a risk factor? Clin J Sport Med 21(4):307–314

Lund J, Arendt J, Hampton S et al (2001) Postprandial hormone and metabolic responses amongst shift workers in Antarctica. J Endocrinol 171:557–564

Magni M, Buscemi G, Zannini L (2018) Cell cycle and apoptosis regulator 2 at the interface between DNA damage response and cell physiology. Mutat Res 776:1–9

Mallis MM, DeRoshia CW (2005) Circadian rhythms, sleep, and performance in space. Aviat Space Environ Med 76(6 Suppl):B94–B107

Marquet P et al (2003) Sleep-related consolidation of visiomotoric skill: brain mechanism as assessed by functional magnetic resonance imaging. J Neurosci 23:1432–1440

Mason IC, Boubekri M, Figueiro MG, Hasler BP, Hattar S, Hill SM, Nelson RJ, Sharkey KM, Wright KP, Boyd WA, Brown MK, Laposky AD, Twery MJ, Zee PC (2018) Circadian health and light: a report on the National Heart, Lung, and Blood Institute's workshop. J Biol Rhythms 1:748730418789506

Matsuoka S, Inoue K, Okuda S, Ishikawa T, Lee HD, Mouri M (1986) EEG polygraphic sleep study in divers under a 31 ATA He–O2 environment with special reference to the automated analysis of sleep stages. J UOEH 8:293–305

McKenna H, van der Horst GTJ, Reiss I, Martin D (2018) Clinical chronobiology: a timely consideration in critical care medicine. Crit Care 22(1):124

McMahon WR, Ftouni S, Drummond SPA, Maruff P, Lockley SW, Rajaratnam SMW, Anderson C (2018) The wake maintenance zone shows task dependent changes in cognitive function following one night without sleep. Sleep 41(10). https://doi.org/10.1093/sleep/zsy148

Mistlberger R, Rusak B (1989) Mechanisms and models of the circadian timekeeping system. In: Kryger M, Roth T, Dement W (eds) Principles and practice of sleep medicine. WB Saunders Company, Philadelphia, PA, pp 141–152

Mizuno K, Asano K, Inoue Y, Shirakawa S (2005) Consecutive monitoring of sleep disturbance for four nights at the top of Mt Fuji (3776 m). Psychiatry Clin Neurosci 59:223–225

Moline ML, Pollak CP, Monk TH et al (1992) Age-related differences in recovery from simulated jet lag. Sleep 15:28–40

Mollicone DJ et al (2007) Optimizing sleep/wake schedules in space: sleep during chronic nocturnal sleep restriction with and without diurnal naps. Acta Astronaut 60:354–361

Montmayeur A, Buguet A (1992) Sleep patterns of European expatriates in a dry tropical climate. J Sleep Res 1:191–196

Moore-Ede MCS, Sulzman FM, Fuller C (1982) The clocks that time us. Harvard University Press, Cambridge

Morgenthaler TI, Lee-Chiong T, Alessi C, Standards of Practice Committee of the American Academy of Sleep Medicine et al. (2007) Practice parameters for the clinical evaluation and treatment of circadian rhythm sleep disorders. An American Academy of Sleep Medicine report. Sleep 30(11):1445–1459

Morris CJ, Purvis TE, Hu K, Scheer FA (2016) Circadian misalignment increases cardiovascular disease risk factors in humans. Proc Natl Acad Sci U S A 113:E1402–E1411

Morrison SA, Mirnik D, Korsic S, Eiken O, Mekjavic IB, Dolenc-Groselj L (2017) Bed rest and hypoxic exposure affect sleep architecture and breathing stability. Front Physiol 8:410

Munafo D, Loewy D, Reuben K, Kavy G, Hevener B (2018) Sleep deprivation and the workplace: prevalence, impact, and solutions. Am J Health Promot 32(7):1644–1646

Mundigler G, Delle-Karth G, Koreny M (2002) Impaired circadian rhythm of melatonin secretion in sedated critically ill patients with severe sepsis. Crit Care Med 30:536–540

Murphy BA, Vick MM, Sessions DR, Cook RF, Fitzgerald BP (2007) Acute systemic inflammation transiently synchronizes clock gene expression in equine peripheral blood. Brain Behav Immun 21(4):467–476

Natvik S, Bjorvatn B, Moen BE, Mageroy N, Sivertsen B, Pallesen S (2011) Personality factors related to shift work tolerance in two- and three-shift workers. Appl Ergon 42(5):719–724

Nesbitt AD, Dijk DJ (2014) Out of synch with society: an update on delayed sleep phase disorder. Curr Opin Pulm Med 20(6):581–587

Nicholson AN, Smith PA, Stone BM, Bradwell AR, Coote JH (1988) Altitude insomnia: studies during an expedition to the Himalayas. Sleep 11:354–361

Nilsson EK, Bostrom AE, Mwinyi J (2016) Epigenomics of total acute sleep deprivation in relation to genome-wide DNA methylation profiles and RNA expression. Omics 20:334–342

Nobel Assembly at Karolinska Institutet (2017) Nobel Price for physiology or medicine. Nobelförsamlingen, Stockholm

Okyar A, Lévi F (2008) Circadian control of cell cycle pathways: relevance for cancer chronotherapeutics. In: Yoshida K (ed) Trends in cell cycle research. Research Signpost, Thiruvananthapuram, pp 293–317

Ortiz-Naretto AE, Pereiro MP, Ernst G, Borsini EE (2018) Sleep respiratory disturbances during the ascent to Mount Aconcagua. Sleep Sci 11(1):20–24

Oster H, Challet E, Ott V, Arvat E, de Kloet ER, Dijk DJ, Lightman S, Vgontzas A, Van Cauter E (2017) The functional and clinical significance of the 24-hour rhythm of circulating glucocorticoids. Endocr Rev 38(1):3–45

Ozaki S, Uchiyama M, Shirakawa S, Okawa M (1996) Prolonged interval from body temperature nadir to sleep offset in patients with delayed sleep phase syndrome. Sleep 19(1):36–40

Paine S, Fink J, Gander P, Warman GR (2014) Identifying advanced and delayed sleep phase disorders in the general population: a national survey of New Zealand adults. Chronobiol Int 31:627–636

Palca JW, Walker JM, Berger RJ (1986) Thermoregulation, metabolism, and stages of sleep in cold-exposed men. J Appl Physiol 61:940–947

Pickett GF, Morris AF (1975) Effects of acute sleep and food deprivation on total body response time and cardiovascular performance. J Sports Med Phys Fitness 15:49–56

Pilcher JJ, Coplen MK (2000) Work/rest cycles in railroad operations: effects of shorter than 24-h shift work schedules and on-call schedules on sleep. Ergonomics 43:573–588

Pilorz V, Helfrich-Förster C, Oster H (2018) The role of the circadian clock system in physiology. Pflugers Arch 470(2):227–239

Plywaczewski R, Wu TY, Wang XQ, Cheng HW, Sliwinski P, Zielinski J (2003) Sleep structure and periodic breathing in Tibetans and Han at simulated altitude of 5000 m. Respir Physiol Neurobiol 136:187–197

Pollard LC, Choy EH, Gonzalez J, Khoshaba B, Scott DL (2006) Fatigue in rheumatoid arthritis reflects pain, not disease activity. Rheumatology (Oxford) 45:885–889

Polyakov V, Lacota NG, Gundel A (2001) Human thermohomeostasis onboard "Mir" and in simulated microgravity studies. Acta Astronaut 49:137–143

Portela LF, Rotenberg L, Waissmann W (2004) Self-reported health and sleep complaints among nursing personnel working under 12 h night and day shifts. Chronobiol Int 21:859–870

Presser HB (1999) Towards a 24-hour economy. Science 284:1778–1779

Prince TM, Abel T (2013) The impact of sleep loss on hippocampal function. Learn Mem 20:558–569

Pritchett D, Reddy AB (2017) No FAD, no CRY: redox and circadian rhythms. Trends Biochem Sci 42(7):497–499

Radomski MW, Boutelier C (1982) Hormone response of normal and intermittent cold-preadapted humans to continuous cold. J Appl Physiol 53:610–616

Rechtschaffen A, Bergmann BM, Everson CA, Kushida CA, Gilliland MA (2002) Sleep deprivation in the rat: X. Integration and discussion of the findings. Sleep 25(1):68–87

Regestein QR, Monk TH (1995) Delayed sleep phase syndrome: a review of its clinical aspects. Am J Psychiatry 152(4):602–608
Reid K, Zee PC (2009) Circadian rhythm disorders. Semin Neurol 29:393–405
Reitz G, Berger T, Bilski P, Facius R, Hajek M, Petrov V, Puchalska M, Zhou D, Bossler J, Akatov Y, Shurshakov V, Olko P, Ptaszkiewicz M, Bergmann R, Fugger M, Vana N, Beaujean R, Burmeister S, Bartlett D, Hager L, Pálfalvi J, Szabó J, O'Sullivan D, Kitamura H, Uchihori Y, Yasuda N, Nagamatsu A, Tawara H, Benton E, Gaza R, McKeever S, Sawakuchi G, Yukihara E, Cucinotta F, Semones E, Zapp N, Miller J, Dettmann J (2009) Astronaut's organ doses inferred from measurements in a human phantom outside the International Space Station. Radiat Res 171:225–235
Reyner LA, Horne JA (2013) Sleep restriction and serving accuracy in performance tennis players, and effects of caffeine. Physiol Behav 120:93–96
Roach GD, Lamond N, Dorrian J, Burgess H, Holmes A, Fletcher A (2005) Changes in the concentration of urinary 6-sulphatoxymelatonin during a week of simulated night work. Ind Health 43:193–196
Rosenthal NE, Joseph-Vanderpool JR, Levendosky AA et al (1990) Phase-shifting effects of bright morning light as treatment for delayed sleep phase syndrome. Sleep 13(4):354–361
Rostain JC, Gardette-Chauffour MC, Naquet R (1997) EEG and sleep disturbances during dives at 450 msw in helium–nitrogen–oxygen mixture. J Appl Physiol 83:575–582
Ruby NF, Brennan TJ, Xie X et al (2002) Role of melanopsin in circadian responses to light. Science 298(5601):2211–2213
Rundfeldt LC, Maggioni MA, Coker RH, Gunga H-C, Riveros-Rivera A, Schalt A, Steinach M (2018) Cardiac autonomic modulations and psychological correlates in the Yukon Arctic ultra: the longest and the coldest ultramarathon. Front Physiol 9:35
Sack RL, Brandes RW, Kendall AR, Lewy AJ (2000) Entrainment of free-running circadian rhythms by melatonin in blind people. N Engl J Med 343(15):1070–1077
Sack RL, Auckley D, Auger RR, Carskadon MA, Wright KP Jr, Vitiello MV, Zhdanova IV, American Academy of Sleep Medicine (2007) Circadian rhythm sleep disorders: part II, advanced sleep phase disorder, delayed sleep phase disorder, free-running disorder, and irregular sleep-wake rhythm. An American Academy of Sleep Medicine review. Sleep 30(11):1484–1501
Salva MAQ, Hartley S, Léger D, Dauvilliers YA (2017) Non-24-hour sleep-wake rhythm disorder in the totally blind: diagnosis and management. Front Neurol 8:686
Santy PA, Kapanka H, Davis JR, Stewart DF (1988) Analysis of sleep on shuttle missions. Aviat Space Environ Med 59(11 Pt 1):1094–1097
Sawka MN, Gonzalez RR, Pandolf KB (1984) Effects of sleep deprivation on thermoregulation during exercise. Am J Physiol 246:R72–R77
Saxvig IW, Wilhelmsen-Langeland A, Pallesen S, Vedaa O, Nordhus IH, Bjorvatn B (2014) A randomized controlled trial with bright light and melatonin for delayed sleep phase disorder: effects on subjective and objective sleep. Chronobiol Int 31(1):72–86
Scheer FA, Hilton MF, Mantzoros CS, Shea SA (2009) Adverse metabolic and cardiovascular consequences of circadian misalignment. Proc Natl Acad Sci U S A 106(11):4453–4458
Scheiermann C, Gibbs J, Ince L, Loudon A (2018) Clocking in to immunity. Nat Rev Immunol 18(7):423–437
Schrader H, Bovim G, Sand T (1993) The prevalence of delayed and advanced sleep phase syndromes. J Sleep Res 2(1):51–55
Schweitzer PK, Randazzo AC, Stone K, Erman M, Walsh JK (2006) Laboratory and field studies of naps and caffeine as practical countermeasures for sleep-wake problems associated with night work. Sleep 29:39–50
Segawa K, Nakazawa S, Tsukamoto Y et al (1987) Peptic ulcer is prevalent among shift workers. Dig Dis Sci 32:449–453
Sewitch DE, Kittrell EMV, Kupfer DJ, Reynolds CF (1986) Body temperature and sleep architecture in response to mild cold stress in women. Physiol Behav 36:951–957
Sharkey KM, Fogg LF, Eastman CI (2001) Effects of melatonin administration on daytime sleep after simulated night shift work. J Sleep Res 10(3):181–192

Shostak A (2017) Circadian clock, cell division, and cancer: from molecules to organism. Int J Mol Sci 18(4):pii:E873

Silva RH, Abílio VC, Takatsu AL, Kameda SR, Grassl C, Chehin AB, Medrano WA, Calzavara MB, Registro S, Andersen ML, Machado RB, Carvalho RC, Ribeiro Rde A, Tufik S, Frussa-Filho R (2004) Role of hippocampal oxidative stress in memory deficits induced by sleep deprivation in mice. Neuropharmacology 46(6):895–903

Simmons E, McGrane O, Wedmore I (2015) Jet lag modification. Curr Sports Med Rep 14(2):123–128

Skein M, Duffield R, Edge J (2011) Intermittent-sprint performance and muscle glycogen after 30 h of sleep deprivation. Med Sci Sports Exerc 43:1301Y11

Skein M, Duffield R, Minett G (2013) The effect of overnight sleep deprivation after competitive rugby league matches on postmatch physiological and perceptual recovery. Int J Sports Physiol Perform 8:556–564

Skornyakov E, Gaddameedhi S, Paech GM, Sparrow AR, Satterfield BC, Shattuck NL, Layton ME, Karatsoreos I, VAN Dongen HPA (2018) Cardiac autonomic activity during simulated shift work. Ind Health 57(1):118–132

Smith MR, Eastman CI (2008) Night shift performance is improved by a compromise circadian phase position: study 3. Circadian phase after 7 night shifts with an intervening weekend off. Sleep 31(12):1639–1645

Smith MR, Cullnan EE, Eastman CI (2008) Shaping the light/dark pattern for circadian adaptation to night shift work. Physiol Behav 95:449–456

Son GH, Cha HK, Chung S, Kim K (2018) Multimodal regulation of circadian glucocorticoid rhythm by central and adrenal clocks. J Endocrinol Soc 2(5):444–459

Spanagel R, Rosenwasser AM, Schumann G, Sarkar DK (2005) Alcohol consumption and the body's biological clock. Alcohol Clin Exp Res 29:1550–1557

Spiegel K, Sheridan JF, Van CE (2002) Effect of sleep deprivation on response to immunization. JAMA 288:1471–1472

Stahn A, Werner A, Opatz O, Maggioni M, Steinach M, Weller von Ahlefeld V, Moore A, Crucian B, Smith S, Zwart S, Schlabs T, Mendt S, Trippel T, Koralewski E, Koch J, Choukèr A, Reitz G, Shang P, Röcker L, Kirsch K, Gunga HC (2017) Increased core body temperature in astronauts during long-duration space missions. Sci Rep 7(1):16180

Steinach M, Kohlberg E, Maggioni MA, Mendt S, Opatz O, Stahn A, Gunga HC (2016) Sleep quality changes during overwintering at the German Antarctic Stations Neumayer II and III: the gender factor. PLoS One 11(2):e0150099

Swanson LM, Arnedt JT, Rosekind MR, Belenky G, Balkin TJ, Drake C (2011) Sleep disorders and work performance: findings from the 2008 National Sleep Foundation Sleep in America poll. J Sleep Res 20:487–494

Takahashi T, Sasaki M, Itoh H et al (2002) Melatonin alleviates jet lag symptoms caused by an 11-hour eastward flight. Psychiatry Clin Neurosci 56:301–302

Thosar SS, Butler MP, Shea SA (2018) Role of the circadian system in cardiovascular disease. J Clin Invest 128(6):2157–2167

Tobler I, Borbély AA (1993) European isolation and confinement study. Twenty-four hour rhythm of rest/activity and sleep/wakefulness: comparison of subjective and objective measures. Adv Space Biol Med 3:163–183

Tononi G, Cirelli C (2014) Sleep and the price of plasticity: from synaptic and cellular homeostasis to memory consolidation and integration. Neuron 81:12–34

Tortorolo F, Farren F, Rada G (2015) Is melatonin useful for jet lag? Medwave 15(Suppl 3):e6343

Toth LA, Tolley EA, Krueger JM (1993) Sleep as a prognostic indicator during infectious disease in rabbits. Proc Soc Exp Biol Med 203:179–192

Tresguerres JA, Ariznavarreta C, Granados B, Martin M, Villanua MA, Golombek DA, Cardinali DP (2001) Circadian urinary 6-sulphatoxymelatonin, cortisol excretion and locomotor activity in airline pilots during transmeridian flights. J Pineal Res 31:16–22

Tseng CH, Lin FC, Chao HS, Tsai HC, Shiao GM, Chang SC (2015) Impact of rapid ascent to high altitude on sleep. Sleep Breath 19(3):819–826

Van Diest R, Hamulyák K, Kop WJ, van Zandvoort C, Appels A (2002) Diurnal variations in coagulation and fibrinolysis in vital exhaustion. Psychosom Med 64(5):787–792
Van Dongen HP (2004) Comparison of mathematical model predictions to experimental data of fatigue and performance. Aviat Space Environ Med 75(3 Suppl):A15–A36. 302
Van Dongen HP, Maislin G, Mullington JM, Dinges DF (2003) The cumulative cost of additional wakefulness: dose-response effects on neurobehavioral functions and sleep physiology from chronic sleep restriction and total sleep deprivation. Sleep 26(2):117–126
Viswanathan AN, Schernhammer ES (2008) Circulating melatonin and the risk of breast and endometrial cancer in women. Cancer Lett 281:1–7
Wang F, Zhang L, Zhang Y, Zhang B, He Y, Xie S, Li M, Miao X, Chan EY, Tang JL, Wong MC, Li Z, Yu IT, Tse LA (2014) Meta-analysis on night shift work and risk of metabolic syndrome. Obes Rev 15(9):709–720
Warman VL, Dijk DJ, Warman GR, Arendt J, Skene DJ (2003) Phase advancing human circadian rhythms with short wavelength light. Neurosci Lett 342(1–2):37–40
Waterhouse J, Reilly T, Atkinson G (1997) Jet-lag. Lancet 350(9091):1611–1616
Waterhouse J, Buckley P, Edwards B, Reilly T (2003) Measurement of, and some reasons for, differences in eating habits between night and dayworkers. Chronobiol Int 20:1075–1092
Waterhouse J, Reilly T, Atkinson G, Edwards B (2007) Jet lag: trends and coping strategies. Lancet 369(9567):1117–1129
Watson AM (2017) Sleep and athletic performance. Curr Sports Med Rep 16(6):413–418
Weber AL, Cary MS, Connor N, Keyes P (1980) Human non-24-hour sleep-wake cycles in an everyday environment. Sleep 2(3):347–354
Weikum ER, Liu X, Ortlund EA (2018) The nuclear receptor superfamily: a structural perspective. Protein Sci 27(11):1876–1892
Wende AR, Young ME, Chatham J, Zhang J, Rajasekaran NS, Darley-Usmar VM (2016) Redox biology and the interface between bioenergetics, autophagy and circadian control of metabolism. Free Radic Biol Med 100:94–107
Whitmire A, Slack K, Locke J, Keeton K (2013) Sleep quality questionnaire shortduration flyers. NASA/TM-2013-217378. Johnson Space Center, Houston, TX
Winter WC, Potenziano BJ, Zhang Z (2011) Chronotype as a predictor of performance in major league baseball batters. Sleep 34:A167–A168
Wittenbrink N, Ananthasubramaniam B, Münch M, Koller B, Maier B, Weschke C, Bes F, de Zeeuw J, Nowozin C, Wahnschaffe A, Wisniewski S, Zaleska M, Bartok O, Ashwal-Fluss R, Lammert H, Herzel H, Hummel M, Kadener S, Kunz D, Kramer A (2018) High-accuracy determination of internal circadian time from a single blood sample. J Clin Invest 128(9):3826–3839
Wotring VE (2015) Medication use by U.S. crewmembers on the International Space Station. FASEB J 29:4417–4423
Wu B, Wang Y, Wu X, Liu D, Xu D, Wang F (2018) On-orbit sleep problems of astronauts and countermeasures. Mil Med Res 5(1):17
Wulff K, Dijk D-J, Middleton B, Foster RG, Joyce EM (2012) Sleep and circadian rhythm disruption in schizophrenia. Br J Psychiatry 200:308–316
Xie L, Kang H, Xu Q, Chen MJ, Liao Y, Thiyagarajan M (2013) Sleep drives metabolite clearance from adult brain. Science 342(6156):373–377
Youngstedt SD (2005) Effects of exercise on sleep. Clin Sports Med 24:355–365
Zhou X, Ferguson SA, Matthews RW, Sargent C, Darwent D, Kennaway DJ (2012) Mismatch between subjective alertness and objective performance under sleep restriction is greatest during the biological night. J Sleep Res 21:40–49
Zhu JL, Hjollund NH, Andersen AM, Olsen J (2004) Shift work, job stress, and late fetal loss: the national birth cohort in Denmark. J Occup Environ Med 46:1144–1149
Zulley J, Wever R, Aschoff J (1981) The dependence of onset and duration of sleep on the circadian rhythm of rectal temperature. Pflugers Arch 391(4):314–318

Endocannabinoids, "New-Old" Mediators of Stress Homeostasis

10

Daniela Hauer, Roland Toth, and Gustav Schelling

10.1 Introduction

The starting point of endocannabinoid research was in 1964 with the discovery of tetrahydrocannabinol, or THC, the main psychoactive chemical in cannabis by Raphael (Mechoulam and Gaoni (1965) in Israel. More than 20 years later, in 1988, Allan Howlett and his group discovered the cannabinoid receptor in the rat brain (Devane et al. 1988). This receptor was called CB1 and binds THC with high affinity. The CB1 receptors are extremely wide spread in the central nervous system and are present in far higher concentrations than any other receptor. CB1 receptors are important in a vast number of central and peripheral functions which include movements, reproduction, emotions, memory, reward, pain, food intake, and nausea and vomiting (Pertwee and Ross 2002). A second type of cannabinoid receptors was discovered in 1993. In contrast to the CB1 system, CB2 receptors are found primarily in the periphery and show high concentrations in the immune system (especially in T- and B-cells) (Sido et al. 2016), GI tract, liver, spleen, kidney, bones, heart (Weis et al. 2010), and peripheral nervous system (Vaughn et al. 2010). There is, however, more recent evidence that CB2 receptors also play a role in the CNS which challenges this traditional view (Chen et al. 2017). When following the well-known principle that receptors usually have endogenous ligands that bind to it, Mechoulam's group discovered the first endogenous can nabinoid, arachidonylethanolamide, or anandamide in 1992. A second endocannabinoid called 2-arachidonylglycerol (2-AG) was revealed by the same investigators in 1995. Cannabinoid receptors, endocannabinoids, and the related enzymes are essential components of what is now called the endocannabinoid system (ECS).

D. Hauer · G. Schelling (✉)
Department of Anaesthesiology, Hospital of the Ludwig-Maximilians-University (LMU), Munich, Germany
e-mail: daniela.hauer@med.uni-muenchen.de; Gustav.schelling@med.uni-muenchen.de

R. Toth
Department of Cardiac Surgery, Semmelweis University, Budapest, Hungary

© Springer Nature Switzerland AG 2020

A. Choukér (ed.), *Stress Challenges and Immunity in Space*,
https://doi.org/10.1007/978-3-030-16996-1_10

10.2 Cannabinoid Receptors and Ligands

10.2.1 Endocannabinoid Receptors

The cannabinoid receptors CB1 and CB2 belong to the superfamily of G protein-coupled receptors, coupling to inhibitory G proteins (Gi/o) and inhibit adenylyl cyclase and activate MAP kinase (Jyotaki et al. 2010) (Fig. 10.1). Furthermore, CB1 receptors inhibit presynaptic N- and P/Q-type calcium channels and activate inwardly rectifying potassium channels. Other possible signaling mechanisms involve focal adhesion kinase, phosphatidylinositol-3-kinase, sphingomyelinase, or nitric oxide synthase (Mechoulam et al. 1997). Recent studies described at least one other endocannabinoid receptor system, the TRP channel which may have an important function in pain perception under inflammatory conditions (Storozhuk and Zholos 2018).

10.2.2 Endogenous Cannabinoids (Endocannabinoids)

Endocannabinoids are lipid signaling molecules that behave differently than circulating hormones. They are not stored but synthesized on demand from membrane phospholipids and released into intracellular space via specific membrane transport systems. From there, endocannabinoids in the brain activate presynaptic cannabinoid receptors leading to an inhibition of neurotransmitter release. Degradation and metabolism takes place after cellular reuptake and is performed by specific hydrolases: anandamide is degraded by fatty acid amide hydrolase (FAAH) and 2-AG mainly via monoacylglycerollipase (MAGL) (Fig. 10.2) (Herrera-Solis et al. 2010).

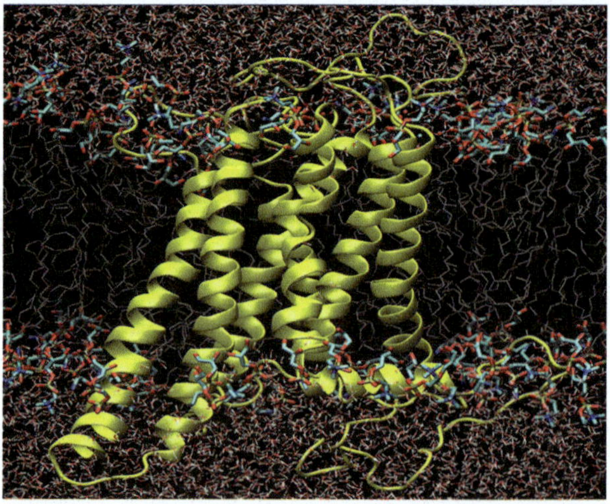

Fig. 10.1 Schematic drawing of a CB1-receptor. The receptor consists of seven transmembrane helices shown in yellow on the image. Courtesy of Patricia H. Reggio, Marie Foscue Rourk Professor, Department of Chemistry and Biochemistry, University of North Carolina, with permission

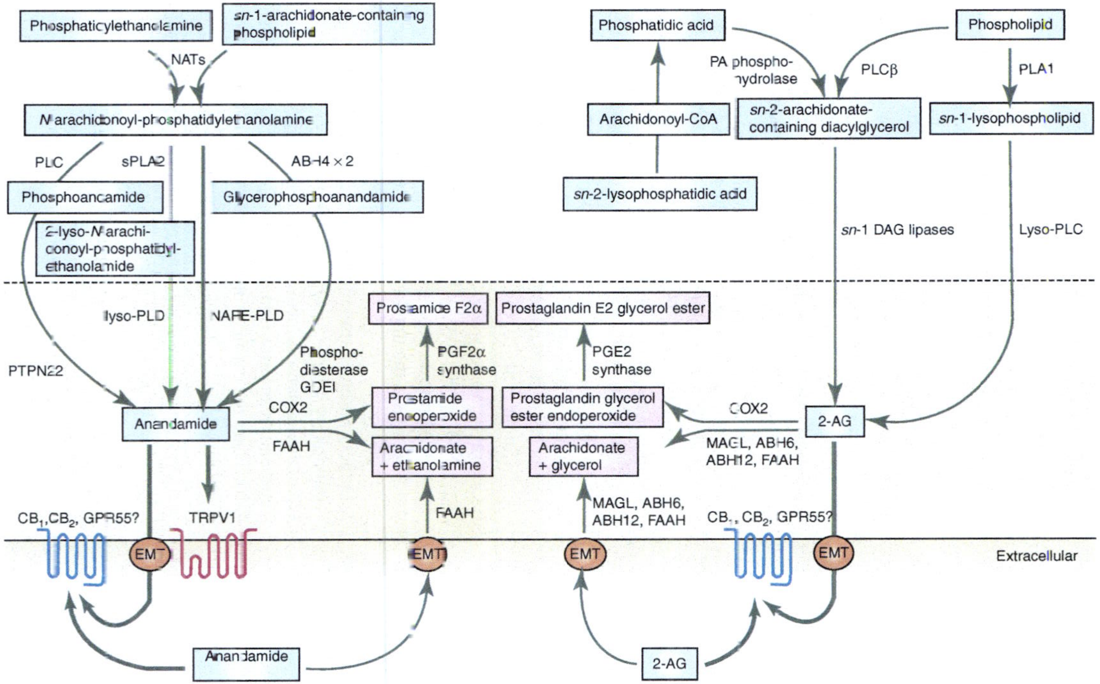

Fig. 10.2 Pathways of synthesis and degradation of the endocannabinoids anandamide and 2-AG. The biosynthetic pathways for anandamide and 2-AG are shown in blue, degradative pathways are shown in pink. Thick arrows denote movement or action. *ABH4/6/12* αβ-hydrolase 4/6/12, *CB1/2* cannabinoid receptor 1/2, *COX2* cyclooxygenase 2, *DAG* diacylglycerol, *EMT* 'endocannabinoid membrane transporter', *FAAH* fatty acid amide hydrolase, *GDE1* glycerophosphodiester phosphodiesterase 1, *GPR55* G protein-coupled receptor 55, *MAGL* monoacylglycerol lipase, *NAPE-PLD N*-acyl-phosphatidylethanolamine-selective phosphodiesterase, *NATs N*-acyltransferases, *PA* phosphatidic acid, *(s)PLA1/2* (soluble) phospholipase A1/2, *PLC* phospholipase C, *PLCβ* phospholipase Cβ, *PLD* phospholipase D, *PTPN22* protein tyrosine phosphatase, nonreceptor type 22, *TRPV1* transient receptor potential, vanilloid subtype 1 receptor. From Di Marzo V: Targeting the endocannabinoid system: to enhance or reduce? Nat. Rev. Drug Discov. 2008; 7: 438–455, with permission

Table 10.1 Names, abbreviations, and presumed biologic functions of several known endocannabinoids

Chemical name	Abbreviation	Biologic functions
2-Arachidonoyl glycerol	2-AG	1. Seems to be the most important endocannabinoid (Tsuboi et al. 2018) 2. Heart, circulation system: • May be related to the pathogenesis of Acute Coronary syndrome (Maeda et al. 2009) • Induces relaxations of bovine coronary arteries by extracellular hydrolysis to arachidonic acid and metabolism to eicosanoids (Gauthier et al. 2005) • The concentration of 2-AG responses to orthostatic stress (Schroeder et al. 2009) 3. Behavior: • Stress exposure evokes a significant increase in circulating concentration in women (Hill et al. 2009) • Basal serum concentrations were significantly reduced in women with major depression (Hill et al. 2009) 4. Induces full platelet activation and aggregation with a non-CB1/CB2 receptor-mediated mechanism (Baldassarri et al. 2008). 5. Central nervous system: • Has neuroprotective properties and restores blood-brain barrier function after traumatic brain injury (Panikashvili et al. 2006; Piro et al. 2018) • Is involved in reducing acute nausea in rodents (Sticht et al. 2016)
2-Arachidonyl glyceryl ether (noladin ether)	2-AG ether	1. Central nervous system: • Neuroprotective by binding to and activation of PPARα (Sun et al. 2007) • Increases the uptake of GABA (Venderova et al. 2005) 2. Induces vascular smooth muscle relaxation in rabbit pulmonary artery (Su and Vo 2007) 3. Attenuates sensory neurotransmission in rat mesenteric arteries (Duncan et al. 2004) 4. Increases the eating motivation (Jones and Kirkham 2012)

(continued)

Table 10.1 (continued)

Chemical name	Abbreviation	Biologic functions
Arachidonoyl ethanolamine (anandamide)	AEA (ANA)	1. Heart, circulation system: • Hypotension (del Carmen García and Celuch 2003) • Has a dose-dependent coronary vasodilatator effect in isolated rat heart (Wagner et al. 2005) • Causes a PPARγ-mediated, time-dependent vasorelaxation in rat isolated aortae (O'Sullivan and Kendall 2009) • Has a negative inotropic response (Ford et al. 2002) • May be related to the pathogenesis of Acute Coronary Syndrome and is produced in infarct-related coronary artery (Maeda et al. 2009) • Generated by circulating monocytes and platelets during cardiogenic shock (Maeda et al. 2009) • Increases the duration of the QRS–EKG complex in rats (Krylatov et al. 2007) 2. Basal serum concentrations were significantly reduced in women with major depression (Hill et al. 2009) 3. Alcohol tolerance may increase accumulation of AEA (Basavarajappa and Hungund 2005) 4. Induces apoptotic body formation and DNA fragmentation in human neuronal and immune cells (Maccarrone et al. 2000) 5. Inflammation: • Inhibits the release of calcitonin gene-related peptide in both skin and spinal cord (Pertwee 2001) • Increases interleukin-6 production by mouse brain cortical astrocytes (Pertwee 2001) 6. Central nervous system: • Increases locomotor activity an influences sleeping periods (Murillo-Rodriguez et al. 1998) • Influences working memory (Mallet and Beninger 1996) and has impact on reward processes (Zona et al. 2017) • Has a role in the neural generation of motivation and pleasure (Stephen et al. 2007)

(continued)

Table 10.1 (continued)

Chemical name	Abbreviation	Biologic functions
Docosatetraenoyl-ethanolamide	DEA	1. Inhibits the norepinephrine-induced migration of colon carcinoma cells (Joseph et al. 2004) 2. Shows an increase of plasma levels after a fasting period of 24 h in rodents (Olatinsu et al. 2017)
N-Arachidonoyl-dopamine	NADA	1. Mediates vasorelaxation in different arteries (Grabiec and Dehghani 2017) 2. induces both pro- and antinociceptive effects in the central and peripheral nervous system (Grabiec and Dehghani 2017) 3. Is a potent inhibitor of early and late activation events in Ag-stimulated human peripheral T cells (Sancho et al. 2004) 4. Exerts neuroprotective effects in states like ischemia (Grabiec and Dehghani 2017)
N-Arachidonyl glycine	NAGly	1. Vascular: • Exerts vasodilatory activity via G protein-coupled receptor 18 activation (Al Suleimani and Al Mahruqi 2017) 2. Pain: • Is a lipid mediator especially in pain modulation and anti-inflammation (Im 2009) • Reduced the mechanical allodynia in rat model (Vuong et al. 2008)
O-Arachidonoyl-ethanolamide (virodhamine)	O-AEA	1. Potently inhibits the large conductance Ca^{2+}-activated K^{+}-channel (Godlewski et al. 2009) 2. In human bronchial epithelial cells induces an additional Ca^{2+} entry (Effimia Gkoumassi et al. 2009) 3. Provides antagonistic effects against the depressive behaviors in mouse model (Hayase 2007, 2008) 4. Induces several arteries (Kozlowska et al. 2008) 5. Has anorectic properties (Tsuboi et al. 2018) 6. Acts as an agonist on monoamine oxidase-B (MAO-B) receptors and could therefore be important in the therapeutic approaches of neurological diseases (Pandey et al. 2009)
N-Oleoyl dopamine	OLDA	1. Is a potent inhibitor of early and late activation events in Ag-stimulated human peripheral T cells (Sancho et al. 2004) 2. Decreases muscle rigidity induced by reserpine in rats (Konieczny et al. 2009) 3. Increases locomotor activity in the rat (Przegalinski et al. 2006) 4. Causes pain sensations via activation of TRPV1 (Bo Tan et al. 2006) 5. Is involved in the modulations of the activity of dopaminergic neurons in the midbrain through interactions with TRPV1, cannabinoid receptors and dopamine uptake (Sergeeva et al. 2017) 6. Induce excitation of histaminergic neurons (De Luca et al. 2018)

Table 10.1 (continued)

Chemical name	Abbreviation	Biologic functions
Oleoyl ethanolamide	OEA	1. Has an analgesic effect in a rat model of neuropathic pain (Jhaveri et al. 2006) 2. Exhibited a significant decline during the stress recovery phase (Hill et al. 2009) 3. Directly inhibits T-cell responses by reducing their production of TNF and IFN-γ (Chiurchiù et al. 2018)
Palmitoyl-ethanolamide	PEA	1. Has anti-inflammatory and anti-nociceptive properties via direct inhibition of T-cell responses (Chiurchiù et al. 2018) 2. Exhibited a significant decline during the stress recovery phase (Hill et al. 2009) 3. Is higher in a highlander population exposed to lower oxygen and maybe associated with a higher hemoglobin (Alarcón-Yaquetto et al. 2017)
Stearoyl-ethanolamide	SEA	1. Exerts anorexic effects in mice via downregulation of liver enzyme activation (Terrazzino et al. 2004) 2. Has nitric oxide -regulated pro-apoptotic activity on C6 glioma cells in rat (Maccarrone et al. 2002)

Apart from the abovementioned and better-investigated endocannabinoids anandamide and 2-AG, a number of newer endocannabinoids have been described which include virodhamide, oleoylethanolamide, and palmitoylamide. Not all of them seem to be biologically active under physiologic conditions but it is believed that every endocannabinoid has its own properties and function under certain conditions (e.g. posttraumatic stress disorder (PTSD); Hauer et al. 2013). Table 10.1 gives an overview on names, chemical structure, and biologic function of a number of known endocannabinoids.

10.2.3 Endocannabinoid Measurement

The source of peripherally and centrally measured endocannabinoids is not entirely clear but is thought to be nucleated blood cells and brain microglia and neurons, respectively (Herrera-Solis et al. 2010). As previously mentioned the endocannabinoids are not stored in vesicles or circulate in the blood stream like hormones but are synthesized on demand and immediately degraded. This makes endocannabinoid measurements in blood and tissues difficult to interpret. In particular, endocannabinoid synthesis in known to be continued ex-vivo in stored blood samples (Fig. 10.3) which makes immediate sample centrifugation and freezing mandatory (Schmidt et al. 2006; Vogeser et al. 2006). The standardization of endocannabinoid extraction from blood samples and tissues is still in progress and not standardized throughout all research groups measuring enocannabinoids. Nevertheless, endocannabinoid measurements can be performed successfully and if standardized

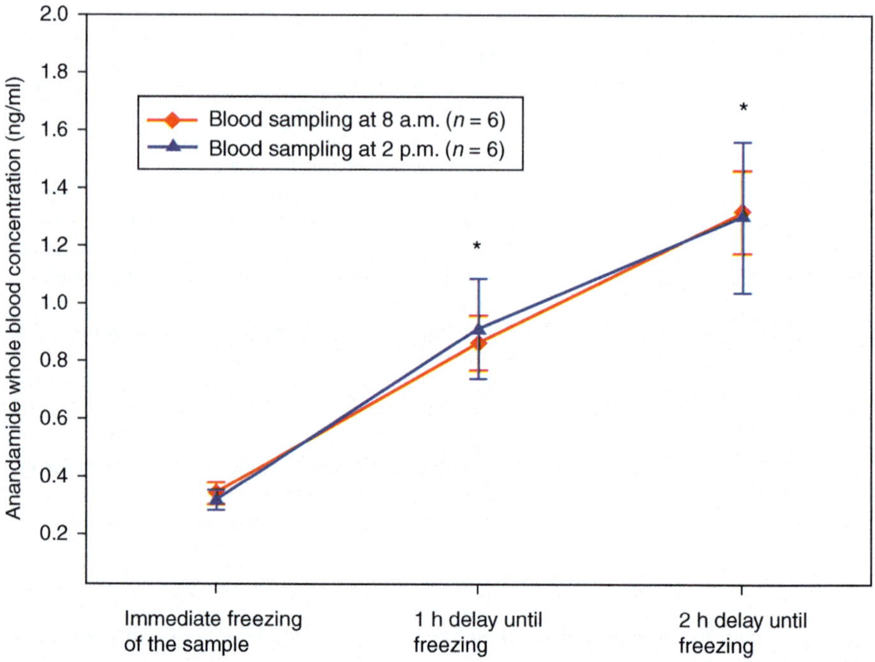

Fig. 10.3 Effect of storage of blood samples on whole blood levels of the endocannabinoid anandamide. Samples were either immediately frozen or kept at room temperature for 1 and 2 h. Anandamide levels increased linearly and significantly over time which indicates that blood samples need to be immediately processed (either centrifuged or frozen) to get meaningful results from endocannabinoid measurements. In order to demonstrate a possible circadian influence on anandamide blood levels and this effect, measurements were performed at 8 am (red lines) and repeated at 2 pm (blue lines) in the same individuals. These measurements did not show any circadian effect although other research groups have reported different results (Vaughn et al. 2010)

conditions are strictly kept throughout the whole process from collection and extraction until measurements, a "snap-shot" of this fast acting system can give important information about endocannabinoid signaling in well-defined situations.

We developed a method to measure endocannabinoids. This method is based on high performance liquid chromatography-tandem mass spectrometry (HPLC-MS/MS) (Vogeser et al. 2006; Hauer et al. 2013). Our method is linear within a range of 0.1–2 ng/ml for anandamide and 0.5–10 ng/ml for 2-AG. The inter-assay coefficient of variation is 34% for a mean anandamide concentration of 0.2 ng/ml. The lower limit of detection of the method (defined as a signal/noise ration >4:1) is 0.025 ng/ml for anandamide and 0.33 ng/ml for 2-AG. In biological matrices, 2-AG (including its deuterated analog) rapidly isomerizes to 1-AG (Vogeser and Schelling 2007). We and other groups therefore tend to quantify 2-AG as the sum of 1- and 2-esterns -of arachidonic acid in consent of the knowledge that both isomers can activate the CB1-receptors. (Vogeser and Schelling 2007).

An alternative to spot-measurements of endocannabinoids in blood samples is the use of hair as a retrospective specimen for the long-term recording of endocannabinoids as a possible indicator of stress-related pathology. This approach allows the simultaneous determination of glucocorticoids from identical hair samples which have extensively been studied (Steudte et al. 2013). An early pilot study of monthly hair analysis for endocannabinoids over a pregnancy cycle in a single individual demonstrated changes in endocannabinoid signals in hair with negative correlations between endocannabinoids and cortisol/cortisone concentrations over time (Krumbholz et al. 2013) (see also Chap. 29). When this analytic technique was used in a sample of rebel war survivors from Uganda, individuals with PTSD had significantly lower hair concentrations of OEA when compared to rebel war survivors without current and lifetime PTSD. Furthermore, there was a strong negative correlation between all measured endocannabinoids and PTSD symptom intensity (Wilker et al. 2016). This suggests that the measurement of endocannabinoids in hair may give additional information of the state of the endocannabinoid system in humans.

10.3 Important Biologic Functions of the Endocannabinoid System to Control Homeostasis in Humans

10.3.1 Immunologic Functions of the Endocannabinoid System

The effects of marijuana smoking or THC on immune cell function were investigated long before cannabinoid receptors and the endocannabinoids were discovered (Klein et al. 2003). The administration of THC to healthy volunteers resulted in a long-lasting suppression endocannabinoid signaling (Thieme et al. 2014). The cannabinoids exhibit complex regulatory effects on the immune system and are involved in immune regulation by the suppression of cell activation, inhibition of pro-inflammatory cytokine production, nuclear-factor kappa B (NF-kB)-dependent apoptosis and modulation of the functions of T helper subsets (Table 10.2) (Pandey et al. 2009).

The immune effects of cannabinoids and the endocannabinoid system provide promising therapeutic implications in a variety of neuroinflammatory, autoimmune, atherosclerotic metabolic and allergic conditions (Tanasescu and Constantinescu 2010). The effects of exogenous cannabinoids and endocannabinoids on the immune system are incompletely understood however, and involve interactions with innate, humoral and cell-mediated immunity. The discussed pathways through which cannabinoids act are apoptosis, inhibition of proliferation, suppression of cytokine and chemokine production, and induction of T-regulatory cells (Braun et al. 2011).

There is evidence that cannabinoid effects on the immune system are particularly pronounced under conditions of stress and that it they may represent a universally acting protective system against tissue damage of multiple and also nonprotein origin (Pacher and Mechoulam 2011).

Indeed there is evidence that the activation of CB2 receptors via 2-AG dampen the immune response and therefore the inflammatory process in mice (Dotsey et al. 2017).

Table 10.2 Effects of endocannabinoids on immune cells

Immune cells	Functions affected	Receptor involved
T-lymphocytes	Proliferation; cell death by apoptosis; Th1/Th2 cytokine secretion, polarization; cell number	CB2
B-lymphocytes	Inhibition of antibody formation; Ig production; Ig isotype switching; proliferation; cell number	CB1 and CB2
Haematopoietic cell line	Cell growth	Non-CB1, CB2
Macrophages	Decreased Inflammatory mediators; antigen presentation; migration; phagocytosis; increased adhesion	CB2
Mast cells	Down modulate mast cell activation; decreased TNF-α; decreased mast cell-dependent angiogenesis	Non-CB1, CB2 CB1, CB2
Dendritic cells	Growth and maturation; apoptosis; recruitment during innate immune response	CB1, CB2
Natural killer cells and neutrophils	Cytolytic activity; chemokines; cytokines	Non-CBl, CB2
Cancer cells	Cell cycle arrest; apoptosis; growth inhibitor	CB1, CB2 receptor. TRPV1, lipid rafts

From: Pandey et al. Endocannabinoids and immune regulation. Pharmacol Res 2009; 60: 85–92, with permission

In contrast, other studies have shown a positive correlation between 2-AG levels and proinflammatory cytokine-levels (interleukin-6) (Knight et al. 2015). Therefore, at present, a close association between the endocannabinoid system und the immune system can be assumed but the direction of the regulatory response is not always clear.

10.3.2 The Endocannabinoid System and Cardiovascular Functions

10.3.2.1 Vascular Functions

Cannabinoid modulation of vascular tone appears to contribute to the pathophysiology of cardiovascular disorders (Batkai et al. 2004). Endocannabinoids are potent vasodilators that cause hypotension in anesthetized animals and biphasic effects (initial hypertension followed by hypotension) in conscious animals (Ledent et al. 1999). Cannabinoids have also been shown to directly induce vasodilatation in isolated blood vessels (Howlett 2005). Cannabinoid-induced vasodilatation in isolated arteries occurs via endothelium-dependent as well as endothelium-independent pathways (Su and Vo 2007). Endothelium-derived hyperpolarizing factor released by endothelial cannabinoid CB1 receptor-coupled pertussis toxin-sensitive G proteins leading to Ca^{2+}-activated K^{+}-channel and membrane hyperpolarization of vascular smooth muscle has been proposed as one mechanism (Sunano et al. 1999; Begg et al. 2003).

Some known effects of different endocannabinoids on the vascular system are listed in the following:

- 2-AG ether induces relaxation in vascular smooth muscle, this effect is completely abolished by cannabinoid CB1 receptor antagonists (Su and Vo 2007)
- 2-AG leads to vascular hyporeactivity to noradrenalin in sepsis due to higher activation and expression of CB1-receptors; this effect can be inhibited by the CB1-receptor-antagonist AM 251 in vitro (Singh et al. 2018). Anandamide can cause vasorelaxation via "classical" CB1 receptors, via a putative endothelial cannabinoid receptor, via sensory nerve activation, or the release of nitric oxide, vasoactive prostanoids or endothelium-derived hyperpolarizing factor (EDHF) and decreases blood pressure in mice (Offertaler et al. 2003; Randall et al. 2004). There is also evidence that anandamide interferes with intracellular calcium release in vascular smooth muscle preparations (White and Hiley 1998). Anandamide increased human skin microcirculatory flow and relaxed isolated human pulmonary artery rings in an experimental set-up (Movahed et al. 2005; Kozlowska et al. 2008).
- Virodhamine (O-AEA) causes a full, slowly developing relaxation of the isolated human pulmonary artery, so it is possible that the endothelial cannabinoid receptor in the human pulmonary artery could be a target for the treatment of pulmonary hypertension (Kozlowska et al. 2008).
- Oleandamide (OEA) causes vasorelaxation partly via activation of sensory nerves and partly via an endothelium-dependent mechanism and cyclooxygenase inhibitors increase the potency of OEA as a vasorelaxant (Wheal et al. 2010; Ho et al. 2008).
- N-arachidonyl glycine (NAGly) caused mesenteric arterial relaxation stimulating endothelial release of nitric oxide and through nitric oxide independent mechanisms, resulting in relaxation of endothelium-denuded vessels (Parmar and Ho 2010).

In addition to modulation of vascular tone, cannabinoids also regulate vascular homeostasis. While cardioprotective properties of cannabinoids have been observed, proapoptotic and anti-angiogenic activities of cannabinoids also have been described (Zhang et al. 2010; Durst et al. 2007). Anandamide mainly behaves as an inhibitor of angiogenesis (Pisanti et al. 2007); 2-AG interacts with endothelin-1 and may play a role in microvascular function (Chen et al. 2000). An activated endocannabinoid system may modulate vascular repair and angiogenesis, these functions could be important in the maintenance of vascular integrity (Zhang et al. 2010).

In animals, intracisternal application of cannabinoids leads to sympatho-adrenergic activation with increased blood pressure and elevated norepinephrine concentrations (Niederhoffer and Szabo 1999; Pfitzer et al. 2005). Moreover, activation of CB1 in the brain modulates baroreflex regulation (Brozoski et al. 2005). In contrast, peripheral endocannabinoid application lowers blood pressure and decreases norepinephrine concentrations, presumably by inhibiting norepinephrine release from presynaptic neurons (Niederhoffer and Szabo 1999; Pfitzer et al. 2005).

Genetic deletion or pharmacological blockade of CB1 receptors has no effect on blood pressure in healthy and normotensive animals, but CB1 receptor blockade increases blood pressure in experimental hypotension, septic shock, and myocardial infarction (Wagner et al. 1997, 2001; Batkai et al. 2004).

10.3.2.2 Cardiac Functions

Administration of anandamide, or synthetic cannabinoids, causes CB1 receptor-mediated hemodynamic changes which are complex, involving phases of both increased and decreased blood pressure as well as changes in heart rate (Randall et al. 2004). The acute intravenous administration of THC to humans resulted in an increase in heart rate and blood pressure which was accompanied by an increase in serum cortisol levels (Thieme et al. 2014). Anandamide exerts its cardiovascular effects in mice and rats through cannabinoid CB1 and CB2 receptors and vanilloid TRPV1 receptors (Pacher et al. 2006). A growing body of evidence suggests that endocannabinoid signaling plays a critical role in the pathogenesis of atherogenesis and its clinical manifestations (Mach and Steffens 2008). Increased endocannabinoid signaling is associated with disease progression and an increased risk for acute thrombotic events (Tanasescu and Constantinescu 2010). Recent data have shown that increased 2-AG levels after acute myocardial infarction could also worsen the cardiac function in mice (Schloss et al. 2019). The presence of CB1 and CB2 receptors in healthy and failing human heart has been shown by our group. In particular, on healthy human left ventricular myocardium, mRNA transcripts of CB1 and CB2 receptors were expressed in an almost equal proportion whereas in patients with chronic heart failure, mRNA expression of CB1 receptors was shown to be down-regulated 0.7-fold, whereas expression of CB2 receptors was upregulated more than 11-fold. These findings indicate that the endocannabinoid system has a possible role in the regulation of cardiac function under both physiologic and pathologic conditions.

10.3.3 The Endocannabinoid System and Bone Metabolism

Bone metabolism can be regarded as a constant modeling/remodeling process which continuously renews the mineralized bone matrix. All components of the ECS appear to be present in human skeleton, and the ECS has been shown to play an important role in regulating this process (Bab et al. 2008). Bone forming (osteoblasts) as well as well as bone resorbing cells (osteoclasts) are able to synthesize anandamide and 2-AG and CB1 receptors have been shown to be present in symphatic nerve endings close to osteoblasts. Noradrenergic signals from the central and peripheral nervous system regulate bone metabolism involving beta-2-adrenergic receptors and results in bone loss, an effect which has been demonstrated in stressed patients with depression (Bab and Yirmiya 2010) and put up as a hypothesis in humans during space flight where loss of bone mass is a highly prevalent problem (Strollo 1999). Activation of CB1 receptors in symphatic nerve endings by 2-AG could inhibit nordadrenergic signaling and thus reduce the symphatically driven downregulation of bone formation. A comparable effect of endocannbinoids has also been postulated in patients with the Complex Regional Pain Syndrome, a symphatically driven chronic pain disorder of the extremities associated with bone loss (Kaufmann et al. 2009). The activation of CB2 receptors expressed by osteoblasts and osteoclasts by stress-activated endocannbinoids could stimulate bone formation and inhibit bone resorption. CB2-receptor deficient knock-out mice exhibit a markedly accelerated bone loss which resembles age-related osteoporosis in humans (Ofek et al. 2006). On the other hand there are recent data that CB1- and CB2-knock-out mice may prevent an age-related bone mass loss due to lower resorption rate (lower number of osteoclasts) (Sophocleous et al. 2017). A single nucleotide polymorphism of the CNR2 gene on human chromosome 1p36 which encodes for the CB2 receptor in women is predictive for low bone mineral density (Karsak et al. 2005). These findings suggest that the endocannabinoid system maybe a promising target to address in order to prevent osteoporosis both on Earth and maybe even in space during long-term exposure to microgravity.

10.3.4 The ECS, Stress, and Control of the HPA Axis

The classic definition of stress as the "body's nonspecific response to a demand placed on it" goes back to the Austrian–Canadian Researcher Hans Seyle who is regarded as the father of modern stress research (Selye 1936). The most important physiologic response to stress is the activation of hypothalamic-pituitary-adrenal (HPA) axis which controls the neuroendocrine response to aversive stimuli (see also Chap. 7). Whereas a fast and timely activation of the HPA—axis is important for adequate functioning and even survival during stress, the rapid shut-down of the glucocorticoid response after termination of the stressful stimuli appears to be equally important. The endocannabinoid system has been shown to play an important role in the limitation of HPA—axis activation during and after stress (Steiner and Wotjak 2008).

Recent studies in animals have demonstrated that the systemic administration of glucocorticoids in the absence of a stressor results in a fast increase in anandamide and 2-AG concentrations in limbic structures of the brain important for the regulation of the stress response (Hill and McEwen 2009a). These effects of glucocorticoids are probably mediated by a nongenomic mechanism (Hill and McEwen 2009b) The exposure to acute stress, on the other hand, led to an increase in 2-AG while anandamide concentrations decreased. Pharmacologic or genetic blockade of the CB1 receptor in animals resulted in an exaggregated response to stress which suggests that stress-induced endocannabinoid signaling in the hypothalamus limits the responsiveness of the HPA—axis to stress.

In order to analyze the relationship between peripheral endocannabinoid signaling and HPA—axis activity in humans exposed to acute stress more closely, we performed a series of parabolic flight experiments (Chouker et al. 2010). In participants with high self-reported stress levels during these experiments (induced by acute motion sickness), anandamide blood concentrations showed an early decrease which was followed by a massive increase in plasma cortisol levels. In contrast, volunteers with low stress exposure who tolerated the experiment well, showed an early increase in anandamide and no activation of the HPA—axis. Interestingly, highly stressed volunteers showed almost no increase in plasma 2-AG levels despite a massive activation of the HPA—axis, whereas the opposite pattern was seen in individuals who tolerated the experiment and reported low levels of stress. Changes in anandamide activity clearly preceded alterations in 2-AG signaling in these experiments. These findings suggest a tonic regulation of HPA—axis responsiveness by anandamide where an early decrease in anandamide activity results in a sensitization of the HPA—axis during the early phase of the stress reaction which is later followed by a strong activation. In contrast, the late increase in 2-AG activity seems to be more related to a limitation of HPA—axis activation which was clearly absent in the participants with a massive stress reaction. Also chronic stress can lead to disrupted stress responses: Individuals being isolated in an artificial environment during a 500 day study simulating trip to the Mars showed decreased 2-AG-levels as well as decreased cortisol levels at the end of the study (Yi et al. 2016).

The biologic bases for these different patterns of endocannabinoid and glucocorticoid reactivity under stress are currently unknown but recent findings suggest that genetic factors involving glucocorticoid receptor sensitivity may play a role (Hauer et al. 2010). It is of interest to note that the activity of both systems appears to be impaired in chronic stress-related disorders such as depression (Hill et al. 2009) or posttraumatic stress disorder (Hauer et al. 2010).

10.3.5 Endocannabinoids as Regulators of (Posttraumatic) Stress Response

The exquisite responsiveness of the endocannabinoid system in humans to acute stress has been demonstrated in several recent studies. Physical stress in trained and physically fit individuals induced by hard exercise during mountaineering or cycling resulted in elevated EC concentrations which returned to baseline after termination of the stressful activity (Feuerecker et al. 2012). Emotional Stress induced by the Trier Social Stress Test resulted in a significant increase in the endocannabinoids anandamide, palmitoylethanolamide (PEA) and oleoylethanolamide (OEA) plasma concentrations (Dlugos et al. 2012). As stated above (Sect. 10.3.4.), exposure of healthy volunteers to kinetic stress gave rise to a significant increase in plasma endocannabinoid concentrations in stress-tolerant participants, whereas highly stressed individuals showed an absent endocannabinoid response and a massive activation of the hypothalamic-pituitary-adrenal (HPA) axis (Chouker et al. 2010). Increased blood concentrations of endocannabinoids were also seen in astronauts spending months on the International Space Station (ISS) (Strewe et al. 2012) or in Antarctic isolation at sea level (Strewe et al. 2018).

These findings suggest that an activated endocannabinoid system is required to maintain homeostasis during stress and enhance the recovery process after a high stress exposure. One could further assume that in chronically stressed individuals with PTSD the endocannabinoid system is either chronically elevated with higher concentrations of circulating endocannabinoids or deficient in signaling with abnormally low levels. Recent studies have in fact pointed to both possibilities. In a population of refugees from the Middle East, Afghanistan, and Africa with multiple experiences of war and torture resulting in PTSD, plasma concentrations of the endocannabinoids anandamide, 2-AG, SEA, OEA, and PEA were significantly elevated when compared to healthy nonstressed individuals or traumatized patients from the same regions who had not developed PTSD (Hauer et al. 2013). In contrast, in a different study population of individuals exposed to the events of 9/11 in New York, there was no difference in anandamide, PEA, and OEA concentrations between individuals with and without PTSD but 2-AG levels in PTSD patients were significantly lower (Hill et al. 2013). A third study evaluated patients from the same geographic area after physical assault or motor vehicle accidents. In this study, anandamide and OEA plasma concentrations were significantly lower in PTSD whereas 2-AG and PEA levels were not different when compared to controls. This study has also demonstrated greater CB_1-receptor availability after administration of a CB_1 selective radiotracer (Neumeister et al. 2013). This supports the notion, that abnormal CB_1 receptor-mediated endocannabinoid signaling may be implicated in PTSD, at least in selected patients and an unclear direction of the effect. These conflicting findings could be due to differences in the patient populations under study. Refugees with PTSD after recent war and torture experiences may react more to the stress of blood sampling in a novel medical environment (Hauer et al. 2013) than individuals from western societies and this could result in higher endocannabinoid plasma levels. The sensitivity of plasma endocannabinoids to short-term increases in stress-levels limits the usefulness of blood endocannabinoids as biomarkers for stress-associated disorders.

10.3.6 The Endocannabinoid System and Food Intake

Exogenous cannabis is known to increase appetite and is therefore used in the therapy of weight loss and anorexia in patients suffering from human immunodeficiency virus (HIV) or cancer. But in what way are endocannabinoids involved in food intake, appetite, and obesity? There are hints that there is a dose-dependent decrease in 2-AG-levels following sugar consumption (Feuerecker et al. 2012) as well as anandamide levels (Di Marzo et al. 2009). If these effects are due to increased insulin levels is still discussed: In one study insulin-sensitive individuals with obesity showed lower 2-AG levels than insulin-resistant obese patients but as insulin resistance was associated with significantly higher body mass index the potential relationship still remains unclear (Abdulnour et al. 2014).

Studies revealed that high endocannabinoid levels are associated with obesity and that obese individuals do not show a decline in anandamide levels after food intake (Monteleone et al. 2016) whereas individuals with a normal body weight showed a postprandial decrease in 2-AG-levels (Gatta-Cherifi et al. 2012). Several studies found a positive correlation between obesity and 2-AG levels as well as anandamide levels. But there is also a difference in the motivation of eating regarding the circulating endocannabinoids: Hedonic eating is positively correlated with higher endocannabinoid plasma levels. After a nonfavourite meal the 2-AG declined whereas the 2-AG concentration increased after a favorite food (Monteleone et al. 2016).

There are also evidence from animal studies that there is a relationship between the endocannabinoid and the glucocorticoid system regarding obesity: Glucocorticoid treated mice showed not only increased body weight and obesity but also higher 2-AG levels. Dyslipidemia was reversed in CB1-knock-out mice and in mice treated with AM251, a CB1 receptor antagonist (Bowles et al. 2015).

10.3.7 The Cannabinoid Effects and Sleep

Some of the more intriguing aspects of endocannabinoids is their important role in the regulation of sleep and stress recovery. The endocannabinoid system appears to modulate stress-related endocrine and behavioral responses in order to restore homeostasis after potentially harm- and stressful situations for the organism (Tasker 2004).

These assumptions are corroborated by the observation that highly stressed humans (e.g. war veterans with posttraumatic stress disorder) show a high incidence of chronic marijuana abuse (Sah 2002). Drowsiness or sleepiness are well-known effects in the later stages of intoxication by marijuana (Freemon 1972). Early experiments in mice suggested that anandamide may be a mediator of sleep induction (Mechoulam et al. 1997) and recent experimental evidence has convincingly demonstrated that endocannabinoids are important for sleep regulation with a particularly pronounced effect on rapid-eye-movement (REM) sleep (Herrera-Solis et al. 2010). Sleep deprivation in human volunteers resulted in a significant increase in cerebrospinal fluid concentrations of oleoylethanolamide, an endogenous lipid messenger that is released after neural injury with neuroprotective and neurotrophic effects (Koethe et al. 2009). In patients with sleep apnea, a disorder characterized by nocturnal sleep deprivation and daytime hypersomnolence, plasma concentrations of oleoylethanolamide were twofold higher than in healthy controls. These findings suggest that oleoylethanolamide may be part of a neuroprotective mechanism against chronic oxidative stress and promote wakefulness after sleep deprivation (Jumpertz et al. 2010). But also 2-AG levels are increased in sleep deprived probands almost parallel to increased hedonic eating in the afternoon assuming endocannabinoid-related activation of reward circuit in the brain (Hanlon et al. 2016; Hillard 2018). The increase of 2-AG levels after sleep deprivation is shifted to later times as a result of a shifted circadian rhythm—the mechanism is not clear but some authors suggests a stress-related effect onto glucocorticoid receptor effects (Hillard 2018).

Additionally there are data suggesting a preventive role of the ECS during hibernating marmots regarding bone-/energy metabolism and immune function (Mulawa et al. 2018).

These observations suggest that the pharmacologic manipulation of endocannabinoid signaling could represent a presumed countermeasure against negative biologic consequences of sleep depriviation during stressful conditions.

10.4 Summary

Recent experiments in animals and humans point to the fact that the ECS is a critical and highly important regulator of adaptation processes to acute and chronic stress. The ECS affects major stress-sensitive and key organ functions and is involved in the maintenance of immune- and cardiovascular functions, as well as in the pathology of stress-associated motion sickness. The ECS in implicated in a complex interaction with other stress-response systems including the sympathoadrenergic and the HPA—axis. Hereby it affects innate and adaptive immune functions. Important roles of the ECS include the regulation of bone turnover with possible important consequences for the treatment of osteoporosis, one of the most prevalent disorders in post-menopausal women. Because as all systems described above (immune, bone, sleep, memory, etc.) are affected by space flight, understanding, and targeting the ECS seems to be of promising relevance for manned space missions. With this regard, the ECS represents a valid and important example of how findings from basic research, preclinical studies in volunteers and clinical investigations in patients can be applied for space research and find their way back to Earth resulting in a better understanding and new treatment options for common disorders in humans. Selected groups of patients may benefit from cannabis use as a drug (e.g. those with PTSD) but methodological difficulties limit the proof of causal relationships. Given the recent changes in medical and legal status in some countries, the number of cannabis user and addicts may grow and further longitudinal studies on adverse and occasionally positive effects of cannabis on memory function and other organ functions are needed.

References

Abdulnour J, Yasari S, Rabasa-Lhoret R, Faraj M, Petrosino S, Piscitelli F, Prud' Homme D, Di Marzo V (2014) Circulating endocannabinoids in insulin sensitive vs. insulin resistant obese postmenopausal women. A MONET group study. Obesity (Silver Spring) 22(1):211–216

Alarcón-Yaquetto DE, Caballero L, Gonzales GF (2017) Association between plasma N-Acylethanolamides and high hemoglobin concentration in Southern Peruvian highlanders. High Alt Med Biol 18(4):322–329

Al Suleimani YM, Al Mahruqi AS (2017) The endogenous lipid N-arachidonoyl glycine is hypotensive and nitric oxide-cGMP-dependent vasorelaxant. Eur J Pharmacol 794:209–215

Bab I, Ofek O, Tam J, Rehnelt J, Zimmer A (2008) Endocannabinoids and the regulation of bone metabolism. J Neuroendocrinol 20(Suppl 1):69–74

Bab IA, Yirmiya R (2010) Depression and bone mass. Ann N Y Acad Sci 1192:170–175

Baldassarri S, Bertoni A, Bagarotti A, Sarasso C, Zanfa M, Catani MV, Avigliano L, Maccarrone M, Torti M, Sinigaglia F (2008) The endocannabinoid 2-arachidonoylglycerol activates human platelets through non-CB1/CB2 receptors. J Thromb Haemost 6:1772–1779

Basavarajappa BS, Hungund BL (2005) Role of the endocannabinoid system in the development of tolerance to alcohol. Alcohol Alcohol 40:15–24

Batkai S, Pacher P, Osei-Hyiaman D, Radaeva S, Liu J, Harvey-White J, Offertaler L, Mackie K, Rudd MA, Bukoski RD, Kunos G (2004) Endocannabinoids acting at cannabinoid-1 receptors regulate cardiovascular function in hypertension. Circulation 110:1996–2002

Begg M, Mo FM, Offertaler L, Batkai S, Pacher P, Razdan RK, Lovinger DM, Kunos G (2003) G protein-coupled endothelial receptor for atypical cannabinoid ligands modulates a Ca^{2+}-dependent K^{+} current. J Biol Chem 278:46188–46194

Bo Tan HBB, Rimmerman N, Srinivasan H, Yu YW, Krey JF, Monn MF, Chen JS-C, Hu SS-J, Pickens SR, Walker JM (2006) Targeted lipidomics: discovery of new fatty acyl amides. AAPS J 8(3):E461–E465

Bowles NP, Karatsoreos IN, Li X, Vemuri VK, Wood JA, Li Z, Tamashiro KL, Schwartz GJ, Makriyannis AM, Kunos G, Hillard CJ, McEwen BS, Hill MN (2015) A peripheral endocannabinoid mechanism contributes to glucocorticoid-mediated metabolic syndrome. Proc Natl Acad Sci U S A 112(1):285–290

Braun A, Engel T, Aguilar-Pimentel JA, Zimmer A, Jakob T, Behrendt H, Mempel M (2011) Beneficial effects of cannabinoids (CB) in a murine model of allergen-induced airway inflammation: Role of CB(1)/CB(2) receptors. Immunobiology 216(4):466–476

Brozoski DT, Dean C, Hopp FA, Seagard JL (2005) Uptake blockade of endocannabinoids in the NTS modulates baroreflex-evoked sympathoinhibition. Brain Res 1059:197–202

Chen DJ, Gao M, Gao FF, Su QX, Wu J (2017) Brain cannabinoid receptor 2: expression, function and modulation. Acta Pharmacol Sin 38:312–316

Chen Y, McCarron RM, Ohara Y, Bembry J, Azzam N, Lenz FA, Shohami E, Mechoulam R, Spatz M (2000) Human brain capillary endothelium: 2-arachidonoglycerol (endocannabinoid) interacts with endothelin-1. Circ Res 87:323–327

Chiurchiù V, van der Stelt M, Centonze D, Maccarrone M (2018) The endocannabinoid system and its therapeutic exploitation in multiple sclerosis: clues for other neuroinflammatory diseases. Prog Neurobiol 160:82–100

Chouker A, Kaufmann I, Kreth S, Hauer D, Feuerecker M, Thieme D, Vogeser M, Thiel M, Schelling G (2010) Motion sickness, stress and the endocannabinoid system. PLoS One 5:e10752

del Carmen García MA-GE, Celuch SM (2003) Hypotensive effect of anandamide through the activation of CB1 and VR1 spinal receptors in urethane-anesthetized rats. Naunyn Schmiedeberg's Arch Pharmacol 368(4):270–276

De Luca R, Mazur K, Kernder A, Suvorava T, Kojda G, Haas HL, Sergeeva OA (2018) Mechanisms of N-oleoyldopamine activation of central histaminergic neurons. Neuropharmacology 143:327–338

Di Marzo V, Verrijken A, Hakkarainen A, Petrosino S, Mertens I, Lundbom N, Piscitelli F, Westerbacka J, Soro-Paavonen A, Matias I, Van Gaal L, Taskinen MR (2009) Role of insulin as a negative regulator of plasma endocannabinoid levels in obese and nonobese subjects. Eur J Endocrinol 161(5):715–722

Dotsey E, Ushach I, Pone E, Nakajima R, Jasinskas A, Argueta DA, Dillon A, DiPatrizio N, Davies H, Zlotnik A, Crompton PD, Felgner PL (2017) Transient cannabinoid Receptor 2 blockade during immunization heightens intensity and breadth of antigen-specific antibody responses in young and aged mice. Sci Rep 7:42584

Devane WA, Dysarz FA 3rd, Johnson MR, Melvin LS, Howlett AC (1988) Determination and characterization of a cannabinoid receptor in rat brain. Mol Pharmacol 34:605–613

Dlugos A, Childs E, Stuhr KL, Hillard CJ, de Wit H (2012) Acute stress increases circulating anandamide and other N-acylethanolamines in healthy humans. Neuropsychopharmacology 37:2416–2427

Duncan M, Millns P, Smart D, Wright JE, Kendall DA, Ralevic V (2004) Noladin ether, a putative endocannabinoid, attenuates sensory neurotransmission in the rat isolated mesenteric arterial bed via a non-CB1/CB2 G(i/o) linked receptor. Br J Pharmacol 142:509–518

Durst R, Danenberg H, Gallily R, Mechoulam R, Meir K, Grad E, Beeri R, Pugatsch T, Tarsish E, Lotan C (2007) Cannabidiol, a nonpsychoactive Cannabis constituent, protects against myocardial ischemic reperfusion injury. Am J Physiol Heart Circ Physiol 293:H3602–H3607

Effimia Gkoumassi BGJDMJD, Elzinga CRS, Hasenbosch RE, Meurs H, Nelemans SA, Schmidt M, Zaagsma J (2009) (Endo)cannabinoids mediate different Ca^{2+} entry mechanisms in human bronchial epithelial cells. Naunyn-Schmied Arch Pharmacol 380(1):67–77

Feuerecker M, Hauer D, Toth R, Demetz F, Holzl J, Thiel M, Kaufmann I, Schelling G, Chouker A (2012) Effects of exercise stress on the endocannabinoid system in humans under field conditions. Eur J Appl Physiol 112:2777–2781

Ford WR, Honan SA, White R, Hiley CR (2002) Evidence of a novel site mediating anandamide-induced negative inotropic and coronary vasodilatator responses in rat isolated hearts. Br J Pharmacol 135:1191–1198

Freemon FR (1972) Effects of marihuana on sleeping states. JAMA 220:1364–1365

Gatta-Cherifi B, Matias I, Vallée M, Tabarin A, Marsicano G, Piazza PV, Cota D (2012) Simultaneous postprandial deregulation of the orexigenic endocannabinoid anandamide and the anorexigenic peptide YY in obesity. Int J Obes (Lond) 36(6):880–885

Gauthier KM, Baewer DV, Hittner S, Hillard CJ, Nithipatikom K, Reddy DS, Falck JR, Campbell WB (2005) Endothelium-derived 2-arachidonylglycerol: an intermediate in vasodilatory eicosanoid release in bovine coronary arteries. Am J Physiol Heart Circ Physiol 288:H1344–H1351

Godlewski G, Offertaler L, Osei-Hyiaman D, Mo FM, Harvey-White J, Liu J, Davis MI, Zhang L, Razdan RK, Milman G, Pacher P, Mukhopadhyay P, Lovinger DM, Kunos G (2009) The endogenous brain constituent N-arachidonoyl L-serine is an activator of large conductance Ca^{2+}-activated K^+ channels. J Pharmacol Exp Ther 328:351–361

Grabiec U, Dehghani F (2017) N-Arachidonoyl dopamine: a novel endocannabinoid and endovanilloid with widespread physiological and pharmacological activities. Cannabis Cannabinoid Res 2(1):183–196. https://doi.org/10.1089/can

Hanlon EC, Tasali E, Leproult R, Stuhr KL, Doncheck E, de Wit H, Hillard CJ, Van Cauter E (2016) Sleep restriction enhances the daily rhythm of circulating levels of endocannabinoid 2-arachidonoylglycerol. Sleep 39(3):653–664

Hauer D, Schelling G, Gola H, Campolongo P, Morath J, Roozendaal B, Hamuni G, Karabatsiakis A, Atsak P, Vogeser M, Kolassa IT (2013) Plasma concentrations of endocannabinoids and related primary fatty acid amides in patients with post-traumatic stress disorder. PLoS One 8:e62741

Hauer D, Weis F, Papassotiropoulos A, Schmoeckel M, Lieke J, Kaufmann I, Kirchhoff F, Vogeser M, Roozendaal B, Briegel J, de Quervain D, Schelling G (2010) Relationship of a common polymorphism of the glucocorticoid receptor gene to traumatic memories and posttraumatic stress disorder in patients after intensive care therapy. Crit Care Med 30:7

Hayase T (2007) Chronologically overlapping occurrences of nicotine-induced anxiety- and depression-related behavioral symptoms: effects of anxiolytic and cannabinoid drugs. BMC Neurosci 8:76

Hayase T (2008) Nicotine (NC)-induced "depressive" behavioral symptoms and effects of antidepressants including cannabinoids (CBs). J Toxicol Sci 33:555–564

Herrera-Solis A, Vasquez KG, Prospero Garcia O (2010) Acute and subchronic administration of anandamide or oleamide increases REM sleep in rats. Pharmacol Biochem Behav 95:106–112

Hill MN, McEwen BS (2009a) Involvement of the endocannabinoid system in the neurobehavioural effects of stress and glucocorticoids. Prog Neuro-Psychopharmacol Biol Psychiatry. https://doi.org/10.1016/j.pnpbp.2009.11.001

Hill MN, McEwen BS (2009b) Endocannabinoids: the silent partner of glucocorticoids in the synapse. Proc Natl Acad Sci U S A 106:4579–4580

Hill MN, Miller GE, Carrier EJ, Gorzalka BB, Hillard CJ (2009) Circulating endocannabinoids and N-acyl ethanolamines are differentially regulated in major depression and following exposure to social stress. Psychoneuroendocrinology 34:1257–1262

Hill MN, Bierer LM, Makotkine I, Golier JA, Galea S, McEwen BS, Hillard CJ, Yehuda R (2013) Reductions in circulating endocannabinoid levels in individuals with post-traumatic stress disorder following exposure to the World Trade Center attacks. Psychoneuroendocrinology 38:2952–2961

Hillard CJ. (2018) Circulating endocannabinoids: from whence do they come and where are they going? Neuropsychopharmacology:155–172

Ho WS, Barrett DA, Randall MD (2008) 'Entourage' effects of N-palmitoylethanolamide and N-oleoylethanolamide on vasorelaxation to anandamide occur through TRPV1 receptors. Br J Pharmacol 155:837–846

Howlett AC (2005) Cannabinoid receptor signaling. Handb Exp Pharmacol 168:53–79

Im DS (2009) New intercellular lipid mediators and their GPCRs: an update. Prostaglandins Other Lipid Mediat 89:53–56

Jhaveri MD, Richardson D, Kendall DA, Barrett DA, Chapman V (2006) Analgesic effects of fatty acid amide hydrolase inhibition in a rat model of neuropathic pain. J Neurosci 26:13318–13327

Joseph J, Niggemann B, Zaenker KS, Entschladen F (2004) Anandamide is an endogenous inhibitor for the migration of tumor cells and T lymphocytes. Cancer Immunol Immunother 53:723–728

Jones EK, Kirkham TC (2012) Noladin ether, a putative endocannabinoid, enhances motivation to eat after acute systemic administration in rats. Br J Pharmacol 166(6):1815–1821

Jumpertz R, Wiesner T, Bluher M, Engeli S, Batkai S, Wirtz H, Bosse-Henck A, Stumvoll M (2010) Circulating endocannabinoids and N-acyl-ethanolamides in patients with sleep apnea-specific role of oleoylethanolamide. Exp Clin Endocrinol Diabetes 118:591–595

Jyotaki M, Shigemura N, Ninomiya Y (2010) Modulation of sweet taste sensitivity by orexigenic and anorexigenic factors. Endocr J 57:467–475

Karsak M, Cohen-Solal M, Freudenberg J, Ostertag A, Morieux C, Kornak U, Essig J, Erxlebe E, Bab I, Kubisch C, de Vernejoul MC, Zimmer A (2005) Cannabinoid receptor type 2 gene is associated with human osteoporosis. Hum Mol Genet 14:3389–3396

Kaufmann I, Hauer D, Huge V, Vogeser M, Campolongo P, Chouker A, Thiel M, Schelling G (2009) Enhanced anandamide plasma levels in patients with complex regional pain syndrome following traumatic injury: a preliminary report. Eur Surg Res 43:325–329

Klein TW, Newton C, Larsen K, Lu L, Perkins I, Nong L, Friedman H (2003) The cannabinoid system and immune modulation. J Leukoc Biol 74:486–496

Knight JM, Szabo A, Zhao S, Lyness JM, Sahler OJ, Liesveld JL, Sander T, Rizzo JD, Hillard CJ, Moynihan JA (2015) Circulating endocannabinoids during hematopoietic stem cell transplantation: a pilot study. Neurobiol Stress 2:44–50

Koethe D, Schreiber D, Giuffrida A, Mauss C, Faulhaber J, Heydenreich B, Hellmich M, Graf R, Klosterkotter J, Piomelli D, Leweke FM (2009) Sleep deprivation increases oleoylethanolamide in human cerebrospinal fluid. J Neural Transm 116:301–305

Konieczny J, Przegalinski E, Pokorski M (2009) N-Oleoyl-dopamine decreases muscle rigidity induced by reserpine in rats. Int J Immunopathol Pharmacol 22:21–28

Kozlowska H, Baranowska M, Schlicker E, Kozlowski M, Laudanski J, Malinowska B (2008) Virodhamine relaxes the human pulmonary artery through the endothelial cannabinoid receptor and indirectly through a COX product. Br J Pharmacol 155:1034–1042

Krumbholz A, Anielski P, Reisch N, Schelling G, Thieme D (2013) Diagnostic value of concentration profiles of glucocorticosteroids and endocannabinoids in hair. Ther Drug Monit 35:600–607

Krylatov AV, Maslov LN, Ermakov S, Lasukova OV, Barzakh EI, Crawford D, Pertwee RG (2007) Significance of cardiac cannabinoid receptors in regulation of cardiac rhythm, myocardial contractility, and electrophysiologic processes in heart. Izv Akad Nauk Ser Biol 1:35–44

Ledent C, Valverde O, Cossu G, Petitet F, Aubert JF, Beslot F, Bohme GA, Imperato A, Pedrazzini T, Roques BP, Vassart G, Fratta W, Parmentier M (1999) Unresponsiveness to cannabinoids and reduced addictive effects of opiates in CB1 receptor knockout mice. Science 283:401–404

Maccarrone M, Pauselli R, Di Rienzo M, Finazzi-Agro A (2002) Binding, degradation and apoptotic activity of stearoylethanolamide in rat C6 glioma cells. Biochem J 366:137–144

Maccarrone M, Lorenzon T, Bari M, Melino G, Finazzi-Agro A (2000) Anandamide induces apoptosis in human cells via vanilloid receptors. Evidence for a protective role of cannabinoid receptors. J Biol Chem 275:31938–31945

Mach F, Steffens S (2008) The role of the endocannabinoid system in atherosclerosis. J Neuroendocrinol 20(Suppl 1):53–57

Maeda N, Osanai T, Kushibiki M, Fujiwara T, Tamura Y, Oowada S, Higuma T, Sasaki S, Yokoyama J, Yoshimachi F, Matsunaga T, Hanada H, Okumura K (2009) Increased serum anandamide level at ruptured plaque site in patients with acute myocardial infarction. Fundam Clin Pharmacol 23:351–357

Mallet PEB, Beninger RJ (1996) The endogenous cannabinoid receptor agonist anandamide impairs memory in rats. Behav Pharmacol 7:261–275

Mechoulam R, Gaoni Y (1965) A total synthesis of Dl-delta-1-tetrahydrocannabinol, the active constituent of hashish. J Am Chem Soc 87:3273–3275

Mechoulam R, Fride E, Hanus L, Sheskin T, Bisogno T, Di Marzo V, Bayewitch M, Vogel Z (1997) Anandamide may mediate sleep induction. Nature 389:25–26

Monteleone AM, Di Marzo V, Monteleone P, Dalle Grave R, Aveta T, Ghoch ME, Piscitelli F, Volpe U, Calugi S, Maj M (2016) Responses of peripheral endocannabinoids and endocannabinoid-related compounds to hedonic eating in obesity. Eur J Nutr 55(4):1799–1805

Movahed P, Evilevitch V, Andersson TL, Jonsson BA, Wollmer P, Zygmunt PM, Hogestatt ED (2005) Vascular effects of anandamide and N-acylvanillylamines in the human forearm and skin microcirculation. Br J Pharmacol 146:171–179

Mulawa EA, Kirkwood JS, Wolfe LM, Wojda SJ, Prenni JE, Florant GL, Donahue SW (2018) Seasonal changes in endocannabinoid concentrations between active and hibernating marmots (marmota flaviventris). J Biol Rhythms 33(4):388–401

Murillo-Rodriguez E, Sanchez-Alavez M, Navarro L, Martinez-Gonzalez D, Drucker-Colin R, Prospero-Garcia O (1998) Anandamide modulates sleep and memory in rats. Brain Res 812:270–274

Neumeister A, Normandin MD, Pietrzak RH, Piomelli D, Zheng MQ, Gujarro-Anton A, Potenza MN, Bailey CR, Lin SF, Najafzadeh S, Ropchan J, Henry S, Corsi-Travali S, Carson RE, Huang Y (2013) Elevated brain cannabinoid CB receptor availability in post-traumatic stress disorder: a positron emission tomography study. Mol Psychiatry 18(9):1034–1040

Niederhoffer N, Szabo B (1999) Effect of the cannabinoid receptor agonist WIN55212-2 on sympathetic cardiovascular regulation. Br J Pharmacol 126:457–466

O'Sullivan SE, Kendall DA (2009) Cannabinoid activation of peroxisome proliferator-activated receptors: potential for modulation of inflammatory disease. Immunobiology 215(8):611–616

Ofek O, Karsak M, Leclerc N, Fogel M, Frenkel B, Wright K, Tam J, Attar-Namdar M, Kram V, Shohami E, Mechoulam R, Zimmer A, Bab I (2006) Peripheral cannabinoid receptor, CB2, regulates bone mass. Proc Natl Acad Sci U S A 103:696–701

Offertaler L, Mo FM, Batkai S, Liu J, Begg M, Razdan RK, Martin BR, Bukoski RD, Kunos G (2003) Selective ligands and cellular effectors of a G protein-coupled endothelial cannabinoid receptor. Mol Pharmacol 63:699–705

Olatinsu AO, Sihag J, Jones PJH (2017) Relationship between circulating fatty acids and fatty acid ethanolamide levels after a single 2-h dietary fat feeding in male Sprague-dawley rats: elevated levels of oleoylethanolamide, palmitoylethanolamide, linoleoylethanolamide, arachidonoylethanolamide and docosahexanoylethanolamide after a single 2 h dietary fat feeding in male Sprague Dawley rats. Lipids 52(11):901–906

Pacher P, Mechoulam R (2011) Is lipid signaling through cannabinoid 2 receptors part of a protective system? Prog Lipid Res 50(2):193–211

Pacher P, Batkai S, Kunos G (2006) The endocannabinoid system as an emerging target of pharmacotherapy. Pharmacol Rev 58:389–462

Pandey R, Mousawy K, Nagarkatti M, Nagarkatti P (2009) Endocannabinoids and immune regulation. Pharmacol Res 60:85–92

Panikashvili D, Shein NA, Mechoulam R, Trembovler V, Kohen R, Alexandrovich A, Shohami E (2006) The endocannabinoid 2-AG protects the blood-brain barrier after closed head injury and inhibits mRNA expression of proinflammatory cytokines. Neurobiol Dis 22(2):257–264

Parmar N, Ho WS (2010) N-Arachidonoyl glycine, an endogenous lipid that acts as a vasorelaxant via nitric oxide and large conductance calcium-activated potassium channels. Br J Pharmacol 160:594–603

Pertwee RG (2001) Cannabinoid receptors and pain. Prog Neurobiol 63:569–611

Pertwee RG, Ross RA (2002) Cannabinoid receptors and their ligands. Prostaglandins Leukot Essent Fatty Acids 66:101–121

Pfitzer T, Niederhoffer N, Szabo B (2005) Search for an endogenous cannabinoid-mediated effect in the sympathetic nervous system. Naunyn Schmiedeberg's Arch Pharmacol 371:9–17
Piro JR, Suidan GL, Quan J, Pi Y, O'Neill SM, Ilardi M, Pozdnyakov N, Lanz TA, Xi H, Bell RD, Samad TA (2018) Inhibition of 2-AG hydrolysis differentially regulates blood brain barrier permeability after injury. J Neuroinflammation 15(1):142
Pisanti S, Borselli C, Oliviero O, Laezza C, Gazzerro P, Bifulco M (2007) Antiangiogenic activity of the endocannabinoid anandamide: correlation to its tumor-suppressor efficacy. J Cell Physiol 211:495–503
Przegalinski E, Filip M, Zajac D, Pokorski M (2006) N-Oleoyl-dopamine increases locomotor activity in the rat. Int J Immunopathol Pharmacol 19:897–904
Randall MD, Kendall DA, O'Sullivan S (2004) The complexities of the cardiovascular actions of cannabinoids. Br J Pharmacol 142:20–26
Sah P (2002) Never fear, cannabinoids are here. Nature 418:488–489
Sancho R, Macho A, de La Vega L, Calzado MA, Fiebich BL, Appendino G, Munoz E (2004) Immunosuppressive activity of endovanilloids: N-arachidonoyl-dopamine inhibits activation of the NF-kappa B, NFAT, and activator protein 1 signaling pathways. J Immunol 172:2341–2351
Schmidt A, Brune K, Hinz B (2006) Determination of the endocannabinoid anandamide in human plasma by high-performance liquid chromatography. Biomed Chromatogr 20:336–342
Schroeder C, Batkai S, Engeli S, Tank J, Diedrich A, Luft FC, Jordan J (2009) Circulating endocannabinoid concentrations during orthostatic stress. Clin Auton Res 19:343–346
Schloss MJ, Horckmans M, Guillamat-Prats R, Hering D, Lauer E, Lenglet S, Weber C, Thomas A, Steffens S (2019) 2-Arachidonoylglycerol mobilizes myeloid cells and worsens heart function after acute myocardial infarction. Cardiovasc Res 115(3):602–613
Selye H (1936) Syndrome produced by diverse nocuous agents. Nature 138:32
Sergeeva OA, De Luca R, Mazur K, Chepkova AN, Haas HL, Bauer A (2017) N-oleoyldopamine modulates activity of midbrain dopaminergic neurons through multiple mechanisms. Neuropharmacology 119:111–122
Sido JM, Nagarkatti PS, Nagarkatti M (2016) Production of endocannabinoids by activated T cells and B cells modulates inflammation associated with delayed-type hypersensitivity. Eur J Immunol 46:1472–1479
Singh P, Sharma P, Nakade UP, Sharma A, Gari M, Choudhury S, Shukla A, Garg SK (2018) Endocannabinoid-mediated modulation of Gq protein-coupled receptor mediates vascular hyporeactivity to nor-adrenaline during polymicrobial sepsis. Pharmacol Rep 70(6):1150–1157
Sophocleous A, Marino S, Kabir D, Ralston SH, Idris AI (2017) Combined deficiency of the Cnr1 and Cnr2 receptors protects against age-related bone loss by osteoclast inhibition. Aging Cell 16(5):1051–1061
Steiner MA, Wotjak CT (2008) Role of the endocannabinoid system in regulation of the hypothalamic-pituitary-adrenocortical axis. Prog Brain Res 170:397–432
Stephen V, Mahler KSS, Berridge KC (2007) Endocannabinoid hedonic hotspot for sensory pleasure: anandamide in nucleus accumbens shell enhances 'liking' of a sweet reward. Neuropsychopharmacology 32(11):2267–2278
Steudte S, Kirschbaum C, Gao W, Alexander N, Schonfeld S, Hoyer J, Stalder T (2013) Hair cortisol as a biomarker of traumatization in healthy individuals and posttraumatic stress disorder patients. Biol Psychiatry 74:639–646
Sticht MA, Limebeer CL, Rafla BR, Abdullah RA, Poklis JL, Ho W, Niphakis MJ, Cravatt BF, Sharkey KA, Lichtman AH, Parker LA (2016) Endocannabinoid regulation of nausea is mediated by 2-arachidonoylglycerol (2-AG) in the rat visceral insular cortex. Neuropharmacology 102:92–102
Storozhuk MV, Zholos AV (2018) TRP channels as novel targets for endogenous ligands: focus on endocannabinoids and nociceptive signalling. Curr Neuropharmacol 16:137–150
Strewe C, Feuerecker M, Nichiporuk I, Kaufmann I, Hauer D, Morukov B, Schelling G, Chouker A (2012) Effects of parabolic flight and spaceflight on the endocannabinoid system in humans. Rev Neurosci 23:673–680

Strewe C, Thieme D, Dangoisse C, Fiedel B, van den Berg F, Bauer H, Salam AP, Gössmann-Lang P, Campolongo P, Moser D, Quintens R, Moreels M, Baatout S, Kohlberg E, Schelling G, Choukèr A, Feuerecker M (2018) Modulations of neuroendocrine stress responses during confinement in antarctica and the role of hypobaric hypoxia. Front Physiol 9:1647
Strollo F (1999) Hormonal changes in humans during spaceflight. Adv Space Biol Med 7:99–129
Su JY, Vo AC (2007) 2-Arachidonylglyceryl ether and abnormal cannabidiol-induced vascular smooth muscle relaxation in rabbit pulmonary arteries via receptor-pertussis toxin sensitive G proteins-ERK1/2 signaling. Eur J Pharmacol 559:189–195
Sun Y, Alexander SP, Garle MJ, Gibson CL, Hewitt K, Murphy SP, Kendall DA, Bennett AJ (2007) Cannabinoid activation of PPAR alpha; a novel neuroprotective mechanism. Br J Pharmacol 152:734–743
Sunano S, Watanabe H, Tanaka S, Sekiguchi F, Shimamura K (1999) Endothelium-derived relaxing, contracting and hyperpolarizing factors of mesenteric arteries of hypertensive and normotensive rats. Br J Pharmacol 126:709–716
Tanasescu R, Constantinescu CS (2010) Cannabinoids and the immune system: an overview. Immunobiology 215:588–597
Tasker J (2004) Endogenous cannabinoids take the edge off neuroendocrine responses to stress. Endocrinology 145:5429–5430
Terrazzino S, Berto F, Dalle Carbonare M, Fabris M, Guiotto A, Bernardini D, Leon A (2004) Stearoylethanolamide exerts anorexic effects in mice via down-regulation of liver stearoyl-coenzyme A desaturase-1 mRNA expression. FASEB J 18:1580–1582
Thieme U, Schelling G, Hauer D, Greif R, Dame T, Laubender RP, Bernhard W, Thieme D, Campolongo P, Theiler L (2014) Quantification of anandamide and 2-arachidonoylglycerol plasma levels to examine potential influences of tetrahydrocannabinol application on the endocannabinoid system in humans. Drug Test Anal 6:17–23
Tsuboi K, Uyama T, Okamoto Y, Ueda N (2018) Endocannabinoids and related N-acylethanolamines: biological activities and metabolism. Inflamm Regen. 38:28
Vaughn LK, Denning G, Stuhr KL, de Wit H, Hill MN, Hillard CJ (2010) Endocannabinoid signalling: has it got rhythm? Br J Pharmacol 160:530–543
Venderova K, Brown TM, Brotchie JM (2005) Differential effects of endocannabinoids on [(3) H]-GABA uptake in the rat globus pallidus. Exp Neurol 194:284–287
Vogeser M, Schelling G (2007) Pitfalls in measuring the endocannabinoid 2-arachidonoyl glycerol in biological samples. Clin Chem Lab Med 45:1023–1025
Vogeser M, Hauer D, Christina Azad S, Huber E, Storr M, Schelling G (2006) Release of anandamide from blood cells. Clin Chem Lab Med 44:488–491
Vuong LA, Mitchell VA, Vaughan CW (2008) Actions of N-arachidonyl-glycine in a rat neuropathic pain model. Neuropharmacology 54:189–193
Wagner JA, Abesser M, Karcher J, Laser M, Kunos G (2005) Coronary vasodilator effects of endogenous cannabinoids in vasopressin-preconstricted unpaced rat isolated hearts. J Cardiovasc Pharmacol 46:348–355
Wagner JA, Varga K, Ellis EF, Rzigalinski BA, Martin BR, Kunos G (1997) Activation of peripheral CB1 cannabinoid receptors in haemorrhagic shock. Nature 390:518–521
Wagner JA, Hu K, Bauersachs J, Karcher J, Wiesler M, Goparaju SK, Kunos G, Ertl G (2001) Endogenous cannabinoids mediate hypotension after experimental myocardial infarction. J Am Coll Cardiol 38:2048–2054
Weis F, Beiras-Fernandez A, Sodian R, Kaczmarek I, Reichart B, Beiras A, Schelling G, Kreth S (2010) Substantially altered expression pattern of cannabinoid receptor 2 and activated endocannabinoid system in patients with severe heart failure. J Mol Cell Cardiol 48:1187–1193
Wheal AJ, Alexander SP, Randall MD (2010) Vasorelaxation to N-oleoylethanolamine in rat isolated arteries: mechanisms of action and modulation via cyclooxygenase activity. Br J Pharmacol 160:701–711
White R, Hiley CR (1998) The actions of some cannabinoid receptor ligands in the rat isolated mesenteric artery. Br J Pharmacol 125:533–541

Wilker S, Pfeiffer A, Elbert T, Ovuga E, Karabatsiakis A, Krumbholz A, Thieme D, Schelling G, Kolassa IT (2016) Endocannabinoid concentrations in hair are associated with PTSD symptom severity. Psychoneuroendocrinology 67:198–206

Yi B, Nichiporuk I, Nicolas M, Schneider S, Feuerecker M, Vassilieva G, Thieme D, Schelling G, Choukèr A (2016) Reductions in circulating endocannabinoid 2-arachidonoylglycerol levels in healthy human subjects exposed to chronic stressors. Prog Neuropsychopharmacol Biol Psychiatry 67:92–97

Zona LC, Fry BR, LaLonde JA, Cromwell HC (2017) Effects of anandamide administration on components of reward processing during free choice. Pharmacol Biochem Behav 158:14–21

Zhang X, Maor Y, Wang JF, Kunos G, Groopman JE (2010) Endocannabinoid-like N-arachidonoyl serine is a novel pro-angiogenic mediator. Br J Pharmacol 160:1583–1594

Immune System in Space: General Introduction and Observations on Stress-Sensitive Regulations

11

Brian Crucian and Alexander Choukér

11.1 General Principles of Immune Functioning

11.1.1 From Physical Barriers to Tailored Immune Responses

The immune system is composed of a wide range of distinct cell types found in peripheral organized tissue where primary immune response occurs (e.g., in spleen, tonsils) and in a vast recirculating pool of cells in the blood and lymph (-nodes) providing the means to deliver the immune-competent cells to sites where they are needed and also to allow a generalized immune response.

This system has been shaped during evolution to protect the organism against potentially pathogenic bacteria and viruses in a rapid and also specific manner. In the human body, the immune system is the largest organ and consists of more than four trillions of cells weighing more than liver and brain altogether. Hence, it is not the mass of an organ anatomist in former times would have considered. Existing at the interface of the environment and the host, the *first line* of host defense is of physical and chemical nature and consists of barriers (e.g., skin) and mucus, enzymes, and acidified thin layers of liquid. Together with the *second line* of defense—the so-called innate immune system—pathogenic germs can be eliminated by phagocytes located in the tissues, e.g., in the lung or in the intestine, at those locations that represent potentially vulnerable areas for germs at the border between the environment and the host. Only higher vertebrates like humans developed the mechanism of reacting to invading germs in a specific manner. This response is targeted exactly to the typical pattern of a

B. Crucian (✉)
NASA-Johnson Space Center, Houston, TX, USA
e-mail: brian.crucian-1@nasa.gov

A. Choukér
Laboratory of Translational Research Stress and Immunity, Department of Anaesthesiology, Hospital of the University of Munich (LMU), Munich, Germany
e-mail: alexander.chouker@med.uni-muenchen.de

© Springer Nature Switzerland AG 2020

A. Choukér (ed.), *Stress Challenges and Immunity in Space*,
https://doi.org/10.1007/978-3-030-16996-1_11

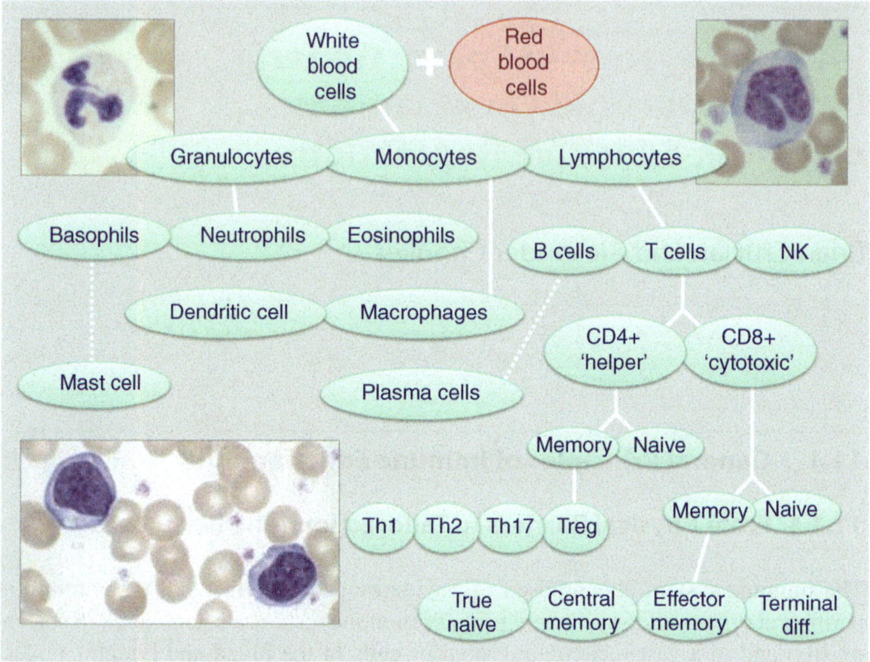

Fig. 11.1 Distribution of primary innate and adaptive immune cell populations in human blood. An analysis of the levels of immune cell subsets can provide information about constitutive immune processes. Populations may increase as cells proliferate during an immune response or contract as cells become sequestered at localized sites of inflammation. Typically, these subsets are quantified by staining specific populations with fluorescent dyes, followed by flow cytometry analysis

germ, allowing also the recognition of the microorganism for a further more rapid and more efficient action in the course of a later, second contact. The immune systems' *third-line* response is based on the action of B- and T-lymphocytes, and is called the adaptive immunity (Choukèr et al. 2008) (Fig. 11.1, see also Chap. 12).

11.1.2 Innate and Adaptive Immunity

Human immunity, after skin and other physical barriers, is actually comprised of cells of two "distinct" yet interconnected systems, the innate and adaptive immune systems. Innate immunity, mediated by neutrophils, monocytes/macrophages, dendritic cells, and NK, is the primary line of cellular defense and consists of an immediate (constitutively present) nonspecific response. An innate immune response does not result in immunologic memory; so subsequent responses to the same pathogen are not aided by the previous exposure. In contrast, adaptive immunity is the secondary line of cellular defense. It consists of a delayed responses, as mediated primarily by lymphocytes, it is antigen-specific and results in immunologic memory. Subsequent reexposures to the same pathogen result in significantly greater and more rapid antigen-specific responses due to the presence of circulating memory B and T cells. There are actually two general components to adaptive

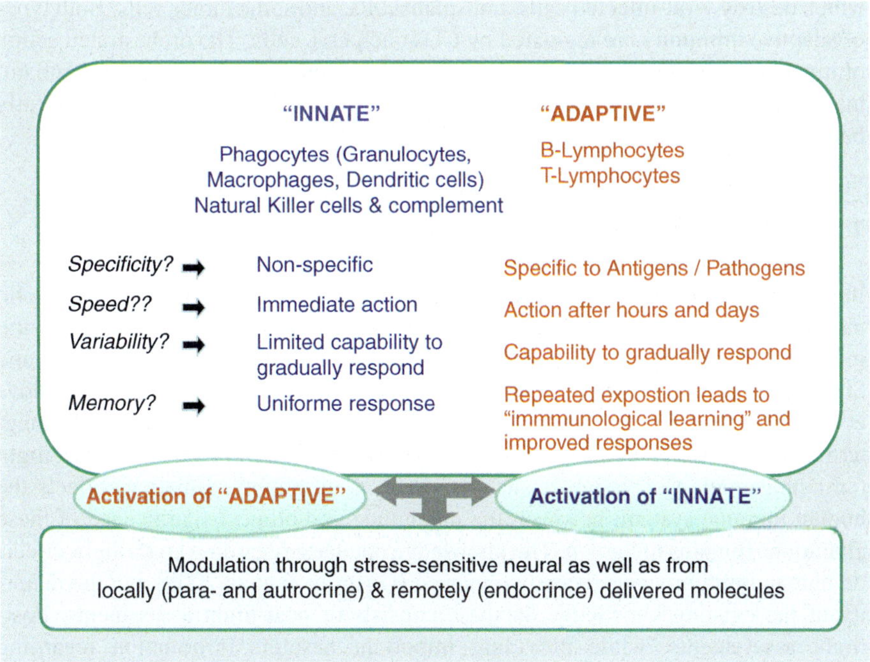

Fig. 11.2 Immunity describes the state of having adequate host defense to protect from infection and disease and is based on a distinct and well-orchestrated interaction of leukocytes (= white blood cells). "Innate": granuolcytes (see phagocytes) represent up to 80% of all blood circulating phagocytes and one of the prime cell types of the innate immunity and small proteins (complement system), which when activated stimulate the phagocytic cells and amplify attraction of further immune cells to clear foreign and damaged material/tissue and to raise inflammatory processes. "Adaptive": leukocytes of the adaptive immune systems consist of antibody-producing B-lymphocytes and T-lymphocytes, thereby allowing to *adapt* the degree of the immune responses to the specific pathogens together with the ability to specifically recognize ("learn") the respective antigens in future. The reciprocal activation of the innate and adaptive immune systems guarantees the mounting of a fast, strong, and efficient attack at the first antigen contact and even more, each subsequent time when this pathogen is encountered again. The control—the activation and limitation of responses—is regulated by a very complex and yet not fully understood local (autocrine/paracrine) and remote (endocrine, nerval control) acting messengers (cytokines/hormones) which altogether modulate the sensitivity and responsiveness of the human immune system to protect the human host against thousands of potentially harmful invading pathogens

immunity: antibody mediated and cell mediated. Antibody (humoral) immunity is mediated by B cells and antibody production. Antibodies bind to specific antigens, which signal phagocytes to engulf, kill, and remove the target. Antibodies also initiate targeted cell killing, via a process termed "antibody-dependant cellular cytotoxicity." In this process, soluble antibodies coat a cellular target which directs cytotoxic lymphocytes to bind and kill that target. The principle of vaccination consists of inoculation of nonvirulent pathogen proteins, against which an antibody response is generated. The soluble antibodies and memory cells then persist for years, conferring protection to the host against the specific pathogen for which the vaccine was generated. Cell-mediated immunity (CMI) is mediated by $CD8^+$ cytotoxic T cells, which destroy viral-infected cells, transplant cells, and some tumor cells. Both types of adaptive immunity are regulated by $CD4^+$ helper T cells. The orchestrated action of immune activation and the inverse action of immune suppression to limit inflammation, both processed through endo-, para-, and autocrine pathways, are currently being defined by active research efforts (see Fig. 11.2).

11.2 The Immune System in Space

Immunological problems of spaceflight have been observed since the first Apollo missions, where more than half of the astronauts suffered from bacterial or viral infections (Hawkins and Zieglschmid 1975). Also in crewmembers of Skylab and Soyuz, a reduced reactivity of blood lymphoid cells has been observed (Konstantinova et al. 1973; Kimzey 1977) indicating a potentially alarming observation for long-term space missions. Four decades of investigation clearly indicate that spaceflight (an amalgamation of variables such as stress, microgravity, radiation) affects the human immune system; however, the magnitude and clinical significance of these alterations remains unknown. The knowledge database regarding spaceflight effects on human immunity was recently reviewed (Gueguinou et al. 2009), but the majority of the existing knowledge database consists of post-flight assessments. Post-flight assessments, while providing important baseline information regarding potential in-flight dysregulation, do not necessarily reflect the in-flight status of the immune system. The physiological stress of reentry and a high-g landing profile, as well as readaptation to unit gravity following prolonged deconditioning, both may skew post-flight data. Post-flight findings include alterations in the distribution of the peripheral blood leukocytes, reductions in the function of specific immune cell subpopulations, and altered stress hormone levels. Fortunately, with the advent of the International Space Station (ISS), assessments of crewmembers participating in long-duration missions became possible, including in-flight sampling. To date, immune assessments of ISS astronauts have included the purification and freezing of blood plasma for subsequent analysis, and the return of ambient living blood for functional cellular assessments. Return of ambient blood, if it was collected near to undocking and hatch closure of a returning vehicle, is available in a terrestrial laboratory 37–48 h after collection. Therefore, immunological assays must be selected which tolerate this delay in processing.

11.2.1 Peripheral Leukocyte Distribution and Spaceflight

The distribution of cells of the immune system (both the adaptive and innate subsections), are usually determined in peripheral blood by flow cytometry, and is a measure of in vivo immune responses. The principle is similar to that of the common "complete blood count" (CBC) measurement (a hematological measurement); however, the number of specific immune cell subpopulations identified by cytometry is much greater. The increased number of subsets are positively identified by staining of cell-specific membrane protein antigens with various fluorescent antibodies, followed by detection and resolution on a flow cytometer. Depending on the specific clinical situation, various subpopulations of immune cells in the peripheral blood (Fig. 11.1) may increase due to clonal expansion/proliferation, or be reduced as cells migrate out of the blood to localized sites of inflammation, die via apoptosis, or become reduced due to a block in cellular maturation processes. Many previous studies have examined the distribution of immune cells following spaceflight, and the data are summarized in Table 11.1. Although specific subset findings may vary among the studies, the most common alterations are increased granulocytes, and reduced lymphocytes and monocytes. Among lymphocyte subsets, T cell and Natural killer (NK) cell percentages are usually reported to be decreased post spaceflight. Reports on B cell percentages are varied, as are reports regarding $CD4^+$ and $CD8^+$ T cell subsets. In addition to these commonly assessed "bulk" subsets, there are a number of "fine" lymphocyte subsets which have been identified, usually by multicolor flow cytometry (see also Chap. 27). Among these "fine" subsets, Gridley et al. (2009) reported increased $CD25^+$ T cells, Crucian et al. reported increased memory T cells and reduced senescent T cells, and Ortega et al. reported activation and lineage differentiation within murine bone marrow cells (Ortega et al. 2009; Crucian et al. 2008). It is useful to interpret these results, especially findings validated by many studies, and utilize peripheral leukocyte subsets as a monitor of in vivo immune changes in crewmembers. Granulocytosis, with concurrent decreases in lymphocyte and monocyte percentages, is most likely related to demargination of the neutrophils to the blood in response to the elevated stress hormone levels following landing. Post-flight findings are very likely to be influenced by the confounding variables of both high-g landing, and readaptation to unit gravity following prolonged deconditioning. Both are significant physiological stressors. It appears that T cell and NK percentages are commonly reduced, whereas B cell percentages are commonly elevated. The CD4:CD8 ratio is unaltered or increased, depending on the study. De Rosa et al. (2001) have indicated that changes in "fine" lymphocyte subsets may have clinical significance, even when changes in "bulk" lymphocyte subsets are not evident. Fine lymphocyte and T cell subsets are unique subsets identified by the presence of multiple markers, such as cytotoxic/effector $CD8^+$ T cells, or central memory T cells. As discussed above, the availability of ISS affords the opportunity to assess peripheral leukocyte distribution in humans during a 6 month orbital spaceflight. Crucian et al. found that most bulk leukocyte and lymphocyte subsets were generally unaltered in the peripheral blood of astronauts, both during short- and long-duration spaceflight (Space Shuttle and ISS crews, respectively) (Crucian et al. 2013, 2015). There was a persistent elevation in the white blood cells (WBC) and granulocyte concentration (absent altered relative percentage alterations), but as discussed by Kunz et al., this may reflect simple dehydration during spaceflight (Kunz et al. 2017). However, there is a persistent alteration in the phenotype of $CD8^+$ T cells, towards a more mature phenotype, which persists during orbital spaceflight (Crucian et al. 2014). This may possibly be associated with the, somewhat ineffective, attempt to control the reactivation of latent herpesviruses (Mehta et al. 2017) as also discussed elsewhere in this volume (Chap. 19).

Table 11.1 Representative studies: peripheral leukocyte distribution during or following spaceflight

			Standard cell subsets										
	Subjects	Flight	WBC	Gran	Lym	Mono	T	B	NK	CD4	CD8	Ratio	Other subsets of interest
Taylor et al. (1986)	Human	Short			↓	↓		↓					Humans, post-flight, short duration.
Allebban et al. (1994)	Rat	Short	↓	↑	↓	↓		nc		nc	nc		Rats, post-flight, peripheral blood.
Ichiki et al. (1996)	Rat	Short	↓	↑	↓	↓	↓	↓		↓	↓		
Chapes et al. (1999a)	Rat	Short		↑	↓	↓							
Chapes et al. (1999b)	Rat	Short			↓	↓							
Stowe et al. (1999)	Human	Short		↑	↓	var							Bands increased, integrin-alpha M (CD11b) decreased, L-selectin increased.
Crucian et al. (2000)	Human	Short		↑	nc	nc	↓	nc	nc			↑	Bulk memory naïve variable, CD14/CD16 monocytes decreased.
Mills et al. (2001)	Human	Short	↑	↑		↑		↑	↓	↑			Post-flight study; mission durations of 1 week correlated with increased levels of sympathetic stress hormones and circulating leukocytes; whereas for mission duration of 2 weeks the sympathetic and leukocyte changes were attenuated.
Stowe et al. (2003)	Human	Short	↑	↑		var			var	↑			Responses vary between 9 and 16 day missions. 9 day missions = sympathetic responses, 16 day missions = glucocorticoid responses (due to deconditioning).

	Subjects	Flight	Standard cell subsets										
			WBC	Gran	Lym	Mono	T	B	NK	CD4	CD8	Ratio	Other subsets of interest
Pecaut et al. (2003)	Mouse	Short											No change in murine circulating leukocytes. Increase in splenic lymphocyte percentages, decrease in granulocytes. Splenic lymphocytes, reduced T cells, increased B cells, reduced CD4+, and reduced ratio. CD34+ population increased. Data indicates shift in marrow populations, not response to changes in periphery.
Gridley et al. (2003)	Mouse	Short											Short-duration flight: elevated CD25+ lymphs, CD25+ T cells, NK1.1+/CD25+ cells. CD71 decreased during flight.
Rykova et al. (2008)	Human	Short/long					nc		↓	nc	nc	nc	
Crucian et al. (2008)	Mouse	Short/long		↑	↓	nc	nc	↑	↓	nc	nc	nc	Short-duration flight: elevation in CD4+/CD45RO+ memory T cells, reduction in senescent CD8+ T cells (CD28−/CD244+). Long-duration flight: elevated percentage of CD45RO+ memory T cells for both the CD4+ and CD8+ subsets.
Gridley et al. (2009)	Mouse	Short						↓	↓	↑			Murine spleen: T cell and B cell counts low, NK1.1+ NK elevated.
Ortega et al. (2009)	Mouse	Short											Bone marrow: assess maturation/activation of granulocytic lineage cells. No composite phenotypic differences following flight, but were some subpopulation differences. Elevation of CD11b in the R2 subpopulation suggests neutrophil activation. Also, decreased expression of other antigens suggests marrow cells were more differentiated.

(continued)

Table 11.1 (continued)

			Standard cell subsets										
	Subjects	Flight	WBC	Gran	Lym	Mono	T	B	NK	CD4	CD8	Ratio	Other subsets of interest
Crucian et al. (2013)	Human	Short	↑[a]	↑[a]							↓[a]		Short-duration—terrestrial analysis of samples collected in-flight near the end of a 7–14 day spaceflight. Short-duration spaceflight was found not to alter the majority of the bulk populations examined, including the WBC, differential, and leukocyte subsets. $CD8^+$ T cell subsets were altered during spaceflight.
Crucian et al. (2015)	Human	Long	↑	↑							#		Long-duration—phenotype of leukocyte, lymphocyte and T cell subsets. Terrestrial analysis of samples collected at 3 points during a 6 month orbital spaceflight. Findings include elevated in-flight WBC, granulocytes and redistribution of $CD8^+$ T cell subsets based on maturation state (#).
Simpson et al. (2016)	Human	Long							nc				Long-duration—phenotype and function of NK assessed by terrestrial analysis of samples collected at 2 points during a 6 month orbital spaceflight. NK function was reduced during flight, corresponding with reductions in perforin and granzyme B. NK expression of NKG2C and NKG2A were unchanged with exception of a crewmember experiencing a CMV reactivation and corresponding expansion of $NKG2C^+$ NK.

[a]Post flight alteration
#Multi-parametric changes in expression patterns

Simpson et al. have investigated both the phenotype and function of NK in ISS astronauts. Their data demonstrated common reductions in NK function with unaltered NK phenotype (NKG2A and NKG2C expression) with the possible exception of astronauts experiencing the reactivation of CMV (Simpson et al. 2016). In summary these alterations in leukocyte distribution (analog, animal, and human) commonly indicate mobilization of the adaptive immune response associated with spaceflight conditions. In an applied assessment such as spaceflight, it cannot necessarily be determined if this mobilization is in response to flight-associated stressors, microgravity, solely the reactivation of latent herpes viruses, or an external infectious agent.

11.2.2 Observations on Immune–Stress Responses in Space

There have been comparatively few in-flight immune studies in humans. Pierson et al. found latent viral reactivation occurs during short-duration flight (Mehta et al. 2000a, b, 2004; Payne et al. 1999; Pierson et al. 2005; Stowe et al. 2001). Delayed type hypersensitivity (DTH) responses, mediated by monocytic infiltration and memory T cells, have been shown to be blunted during long-duration space flight (Cogoli 1993; Gmünder et al. 1994). More recently, blood analysis of ISS astronauts has revealed reduced adaptive immune cell function and the persistent reactivation of latent herpesviruses (Crucian et al. 2014; Simpson et al. 2016; Mehta et al. 2014, 2017). Potential causes for this phenomenon include crewmember stress, isolation, disrupted circadian rhythms, microgravity, and radiation, and more detailed discussion of the findings, are expounded upon the chapters of Part III of this volume. In fact, it is likely that a synergy of these factors is responsible for the observations.

Spaceflight is an applied situation. For humans, there is no way to separate the variables to ascertain mechanistic causes from among the various stressors. Therefore, to prepare for space flight, the knowledge gap on the role of one or more of the distinct stressors in space can be bridged or aided by investigations which address selected stressors—e.g., only microgravity conditions, or confinement without the effects of low gravity—in appropriate Earth analogue conditions. Analog experiments are of critical importance to simulate and to investigate the effect of environmental factors likely to affect immunity in space as well. These factors include: long-term bed rest in 6° head-down tilt (to simulate weightlessness) as well as confinement; (Ant-)Arctic overwintering; isolation chambers (Mars500, see Chap. 37); and underwater habitats (NEEMO, see Chap. 36). Such ground analogs can serve as a suitable Earth-bound simulation experiment to mirror some mission-related stressors, but in a terrestrial location. Although a full comparability of the studies as mentioned below could not have been achieved due to different models, study protocols, and methods, these studies revealed that stress due to confinement induces various changes in neuroendocrine and immune responsiveness.

To date we have learned that stressful situations shown to result in a nerval (e.g., sympathetic) and/or endocrine or auto/paracrine mediated effects on immune responses (Fig. 11.2). This has been exemplarily indicated by a glucocorticoid-mediated immune downregulation which was also associated with reactivation of mostly dormant virus (see Chap. 19). In accordance to these observations, other groups demonstrated dynamics and severity of immune (dys-) function in long-term as well as short-term approaches, respectively. For instance, the consequences of confinement and stress-associated neuroendocrine and immune change have been investigated for different durations during bed rest, e.g. under the conditions of 120 days 6° head-downtilt (HDT) (Choukèr et al. 2001) or in confinement from 10 days (Shimamiya et al. 2004) to 60 days (Hennig and Netter 1996), to 110 days up to 240 days (Choukèr et al. 2002), and T cell dysfunction became evident within the peripheral blood mononuclear cell compartment, including a paradoxical atypical monocytosis associated with altered production of inflammatory cytokines (Tingate et al. 1997). Moreover, more specific information on changes of the human immune system have been described very recently from a month-long study in the Canadian arctic (Crucian et al. 2007) or in the Concordia base in continental Antarctic environment (Crucian et al. 2011; Feuerecker et al. 2018) indicating that several stress responses and specific immune changes were quite similar to those observed in astronauts after space flight. It was found that stressful conditions of psychological or physical nature under such space (-analogous) conditions can modulate innate and adaptive immune responses, respectively. An important role was ascribed to the catecholaminergic- and glucocorticoid-system to affect the host's immune responses together with direct nerval pathways (Tracey 2002), although other not-yet-so-well-investigated stress-response systems along with the endocannabinoid-system are also found to play a role. The latter has been shown to be activated under the conditions of stress (Choukèr et al. 2010) and to modulate a variety of immune cell functions in humans and animals, including T-helper cell development and chemotaxis. It hence appears that the "immunocannabinoid" system is involved in regulating the brain-gut-immune axis (Klein et al. 2003; Acharya et al. 2017) which exerts an important pathophysiological control of inflammation (D'argenio et al. 2006) (see also Chap. 10). This strong interaction between nerval, neuroendocrine mediators, and the immune system (see Chaps. 4 and 6) has been also reviewed by Macho et al. (2001) and Cacioppo et al. (2002) is to be extended by immunotropic metabolic factors to regulate in a way similar to auto and paracrine regulation, the immune cells (mitochondrial) functions under resting and activating status (see Chaps. 4, 16 and 17).

Some first and confirmative interaction of space flight stress and the consequences on immune functions have been recently published from rodent research in C57BL/6NT mice flown on a 13-day space mission. When a battery of functional, enumerative, and genetic analyses data from space-flown mice were compared to ground control, immune cells' responsiveness was significantly found to be affected. The Phytohemagglutinin (PHA)-induced splenocyte DNA synthesis was highly reduced, as it was observed for the interleukin-2 (IL-2) production after activation of spleen cells with anti-CD3 monoclonal antibody. In accordance to that interleukin 10 levels, a cytokine shown to be involved in the limitation of pro-inflammatory processes was elevated. Interestingly, however, interferon (IFN)-gamma and macrophage inflammatory protein-1-alpha were increased in flight mice. Moreover, Gridley and coworkers observed further that T lymphocytes ($CD3^+$) and B lymphocytes ($CD19^+$) numbers were low in the spleens of mice flown on in this mission with higher percentage of NK (Gridley et al. 2009). Interestingly, it was shown also that space flight resulted in significant changes of thymic mRNA expression in T cell signaling regulating genes, including: cytotoxic T lymphocyte antigen-4, IFN-alpha2a (up), and CD44 (down) and the genes that regulate stress and glucocorticoid receptor metabolism (Lebsack et al. 2010). Along the line of these observations indicating dysbalanced immune functional properties under conditions of stress, it was also observed that cancer-related gene expression patterns were significantly modified after being subjected to the stressful spaceflight environment (Gridley et al. 2009).

11.2.3 The Immune-Dysregulation and Hypersensitivity: From In Vitro Observation and Clinical Incidences

However, it is yet unknown to what degree the alterations during orbital flight would persist, or worsen, for the duration of exploration-class deep space missions. The question remains open, if, why, when and to which degree adaptation processes, in a bidirectional sense also of conditioning and deconditioning, can occur for immune functions as suggested for other organ systems (Nicogossian et al. 1994). With a prolonged immune dysregulation during exploration-class deep space missions, several potential clinical risks to crewmembers, including hypersensitivities, autoimmunity, infectious disease, and malignancies, can endanger mission success. Indeed, clinical incidence onboard ISS would be a very relevant predictor of, should the observed immune dysregulation persist, clinical risks during deep space/high radiation missions of exploration. Such "incidence" data for ISS has been historically lacking, as crewmembers private medical data is not open for researchers. However, two recent studies reveal insight into the clinical situation onboard ISS. Summary incidence numbers indicate both infectious disease and atypical allergy/atopic dermatitis are the most reported clinical events onboard ISS (Crucian et al. 2016b). It should be noted however, that since disease cannot be confirmed, these findings are more a tabulation of symptoms than standard epidemiology. Further, some ISS crewmembers experience a persistent skin rash phenomenon, as captured in a recent case study of an ISS astronaut (Crucian et al. 2016a). It is debatable what the incidence "'should" be; given there is no terrestrial correlation for ISS crewmembers: extremely fit and well-trained individuals who undergo a pre-flight medical screening; living in an isolated yet extreme environment with unique stressors. However, as these stressors will only increase, and clinical care capability decrease, during upcoming deep space missions these phenomenon bear both investigation and the development of relevant countermeasures.

Here ground analoges can help reflecting long-duration exploration class mission and allow further guiding the research to understand the impact of harsh conditions of life on the occurrence of immune dysfunctional states. The Antarctic environment has helped us to understand human adaption to living under duress, but also to follow the time slopes of their development, e.g. of a Type IV hypersensitivity reaction pattern ex vivo upon stimulation with recall antigen (Feuerecker et al. 2018). The occurrence of a pathology to newly developed allergies—either during or after overwintering in Antarctic stations—has been reported and hence a more stringent investigation on the incidences is now under a systematic epidemiologic evaluation run at several Antarctic bases.

11.3 Limitations and New Approaches for Immune-Directed Research in Space

Space flight experimental opportunities are infrequent and expensive to conduct and there are many constraints to performing science experiments in space (e.g. hazard aspect, blood handling, up/download mass limitations, and conditioned storage). During spaceflight missions, collection of biological samples (e.g. blood, saliva, urine) is limited due to constraints on crewmember time and lack of resources such as refrigeration and many of the instruments typically used in medical experiments do not function properly in microgravity. Thus, the majority of space flight data has, to date, still been derived from pre- and post-flight, or animal/analog studies. Although valuable information has been derived from the summary spaceflight-immunology research, there are also limitations that must be acknowledged. These include a several hour delay between landing and in-flight bio-specimen collection, the effects of reentry stress on biological and biochemical measures, and differences in mission duration. Moreover, some discrepancies can be attributed to methodologic differences, adding to the variety of types and extent of stressors encountered during space missions, altogether resulting in the variability of immune responses assessed during and after space flight (Borchers et al. 2002).

In recent years the scientific community has became aware that some of these limitations can be also overcome by evolved methods, extensively tested and standardized hardware, and by the use of safe reagents and "streamlined" procedures that help to gather reliable data. Also, with improved post-flight logistics, it is now possible to return living blood samples from the ISS, which allows a functional determination of immune status during flight, without flying significant analytical hardware to space (Fig. 11.3). As ISS has aged, utilization has matured too. More samples are being collected, and are of various types and/or sample processing. Instruments such as flow cytometers and biomolecule sequencers have been triaged onboard ISS. The techniques required for "omics" analysis, such as the purification of cell populations during spaceflight has been validated (Rizzardi et al. 2016). Moreover, the combination of knowledge and sharing specimen, to address also the combined action of multi-factorial sources of stress conditions during space flight, will help to more comprehensively explore the consequences on the immune system.

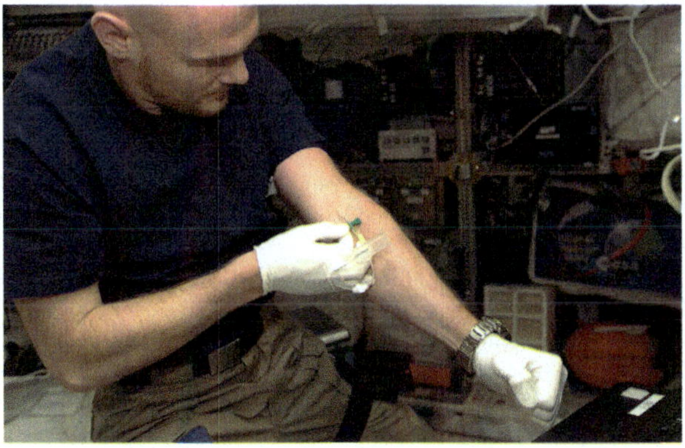

Fig. 11.3 ESA Astronaut Alexander Gerst collects a blood sample onboard the International Space Station. For studies to assess immune function, stress, and viral reactivation during spaceflight, blood samples are typically collected near to undocking of a return vehicle such that live whole blood may be returned to Earth for subsequent processing and analysis. Image credit: ESA/NASA

11.4 Summary

Because of the nature of the immune system in protecting the host and maintaining health in changing and challenging environment, it seems to be a 'natural' and homeodynamic situation that functions of the immune system are affected during the course of space flight. However, long-duration missions in space and especially interplanetary missions require distinct knowledge on how the humans' immune system will cope with these extreme environmental conditions and to which extent an adaptation can occur. Recent investigations on crewmembers of long-duration space missions have revealed the potential development of immune dysfunctions into two directions: immune hyper-activity which may result in risks such as hypersensitivity or autoimmunity and immune hypo-reactivity, which means an anticipated increased risk for infectious diseases and viral reactivation. Such alterations, should they persist during prolonged interplanetary space missions, and habitation on moon or Mars, could lead to diseases associated with immune imbalance such as hypersensitivity reactions, allergies, autoimmune diseases, chronic inflammation and other inflammation-related and infectious diseases, also when space crews return back to Earth. Therefore, continued fundamental and clinical immunology research is critically needed to allow in a multidisciplinary approach a better understanding of how microgravity, cosmic radiation, and other stressful mission-associated risk factors can influence the adequate functioning of the immune system. These investigations will also help patients on Earth that will clearly benefit from advances in research in clinical immunology to better understand immune function in nominal and off-nominal conditions of life.

Acknowledgments We are thankful to R. Stowe and O. Ullrich as well as to J.-P. Frippiat, M. Feuerecker, M. Rykova, and C. Sams for their support and for kindly providing selective information to this chapter. We extend our thanks to Sandra Matzel for the support in the preparation of the figures.

References

Acharya N, Penukonda S, Shcheglova T, Hagymasi AT, Basu S, Srivastava PK (2017) Endocannabinoid system acts as a regulator of immune homeostasis in the gut. Proc Natl Acad Sci U S A 114(19):5005–5501

Allebban Z, Ichiki AT, Gibson LA, Jones JB, Congdon CC, Lange RD (1994) Effects of spaceflight on the number of rat peripheral blood leukocytes and lymphocyte subsets. J Leukoc Biol 55(2):209–213

Borchers AT, Keen CL, Gershwin ME (2002) Microgravity and immune responsiveness: implications for space travel. Nutrition 18:889–898

Cacioppo JT, Kiecolt-Glaser JK, Malarkey WB et al (2002) Autonomic and glucocorticoid associations with the steady-state expression of latent Epstein-Barr virus. Horm Behav 42:32–41

Chapes SK, Simske SJ, Forsman AD, Bateman TA, Zimmerman RJ (1999a) Effects of space flight and IGF-1 on immune function. Adv Space Res 23(12):1955–1964

Chapes SK, Simske SJ, Sonnenfeld G, Miller ES, Zimmerman RJ (1999b) Effects of spaceflight and PEG-IL-2 on rat physiological and immunological responses. J Appl Physiol 86(6):2065–2076

Choukèr A, Thiel M, Baranov V et al (2001) Simulated microgravity, psychic stress, and immune cells in men: observations during 120-day 6 degrees HDT. J Appl Physiol 90:1736–1743

Choukèr A, Smith L, Christ F et al (2002) Effects of confinement (110 and 240 days) on neuroendocrine stress response and changes of immune cells in men. J Appl Physiol 92:1619–1627

Choukèr A, Morukov B, Sams C (2008) Clinical immunology in new frontiers. Scientific American presents: looking up, Europe's quiet revolution in microgravity research. Sci Am J 2008:24–31

Choukèr A, Kaufmann I, Kreth S, Hauer D, Feuerecker M, Thieme D, Vogeser M, Thiel M, Schelling G (2010) Motion sickness, stress and the endocannabinoid system. PLoS One 5:e10752

Cogoli A (1993) The effect of space flight on human cellular immunity. Environ Med 37(2):107–116

Crucian BE, Cubbage ML, Sams CF (2000) Altered cytokine production by specific human peripheral blood cell subsets immediately following space flight. J Interf Cytokine Res 20(6):547–556

Crucian B, Lee P, Stowe R et al (2007) Immune system changes during simulated planetary exploration on Devon Island, high arctic. BMC Immunol 8:7

Crucian BE, Stowe RP, Pierson DL, Sams CF (2008) Immune system dysregulation following short- vs long-duration spaceflight. Aviat Space Environ Med 79(9):835–843

Crucian BE, Feuerecker M, Salam AP, Rybka A, Stowe RP, Morrels M, Mehta SK, Quiriarte H, Quintens R, Thieme U, Kaufmann I, Baatout DS, Pierson DL, Sams CF, Choukèr A (2011) The ESA-NASA 'CHOICE' study: winterover at Concordia station, interior Antarctica, as an analog for spaceflight-associated immune dysregulation. In: 18th IAA humans in space symposium, Houston, Texas, 11–15 April 2011

Crucian B, Stowe R, Mehta S, Uchakin P, Quiriarte H, Pierson D, Sams C (2013) Immune system dysregulation occurs during short duration spaceflight on board the space shuttle. J Clin Immunol 33(2):456–465. https://doi.org/10.1007/s10875-012-9824-7

Crucian BE, Zwart SR, Mehta S, Uchakin P, Quiriarte HD, Pierson D, Sams CF, Smith SM (2014) Plasma cytokine concentrations indicate that in vivo hormonal regulation of immunity is altered during long-duration spaceflight. J Interf Cytokine Res 34(10):778–786. https://doi.org/10.1089/jir.2013.0129

Crucian B, Stowe RP, Mehta S, Quiriarte H, Pierson D, Sams C (2015) Alterations in adaptive immunity persist during long-duration spaceflight. NPJ Microgravity 1:15013. https://doi.org/10.1038/npjmgrav.2015.13

Crucian B, Johnston S, Mehta S, Stowe R, Uchakin P, Quiriarte H, Pierson D, Laudenslager ML, Sams C (2016a) A case of persistent skin rash and rhinitis with immune system dysregulation onboard the International Space Station. J Allergy Clin Immunol Pract 4(4):759–762.e8. https://doi.org/10.1016/j.jaip.2015.12.021

Crucian B, Babiak-Vazquez A, Johnston S, Pierson DL, Ott CM, Sams C (2016b) Incidence of clinical symptoms during long-duration orbital spaceflight. Int J Gen Med 9:383–391. eCollection 2016

D'argenio G, Valenti M, Scaglione G et al (2006) Up-regulation of anandamide levels as an endogenous mechanism and a pharmacological strategy to limit colon inflammation. Gastroenterology 130:A348

De Rosa SC, Herzenberg LA, Roederer M (2001) 11-color, 13-parameter flow cytometry: identification of human naive T cells by phenotype, function, and T-cell receptor diversity. Nat Med 7(2):245–248

Feuerecker M, Crucian BE, Quintens R, Pagel J-I, Salam AP, Rybka A, Moreels M, Strewe C, Stowe R, Mehta S, Schelling G, Thiel M, Baatout S, Sams C, Choukèr A (2018) Immune sensitization during one year in the Antarctic high altitude Concordia Environment. Allergy. https://doi.org/10.1111/all.13545

Gmünder FK, Konstantinova I, Cogoli A, Lesnyak A, Bogomolov W, Grachov AW (1994) Cellular immunity in cosmonauts during long duration spaceflight on board the orbital MIR station. Aviat Space Environ Med 65(5):419–423

Gridley DS, Nelson GA, Peters LL, Kostenuik PJ, Bateman TA, Morony S et al (2003) Genetic models in applied physiology: selected contribution: effects of spaceflight on immunity in the C57BL/6 mouse. II. Activation, cytokines, erythrocytes, and platelets. J Appl Physiol 94(5):2095–2103

Gridley DS, Slater JM, Luo-Owen X, Rizvi A, Chapes SK, Stodieck LS et al (2009) Spaceflight effects on T lymphocyte distribution, function and gene expression. J Appl Physiol 106(1):194–202

Gueguinou N, Huin-Schohn C, Bascove M, Bueb JL, Tschirhart E, Legrand-Frossi C et al (2009) Could spaceflight-associated immune system weakening preclude the expansion of human presence beyond Earth's orbit? J Leukoc Biol 86(5):1027–1038

Hawkins W, Zieglschmid J (1975) Clinical aspects of crew health. In: Johnston R, Dietlein L, Berry C (eds) Biomedical results of Apollo. NASA, Washington, DC, pp 43–81

Hennig J, Netter P (1996) Local immunocompetence and salivary cortisol in confinement. Adv Space Biol Med 5:115–132

Ichiki AT, Gibson LA, Jago TL, Strickland KM, Johnson DL, Lange RD et al (1996) Effects of spaceflight on rat peripheral blood leukocytes and bone marrow progenitor cells. J Leukoc Biol 60(1):37–43

Kimzey SL (1977) Hematology and immunology studies. In: Biomedical results from Skylab. NASA-SP-377. National Aeronautics and Space Administration, U.S. Goverment Printing Office, Washington, DC, pp 249–282

Klein TW, Newton C, Larsen K et al (2003) The cannabinoid system and immune modulation. J Leukoc Biol 74:486–496

Konstantinova IV, Antropova YN, Legenkov VI, Zazhirey VD (1973) Study of reactivity of blood lymphoid cells in crew members of the Soyuz-6, Soyuz-7 and Soyuz-8 spaceships before and after flight. Space Biol Med 7:48–55

Kunz H, Quiriarte H, Simpson RJ, Ploutz-Snyder R, McMonigal K, Sams C, Crucian B (2017) Alterations in hematologic indices during long-duration spaceflight. BMC Hematol 17:12. https://doi.org/10.1186/s12878-017-0083-y. eCollection 2017

Lebsack TW, Fa V, Woods CC, Gruener R, Manziello AM, Pecaut MJ, Gridley DS, Stodieck LS, Ferguson VL, Deluca DJ (2010) Microarray analysis of spaceflown murine thymus tissue reveals changes in gene expression regulating stress and glucocorticoid receptors. J Cell Biochem 110(2):372–381

Macho L, Kvetnansky R, Fickova M et al (2001) Endocrine responses to space flights. J Gravit Physiol 8:117–120

Mehta SK, Stowe RP, Feiveson AH, Tyring SK, Pierson DL (2000a) Reactivation and shedding of cytomegalovirus in astronauts during spaceflight. J Infect Dis 182(6):1761–1764

Mehta SK, Pierson DL, Cooley H, Dubow R, Lugg D (2000b) Epstein-Barr virus reactivation associated with diminished cell-mediated immunity in Antarctic expeditioners. J Med Virol 61(2):235–240

Mehta SK, Cohrs RJ, Forghani B, Zerbe G, Gilden DH, Pierson DL (2004) Stress-induced subclinical reactivation of varicella zoster virus in astronauts. J Med Virol 72(1):174–179

Mehta SK, Laudenslager ML, Stowe RP, Crucian BE, Sams CF, Pierson DL (2014) Multiple latent viruses reactivate in astronauts during Space Shuttle missions. Brain Behav Immun 41:210–217. https://doi.org/10.1016/j.bbi.2014.05.014

Mehta SK, Laudenslager ML, Stowe RP, Crucian BE, Feiveson AH, Sams CF, Pierson DL (2017) Latent virus reactivation in astronauts on the international space station. NPJ Microgravity 3:11. https://doi.org/10.1038/s41526-017-0015-y. eCollection 2017

Mills PJ, Meck JV, Waters WW, D'Aunno D, Ziegler MG (2001) Peripheral leukocyte subpopulations and catecholamine levels in astronauts as a function of mission duration. Psychosom Med 63(6):886–890

Nicogossian A, Sawin C, Huntoon C (1994) Overall physiologic response to spaceflight. In: Nicogossian A, Huntoon C, Pool S (eds) Space physiology and medicine, 3rd edn. Lea and Febiger, Philadelphia

Ortega MT, Pecaut MJ, Gridley DS, Stodieck LS, Ferguson V, Chapes SK (2009) Shifts in bone marrow cell phenotypes caused by spaceflight. J Appl Physiol 106(2):548–555

Payne DA, Mehta SK, Tyring SK, Stowe RP, Pierson DL (1999) Incidence of Epstein-Barr virus in astronaut saliva during spaceflight. Aviat Space Environ Med 70(12):1211–1213

Pecaut MJ, Nelson GA, Peters LL, Kostenuik PJ, Bateman TA, Morony S et al (2003) Genetic models in applied physiology: selected contribution: effects of spaceflight on immunity in the C57BL/6 mouse. I. Immune population distributions. J Appl Physiol 94(5):2085–2094

Pierson DL, Stowe RP, Phillips TM, Lugg DJ, Mehta SK (2005) Epstein-Barr virus shedding by astronauts during space flight. Brain Behav Immun 19(3):235–242

Rizzardi LF, Kunz H, Rubins K, Chouker A, Quiriarte H, Sams C, Crucian BE, Feinberg AP (2016) Evaluation of techniques for performing cellular isolation and preservation during microgravity conditions. NPJ Microgravity 2:16025. https://doi.org/10.1038/npjmgrav.2016.25. eCollection 2016

Rykova MP, Antropova EN, Larina IM, Morukov BV (2008) Humoral and cellular immunity in cosmonauts after the ISS missions. Acta Astronaut 63(7–10):697–705

Shimamiya T, Terada N, Hiejima Y, Wakabayashi S, Kasai H, Mohri M (2004) Effects of 10-day confinement on the immune system and psychological aspects in humans. J Appl Physiol 97:920–924

Simpson RJ, Bigley AB, Spielmann G, Kunz HE, Agha N, Baker F, Rooney B, Mylabathula PL, Graff RM, Crucian BE, Laughlin M, Mehta SK, Pierson DL (2016) Long duration spaceflight impairs NK-cell function in astronauts. Med Sci Sports Exerc 48(5 Suppl 1):87

Stowe RP, Sams CF, Mehta SK, Kaur I, Jones ML, Feeback DL et al (1999) Leukocyte subsets and neutrophil function after short-term spaceflight. J Leukoc Biol 65(2):179–186

Stowe RP, Mehta SK, Ferrando AA, Feeback DL, Pierson DL (2001) Immune responses and latent herpesvirus reactivation in spaceflight. Aviat Space Environ Med 72(10):884–891

Stowe RP, Sams CF, Pierson DL (2003) Effects of mission duration on neuroimmune responses in astronauts. Aviat Space Environ Med 74(12):1281–1284

Taylor GR, Neale LS, Dardano JR (1986) Immunological analyses of U.S. space shuttle crewmembers. Aviat Space Environ Med 57(3):213–217

Tingate TR, Lugg DJ, Muller HK, Stowe RP, Pierson DL (1997) Antarctic isolation: immune and viral studies. Immunol Cell Biol 75:275–283

Tracey KJ (2002) The inflammatory reflex. Nature 420:853–859

Innate Immunity Under the Exposome of Space Flight 12

Judith-Irina Buchheim, Matthias Feuerecker, and Alexander Choukér

12.1 Introduction

During evolution, life developed strategies for survival. Adaptation to the environment requires the formation of an efficient immune defense. Five hundred million years ago, jawed vertebrates developed the capability of genetic rearrangement and as such the adaptive immune system, a specific form of defense with the capacity to build-up a strong immunologic memory. Two hundred million years earlier, sponges and higher aquatic invertebrates (e.g. starfish) developed phagocytosis as well as the production and release of cytokines, the two most ancient functions belonging to innate immunity (Kvell et al. 2007). Whereas adaptive immune functions are found mostly in vertebrates, features of the innate immune system can be identified in all life on Earth. Its strategy lies in a standardized, less-specific but immediate reply to invading microorganisms without the set-up of a long-term memory (Murphy et al. 2012). The multiverse portfolio ranges from preformed surface proteins at the anatomical border to the environment, a fast detection of typical pathogenic cell wall components to short response times with preformed proteins and phagocytic cells. All these mechanisms are the result of constant learning and adaptation to the environment. As humankind moves on to explore the universe, it will also continue to develop "the human factor" of space exploration. We will aim for permanently manned space orbiter like the International Space Station (ISS) or the planned "Gateway" as well as further habitat missions on the moon and even Mars. As a likely consequence, more humans will become astronauts and exposed to this special living conditions aboard a space station that can be summarized as space "exposome" (Crucian et al. 2018; Wild 2012). Strong stressors of physical (radiation,

J.-I. Buchheim (✉) · M. Feuerecker · A. Choukér
Laboratory of Translational Research Stress and Immunity, Department of Anaesthesiology, Hospital of the University of Munich (LMU), Munich, Germany
e-mail: Judith-Irina.Buchheim@med.uni-muenchen.de

© Springer Nature Switzerland AG 2020

A. Choukér (ed.), *Stress Challenges and Immunity in Space*,
https://doi.org/10.1007/978-3-030-16996-1_12

microgravity), biological (bone and muscle loss, altered circadian rhythm and sleep), microbial (increased virulence of pathogens, allergic diathesis), or psychological (isolation, confinement) nature merge and influence health, mood, and performance.

During the Apollo and Skylab Missions, nearly all astronauts showed signs of infections on board or immediately following space flight (Johnston et al. 1975a, b; Michel et al. 1976). More than 40 years after these early findings it is clear that the exposure to space flight affects the immune system in many ways (Cogoli 1993; Gueguinou et al. 2009; Konstantinova et al. 1993; Sonnenfeld 1994). Studies have been conducted not only aboard space stations but also using terrestrial space analogs such as bed rest studies, parabolic flight, or Antarctic winter-over campaigns. Yet at the moment it remains unclear why these observed changes occur, how they are triggered mechanistically or how they can be prevented. This chapter will focus on the key players of human innate immunity, the known immune alterations derived from space analog research and space flight studies, and will identify promising candidates for future research.

12.2 The Armory of Innate Immunity

The well-preserved innate immune system is constitutively active and exerts an immediate, nonspecific immune reaction directly at the anatomical border to the environment. Under constant exposure to potential pathogens, tissue-specific mechanical (epithelial tight junctions, tears, longitudinal air flow), chemical (low pH, pulmonary surfactant) and microbial (commensal microbiota) barriers have developed. Specialized epithelia such as in skin, lungs or gut support the barrier. The mucosal epithelium of the respiratory tract carries specialized cilia to facilitate the transport of mucus, the viscous substance secreted by the epithelial cells, which contains various glycoproteins (mucins). They help preventing the adherence of bacteria and the movement of the cilia leads to an outward transport of pathogens engulfed in mucus. The low pH inside the stomach hinders bacteria from entering the abdomen and the microbiome of the gut is a fertile matrix for the diversity of beneficial commensal bacteria. In case of a skin wound, a coordinated process of healing will be started immediately by resident leukocytes to repair the defect. Preformed antimicrobial proteins and enzymes produced by epithelial cells help to disintegrate the cell wall of pathogens. Defensins are capable of instantly destroying the cellular wall of bacteria and fungi and can also be found in plants and insects. Phagocytes (e.g. macrophages) produce lysozyme, which is found in tears, in the small intestine and on the skin. Lysozyme specifically breaks polymers of glucose (glucans) found in the cellular wall of gram-positive bacteria (Murphy et al. 2012). After overcoming the epithelial barriers, the first line of defense, pathogens most likely encounter the complement system, plasma proteins that very efficiently mark (opsonize) microbes for phagocytes or eliminate bacteria directly through pore forming complexes (Heesterbeek et al. 2018). Interestingly, complement is also capable to regulate B and T cell function from inside the cell, a more recent discovery (West et al. 2018). As such it is responsible for cell homeostasis and regulation of apoptosis. To emphasize the growing role of complement, based on the recent discoveries, the variety of functions has led to the development of the term "complosome" (West et al. 2018). Special features of the cellular membrane of pathogens are, however, not only discovered and recognized by complement. These so-called pathogen-associated molecular patterns (PAMPs) and damage-associated molecular patterns (DAMPs) can be recognized by various pattern recognition receptors (PRRs) expressed on the surface of immune cells or secreted (Zhang et al. 2010). Secreted PRRs such as peptidoglycan recognition protein short (PGRP-S) and Pentraxin 3 (PTX-3), are soluble compounds stored in the granules of neutrophils that act as antibody-like molecules when secreted. PTX-3 is critical for resistance to *Aspergillus fumigatus* infection through its opsonic properties and activation of the complement cascade (Bottazzi et al. 2010). The large group of PRRs also count many well-studied receptor families such as the toll like receptors (TLRs) or the nucleotide-binding oligomerization domain (NOD)-like receptors (NLRs). The distinct TLRs detect different types of molecular patterns. They reside on the surface (e.g. for bacterial PAMPs) but also inside the cell to detect viral double stranded RNA (dsRNA). Specialized phagocytic receptors such as Dectin-1, highly expressed on the surface of all phagocytosing cells (see below), are capable to detect polymers of glucose. Dectin-1 recognizes β-1,3-linked glucans, a common component of most fungi (Murphy et al. 2012). Apart from phagocytosis, activation of Dectin-1 on the surface of neutrophils and macrophages triggers a cytokine response involving spleen tyrosine kinase (SYK), protein kinase C delta (PKC delta) and caspase activation and recruitment domain containing 9 (CARD9) (Hopke et al. 2016; Roth et al. 2016). Cytokines and chemokines are small proteins (~25 kDa) such as tumor necrosis factor (TNF), bradykinins or prostaglandins. When released, they induce a local increase in vessel permeability (edema formation) and further recruitment of leukocytes to the site of infection. The core body temperature, which is tightly regulated at 37°C by the hypothalamus, rises to facilitate microbe elimination (fever). This is mediated by so called endogenous pyrogens (e.g. TNF, interleukin 1β) that stimulate the production of prostaglandin E2 acting on the hypothalamus, which in turn enhances heat production of brown fat tissue and vasoconstriction (Murphy et al. 2012). The humoral arm of innate immunity in addition includes neutrophil extracellular traps (NETs). NETs are formed within minutes upon stimulation of phagocytes or hours upon cytolysis of phagocytes at sites of inflammation (Kumar and Sharma 2010), and are networks of extracellular structures that bind and kill microorganisms. NETs consist of nuclear contents such as neutrophil DNA and histones "decorated" with granules that exert various functions such as microbial recognition (e.g. PTX-3, PGRP-S) and antimicrobial activity (Mantovani et al. 2011). NETs have been shown to trap and kill microorganisms such as *Candida albicans*.

Specialized cells form the other part of innate immunity. There are different myeloid phagocytes known: granulocytes, macrophages, and dendritic cells (DC). Phagocytes can be tissue-specific residents e.g. tissue macrophages in lung, liver or intestines, circulating as precursors (monocytes) or patrolling non classical monocytes that get chemically attracted and migrate to the area of interest where they differentiate (Buscher et al. 2017; Murphy et al. 2012). In general, the key steps include the interaction of leukocytes with vascular endothelial cells through adhesion molecules such as β2-integrin and selectins, the recruitment to the perivascular space via transmigration and the recognition of microorganisms (Marki et al. 2015; Zarbock and Ley 2009) (Fig. 12.1).

The elimination of pathogens is achieved via several toxic chemicals (e.g. reactive oxygen species) and the activation of the complement system which results in opsonization of pathogens that facilitate the engulfment and subsequent destruction of microorganisms by phagocytes (see above). Granulocytes also called polymorphonuclear leukocytes (PMNs) due to their irregularly shaped nuclei, store different types of antimicrobial proteins inside their cytoplasmic granules. They are divided into neutrophils, eosinophils, and basophils according to their staining patterns. The strongest phagocytic activity is found in neutrophils where also the oxidative burst reaction, the generation of reactive oxygen species (ROS) for the elimination of pathogens takes place. PMNs and macrophages are capable to scavenge microbes also independently without prior opsonization. PMNs hold a G-protein coupled fMet-Leu-Phe (fMLP) receptor with high affinity for bacterial proteins starting with *N*-formyl methionine (fMet) residues. Upon activation, hydrolytic enzymes and reactive oxygen species (ROS) such as superoxide (O_2^-), H_2O_2 and many other highly toxic molecules are generated. These reactive compounds cause considerable damage not only to engulfed pathogens inside the cell but also to the tissue if they enter the extracellular space. The innate immune system is, in general, not solely involved in combating microbes but also responsible for eliminating dysfunctional host tissues and cells. Such cytotoxic functions, however, can lead under certain conditions to uncontrolled inflammation and tissue damage. This dual and sometimes conflicting role of innate immunity—to protect the host from infection as well as having the potential to be highly toxic to the host—can be described as a Janus-faced role of innate immunity (Nussler et al. 1999). (Fig. 12.2).

Interestingly, neutrophils are capable to sense the size of their microbial opponent and use NETs (see above) also to attack pathogens of bigger size that would be too big for phagocytosis as a whole (Branzk et al. 2014). *Candida*, *Aspergillus fumigatus* and other fungi are capable to shield β-glucans in their cellular wall from detection by dectin-1 (Carrion Sde et al. 2013). NETs are then specifically used by neutrophils to uncover β-glucans in fungi and initiate the immune response. Other leukocytes involved in innate immunity are natural killer (NK) cells. Especially during a viral host infection, cytokines like interferon (IFN) β that are released by macrophages or dendritic cells activate NK. They express invariant receptors on their surface and carry cytotoxic granules containing the pore forming protein perforin, which enables the deposition of proteases inside the target cell that induces programmed cell death (apoptosis). Almost all cells except red blood cells possess tissue-specific major histocompatibility cluster (MHC) I, a set of glycoproteins expressed on their surface. NK are capable to scan tissue cells for atypical or missing MHC I expression. An NK inhibitory signal is activated if a tissue cell expresses MHC I correctly. If this negative signal cannot be triggered by the tissue

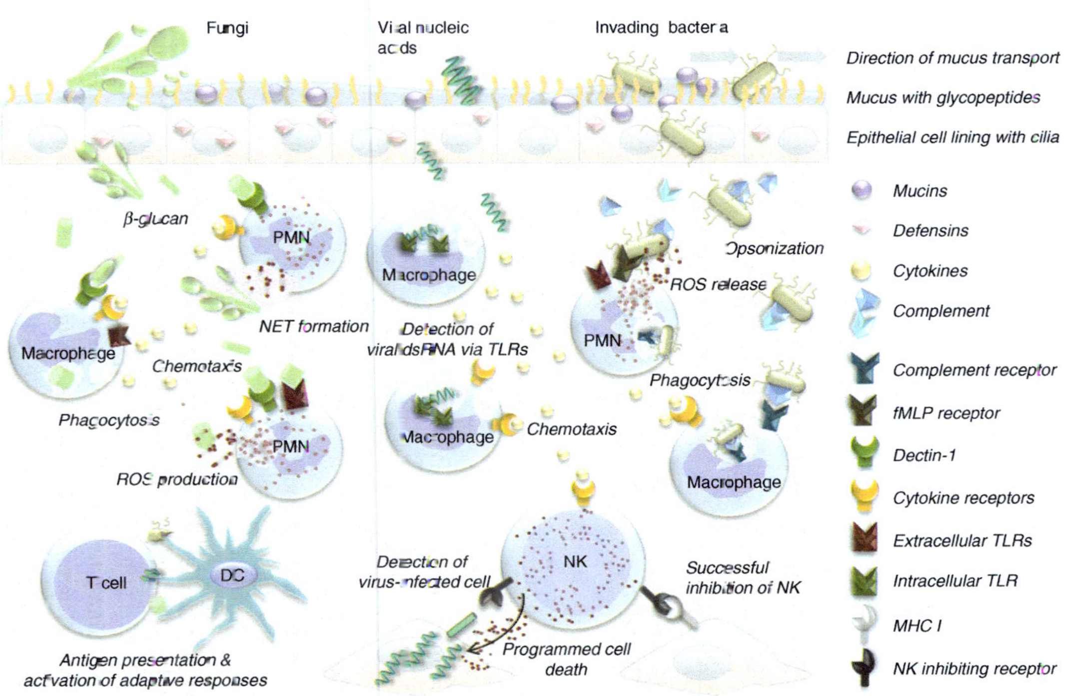

Fig. 12.1 Principles of innate immune responses against fungal, viral and bacterial pathogens: Exemplary lung epithelium with cilia and mucus transport as physical barrier. Preformed enzymes and peptides support the natural barrier function. The inflammatory response raised by pathogens include the recognition of pathogen structures by pattern recognition receptors and complement activation. Dectin-1 and Toll-like receptors (TLR) are innate immune-cell bound receptors. TLR recognize typical cell wall components of microbes (PAMPs) that are different from host molecules. Cell to cell interactions, release of cytokines promote cell recruitment and, together with complement proteins, phagocytosis and antigen presentation. Antigen presenting cells (APCs) such as dendritic cells (DC) are bridging the gap to activate adapted immune responses

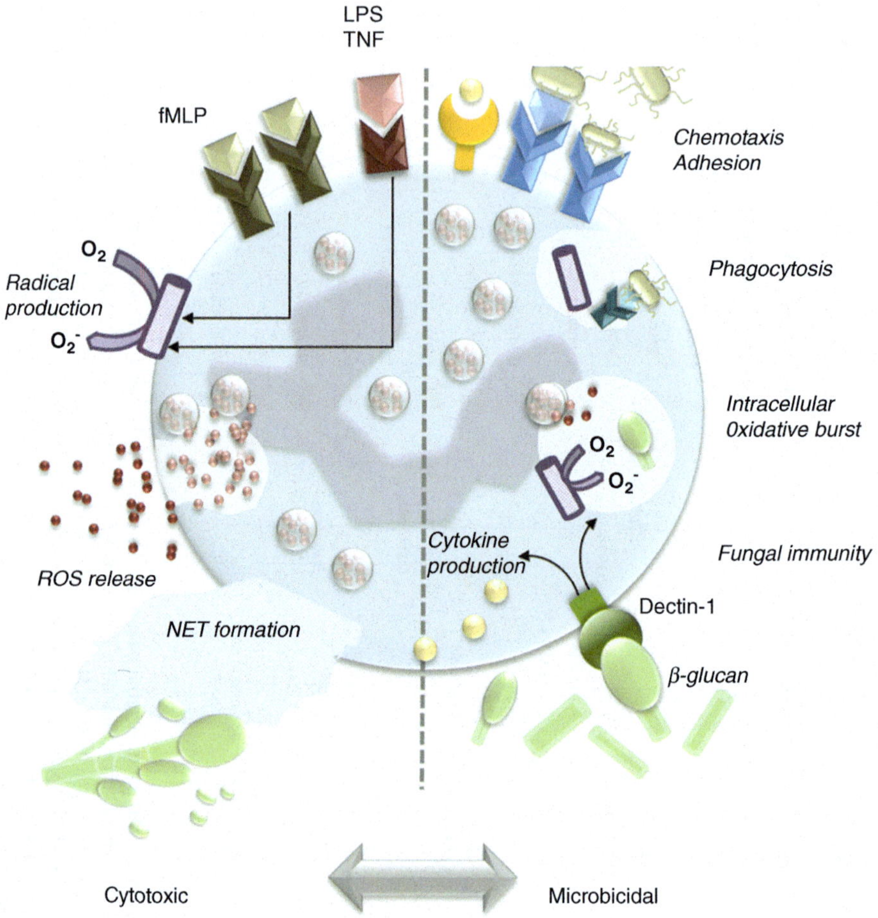

Fig. 12.2 Cytotoxic and microbicidal functions of PMNs. Tumor necrosis factor (TNF), *N*-formyl methionyl-leucyl-phenylalanine (fMLP), Neutrophil extracellular trap (NET)

cell e.g. during a viral infection, the NK is activated, and the elimination program started. Dendritic cells (DC), another group of phagocytes, are disintegrating pathogen components and presenting their foreign proteins on the surface in order to activate T cells, thus bridging the gap and transmitting information between innate and adaptive immunity. Therefore, they are also counted to the group of antigen presenting cells (APC) and are maybe the most important member.

The efficiency of innate immune responses is founded in combining different independent and at the same time communicating strategies. It is therefore not a "stand-alone" program within the human immune system but is involved in cross-talk with the adaptive immune system in particular, the activation, recruitment, differentiation, and proliferation of lymphocytes. This process is mediated principally through DC, which process microbial antigens and present them to naïve lymphocytes. This step is considered one of the most important activators of the adaptive immune response (see Chaps. 13–15). In turn, the adaptive immune system has reciprocal actions on cells of the innate immune system (Murphy et al. 2012).

12.3 The Innate Immune System in Space and Earth-Bound Space Analogs

Given the key role the innate immune system plays in (1) the immediate defense against invading pathogens, (2) activation of the adaptive immune system and (3) tissue inflammation, both downregulation and upregulation of the innate immune system could have significant adverse effects on the health of astronauts, and potentially disastrous consequences during deep space missions. Therefore, two critical questions that need to be addressed in preparation for future long-term missions are (1) whether the innate immune system of healthy individuals becomes dysfunctional during the evolutionary unforeseen and highly physically and psychologically stressful conditions of spaceflight and (2) whether the modulation of innate immune cell functions under conditions of spaceflight affects adaptive immunity?

12.3.1 Space Missions

Human space exploration involves short and long-term stays in an extreme, life-endangering environment triggering distinct immune responses. We base our knowledge mostly on data from pre- and postflight studies. This is due to the complex research environment, the restricted technical possibilities, limited man power and crew time as a cost factor as well as the considerable efforts necessary for conducting an in-flight study. But humans are not the only species to have ventured into space. Since the innate immune system originally evolved in lower order organisms, they serve as useful models for innate immunity research (e.g. fruit fly *Drosophila melanogaster*) both on Earth and in space (Taylor et al. 2014). Organisms such as rodents (Allebban et al. 1994; Baqai et al. 2009; Gridley et al. 2009; Pecaut et al. 2003; Ward et al. 2018) and amphibians (Bascove et al. 2011) have also been used to investigate certain aspects of immunity during and post spaceflight. The use of animal models allows for better control of experimental conditions and the need for only relatively small laboratory volumes onboard a space module. Clearly, however, animal models simplify many of the complex and variable physiological and emotional conditions associated with human spaceflight. Stress and associated neuroendocrine responses in animals, especially in lower order organisms, are unlikely to accurately reflect those that occur in humans. Furthermore, it has been observed that cellular gene expression can differ decisively between humans and rodents in response to different models of microgravity (Paulsen et al. 2015). Therefore, this chapter concentrates on studies undertaken in humans alone.

Since April 1961 more than 555 people have flown to space and all have experienced the conglomerate of psychological and physical stressors. Increased susceptibility to infection in astronauts was already reported during the Apollo era, when a surprisingly high incidence of infectious diseases or inflammation related symptoms on board or after spaceflight were reported (Johnston et al. 1975a). Many studies have tackled this issue and data from space analog research has helped greatly to understand single aspects of the multitude of stressors encountered in space (Pagel and Chouker 2016).

Only 24 astronauts have travelled beyond low-Earth orbit. During low-Earth orbit our home planet remains always in sight, there is live communication with ground control and relatives, the possibility of emergency evacuation, and shielding against solar and cosmic radiation by the Earth's magnetic field. None of these factors will be present during deep-space missions and the adverse physical and psychological effects when travelling beyond our "fore-garden," for example to Mars, are likely to be marked.

The *core body temperature* is kept within strict margins and controlled in the hypothalamus. Higher temperatures are observed during intense physical exercise or fever in relation to an inflammatory response (see above). A recent study showed a persistently elevated core body temperature in astronauts during long-term spaceflight and a faster temperature rise during physical exercise. Concomitantly, interleukin-1 receptor antagonist (IL-1ra) was elevated inflight suggesting a persistent low-grade pro-inflammatory state in response to the exposome inflight (Stahn et al. 2017). The effects in missions longer than 6 months remain unknown (see Chap. 26).

Neutrophils comprise the majority of circulating leukocytes (50–80%) and are amongst the first cells involved in the immune response to infection. Together with monocytes, they are the focus of this chapter, whilst the effect of spaceflight on NK is reviewed in Chap. 13. Analysis of peripheral leukocyte subsets immediately post spaceflight reveals, in general, an increase in neutrophil counts. However, most of this data comes from short term (less than 3 weeks) spaceflight experiments (Stowe et al. 1999; Taylor et al. 1986). More recent work showed that over the time course of a long-term mission on the ISS (6 months) a slight increase in the absolute neutrophil count occurred. Nevertheless, these counts are found doubled on day 1 after return, that quickly return to normal numbers (Crucian et al. 2015). This observed effect is likely triggered by the landing process, which can apply dramatic acute physical stress to the human body due to the microgravity, hypergravity and fierce vibration. Because cell counts do not necessarily correlate with cell function, Kaur et al. (2004) examined the oxidative burst capacity and the phagocytic abilities of neutrophils drawn from astronauts. There were no significant changes during 5 days missions whereas 9–11 days missions resulted in significantly reduced oxidative burst capacity upon activation and reduced ability to phagocytose *Escherichia coli* (*E. coli*). Preliminary results from the ISS-IMMUNO study have also shown that ROS production following stimulation with fMLP is preserved in neutrophils after 6 months in orbit but that neutrophils were found highly activated upon return (Buchheim et al. 2019). fMLP is a synthetic peptide that mimics bacterially derived peptides and binds to the fMLP receptor on neutrophils leading to their activation (see above).

Monocytes form the other major cell types of the innate immune system. One of their functions is to migrate rapidly to sites of inflammation where they then differentiate into specialized phagocytic cells. They primarily phagocytose not only pathogens but also apoptotic cells and debris. They also function as antigen presenting cells and promote the activation, proliferation and differentiation of naïve T-cells similar to dendritic cells that mainly function as antigen presenting cells. In contrast to the consistent changes seen in neutrophil counts, monocyte counts show conflicting results during spaceflight experiments. As for neutrophils, Kaur (Kaur et al. 2005) demonstrated a reduction in the ability of monocytes to phagocytose *E. coli* after short-duration spaceflight. In addition, the response of monocytes challenged with the gram-negative toxin lipopolysaccharide (LPS) was differentially modulated by spaceflight (Kaur et al. 2008). On the one hand, there was a reduction in intracellular pro-inflammatory cytokines such as interleukin-6 (IL-6) and interleukin-1 (IL-1) whilst on the other hand there was an increase in intracellular IL-8, a chemoattractant of the CXC-family. A larger study measuring 22 cytokines in astronauts during long-term flight showed increased levels of pro-inflammatory plasmatic cytokines such as TNF, IL-8, and IL-1ra (Crucian et al. 2014) that corroborate the findings of the increase in body temperature (Stahn et al. 2017). On the other hand, chronic exposure to stress modifies immunity as delayed mucosal wound healing and a decrease in the key cytokine IL-1β was previously reported during chronic stress (Marucha et al. 1998). It seems that the multitude of stressors during long-term spaceflight modify classic concepts of immunity and results from cell-based assays should be interpreted cautiously. Whilst the investigation of distinct cellular responses is essential to our understanding of basic immunology, there is a need to understand the clinical consequences of stressors on immunity in an integrated and comprehensive manner, including the crosstalk between the innate and adaptive systems. It is furthermore important to imply functional immune tests that better reflect immune function rather than blood levels of distinct cytokines. The classic stress hormone cortisol driven by the hypothalamic-pituitary-adrenal (HPA) axis or the catecholamines norepinephrine and epinephrine are potent modulators of immune responses (Sorrells and Sapolsky 2007). Therefore, a stressful environment can impair ROS production and monocytic activation leading to a defective adaptive immune response towards *Candida* (Buchheim et al. 2018). Macrophages are specifically perceptive of metabolic, endocrine or dietary modulations of the host environment as they express a tremendous amount of neuroendocrine surface receptors (reviewed in (Jurberg et al. 2018)) and show distinct functional phenotypes. The pro-inflammatory (classically activated) phenotype is triggered by IL-12 and IFN-γ and participating in immune defense or tissue repair, whereas the heterogenous group of anti-inflammatory type macrophages are triggered by Th2 cytokines such as IL-4, IL-10 promoting eosinophilic responses and atopy or glucocorticoids and TGF-β supporting healing processes (Jurberg et al. 2018). Recent reports highlight the higher incidence of dermatologic symptoms with allergies and hypersensitivities as well as the occurrence of persistent skin rashes aboard the ISS (Crucian et al. 2016a, b). The delayed type hypersensitivity skin test (DTH, Biomerieux) served as useful tool for monitoring the integrated cellular immune response. In this test, bacterial and fungal antigens are injected intracutaneously and the degree of skin erythema in response to the antigen challenge measured after 48 h. Spaceflight has two aspects of stress: acute versus chronic. A strong acute stress reaction because of a life-threatening event, such as the return to Earth after a longer stay aboard the ISS seems to be a potent immune activator that mobilizes immune cells regardless of the overall immune status. Chronic exposure to stress however leads to an activation of similar mechanisms over a longer time. Activation of the HPA axis and cortisol levels limit macrophage migration on the long-term but a short spike of cortisol may even stimulate it (Yeager et al. 2016). Similarly, T cells are redistributed to areas of action such as the blood stream or the skin during an acute stress response, since skin damage is a likely event during a fight or flight situation. Long-term activation, however, causes that fewer T cells respond leading to a reduced number of T cells in the skin. Gmünder (Gmunder et al. 1994) and others (Dhabhar 2002; Dhabhar and McEwen 1997) concomitantly observed a reduction in DTH skin reactivity during long term space flight. The cutaneous DTH test was phased out in 2002 due to both the risk of sensitization to the antigens and licensing issues. As such, a DTH-like in vitro test has been developed (Immumed Inc., Germany) and consists of the incubation of whole blood with antigens for a maximum of 48 h and the subsequent characterization of the cytokine response. Preliminary data using this assay performed with blood taken from astronauts after a 6 month mission (Buchheim et al. 2019) appear to strongly corroborate the reports from Stowe (Stowe et al. 2001) of significant attenuation of anti-viral immunity as demonstrated by viral reactivation during spaceflight (Chap. 19). Interestingly however, the whole blood immune response (TNF and IL-1b) following stimulation with e.g. fungal antigen mixture was higher post-flight compared to baseline and quantitatively higher than any other response (Fig. 12.3).

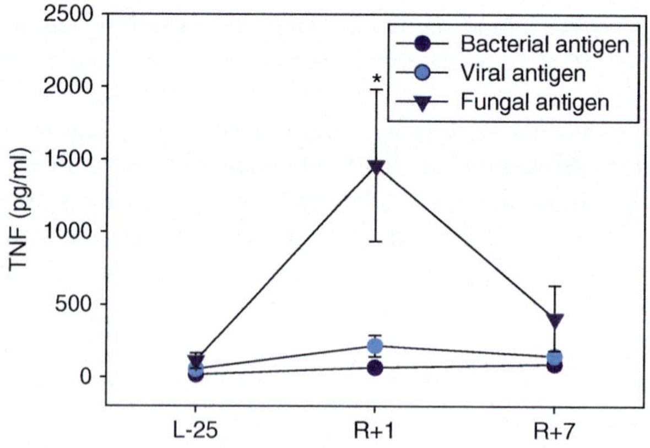

Fig. 12.3 DTH-like in vitro cytokine release assay after space flight. Whole blood was stimulated 1 month before launch (L − 25), on day 1 (R + 1) and day 7 after return (R + 7) with either bacterial, viral, or fungal antigens. After 48 h incubation at 37°C determination of quantitative TNF levels (pg/ml) was carried out in the supernatant. Data are presented as means ±SEM, $n = 10$, $*p < 0.05$ vs. baseline and viral or bacterial antigen. "Bacterial": Tetanus-, Diphtheria- and Pertussis-toxoid, "Viral": CMV-EBV-lysate, Influenza protein; "Fungal": *Candida* and *Trichophyton* lysate

Cytokine plasma concentrations of IL-1β were also found significantly increased in flight as seen before (Crucian et al. 2014) but showed baseline concentrations already on day 1 after return and could therefore not account for the higher amounts detected on this timepoint after stimulation (Buchheim et al. 2019). It remains unclear if this increase in fungal immune responses is due to (1) priming of neutrophils and monocytes, to (2) the higher activation of phagocytes in response to β-glucans in general, or due to (3) enhanced crosstalk between the adaptive and innate immune systems. The degree and type of microbial, especially fungal, colonization on the ISS ((Novikova et al. 2006) see also Chap. 25) as well as their strikingly more aggressive features (Knox et al. 2016; Romsdahl et al. 2018) (see Chap. 18) might be an additional contributing factor for the observed upregulation of fungal immune responses. Being exposed to a stressful and antigen-loaded environment together with a constant inflammatory response is not only threatening health on the short term. If exposure becomes "routine," a pathogenic immune phenotype can occur showing signs of premature immune aging (*Inflammaging* (Franceschi et al. 2000)) characterized by latent viral infections, allergic or autoimmune disease. While it remains completely unknown, whether long-term spaceflight might further enhance the incidence of this pro-inflammatory state, some typical features have been found in astronauts (Crucian et al. 2016a, b; Mehta et al. 2017a, b) and it was quantified that changes in thymopoiesis occur in association with stress in space (Benjamin et al. 2016) suggesting immunosenescence similar to the characteristics as seen with elderly patients (Bauer and Fuente Mde 2016; Pawelec and Gouttefangeas 2006; Sanada et al. 2018). This pro-inflammatory state together with risks of auto-immune disease will be the topic of future research. It is also an excellent way of enhancing knowledge transfer back to Earth as understanding the mechanistic pathogenesis will help astronauts on missions but also the aging population on Earth.

12.3.2 Space Analogs

Because of the high logistical challenges and operational costs associated with space missions, Earth-bound models with characteristics—mimicking specific spaceflight conditions—can serve as useful support research platforms for immune system changes (Chap. 36). Space relevant stressors such as microgravity, confinement, and isolation can be addressed individually or in their interactions. Such research platforms have been successfully used throughout the last few decades and allow for greater participant numbers as well as control groups.

A useful model to simulate conditions of short-lived micro-gravity is *parabolic flight*. A specially equipped airplane performs repeated parabolas resulting in short cycles (about 22 s each) of zero gravity followed by hypergravity (1.8 g). The ability of neutrophils to generate reactive oxygen species (ROS) both spontaneously and in response to activation with fMLP ± TNF was studied (Kaufmann et al. 2009). There was an increase in ROS generation by neutrophils in response to fMLP ± TNF after parabolic flight when compared to pre-flight. The spontaneous generation of ROS, however, remained unchanged. The ability of adenosine, an important suppressor of innate immune cytotoxic functions (Sitkovsky et al. 2004), to reduce ROS generation was augmented by parabolic flight. This response was modulated by an upregulation in Adenosine-2A receptor function. This "STOP" signal on ROS generation might be an evolved mechanism to limit uncontrolled cytotoxic devastation (Kaufmann et al. 2011). The stress hormone cortisol is capable to trigger the release of members of the endocannabinoid system, an important stress response system next to the HPA axis with multiple roles physiological processes of stress (Chap. 10, (Hill et al. 2010)). They are also potent immune modulators and shown to be powerful suppressors of innate and adaptive immunity (Buchheim et al. 2018). The anandamide *N*-arachidonoylethanolamide/anandamide (AEA) and 2-arachidonoylglycerol (2-AG) were found rapidly increased during parabolic flight (Chouker et al. 2010; Strewe et al. 2012). It remains to be seen whether the endocannabinoid system is involved in the immune alterations observed aboard the ISS but higher blood levels found in astronauts suggest that this is the case (Strewe et al. 2012).

A further model for gravitational deconditioning in space is 6° head down tilt (HDT) bed rest on Earth, in which the effects of microgravity can be mimicked. Unlike parabolic flight, bed rest allows for the investigation of the long-term effects of simulated microgravity on physiological systems. Head down tilt bed rest results in changes not only of the cardiovascular and musculoskeletal systems, but also the neuroendocrine and hematopoietic systems. One hundred and twenty days of 6° HDT bed rest led to increased levels of psychic stress and alteration of cortisol circadianicity. HDT bed rest also led to an increase in phagocyte and natural killer cell counts. Upregulation of the adhesion molecule β2-integrin on the surface of neutrophils and increased levels of IL-6 suggested a pro-inflammatory state (Chouker et al. 2001). β2-integrin allows neutrophils to attach to and roll along the vascular endothelium of inflamed sites. It is a key factor in neutrophil host defense functions. CD62L or L-selectin facilitates leukocyte recruitment into the perivascular space via binding to P-selectin glycoprotein ligand 1 (PSGL-1) on the surface of endothelial cells. Five days of HDT induced a noninflammatory shedding of L-selectin in granulocytes. No other functional immune effects were observed, suggesting a more mechanical shedding due to fluid shifts (Feuerecker et al. 2013). Other HDT bed rest studies have confirmed alterations in leukocyte subsets (Stowe et al. 2008), in addition to increased levels of the pro-inflammatory cytokine TNF (Shearer et al. 2009). The effect of bed rest on immunity is controversial and is likely to be dependent on different durations and conditions of bed rest (Crucian et al. 2009; Mehta et al. 2007).

The relationship between purinergic pathways and immunity and inflammation has not been explored in space so far but could reveal potential pharmacological targets to reduce uncontrolled cytotoxic responses by immune cells primed by stress and microgravity. Such a pharmacological approach could be beneficial as a "parachute against uncontrolled hyperinflammation" (Kaufmann et al. 2011). In the context of the *Planetary Habitat Simulation Study* (PlanHab), immobilization and hypoxia of healthy subjects was tested at the Olympic Sport Centre Planica

(Ratece, Slovenia), a model for future extra-terrestrial habitat design under reduced gravity. Interestingly, adenosine release was not only triggered under hypoxic conditions as hypothesized (Chap. 16) but was also detected under normoxic conditions although in lower concentrations. The reason for this increase in all groups remains to be elucidated; it might relate to dietary restrictions or muscle break down (Strewe et al. 2018).

The lack of gravitational force is only one of many stressors affecting humans in space. *Isolation and confinement* are powerful stressors that can have profound effects on the human psyche. Several studies have been performed in the last few decades testing the effects of variables such as habitat volume, mission duration and crew size and composition during group confinement on stress responses (reviewed in (Pagel and Chouker 2016)). For instance, the SFINCSS-99 (Simulation of Flight of International Crew on Space Station) study examined the effects of confinement on four healthy men living in a low-volume containment chamber for 110 and 240 days with limited contact with the outside world. The innate immune system showed a trend towards activation as demonstrated by an increase in neutrophil counts together with an upregulation of β2-integrin on the surface of neutrophils (Chouker et al. 2002). The use of the same analog facility continued with two studies that tested for the psychological and physiological, including immunological, consequences of extended duration of isolation and confinement. A short pilot study (Mars105) lasting for 105 days in 2009 was followed by a 520-day simulated Mars mission (Mars500, see also Chap. 37) in 2011. Due to the shorter isolation duration, many studies performed during Mars105 drew the conclusion that the simulation was not as stressful as expected on participants; the observed effects were therefore limited (Gemignani et al. 2014; Strewe et al. 2015). However, PMN related ROS production and cellular activation was increased in volunteers (Strewe et al. 2015). On the contrary, stress response systems (endocannabinoid system and salivary cortisol) were found activated during 520 days of isolation (Jacubowski et al. 2015; Yi et al. 2014, 2015), and could be triggered to higher levels shortly after the isolation period compared to control subjects when both groups were subjected to a parabolic flight experiment (Yi et al. 2015). Currently, this facility is being used to perform a terrestrial isolation experiment ("SIRIUS"), which simulates a lunar space mission and presence on a lunar station. This work will be relevant for the upcoming "Gateway" project (deployment of a cis-Lunar space station) for physiological research and the evaluation of potential countermeasures.

The consequences of prolonged isolation and confinement on the health of the crew members allows us to judge the risks resulting from a real Mars mission and develop appropriate mitigation strategies. Over the last decade, Antarctica has also been identified as a suitable platform to investigate the effects of "real-life" hostile confinement on physiology and is considered a highly useful analog for lunar and interplanetary missions. In addition, Antarctic bases at high altitude, such as Concordia station at ~3200 m, serve as useful platforms to test for additional space-relevant stressors such as hypoxia in combination with isolation and confinement (Chaps. 16 and 38).

12.4 Immune Crosstalk

During the evolution of the immune system, adaptive immunity becomes apparent in vertebrates, which, contrarily to innate immunity, is characterized by a more specific response with a certain time delay and excellent memory function. As our knowledge on the phylogeny of both systems and their features expands, we have come to realize that one system depends on the other, communication between both is critical and cell families that we defined as adaptive have members with innate features that shift them to the innate system and vice versa (e.g. innate lymphoid cells (ILCs)). Also, distinct features of adaptive immunity have been found in non-vertebrates. We must accept that separating the two systems is rather artificial and borders become more and more blurred (Criscitiello and de Figueiredo 2013; Kvell et al. 2007). However, in lack of a better systematic, we will have to continue to apply the two-system classification.

Communication between cells, tissues, and organs can occur through different mechanisms. The most evident is cell–cell interaction via the use of surface proteins such as receptors or the release of signaling cytokines for recruitment and chemotaxis. One of the most important intercellular communication scenarios is the antigen presentation by dendritic cells to T cells leading to their activation and proliferation and the initiation of adaptive responses (Chaps. 11 and 14). Multiple signals are necessary to trigger a T cell response and the subsequent differentiation of T cell subsets (Murphy et al. 2012). PMNs were shown to modulate the maturation, activation and survival of NK, as well as their cytotoxic and their pro-inflammatory capacity. The latter is predominantly mediated through the production of IFN-γ and the direct interaction of PMNs with NK through the intercellular adhesion molecules ICAM-3 and CD18-CD11b. The relationship between neutrophils and NK is not unidirectional however, as NK are capable of priming neutrophils to produce reactive oxygen species and pro-inflammatory cytokines, resulting overall in an efficient "defensive alliance" (Costantini and Cassatella 2011). The interplay between the adaptive and the innate immune systems can be seen in the efficient co-recruitment of cells of the innate and the adaptive immune system to sites of infection. Distinct T cell populations (e.g. regulatory T cells, Th17 cells) are capable of modifying the activation status and survival of PMNs, predominantly through the release of granulocyte monocyte colony stimulating factor (GM-CSF) and TNF (Mantovani et al. 2011). Kaufmann et al. (2009) demonstrated that elevated blood concentrations of GM-CSF and TNF were associated with priming of neutrophils after gravitational stress. Reciprocally, T cells are guided by activated PMNs to sites of inflammation by a battery of CC and CXC (Rot and von Andrian 2004). Furthermore, activated neutrophils serve as a major source of mediators that affect B cell differentiation and functions e.g. through B cell activation factor (BAFF) (Scapini et al. 2008). More recently, inter-cellular communication via immune cell derived extracellular vesicles (EVs) have come into focus. EVs are a heterogenic group of vesicles characterized by submicron size, lipid bilayer structure, and transport of proteins and nucleic acids (Groot Kormelink et al. 2018). Smaller vesicles so called exosomes range in a size of 30–100 nm. They have been shown to contribute to innate immunity and participate in antigen recognition through PRRs (Kesimer et al. 2009; Kouwaki et al. 2017). The cargo of exosomes is characteristically different depending on the tissue or cell of origin. Frequently, exosomes deliver host and viral mRNA or microRNAs (miRNA) to target cells. As such, virus-infected cells were shown to deliver viral RNA to dendritic cells and macrophages where an immune response is triggered (Kouwaki et al. 2017). This form of communication, if unintentionally sustained, can lead to a hyperinflammatory state and trigger autoimmunity. As such, a failed regulation and therefore constant maintenance of an IFN-γ signal via host miRNAs in secreting DCs was found to promote autoimmune disease (Salvi et al. 2018). This exciting new concept of communication remains to be explored under spaceflight conditions. But given the assumption of a dysregulated and maybe hyperinflammatory state in astronauts, it is likely that future research will take a closer look.

Besides, immune modulation is possible due to the influence of other systems such as stress hormones, epinephrine and norepinephrine or endocannabinoids and their acting receptors (Buchheim et al. 2018; Jurberg et al. 2018). Many of these neuroendocrine receptors are found on the surface of macrophages that respond with shifts in phenotype and distinct functions. Although modulation of basic phagocyte functions has been shown to occur as a result of spaceflight, further research is needed to evaluate to what degree some of the more complex armory of the innate immune system and its relationships and crosstalk with cells of the adaptive immune system are altered.

12.5 Summary

Humanity always strove to explore unknown territory and to tap out its full potential. Long-duration missions to the Earth's moon and Mars are the next goal for space exploration. One of the major concerns is the effect of the biophysical interplay of mechanistic stimuli, psychoneuroendocrine burdens and microbial attack, the so-called space exposome, on the innate human immune system. What is known today is that spaceflight induces shifts in percental populations and absolute numbers of peripheral leukocytes, hampers leukocyte function and pushes cytokine profiles to an inflammatory state together with a higher core body temperature. Viral shedding correlates with plasma cytokine alterations and functional impairment of immune responses after antigen stimulation. Astronauts are prone for fungal infections, since high atmospheric load with potentially pathogenic fungi meet with incapacitated or dysregulated immune functions. The changes in innate immune responses may also result from a complex crosstalk with adaptive immune responses. To what degree upregulation of certain aspects of innate immunity can be considered a compensatory mechanism for adaptive immune suppression, and vice versa, in an effort to maintain immunological homeostasis under conditions of stress is unclear and will need a more integrative approach to immunology research in space. Furthermore, the infrastructure aboard needs to be enhanced as such that test platforms and newer monitoring tools can become a standard resource for astronauts.

This can and of course must initially be tested in the context of research but should follow down the path to a standard medical and diagnosing application. Long-term space missions away from Earth will inevitably cause a rise in clinical symptoms and need for specific treatments due to longer mission duration and limited resources. Questionnaires that allow the anonymous reporting of incidence, persistence and duration of clinical symptoms should be used on a general scale. Little do researchers know about the frequency of symptoms related to clinical disease although they are the basis for designing studies and appropriate countermeasures. Medical confidentiality is of course of central importance and given the small population of active astronauts worldwide, it seems logical that underreporting and unavailability of incidence numbers is a fact. Terrestrial spaceflight analog platforms can help to increase number of participants, generate data about incidence and severity of symptoms and facilitate the implementation of new procedures or devices before going to space. Nevertheless, all gathered information and every promising countermeasure needs verification in the unique space environment. Only then we can make sure that lessons are learned, and we are ready to travel beyond our current horizon.

Acknowledgments This work has been supported by the DLR on behalf of the Federal Ministry of Economics and Technology (BMWi 50WB0719 and 50WB0919, 50WB1319), the European Space Agency (ESA's ELIPS 3 and 4 program, and E3P), the Russian Space Agency (Roscosmos), and the program of fundamental research (theme 65.1) of the Institute for Biomedical Problems (IBMP). We explicitly thank all the helping hands, operators, scientists, and administrators at Roscosmos, IBMP, TsNIIMash in Russia, at ESA, CNES, and DLR as well as the NASA Kennedy Space Centre and the Johnsons' Space Centre, who made this project possible. Our highest appreciation is expressed to the ISS crews who have participated with motivation and who realized these studies with outstanding professionalism.

In honor and memory of Cosmonaut Dr. Boris Morukov former vice director of IBMP and mission director of Mars500 who passed away in 2015.

References

Allebban Z, Ichiki AT, Gibson LA, Jones JB, Congdon CC, Lange RD (1994) Effects of spaceflight on the number of rat peripheral blood leukocytes and lymphocyte subsets. J Leukoc Biol 55:209–213

Baqai FP et al (2009) Effects of spaceflight on innate immune function and antioxidant gene expression. J Appl Physiol 106:1935–1942. https://doi.org/10.1152/japplphysiol.91361.2008

Bascove M, Gueguinou N, Schaerlinger B, Gauquelin-Koch G, Frippiat JP (2011) Decrease in antibody somatic hypermutation frequency under extreme, extended spaceflight conditions. FASEB J 25:2947–2955. https://doi.org/10.1096/fj.11-185215

Bauer ME, Fuente Mde L (2016) The role of oxidative and inflammatory stress and persistent viral infections in immunosenescence. Mech Ageing Dev 158:27–37. https://doi.org/10.1016/j.mad.2016.01.001

Benjamin CL et al (2016) Decreases in thymopoiesis of astronauts returning from space flight. JCI Insight 1:e88787. https://doi.org/10.1172/jci.insight.88787

Bottazzi B, Doni A, Garlanda C, Mantovani A (2010) An integrated view of humoral innate immunity: pentraxins as a paradigm. Annu Rev Immunol 28:157–183. https://doi.org/10.1146/annurev-immunol-030409-101305

Branzk N, Lubojemska A, Hardison SE, Wang Q, Gutierrez MG, Brown GD, Papayannopoulos V (2014) Neutrophils sense microbe size and selectively release neutrophil extracellular traps in response to large pathogens. Nat Immunol 15:1017–1025. https://doi.org/10.1038/ni.2987

Buchheim J-I et al (2018) Oxidative burst and Dectin-1-triggered phagocytosis affected by norepinephrine and endocannabinoids: implications for fungal clearance under stress. Int Immunol 30:79–89. https://doi.org/10.1093/intimm/dxy001

Buchheim J-I et al (2019) Stress related shift toward inflammaging in cosmonauts after long-duration space flight. Front Physiol 10:85. https://doi.org/10.3389/fphys.2019.00085

Buscher K, Marcovecchio P, Hedrick CC, Ley K (2017) Patrolling mechanics of non-classical monocytes in vascular inflammation. Front Cardiovasc Med 4:80. https://doi.org/10.3389/fcvm.2017.00080

Carrion Sde J, Leal SM Jr, Ghannoum MA, Aimanianda V, Latge JP, Pearlman E (2013) The RodA hydrophobin on Aspergillus fumigatus spores masks dectin-1- and dectin-2-dependent responses and enhances fungal survival in vivo. J Immunol 191:2581–2588. https://doi.org/10.4049/jimmunol.1300748

Chouker A et al (2001) Simulated microgravity, psychic stress, and immune cells in men: observations during 120-day 6 degrees HDT. J Appl Physiol 90:1736–1743. https://doi.org/10.1152/jappl.2001.90.5.1736

Chouker A et al (2002) Effects of confinement (110 and 240 days) on neuroendocrine stress response and changes of immune cells in men. J Appl Physiol 92:1619–1627. https://doi.org/10.1152/japplphysiol.00732.2001

Chouker A et al (2010) Motion sickness, stress and the endocannabinoid system. PLoS One 5:e10752. https://doi.org/10.1371/journal.pone.0010752

Cogoli A (1993) The effect of space flight on human cellular immunity. Environ Med 37: 107–116

Costantini C, Cassatella MA (2011) The defensive alliance between neutrophils and NK cells as a novel arm of innate immunity. J Leukoc Biol 89:221–233. https://doi.org/10.1189/jlb.0510250

Criscitiello MF, de Figueiredo P (2013) Fifty shades of immune defense. PLoS Pathog 9:e1003110. https://doi.org/10.1371/journal.ppat.1003110

Crucian BE et al (2009) Immune status, latent viral reactivation, and stress during long-duration head-down bed rest. Aviat Space Environ Med 80:A37–A44

Crucian BE et al (2014) Plasma cytokine concentrations indicate that in vivo hormonal regulation of immunity is altered during long-duration spaceflight. J Interferon Cytokine Res 34:778–786. https://doi.org/10.1089/jir.2013.0129

Crucian B, Stowe RP, Mehta S, Quiriarte H, Pierson D, Sams C (2015) Alterations in adaptive immunity persist during long-duration spaceflight. NPJ Microgravity 1:15013. https://doi.org/10.1038/npjmgrav.2015.13

Crucian B, Babiak-Vazquez A, Johnston S, Pierson DL, Ott CM, Sams C (2016a) Incidence of clinical symptoms during long-duration orbital spaceflight. Int J Gen Med 9:383–391. https://doi.org/10.2147/IJGM.S114188

Crucian B et al (2016b) A case of persistent skin rash and rhinitis with immune system dysregulation onboard the International Space Station. J Allergy Clin Immunol Pract 4:759–762.e758. https://doi.org/10.1016/j.jaip.2015.12.021

Crucian BE et al (2018) Immune system dysregulation during spaceflight: potential countermeasures for deep space exploration missions. Front Immunol 9:1437. https://doi.org/10.3389/fimmu.2018.01437

Dhabhar FS (2002) Stress-induced augmentation of immune function—the role of stress hormones, leukocyte trafficking, and cytokines. Brain Behav Immun 16:785–798

Dhabhar FS, McEwen BS (1997) Acute stress enhances while chronic stress suppresses cell-mediated immunity in vivo: a potential role for leukocyte trafficking. Brain Behav Immun 11:286–306. https://doi.org/10.1006/brbi.1997.0508

Feuerecker M et al (2013) Five days of head-down-tilt bed rest induces noninflammatory shedding of L-selectin. J Appl Physiol 115:235–242. https://doi.org/10.1152/japplphysiol.00381.2013

Franceschi C, Bonafe M, Valensin S, Olivieri F, De Luca M, Ottaviani E, De Benedictis G (2000) Inflamm-aging. An evolutionary perspective on immunosenescence. Ann N Y Acad Sci 908:244–254

Gemignani A et al (2014) How stressful are 105 days of isolation? Sleep EEG patterns and tonic cortisol in healthy volunteers simulating manned flight to Mars. Int J Psychophysiol 93:211–219. https://doi.org/10.1016/j.ijpsycho.2014.04.008

Gmunder FK, Konstantinova I, Cogoli A, Lesnyak A, Bogomolov W, Grachov AW (1994) Cellular immunity in cosmonauts during long duration spaceflight on board the orbital MIR station. Aviat Space Environ Med 65:419–423

Gridley DS et al (2009) Spaceflight effects on T lymphocyte distribution, function and gene expression. J Appl Physiol 106:194–202. https://doi.org/10.1152/japplphysiol.91126.2008

Groot Kormelink T, Mol S, de Jong EC, Wauben MHM (2018) The role of extracellular vesicles when innate meets adaptive. Semin Immunopathol. https://doi.org/10.1007/s00281-018-0681-1

Gueguinou N, Huin-Schohn C, Bascove M, Bueb JL, Tschirhart E, Legrand-Frossi C, Frippiat JP (2009) Could spaceflight-associated immune system weakening preclude the expansion of human presence beyond Earth's orbit? J Leukoc Biol 86:1027–1038. https://doi.org/10.1189/jlb.0309167

Heesterbeek DAC, Angelier ML, Harrison RA, Rooijakkers SHM (2018) Complement and bacterial infections: from molecular mechanisms to therapeutic applications. J Innate Immun 10:455–464. https://doi.org/10.1159/000491439

Hill MN, Karatsoreos IN, Hillard CJ, McEwen BS (2010) Rapid elevations in limbic endocannabinoid content by glucocorticoid hormones in vivo. Psychoneuroendocrinology 35:1333–1338. https://doi.org/10.1016/j.psyneuen.2010.03.005

Hopke A, Nicke N, Hidu EE, Degani G, Popolo L, Wheeler RT (2016) Neutrophil attack triggers extracellular trap-dependent Candida cell wall remodeling and altered immune recognition. PLoS Pathog 12:e1005644. https://doi.org/10.1371/journal.ppat.1005644

Jacubowski A et al (2015) The impact of long-term confinement and exercise on central and peripheral stress markers. Physiol Behav 152:106–111. https://doi.org/10.1016/j.physbeh.2015.09.017

Johnston RS, Dietlein LF, Berry CA (1975a) Biomedical results of Apollo. Nasa Sp 368. Scientific and Technical Information Office, National Aeronautics and Space Administration: for sale by the Supt. of Docs., U.S. Govt. Printing Office, Washington, DC

Johnston RS, Dietlein LF, Charles AB (1975b) Biomedical results of Apollo. NASA, Washington, DC

Jurberg AD, Cotta-de-Almeida V, Temerozo JR, Savino W, Bou-Habib DC, Riederer I (2018) Neuroendocrine control of macrophage development and function. Front Immunol 9:1440. https://doi.org/10.3389/fimmu.2018.01440

Kaufmann I, Schachtner T, Feuerecker M, Schelling G, Thiel M, Chouker A (2009) Parabolic flight primes cytotoxic capabilities of polymorphonuclear leucocytes in humans. Eur J Clin Investig 39:723–728. https://doi.org/10.1111/j.1365-2362.2009.02136.x

Kaufmann I, Feuerecker M, Salam A, Schelling G, Thiel M, Chouker A (2011) Adenosine A2(A) receptor modulates the oxidative stress response of primed polymorphonuclear leukocytes after parabolic flight. Hum Immunol 72:547–552. https://doi.org/10.1016/j.humimm.2011.03.021

Kaur I, Simons ER, Castro VA, Mark Ott C, Pierson DL (2004) Changes in neutrophil functions in astronauts. Brain Behav Immun 18:443–450. https://doi.org/10.1016/j.bbi.2003.10.005

Kaur I, Simons ER, Castro VA, Ott CM, Pierson DL (2005) Changes in monocyte functions of astronauts. Brain Behav Immun 19:547–554. https://doi.org/10.1016/j.bbi.2004.12.006

Kaur I, Simons ER, Kapadia AS, Ott CM, Pierson DL (2008) Effect of spaceflight on ability of monocytes to respond to endotoxins of gram-negative bacteria. Clin Vaccine Immunol 15:1523–1528. https://doi.org/10.1128/CVI.00065-08

Kesimer M et al (2009) Characterization of exosome-like vesicles released from human tracheobronchial ciliated epithelium: a possible role in innate defense. FASEB J 23:1858–1868. https://doi.org/10.1096/fj.08-119131

Knox BP et al (2016) Characterization of Aspergillus fumigatus isolates from air and surfaces of the International Space Station. mSphere 1. https://doi.org/10.1128/mSphere.00227-16

Konstantinova IV, Rykova MP, Lesnyak AT, Antropova EA (1993) Immune changes during long-duration missions. J Leukoc Biol 54:189–201

Kouwaki T, Okamoto M, Tsukamoto H, Fukushima Y, Oshiumi H (2017) Extracellular vesicles deliver host and virus RNA and regulate innate immune response. Int J Mol Sci 18. https://doi.org/10.3390/ijms18030666

Kumar V, Sharma A (2010) Neutrophils: Cinderella of innate immune system. Int Immunopharmacol 10:1325–1334. https://doi.org/10.1016/j.intimp.2010.08.012

Kvell K, Cooper EL, Engelmann P, Bovari J, Nemeth P (2007) Blurring borders: innate immunity with adaptive features. Clin Dev Immunol 2007:83671. https://doi.org/10.1155/2007/83671

Mantovani A, Cassatella MA, Costantini C, Jaillon S (2011) Neutrophils in the activation and regulation of innate and adaptive immunity. Nat Rev Immunol 11:519–531. https://doi.org/10.1038/nri3024

Marki A, Esko JD, Pries AR, Ley K (2015) Role of the endothelial surface layer in neutrophil recruitment. J Leukoc Biol 98:503–515. https://doi.org/10.1189/jlb.3MR0115-011R

Marucha PT, Kiecolt-Glaser JK, Favagehi M (1998) Mucosal wound healing is impaired by examination stress. Psychosom Med 60:362–365

Mehta SK, Crucian B, Pierson DL, Sams C, Stowe RP (2007) Monitoring immune system function and reactivation of latent viruses in the Artificial Gravity Pilot Study. J Gravit Physiol 14:P21–P25

Mehta SK, Laudenslager ML, Stowe RP, Crucian BE, Feiveson AH, Sams CF, Pierson DL (2017a) Latent virus reactivation in astronauts on the international space station. NPJ Microgravity 3:11. https://doi.org/10.1038/s41526-017-0015-y

Mehta SK et al (2017b) Localization of VZV in saliva of zoster patients. J Med Virol 89:1686–1689. https://doi.org/10.1002/jmv.24807

Michel EL, Johnston RS, Dietlein LF (1976) Biomedical results of the Skylab Program. Life Sci Space Res 14:3–18

Murphy K, Travers P, Walport M, Janeway C (2012) Janeway's immunobiology, 8th edn. Garland Science, New York

Novikova N et al (2006) Survey of environmental biocontamination on board the International Space Station. Res Microbiol 157:5–12. https://doi.org/10.1016/j.resmic.2005.07.010

Nussler AK, Wittel UA, Nussler NC, Beger HG (1999) Leukocytes, the Janus cells in inflammatory disease. Langenbeck's Arch Surg 384:222–232

Pagel JI, Chouker A (2016) Effects of isolation and confinement on humans-implications for manned space explorations. J Appl Physiol 120:1449–1457. https://doi.org/10.1152/japplphysiol.00928.2015

Paulsen K et al (2015) Regulation of ICAM-1 in cells of the monocyte/macrophage system in microgravity. Biomed Res Int 2015:538786. https://doi.org/10.1155/2015/538786

Pawelec G, Gouttefangeas C (2006) T-cell dysregulation caused by chronic antigenic stress: the role of CMV in immunosenescence? Aging Clin Exp Res 18:171–173

Pecaut MJ et al (2003) Genetic models in applied physiology: selected contribution: effects of spaceflight on immunity in the C57BL/6 mouse. I. Immune population distributions. J Appl Physiol 94:2085–2094. https://doi.org/10.1152/japplphysiol.01052.2002

Romsdahl J et al (2018) Characterization of Aspergillus niger Isolated from the International Space Station. mSystems 3. https://doi.org/10.1128/mSystems.00112-18

Rot A, von Andrian UH (2004) Chemokines in innate and adaptive host defense: basic chemokinese grammar for immune cells. Annu Rev Immunol 22:891–928. https://doi.org/10.1146/annurev.immunol.22.012703.104543

Roth S et al (2016) Vav proteins are key regulators of Card9 signaling for innate antifungal immunity. Cell Rep 17:2572–2583. https://doi.org/10.1016/j.celrep.2016.11.018

Salvi V et al (2018) Exosome-delivered microRNAs promote IFN-alpha secretion by human plasmacytoid DCs via TLR7. JCI Insight 3. https://doi.org/10.1172/jci.insight.98204

Sanada F, Taniyama Y, Muratsu J, Otsu R, Shimizu H, Rakugi H, Morishita R (2018) Source of chronic inflammation in aging. Front Cardiovasc Med 5:12. https://doi.org/10.3389/fcvm.2018.00012

Scapini P, Bazzoni F, Cassatella MA (2008) Regulation of B-cell-activating factor (BAFF)/B lymphocyte stimulator (BLyS) expression in human neutrophils. Immunol Lett 116:1–6. https://doi.org/10.1016/j.imlet.2007.11.009
Shearer WT et al (2009) Immune responses in adult female volunteers during the bed-rest model of spaceflight: antibodies and cytokines. J Allergy Clin Immunol 123:900–905. https://doi.org/10.1016/j.jaci.2008.12.016
Sitkovsky MV et al (2004) Physiological control of immune response and inflammatory tissue damage by hypoxia-inducible factors and adenosine A2A receptors. Annu Rev Immunol 22:657–682. https://doi.org/10.1146/annurev.immunol.22.012703.104731
Sonnenfeld G (1994) Effect of space flight on cytokine production. Acta Astronaut 33:143–147
Sorrells SF, Sapolsky RM (2007) An inflammatory review of glucocorticoid actions in the CNS. Brain Behav Immun 21:259–272. https://doi.org/10.1016/j.bbi.2006.11.006
Stahn AC et al (2017) Increased core body temperature in astronauts during long-duration space missions. Sci Rep 7:16180. https://doi.org/10.1038/s41598-017-15560-w
Stowe RP, Sams CF, Mehta SK, Kaur I, Jones ML, Feeback DL, Pierson DL (1999) Leukocyte subsets and neutrophil function after short-term spaceflight. J Leukoc Biol 65:179–186
Stowe RP, Mehta SK, Ferrando AA, Feeback DL, Pierson DL (2001) Immune responses and latent herpesvirus reactivation in spaceflight. Aviat Space Environ Med 72:884–891
Stowe RP, Yetman DL, Storm WF, Sams CF, Pierson DL (2008) Neuroendocrine and immune responses to 16-day bed rest with realistic launch and landing G profiles. Aviat Space Environ Med 79:117–122
Strewe C et al (2012) Effects of parabolic flight and spaceflight on the endocannabinoid system in humans. Rev Neurosci 23:673–680. https://doi.org/10.1515/revneuro-2012-0057
Strewe C et al (2015) Functional changes in neutrophils and psychoneuroendocrine responses during 105 days of confinement. J Appl Physiol 118:1122–1127. https://doi.org/10.1152/japplphysiol.00755.2014
Strewe C et al (2018) PlanHab study: consequences of combined normobaric hypoxia and bed rest on adenosine kinetics. Sci Rep 8:1762. https://doi.org/10.1038/s41598-018-20045-5
Taylor GR, Neale LS, Dardano JR (1986) Immunological analyses of U.S. Space Shuttle crewmembers. Aviat Space Environ Med 57:213–217
Taylor K et al (2014) Toll mediated infection response is altered by gravity and spaceflight in Drosophila. PLoS One 9:e86485. https://doi.org/10.1371/journal.pone.0086485
Ward C, Rettig TA, Hlavacek S, Bye BA, Pecaut MJ, Chapes SK (2018) Effects of spaceflight on the immunoglobulin repertoire of unimmunized C57BL/6 mice. Life Sci Space Res 16:63–75. https://doi.org/10.1016/j.lssr.2017.11.003
West EE, Kolev M, Kemper C (2018) Complement and the regulation of T cell responses. Annu Rev Immunol 36:309–338. https://doi.org/10.1146/annurev-immunol-042617-053245
Wild CP (2012) The exposome: from concept to utility. Int J Epidemiol 41:24–32. https://doi.org/10.1093/ije/dyr236
Yeager MP, Pioli PA, Collins J, Barr F, Metzler S, Sites BD, Guyre PM (2016) Glucocorticoids enhance the in vivo migratory response of human monocytes. Brain Behav Immun 54:86–94. https://doi.org/10.1016/j.bbi.2016.01.004
Yi B et al (2014) 520-d Isolation and confinement simulating a flight to Mars reveals heightened immune responses and alterations of leukocyte phenotype. Brain Behav Immun 40:203–210. https://doi.org/10.1016/j.bbi.2014.03.018
Yi B et al (2015) The impact of chronic stress burden of 520-d isolation and confinement on the physiological response to subsequent acute stress challenge. Behav Brain Res 281:111–115. https://doi.org/10.1016/j.bbr.2014.12.011
Zarbock A, Ley K (2009) Neutrophil adhesion and activation under flow. Microcirculation 16:31–42. https://doi.org/10.1080/10739680802350104
Zhang Q et al (2010) Circulating mitochondrial DAMPs cause inflammatory responses to injury. Nature 464:104–107. https://doi.org/10.1038/nature08780

NK Cell Assessments: A 40-Years-Old History of Immune–Stress Interaction in Space with a Promising Future

13

Marina Rykova, Eugenia Antropova, Boris Morukov†, and Sergey Ponomarev

Natural killer cells (NK) are special cytotoxic lymphocytes which constitute a major component of the innate immune system. NK were first described in the early 1970s on a functional basis according to their ability to lyse tumor cells in the absence of prior stimulation (Herberman 1974). Today, NK are widely viewed as efficient non-MHC-restricted effector cells which are capable of patrolling the host and getting involved in the defensive action against infectious agents, tumour cells, and certain immature normal cells. In addition, recent research highlights the fact that NK are also regulatory cells engaged in reciprocal interactions with dendritic cells, macrophages, T cells, and B cells. Thus NK may serve as a bridge in an interactive loop between innate and adaptive immunity. Thus, dendritic cells stimulate NK which then deliver a co-stimulatory signal to T or B cells allowing for an optimal immune response (Blanca et al. 2001; Vivier et al. 2008). NK are able to response immediately after recognising specific signals, including stress signals, "danger" signals or signals from molecules of foreign origin (Fig. 13.1). Suppression of NK activity and/or reduction in the absolute number of NK circulating in peripheral blood has been found to be associated with the development and progression of cancer, with acute and chronic viral infections, chronic fatigue syndrome, psychiatric depression, various immunodeficiency syndromes, and certain autoimmune diseases (Mandal and Viswanathan 2015). In this context, intact innate immunity, as evidenced by normal levels of NK activity and numbers of circulating NK, appears to play an important role in health (Vivier et al. 2011). On the contrary, low or absent NK activity is often a sign of disease or an early predictor of susceptibility to

†Boris Morukov was deceased in 2015.

M. Rykova · E. Antropova · B. Morukov (deceased) · S. Ponomarev (✉)
State Scientific Center of Russian Federation, Institute of Biomedical Problems, Russian Academy of Sciences, Russian Federation, Moscow, Russia
e-mail: dr.grey@bk.ru

© Springer Nature Switzerland AG 2020

A. Choukér (ed.), *Stress Challenges and Immunity in Space*,
https://doi.org/10.1007/978-3-030-16996-1_13

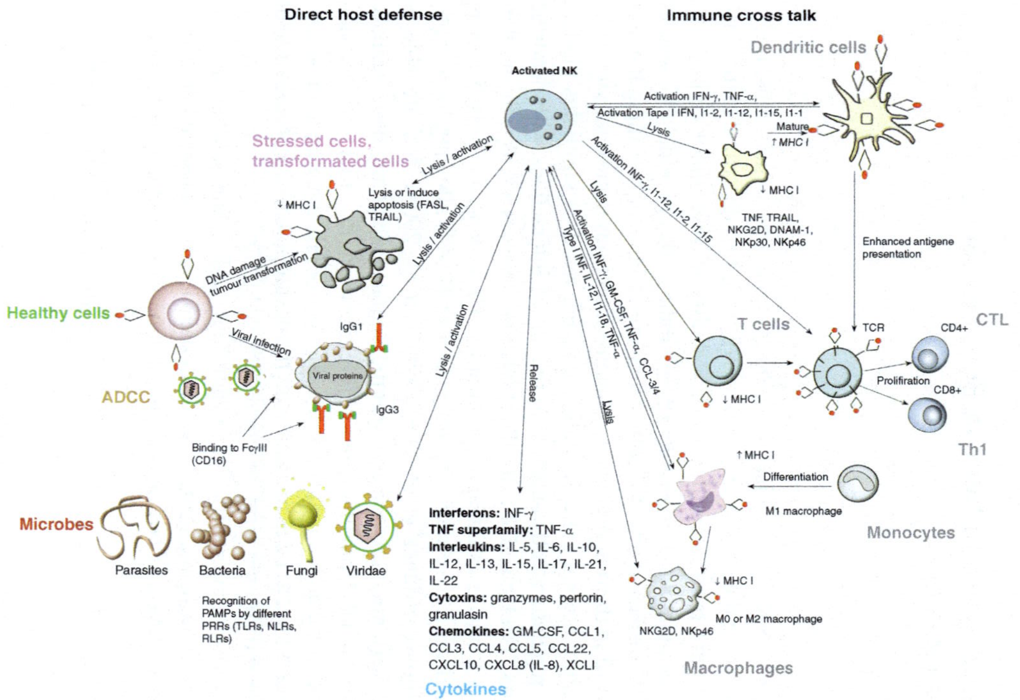

Fig. 13.1 Principle role of NK in adaptive and innate immunity response. NK provide rapid direct responses to viral-infected or tumor cells. Typically NK detect major histocompatibility complex (MHC) presented on infected cell surfaces, triggering cytokine release, initiating apoptosis, or lysis. However NK have the ability to recognize stressed cells in the absence of antibodies and MHC that differentiates them from T-cells of adaptive immunity and allows to perform a fast reaction before the adaptive response activation. Synthesizing cytokines NK influence directly on viruses and activate different cells of both adaptive and innate immunity. NK could also be activated by cytokines produced by other immune cells

disease. Because NK appear to be involved in multiple effector, regulatory, and development steps of the immune system, investigation of NK may serve as an important marker for clinical risk associated with prolonged stressful environmental conditions, including stress factors uniquely associated with space travel. The intent of this chapter is to provide a summary of the effects of stress factors during extended spaceflights on NK, providing—together with recent terrestrial progress in this field—an outlook on future research and application in the light of NK immunology and immunotherapy.

13.1 Review on NK Activities After Long-Duration Missions: From Salyut to ISS

13.1.1 NK Cytotoxicity

NK total cytotoxic activity has been tested on the basis of the amount of nondegraded [^{3}H]RNA that remains in target K562 cells after contact with NK (Rykova et al. 1981). Since the first investigation of the NK functional activity in one Russian cosmonaut after 185-day mission on Salyut-6 in 1980, the cytotoxic activity of NK was examined after long-duration flights on Salyut-6,7, Mir, and ISS in 88 crewmembers (Konstantinova and Fuchs 1991; Konstantinova et al. 1993; Rykova et al. 2001, 2008; Rykova 2013). Flight durations varied from 65 to 438 days. The studies showed a high degree of individual variability of NK cytotoxic activity. Thus, more than 60% of crewmembers had reduced NK cytotoxic activity from preflight values after long-duration missions (Fig. 13.2). On the first day after landing, these reductions in cytotoxicity were moderate in 15 of the 74 cosmonauts (decline by 20–50%), in the middle range in 17 cosmonauts (decline by 51–90%), and in the significant

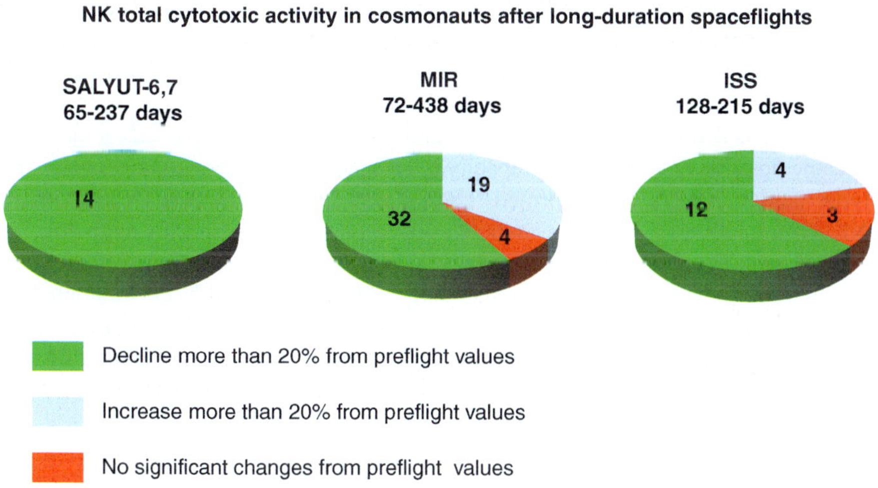

Fig. 13.2 NK total cytotoxic activity in cosmonauts after long-duration spaceflights

range in 7 cosmonauts (cytotoxicity close to zero). At day 7–9 after landing, NK cytotoxic activity of only 4 crewmembers increased toward their preflight values, but in 16 cosmonauts NK cytotoxic activity continued to decrease. Examination of cosmonauts at days 14–19 postflight demonstrated that NK cytotoxic activity remained decreased in almost 30% of all cosmonauts.

In contrast to the Salyut missions of equal duration, in which 100% of the crew members showed lower NK activities postflight, not all cosmonauts from Mir and ISS displayed NK suppression at the first day after long-duration missions. Furthermore, compared to preflight values 36% and 21% of the crew members, respectively, demonstrated even a significant increase in NK cytotoxic activity which tended back to baseline by the last recovery data point (within 7–14 days after landing).

Results obtained from investigations of cosmonauts allowed to identify four different types of NK reactions to long-duration spaceflight conditions: first, a decrease in cytotoxicity on the first day after landing with an increase towards preflight values within 1 week; second, the absence of a decrease in cytotoxicity on the first day after landing and a significant decrease in cytotoxicity on the seventh day; third, a decrease in cytotoxicity at days 1–7 postflight; fourth, an increase in cytotoxicity on the first day after landing with a trend towards preflight values upon examination at later periods.

13.1.2 NK Activities, Subpopulations and Counts: Results and Considerations from More Recent Long-Duration Missions

NK execute their cytotoxicity in a sequential manner, which involves (1) formation of a conjugate between NK and target cell (2) delivery of lethal hit and (3) disassociation of target cell and NK. A disassociated NK can restart and bind to new targets and finally eliminate them, commonly referred to a recycling capacity or killing frequency (Ullberg and Jondal 1981). In order to define the nature of changes in NK activity after long-duration flights single cell-in-agarose assays with mononuclear cells were performed (Grimm and Bonavida 1979). By this method, it became possible to estimate the capacity of effector cells binding to target cells, the lytic activity of NK in formed conjugates and the percentage of active NK. Examination of 27 cosmonauts who lived on board the space station Mir for 2–12 months revealed a significant decrease of the proportion of lymphocytes that bound target cells and the percentage of active NK in peripheral blood mononuclear cells (PBMCs) within 7 days of landing (Meshkov and Rykova 1995). In most cases these changes correlated with decreased NK total cytotoxic activity in PBMCs. However, the percentage of cytotoxic cells in conjugates after flights did not differ from the data observed before launch.

The induction of many NK effector functions, including cytotoxicity, requires that the NK contacts its target cell. Therefore, a likely explanation for the finding of a decrease in the cytotoxic activity of NK in crew members after long-duration space missions might be a defect in the initial step of formation of the lytic NK immunological synapse. Because early steps in NK synapse formation requires adhesion by the integrin family of adhesion molecules (Orange 2008), it seems possible that NK from cosmonauts are not able to adhere to their target cells, which results in a defective cytotoxicity based on defective immunological synapses. This hypothesis is corroborated by the observation of a marked decrease in expression of the adhesion molecules CD11b on NK after landing in comparison to preflight values.

In contrast, spaceflight did not induce significant changes in events which occur following the interaction between a cytolytic cell and its target cell resulting in the directed secretion of lytic granules. At the same time, after a 179-day mission on board the Mir station NK total cytotoxic activity was significantly decreased in one cosmonaut, but the capacity of lymphocytes to bind target cells was relatively unchanged. In this case the inhibition of the recycling function of NK, which have already participated in serial killing, may underlay this observation.

NK are a heterogeneous subset of large granular lymphocytes. Recent studies have identified at least 48 distinct NK subsets, whose significance and function are largely uncertain (Jonges et al. 2001). A few NK subsets, however, have been well defined. NK do not express T cell receptors (CD3) on their surface but express antigens CD16 and CD56. There has been debate over the past years whether changes in cellular distribution reflect changes in NK total cytotoxic activity. It appears that changes in circulating NK number account for most of the change in total NK cytotoxic activity after long-duration flights. Thus, in most crew members a reduction in total NK cytotoxic activity after landing was accompanied by a similar reduction in the percentage of circulating $CD3^{-}CD56^{+}CD16^{+}$ NK. However, not all of the effects of long-duration space missions on total NK cytotoxic activity can be attributed solely to changes in NK number. In particular, two variants of changes in NK number were observed. First, the percentage of $CD3^{-}CD56^{+}CD16^{+}$ NK was unchanged or even elevated (as compared to preflight baseline values), but total NK cytotoxic activity was reduced. Second, the percentage of $CD3^{-}CD56^{+}CD16^{+}$ NK was significantly decreased, but total NK cytotoxic activity was significantly increased. These results showed that investigations of the effects of spaceflight factors on NK should include examination of NK activity and NK numbers (Rykova 2013).

In the recent years NK have been categorized into major groups based on the level of CD56 expression (CD56 bright and dim) (Cooper et al. 2001). These $CD56^{bright}$ and $CD56^{dim}$ subsets are tough to differ in their tissue homing properties due to differential patterns of expression of chemokine receptors and adhesion molecules. $CD56^{dim}$ NK comprise the majority of peripheral blood NK (approximately 95%) and express high levels of the Fc-γIII receptor CD16. Ex vivo $CD56^{dim}$ cells contain detectable perforin and are thus expected to mediate cytotoxicity. In contrast, the $CD56^{bright}$ NK express low levels of perforin but high levels of cytokines and are thought to be an important inflammatory or regulatory subset (Orange and Ballas 2006). It is surprising that only a few attempts have been made to further characterize NK subsets distribution during stress. Thus Bosch and coworkers (2005) reported that acute psychologic stress increased the number of cytotoxic

CD56^{dim} NK without altering the regulatory CD56^{bright} NK counts. Conversely, chronic stressors (as well as 1 month of intensive competitive sports training) caused an increase of the numbers of CD56^{bright} NK but no change in the numbers of CD56^{dim} NK (Suzui et al. 2004). At the same time the examination of nine cosmonauts who flew 187–196-day missions aboard the ISS showed that the number of CD56^{dim} and CD56^{bright} NK (among CD16^{+}CD56^{+} and CD16^{-}CD56^{+} NK) was decreased during the first 7 days after landing. These results suggest that space mission associated factors can result not only in decreased levels of circulating NK with a high cytotoxicity but also decreased level of the major cytokine producing subset of NK.

It is well known that interleukin 2 (IL-2) activation of NK rapidly induces the expression of CD69 (Lanier et al. 1988). This activation antigen represents a triggering surface molecule on NK clones as its stimulation triggers the cytolytic machinery of these cells (Borrego et al. 1993). Beyond its role in NK cytotoxicity, CD69 is involved in regulating other NK functions such as TNF production and the expression of other, functionally relevant, activation antigens, such as ICAM-1 by a mechanism that involves Ca^{2+} mobilization (Borrego et al. 1999). Expression of cell surface marker CD69 on NK, induced by IL-2, was assessed by performing whole blood cultures for 18 h in ten cosmonauts before and after long-duration missions aboard the ISS. A marked decrease in the content of CD56^{+} NK expressing CD69 was observed on the first and seventh days after landing. Such effect may reflect in part a decrease in the functional potential of NK.

13.1.3 Does NK Cytotoxicity Decline Proportionally to Mission Duration?

It was of interest to clarify whether the magnitude of the reduction in cytotoxic activity of NK correlates with the length of time an individual stays in microgravity associated with spaceflight. The data from all cosmonauts examined after prolonged flights were divided into four groups by the duration on board the orbital station (65–131 days, 145–199 days, 208–241 days and 312–438 days). No future decline was noted when flight duration increased from a few months to 14 months. The most of nine cosmonauts, participating in two or three missions each lasting 65–131 and 145–199 days, had similar changes in NK cytotoxicity after all flights. For one of these cosmonauts, a decrease of cytotoxic activity was significant after the first spaceflight lasting 65 days aboard the Salyut 7 station, but a moderate reduction occurred after two missions lasting 152 and 175 days aboard the Mir station. It is to note that examination of 17 crew members who flew for 145–199 days demonstrated in some cases a considerable magnitude of changes in NK cytotoxicity after the first flight and subsequent missions. Comparison of the depth and direction of changes after such repeat space missions let assume that revealed discrepancy could have been associated with some specific features of each flight or differences in the countermeasures that were used. It is important to note, that some crew members had, even after 12 or 15 months in flight, a rather high NK activity—at least at all measurement periods after landing. These results suggest that the immune system (or some of its components) could be adapting to prolonged exposure to microgravity and to the combination of stressors affecting the immune homeostasis.

13.2 NK Cytotoxicity Is Affected by Countermeasures

Results from the Mir station and ISS showed that lower NK suppression and even increase of activation are of no surprise when taking into account the countermeasures that were put in place compared to Salyut. The system of countermeasures used by Russian cosmonauts in spaceflights on board of Mir and ISS included as primary components: physical methods aimed to maintain the distribution of fluids at levels close to those experienced on Earth; physical exercises and loading suits aimed to load the musculoskeletal and the cardiovascular systems; measures that prevent the loss of fluids, mainly, water–salt additives which help to maintain orthostatic tolerance and endurance to gravitational overloads during the return to Earth; well-balanced diet and medications directed to correct possible negative reactions of the body to weightlessness. Fulfilment of countermeasure's protocols in flight was thoroughly controlled (Kozlovskaya and Grigoriev 2004; Kozlovskaya et al. 1995). Recent scientific research has shown that NK are highly influenced by physical exercise (Walsh et al. 2011). The possible important mechanisms behind exercise-induced changes in NK function are cytokines, hyperthermia, and stress hormones, including catecholamines, growth hormone, cortisol, and beta-endorphins. Infusion studies mimicking stress hormone levels in blood during exercise indicate that increased plasma-adrenaline accounts for at least part of the exercise-induced modulation of NK function (Pedersen and Ullum 1994). Many studies have documented that severe exercise is followed by immunodepression, but moderate regular exercise has a beneficial effect on NK function (Mackinnon 1999). In this connection, elevated NK cytotoxicity after long-duration space missions in some cosmonauts could have been associated not only with the constitutional characteristics of the individuals but also with the performance by the crew members of a set of recommended prophylactic measures, first of all moderate regular physical exercise.

13.3 Is NK Cytotoxicity Associated to Viral Reactivation?

An alternative explanation of an increase in NK cytotoxicity in some cosmonauts after long duration space missions is that enhancement of NK cytotoxicity was associated with latent viral reactivation during spaceflight which was observed during short- and long-duration flight (Mehta et al. 2000, 2005; Stowe et al. 2001a, b; Pierson et al. 2005; Cohrs et al. 2008; Mehta et al. 2014, 2017). Indeed, in the 1970s, it was reported that a normal NK response to a viral infection should be an initial rise in activity, followed by elevated activity during the course of infection, and then a gradual return to baseline, while NK numbers can be reduced during infectious episodes (Whiteside and Herberman 1989). More recent studies have shown that NK activation by viruses occurs predominantly via the indirect pathway—involving cytokines and cell contact-dependent signals from accessory cells. While there are many reports of direct NK activation by binding of pathogen-derived molecules to NK surface receptors, in only very few of these cases (influenza virus activation of human NK) have both the pathogen ligand and the NK receptor been defined at the gene and protein levels (Horowitz et al. 2012). At the same time there is increasing evidence indicates that most, if not all, members of the herpesvirus family suppress NK activity. On the other hand it has been demonstrated that some viruses have the capacity to inhibit activating pathways of NK through a multifaceted strategies, including indirect and direct ones. First, NK activation is inhibited by inhibitory signals provided through interaction of receptors on NK with self-MHC class I products. Second, several viruses encode cytokine mimicry to escape host immune surveillance. Third, it is a universal strategy for viruses to escape NK recognition by encoding viral proteins to disturb NK receptor–ligand interaction. Moreover, viruses can directly infect NK, which results in the impairments of NK function and also induces marked apoptosis of NK (Wang et al. 2013; De Pelsmaeker et al. 2018). However, the relationship of the signaling balance of activating and inhibitory receptors of NK and latent viral reactivation in spaceflight has not been fully investigated, but the establishment of such a relationship could lead to the development of countermeasures that could prevent any compromises in resistance to infection resulting from exposure to flight conditions.

13.4 NK and Antitumor Effects

Clearly, the combined exposure to microgravity and solar energetic particles (SEP) and galactic cosmic ray (GCR) radiation that would occur during extended deep space missions, such as to Mars, could potentially increase risk of cancer. The scientific literature over the past 50 years has provided strong support to the cancer immunosurveillance hypothesis. There have been published several excellent reviews recently regarding the role of NK in protection against cancer (Guillerey and Smyth 2015; Pahl and Cerwenka 2017; Sharma et al. 2017). In general, NK recognize and respond to malignant cells—in two ways. First, they bear natural cytotoxicity receptors (NCRs) that detect the altered expression of ligands on the surface of tumor cells, which ultimately triggers NK activation. Second, they bind antibody coated targets via immunoglobulin receptors leading to antibody-dependent cellular cytotoxicity (ADCC). Clinical and experimental evidence demonstrate that patients with a variety of solid malignancies and large tumor burdens have decreased NK activity in the circulation, and that this low NK activity may be significantly associated with the development of distant metastases. Furthermore, in patients treated for metastatic disease, the survival time correlates directly with the levels of NK activity. In patients with hematologic malignancies, there appears to be a correlation between NK activity and the status of disease; the more advanced the disease, the lower the NK activity. Decreased NK activity may also be an important risk factor for the development of malignancy in humans. The prognostic significance of low NK activity in patients with cancer has been recently emphasized; thus, low NK activity may have prognostic value in predicting relapses, poor responses to treatment and, especially, decreased survival time (Levy et al. 2011; Berrien-Elliot et al. 2015).

It is possible that the patient's NK are able to contribute to the host's anti-tumor immune responses, either by eliminating circulating tumor cells or by secreting the appropriate, perhaps Th1-stimulating, cytokines during cross talk with dendritic cells in the periphery and/or in the secondary lymphoid tissues. NK can directly kill target tumor cells in two ways. First, they bear natural cytotoxicity receptors (NCRs) that detect the altered expression of ligands on the surface of tumor cells, which ultimately triggers NK activation. These NK can kill target tumor cells by releasing cytoplasmic granules containing perforin (membrane pore-forming proteins) and granzymes (serine proteases) that leads to tumor-cell apoptosis (programmed cell death) by caspase-dependent and -independent pathways. Furthermore, the expression of members of the tumor necrosis factor (TNF)-family like FAS ligand (FASL), TNF, and TNF-related apoptosis inducing ligand (TRAIL) are able to induce tumor-cell apoptosis upon formation of immune synapses (Leischner et al. 2016).

Another way of direct killing is through in antibody-dependent cell-mediated cytotoxicity (ADCC). NK can express FcγRIIIA and/or FcγRIIC, which can bind to the Fc portion of immunoglobulins, transmitting activating signals within NK. Once activated through Fc receptors by antibodies bound to target cells, NK are able to lyse target cells without priming, and secrete cytokines like IFNγ to recruit adaptive immune cells. This ADCC of tumor cells is utilized in the treatment of various cancers overexpressing unique antigens, such as neuroblastoma, breast cancer, B cell lymphoma, and others. NK also express a family of receptors called killer immunoglobulin-like receptors (KIRs), which regulate the function and response of NK toward target cells through their interaction with their cognate ligands that are expressed on tumor cells (Wang et al. 2015).

Acting as regulatory cells, NK interact with dendritic cells, macrophages, T-cells, and endothelial cells through production of various cytokines such as INF-γ, TNF, and IL-10 as well as other chemokines and growth factors. By releasing INF-γ, these cells induce $CD8^+$ cytotoxic cells to differentiate into CTLs. They also help in differentiation of CTLs via induction of $CD4^+$ cells. Moreover, in course of killing of tumor cells by NK, antigens produced, are taken up by antigen presenting cells (APC's) for induction of adaptive immune response (Wang et al. 2015). Unfortunately, no studies are currently done to establish the intimate mechanism of tumor development and interactions occurring between cancer cells and NK under spaceflight conditions.

13.5 NK in Autoimmunity and Allergy

Although the existence of clinical risks related to spaceflight-associated dysfunction of immune surveillance has not been firmly concluded, concerns of the occurrence of autoimmune and *allergic* diseases in long-duration spaceflights are growing.

Recent studies have documented that innate immune cells and, in particular, NK play a key role in the pathogenesis of autoimmune diseases (Gianchecchi et al. 2018; Zhang and Tian 2017). The data obtained showed the correlation between NK number and/or functional alterations, such as a defective cytotoxic activity and several autoimmune conditions such as multiple sclerosis, rheumatoid arthritis, systemic lupus erythematosus, Sjögren's syndrome, and type I diabetes mellitus (Fogel et al. 2013). NK act at all the phases of autoimmune diseases including self-antigen-releasing, T cell priming in secondary lymphoid organs, and finally trafficking to the target organ for tissue destruction (Flodstrom-Tullberg et al. 2009). However, among the different autoimmune pathologies and even within the same disease, NK function is significantly different either promoting or even protecting against the onset of the autoimmune condition, in other words, NK act as a two-edged weapon and play opposite roles with both regulatory and inducer activity in autoimmune diseases (Tian et al. 2012). Although the precise mechanism for the opposite effects of NK was not fully elucidated, the importance of NK in autoimmunity might be associated with different NK subsets, different tissue microenvironment and different stages of corresponding diseases (Peng and Tian 2014).

Strong evidence that NK contribute to the pathogenesis of human autoimmune disorders was given by genetic association studies in a variety of autoimmune disease. Thus, in particular, several findings have pointed out associations between risk of systemic or organ-specific autoimmune diseases and KIR/HLA genotypes, which indicate that self-tolerance may be broken with inappropriate receptor and ligand pairs or with the interrupted signal balance. For example, the presence of an activating receptor for HLA-I associated with the lack or reduction of inhibitory pairs has been shown in such autoimmune diseases as Behçet's disease, type I diabetes mellitus, systemic lupus erythematosus, multiple sclerosis, psoriasis/psoriatic arthritis, ankylosing spondylitis and rheumatoid arthritis (Fogel et al. 2013; Poggi and Zocchi 2014).

Unfortunately not much is known about the role of NK in allergy. As is well known the first step in the sequence of events leading to atopic diseases, including atopic dermatitis, food allergy, allergic rhinitis, asthma, and anaphylaxis asthma occurs at the site of sensitization. At this stage, activated NK may regulate the extent of DC maturation and thus the magnitude of antigen presentation to naïve CD4 T cells. The production of inflammatory cytokines by NK can subsequently affect the development of allergen-specific Th2 cells by modulating the immune microenvironment. Cytokines such as IFN-γ produced by NK1 may potentially inhibit the development of Th2 cells. At the same time NK2 can produce cytokines such as IL-4, which may enhance Th2 development. During allergen challenge, activated NK may further potentate the response by producing cytokines such as IL-5, IL-13, and IFN-γ, which can enhance or dampen inflammation (Deniz et al. 2013; Mathias 2015; Kim and Jang 2018). Finally, IgE, acting through FcγRIII, can activate NK resulting in cytokine/chemokine production (Arase et al. 2003).

13.6 NK as a Potential Therapeutic Target

NK constitute our bodies' frontline defense system. Now it is clear that NK have predominant role in several disease conditions. In this connection they are the potential target for immunotherapy, due to their high plasticity and ability to exert a potent and extremely rapid response, which can influence the outcome of the adaptive immune response. The therapeutic role of NK has been studied in several diseases such as cancer, asthma, multiple sclerosis, diabetes, and arthritis (Mandal and Viswanathan 2015; Schäfer et al. 2017; Schmidt et al. 2018).

The current strategies and efforts on enhancing NK activity include infusion of NK-activating cytokines or vaccines (to activate the impaired endogenous NK) (Romee et al. 2014) and infusion of autologous or allogeneic NK (to replace the impaired endogenous NK) (Veluchamy et al. 2017; Sanchez-Martinez et al. 2018). An interesting alternative to allogeneic NK is KIR/KIRL blocking antibodies that activate endogenous NK (Vey et al. 2012).

Recent efforts to improve the clinical efficacy of NK immunotherapy has led to the development of genetically engineered NK that express a chimeric antigen receptor (CAR). Primary NK and NK lines can be engineered to express CARs which redirect the anti-tumor specificity of NK on an antigen-dependent basis. Through the manipulation of signaling motifs critical for lymphocyte activation, CARs are also designed to utilize specific intracellular signaling molecules which can further refine NK function and optimize their therapeutic potential. The use of a clonal cell line derived from a human NK leukemia, known as NK-92, has been genetically modified to express fully functional CARs and these cells have shown great promise with regards to their safety and efficacy in recent clinical trials. Moreover, the use of irradiated cell lines may provide a fast and affordable off-the-shelf option for a personalized cellular immunotherapy treatment (Abel et al. 2018; Daher and Rezvani 2018).

In addition, one of the strategies to improve NK activity is based on the use of immunomodulatory drugs (IMiDs) such as thalidomide, lenalidomide, or pomalidomide. This could be due to several mechanisms. First, IMiDs could suppress the production of inhibitory cytokines and promote the production of IL-2, an NK growth factor, by T cells. The latter effect could be the result of induced degradation of Ikaros and Aiolos, two transcription factors repressing IL-2 production in T cells. Moreover, IMiDs can enhance the expression of the NKG2D and DNAM-1 activating receptor ligands MICA and PVR/CD155 also through the downregulation of Ikaros and Aiolos (Besson et al. 2018).

Especially in the last decade, it was established many molecular connections between dietary compounds and their influence on immune cells. Several natural compounds, in particular, the most extensively studied stilbene resveratrol exhibit stimulating properties on NK and their killing ability on different levels. In line with the immune-modulating properties of nutrition-derived compounds resveratrol has been described to affect NK function directly via alteration of the expression of activating cell surface receptors like NKG2D and stimulation of JNK and ERK-1/2 MAP kinase activity and perforin expression (Leischner et al. 2016).

Undoubtedly further studies clarifying the potential of targeting or using of NK in the immunotherapy during the duration of an exploration-class deep space mission have to be initiated.

13.7 Are the Changes Observed in the NK Population of Astronauts Specifically Due to Spaceflight?

Although studies such as the one described above provide useful information on NK function after long-duration space missions, the main question, whether the altered NK were already present during flight or if this alteration was caused solely by landing effects, remains. The nature of immune dysregulation of NK during flight is currently being investigated, and studies to date have demonstrated a significant reduction in NK activities against all tumor target cells (K562, U266, 221.AEH and 721.221), perforin and granzyme b levels during the in-flight phase (FD90) of the mission. Interestingly, in some crewmembers, NK activities remained suppressed at day 1–66 after landing (Simpson et al. 2016). Moreover, comparing results from brief and long flights may facilitate understanding of the differences and interrelationships between the multiple stresses associated with spaceflight and landing stressor. Despite of substantial differences in methodology used, the results from American and Russian studies conducted before and after short-duration missions (4–16 days) showed decreased NK cytotoxicity, diminished capacity of lymphocytes to bind target cells and reduced numbers of NK after landing (Konstantinova and Fuchs 1991; Meshkov and Rykova 1995; Tipton et al. 1996; Mehta et al. 2001; Mills et al. 2001; Rykova et al. 2001, 2008; Borchers et al. 2002; Sonnenfeld and Shearer 2002; Crucian et al. 2008). The data suggest that landing stress (i.e. an acute response to reentry and readaptation to unit gravity) may be a major mediating factor in the impaired NK function of cosmonauts. On the other hand, in ground-based space analog studies there was a significant decrease in total NK cytotoxicity during and after 120- and 360-day bed rest with head-down tilt (Konstantinova and Fuchs 1991).

Data of the natural cytotoxicity system in subjects who participated in the 520-day isolation project ("Mars-500", see also Chap. 37) with artificial environment showed that despite of physical training starting at the beginning of a 12-month stay in isolation and lasting up to recovery period of 7 days, there was a significant decrease not only in the relative but also in the absolute content of the peripheral blood circulating mature $CD3^{-}CD16^{+}CD56^{+}$-lymphocyte. Additionally, the study of phenotypic features of NK showed that the observed changes were associated with a decrease in the content in the peripheral blood of the following subclasses of NK: lymphocytes with membrane surface molecule CD16 expression and lymphocytes which expressed membrane surface molecule CD56. It should be noted that at the final stage of the experiment, there was a decrease in lymphocyte number in most subjects which belonged to the subtype of NK with high cytolytic activity ($CD56^{+}CD16^{+}$-NK), rather than to the subtype of NK with moderate and low cytolytic activity ($CD16^{-}CD56^{+}$ and $CD16^{+}CD56^{-}$-NK). Studying the expression of the early CD69 activation marker on CD56⁺ lymphocytes in a 18 h culture (nonstimulated and stimulated with IL-2) showed, that there was a decrease in the expression of CD69 on the NK lymphocytes in the absence of stimulation, as well as when cultured in the presence of IL-2 in all subjects by the fourth month of the experiment. This phenomenon may be due not only to a decrease in the functional potential of NK but also to the depletion of the reserve capacities of the system of natural cytotoxicity (Morukov et al. 2013).

This observation indicates that functionally deficient NK may not result exclusively from landing stressors but could represent alterations in spaceflight. In addition, since experimental conditions of bed rest were the same for all subjects, the magnitude of changes differed dramatically between individuals. This data underscores the importance of individual differences in responsiveness.

In brief, the mechanisms responsible for changes in NK activity during or after long-duration missions are not clear and may be the results of a combination of factors, including complexes of multiple neuropeptide hormones and catecholamines (particularly glucocorticoids) (Meehan et al. 1993; Stowe et al. 2003). Future studies of the role of psychoneuroendocrine factors on spaceflight-induced alterations in NK as well as the interaction with the microbial load on-board will expand significantly our understanding of the mechanisms of immune reactions as well as provide novel information about relationship of the immune system with other functional systems participating in adaptation of organism to altering exogenous and endogenous factors. Also we can't exclude the direct influence of spaceflight factors on the gene expression which plays an important role in the immune response.

13.8 Conclusion

The beginning of the twenty-first century is marked by the growing interest of the international community in interplanetary flights. This is illustrated by the intense development of projects related to the possibility of manned flights to Mars, which seems to be of the highest scientific interest among the planets of the solar system. Surely, manned flights to Mars require the solution of a set of problems related to the preservation of the health and working capacity of the crew at all stages of the mission and after its termination. Disorder of the NK which are one of the components of immune system may limit an increase in the duration of a human's stay under the conditions of a spaceflight, because the deviation taking place may be regarded in certain cases as formation of an immune deficiency state that represents an increase degree of risk of the occurrence of viral, autoimmune/allergic diseases and tumor during space missions. In view of this, one of the key challenges facing space immunology is to create a comprehensive system to control and predict possible adverse changes in the functioning of NK and the development of adequate methods of their prevention and relief. Naturally, the solution to this common problem is not possible without solving a number of particular problems. The latter are primarily associated with the creation of an automated clinical immunological laboratory, which will make it possible not only to evaluate the NK at the preflight and postflight stages but also to constantly control the state of immunological resistance throughout the flight, and likely also in conjunction with other organ systems (see also Chap. 18). A special task in the near future can be the development of one of the main directions of future medicine—identification of the predisease stage, at which it is easy to prevent the development of a disease based on the systematic and close monitoring of a large group of healthy people and identifying immunological changes prior to the emergence of certain diseases. An equally important task is the development of the concept of the genetic prediction of the probability of negative deviations in NK system, which may have a detrimental effect during spaceflight, at the stage of selection of crew members of interplanetary missions.

References

Abel AM, Yang C, Thakar MS, Malarkannan S (2018) Natural killer cells: development, maturation, and clinical utilization. Front Immunol 9:1869. https://doi.org/10.3389/fimmu.2018.01869

Arase N, Arase H, Hirano S et al (2003) IgE-mediated activation of NK cells through Fc gamma RIII. J Immunol 170(6):3054–3058

Berrien-Elliot MM, Romee R, Fehniger TA (2015) Improving natural killer cell cancer immunotherapy. Curr Opin Organ Transplant 20(6):671–680

Besson L, Charrier E, Karlin L et al (2018) One-year follow-up of natural killer cell activity in multiple myeloma patients treated with adjuvant lenalidomide therapy. Front Immunol 9:704

Blanca IR, Bere EW, Youn HA, Ortaldo JR (2001) Human B cell activation by autologous NK cells is regulated by CD40-CD40 ligand interaction: role of memory B cells and CD5+ B cells. J Immunol 167:6132–6139

Borchers AT, Keen CL, Gershwin ME (2002) Microgravity and immune responsiveness: implications for space travel. Nutrition 18(10):889–898

Borrego F, Pena J, Solana R (1993) Regulation of CD69 expression on human natural killer cells: differential involvement of protein kinase C and protein tyrosine kinases. Eur J Immunol 23(5):1039–1043

Borrego F, Robertson MJ, Ritz J et al (1999) CD69 is a stimulatory receptor for natural killer cell and its cytotoxic effect is blocked by CD94 inhibitory receptor. Immunology 97(1):159–165

Bosch JA, Berntson GG, Cacioppo JT et al (2005) Differential mobilization of functionally distinct NK subsets during acute psychologic stress. Psychosom Med 67:366–375

Cohrs RJ, Mehta SK, Schmid DS et al (2008) Asymptomatic reactivation and shed of infectious varicella zoster virus in astronauts. J Med Virol 80:1116–1122

Cooper MA, Fehniger TA, Caligiuri MA (2001) The biology of human natural killer-cell subsets. Trends Immunol 22:633–664

Crucian BE, Stowe RP, Pierson DL, Sams CF (2008) Immune system dysregulation following short- vs long-duration spaceflight. Aviat Space Environ Med 72(9):835–843

Daher M, Rezvani K (2018) Next generation natural killer cells for cancer immunotherapy: the promise of genetic engineering. Curr Opin Immunol 51:146–153. https://doi.org/10.1016/j.coi.2018.03.013. Epub 2018 Mar 30

De Pelsmaeker S, Romero N, Vitale M, Favoreel HW (2018) Herpesvirus evasion of natural killer cells. J Virol 92(11):e02105–e02117

Deniz G, van de Veen W, Akdis M (2013) Natural killer cells in patients with allergic diseases. J Allergy Clin Immunol 132(3):527–535

Flodstrom-Tullberg M, Bryceson YT, Shi FD, Hoglund P, Ljunggren HG (2009) Natural killer cells in human autoimmunity. Curr Opin Immunol 21:634–640

Fogel LA, Wayne M, Yokoyama WM, French AR (2013) Natural killer cells in human autoimmune disorders. Arthritis Res Ther 15:216

Gianchecchi E, Delfino DV, Fierabracci A (2018) NK cells in autoimmune diseases: linking innate and adaptive immune responses. Autoimmun Rev 17(2):142–154

Grimm E, Bonavida B (1979) Mechanism of cell mediated cytotoxicity at the single cell level. I. Estimation of cytotoxic T lymphocyte frequency and relative lytic efficiency. J Immunol 123:2861–2868

Guillerey C, Smyth MJ (2015) NK cells and cancer immunoediting. In: Vivier E, Di Santo J, Moretta A (eds) Natural killer cells, Current topics in microbiology and immunology, vol 395. Springer, Cham

Herberman RB (1974) Cell-mediated immunity to tumor cells. In: Klein G, Weinhouse S (eds) Advances in cancer research, vol 19. Academic, New York, pp 107–263

Horowitz A, Stegman KA, Rile EM (2012) Activation of natural killer cells during microbial infections. Front Immun 2:88

Jonges LE, Albertsson P, van Vlierberghe RL et al (2001) The phenotypic heterogeneity of human natural killer cells: presence of at least 48 different subsets in the peripheral blood. Scand J Immunol 53:103–110

Kim JH, Jang YJ (2018) Role of natural killer cells in airway inflammation. Allergy Asthma Immunol Res 10(5):448–456

Konstantinova IV, Fuchs BB (1991) The immune system in Space and other extreme conditions. Harwood Academic Publishers, Reading, Berkshire

Konstantinova IV, Rykova MP, Lesnyak AT, Antropova EN (1993) Immune changes during long-duration missions. J Leukoc Biol 54(3):189–201

Kozlovskaya IB, Grigoriev AI (2004) Russian system of countermeasures on board of the International Space Station (ISS): the first results. Acta Astronaut 55:233–237

Kozlovskaya IB, Grigoriev AI, Stepantzov VI (1995) Countermeasure of the negative effects of weightlessness on physical systems in long-term space flights. Acta Astronaut 36:661–668

Lanier LL, Buck DW, Rhodes L et al (1988) Interleukin 2 activation of natural killer cells rapidly induces the expression and phosphorylation of the Leu-23 activation antigen. J Exp Med 167(5):1572–1585

Leischner C, Burkard M, Pfeiffer MM et al (2016) Nutritional immunology: function of natural killer cells and their modulation by resveratrol for cancer prevention and treatment. Nutr J 15(1):47

Levy EM, Roberti MP, Mordoh J (2011) Natural killer cells in human cancer: from biological functions to clinical applications. J Biomed Biotechnol 2011:676198

Mackinnon LT (1999) Advances in exercise immunology. Human Kinetics, Champaign, IL

Mandal A, Viswanathan C (2015) Natural killer cells: in health and disease. Hematol Oncol Stem Cell Ther 8:47–55

Mathias CB (2015) Natural killer cells in the development of asthma. Curr Allergy Asthma Rep 15(2):500

Meehan R, Whitson P, Sams C (1993) The role of psychoneuroendocrine factors on spaceflight-induced immunological alterations. J Leukoc Biol 54:236–244

Mehta SK, Stowe RP, Feiveson AH et al (2000) Reactivation and shedding of cytomegalovirus in astronauts during spaceflight. J Infect Dis 182:1761–1764

Mehta SK, Kaur I, Grimm EA et al (2001) Decreased non-MHC-restricted ($CD56^+$) killer cell cytotoxicity after spaceflight. J Appl Physiol 91:1814–1818

Mehta SK, Cohrs RJ, Forghani B et al (2005) Stress-induced subclinical reactivation of varicella zoster virus in astronauts. J Med Virol 72:174–179

Mehta SK, Laundenslager ML, Stowe RP et al (2014) Multiple latent viruses reactivate in astronauts during space shuttle missions. Brain Behav Immun 41:210–217

Mehta SK, Laundenslager ML, Stowe RP et al (2017) Latent virus reactivation in astronauts on the international space station. NPJ Microgravity 3:11. https://doi.org/10.1038/s41526-017-0015-y

Meshkov D, Rykova M (1995) The natural cytotoxicity in cosmonauts on board space stations. Acta Astronaut 36:719–726

Mills PJ, Meck JV, Waters WW et al (2001) Peripheral leukocyte subpopulations and catecholamine levels in astronauts as a function of mission duration. Psychosom Med 63:886–890
Morukov BV, Rykova MP, Antropova EN et al (2013) Immunological aspects of a space flight to mars. Hum Physiol 39:126–135
Orange JS (2008) Formation and function of the lytic of the NK-cell immunological synapse. Nat Rev Immunol 8:713–725
Orange JS, Ballas ZK (2006) Natural killer cells in human health and disease. Clin Immunol 118:1–10
Pahl J, Cerwenka A (2017) Tricking the balance: NK cells in anti-cancer immunity. Immunobiology 222:11–20
Pedersen BK, Ullum H (1994) NK cell response to physical activity: possible mechanisms of action. Med Sci Sports Exerc 26(2):140–146
Peng H, Tian Z (2014) NK cell trafficking in health and autoimmunity: a comprehensive review. Clin Rev Allergy Immunol 47:119–127
Pierson DL, Stowe RP, Phillips TM et al (2005) Epstein-Barr virus shedding by astronauts during space flight. Brain Behav Immun 19:235–242
Poggi A, Zocchi MR (2014) NK cell autoreactivity and autoimmune diseases. FrontImmunol 5:27
Romee R, Leong JW, Fehniger TA (2014) Utilizing cytokines to function-enable human NK cells for the immunotherapy of cancer. Scientifica. https://doi.org/10.1155/2014/205796
Rykova MP (2013) Immune system of Russian cosmonauts after orbital space flights. Hum Physiol 39:557–566
Rykova MP, Spirande IV, Zedgenidze MS et al (1981) New high sensitive technique for testing natural killers. Immunologiya N3:88–90. (in Russian)
Rykova MP, Antropova EN, Meshkov DO (2001) Results of immunological studies. In: Orbital'naya stantsiya, vol 2. MIR (Mir Space Station), Moscow, p 615
Rykova MP, Antropova EN, Larina IM, Morukov BV (2008) Humoral and cellular immunity in cosmonauts after the ISS missions. Acta Astronaut 63:697–705
Sanchez-Martinez D, Allende-Vega N, Orecchioni S et al (2018) Expansion of allogeneic NK cells with efficient antibody-dependent cell cytotoxicity against multiple tumors. Theranostics 8(14):3856–3869
Schäfer C, Ascui G, Ribeiro CH, López M, Prados-Rosales R, González PA, Bueno SM, Riedel CA, Baena A, Kalergis AM, Carreño LJ (2017) Innate immune cells for immunotherapy of autoimmune and cancer disorders. Int Rev Immunol 36:1–23
Schmidt S, Tramsen L, Rais B et al (2018) Natural killer cells as a therapeutic tool for infectious diseases - current status and future perspectives. Oncotarget 9. https://doi.org/10.18632/oncotarget.25058
Sharma P, Kumar P, Sharma R (2017) Natural killer cells - their role in tumour immunosurveillance. J Clin Diagn Res 11(8):BE01–BE05
Simpson RJ, Bigley AB, Spielmann G, Kunz HE, Agha N, Baker F et al (2016) Long duration spaceflight impairs NK-cell function in astronauts. Med Sci Sports Exerc 48(5 Suppl 1):87
Sonnenfeld G, Shearer WT (2002) Immune function during space flight. Nutrition 18(10):899–903
Stowe RP, Mehta SK, Ferrando AA et al (2001a) Immune responses and latent herpesvirus reactivation in spaceflight. Aviat Space Environ Med 72:884–891
Stowe RP, Pierson DL, Barrett AD (2001b) Elevated stress hormone levels relate to Epstein-Barr virus reactivation in astronauts. Psychosom Med 63:891–895
Stowe RP, Sams CF, Pierson DL (2003) Effects of mission duration on neuroimmune responses in astronauts. Aviat Space Environ Med 74:1281–1284
Suzui M, Kawai T, Kimura H, Takeda K et al (2004) Natural killer cell lytic activity and CD56dim and CD56bright cell distributions during and after intensive training. J Appl Physiol 96(6):2167–2173
Tian Z, Gershwin ME, Zhang C (2012) Regulatory NK cells in autoimmune disease. J Autoimmun 39(3):206–215
Tipton CM, Greenleaf JE, Jackson CG (1996) Neuroendocrine and immune system responses with spaceflights. Med Sci Sports Exerc 28:988–998

Ullberg M, Jondal M (1981) Recycling and target binding capacity of human natural killer cells. J Exp Med 153:615–628

Veluchamy JP, Kok N, van der Vliet HJ et al (2017) The rise of allogeneic natural killer cells as a platform for cancer immunotherapy: recent innovations and future developments. Front Immunol 8:631

Vey N, Bourhis JH, Boissel N et al (2012) A phase 1 trial of the anti-inhibitory KIR mAb IPH2101 for AML in complete remission. Blood 120:4317–4323

Vivier E, Tomasello E, Baratin M et al (2008) Functions of natural killer cells. NatImmunol 9(5):503–510

Vivier E, Raulet DH, Moretta A (2011) Innate or adaptive immunity? The example of natural killer cells. Science 331(6013):44–49

Walsh NP, Gleeson M, Shephard RJ, Gleeson M (2011) Position statement. Part one: Immune function and exercise. Exerc Immunol Rev 17:6–63

Wang D, Ma Y, Wang J, Liu X, Fang M (2013) Natural killer cells in innate defense against infective pathogens. J Clin Cell Immunol S13:006

Wang W, Erbe AK, Hank JA et al (2015) NK cell-mediated antibody-dependent cellular cytotoxicity in cancer immunotherapy. Front Immunol 6:368

Whiteside TL, Herberman RB (1989) The role of natural killer cells in human disease. Clin Immunol Immunopathol 53:227–228

Zhang C, Tian Z (2017) NK cell subsets in autoimmune diseases. Autoimmunity 83:22–30

Adaptive Immunity and Spaceflight

14

Brian Crucian, George Makedonas, and Clarence Sams

14.1 Adaptive Immune Cell Function

The distribution of the various immune cell subsets and their functional capacity are mutually exclusive considerations. Cellular distribution may be dramatically altered while cell functional capacity is normal. Alternatively, the distribution of peripheral immune cells may be unaltered, when functional capacity is reduced. Either a loss in cell numbers or a reduction in cell function may each separately result in a clinical immunodeficiency. Considering this, any assessment of immune status during spaceflight must include assays of both cell number and measures of cell function. There are many distinct assays to measure immune cell function. Some assays may be generalized and apply to several cell types, whereas others may be highly specific and apply only to a particular cell subset. Most functional assays consist of cell culture in the presence of a stimulus, after which any event in the normal activation process may be monitored. For T cells, a specific sequence of cellular events, beginning within seconds and progressing over the course of days, leads to full activation and execution of an in vivo response. Very early events (within minutes) include transmission of intracellular signals to the nucleus, protein kinase C (PKC) phosphorylation, and cytoskeleton reorganization. Over the course of hours there is expression of new mRNA and translation of new protein. Activation markers are expressed on the T cell surface at specific times post-activation, each with a specific purpose. CD69, preformed in the cytoplasm, is expressed within an hour, whereas CD25 (Interleukin (IL) 2 receptor) is expressed at about 24 h post-activation. A

B. Crucian (✉) · C. Sams
NASA-Johnson Space Center, Houston, TX, USA
e-mail: brian.crucian-1@nasa.gov

G. Makedonas
NASA-Johnson Space Center, Houston, TX, USA

JES Tech, Houston, TX, USA

© Springer Nature Switzerland AG 2020

A. Choukér (ed.), *Stress Challenges and Immunity in Space*,
https://doi.org/10.1007/978-3-030-16996-1_14

third marker of activation, HLA-DR (a MHC class II molecule), is expressed between 48 and 72 h post-activation. IL-2 is secreted, activating other T cells as well as the source cells (autocrine stimulation). Depending on the local cytokine environment, antigen type/dose, and other factors, naive $CD4^+$ T-helper cells will proliferate and develop into a variety of Th1, Th2, or Th17 cells. These cells will then produce additional effector cytokines to drive the adaptive immune response toward a particular bias, specific for the initiating pathogen or antigen. Cell division and expansion (proliferation) usually begins at around 72 h. Although any of the cellular events involved in T cell activation may be monitored as a functional assay, the expression of activation markers and production of cytokines are the most commonly used. These assays are relatively simple, and indicate if T cell activation has initiated, progressed, and resulted in effector function (cytokine production, etc.). Representative studies demonstrating specific adaptive immune functional alterations that have been observed during or immediately following spaceflight are summarized in Table 14.1. Several groups have observed that T cells in culture do not activate during spaceflight (or simulated microgravity) conditions. Specific findings include alterations in T cell activation characteristics, intracellular signaling, and other observable changes at the single-cell level (Cogoli 1997; Hashemi et al. 1999; Hughes-Fulford 2003).

Microgravity culture was recently demonstrated to affect negatively several T cell intracellular signaling pathways at points upstream of PKC, a family of enzymes involved in transducing cellular signals by controlling the function of other proteins. However, phosphorylation of PKC was not downregulated in T cells incubated in the presence of Concanavalin A (Con A) and anti-CD28 antibodies in simulated microgravity (Boonyaratanakornkit et al. 2005), and T cell stimulation with phorbol myristate acetate (PMA) + ionomycin (direct PKC activation) was able to fully restore T cell activation during simulated microgravity (Cooper and Pellis 1998). This suggested that signaling pathways "upstream" of PKC might be responsible for the "microgravity defect" in normal T cell activation observed in microgravity. Recently, using parabolic flight and the 2D clinostat system to represent the microgravity environment, altered gravity depresses the cell surface expression of CD3 and CD25, the receptor required for IL-2 recognition. In addition, these conditions impaired LAT phosphorylation and the global levels of ZAP70, elements critical to initiate the signaling cascade of T cells (Tauber et al. 2015). Hughes-Fulford et al. analyzed differential gene expression following T cell activation comparing normal gravity to simulated microgravity conditions, to identify gravity-sensitive intracellular signaling pathways. The study found that 99 genes were significantly upregulated during early T cell activation in normal gravity; however during simulated microgravity the majority of these showed no mitogen-induced gene expression (Boonyaratanakornkit et al. 2005). Most of the gravity-sensitive genes were regulated by transcriptional factors NF-kB, CREB, Ets-like protein, AP-1, or STAT1. The authors concluded that gravity is a key regulator of cell-mediated immunity, and its absence slows or impedes signaling pathways required for early T cell activation. This phenomenon was validated onboard the International Space Station (ISS): in response to mitogenic stimulation, human leukocytes exhibited significant

Table 14.1 Representative studies: adaptive immune function and spaceflight

Category	Author	Flight phase	Findings
T cell intracellular signaling	Cogoli (1997)	Review	T cells during reduced gravity, expression of IL-2 receptor is inhibited, cytoskeleton interactions with rap proteins are disturbed, the function of PKC is altered
	Hashemi et al. (1999)	In	TCR stimulation during spaceflight: engagement and internalization, but T cells failed to express CD25. T cell intracellular signaling (likely PKC pathway) within first few hours of activation is disrupted
	Hatton et al. (2002)	In	Microgravity differentially modifies protein kinase C isoform translocation in monocytes and T cells
	Tauber et al. (2015)	Parabolic flight and 2D Clinostat	Altered gravity conditions suppressed CD3 and IL-2 receptor surface expression, and reduced levels of ZAP70 and phosphorylated LAT
T cell cytoskeletal rearrangement/ MTOC orientation	Lewis et al. (1998)	In	Jurkat cells stimulated during spaceflight: diffuse shortened microtubules, poorly defined microtubule organizing centers (MTOC), and increased percentage of FAS/APO-1 apoptotic cells. Cytoskeletal changes potential cause of poor T cell activation during flight
	Hughes-Fulford (2003)	In	Collapse of actin cytoskeleton, loss of focal adhesions, increase in PGE2 during microgravity. Actin and microtubule modification in microgravity, alteration in anabolic signal transduction, probably through growth factor receptors and associated kinase pathways connected to cytoskeleton

(continued)

Table 14.1 (continued)

Category	Author	Flight phase	Findings
Leukocyte proliferation or blastogenesis	Konstantinova et al. (1973)	Post	Human lymphocytes have decreased responses to mitogens following spaceflight
	Kimzey et al. (1975)	Post	Marked decreased in mitogen-stimulated RNA and DNA synthesis by lymphocytes following both 28 and 59 days spaceflight
	Cogoli et al. (1984)	In	Mitogenic stimulation of cultured lymphocytes during spaceflight shows almost no proliferative/DNA synthesis response
	Nash and Mastro (1992)	Post	Following 4 day spaceflight, proliferation of lymph node lymphocytes to ConA or PMI-I depressed, not restored by exogenous IL-1 or IL-2. Response to PHA not decreased. Proliferation of splenocytes was not depressed
	Cogoli (1993)	In	T lymphocyte response to mitogens depressed by average of 56% in humans during and following spaceflight ($n = 129$)
	Grove et al. (1995)	Post	Following 10 day spaceflight, lymph node lymphocytes and splenocytes showed decreased IL-2 production in response to T cell receptor (TCR)-independent mitogens, but unchanged following TCR stimulation
	Pippia et al. (1996)	In	Addition of exogenous IL-1 does not restore loss of lymphocyte proliferative response during microgravity
	Sonnenfeld et al. (1998)	Post	Flown rats: blastogenesis of spleen cells in responses to mitogen inhibited for dams, but not for pups
	Gridley et al. (2009)	Post	Mice flown onboard Space Shuttle: splenocyte DNA synthesis significantly reduced in response to PHA stimulation
	Thiel et al. (2012)	2D Clinostat and parabolic flight	Functional weightlessness conditions disturbed in T lymphocytes the appropriate expression of several cell cycle regulatory proteins

Category	Author	Flight phase	Findings
T cell function (early blastogenesis)	Hashemi et al. (1999)	In	Cell cultures performed during spaceflight: TCR stimulation does not result in early blastogenesis (CD69/CD25), phorbol ester bypass only partially restores an activity. TCR internationalization and monocyte function appear not to be responsible for T cell deficit
	Crucian et al. (2008)	Post	T cell early blastogenesis depressed following long-duration spaceflight. Percentage of T cell subsets capable of being stimulated to secrete cytokines depressed following both short and long-duration spaceflight
	Bradley et al (2017)	Simulated Microgravity (sMG)	T cells in long-term sMG culture produce less IL-2 in response to DC+ cognate peptide stimulation than control static T cells, and express higher levels of CTLA-4
	Hughes-Fulford et al. (2015)	In	Human leukocytes stimulated with mitogens onboard ISS. miR-21 gene expression suppressed in microgravity compared to normal gravity. Global microarray proved significant suppression of 85 genes in microgravity conditions compared to normal gravity samples
	Martinez et al. (2015)	In, sMG	Compared true microgravity condition with lab models of microgravity for murine T cell activation gene expression; sMG conditions replicated true microgravity for suppression of a panel of activation genes
Delayed type hypersensitivity	Taylor and Janney (1992)		DTH responses diminished during short-duration spaceflight. $N = 10$; antigens administered 46 h before landing, reactions read 2 h following landing
	Gmuncer et al (1994)		DTH responses diminished during or following long-duration spaceflight onboard MIR space station in 4 of 5 subjects

(continued)

Table 14.1 (continued)

Category	Author	Flight phase	Findings
Virus-specific T cell function	Stowe (2003)		Levels of EBV viral-specific T cells elevated prior to flight, their function reduced at landing
	Stowe (2009)		Samples collected during short-duration spaceflight: EBV-specific T cell levels increased, their function is reduced
	Mehta et al. (2014)	In	Indirect evidence of T cell dysfunction: during short-duration Shuttle missions, 14/17 astronauts shed Epstein Barr virus (EBV) in their saliva; 7 of these shed Varicella-Zoster virus (VZV) and 8 shed Cytomegalovirus (CMV)
	Crucian et al. (2015)	In	Astronaut immunity assessed at three intervals during 6-month mission. Significant reduction in mitogen-stimulated T cell function
	Mehta et al. (2017)	In	Indirect evidence of T cell dysfunction: compared to healthy ground-based control subjects, astronauts undertaking long-duration missions on ISS experienced a higher frequency of EBV, VZV, and CMV reactivation (8/23 vs 0/20)
NK function	Konstantinova et al. (1995)		Postflight, NK binding and lysis of target cells reduced, production of NK decreased
	Meshkov and Rykova (1995)		Decrease in percentage of lymphocytes that can bind targets leads to reduction in total NK activity
	Buravkova et al. (2004)		During cell culture onboard the ISS, NK binding and cytotoxicity to K562 target cells was unchanged
	Simpson et al. (2016)	In	During in-flight time points, NK from all 6 crewmembers compared to those from ground control subjects possessed less perforin and less Granzyme B; correlated with reduced cytotoxic responses against tumor targets

depression of 85 genes compared to their counterparts in normal gravity. Furthermore, miR-21 gene expression was lower, suggesting T cell activation may be regulated by micro-RNAs (Hughes-Fulford et al. 2015).

While these microgravity cell culture measurements represent solid findings, it is unknown if they reflect a true clinical risk for crewmembers. Since T cells from otherwise normal and healthy test subjects also do not activate during microgravity culture, these findings may potentially represent some gravity-sensing "artifact" associated with the unique culture conditions. The in vivo situation is quite different from pure microgravity culture, with the addition of adhesion, tissue environment, pressure, motion, and shear flow variables. It is also possible that microgravity does indeed inhibit T cell function to some degree in vivo, thus explaining the in-flight DTH response data. However, additional crewmember variables such as stress and isolation make such mechanistic distinctions problematic.

Removing and culturing cells from crewmembers or animals during or following spaceflight, and performing cell function assays using terrestrial cell culture, provides subject-specific clinical information. Almost universally, activation, blastogenesis, and proliferation are altered in crewmembers or animals postflight (Fig. 14.1). Since these cultures are free of the microgravity cell culture "artifact," this must represent legitimate in vivo immunosuppression. It is likely that multiple factors may explain these observations. Certainly stress hormone levels associated with the high-g reentry and readaptation to unit gravity play a significant role. Elevated stress hormone levels are commonly observed postflight (Gmunder et al. 1994; Meehan et al. 1992; Mills et al. 2001; Stowe et al. 2003), and corticosteroids are known to suppress immune function. It is also possible that any microgravity effect on T cell function may linger, and thus require some "recovery" time period.

Published this last decade, there have been a few seminal studies of crewmember whole blood and lymphocyte populations extracted thereof from time points during a space mission to extend our understanding of the human response to space voyage. To distinguish stress-induced immune perturbation from true immunodeficiency due to exposure to the microgravity environment, we performed a comprehensive investigation of the immune system of astronauts during short-duration spaceflight onboard the Space Shuttle (Crucian et al. 2013), as well as during long-duration missions at the ISS (Crucian et al. 2015). In the former project, blood samples from 19 astronauts were drawn immediately prior to return, to eliminate the variable of stress caused by landing, and compared to a post-landing time point. In terms of T cell function, responses to staphylococcal enterotoxin were reduced during and following spaceflight, as was virus-specific T cell responses. Cytokine production profiles following mitogenic stimulation were significantly altered both during and following spaceflight: production of IFNγ, IL-17 and IL-10 were reduced in flight, but production of TNF and IL-8 were elevated. The long-duration project was more comprehensive, sampling blood from 23 astronauts at time points before, during, and after their 6-month ISS expedition. In-flight samples were returned to Earth within 48 h of collection for immediate analysis. A reduction in both CD4 and CD8 T cell function persisted for the duration of the 6-month spaceflights, with differential responses between mitogens suggesting an activation

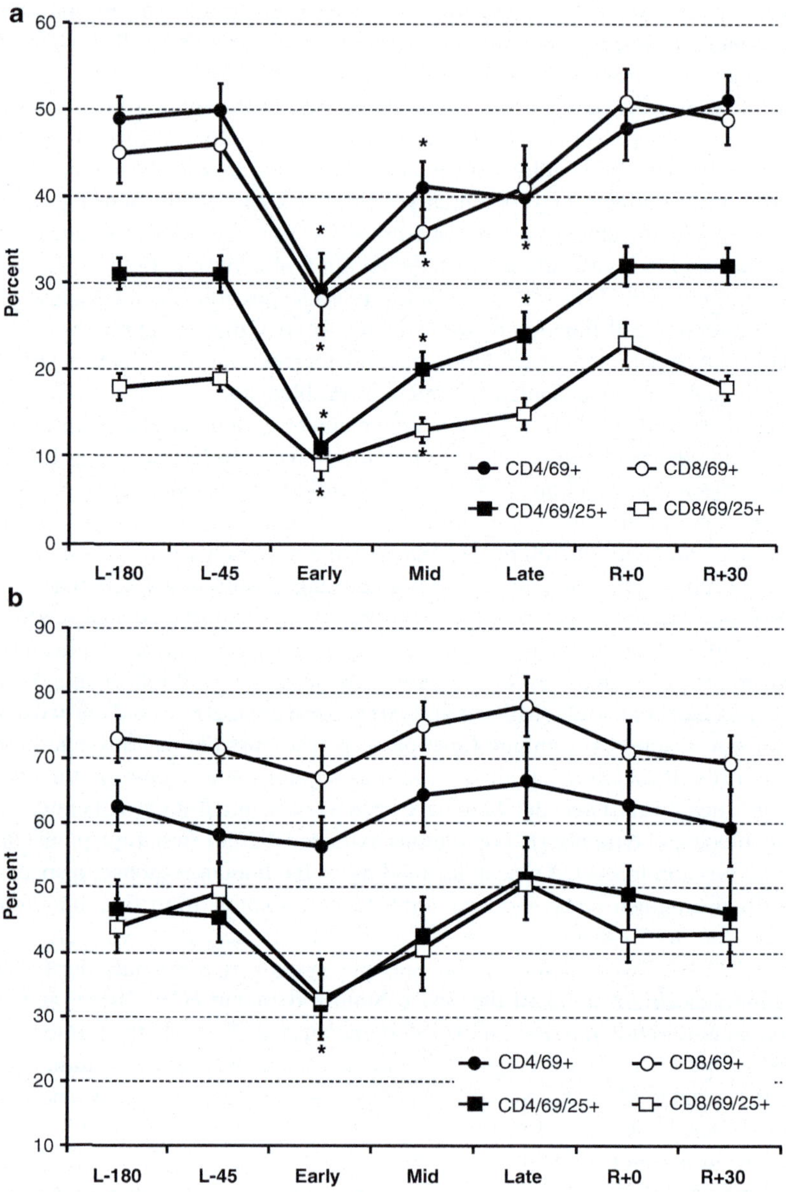

Fig. 14.1 (**a**) T-cell function (early blastogenesis) data: expression of either CD69 or CD69/CD25 following 24-h culture in the presence of (**a**) staphylococcal enterotoxin (**a** and **b**); or (**b**) antibodies to CD3 and CD28. Data are presented as mean ± standard error. Significance was evaluated via a Student's paired *t*-test by comparing all other data points to L-180 baseline data. Significant differences ($P \leq 0.05$) are indicated (*). Sample size for all data is 23 ISS astronaut subjects. *ISS* International Space Station. Adapted from Crucian B, et al. Alterations in adaptive immunity persist during long-duration spaceflight. NPJ Microgravity. 2015

threshold shift. Significant reductions in mitogen-stimulated IFNγ, IL-10, IL-5, TNF, and IL-6 persisted during spaceflight. Together, these studies form a profound argument that exposure to a microgravity environment imparts a true immune deficiency to humans that is not simply due to the stress of the voyage. This phenomenon begs the question: does space-induced immunodeficiency pose a clinical risk to the astronauts? A manifestation of impaired adaptive immunity is susceptibility to opportunistic infections and/or a failure to control established viral infections. Indeed, latent herpes virus reactivation, occurs to high levels during spaceflight (see Table 14.1; discussed in detail in Chap. 19).

14.2 Alterations in Cytokine Production Profiles

In addition to the generalized functional measures described above, a primary end point of effector cell function following stimulation is the production of cytokines. Cytokines are the signaling molecules of the immune system, and are responsible for cellular cross talk, recruitment, and activation among immune cell subsets. They regulate the type and magnitude of any immune response. Cytokines generally possess a short half-life, and thus function at low concentrations in a localized environment restricting the reaction to the site of pathology. Upon binding their receptor on the surface of their target cells, cytokines may induce the expression of activation genes, trigger new protein synthesis, and initiate proliferation and cellular effector function. CD4 T cells are the primary cytokine producing cells within adaptive immunity, and essentially regulate adaptive immune responses between cellular and humoral, as well as pro- and anti-inflammatory biases. Recent advances in cytokine biology have identified several specific categories of T cell immune responses based on specific patterns of cytokine expression. These categories are Th1, Th2, and Th17 T cell responses, and generally result in monocytic, mast cell/basophilic, or neutrophilic patterns of inflammation, respectively. These categories of cytokines generally inhibit each other's production, so only a single immune bias may be realized. Proper T cell cytokine regulation is important for immune health, and inappropriate biases (either hypo or hyper) may result in clinical disease. Rheumatoid arthritis is a Th17-mediated disease, whereas excessive Th2 responses can result in allergies and hypersensitivities (Jager and Kuchroo 2010).

Much may be learned about immune status and function by measuring cytokine levels following immune cell stimulation during cell culture. There have been many studies that investigated cytokine production profiles associated with spaceflight. A summation of representative data regarding cytokines and spaceflight is presented in Table 14.2. As might be expected the spaceflight data vary considerably based on human vs animal, vehicle/duration, mitogen used, culture system used, and other experimental variables. There are however, some generally common findings. The adaptive immune cytokine IFNγ is found to be suppressed during/following spaceflight (Gould et al. 1987; Sonnenfeld et al. 1988; Crucian et al. 2000, 2008), whereas production of inflammatory cytokines such as IL-1, IL-6, and IL-8 seem to vary associated with spaceflight (Chapes et al. 1994; Miller et al. 1995; Sonnenfeld et al.

Table 14.2 Representative studies: adaptive immune cytokines and spaceflight

Author	Flight phase	Human/ animal	Cells	Mitogen	Findings
Gould et al. (1987)	Post, short	Rats	Splenocytes	Concanavalin A (Con-A)	IFNγ production severely inhibited, IL-3 production unaffected
Sonnenfeld et al. (1988)	Post, short	Rats	Splenocytes	Con-A	IFNγ production severely inhibited, IL-3 production normal
Chapes et al. (1994)	In, short		Marrow macrophage line	LPS	Production of TNF, IL-1, IFNβ, IFNγ elevated during spaceflight
Miller et al. (1995)	Post, short	Rats	Splenocytes, thymocytes	Con-A or LPS	Both spleen and thymus cells secreted elevated IL-2, but only thymus cells secreted elevated IL-6
Sonnenfeld et al. (1996)	Post, short	Rhesus monkey	Leukocytes		Production of IL-1 and expression of IL-2 receptor decreased after spaceflight
Crucian et al. (2000)	Post, short	Human	Peripheral blood mononuclear cells (PBMC)	PMA/I	Intracellular production of IL-2 reduced in CD4 and CD8-positive T cells; production of IFNγ reduced in CD4 T cells only
Kaur et al. (2008)	Post, short	Human	Monocytes	LPS	Postflight, crew monocytes produced lower levels of IL-6 and Il-1β, and higher levels of IL-1ra and IL-8 compared to control subjects
Crucian et al. (2008)	Post, short and long	Human	Whole blood	Anti-CD3, PMA-I	Short duration: percentage of T cells secreting IL-2 and IFNγ decreased, IFNγ:IL-10 ratio decreased. Long duration: percentage of T cells secreting IL-2 reduced, secreting IFNγ unchanged, IFNγ:IL-10 ratio reduced
Gridley et al. (2009)	Post, short	Mouse	Splenocytes	Anti-CD3	Decreased production of IL-2, increased production of IL-10, IFNγ, MIP-1a following short-duration spaceflight
Fitzgerald et al. (2009)	In	Human	Lymphoid tissue	Antigenic polyclonal	Cell culture on board ISS, lymphoid cells did not respond to antigenic or polyclonal stimulation, losing ability to produce antibodies or cytokines (cells stimulated before launch maintained their ability onboard ISS)

Author	Flight phase	Human/animal	Cells	Mitogen	Findings
Gridley et al. (2009)	Post, short	Mouse	Splenocytes	LPS	Secretion of IL-6 and IL-10, but not TNF, increased in flown mice
Morukov et al. (2010)	Post, short	Humans	PBMC	LPS + PHA	Alterations in Th1:Th2 cytokine production following spaceflight indicating a Th2 shift
Mehta et al. (2013)	Post, long	Humans	Plasma	None	9/17 astronauts shed at least 1 latent herpes virus. This cohort had elevations in 10 plasma cytokines at return over baseline values
Crucian et al. (2013)	In, long	Humans	Plasma	None	Plasma levels of IFNα, IFNγ, IL-1β, IL-4, IL-10, IL-12, and TNF were significantly elevated in-flight
Crucian et al. (2015)	In, long	Human	PBMC	Anti-CD3, PMA+I	Blood samples collected from 22 ISS astronauts were stimulated to assess cytokine production profiles. Reductions were observed in the production of various Th1, Th2, and Th17 cytokines
Chang et al. (2015)	In, short	Mouse	OT-II CD4 Transgenic cells	OVA	Compared to ground mice controls, in flight mice with OT-II cells produced fivefold more IFN-γ and tenfold more IL-17; immune tolerance may be impaired in spaceflight, leading to excessive inflammatory responses

1996; Kaur et al. 2008). There is a postflight decrease in the IFNγ:IL-10 ratio and a Th1:Th2 shift during spaceflight (Crucian et al. 2008). A Th1:Th2 shift would seem to be supported by other evidence, such as diminished cell-mediated immunity during spaceflight (Taylor and Janney 1992; Gmunder et al. 1994), yet unaltered humoral immunity during spaceflight (Fuchs and Medvedev 1993; Stowe et al. 1999). In fact, Sonnenfeld summarized that cytokine changes in astronauts during spaceflight would not involve a general shutdown of the cytokine network, but rather involve selective alterations of specific cytokine functions (Sonnenfeld and Miller 1993). In support of this concept, plasma samples from 19 astronauts aboard Space Shuttle revealed higher levels of IFNα, IFNγ, IL-1β, IL-4, IL-10, IL-12, and TNF compared to postflight samples. Furthermore, PBMC responses to mitogenic stimulation included decreases in IFN-γ, IL-17, and IL-10 production, but increases in TNF and IL-8, during spaceflight compared with post-landing samples. More in-flight assessments, or in-flight sampling followed by terrestrial analysis, using comprehensive cytokine arrays will be required to completely understand cytokine dysregulation in-flight, and derive associated clinical risks. Moreover, the role of temperature dysregulation and increase in the core body temperature as observed in space crew on the ISS could either be induced by the immune changes or could be an independently occurring homeostatic change that causes immune dysfunctional states (see Chap. 26).

14.3 Humoral Immunity

Humoral immunity has not been investigated to the same extent as the cellular aspects of adaptive immunity. Those studies which have been performed indicate humoral immunity may not suffer as much during spaceflight as cellular immunity. Following spaceflight, no changes were observed in total immunoglobulins and Ig isotypes for Russian cosmonauts (Fuchs and Medvedev 1993). Similar results were reported following Space Shuttle flights for US astronauts (Stowe et al. 1999), and in Russian cosmonauts following long-duration flight onboard the ISS (Rykova et al. 2006; 2008). Ongoing studies, such as the vaccination of 1-year crewmembers during the NASA “Twins” investigation, and an ongoing assessment of antibody-free light chains performed at Lousiana State University, should soon provide further insight regarding the effects of spaceflight on humoral immunity.

14.4 Ground-Analog Studies

Despite the great success of the Space Shuttle, MIR and ISS programs, historically there has been extremely limited access to in-flight resources to support on-orbit studies. Limitations on up/down mass capability, crew time, power, volume, and microgravity-compatible instrumentation have all precluded large-scale or complicated studies as may be readily performed on Earth. Some aspects of human spaceflight may be replicated to high fidelity using "ground-based space analogs." It is thus very advantageous when possible to advance spaceflight knowledge using these available analogs. Choice of analog depends highly on the physiological system of interest. For example, prolonged head-down tilt (HDT) bed rest is an appropriate analog for bone loss and muscle deconditioning. However, HDT bed rest does not replicate the likely causal factors of spaceflight-associated immune dysregulation: mission stress, isolation, confinement, disrupted circadian rhythm, microgravity, radiation, etc. The best likely terrestrial analog for these particular space factors would include extreme environment deployment, prolonged isolation, legitimate subject risk, and operational mission activities. For immune studies, NASA has investigated deployment to the Canadian arctic and undersea station deployment (NEEMO) as human analogs for immune dysregulation. The Russian Space Agency had completed the Mars500 project, a high-fidelity simulation for the confinement and duration of a projected Mars mission (see Chap. 37), which include an immune surveillance component that indicated significant immune dysregulation with indication of immune hypersensitive pattern. Interestingly, most pronounced changes were observed at the transition phases which reflected to a certain extent the stress related to mission stress and environmental changes (Yi et al. 2014, 2015). Some of the observations could be reconfirmed during the European Space Agency (ESA-) CHOICE investigation at the French-Italian "Concordia Station" at Dome C on the high Antarctic plateau confirming incrementing immune dysregulation and ex vivo assessed Type IV hypersensitivity pattern (Feuerecker et al. 2018). The Antarctic analog, with prolonged 1-year mission deployment, extreme environment and risk, long travel, subject isolation, disrupted circadian rhythm, mission objectives, and station environment, likely represents one of the best human space analogs for immune dysregulation on Earth. This ESA study is ongoing, but deepen the knowledge on mechanisms on immune dysregulation to hereby validate life at Concordia as a ground analog for spaceflight immune dysregulation (see also Chaps. 16, 25, 36, and 38). Any validated analog could then be used for further mechanistic studies or evaluation of countermeasures. Preliminary data indicate that winterover at Concordia station, at significant altitude and in conditions of hypobaric hypoxia, may be somewhat distinctive to current spaceflight condition but could well reflect a future space exploration scenario referring to habitat atmosphere. Hypoxia is known to influence immunity, largely in a stimulatory fashion. Therefore, to allow comparative effects on the lead of the environmental stressor during life in harsh conditions during winterover, coastal Antarctica stations, although somewhat less extreme in environment than interior stations, has been implemented and may afford a closer approximation to spaceflight from an immunological perspective. Specifically, an ongoing NASA-German study is currently assessing winterover at the German "Neumayer III" station—the CHOICE-coastal study—using assays similar to those which define altered immunity in astronauts.

Other nonhuman analogs exist which may have significant utility. Single cell studies may be performed in "clinostat" or "bioreactor" devices, which simulate microgravity during cell culture by slowly spinning the culture vessel, resulting in a constantly randomized gravity vector. Studies have indicated similar dysregulation of T cell activation during bioreactor culture as has been observed during on-orbit culture (Hashemi et al. 1999; Thiel et al. 2012; Bradley et al. 2017). Animal studies, offering a greater variety of sampling opportunities than human studies, may be performed using various stress models such as hind-limb suspension or confinement situations. As a proof-of-concept, Martinez et al. (2015) compared a couple of laboratory models of microgravity to true microgravity; the simulated environmental conditions replicated the silencing of a plethora of T cell activation genes observed in mice subjected to spaceflight.

14.5 Conclusion

This chapter has summarized the current understanding of the relationship between spaceflight and the human adaptive immune system. Although some of the return findings are almost certainly related to landing and readaptation, it is likely that at least some postflight observations may reflect the impact of in-flight immune alterations. In fact, the very limited in-flight studies that have been performed confirm an adaptive immune system decrement during both short- and long-duration spaceflight. What remains in need is a comprehensive in-flight immune assessment during long-duration expeditions, comparable to what has been done in some ground-based space analog environments. This, together with ongoing and new Earth-bound and clinical studies, is required to define the impact on the human immune system of the "Space-flight Exposome"—the sum of all the environmental stressors with which crewmembers must contend: radiation, microgravity, isolation and confinement, temperature and circadian misalignment, altered nutrition, etc... (Chouker et al. 2008). If we determine a severe dysregulation persists during long-duration missions—as preliminary data from ongoing projects on the ISS indicate ("Functional/Integrated Immune", "IMMUNO-2", and IBMP-Immune studies)—and hence are not transiently related to either launch or landing, then the clinical risks of spaceflight to astronauts increase significantly (Crucian and Sams 2009; Crucian et al. 2016a, b). Thus, a strategy for both robust immune surveillance and countermeasures validation must be devised and operational prior to the initiation of deep space exploration missions (Frippiat et al. 2016).

References

Boonyaratanakornkit JB, Cogoli A, Li CF, Schopper T, Pippia P, Galleri G et al (2005) Key gravity-sensitive signaling pathways drive T cell activation. FASEB J 19(14):2020–2022

Bradley JH, Stein R, Randolph B, Molina E, Arnold JP, Gregg RK (2017) T cell resistance to activation by dendritic cells requires long-term culture in simulated microgravity. Life Sci Space Res (Amst) 15:55–61. https://doi.org/10.1016/j.lssr.2017.08.002

Buravkova LB, Rykova MP, Grigorieva V, Antropova EN (2004) Cell interactions in microgravity: cytotoxic effects of natural killer cells in vitro. J Gravit Physiol 11(2):P177–P180

Chang TT, Spurlock SM, Candelario TL, Grenon SM, Hughes-Fulford M (2015) Spaceflight impairs antigen-specific tolerance induction in vivo and increases inflammatory cytokines. FASEB J 29(10):4122–4132. https://doi.org/10.1096/fj.15-275073. Epub 2015 Jun 17

Chapes SK, Morrison DR, Guikema JA, Lewis ML, Spooner BS (1994) Production and action of cytokines in space. Adv Space Res 14(8):5–9

Chouker A, Morukov B, Sams C (2008) Clinical immunology in new frontiers. Scientific American Presents: Looking up, Europe's quiet revolution in microgravity research. Scientific American, New York, pp 24–31

Cogoli A (1993) The effect of space flight on human cellular immunity. Environ Med 37(2):107–116

Cogoli A (1997) Signal transduction in T lymphocytes in microgravity. Gravit Space Biol Bull 10(2):5–16

Cogoli A, Tschopp A, Fuchs-Bislin P (1984) Cell sensitivity to gravity. Science 225(4658):228–230

Cooper D, Pellis NR (1998) Suppressed PHA activation of T lymphocytes in simulated microgravity is restored by direct activation of protein kinase C. J Leukoc Biol 63(5):550–562

Crucian B, Sams C (2009) Immune system dysregulation during spaceflight: clinical risk for exploration-class missions. J Leukoc Biol 86(5):1017–1018

Crucian BE, Cubbage ML, Sams CF (2000) Altered cytokine production by specific human peripheral blood cell subsets immediately following space flight. J Interf Cytokine Res 20(6):547–556

Crucian BE, Stowe RP, Pierson DL, Sams CF (2008) Immune system dysregulation following short- vs long-duration spaceflight. Aviat Space Environ Med 79(9):835–843

Crucian B, Stowe R, Mehta S, Uchakin P, Quiriarte H, Pierson D, Sams C (2013) Immune system dysregulation occurs during short duration spaceflight on board the space shuttle. J Clin Immunol 33(2):456–465. https://doi.org/10.1007/s10875-012-9824-7

Crucian B, Stowe RP, Mehta S, Quiriarte H, Pierson D, Sams C (2015) Alterations in adaptive immunity persist during long-duration spaceflight. NPJ Microgravity. Sep 3;1:15013. doi: https://doi.org/10.1038/npjmgrav.2015.13

Crucian B, Johnston S, Mehta S, Stowe R, Uchakin P, Quiriarte H, Pierson D, Laudenslager ML, Sams C (2016a) A case of persistent skin rash and rhinitis with immune system dysregulation onboard the International Space Station. J Allergy Clin Immunol Pract 4(4):759–762.e8. https://doi.org/10.1016/j.jaip.2015.12.021

Crucian B, Babiak-Vazquez A, Johnston S, Pierson DL, Ott CM, Sams C (2016b) Incidence of clinical symptoms during long-duration orbital spaceflight. Int J Gen Med 9:383–391. eCollection 2016

Feuerecker M, Crucian BE, Quintens R, Pagel J-I, Salam AP, Rybka A, Moreels M, Strewe C, Stowe R, Mehta S, Schelling G, Thiel M, Baatout S, Sams C, Choukèr A (2018) Immune sensitization during one year in the Antarctic high altitude Concordia Environment. Allergy. https://doi.org/10.1111/all.13545

Fitzgerald W, Chen S, Walz C, Zimmerberg J, Margolis L, Grivel JC (2009) Immune suppression of human lymphoid tissues and cells in rotating suspension culture and onboard the International Space Station. In Vitro Cell Dev Biol Anim 45(10):622–632

Frippiat JP, Crucian BE, de Quervain DJ, Grimm D, Montano N, Praun S, Roozendaal B, Schelling G, Thiel M, Ullrich O, Choukèr A (2016) Towards human exploration of space: the THESEUS review series on immunology research priorities. NPJ Microgravity 2:16040. https://doi.org/10.1038/npjmgrav.2016.40

Fuchs BB, Medvedev AE (1993) Countermeasures for ameliorating in-flight immune dysfunction. J Leukoc Biol 54(3):245–252

Gmunder FK, Konstantinova I, Cogoli A, Lesnyak A, Bogomolov W, Grachov AW (1994) Cellular immunity in cosmonauts during long duration spaceflight on board the orbital MIR station. Aviat Space Environ Med 65(5):419–423

Gould CL, Lyte M, Williams J, Mandel AD, Sonnenfeld G (1987) Inhibited interferon-gamma but normal interleukin-3 production from rats flown on the space shuttle. Aviat Space Environ Med 58(10):983–986

Gridley DS, Slater JM, Luo-Owen X, Rizvi A, Chapes SK, Stodieck LS et al (2009) Spaceflight effects on T lymphocyte distribution, function and gene expression. J Appl Physiol 106(1):194–202

Grove DS, Pishak SA, Mastro AM (1995) The effect of a 10-day space flight on the function, phenotype, and adhesion molecule expression of splenocytes and lymph node lymphocytes. Exp Cell Res 219(1):102–109

Hashemi BB, Penkala JE, Vens C, Huls H, Cubbage M, Sams CF (1999) T cell activation responses are differentially regulated during clinorotation and in spaceflight. FASEB J 13(14):2071–2082

Hatton JP, Gaubert F, Cazenave JP, Schmitt D (2002) Microgravity modifies protein kinase C isoform translocation in the human monocytic cell line U937 and human peripheral blood T-cells. J Cell Biochem 87(1):39–50

Hughes-Fulford M (2003) Function of the cytoskeleton in gravisensing during spaceflight. Adv Space Res 32(8):1585–1593

Hughes-Fulford M, Chang TT, Martinez EM, Li CF (2015) Spaceflight alters expression of microRNA during T-cell activation. FASEB J 29(12):4893–4900. https://doi.org/10.1096/fj.15-277392

Jager A, Kuchroo VK (2010) Effector and regulatory T-cell subsets in autoimmunity and tissue inflammation. Scand J Immunol 72(3):173–184

Kaur I, Simons ER, Kapadia AS, Ott CM, Pierson DL (2008) Effect of spaceflight on ability of monocytes to respond to endotoxins of gram-negative bacteria. Clin Vaccine Immunol 15(10):1523–1528

Kimzey SL, Ritzmann SE, Mengel CE, Fischer CL (1975) Skylab experiment results: hematology studies. Acta Astronaut 2(1–2):141–154

Konstantinova IV, Antropova EN, Legen'kov VI, Zazhirei VD (1973) Reactivity of lymphoid blood cells in the crew of "Soiuz-6", "Soiuz-7" and "Soiuz-8" spacecraft before and after flight. Kosm Biol Med 7(6):35–40

Konstantinova IV, Rykova M, Meshkov D, Peres C, Husson D, Schmitt DA (1995) Natural killer cells after ALTAIR mission. Acta Astronaut 36(8–12):713–718

Lewis ML, Reynolds JL, Cubano LA, Hatton JP, Lawless BD, Piepmeier EH (1998) Spaceflight alters microtubules and increases apoptosis in human lymphocytes (Jurkat). FASEB J 12(11):1007–1018

Martinez EM, Yoshida MC, Candelario TL, Hughes-Fulford M (2015) Spaceflight and simulated microgravity cause a significant reduction of key gene expression in early T-cell activation. Am J Physiol Regul Integr Comp Physiol 308(6):R480–R488. https://doi.org/10.1152/ajpregu.00449.2014. Epub 2015 Jan 7

Meehan RT, Neale LS, Kraus ET, Stuart CA, Smith ML, Cintron NM et al (1992) Alteration in human mononuclear leucocytes following space flight. Immunology 76(3):491–497

Mehta SK, Crucian BE, Stowe RP, Simpson RJ, Ott CM, Sams CF, Pierson DL (2013) Reactivation of latent viruses is associated with increased plasma cytokines in astronauts. Cytokine 61(1):205–209. https://doi.org/10.1016/j.cyto.2012.09.019.PubMed

Mehta SK, Laudenslager ML, Stowe RP, Crucian BE, Sams CF, Pierson DL (2014) Multiple latent viruses reactivate in astronauts during Space Shuttle missions. Brain Behav Immun 41:210–217. https://doi.org/10.1016/j.bbi.2014.05.014

Mehta SK, Laudenslager ML, Stowe RP, Crucian BE, Feiveson AH, Sams CF, Pierson DL (2017) Latent virus reactivation in astronauts on the international space station. NPJ Microgravity 2017Apr 3:11. doi: https://doi.org/10.1038/s41526-017-0015-y. eCollection 2017

Meshkov D, Rykova M (1995) The natural cytotoxicity in cosmonauts on board space stations. Acta Astronaut 36(8–12):719–726

Miller ES, Koebel DA, Sonnenfeld G (1995) Influence of spaceflight on the production of interleukin-3 and interleukin-6 by rat spleen and thymus cells. J Appl Physiol 78(3):810–813

Mills PJ, Meck JV, Waters WW, D'Aunno D, Ziegler MG (2001) Peripheral leukocyte subpopulations and catecholamine levels in astronauts as a function of mission duration. Psychosom Med 63(6):886–890

Morukov VB, Rykova M, Antropova EN, Berendeeva TA, Ponomarev SA, Larina IM (2010) Indicators of innate and adaptive immunity of cosmonauts after long-term space flight to international space station. Fiziol Cheloveka 36(3):19–30

Nash PV, Mastro AM (1992) Variable lymphocyte responses in rats after space flight. Exp Cell Res 202(1):125–131

Pippia P, Sciola L, Cogoli-Greuter M, Meloni MA, Spano A, Cogoli A (1996) Activation signals of T lymphocytes in microgravity. J Biotechnol 47(2–3):215–222

Rykova MP, Gertsik Iu G, Antropova EN, Buravkova LB (2006) Immunoglobulin e and allergen-specific IgE antibodies in cosmonauts before and after long-duration missions on the International Space Station. Aviakosm Ekolog Med 40(2):19–22

Rykova MP, Antropova EN, Larina IM, Morukov BV (2008) Humoral and cellular immunity in cosmonauts after the ISS missions. Acta Astronaut 63(7–10):697–705

Simpson RJ, Bigley AB, Spielmann G, Kunz HE, Agha N, Baker F, Rooney B, Mylabathula PL, Graff RM, Crucian BE, Laughlin M, Mehta SK, Pierson DL (2016) Long duration spaceflight impairs NK-cell function in astronauts. Med Sci Sports Exerc 48(5 Suppl 1):87

Sonnenfeld G, Miller ES (1993) The role of cytokines in immune changes induced by spaceflight. J Leukoc Biol 54(3):253–258

Sonnenfeld G, Gould CL, Williams J, Mandel AD (1988) Inhibited interferon production after space flight. Acta Microbiol Hung 35(4):411–416

Sonnenfeld G, Davis S, Taylor GR, Mandel AD, Konstantinova IV, Lesnyak A et al (1996) Effect of space flight on cytokine production and other immunologic parameters of rhesus monkeys. J Interf Cytokine Res 16(5):409–415

Sonnenfeld G, Foster M, Morton D, Bailliard F, Fowler NA, Hakenewerth AM et al (1998) Spaceflight and development of immune responses. J Appl Physiol 85(4):1429–1433

Stowe RP (2003) Impaired effector function in virus-specific T cells in astronauts. NASA Investigators Workshop, Houston, 2003

Stowe RP (2009) Validation of procedures for monitoring crewmember immune function. NASA Investigators Workshop, Houston, 2009

Stowe RP, Sams CF, Mehta SK, Kaur I, Jones ML, Feeback DL et al (1999) Leukocyte subsets and neutrophil function after short-term spaceflight. J Leukoc Biol 65(2):179–186

Stowe RP, Sams CF, Pierson DL (2003) Effects of mission duration on neuroimmune responses in astronauts. Aviat Space Environ Med 74(12):1281–1284

Tauber S, Hauschild S, Paulsen K, Gutewort A, Raig C, Hürlimann E, Biskup J, Philpot C, Lier H, Engelmann F, Pantaleo A, Cogoli A, Pippia P, Layer LE, Thiel CS, Ullrich O (2015) Signal transduction in primary human T lymphocytes in altered gravity during parabolic flight and clinostat experiments. Cell Physiol Biochem 35(3):1034–1051. https://doi.org/10.1159/000373930

Taylor GR, Janney RP (1992) In vivo testing confirms a blunting of the human cell-mediated immune mechanism during space flight. J Leukoc Biol 51(2):129–132

Thiel CS, Paulsen K, Bradacs G, Lust K, Tauber S, Dumrese C, Hilliger A, Schoppmann K, Biskup J, Gölz N, Sang C, Ziegler U, Grote KH, Zipp F, Zhuang F, Engelmann F, Hemmersbach R, Cogoli A, Ullrich O (2012) Rapid alterations of cell cycle control proteins in human T lymphocytes in microgravity. Cell Commun Signal 10(1):1. https://doi.org/10.1186/1478-811X-10-1

Yi B, Rykova M, Feuerecker M, Jäger B, Ladinig C, Basner M, Hörl M, Matzel S, Kaufmann I, Strewe C, Nichiporuk I, Vassilieva G, Rinas K, Baatout S, Schelling G, Thiel M, Dinges DF, Morukov B, Choukèr A (2014) 520 d Isolation and confinement simulating a flight to Mars reveals heightened immune responses and alterations of leukocyte phenotype. Brain Behav Immun 40:203–210. https://doi.org/10.1016/j.bbi.2014.03.018

Yi B, Rykova M, Jäger G, Feuerecker M, Hörl M, Matzel S, Ponomarev S, Vassilieva G, Nichiporuk I, Choukèr A (2015) Influences of large sets of environmental exposures on immune responses in healthy adult men. Sci Rep 5:13367. https://doi.org/10.1038/srep13367

B-Cell Immunology in Space

15

Coralie Fonte and Jean-Pol Frippiat

15.1 Introduction

Spaceflight corresponds to a unique allostatic (over-) load (see Chap. 4) as by combination of stressors that can be classified in two main categories: chronic stressors (e.g. microgravity, confinement, isolation, circadian rhythm misalignment) and acute stressors encountered during periods of intense activity, such as dockings and extravehicular activities, but also during take-off and landing phases (hypergravity exposure). While acute stress has been described as beneficial (mobilization of individual's defense capabilities), chronic stress appears to have deleterious effects as several studies have shown that it contributes to many pathologies (reviewed in Godbout and Glaser 2006, see also Chaps. 4 and 6). Studies performed on humans and animals returning from space travels have shown that this extreme environment has a negative impact on almost all physiological systems. It causes muscle atrophy, bone demineralization, cardiovascular and metabolic dysfunctions, alters cognitive processes, and reduces immunological competence. Regarding this last point, it was shown that 15 of the 29 astronauts involved in Apollo missions developed bacterial or viral infections during, immediately after, or within 1 week of landing (Kimzey 1977). In addition, the first epidemiological study based on medical data collected on 46 astronauts who spent 6 months onboard the International Space Station (ISS), showed that 46% of them had to face immunological problems (Crucian et al. 2016). These observations demonstrate that, on average, spaceflight affects the immune system of 50% of the astronauts and that immune dysregulations occur during spaceflight and not only after landing. They are therefore not exclusively due to the hypergravity phase associated to landing or to Earth gravity rehabilitation (Crucian et al. 2016).

C. Fonte · J.-P. Frippiat (✉)
Stress Immunity Pathogens Laboratory, EA 7300, Faculty of Medicine,
Lorraine University, Vandœuvre-lès-Nancy, France
e-mail: coralie-fonte@univ-lorraine.fr; jean-pol.frippiat@univ-lorraine.fr

© Springer Nature Switzerland AG 2020
A. Choukér (ed.), *Stress Challenges and Immunity in Space*,
https://doi.org/10.1007/978-3-030-16996-1_15

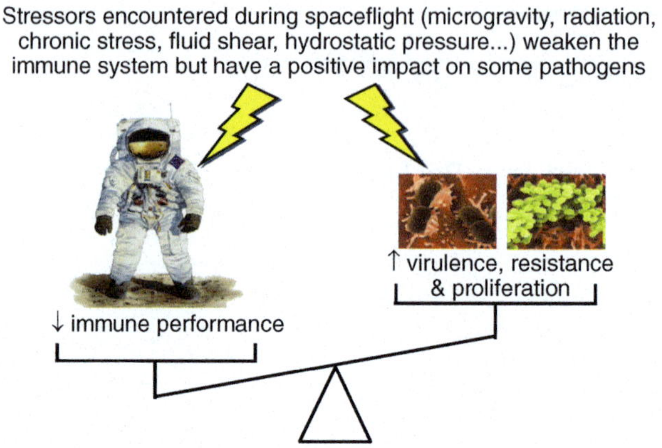

Fig. 15.1 Environmental changes (stressors) encountered during spaceflight weaken the immune system but can also increase the virulence, resistance and proliferation of some pathogens. This imbalance likely contributes to explain increased susceptibility to infections. Reprinted from Frippiat et al. (2016) with the permission of Nature Publishing Group

In parallel of these immunological modifications, spaceflight-induced changes in microbial physiology have to be considered because some bacteria can take advantage of this extreme environment (see Chap. 18). For example, *Salmonella* cultured under real or simulated spaceflight conditions exhibited increased virulence, resistance, and survival in macrophages (Nickerson et al. 2000). In addition, reduced activity of some antibiotics associated with variations in pharmacokinetics have been reported (Taylor and Sommer 2005; Klaus and Howard 2006). These observations, coupled with the weakening of the immune system, certainly contribute to explain why nearly 50% of the astronauts exhibit in-flight immunological alterations (Fig. 15.1).

In this chapter, we will focus on the effects of spaceflight on the development and the responses of B lymphocytes (Guéguinou et al. 2009; Frippiat et al. 2016).

15.2 Effects of Spaceflight Conditions on B-Cell Development

Cells that make up the immune system derive from hematopoietic stem cells that give rise to common myeloid progenitors (CMP) and common lymphoid progenitors (CLP). After several steps of differentiation, CMP generate myeloid cells (granulocytes, monocytes, macrophages, dendritic cells) while CLP generate lymphoid cells (B and T lymphocytes and Natural Killer cells (NK)).

Different studies have shown that spaceflight conditions affect lymphopoiesis (the development of lymphocytes) thanks to the use of different animal models: mouse and the urodele amphibian *Pleurodeles waltl*. The latter lends itself well to the constraints associated with space experiments and has all the cardinal elements of the mammalian immune system (Frippiat 2013). It can therefore be used to improve our understanding of the immunosuppressive effects of spaceflight. Using that animal model, it was shown that the expression of IgM immunoglobulin heavy chain, of the lymphoid-determining transcription factor Ikaros, and of NFκ-B members that play a crucial role in lymphocyte development (Vallabhapurapu and Karin 2009) are modified when amphibian embryos are subjected to gravity changes, suggesting a modification in B lymphopoiesis (Huin-Schohn et al. 2013) (Fig. 15.2). This hypothesis was then confirmed in C57/BL6 mice subjected to an anti-orthostatic suspension (HU for "Hindlimb Unloading," a ground-based model frequently used to simulate the effects of spaceflight on rodents) for 21 days to mimic a long-duration mission at the human-scale. Indeed, it was shown that HU induces a decrease in both bone microstructure and the frequency of B-cell progenitors in the bone marrow. While multipotent hematopoietic progenitors were not affected by HU, a decrease in B lymphopoiesis was observed in femoral bone marrow as of the CLP stage with a major block at the pro-B to pre-B cell transition (five- to tenfold decrease) (Fig. 15.3) (Lescale et al. 2015). These reductions persisted during the 21-days period of suspension and were not due to stress because biochemical and behavioral studies have shown that mice adapt to HU in less than 3 days. Various studies have subsequently identified several

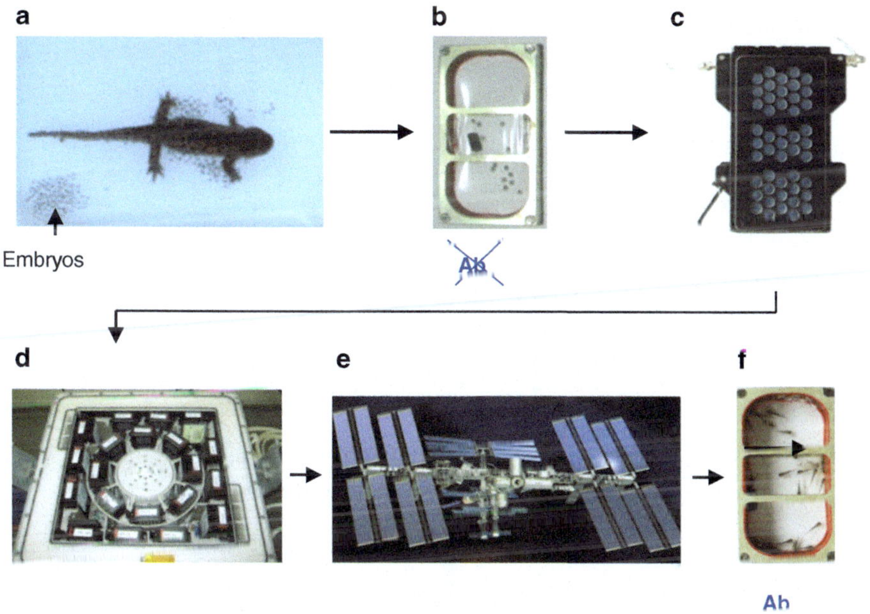

Fig. 15.2 Schematic presentation of the AMPHIBODY experiment performed onboard the ISS in 2006. Embryos of the urodele Amphian *P. waltl*, which do not yet produce antibodies (**a**), were placed in miniaquaria (**b**) that were inserted into Type 1 containers (**c**). These containers were loaded into the Kubik incubator (**d**) that was transported to the ISS (**e**). Embryos developed over a 10 day period in the space station. At landing, antibody-producing larvae were recovered (**f**). "Ab" stands for antibody. Reprinted from Frippiat (2013) with permission from Elsevier

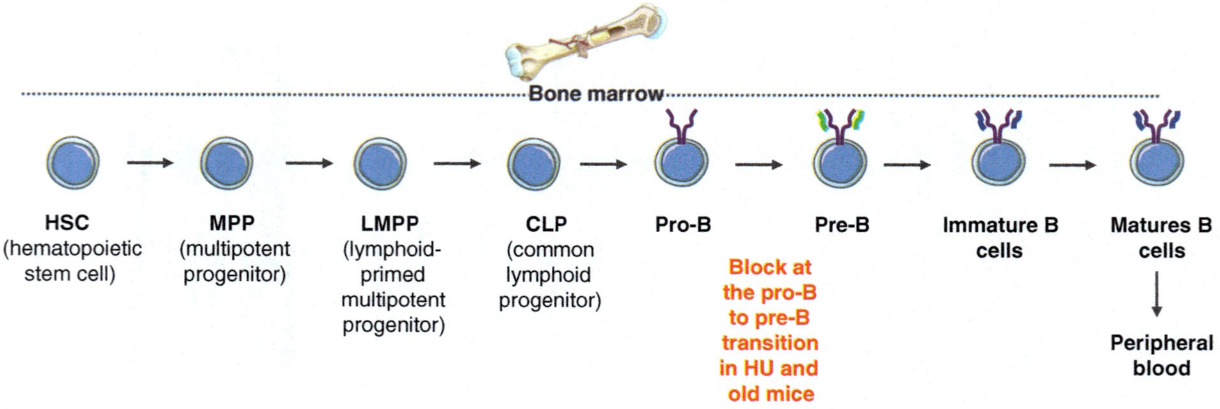

Fig. 15.3 Schematic presentation of B-cell development in the bone marrow (B lymphopoiesis). Antibody heavy chains appear at the pro-B stage (purple). At the pre-B stage, this chain is associated with a pseudo light chain (green) that will be replaced by a light chain (blue) in immature B-cells. Microgravity simulated by anti-orthostatic suspension (HU) and ageing weaken B lymphopoiesis as of the CLP stage with a major block at the pro-B to pre-B cell transition (Lescale et al. 2015)

causes of this B lymphopoiesis weakening such as a reduced expression of B-cell transcription factors (EBF and Pax5) and an alteration in STAT5-mediated IL-7 signaling, a signaling pathway required for the differentiation of B lymphocytes. This decrease in B lymphopoiesis is most certainly linked to suspension-induced osteoporosis because immune-competent B-cells derive from hematopoietic stem cells that reside in the bone marrow in specialized niches made up of bone and vascular structures, including bone-forming osteoblasts and bone-resorbing osteoclasts (Mercier et al. 2011). This hypothesis is supported by the fact that a doubling of the level of progenitors and stem cells was observed in the blood of HU mice suggesting that the interactions between these cells and their niches are disrupted. This example highlight the importance of keeping in mind that all physiological systems interact within an organism and that consequently, interconnections between systems must always be considered.

Some of these observations were confirmed in C57/BL6 mice subjected to real microgravity during a 30-day flight onboard the Bion-M 1 biosatellite (April 19–May 19, 2013). Indeed, osteocyte death, which may trigger bone resorption and result in bone mass and microstructural deterioration, was noted at landing (Gerbaix et al. 2017). Furthermore, a decrease in the expression of proteins necessary for the maturation of immune cells was noted in these mice (Tascher et al. 2019).

15.3 Effects of Spaceflight Conditions on B-Cell Responses

To determine whether the humoral immune response is affected by spaceflight conditions, adult *P. waltl* were immunized during a 5-month stay onboard the Mir space station (Fig. 15.4). These animals are, until now, the only ones to have been immunized in space. At the same time, other *P. waltl* of the same age and gender were reared in the laboratory under the same conditions as those found in the space station and immunized with the same antigen. The quantification of IgY (the physiological counterpart of human IgA—Schaerlinger et al. 2008) and IgM heavy-chain mRNAs in the spleen of these animals revealed that the level of IgM transcription was unaffected in flown animals, while the level of IgY transcription was at least three times higher in the spleen of animals immunized onboard Mir (Boxio et al. 2005). These data support the increase of IgA previously noted in astronauts by Konstantinova et al. (1993). This increase could result from a change in the distribution of IgY-producing B-cells because it has been reported that spaceflight can modify the distribution of leukocytes. More detailed studies of these amphibian humoral responses revealed spaceflight-induced changes in the expression of the VHII and VHVI gene segment subgroups used to build specific antibodies in response to the antigenic challenge. VHII and VHVI subgroups were found in 28% and 58% of IgM heavy-chains from animals immunized on Earth, respectively, and in 61% and 24% of IgM heavy-chains from animals immunized onboard Mir. More precisely, it was shown that *P. waltl* that were immunized on Earth or in space used the same VHII gene segment to build their IgM heavy-chains but that the expression of this

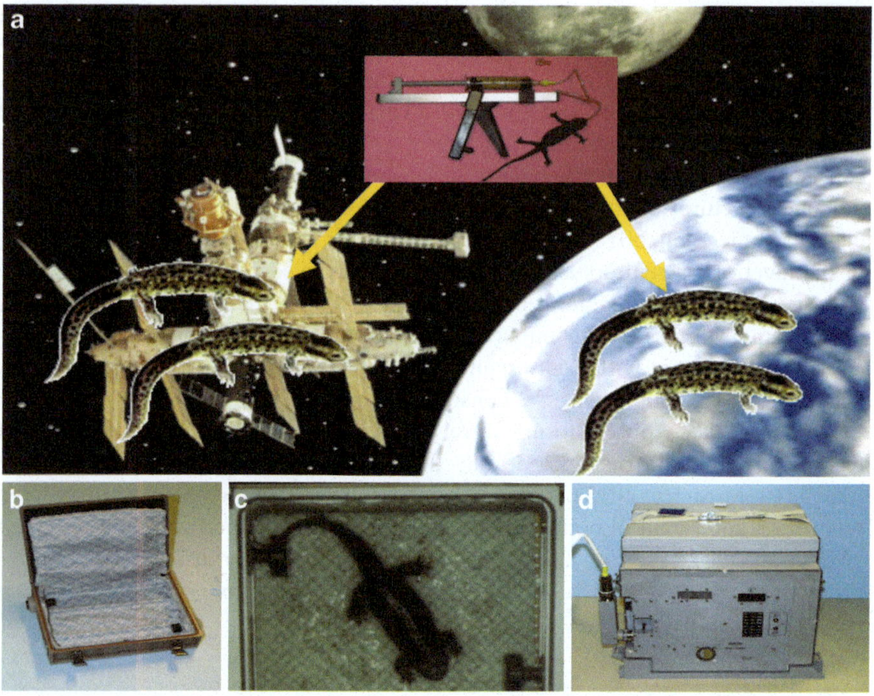

Fig. 15.4 Schematic presentation of the Genesis experiment performed in 1999. During that experiment adult *P. waltl* were immunized and reared during 5 months onboard the Mir space station (**a**). During their stay in the station, animals were force-fed using a syringe with a catheter. As controls, adult *P. waltl* were immunized and reared on Earth in the same conditions as onboard Mir. During that experiment, animals were kept in boxes on damp towels (**b**, **c**) that were placed inside the CTA ("Container de Transport Aller") (**d**) to regulate the temperature. Reprinted from Frippiat (2013) with permission from Elsevier

segment was two times higher under spaceflight conditions. It was also shown that animals immunized on Earth used four different VHVI gene segments to build their IgM heavy-chains, while animals immunized onboard Mir used only two of them. Strong decreases in the expression of IgM heavy-chain mRNAs encoded by the VHVI.A and VHVI.B segments were noted in flown animals (Bascove et al. 2009). These data indicate that flown animals expressed different Ig heavy-chain mRNAs by comparison to animals immunized on Earth. Finally, somatic hypermutations in heavy-chain variable domains (VH) of IgM antibodies were analyzed. Somatic hypermutations are important because they enable B lymphocytes to improve their antibody binding sites. To calculate the frequency of these mutations, cDNA sequences of IgM heavy-chain transcripts encoding VHII domains were aligned with the corresponding VHII germline sequence for each of the studied animals. This study revealed that somatic hypermutation occurs in space following immunization but at frequency two times lower than on Earth (Bascove et al. 2011). This observation suggests that antibody affinity maturation could be less efficient in space, thereby confirming the disruption of the humoral immune response under spaceflight conditions. This decrease in somatic hypermutation frequency likely results from the combination of several spaceflight-associated changes, such as the severe reduction in T cell activation, changes in cell–cell contacts, signal transduction, gene transcription, cytokine production, and cytoskeleton organization. Of all these changes, those concerning the cytoskeleton are of particular interest because this structure is involved in many functions. It participates in cell mobility, signal transduction, gene expression and cell cycle. The cytoskeleton could therefore be one of the structures through which a cell detects a change in gravitational force (see Chap. 17).

In complement of these in-flight studies, the impacts of gravity changes on the immune system of adult mice were studied. Eight-week-old male C57BL/6 mice were centrifuged for 21 days at 2 or 3 g. A substantial increase in serum corticosterone (a stress hormone) concentration was observed at the end of the centrifugation in 3 g mice, but not in 2 g mice. In the same way, a significant increase in the level of anxiety, persisting for more than two weeks after the return to normogravity, was noted in 3 g mice. A significant reduction in the proliferative response of B lymphocytes from 2 g mice, but not from 3 g mice, was associated to these changes. T lymphocyte proliferative responses fell at 2 and 3 g but less sharply than for B cells. Together, these results suggest a breakdown in adaptation at 3 g which takes the form of an increase in serum corticosterone, lasting anxiety, and a reduction of the proliferative response of T lymphocytes, whereas after 21 days at 2 g, mice only show a lower response of their B and T lymphocytes. It therefore seems that at 2 g, hypergravity on its own can affect the response of these two cell types (Guéguinou et al. 2012). On the other hand, when mice were subjected to HU during 21 days to simulate physiological changes observed in astronauts, no differences in serum corticosterone concentrations or in thymus and spleen masses were noted indicating adaptation. However, it appeared that the anti-orthostatic position led, in the absence of stress, to an inversion of the B/T lymphocyte ratio in mice's spleen and to a 40% weakening of B lymphocytes proliferative response (Gaignier et al. 2014) as was shown for 2 g mice. Taken together, these results show that B lymphocytes are sensitive to changes in gravity, which could explain the greater sensitivity to pathogens previously reported for HU mice (Aviles et al. 2003).

To understand why B lymphocytes from mice exposed to gravitational stress have a poorer response to mitogen stimulation, it is of interest to look at neurohormone receptors present on these cells. Glucocorticoid receptors have been reported on B-cells (Gruver-Yates et al. 2014) and very recently corticotropin-releasing hormone receptors 2, CRHR2 (Harlé et al. 2018). These receptors render B-cells directly sensitive to glucocorticoids and CRH produced when the hypothalamic-pituitary-adrenal (HPA) axis is activated by stress.

15.4 Similarities Between Spaceflight-Induced Modifications and Aging

Aging profoundly impairs the immune system and is characterized by changes in hematopoiesis, adaptive, and innate immunity. Thymus atrophy and the subsequent reduction in T-cell production are the most noticeable age-related events. Accelerated immune aging is also observed among individuals with high stress (Casaletto et al. 2018).

Interestingly, some of these age-associated changes are observed in astronauts who are subjected to numerous stressors during space missions. Indeed, increased susceptibility to infection and decreased response to vaccine (Sasaki et al. 2011), that may be correlated with impaired development and functions of B and T lymphocytes, have been observed in the elderly (Shaw et al. 2013). The development of these cells and their responses are also reduced when the gravitational force is modified (see above). HU rapidly induced modifications presenting many similarities with the changes observed in old mice both with regard to trabecular bone microarchitecture and to the cellular and molecular changes accompanying reduced B-cell differentiation (Lescale et al. 2015). Aging is accompanied by a decrease in antibody affinity (Sasaki et al. 2011) as shown in in-flight immunized *P. waltl* (Bascove et al. 2011). The use of antibody VH gene segments is modified during aging (Gibson et al. 2009) and in space (Bascove et al. 2009) which affects the diversity of the antibody repertoire. Taken together, these observations suggest that stressors encountered during space missions could lead to premature aging of the immune system.

15.5 Conclusion and Perspectives

On average, 50% of the astronauts are confronted to immunological problems during space missions. Our understanding of these dysfunctions has greatly progressed. However, the impact of long-term missions remains to be clarified. Indeed, most of our current knowledge has derived from space missions whose duration did not exceed six months. Moreover, since the homeostasis of the immune system is linked to that of other systems that are also affected by spatial conditions such as the musculoskeletal, nervous, and cardiovascular systems, it will be necessary to study the impact of spaceflight conditions on these interconnections through interdisciplinary approaches.

Knowledge that will emerge from these researches will be necessary to, on the one hand, better understand the risks incurred during future long-term space missions, where the crew will be left on its own without the possibility of a quick return to Earth and, on the other hand, to develop pharmacological or nutritional countermeasures to restore the immune system to prevent the development or aggravation of pathologies. These researches are also of interest for people living on Earth, since they could open therapeutic ways to treat immunosenescence and chronic stress-induced immune dysfunctional states.

Acknowledgments J.P.F. and his team were supported by the French National Space Agency (CNES), the European Space Agency (ESA), the French Ministry of Higher Education and Research, the Lorraine University, and the Lorraine area.

References

Aviles H, Belay T, Fountain K, Vance M, Sonnenfeld G (2003) Increased susceptibility to Pseudomonas aeruginosa infection under hindlimb-unloading conditions. J Appl Physiol 95:73–80

Bascove M, Huin-Schohn C, Guéguinou N, Tschirhart E, Frippiat JP (2009) Spaceflight-associated changes in immunoglobulin VH gene expression in the amphibian *Pleurodeles waltl*. FASEB J 23:1607–1615. https://doi.org/10.1096/fj.08-121327

Bascove M, Guéguinou N, Schaerlinger B, Gauquelin-Koch G, Frippiat JP (2011) Decrease in antibody somatic hypermutation frequency under extreme, extended spaceflight conditions. FASEB J 25:2947–2955. https://doi.org/10.1096/fj.11-185215

Boxio R, Dournon C, Frippiat JP (2005) Effets of a long-term spaceflight on immunoglobulin heavy chains of the urodele amphibian *Pleurodeles waltl*. J Appl Physiol 98:905–910

Casaletto KB, Staffaroni AM, Elahi F, Fox E, Crittenden PA, You M, Neuhaus J, Glymour M, Bettcher BM, Yaffe K, Kramer JH (2018) Perceived stress is associated with accelerated monocyte/macrophage aging trajectories in clinically normal adults. Am J Geriatr Psychiatry 26(9):952–963. https://doi.org/10.1016/j.jagp.2018.05.004

Crucian B, Babiak-Vazquez A, Johnston S, Pierson DL, Ott CM, Sams C (2016) Incidence of clinical symptoms during long-duration orbital spaceflight. Int J Gen Med 9:383–391. https://doi.org/10.2147/IJGM.S114188

Frippiat JP (2013) Contribution of the urodele amphibian *Pleurodeles waltl* to the analysis of spaceflight-associated immune system deregulation. Mol Immunol 56:434–441. https://doi.org/10.1016/j.molimm.2013.06.011

Frippiat JP, Crucian BE, de Quervain DJF, Grimm D, Montano N, Praun S et al (2016) Towards human exploration of space: the THESEUS review series on immunology research priorities. NPJ Microgravity 2:16040. https://doi.org/10.1038/npjmgrav.2016.40

Gaignier F, Schenten V, De Carvalho Bittencourt M, Gauquelin-Koch G, Frippiat JP, Legrand-Frossi C (2014) Three weeks of murine hindlimb unloading induces shifts from B to T and from Th to Tc splenic lymphocytes in absence of stress and differentially reduces cell-specific mitogenic responses. PLoS One 9:e92664. https://doi.org/10.1371/journal.pone.0092664

Gerbaix M, Gnyubkin V, Farlay D, Olivier C, Ammann P, Courbon G et al (2017) One-month spaceflight compromises the bone microstructure, tissue-level mechanical properties, osteocyte survival and lacunae volume in mature mice skeletons. Sci Rep 7:2659. https://doi.org/10.1038/s41598-017-03014-2

Gibson KL, Wu YC, Barnett Y, Duggan O, Vaughan R, Kondeatis E et al (2009) B-cell diversity decreases in old age and is correlated with poor health status. Aging Cell 8:18–25. https://doi.org/10.1111/j.1474-9726.2008.00443.x

Godbout JP, Glaser R (2006) Stress-induced immune dysregulation: implications for wound healing, infectious disease and cancer. J Neuroimmune Pharmacol 1:421–427. https://doi.org/10.1007/s11481-006-9036-0

Gruver-Yates AL, Quinn MA, Cidlowski JA (2014) Analysis of glucocorticoid receptors and their apoptotic response to dexamethasone in male murine B cells during development. Endocrinology 155:463–474. https://doi.org/10.1210/en.2013-1473

Guéguinou N, Huin-Schohn C, Bascove M, Bueb JL, Tschirhart E, Legrand-Frossi C et al (2009) Could spaceflight-associated immune system weakening preclude the expansion of human presence beyond Earth's orbit? J Leukoc Biol 86:1027–1038. https://doi.org/10.1189/jlb.0309167

Guéguinou N, Bojados M, Jamon M, Derradji H, Baatout S, Tschirhart E et al (2012) Stress response and humoral immune system alterations related to chronic hypergravity in mice. Psychoneuroendocrinology 37:137–147. https://doi.org/10.1016/j.psyneuen.2011.05.015

Harlé G, Kaminski S, Dubayle D, Frippiat JP, Ropars A (2018) Murine splenic B cells express corticotropin-releasing hormone receptor 2 that affect their viability during a stress response. Sci Rep 9:143. https://doi.org/10.1038/s41598-017-18401-y

Huin-Schohn C, Guéguinou N, Schenten V, Bascove M, Gauquelin-Koch G, Baatout S et al (2013) Gravity changes during animal development affect IgM heavy-chain transcription and probably lymphopoiesis. FASEB J 27:333–341. https://doi.org/10.1096/fj.12-217547

Kimzey SL (1977) Hematology and immunology studies. In: Johnson RS, Dietlein LF (eds) Biomedical results from Skylab. National Aeronautics and Space Administration, U.S. Goverment Printing Office, Washington, DC, pp 249–282

Klaus DM, Howard HN (2006) Antibiotic efficacy and microbial virulence during space flight. Trends Biotechnol 24:131–136. https://doi.org/10.1016/j.tibtech.2006.01.008

Konstantinova IV, Rykova MP, Lesnyak AT, Antropova EA (1993) Immune changes during long-duration missions. J Leukoc Biol 54:189–201

Lescale C, Schenten V, Djeghloul D, Bennabi M, Gaignier F, Vandamme K et al (2015) Hind limb unloading, a model of spaceflight conditions, leads to decreased B lymphopoiesis similar to aging. FASEB J 29:455–463. https://doi.org/10.1096/fj.14-259770

Mercier FE, Ragu C, Scadden DT (2011) The bone marrow at the crossroads of blood and immunity. Nat Rev Immunol 12:49–60. https://doi.org/10.1038/nri3132

Nickerson CA, Ott CM, Mister SJ, Morrow BJ, Burns-Keliher L, Pierson DL (2000) Microgravity as a novel environmental signal affecting Salmonella enterica serovar Typhimurium virulence. Infect Immun 68:3147–3152

Sasaki S, Sullivan M, Narvaez CF, Holmes TH, Furman D, Zheng NY et al (2011) Limited efficacy of inactivated influenza vaccine in elderly individuals is associated with decreased production of vaccine-specific antibodies. J Clin Invest 121:3109–3119. https://doi.org/10.1172/JCI57834

Schaerlinger B, Bascove M, Frippiat JP (2008) A new isotype of immunoglobulin heavy chain in the urodele amphibian *Pleurodeles waltl* predominantly expressed in larvae. Mol Immunol 45:776–786. https://doi.org/10.1016/j.molimm.2007.06.356

Shaw AC, Goldstein DR, Montgomery RR (2013) Age-dependent dysregulation of innate immunity. Nat Rev Immunol 13:875–887. https://doi.org/10.1038/nri3547

Tascher G, Gerbaix M, Maes P, Chazarin B, Ghislin S, Antropova E et al (2019) Analysis of femurs from mice embarked on board BION-M1 biosatellite reveals a decrease in immune cell development, including B cells, after 1 wk of recovery on Earth. FASEB J 33(3):3772–3783. https://doi.org/10.1096/fj.201801463R

Taylor PW, Sommer AP (2005) Towards rational treatment of bacterial infections during extended space travel. Int J Antimicrob Agents 26:183–187. https://doi.org/10.1016/j.ijantimicag.2005.06.002

Vallabhapurapu S, Karin M (2009) Regulation and function of NF-B transcription factors in the immune system. Annu Rev Immunol 27:693–733. https://doi.org/10.1146/annurev.immunol.021908.132641

16 Stress, Hypoxia, and Immune Responses

Claudia Strewe, Manfred Thiel, Michail Sitkovsky, Alexander Choukér, and Matthias Feuerecker

16.1 Introduction

Immune cells are unique in their capacity to distinguish between self and nonself. This ability is more or less pronounced with respect to cells of the adaptive and the innate part of the immune system. Although specificity of immune cells in targeting foreign microbial structures is well developed overall, damage to normal tissue may occur in excessive inflammatory reactions (Fig. 16.1). Under such conditions, overactive immune cells release a large spectrum of cytotoxic mediators causing damage to vascular endothelial cells, especially at the level of microcirculation. As a result, perfusion-dependent oxygen delivery to tissues—in addition to the physiological oxygenation gradient—gets compromised and severe tissue hypoxia ensues. Hypoxia in turn gives rise to the stabilization of hypoxia-inducible factors and leads to the enhanced degradation of adenine nucleotides to adenosine. In search of physiologic mechanisms directed at the limitation of inflammatory collateral tissue damage during ongoing immune cell-mediated destruction of pathogens, a critical role for hypoxia in protecting tissues was revealed. Collateral tissue damage-associated deep hypoxia, hypoxia-inducible factors (HIF), and hypoxia-induced accumulation of adenosine may represent one of the most fundamental and immediate tissue-protecting mechanisms, with HIF-1α, adenosine $A2_A$, and $A2_B$ receptors triggering

C. Strewe (✉) · A. Choukér · M. Feuerecker
Laboratory of Translational Research Stress and Immunity, Department of Anaesthesiology, Hospital of the University of Munich (LMU), Munich, Germany
e-mail: claudia.strewe@med.uni-muenchen.de; matthias.feuerecker@med.uni-muenchen.de

M. Thiel
Department of Anesthesiology, Medical Faculty of Mannheim, University of Heidelberg, Heidelberg, Germany

M. Sitkovsky
New England Inflammation and Tissue Protection Institute, Northeastern University, Boston, MA, USA

© Springer Nature Switzerland AG 2020

A. Choukér (ed.), *Stress Challenges and Immunity in Space*,
https://doi.org/10.1007/978-3-030-16996-1_16

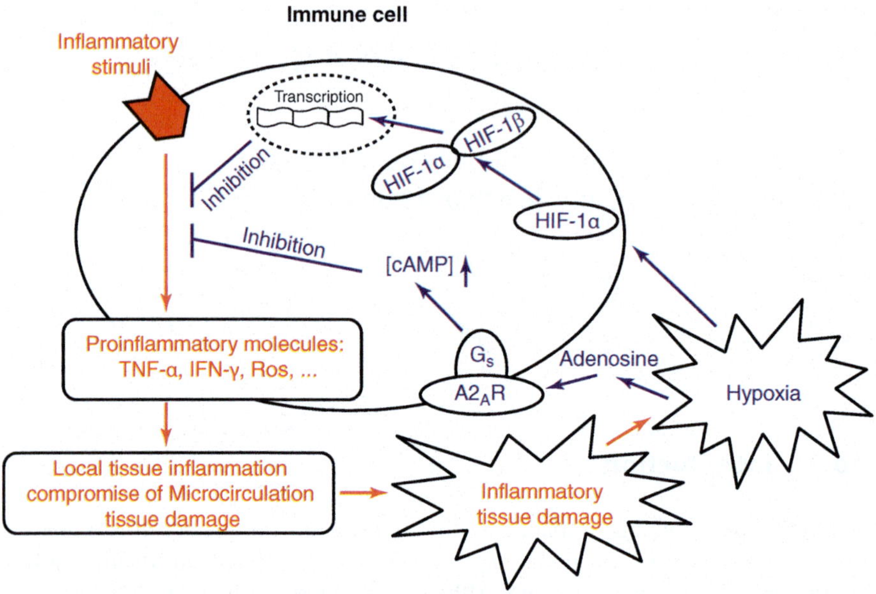

Fig. 16.1 Regulation of immune cells by inflammatory hypoxia: Under conditions of inflammation, immune cells may get overactivated and cause collateral tissue injury ending up in tissue hypoxia. Hypoxia triggers the breakdown of adenosine triphosphate (ATP) to form adenosine which acts in a delayed negative feedback manner. In this model, the extracellular adenosine serves both as a "signal" and the adenosine $A2_A$ receptors as "sensors" of excessive inflammatory collateral tissue damage. Adenosine $A2_A$ receptor activation will stimulate formation of intracellular cyclic adenosine monophosphate (cAMP), a well-known inhibitor of immune cell effector functions. In parallel, hypoxia will stabilize hypoxia-inducible factor (HIF-1α), also shown to counteract immune cell activation. These hypoxia-dependent signaling pathways help to counteract further immune cell activation and hence limit inflammatory collateral tissue injury. Although good in the short term, prolonged activation of these hypoxic mechanisms will lead to severe immunosuppression in the long term. (Modified from Sitkovsky et al. 2004)

"OFF" signals in activated immune cells. In these regulatory mechanisms, oxygen deprivation and extracellular adenosine accumulation serve as "reporters," while HIF-1α, adenosine $A2_A$ and $A2_B$ receptors serve as "sensors" of excessive tissue injury. HIF-1α by not yet characterized signaling pathways and adenosine $A2_A$/$A2_B$ receptor-triggered generation of intracellular cAMP strongly inhibit activated immune cells in a delayed negative feedback manner to prevent additional tissue damage (Sitkovsky et al. 2004; Thiel et al. 2007).

Thus, short-term activation of hypoxia-dependent anti-inflammatory signaling mechanisms is considered to be beneficial in tissue protection from immune cells-mediated inflammatory collateral damage. It represents an evolutionary well preserved axis that protects and regulates and thus maintains human physiological processes when exposed to acute stress of different nature (Dhabhar 2018). In contrast, long-term activation of these anti-inflammatory mechanisms by prolonged stressors and chronic hypoxia may cause severe immunosuppression, predisposing an individual to microbial infections. However, research suggests that the phenotype of hypoxia-derived immune consequences is dependent on the degree of hypoxia and the duration of exposition as it is dependent on the complimentary effects of other stressors (e.g. radiation, metabolic effects and unique other stressors in space).

This brief review gives (1) a short overview on hypoxia and its immune modulating functions. Furthermore, it focuses on the questions whether stress encountered by astronauts/cosmonauts (2) may trigger neurohumoral effector mechanisms leading to tissue hypoxia and/or (3) causes upregulation of the same hypoxia signaling-dependent anti-inflammatory pathways even in the absence of low tissue pO_2 tension, i.e. also under normoxic conditions. Additionally, it questions (4) to which degree this is dependent on the duration of exposition.

16.2 Evidence for Hypoxia to Cause Immunosuppression

The hypoxia → adenosine → adenosine-receptor pathway is an evolutionary old mechanism to control inflammation. In this pathway adenosine acts by binding to widely distributed adenosine $A2_A$ and $A2_B$ receptors located on the cell surface (Linden 2001). Adenosine A2 receptors activate the enzyme adenylyl cylase leading to the enhanced production of cyclic adenosine-monophosphate (cAMP). cAMP itself represents a very potent "OFF" signal in activated immune cells. Although this sequence of events is part of a delayed negative feedback manner mechanism directed at the protection of tissues from uncontrolled ongoing excessive inflammation, downregulation of immune cells may cause severe immunosuppression (reviewed in Linden 2006). Indeed there is a lot of in vivo and in vitro evidence for hypoxic-adenosinergic suppression of cells of innate and specific immunity. Although there are some reports on activation of distinct functions of phagocytes (e.g. upregulation of β2 integrins) upon short-term exposure to hypoxia (Scannell et al. 1995) or moderate hypobaric hypoxia (Hitomi et al. 2003), capacity of phagocytes to kill germs is severely impaired (Segal et al. 1981; Leeper-Woodford and Mills 1992). Similarly, production by macrophages—and T lymphocytes of chemotactic and activating factor or monocyte chemoattractant protein-1 (MCP-1) was inhibited during hypoxia (Bosco et al. 2004).

While T cell-mediated immune responses are suppressed, B-lymphocytes have been shown to still produce immunoglobulins under hypoxic conditions (Sitkovsky et al. 2004). Moreover, hypoxia resulted in a shift of T-helper cell responses from a Th1 type to a Th2 type. This effect is discussed to be mediated by adenosine leading to the inhibition of the differentiation of naïve CD4 T-cells into Th2 cells via adenosine A2 receptors (Panther et al. 2003). As a result of a lack of oxygen, activity of cells especially of those of the specific immune system is severely suppressed (Sitkovsky et al. 2004).

The effects of hypoxia on lymphoid and nonlymphoid tissues can be further aggravated by breathing air with an oxygen content of lower than 20.9 vol.%, because this will enhance the accumulation of extracellular adenosine. Accordingly, we and others (Decking et al. 1997) showed in in vivo experiments that adenosine blood concentrations significantly increased upon breathing 10–12% oxygen (Chouker et al. 2012). These results have recently been confirmed in humans by a cross-over designed study demonstrating that the combined effect of normobaric hypoxia (14% FiO_2) and bed rest (simulated microgravity) triggers highest extracellular adenosine levels but that also bed rest per se induces adenosine release (Strewe et al. 2018b). Altogether, adenosine signaling via $A2_AR$ and $A2_BR$ and accumulation of immunosuppressive intracellular cAMP in activated immune cells (Sitkovsky 2003) inhibits intracellular pathways required for synthesis and secretion of pro-inflammatory and cytotoxic mediators by immune cells, effects which will terminate the immune cells' effector functions and thereby downregulate the immune response. In addition, some beneficial effects of exposition to moderate oxygen tension are also confirmed in clinical studies in patients with inflammatory compromised states (Girardis et al. 2016; Asfar et al. 2017).

16.3 Life Aboard the ISS: Are Astronauts at Risk to Develop Tissue Hypoxia?

Concerning this question, it is of interest to note that several stressful conditions encountered by astronauts/cosmonauts during spaceflight might cause hypoxia leading to the upregulation of hypoxia-related immunosuppressive signaling pathways. During spaceflight multiple stressors take effect on the body's psychological and physiological stress response systems. These systems include neurohumoral control mechanisms (Fig. 16.2) which are old in nature and have been shaped by a long history of evolutionary processes on Earth. As it was not foreseen by evolution

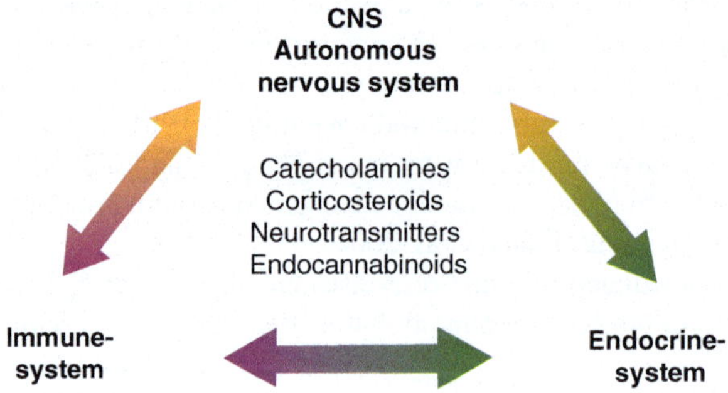

Fig. 16.2 Stress Supersystems: The major stress supersystems of the body comprise the sympathetic and parasympathetic parts of the autonomous nervous system, the endocrinological hypothalamic-pituitary-adrenal gland axis and the immune system which mutually interact in a highly compartmentalized and timely coordinated fashion to cope with psychological and physical challenges

for man to leave Earth, it is hard to predict how these endogenous stress response supersystems may interact during long-duration spaceflights (LDMs) and what the long-term effects on the astronauts'/cosmonauts' health status may look like.

Despite a lack of data collected during the course of spaceflight missions, there is some evidence from Earth-bound models of simulated microgravity for the development of tissue hypoxia or hypoxia-like disturbances in response to stress especially caused by microgravity and/or radiation.

The loss in gravity is associated with an upward shift of the body's blood volume toward both the thoracic cavity and the head, respectively. Such a state of central hypervolemia will elicit a multitude of adaptive responses at the macro- and microcirculatory level (Vorob'ev 2004). Additionally, low or zero g conditions will significantly change airway breathing characteristics. Both altered central blood volume distribution and mechanical airway characteristics might end up in pulmonary ventilation/perfusion mismatch thereby compromising lung gas exchange. Such simple mechanistic effects are likely to be even more pronounced during sleep when physical activity and consciousness are decreased and control of airway patency is impaired (Prisk 1998).

At the microcirculatory level, oxygen delivery may be further limited by peripheral vasoconstriction due to elevated blood levels of stress hormones like vasopressin, adrenaline, and noradrenaline, with the action of the catecholamines being potentiated by vasoconstrictor permissive effects of increased blood concentrations of cortisol (Yang and Zhang 2004). Tissue malperfusion may additionally be aggravated by partial obstruction of capillaries due to catecholaminergic activation and aggregation of blood platelets, known to be induced upon stimulation of $\alpha 2_{\alpha}$ adrenergic receptors located on the surface of thrombocytes (Tschuor et al. 2008.).

In this scenario, a decrease in rheological properties of erythrocytes due to decreased cell membrane fluidity and increased cell stiffness will further add to tissue malperfusion. Such effects are even more pronounced by increased hematocrit values which are frequently observed in astronauts/cosmonauts during LDMs. The rise in hematocrit values, however, is not due to an increase in red cell mass, but results from its decrease which is paralleled by a much stronger restriction of the plasma volume. In consequence, during LDMs circulating blood volume is in the end far below from normal in astronauts/cosmonauts which may predispose them to substantially increased risk for tissue hypoxia in case of unexpected cardiovascular or pulmonary challenges (Talbot and Fisher 1986).

Pathophysiological hypoxia elicited by microgravity is further aggravated in astronauts/cosmonauts by breathing even normoxic or hypoxic air mixtures at hypobaric or normobaric pressures, respectively. Such environmental stressors are likely to be encountered during space exploration missions when air pressure and oxygen content will be reduced in lunar or Martian living habitats to reduce risks of fire or to facilitate extrahabitat activities in hypobaric space suits.

16.4 Are There Environmental Stress Conditions Which Are Able to Upregulate Immunosuppressive Hypoxic Signaling Pathways Irrespective of Tissue Hypoxia?

Little is known about factors able to upregulate or to mimic hypoxic signaling pathways independently from a lack in oxygen, i.e. under conditions of normoxia. A broad review on the different regulatory mechanisms of HIF-1 activity was published by Koyasu et al. (2018). In this regard stress caused by cosmic radiation (see Chaps. 20 and 28) likely represents an important factor encountered by astronauts/cosmonauts during LDMs.

For instance, tumors respond to radiation especially upon reoxygenation in the postradiation phase by the nuclear accumulation of HIF-1α. Enhanced translation of HIF-1α-regulated transcripts in the tumor microenvironment has been demonstrated to modify the expression of multiples immune regulatory components such as CD47, PD-L1, HLA-G (Li et al. 2018) and thus leading to evasion of hypoxic cancer cells from immunity. Furthermore, evidence exists that HIF-1α protects tumor cells from apoptosis by radiation under hypoxia (Fu et al. 2015). Additionally, it has been demonstrated that HIF-1 regulatory mechanisms may not only become active and induce tumor radioresistance under hypoxic but also under normoxic conditions through cancer-specific genetic alterations (Harada 2016). Therefore, all these pathways represent targets to enhance efficacy of antitumor therapies by developing novel cancer immunotherapies. These data describe furtherly explored pathways contributing significantly to our understanding of HIF-1α regulation by radiation which may be one of the reasons for the development of immunosuppression in astronauts/cosmonauts during LDMs.

Accordingly, HIF-1α was shown not only to directly suppress immune-stimulatory cytokine production and to lead via T regulatory cells to dysfunction but also to modulate the acid–base regulation in the tumor microenvironment and the production of immuno-modulatory lactate (Li et al. 2018). Moreover, HIF-1α interacts with key enzymes for the generation of immunosuppressive adenosine. For instance, HIF-1α upregulates the AMP degrading and adenosine forming enzyme ecto 5′-NT (CD73) (Synnestvedt et al. 2002) and increases adenosine $A2_B$ receptor expression (Kong et al. 2006) due to the location of hypoxia-responsive elements (HRE) in the promoter region of these genes. Moreover, HIF-1α by inhibiting adenosine kinase expression (Morote-Garcia et al. 2008) strongly attenuates the removal of formed adenosine via inhibition of its rephosphorylation to AMP. This will greatly increase levels of adenosine upon the release of ATP into the extracellular space and strengthen immunosuppressive signaling via A2 receptor expression (Sitkovsky et al. 2004). Increasing research and knowledge about this signaling cascade ensues the search to counteract these modulations in order to strengthen antitumor therapy and its effects. Hatfield and Sitkovsky (2016) demonstrated that A2A adenosine receptor antagonists are able to weaken immunosuppression that is mediated this way. Such immunosuppressive mechanisms may be further augmented by the ability of radiation to cause direct release of ATP, the mother molecule for the generation of adenosine. A hypothesis under investigation is that opening of multidrug resistance protein P-glycoprotein 1, a MDR ATP/chloride channel, in response to increased local concentrations of peroxides may be one initial response to ionizing radiation. Notably, a rapid release of ATP was detected from cells irradiated with doses as low as 10 cGy (Abraham et al. 1994). As the magnitude of ATP release correlated with the level of MDR1 expression, ATP release was rather mediated by the specific P-glycoprotein transmembrane channel than by nonspecific membrane leakage.

The direct effects of radiation in terms of enhanced ATP release and immunosuppressive adenosine formation are likely to be strengthened by other stressors known to cause ATP release from cells via activation of ATP-permeable channels and/or exocytosis of ATP-containing vesicles. Accordingly, ATP is set free from the intracellular pool upon challenge of cells by numerous stressors, like oxidative stress (Ahmad et al. 2006), mechanical stress (Qin et al. 2008), heat, and osmotic stress (Kimura et al. 2000).

Most importantly, ATP is released into the synaptic cleft via exocytosis as a chemical cotransmitter of noradrenaline or acetylcholine neurotransmission following activation of the autonomous sympathetic and parasympathetic nervous system (Abbracchio et al. 2009). This is of potential immunomodulatory, immunosuppressive significance and may represent a direct link between psychic stress, activation of the autonomous nervous system, and direct immunosuppressive action of adenosine, since primary as well as secondary lymphoid tissues are directly innervated by sympathetic and cholinergic nerve fibers (Bellinger et al. 2008).

16.5 Testing for the Effects of Hypoxia on Humans' Immunity in Ground Space Analogues: Surprising News for Planning Deep Space Exploration?

Despite the clearly identified risk for the development of tissue hypoxia during LDM, to the best of our knowledge there are no space in-flight data available regarding tissue oxygenation in astronauts/cosmonauts. As a consequence it is also not known whether LDMs induce the aforementioned anti-inflammatory hypoxic signaling pathways which were shown by us to significantly contribute to immunosuppression. Most studies on tissue oxygenation therefore arise from Earth-bound models.

For instance, indirect evidence for local circulatory hypoxia has been reported in individuals subjected to head-down tilt to mimic microgravity-induced cardiovascular changes on central blood volume (Vorob'ev 2004). Accordingly, after 1 week in head-down tilt, local O_2 utilization, for example, by the arm was found to be increased as O_2 content was lowered and lactate increased in peripheral venous blood returning from the periphery, while respective parameters and mean pressure were unchanged in arterial blood. Furthermore, direct consequences of hypoxic conditions on human physiology were evidenced during the PlanHab Study, a crossover designed experiment carried out in Slovenia (see also Chap. 36) and exposing subjects to bed rest and/or normobaric hypoxia for 21 days (Strewe et al. 2018b).

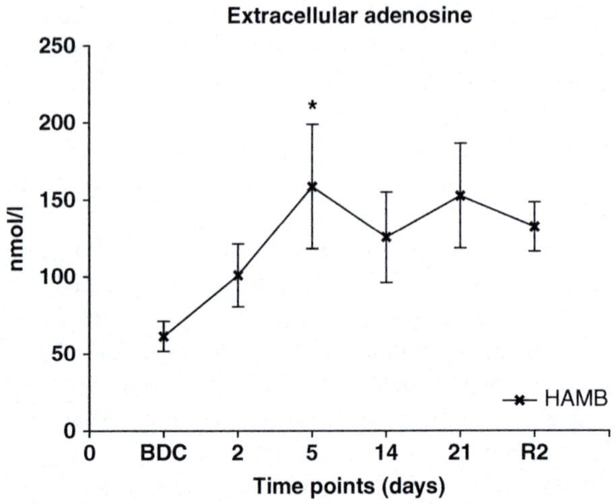

Fig. 16.3 Extracellular adenosine concentrations in plasma; data are means ± SEM; units are nmol/l; HAMB = hypoxic ambulation ($n = 12$); BDC = Baseline Data Collection; R2 = 2 days after the end of condition; $*$ significant difference to BDC in HAMB ($p < 0.05$). Hypoxia in combination with bed rest results in even higher adenosine plasma concentrations than stated in HAMB. Normoxic bed rest causes similar results as HAMB

Its results demonstrated a significant increase of adenosine under hypoxia but unexpectedly also bed rest under normoxic conditions evoked such an increase (Strewe et al. 2018b) (Fig. 16.3).

Thus, hypoxic mechanisms cannot be the sole explanation and further investigations should elucidate these findings. Ongoing genetic analyses are expected to give better insights and further help to analyze more precisely the hypoxia-dependent signaling pathways under such combined influences.

Besides such short- and mid-term experiments and other numerous observational studies strongly suggesting that astronauts/cosmonauts are prone to hypoxia (Grigor'ev et al. 2008) and immunosuppression (Guèguinou et al. 2009) during long-term spaceflight missions, long-term isolation experiments have been run in the Antarctic region to determine how hypobaric hypoxia might affect the complex interaction between the body's stress supersystems at the psychoneuroendocrinological and immunological level (results as summarized in Fig. 16.4). Furthermore, the stress effects caused by long-term confinement will be delineated from those elicited by hypobaric hypoxia. This can be achieved by the comparison of results obtained either at the Concordia base located at high altitude to the data gathered from crews at the Antarctic Neumayer III Station at sea level in the context of the *CHOICE*-Study (*C*onsequences of long-term-*C*onfinement and *H*ypobaric Hyp*O*xia on *I*mmunity in the Antarctic *C*oncordia *E*nvironment) (see also Chaps. 36 and 38). Analyses of recent data from this study revealed changing patterns in the neuroendocrine stress response due to the exceptional Antarctic environment and specifically to the condition of hypobaric hypoxia. The activity of the endocannabinoid system that regulates and takes part in multiple physiological processes from stress (Chouker

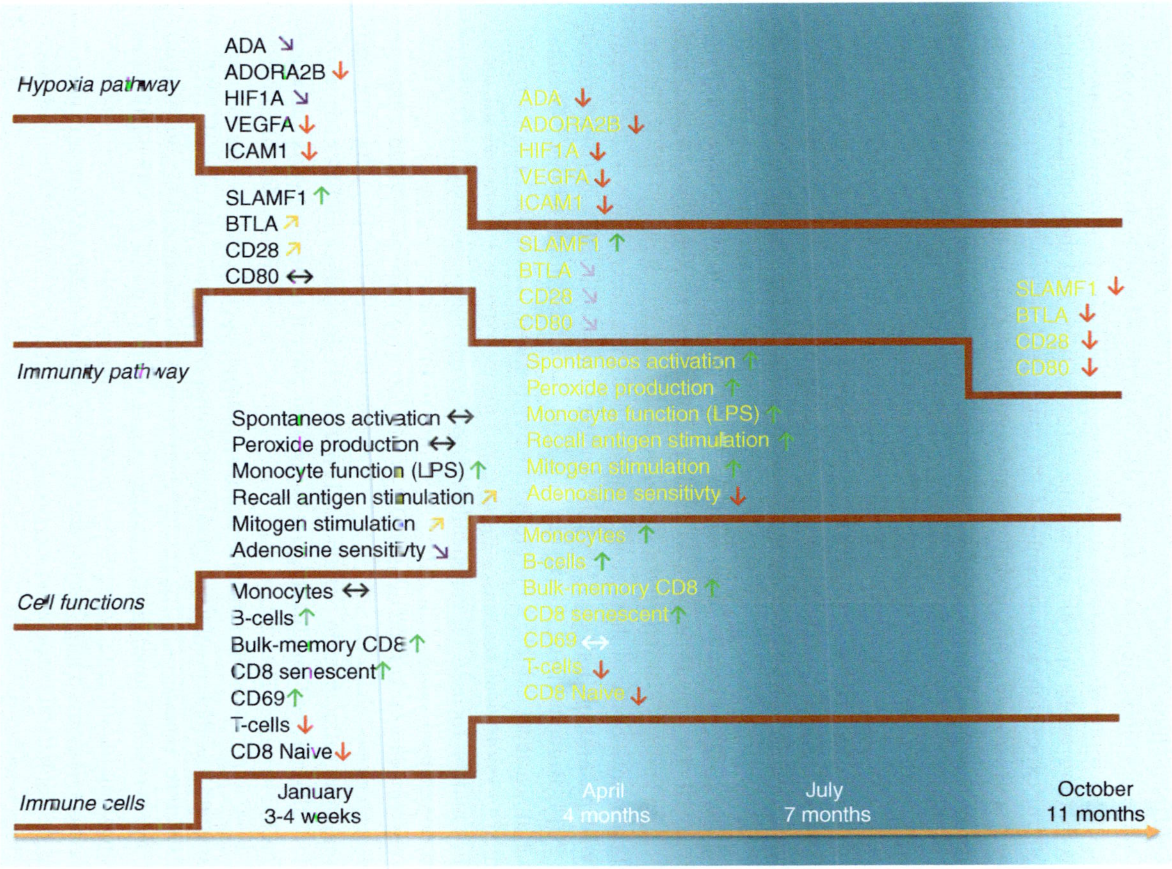

Fig. 16.4 Overview of immune changes during 1 year at Concordia station; adenosine deaminase (ADA), adenosine receptor 2 subtype B (ADORA2B), hypoxia inducible factor 1 alpha (HIF1-α), vascular endothelial growth factor A (VEGF-A), intercellular Adhesion Molecule 1 (ICAM-1), signaling lymphocytic activation molecule family 1 (SLAMF-1), B- and T-lymphocyte attenuator (BTLA) (modified after Feuerecker et al. 2019)

et al. 2010) over memory consolidation (Campolongo et al. 2013) to physical activity (Feuerecker et al. 2012) seems to be generally enhanced in such a surrounding. However, the influence of hypobaric hypoxia as experienced at Concordia station seems to downregulate this activity by an increased sympathoadrenal answer (Strewe et al. 2018a). Thus, the two neuroendocrine systems show a negative correlation.

On the immunological level, former data from the CHOICE study of two overwintering seasons at Concordia in 2009 and in 2010 also demonstrated modulations that mirror the specific environmental conditions and their influence under long-term exposition. Components of the innate and adaptive immune response showed a sensitization and a shift towards pro-inflammation was observed by an altered cytokine expression and a heightened activation of innate cells residing in a sort of alert status (priming). Furthermore, T-cell-mediated immune answers were enhanced when stimulated with recall antigens or other stimuli displaying a further immune imbalance promoting inflammation. Hypoxia-dependent signaling pathways (adenosine, HIF-1α) known to exert a immunosuppressive function and to limit the inflammatory response were less effective thus favoring a disinhibition of immune functions with subsequent possible enhanced inflammation (Feuerecker et al. 2019) (Fig. 16.4). Therefore, it seems that long-term, chronic exposure to hypobaric hypoxia results in an adaptive activation pattern of regulator proteins such as HIF-1α. This is supported by other publications on long-term hypoxia and HIF-1α in humans that reported a short-lived transient activation of HIF-1α dependent pathways. Some of these studies investigated the effects of hypoxia in genetically adapted populations e.g. Tibetans, Andeans on the HIF pathway (Bigham and Lee 2014; Bigham 2016). Petousi et al. (2014) were able to show that "Tibetans living at sea level exhibit a hyporesponsive hypoxia-inducible factor system and blunted physiological responses upon exposure to hypoxia." However, all these studies mainly focused on genetically adapted populations and may hence differ from nonadapted subjects. A situation similar to the one represented in CHOICE was described by Goyal and Longo (2014). The latter group demonstrated, although in a sheep model, that exposition to ~3800 m of altitude for 110 days does not lead to significant differences in protein levels of HIF-1α. In another animal model Baze et al. (2010) showed an altered gene expression pattern in murine livers subjected to chronic hypoxia. They found that "after 32 days, the previously described stress pathways, and in particular the HIF pathway, are no longer significant factors in regulating and maintaining physiological acclimation to hypoxia." (Baze et al. 2010).

Therefore, long-term in contrast to short-term confinement and in combination with hypoxia seems to inhibit the regulatory role of the hypoxia-signaling pathways emphasizing the different impacts on human immunity of the respective condition. In the light of planned long-term space missions such as flights to Mars or residence in space habitats more detailed knowledge of the impacts on the human immune system is mandatory.

In summary, the knowledge on stress-triggered, hypoxia-signaling-dependent anti-inflammatory and immunosuppressive mechanisms can be of critical importance for long-duration space mission and part of unfavorable impact on immune functions. The hypoxic and normoxic pathways of the neurohumoral stress response may synergistically end up in clinically relevant immune modulation. This knowledge about the interactions of stress- and hypoxia-sensitive pathways, as has been gathered in different condition, in space and in space analog studies, as well as from clinical and experimental models, will greatly benefit patients on Earth in the fields of surgery, emergency, and intensive care medicine. This all the more as in the hostile environment of an Intensive Care Unit critically ill patients experience substantial amounts of psychological and physical stress frequently complicated by tissue hypoxia.

Acknowledgments Some of the investigations described have been supported by the National Institutes of Health (awarded Fogarty grants), the German Research Council (Th733/4-1), the European Space Agency (ESA, ELIPS 3 and 4 program), the German National Space Program (DLR, BMWi 50WB0719 and 50WB1319), the French (IPEV), Italian (PNRA), and German (AWI) polar institutes. The authors want to express their appreciation to the overwintering crews and the operators Drs. F. van den Berg, C. Dangoisse, A. Salam, A. Rybka, H. Geißler, B. Fiedel, H. Bauer, P. Gössmann-Lang, L. Dunker, T. Heitland, and B. Gropp who have been altogether supporting the CHOICE study with great professionalism.

Furthermore, we would like to thank also all participants and co-researchers of the PlanHab study for their enthusiasm and professionalism, especially Prof. I. Mekjavic and Dr. T. Debevec.

References

Abbracchio MP, Burnstock G, Verkhratsky A et al (2009) Purinergic signalling in the nervous system: an overview. Trends Neurosci 32:19–29

Abraham EHA, Steiglitz K, Herscher L et al (1994) ATP electrochemical gradient drives P-glycoprotein mediated drug fluxes. Proc AM Assoc Cancer Res 353

Ahmad S, Ahmad A, White CW (2006) Purinergic signaling and kinase activation for survival in pulmonary oxidative stress and disease. Free Radic Biol Med 41:29–40

Asfar P, Schortgen F, Boisrame-Helms J, Charpentier J, Guerot E, Megarbane B, Grimaldi D, Grelon F, Anguel N, Lasocki S, Henry-Lagarrigue M, Gonzalez F, Legay F, Guitton C, Schenck M, Doise JM, Devaquet J, Van Der Linden T, Chatellier D, Rigaud JP, Dellamonica J, Tamion F, Meziani F, Mercat A, Dreyfuss D, Seegers V, Radermacher P, HYPER2S Investigators; REVA Research Network (2017) Hyperoxia and hypertonic saline in patients with septic shock (HYPERS2S): a two-by-two factorial, multicentre, randomised, clinical trial. Lancet Respir Med 5(3):180–190

Baze MM, Schlauch K, Hayes JP (2010) Gene expression of the liver in response to chronic hypoxia. Physiol Genomics 41(3):275–288

Bellinger DL, Millar BA, Perez S et al (2008) Sympathetic modulation of immunity: relevance to disease. Cell Immunol 252:27–56

Bigham AW (2016) Genetics of human origin and evolution: high-altitude adaptations. Curr Opin Genet Dev 41:8–13

Bigham AW, Lee FS (2014) Human high-altitude adaptation: forward genetics meets the HIF pathway. Genes Dev 28(20):2189–2204

Bosco MC, Puppo M, Pastorino S et al (2004) Hypoxia selectively inhibits monocyte chemoattractant protein-1 production by macrophages. J Immunol 172:1681–1690

Campolongo P, Morena M, Scaccianoce S, Trezza V, Chiarotti F, Schelling G, Cuomo V, Roozendaal B (2013) Novelty-induced emotional arousal modulates cannabinoid effects on recognition memory and adrenocortical activity. Neuropsychopharmacology 38(7):1276–1286

Chouker A, Kaufmann I, Kreth S, Hauer D, Feuerecker M, Thieme D, Vogeser M, Thiel M, Schelling G (2010) Motion sickness, stress and the endocannabinoid system. PLoS One 5(5):e10752

Chouker A, Ohta A, Martignoni A, Lukashev D, Zacharia LC, Jackson EK, Schnermann J, Ward JM, Kaufmann I, Klaunberg B, Sitkovsky MV, Thiel M (2012) In vivo hypoxic preconditioning protects from warm liver ischemia-reperfusion injury through the adenosine A2B receptor. Transplantation 94(9):894–902

Decking UK, Schlieper G, Kroll K et al (1997) Hypoxia-induced inhibition of adenosine kinase potentiates cardiac adenosine release. Circ Res 81:154–164

Dhabhar FS (2018) The short-term stress response - mother nature's mechanism for enhancing protection and performance under conditions of threat, challenge, and opportunity. Front Neuroendocrinol 49:175–192

Feuerecker M, Hauer D, Toth R, Demetz F, Holzl J, Thiel M, Kaufmann I, Schelling G, Chouker A (2012) Effects of exercise stress on the endocannabinoid system in humans under field conditions. Eur J Appl Physiol 112(7):2777–2781

Feuerecker M, Crucian BE, Quintens R, Buchheim JI, Salam AP, Rybka A, Moreels M, Strewe C, Stowe R, Mehta S, Schelling G, Thiel M, Baatout S, Sams C, Chouker A (2019) Immune sensitization during one year in the Antarctic high altitude Concordia Environment. Allergy 74(1):64–77

Fu Z, Chen D, Cheng H, Wang F (2015) Hypoxia-inducible factor-1alpha protects cervical carcinoma cells from apoptosis induced by radiation via modulation of vascular endothelial growth factor and p53 under hypoxia. Med Sci Monit 21:318–325

Girardis M, Busani S, Damiani E, Donati A, Rinaldi L, Marudi A, Morelli A, Antonelli M, Singer M (2016) Effect of conservative vs conventional oxygen therapy on mortality among patients in an intensive care unit: the oxygen-ICU randomized clinical trial. JAMA 316(15):1583–1589

Goyal R, Longo LD (2014) Acclimatization to long-term hypoxia: gene expression in ovine carotid arteries. Physiol Genomics 46(19):725–734

Grigor'ev AI, Ivanova SM, Morukov BV et al (2008) Development of cell hypoxia induced by factors of long-term spaceflight. Dokl Biochem Biophys 422:308–311

Guèguinou N, Huin-Schohn C, Bascove M et al (2009) Could spaceflight-associated immune system weakening preclude the expansion of human presence beyond Earth's orbit? J Leukoc Biol 86:1027–1038

Harada H (2016) Hypoxia-inducible factor 1-mediated characteristic features of cancer cells for tumor radioresistance. J Radiat Res 57(Suppl 1):i99–i105

Hatfield SM, Sitkovsky M (2016) A2A adenosine receptor antagonists to weaken the hypoxia-HIF-1alpha driven immunosuppression and improve immunotherapies of cancer. Curr Opin Pharmacol 29:90–96

Hitomi Y, Miyamura M, Mori S et al (2003) Intermittent hypobaric hypoxia increases the ability of neutrophils to generate superoxide anion in humans. Clin Exp Pharmacol Physiol 30:659–664

Kimura C, Koyama T, Oike M et al (2000) Hypotonic stress-induced NO production in endothelium depends on endogenous ATP. Biochem Biophys Res Commun 274:736–740

Kong T, Westerman KA, Faigle M et al (2006) HIF-dependent induction of adenosine A2B receptor in hypoxia. FASEB J 20:2242–2250

Koyasu S, Kobayashi M, Goto Y, Hiraoka M, Harada H (2018) Regulatory mechanisms of hypoxia-inducible factor 1 activity: two decades of knowledge. Cancer Sci 109(3):560–571

Leeper-Woodford SK, Mills JW (1992) Phagocytosis and ATP levels in alveolar macrophages during acute hypoxia. Am J Respir Cell Mol Biol 6:326–334

Li Y, Patel SP, Roszik J, Qin Y (2018) Hypoxia-driven immunosuppressive metabolites in the tumor microenvironment: new approaches for combinational immunotherapy. Front Immunol 9:1591

Linden J (2001) Molecular approach to adenosine receptors: receptor-mediated mechanisms of tissue protection. Annu Rev Pharmacol Toxicol 41:775–787

Linden J (2006) New insights into the regulation of inflammation by adenosine. J Clin Invest 116:1835–1837

Morote-Garcia JC, Rosenberger P, Kuhlicke J et al (2008) HIF-1-dependent repression of adenosine kinase attenuates hypoxia-induced vascular leak. Blood 111:5571–5580

Panther E, Corinti S, Idzko M et al (2003) Adenosine affects expression of membrane molecules, cytokine and chemokine release, and the T-cell stimulatory capacity of human dendritic cells. Blood 101:3985–3990

Petousi N, Croft QP, Cavalleri GL, Cheng HY, Formenti F, Ishida K, Lunn D, McCormack M, Shianna KV, Talbot NP, Ratcliffe PJ, Robbins PA (2014) Tibetans living at sea level have a hyporesponsive hypoxia-inducible factor system and blunted physiological responses to hypoxia. J Appl Physiol (1985) 116(7):893–904

Prisk GK (1998) Sleep and respiration in microgravity. Neurosci News 1:39–45

Qin KR, Xiang C, Xu Z et al (2008) Dynamic modeling for shear stress induced ATP release from vascular endothelial cells. Biomech Model Mechanobiol 7:345–353

Scannell G, Waxman K, Vaziri ND et al (1995) Effects of trauma on leukocyte intercellular adhesion molecule-1, CD11b, and CD18 expressions. J Trauma 39:641–644

Segal AW, Geisow M, Garcia R et al (1981) The respiratory burst of phagocytic cells is associated with a rise in vacuolar pH. Nature 290:406–409

Sitkovsky MV (2003) Use of the A(2A) adenosine receptor as a physiological immunosuppressor and to engineer inflammation in vivo. Biochem Pharmacol 65:493–501

Sitkovsky MV, Lukashev D, Apasov S et al (2004) Physiological control of immune response and inflammatory tissue damage by hypoxia-inducible factors and adenosine A2A receptors. Annu Rev Immunol 22:657–682

Strewe C, Thieme D, Dangoisse C, Fiedel B, van den Berg F, Bauer H, Salam AP, Gossmann-Lang P, Campolongo P, Moser D, Quintens R, Moreels M, Baatout S, Kohlberg E, Schelling G, Chouker A, Feuerecker M (2018a) Modulations of neuroendocrine stress responses during confinement in Antarctica and the role of hypobaric hypoxia. Front Physiol 9:1647

Strewe C, Zeller R, Feuerecker M, Hoerl M, Matzel S, Kumprej I, Crispin A, Johannes B, Debevec T, Mekjavic IB, Eiken O, Thiel M, Schelling G, Chouker A (2018b) PlanHab Study: Consequences of combined normobaric hypoxia and bed rest on adenosine kinetics. Sci Rep 8(1):1762

Synnestvedt K, Furuta GT, Comerford KM et al (2002) Ecto-5′-nucleotidase (CD73) regulation by hypoxia-inducible factor-1 mediates permeability changes in intestinal epithelia HIF-dependent induction of adenosine A2B receptor in hypoxia. J Clin Invest 110:993–1002

Talbot JM, Fisher KD (1986) Influence of space flight on red blood cells. Fed Proc 45:2285–2290

Thiel M, Caldwell CC, Kreth S et al (2007) Targeted deletion of HIF-1alpha gene in T cells prevents their inhibition in hypoxic inflamed tissues and improves septic mice survival. PLoS One 2:e853

Tschuor C, Asmis LM, Lenzlinger PM et al (2008) In vitro norepinephrine significantly activates isolated platelets from healthy volunteers and critically ill patients following severe traumatic brain injury. Crit Care 12:R80

Vorob'ev VE (2004) Changes in oxygen supply and utilization in head-down tilted humans. Aviakosm Ekolog Med 38:48–52

Yang S, Zhang L (2004) Glucocorticoids and vascular reactivity. Curr Vasc Pharmacol 2:1–12

Cellular and Molecular Responses to Gravitational Force-Triggered Stress in Cells of the Immune System

17

Oliver Ullrich and Cora S. Thiel

17.1 The "Immune Problem" in Space as a Cellular "Gravity Problem"

Since the early days of human spaceflight, an enhanced susceptibility to bacterial and viral infections has been observed during the Apollo missions (NASA 1967, 1968). First evidence suggesting disturbed cellular function arose from investigations of lymphocytes from astronauts of the Soyuz and Skylab missions, that showed a considerably decreased response to mitogenic stimulation during and after flight (Konstantinova et al. 1973; Hawkins and Zieglschmid 1975; Kimzey 1977). Then, during the first Spacelab mission, in vitro experiments confirmed a strongly impaired response of lymphocytes to proliferative stimuli under space conditions (Cogoli et al. 1984). Whereby several types of cultured cells are sensitive to gravity (Ingber 1999; Vorselen et al. 2014), the immune system belongs to the most affected systems during spaceflight (reviewed in Comet 2001; Frippiat et al. 2016; Choukèr and Ullrich 2016). Sensitivity of cells of the human immune system to reduced gravity has been confirmed in numerous studies in real and simulated microgravity (reviewed in Choukèr and Ullrich 2016; Thiel et al. 2017a). The very first space experiment which investigated the effects of altered gravity in isolated cells of the human immune system identified a severe inhibition of lymphocyte proliferation in microgravity and enhanced proliferation in hypergravity, demonstrating that cells are in principle sensitive to gravity (Cogoli et al. 1984; Cogoli 1996).

O. Ullrich (✉) · C. S. Thiel
Institute of Anatomy, Faculty of Medicine, University of Zurich, Zurich, Switzerland

Department of Machine Design, Engineering Design and Product Development, Institute of Mechanical Engineering, Otto-von-Guericke-University Magdeburg, Magdeburg, Germany

UZH Space Hub, University of Zurich, Zurich, Switzerland

Space Life Sciences Laboratory (SLSL), Kennedy Space Center, Exploration Park, FL, USA
e-mail: oliver.ullrich@anatomy.uzh.ch, oliver.ullrich@uzh.ch, oliver.ullrich@ovgu.de

© Springer Nature Switzerland AG 2020
A. Choukér (ed.), *Stress Challenges and Immunity in Space*,
https://doi.org/10.1007/978-3-030-16996-1_17

Since then, a plethora of in vitro studies have been conducted and reconfirmed cellular sensitivity to gravity in different immune cell systems, platforms, and affecting many cellular and molecular functions (reviewed in Thiel et al. 2017a): Cells of the monocyte–macrophage system (MMS) demonstrated disturbed cytokine release (Limouse et al. 1991; Schmitt et al. 1996), reduced oxidative burst (Armstrong et al. 1995; Adrian et al. 2013), alteration of the cytoskeleton (Meloni et al. 2006), significant changes in gene expression associated with macrophageal differentiation (Hughes-Fulford et al. 2008), and activation of syk signaling pathway (Brungs et al. 2015) in microgravity. Experiments with T lymphocytes and T cell lines, one of the best investigated cells in microgravity, revealed that a remarkable sensitivity to gravity with respect to cell cycle regulation (Thiel et al. 2012), epigenetic (Singh et al. 2010) and chromatin regulation (Paulsen et al. 2010), gene expression (Chang et al. 2012), micro RNA expression (Mangala et al. 2011), regulation of apoptosis (Lewis et al. 1998; Battista et al. 2012), and expression of cytokines IL-2 and IFN-γ was changed in microgravity (Cogoli and Cogoli-Greuter 1997).

Recent studies described a reactivation of the varicella zoster virus (VZV), a latent nervous system virus, in astronauts (Cohrs et al. 2008; Mehta et al. 2004), an alarming observation for long-term space missions. Because of the obvious and severe effects on the human immune system, serious concerns arose whether spaceflight-associated immune system weakening ultimately precludes the expansion of human presence beyond Earth's orbit (Guéguinou et al. 2009) (see also Chaps. 11–16, and 19). Therefore, immune system deterioration during long-term spaceflights could contribute to an increased susceptibility to infection, autoimmunity and cancer during exploration class missions. Thus, it is an urgent need to understand the cellular and molecular mechanisms by which altered gravity changes the functions of cells of the immune system and to assess the cellular capacity and mechanisms for adaptation to a new gravitational environment. Knowing the molecular and genetic basis of cellular response to altered gravity will provide key information for appropriate risk management, efficient monitoring, and countermeasures against existing limiting factors for human health and performance in altered gravity during manned exploration of the solar system.

The gravity field may act on a cell directly or indirectly. In the experimental systems used, direct influences of altered gravity are more pronounced in vitro, while indirect influences are more apparent in the living organisms (Tairbekov 1996). Therefore, in vitro experiments with living human immune cells in microgravity conditions, such as on board of parabolic flights, sounding rockets, satellites, or the International Space Station (ISS), are providing an ideal platform to elucidate the underlying cellular and molecular mechanisms. In contrast to the logistic limitations of the ISS and other space-based research platforms, parabolic flights provide frequent and repeated access to microgravity and therefore allow replication and modification of experiments within a reasonable time frame, which are not only characteristics, but rather requirements, of modern biomedical research. Thus, access to space is an instrument to elucidate long-term and functional effects of microgravity, whereas ultrashort, initial and primary effects and mechanisms are amenable in short-term-microgravity provided by parabolic flight maneuvers. In vivo and in vitro experiments in microgravity can be performed on board of an aircraft flying on a Keplarian trajectory, described as an unpropelled body in ideally frictionless space subjected to a centrally symmetric gravitational field (Pletser 2016; Studer et al. 2010). Due to the fact that cells of the immune system are obviously influenced by altered gravity, its gravisensitive nature render these cells also an ideal biological model in the search for general gravisensitive mechanisms in mammalian cells (Fig. 17.1).

Whereas it has been supposed that the most of the cellular effects of microgravity or of simulated weightlessness by the clinostat model may be attributed to the generalized unspecific reaction of a cell to external influence (Kondrachuk and Sirenko 1996), recent findings support a more specific nature of the cellular reactions (Paulsen et al. 2010; Thiel et al. 2017c, d). The ability of cells to transduce, to process and to respond to mechanical forces are increasingly recognized as playing predominant roles in a variety of relevant cellular functions (Orr et al. 2006; Ingber 2003a) such as cell migration, growth, mitosis, and differentiation (Engler et al. 2006). The mechanisms by which the cells react to mechanical forces could lead to

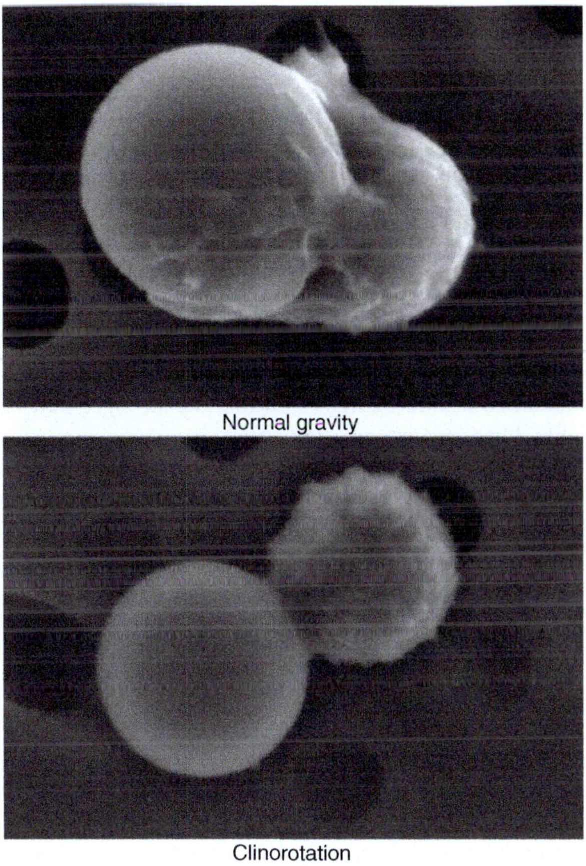

Fig. 17.1 T-cell activation in 1 g (static) culture versus simulated microgravity (clinorotation); Photo credit to Clarence Sams (NASA Johnson Space Centre) and Mayra Nelman-Gonzalez (Wyle)

an idea how cells may transduce gravitational forces. In the last decade, much progress has been made in understanding the response of cells to physical forces (Vogel and Sheetz 2006; Hoffman and Crocker 2009; Wang et al. 2009; Tajik et al. 2016; Athirasala et al. 2017; Uhler and Shivashankar 2018). However, a clear concept how human cells are sensing gravity is missing and research trying to identify and to understand gravity-sensitive mechanisms in human cells is still at the beginning of a long road.

17.2 Gravisensitivity in Cells of the Immune System

Several investigations provide evidence of alterations in signal transduction in lymphocytes. In this type of cells, enhanced phosphorylation of the MAP kinases ERK-1/2, MEK, and p38 and inhibition of nuclear translocation of NF-kB were the predominant responses to simulated weightlessness (Paulsen et al. 2010), whereas p44/42-MAPK-phosphorylation was reduced in microgravity during the MASER-12 sounding rocket experiment (Tauber et al. 2013). mRNA expression of p21, cyclin-dependent kinase 2-associated protein 2 (CDK2AP2, p14), cyclin-dependent kinase 4 (CDK4), cyclin-dependent kinase inhibitor 3 (CDKN3), and cyclin D3 (CCND3) changed rapidly and reversibly in microgravity compared to normogravity or hypergravity (Thiel et al. 2012), whereby microgravity-induced p21 mRNA expression inhibition was dependent on histone acetylation/deacetylation activity (Thiel et al. 2012). Microgravity affected the protein kinase C (PKC) (Hatton et al. 2002; Schmitt et al. 1996), whereas delivery of first activation signal, patching, and capping of conA-binding membrane proteins occurred normally (Cogoli et al. 1992). These findings suggest the existence of gravisensitive cellular targets upstream from PKC and downstream from the T cell receptor (TCR)/CD3, where the lipid-raft-associated membrane-proximal signalosome complex is located. However, experiments with primary human T cells during the MASER-12 experiment suggested that key proteins of T cell signal modules are not severely disturbed in microgravity and that the strong T cell inhibiting signal occurs probably downstream from membrane proximal signaling, such as at the transcriptional level (Tauber et al. 2013). In this context, gene expression analysis of T cells subjected to simulated microgravity revealed an alteration of several signal modules, in particular NF-kB and MAPK-signaling (Boonyaratanakornkit et al. 2005). Also the expression of the early oncogenes c-fos, c-myc, and c-jun is inhibited (summarized in Braeucker et al. 2002). Gravisensitive mechanisms at the chromatin and epigenetic level have been suggested (Paulsen et al. 2010; Singh et al. 2010)

Recently, the dynamics of gene expression response to different gravitational environments in human Jurkat T lymphocytic cells and human myelomonocytic U937 were investigated through the combination of parabolic flight with suborbital ballistic rocket and 2D clinostat and centrifuge experiments, using strict controls for excluding all possible other factors of influence (Thiel et al. 2017c, d, 2018). Human T cells rapidly responded to altered gravity in the time frame of 20 s and 5 min, involving regulatory RNAs as major response (Thiel et al. 2017c). A recent

ISS study suggested that T-cell activation itself may induce a sequence of gene expressions that is self-limited by miR-21 (Hughes-Fulford et al. 2015). Although three gravity-regulated genes (ATP6V1A/D, IGHD3-3/IGHD3-10 and LINC00837) could be identified by cross-validation in completely independent experiment missions (Thiel et al. 2017c), an overall high stability of gene expression in microgravity was found, with olfactory gene expression in the chromosomal region 11p15.4 as particularly robust to altered gravity (Thiel et al. 2017d).

In other studies, gravisensitivity of pro- and anti-apoptotic pathways has been reported in human mononuclear cells (Bakos et al. 2001), human ML-1 thyroid-carcinoma cells (Kossmehl et al. 2002), and astrocytes (Uva et al. 2002a). On the molecular level, microgravity induced Fas, p53, and Bax and reduced Bcl-2 (Kossmehl et al. 2002; Nakamura et al. 2003; Ohnishi et al. 1999). Interestingly, the expression of Fas was elevated in Jurkat-T-cells during spaceflights of the shuttle missions STS (Space Transportation System)-80 and STS-95 (Cubano and Lewis 2000), suggesting an enhanced Fas-FasL-mediated apoptosis of immune cells. During a 14-day spaceflight (SLS-2-mission) an accumulation of p53 has been found in keratinocytes and myocytes, indicating that central regulatory molecules of nuclear signal transduction and cell cycle are influenced by gravity (Ohnishi et al. 1999). In fact, p53 protein was phosphorylated in Jurkat T cells after 20 s in real microgravity (Paulsen et al. 2010). The diminished proliferative response of T cells upon stimulation during microgravity could also be caused by a reduced expression of IL-2 receptor (Schwarzenberg et al. 1999; Walther et al. 1998; Tauber et al. 2013, 2015), resulting in an impairment of positive regulatory feedback loops. Overall, a decreased capacity of T-cells for the production of cytokines is a prominent effect of microgravity on leukocytes (Cogoli and Cogoli-Greuter 1997).

Microgravity also impaired monocyte function: During the spacelab-mission SLS-1 monocytes lost their capability of secreting IL-1 (Cogoli 1993) and of expressing IL-2-receptor (Hashemi et al. 1999). However, the molecular mechanisms are not identified. Examination of gene expression of monocytes under real microgravity demonstrated significant changes in gene induction associated with differentiation of monocytes into macrophages (Hughes-Fulford et al. 2008). Kaur et al. (2005) investigated monocytes isolated from astronauts before and after a mission and compared the results with control groups. They found a reduction of phagocytosis and a reduced oxidative burst- and degranulation-capacity. Meloni et al. (2006) recently demonstrated that simulated weightlessness leads to massive alterations in the cytoskeleton of J-111 cells, which in turn influences motility and recently revealed during an ISS experiment a severe reduction in the locomotion ability of monocytic cells in microgravity (Meloni et al. 2008) (see also Chap. 14). However, primary human macrophages exhibited neither quantitative nor structural changes of the cytoskeleton after 11 days in microgravity during the CELLBOX ISS experiment, and only minor alterations in the metabolite spectrum (Tauber et al. 2017). Lymphocyte function-associated antigen 1 (LFA-1) and intercellular adhesion molecule-1 (ICAM-1) adhesion protein expression—important to tether the antigen presenting cell (APC) to the lymphocyte that is activated and that proliferates in response to the antigen presented and other (co-) activation pathways—seemed also to be sensitive to microgravity (Meloni et al. 2008; Paulsen et al. 2014, 2015; Tauber et al. 2017), whereas their interaction is not altered (Meloni et al. 2008). Surprisingly, the ISS experiment TRIPLE LUX A provided direct evidence of cellular sensitivity in macrophages within seconds and a subsequent ultrafast adaptation in only 42 s to microgravity, through real-time on orbit measurements (Thiel et al. 2017b). Therefore, the ultrafast adaptation response of the oxidative burst, the extensive and rapid adaptation of gene expression, and the cytoskeletal stability after long-term microgravity exposure suggest the existence of rapid force transduction and transmission into a cellular response, followed by adaptation processes.

17.3 Cell Migration and Cytoskeletal Architecture in Altered Gravity

Cell migration is an essential characteristic of life. Multicellular organisms must be motile to obtain nourishment, evade being eaten in their own right, respond to environmental changes, and reproduce. Likewise, unicellular organisms such as *Paramecium* or *Loxodes* must dynamically respond to fluctuations in ever-changing surroundings to assure survival. However, cell migration is also an essential characteristic of many normal and abnormal biological processes within the human organism including embryonic development, defense against infections, wound healing, and tumor metastasis (Lauffenburger and Horwitz 1996; Horwitz and Parsons 1999). Especially cells of the immune system need to move over long distances in the human body in only a few hours. Their high plasticity is a requirement to adapt to the changing environment (Vargas et al. 2017). An example are neutrophil granulocytes that demonstrate the body's first line of host defense by recognizing and killing microorganisms. Neutrophil locomotion is integral for immune effector function, because the cells have to leave the blood vessels and navigate to places of infection and injury to fulfil their main task of phagocytosis.They have the capability to pass through gaps smaller than 1.5 μm by varying the organization of the cytoskeleton in different environments (Thiam et al. 2016). Returning astronauts of spaceflight missions exhibited a strong increase of neutrophil granulocytes immediately after landing (Kaur et al. 2004) and neutrophil chemotactic assays showed a tenfold decrease in the optimal dose–response after the return to Earth (Stowe et al. 1999). In previous studies, changes in gravity demonstrated an inhibition of lymphocyte locomotion through type I collagen (Pellis et al. 1997; Sundaresan et al. 2002), and culture of human bone marrow CD34+ cells using NASA's rotating wall vessels resulted in a decreased migration potential (Plett et al. 2004). An altered movement in microgravity was shown for leukocytes and Jurkat T cells, too (Cogoli-Greuter et al. 1996; Sciola et al. 1999), whereas the underlying signal transduction mechanisms are still illusive. On the other side, T cells become more motile after being cultured in 10 g hypergravity (Galimberti et al. 2006).

The cells in the human body have to respond to different environmental conditions. They sense mechanical forces and signals and react to these changes. So far, it is still controversially discussed if the forces generated by Earth gravitation are too small to generate a cellular response (Vorselen et al. 2014). However, multiple investigators have reported a cellular sensitivity to gravity (Najrana and Sanchez-Esteban 2016). The cytoskeleton of cells is discussed to be a candidate to sense and promote mechanical forces. The mechanism of signal conversion into an intracellular response is termed tensegrity (Ingber 1998).

The cytoskeleton is responsible for giving a cell its shape and for generating the forces required for cell motility. It is an internal network of at least three types of cytosolic fibers: actin filaments, microtubules, and intermediate filaments. Actin, one of the most highly conserved and abundant eukaryotic proteins, is constantly polymerized and depolymerized within cells to invoke cellular motility, tissue formation, and repair (Feldner and Brandt 2002; Lee and Gotlieb 2002). Actin dynamics are considered to be the major component of the cytoskeleton responsible for cell motility. It has been shown to be essential for the migration of T lymphocytes as well as neutrophil granulocyte migration, a conclusion readily assumed as actin-depolymerizing drugs inhibit cellular motility (Hofman et al. 1999; Verschueren et al. 1995). In contrast, an intact microtubule network does not appear to be required for neutrophil migration, because microtubule-disrupting drugs such as colchicine even induce the migration of neutrophils (Niggli 2003), probably by inducing changes in the actin network. In the tensegrity model, all three types of cytoskeletal fibers are involved in the cellular response (Maniotis et al. 1997; Eckes et al. 1998; Hubmayr et al. 1996; Pourati et al. 1998). This model suggests that all nonspecialized cells are able to sense changes in gravity by differences in the balance of force distribution.

Multiple investigators have reported that this complex network of fibers is sensitive to environmental factors such as microgravity and altered gravitational forces (Schatten et al. 2001). The self-organization of microtubules is e.g. gravity-dependent and affected by exposure to altered gravity (Papaseit et al. 2000; Tabony et al. 2007). Furthermore, several studies demonstrate modifications of the actin and microtubule cytoskeleton in microgravity. Already a few minutes of weightlessness affected the cytoskeleton of lymphocytes, astrocytes, neurons, and glial cells, disorganizing microtubules, intermediate filaments, and microfilaments (Uva et al. 2002b, 2005; Roesner et al. 2006). Morphological differences of both the microtubule and actin components of the cytoskeleton have been observed in cells grown in real and simulated microgravity (Uva et al. 2002b; Lewis et al. 1998). Hughes-Fulford (2003) reported that actin reorganization responded to the gravity level. However, cells and tissues should be considered as highly dynamic systems which constantly adapt to new environmental situations. Cellular changes that are visible after short-term altered gravity like seconds or minutes could be compensated after hours or days of exposure to the new stimulus. For instance, primary human macrophages exposed to five minutes of microgravity during a suborbital rocket flight showed changes in the actin cytoskeleton, while after exposure of 11 days of real microgravity, the cells showed neither quantitative nor structural changes of the actin and vimentin cytoskeleton (Tauber et al. 2017; Thiel et al. 2019).

17.4 Graviperception: Lessons Learned from Unicellular Systems

All living organisms on Earth developed under a constant gravitational force (Barlow 1995; Volkmann and Baluska 2006). Therefore it can be anticipated that similar strategies for graviperception has been emerged during evolution of life. Unicellular systems were and are frequently used as model systems to analyze and understand the influence of gravitational forces on the cellular level. Common subjects of study are ciliates like *Paramecium* and *Loxodes*, and flagellates like the algae *Euglena* (reviewed in Hemmersbach and Braeucker 2002; Haeder et al. 2005; Häder and Hemmersbach 1997; Häder et al. 2017). Ciliates and flagellates are particularly interesting, because they can show positive and negative gravitaxis i.e. a movement in the same or opposite direction as the gravity vector, and gravikinesis, an altered swimming velocity, and have the advantage of convenient experimental handling and observation as well as short population doubling times (Planel et al. 1981, 1982; Machemer et al. 1991; Haeder et al. 1996; Hemmersbach et al. 1998; Hemmersbach and Haeder 1999). In these model organisms two different mechanisms of graviperception evolved: (1) in the ciliate *Loxodes* gravity is sensed by specific organelles called Müller vesicles. They consist of a statolith composed of barium sulfate (Penard 1917; Rieder 1977; Fenchel and Finlay 1986) within a vacuole attached to a ciliary stick. The experimental disconnection of the statolith and the ciliary stick led to the loss of gravitaxis, indicating that the Müller vesicles are the cellular gravireceptors (Hemmersbach et al. 1998); (2) in the flagellate *Euglena* and the ciliate *Paramecium* density differences between cytoplasm and extracellular medium activate mechanosensitive ion channels (Hemmersbach et al. 1998, 1999; Lebert and Haeder 1996; Lebert et al. 1997). In *Euglena*, experiments in real microgravity showed that the cells react to gravity and gravity changes (Häder et al. 2006). The entire cell content which is heavier than the medium surrounding the cell, presses on the lower cell membrane and is used for the perception of gravity (Lebert et al. 1997). Lebert and colleagues calculated that under terrestrial conditions the force acting on the lower membrane is between 0.57 and 1.13 pN (Lebert et al. 1999). On the molecular level, it could be shown that the gravireceptor is a specific transient receptor potential (TRP) channel (Elbashir et al. 2001). This type of channel is involved in many cellular processes like photoreception, nociception, and mechanoperception (Barritt and Rychkov 2005). The activation of the TRP channel induces different signal transduction cascades where calcium, cAMP, calmodulin, and phosphorylation processes play an important role (Hemmersbach and Haeder 1999; Häder and Lebert 2001; Schwer et al. 2013; Tahedl et al. 1997; Haeder et al. 2005; Streb et al. 2002). The sensitivity for gravity differs in *Paramecium* (threshold <0.35 g; Hemmersbach et al. 1996, 1998), *Euglena* (threshold ≤ 0.16 g; Haeder et al. 1996, 1997), and *Loxodes* (threshold <0.15 g; Hemmersbach et al. 1998). This was shown by using a slow rotating centrifuge microscope (NIZEMI) in microgravity and identifying the acceleration threshold inducing graviresponse. Below these thresholds, protists were unable to perceive gravity and lost their typical gravity-based directed movements (Hemmersbach-Krause et al. 1993). These results were independent of the previous exposure to microgravity up to 12 days, although cells underwent several division cycles. Besides the described effects on gravity-based orientation, other effects can be observed due to microgravity in protists like *Paramecium*, e.g. an increased cell growth rate, increase in cell volume, decrease in total cell protein content, and lower cell calcium content (Planel et al. 1981, 1982; Planel 2004). The research on effects of microgravity on microorganisms will be of common future interest, since they represent an essential component of biological life support systems during long-term spaceflights.

17.5 Sensing Gravitational Forces in Human Cells

The gravitational force acting on a cell or a subcellular or molecular structure is equal to the mass of the structure and the acceleration toward the center of Earth. One of the first theoretical studies about the physical background of cellular microgravity effects evaluated phenomena like temperature, Brownian movement, convection, hydrostatic forces, and cell membrane stress, concluding that cells with a diameter of 10 μm and more would experience gravity (Pollard 1965). However, because the weight of single normal-sized cells of 10 μm is too small compared with other cellular forces, direct "sensing" of the gravity vector, e.g. to distinct between up and down, seems unlikely (Albrecht-Buehler 1991; Klopp et al. 2002). But, due to the much larger weight of the surrounding tissues and fluids, cells are constantly subjected to environmental force changes caused by gravity and may thus "sense" gravity indirectly at least as the magnitude of gravity-induced weight forces (Albrecht-Buehler 1991), e.g. through the hydrostatic pressure in a fluidic environment. Since decades, it has been asked for mechanism by which the gravitational force becomes transmitted into a biological process (Brown 1991), but was rarely addressed in experimental approaches. In a very recent review about perception of gravity in eukaryotes (Häder et al. 2017), gravity-perception by cytoskeletal processes and mechanosensitive ion channel have been discussed, but the presented theories about gravitational force—transmission in mammalian cells remained speculative and not substantiated by thorough experiments. Therefore, until today, no theory of gravitational force transmission into a biological process in mammalian cells has been experimentally validated so far.

17.5.1 Sensing Stress as Changes in the Gravitational Environment

The environment, which is sensed locally (Choquet et al. 1997), consists of cells and of the intercellular matrix. In microgravity, force-induced breakage of cell–cell- or cell–matrix-adhesion sites could be reduced. The mechanical properties of the matrix have predominant effects on different cell functions (Discher et al. 2005). The forces sufficient to activate such kind of cellular response are very low (10 pN for 1 s, Jiang et al. 2006). Adhesion-mediated signaling provides cells with information about multiple parameters of their microenvironment, including mechanical characteristics (Bershadsky et al. 2006). During the last decade, several molecules and mechanisms of adhesion-mediated signaling have been identified, mostly in endothelial cells, where the response to fluid shear stress alters and regulates several cellular functions (Engler et al. 2006). In endothelial cells, integrins are crucially involved in endothelial mechanosensing of shear stress (Shyy and Chien 2002). Immunoglobulin family adhesion receptor platelet endothelial cell adhesion molecule (PECAM-1) directly transmits mechanical force and, in cooperation with vascular endothelial (VE)-cadherin and vascular endothelial growth factor receptor (VEGFR)-2, mediates the response of confluent endothelial cells to shear flow (Tzima et al. 2005). Interestingly, modulation of the expression of surface adhesion molecules such as ICAM-1 has been reported as the consequence of long-term microgravity (Tauber et al. 2017; Buravkova et al. 2005; Romanov et al. 2001). Additionally, early molecular mechanisms responsible for gravity sensing of endothelial cells involve caveolae and Caveolin-1 phosphorylation (Spisni et al. 2006). Importantly, the appropriate formation and function of the immunological synapse between T cells and antigen-presenting cells require a well-defined spatial orientation of membrane adhesion molecules ICAM-1 and LFA-1 (Mossman et al. 2005). Therefore it is possible that integrin-mediated force transduction renders the immunological synapse a gravisensitive site.

17.5.2 Cellular Mechanosensory System

If the cells are able to sense gravitational forces by adhesion to the extracellular matrix, the question about the responsible mechanosensory system arises. Paradigms of cellular mechanosensing have been reviewed by Orr et al. (2006). One possibility is that the entire actin network serves as a mechanosensor. The folding state of cytoskeleton-associated proteins, which creates or masks binding sites for other proteins, depends on the strains in the actin network (reviewed in Vogel and Sheetz 2006). Forces applied to or taken from the actin network could be therefore transduced in altered binding of signal proteins to the cytoskeleton or the gain or loss of enzyme function. Consequently, microgravity may reduce the force inside the actin network, which could be then transduced into a certain biochemical signal by cytoskeleton-associated proteins. According to the tensegrity model (Fuller 1961), the whole cell is a prestressed structure (Ingber 1993, 2003b), with tensions generated by the actin–myosin network, by cellular force through focal adhesions (Choquet et al. 1997; Tamada et al. 2004), by cell–cell adhesions and polymerization of cytoskeletal elements (Wang et al. 2001, 2002; Wang and Stamenovic 2000; Stamenovic et al. 2002). A cell would not maintain its shape stability under load without a preexisting stress or prestress in the mechanical elements (Ingber 2003b). During gravitational unloading of the cell, intracellular forces induced by the prestress could be altered and transduced into a biochemical response. Force-induced changes of protein conformation and exposure of cryptic binding sites for signal proteins have been described as a possible method of mechanotransduction (reviewed in Vogel and Sheetz 2006). Unfolding of proteins with tandem-repeat domains can occur with applied forces on the order of 50–200 pN, as demonstrated by single-molecule experiments with actinin (Rief et al. 1999), filamin (Furuike et al. 2001), and spectrin (Rief et al. 1999). Tandem-repeat sequences are found in most extracellular matrix (ECM) proteins and many proteins that link the integrins to the cytoskeleton. Therefore, cells could hypothetically "measure" strain by integrating the number of unfolded domains (Hoffman et al. 2007). Many molecules are stabilized by disulphide bonds and their redox state can be therefore sensitive to force (Vogel and Sheetz 2006). In cross-linked molecules, only a small alteration in force could readily cause extensive conformational changes, suggesting binding partners of cross-linking proteins as potential signal proteins in mechanotransduction, such as including heat shock proteins (HSPs), PKC, ras-related protein A (Ral A), Phosphatidylinositol 4,5-bisphosphate (PIP2), Phosphatidylinositol (3,4,5)-trisphosphate (PIP 3), Phosphatidylinositol 3 (PI3)-kinase, MEKK1 mitogen-activated protein kinase 1 (as reviewed in Otey and Carpen 2004; Stossel et al. 2001), and regulatory proteins for the Rho GTPases (Mammoto et al. 2007; Ohta et al. 2006). Interestingly, Rho kinase has been found to regulate the intracellular micromechanical response of adherent cells (Kole et al. 2004) and small G proteins are discussed having a significant role in mechanotransduction (Burridge and Wennerberg 2004). A special class of mechanosensing mechanisms (Orr et al. 2006) is represented by myosins. In most myosin classes, mechanical load alters ADP release. Myosin mechanosensing is best demonstrated by myoIC during adaptation in hearing (LeMasurier and Gillespie 2005). Enzymatic activity can be regulated in response to mechanical force by opening of enzymatic cleavage sites. Fibronectin, for example, has a partially cryptic disulphide isomerase (Langenbach and Sottile 1999) and a cryptic metalloprotease activity (Schnepel and Tschesche 2000), but it is not known whether these activities can be regulated by force (Vogel and Sheetz 2006). It has been demonstrated that application of a force to fibronectin binding induced rapid local src activation forming a directional wave propagated away from the stimulation site along the plasma membrane (Wang et al. 2005). Structural motifs that could change conformation over a range of mechanical forces and could therefore exhibit mechanosensory functions are diverse (Bershadsky et al. 2006; Martinac 2004; Vogel and Sheetz 2006). Considering the myriad of multidomain cross-linking proteins in different cells and subcellular localizations, it is likely that mechanoresponse is a highly specific and specialized process and not a general and nonspecific phenomenon.

Mechanotransduction occurs through sensing of the major matrix signals through focal adhesion proteins and the cytoskeleton (Chan et al. 2009; Riveline et al. 2001; Iskratsch et al. 2014), contributing to the activation of a variety of transcription factors such as YAP/TAZ (Miralles et al. 2003; Dupont et al. 2011; Halder et al. 2012; Miroshnikova et al. 2017), SRF/MRTF-A (Vartiainen et al. 2007; Morita et al. 2007), NF-κB (Hayden and Ghosh 2008), and JMY (Campellone and Welch 2010) and transcription co-activators (Sathe et al. 2016; Wang and Gilmore 2003) to regulate gene expression. In addition, cytoskeleton-regulated Rho GTPase further mediates cell-shape-dependent changes in cell-cycle progression (Mammoto et al. 2007, 2004). Interestingly, SRF binding sites are present in the promoters of a majority of genes inhibited in T cells activated in microgravity (Chang et al. 2012). Given that Ras/MAPK and Rho/actin pathways activate SRF, these cytoskeleton-regulating small GTPases could be disrupted in microgravity explaining the reduced T cell activation (Katsch et al. 2012).

17.5.3 Force Transduction into the Nucleus

Because transcriptome alterations could be detected as early as 20 s after the onset of altered gravitational force two independent cell models and independent experiment campaigns (Thiel et al. 2015, 2017c, d, 2018), fast-reacting transduction cascades between the gravitational force and the regulation of the transcriptome have to be assumed. In summary, mechanosensation at the plasma membrane leads to downstream nucleocytoplasmic shuttling of various transcription regulators (Thorpe and Lee 2017). Whereas significant transcription-caused transcriptome alterations are theoretically possible after 20 s (Callegari 2016; Danko et al. 2013; Darzacq et al. 2007; Maiuri et al. 2011), the preceding signal cascade has to occur within seconds or faster, which has been demonstrated for direct mechanotransduction into the nucleus previously (Wang et al. 2009; Thorpe and Lee 2017; Athirasala et al. 2017), in contrast to biochemical signal transduction cascades into the nucleus, requiring 10 min or more (Karin and Hunter 1995; Tsang et al. 2008).

The physical link between the nucleus and the cytoskeletal compartments is represented by the “linker of nucleoskeleton and cytoskeleton (LINC)” complex, composed of inner and outer nuclear membrane proteins with diverse functions in nuclear positioning, nuclear shaping, chromatin organization, and mechanotransduction (Chang et al. 2015) (see also Fig. 17.2). The LINC complex, formed by the interaction of nesprins and SUN proteins at the nuclear envelope (Lombardi et al. 2011) binds to nuclear lamin proteins, whereby lamins interact with chromatin either directly or through histones and other lamin-associated proteins including emerin, lamin B receptor (LBR), heterochromatin protein 1 (HP1), barrier-to-autointegration factor (BAF), LEM domain-containing protein 3 (LEMD3), and several lamin-associated polypeptide-2 (LAP2) isoforms (Wilson and Foisner 2010). Lamins undergo conformational change under cytoskeletal forces (Ihalainen et al. 2015) and also impact gene expression through their interaction with transcription factors affecting proliferation, differentiation, and apoptosis (Wilson and Foisner 2010; Dreuillet et al. 2002). The inner nuclear membrane (INM) is enriched in a wide range of proteins that are required for nuclear structure, chromosome organization, DNA repair machinery, and transcriptional regulation (Katta et al. 2014). At least 60 proteins have been identified to associate with the INM, but their functions are mostly unknown (Hetzer 2010). Tethering of peripheral chromatin to the nuclear lamina occurs in specific genomic regions termed lamin-associated domains (LADs) of “repressive heterochromatin” which reduces transcription factor accessibility resulting in low gene expression levels (Guelen et al. 2008). Cyclic stretching force induces redistribution of emerin from the inner nuclear membrane (INM) to the outer nuclear

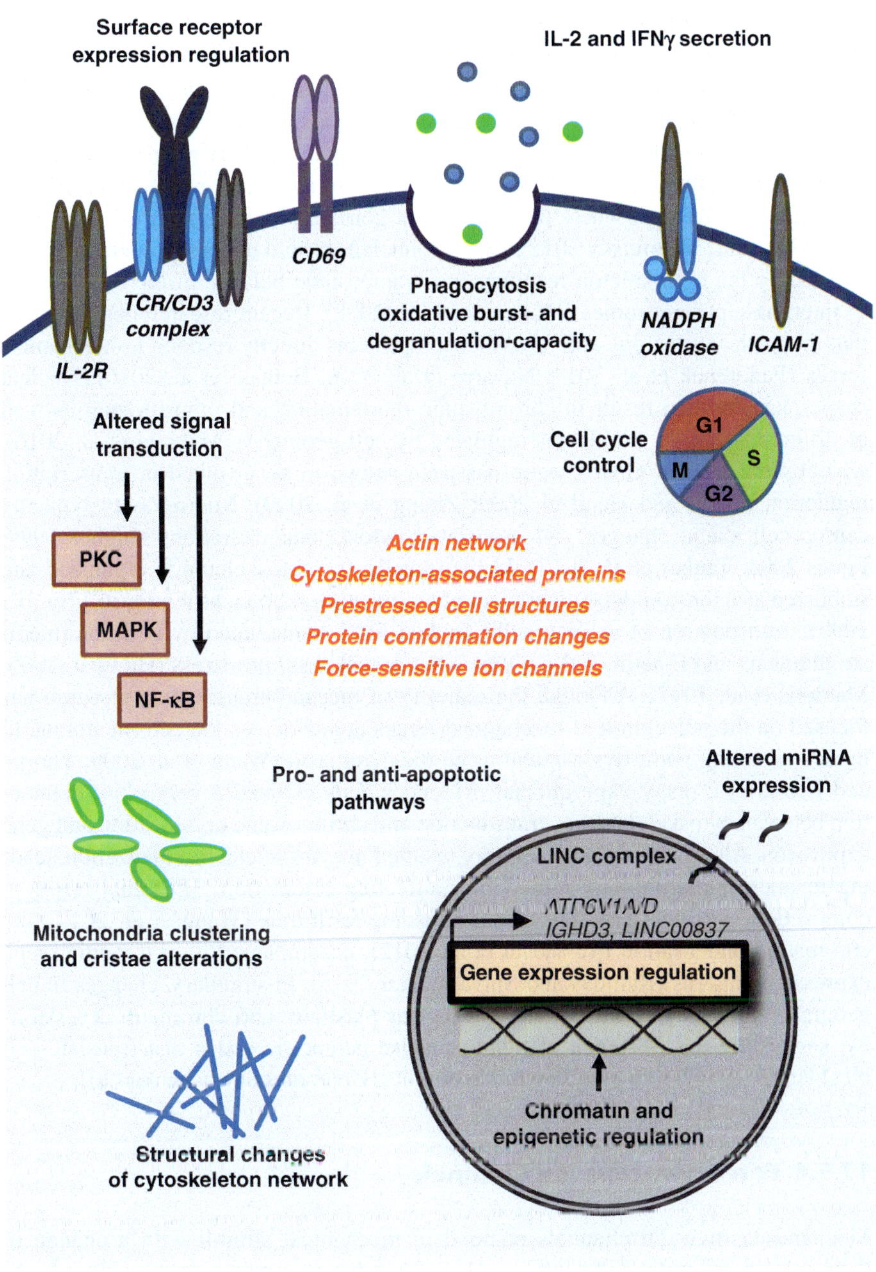

Fig. 17.2 Schematic summary of gravity-sensitive molecular alterations in cells of the immune systems. Red: Potential gravity-sensing systems. All depicted alterations could represent either transient changes during homeostatic regulations or a new steady state in the microgravity environment

membrane (ONM), which results in the detachment of heterochromatin from the nuclear lamina transcription attenuation (Le et al. 2016). In mammals, genome-wide analysis showed that lamin-bound heterochromatin builds up to 40% of the genome (Solovei et al. 2013; Ihalainen et al. 2015), organized in LADs. Consequently, there is accumulating evidence that the nuclear periphery is a transcriptionally repressive compartment with a large genome fraction either in transient or permanent contact with nuclear envelope, where the majority of genes are maintained in a silent state (Shevelyov and Nurminsky 2012). Also the nucleus is held under a pre-stress tension by links to the cytoskeleton resulting in a homeostatic balance of the organization of interphase chromosomes (Mazumder et al. 2008). Recent research demonstrated that the nuclear envelope and its associated proteins directly respond to mechanical forces (Fedorchak et al. 2014; Navarro et al. 2016; Belaadi et al. 2016). Nuclear shape changes then result in chromosome reorientation and repositioning as well as differential lamin A/C levels regulated by cell geometry (Makhija et al. 2016), whereby the nuclear G or F actin has been shown to be involved in transcription regulation (Gieni and Hendzel 2009; Zheng et al. 2009). Microgravity typically causes cell shape changes and associated cytoskeletal alterations (Ingber 1999; Lewis 2004; Tauber et al. 2017). In living cells, the cytoskeleton is organized and stabilized as a tension-dependent form of architecture (known as tensegrity) (Ingber 1999). Transduction of an externally applied force is mediated by both intermediate filaments and F-actin, and requires some cytoskeletal pre-stress (Hu et al. 2005; Maniotis et al. 1997). Although the majority of mechanotransduction research has focused on the perception of mechanical forces at and across the cell membrane to induce signaling pathways originating in the cytoplasm (Wang et al. 2009; Thorpe and Lee 2017), many experimental evidences demonstrated a very close connection between cytoskeletal force transduction and chromosome organization and gene expression: Alterations in cell geometry resulted in cytoskeletal reorganization, leading to nuclear morphology remodeling, affecting orientation, 3D radial position, compaction, and intermingling of chromosome territories (Wang et al. 2017) and chromatin condensation (Versaevel et al. 2012), accompanied by differential gene expression patterns (Ramdas and Shivashankar 2015). In summary, changes in cell geometry obviously results to altered nuclear plasticity and chromatin accessibility, suggesting the existence of highly precise geometric codes that translate cell mechanical signals into precise genetic outputs (Uhler and Shivashankar 2016).

17.5.4 Force-Sensitive Ion Channels

Mechanosensitive ion channels respond to mechanical stimuli with a change in their conductive state. Either the mechanosensitive channel senses directly alterations of the lipid bilayer or transmit forces of the cytoskeleton or the extracellular matrix via a physical connection. Ion-channel involvement in mechanosensing is well described in prokaryotic systems (Martinac 2004). However, prokaryotic ion channels are relatively force-insensitive and only open at high tensions that are approaching the lytic tensions for the lipid bilayer (Vogel and Sheetz 2006), whereas typical membrane tensions in animal cells are a thousand-fold lower than the activating bacterial tensions (Sheetz 2001). Membrane channels, for example, the PKD2 Ca^{2+} channel and inner rectifier K^{+} channel in eukaryotes, are activated upon stretching (Martinac 2004), leading to channel opening, ion flux, and most probably to the recruitment and activation of downstream signaling molecules. In this context, cellular responses to forces in bone and cartilage are probably the consequence of out-of-plane forces on channel–cytoskeleton linkages (Haut Donahue et al. 2004). Force-sensitive ion channels have been discussed as trigger point of mechanotransduction (Ingber 2006, 2008; Goldermann and Hanke 2001; Meissner and Hanke 2005) into complex cellular reactions such as gene expression. The transient receptor potential cation channel TRPC1 is known to be activated by stretch (Maroto et al. 2005; Garrison et al. 2012), it represents a candidate for transduction of gravitational forces in cells of the MMS. This hypothesis was corroborated by the finding that in the unicellular photosynthetic flagellate *Euglena gracilis*, knockdown of a putative TRP channel abolished gravitaxis (Häder et al. 2009).

Possible cellular gravisensing mechanisms and gravisensitive cellular molecules and functions are summarized in Table 17.1.

Table 17.1 Cells of the immune system are exceptionally sensitive to microgravity

Possible cellular gravisensing systems
Actin network/folding state of cytoskeleton-associated proteins
Prestressed structure of the cell (tensegrity model)
Force-induced changes of protein conformation
Force-sensitive ion channels (e.g. TRCP1)
Gravisensitive cellular molecules and mechanisms
Protein kinase C
NF-kB
MAPK-signaling
c-fos, c-myc, and c-jun LINC complex
Chromatin and epigenetic regulation
Pro- and anti-apoptotic pathways
Cell cycle
IL-1 secretion
IL-2-receptor expression
Phagocytosis
Oxidative burst- and degranulation-capacity
Cytoskeleton
Locomotion ability
LFA-1
ICAM-1 ATP6V1A/D IGHD3-3/IGHD3-10 LINC00837

Gravitational forces may be sensed by individual cells in the context of altered extracellular matrix mechanics, cytoskeletal organization, or internal pre-stress in the cell–tissue matrix and transduced into specific molecular alterations which in turn contributes to a complex disturbance of immune cell reactions and interactions

17.6 Conclusion

Taken together, cellular gravisensing may not result from a direct activation of a single gravisensing molecule. Instead, gravitational stress and forces may be sensed by an individual cell in the context of altered extracellular matrix mechanics, cell shape, cytoskeletal organization, or internal pre-stress in the cell–tissue matrix (Ingber 1999). For understanding the force transduction system, it seems not reasonable to focus on any single signaling mechanism in isolation (Strohman 1997; Coffey 1998). The reality is, that normally multiple simultaneous inputs are transduced and integrated within the structural complexity of the living cell (Ingber 1999). Particularly, the gravitational force acts everywhere in and around the cell, and is obviously transduced to a highly time- and cell type-specific pattern of altered transcriptome response (Thiel et al. 2017c, d, 2018).

The development of cellular mechanosensitivity and mechanosensitive signal transduction was probably an evolutionary requirement to enable our cells to sense their extracellular matrix and their individual microenvironment. However, mechanosensitive mechanisms were designed to work under the condition of 1 g, but never had the possibility to adapt and adjust their reaction to conditions below 1 g. Therefore it is possible that the same mechanisms, which enable human cells to sense and to cope with mechanical stress, are potentially dangerous in microgravity. It is a major challenge to find out if our cellular machinery is able to live and to work without gravity force or if our cellular architecture will keep us dependent on the gravity field of Earth. With the completion and utilization of the International Space Station and with mission plans to moon and Mars during the first half of our century, astronautics has entered the era of long-term space exploration class missions. Such long-term missions represent a challenge never experienced before: Small or even marginal medical problems could easily evolve to substantial challenges, which could possibly endanger the entire mission. Since crew performance is the crucial factor during space missions and since evacuation or exchange of the crew is impossible during interplanetary flights, to elucidate the underlying mechanism of limiting factors for human health and performance in microgravity, such as for the immune system, and to identify and test potential counteractive interventions is an urgent need. Therefore, identification of gravisensitive cellular reactions will also help to understand the molecular mechanisms of disturbed immune cell function in space in order to identify, to test, and to provide new targets for therapeutic or preventive intervention related to the immune system of astronauts during long-term space missions.

Acknowledgments We gratefully acknowledge the support by the German Aerospace Center (DLR), Space Agency (grant no. 50WB0613, no. 50WB0912, no. 50WB1219, no. 50WB1519), and ESA (ESTEC Contract nr 20562/07/NL/VJ ESA-CORA-GBF-2005-005). We also gratefully thank our collaboration partners DLR, ESA, NASA, Spaceflorida, Airbus Defense and Space, Novespace, the Swiss Air Force, the Swiss International Airlines, and the Deutsche Lufthansa.

References

Adrian A, Schoppmann K, Sromicki J et al (2013) The oxidative burst reaction in mammalian cells depends on gravity. Cell Commun Signal 11:98

Albrecht-Buehler G (1991) Possible mechanisms of indirect gravity sensing by cells. ASGSB Bull 4:25–34

Armstrong JW, Gerren RA, Chapes SK (1995) The effect of space and parabolic flight on macrophage hematopoiesis and function. Exp Cell Res 216(1):160–168

Athirasala A, Hirsch N, Buxboim A (2017) Nuclear mechanotransduction: sensing the force from within. Curr Opin Cell Biol 46:119–127

Bakos A, Varkonyi A, Minarovits J, Batkai L (2001) Effect of simulated microgravity on human lymphocytes. J Gravit Physiol 8:69–70

Barlow PW (1995) Gravity perception in plants: a multiplicity of systems derived by evolution? Plant Cell Environ 18:951–962

Barritt G, Rychkov G (2005) TRPs as mechanosensitive channels. Nat Cell Biol 7:105–107

Battista N, Meloni MA, Bari M et al (2012) 5-Lipoxygenase-dependent apoptosis of human lymphocytes in the International Space Station: data from the ROALD experiment. FASEB J 26:1791–1798

Belaadi N, Aureille J, Guilluy C (2016) Under pressure: mechanical stress management in the nucleus. Cell 5:27

Bershadsky A, Kozlov M, Geiger B (2006) Adhesion-mediated mechanosensitivity: a time to experiment, and a time to theorize. Curr Opin Cell Biol 18:472–481

Boonyaratanakornkit JB, Cogoli A, Li CF et al (2005) Key gravity-sensitive signaling pathways drive T cell activation. FASEB J 19:2020–2022

Braeucker R, Cogoli A, Hemmersbach R (2002) Graviperception and graviresponse at the cellular level. In: Horneck G, Baumstark-Khan C (eds) Astrobiology the quest for the conditions of life. Springer, Berlin, pp 287–333

Brown AH (1991) From gravity and the organism to gravity and the cell. ASGSB Bull 4:7–18

Brungs S, Kolanus W, Hemmersbach R (2015) Syk phosphorylation–a gravisensitive step in macrophage signalling. Cell Commun Signal 13:9

Buravkova L, Romanov Y, Rykova M et al (2005) Cell-to-cell interactions in changed gravity: ground-based and flight experiments. Acta Astronaut 57:67–74

Burridge K, Wennerberg K (2004) Rho and Rac take center stage. Cell 116:167–179

Callegari A (2016) Eukaryotic transcription factor binding kinetics - a single-molecule and functional study. Thesis. EPFL, Switzerland. https://doi.org/10.5075/epfl-thesis-7267

Campellone KG, Welch MD (2010) A nucleator arms race: cellular control of actin assembly. Nat Rev Mol Cell Biol 11:237

Chan MW, Arora PD, Bozavikov P et al (2009) FAK, PIP5KIγ and gelsolin cooperatively mediate force-induced expression of α-smooth muscle actin. J Cell Sci 122:2769–2781

Chang TT, Walther I, Li CF et al (2012) The Rel/NF-κB pathway and transcription of immediate early genes in T cell activation are inhibited by microgravity. J Leukoc Biol 92:1133–1145

Chang W, Worman HJ, Gundersen GG (2015) Accessorizing and anchoring the LINC complex for multifunctionality. J Cell Biol 208:11–22

Choquet D, Felsenfeld DP, Sheetz MP (1997) Extracellular matrix rigidity causes strengthening of integrin–cytoskeleton linkages. Cell 88:39–48

Chouker A, Ullrich O (2016) The immune system in space: are we prepared. Springer International Publishing, New York, pp 123–127

Coffey DS (1998) Self-organization, complexity and chaos: the new biology for medicine. Nat Med 4:882–885

Cogoli A (1993) The effect of hypogravity and hypergravity on cells of the immune system. J Leukoc Biol 54:259–268

Cogoli A (1996) Gravitational physiology of human immune cells: a review of in vivo, ex vivo and in vitro studies. J Gravit Physiol 3:1–9
Cogoli A, Cogoli-Greuter M (1997) Activation and proliferation of lymphocytes and other mammalian cells in microgravity. Adv Space Biol Med 6:33–79
Cogoli A, Tschopp A, Fuchs-Bislin P (1984) Cell sensitivity to gravity. Science 225:228–230
Cogoli M, Bechler B, Cogoli A et al (1992) Lymphocytes on sounding rockets. Adv Space Res 12:141–144
Cogoli-Greuter M, Meloni MA, Sciola L et al (1996) Movements and interactions of leukocytes in microgravity. J Biotechnol 47:279–287
Cohrs RJ, Mehta SK, Schmid DS et al (2008) Asymptomatic reactivation and shed of infectious varicella zoster virus in astronauts. J Med Virol 80:1116–1122
Comet B (2001) Limiting factors for human health and performance: microgravity and reduced gravity. HUMEX-TN-002 Study on the survivability and adaptation of humans to long-duration interplanetary and planetary environments. Technical Note 2: Critical assessments of the limiting factors for human health and performance and recommendation of countermeasures
Cubano LA, Lewis ML (2000) Fas/APO-1 protein is increased in spaceflown lymphocytes (Jurkat). Exp Gerontol 35:389–400
Danko CG, Hah N, Luo X et al (2013) Signaling pathways differentially affect RNA polymerase II initiation, pausing, and elongation rate in cells. Mol Cell 50:212–222
Darzacq X, Shav-Tal Y, De Turris V et al (2007) In vivo dynamics of RNA polymerase II transcription. Nat Struct Mol Biol 14:796–806
Discher DE, Janmey P, Wang YL (2005) Tissue cells feel and respond to the stiffness of their substrate. Science 310:1139–1143
Dreuillet C, Tillit J, Kress M et al (2002) In vivo and in vitro interaction between human transcription factor MOK2 and nuclear lamin A/C. Nucleic Acids Res 30:4634–4642
Dupont S, Morsut L, Aragona M et al (2011) Role of YAP/TAZ in mechanotransduction. Nature 474:179
Eckes B, Dogic D, Colucci-Guyon E et al (1998) Impaired mechanical stability, migration and contractile capacity in vimen-tin-deficient fibroblasts. J Cell Sci 111:1897–1907
Elbashir SM, Lendeckel W, Tuschl T (2001) RNA interference is mediated by 21- and 22-nucleotide RNAs. Genes Dev 15:188–200
Engler AJ, Sen S, Sweeney HL et al (2006) Matrix elasticity directs stem cell lineage specification. Cell 126:677–689
Fedorchak GR, Kaminski A, Lammerding J (2014) Cellular mechanosensing: getting to the nucleus of it all. Prog Biophys Mol Biol 115:76–92
Feldner JC, Brandt BH (2002) Cancer cell motility–on the road from c-erbB-2 receptor steered signaling to actin reorganization. Exp Cell Res 272:93–108
Fenchel T, Finlay BJ (1986) The structure and function of Müller vesicles in loxodid ciliates. J Protozool 33:69–76
Frippiat JP, Crucian BE, De Quervain DJ et al (2016) Towards human exploration of space: the THESEUS review series on immunology research priorities. NPJ Microgravity 2:16040
Fuller B (1961) Tensegrity. Portfolio Artnews Annu 4:112–127
Furuike S, Ito T, Yamazaki M (2001) Mechanical unfolding of single filamin A (ABP-280) molecules detected by atomic force microscopy. FEBS Lett 498:72–75
Galimberti M, Tolic-Norrelykke IM, Favillini R et al (2006) Hypergravity speeds up the development of T-lymphocyte motility. Eur Biophys J 35:393–400
Garrison SR, Dietrich A, Stucky CL (2012) TRPC1 contributes to light-touch sensation and mechanical responses in low-threshold cutaneous sensory neurons. J Neurophysiol 107:913–922
Gieni RS, Hendzel MJ (2009) Actin dynamics and functions in the interphase nucleus: moving toward an understanding of nuclear polymeric actin. Biochem Cell Biol 87:283–306
Goldermann M, Hanke W (2001) Ion channel are sensitive to gravity changes. Microgravity Sci Technol 13:35

Guéguinou N, Huin-Schohn C, Bascove M et al (2009) Could spaceflight-associated immune system weakening preclude the expansion of human presence beyond earth's orbit? J Leukoc Biol 86:1027–1038

Guelen L, Pagie L, Brasset E et al (2008) Domain organization of human chromosomes revealed by mapping of nuclear lamina interactions. Nature 453:948

Häder DP, Hemmersbach R (1997) Graviperception and graviorientation in flagellates. Planta 203:7–10

Häder DP, Lebert M (2001) Graviperception and gravitaxis in algae. Adv Space Res 27:861–870

Häder DP, Richter PR, Strauch SM et al (2006) Aquacells – flagellates under long-term microgravity and potential usage for life support systems. Microgravit Sci Technol 18:210–214

Häder DP, Richter PR, Schuster M et al (2009) Molecular analysis of the graviperception signal transduction in the flagellate Euglena gracilis: involvement of a transient receptor potential-like channel and a calmodulin. Adv Space Res 43:1179–1184

Häder DP, Braun M, Grimm D et al (2017) Gravireceptors in eukaryotes—a comparison of case studies on the cellular level. NPJ Microgravity 3:13

Haeder DP, Rosum A, Schaefer J et al (1996) Graviperception in the flagellate Euglena gracilis during a shuttle spaceflight. J Biotechnol 47:261–269

Haeder DP, Porst M, Tahedl H et al (1997) Gravitactic orientation in the flagellate Euglena gracilis. Microgravity Sci Technol 10:53–57

Haeder DP, Hemmersbach R, Lebert M (2005) Gravity and the behaviour of unicellular organisms. Cambridge University Press, Cambridge

Halder G, Dupont S, Piccolo S (2012) Transduction of mechanical and cytoskeletal cues by YAP and TAZ. Nat Rev Mol Cell Biol 13:591

Hashemi BB, Penkala JE, Vens C et al (1999) T cell activation responses are differentially regulated during clinorotation and in spaceflight. FASEB J 13:2071–2082

Hatton JP, Gaubert F, Cazenave JP et al (2002) Microgravity modifies protein kinase C isoform translocation in the human monocytic cell line U937 and human peripheral blood T-cells. J Cell Biochem 87:39–50

Haut Donahue TL, Genetos DC et al (2004) Annexin V disruption impairs mechanically induced calcium signaling in osteoblastic cells. Bone 35:656–663

Hawkins W, Zieglschmid J (1975) Clinical aspects of crew health. In: Johnston R, Dietlein L, Berry C (eds) Biomedical results of Apollo. NASA, Washington, DC, pp 43–81

Hayden MS, Ghosh S (2008) Shared principles in NF-κB signaling. Cell 132:344–362

Hemmersbach R, Braeucker R (2002) Gravity-related behaviour in ciliates and flagellates. In: Cogoli A (ed) Cell biology and biotechnology in space, advances in space biology and medicine, vol 8. Elsevier, Amsterdam, pp 59–75

Hemmersbach R, Haeder DP (1999) Graviresponses of certain ciliates and flagellates. FASEB J 13:S69–S75

Hemmersbach R, Voormanns R, Briegleb W et al (1996) Influence of accelerations on the spatial orientation of Loxodes and Paramecium. J Biotechnol 47:271–278

Hemmersbach R, Voormanns R, Bromeis B et al (1998) Comparative studies of the graviresponses of Paramecium and Loxodes. Adv Space Res 21:1285–1289

Hemmersbach R, Volkmann D, Haeder DP (1999) Graviorientation in protists and plants. J Plant Physiol 154:1–15

Hemmersbach-Krause R, Briegleb W, Haeder DP et al (1993) Orientation of Paramecium under the conditions of microgravity. J Eukaryot Microbiol 40:439–446

Hetzer MW (2010) The nuclear envelope. Cold Spring Harb Perspect Biol 2:a000539

Hoffman BD, Crocker JC (2009) Cell mechanics: dissecting the physical responses of cells to force. Annu Rev Biomed Eng 11:259–288

Hoffman BD, Massiera G, Crocker JC (2007) Fragility and mechanosensing in a thermalized cytoskeleton model with forced protein unfolding. Phys Rev E Stat Nonlinear Soft Matter Phys 76:051906

Hofman P, d'Andrea L, Guzman E et al (1999) Neutrophil F-actin and myosin but not microtubules functionally regulate transepithelial migration induced by interleukin 8 across a cultured intestinal epithelial monolayer. Eur Cytokine Netw 10:227–236
Horwitz AR, Parsons JT (1999) Cell migration–movin' on. Science 286:1102–1103
Hu S, Chen J, Butler JP et al (2005) Prestress mediates force propagation into the nucleus. Biochem Biophys Res Commun 329:423–428
Hubmayr RD, Shore SA, Fredberg JJ et al (1996) Pharmacological activation changes stiffness of cultured human airway smooth muscle cells. Am J Phys 271:C1660–C1668
Hughes-Fulford M (2003) Function of the cytoskeleton in gravisensing during spaceflight. Adv Space Res 32:1585–1593
Hughes-Fulford M, Chang T, Li CF (2008) Effect of gravity on monocyte differentiation. Paper presented at the 10th ESA Life Sciences Symposium/29th Annual ISGP Meeting/24th Annual ASGSB Meeting/ELGRA Symposium "Life in Space for Life on Earth", 22–27 June 2008 Angers, France
Hughes-Fulford M, Chang TT, Martinez EM et al (2015) Spaceflight alters expression of microRNA during T-cell activation. FASEB J 29(12):4893–4900
Ihalainen TO, Aires L, Herzog FA et al (2015) Differential basal-to-apical accessibility of lamin A/C epitopes in the nuclear lamina regulated by changes in cytoskeletal tension. Nat Mater 14:1252
Ingber DE (1993) Cellular tensegrity: defining new rules of biological design that govern the cytoskeleton. J Cell Sci 104:613–627
Ingber DE (1998) The architecture of life. Sci Am 278:48–57
Ingber DE (1999) How cells (might) sense microgravity. FASEB J 13:S3–S15
Ingber DE (2003a) Mechanobiology and diseases of mechanotransduction. Ann Med 35:564–577
Ingber DE (2003b) Tensegrity I: cell structure and hierarchical systems biology. J Cell Sci 116:1157–1173
Ingber DE (2006) Cellular mechanotransduction: putting all the pieces together again. FASEB J 20:811–827
Ingber DE (2008) Tensegrity-based mechanosensing from macro to micro. Prog Biophys Mol Biol 97:163–179
Iskratsch T, Wolfenson H, Sheetz MP (2014) Appreciating force and shape—the rise of mechanotransduction in cell biology. Nat Rev Mol Cell Biol 15:825
Jiang G, Huang AH, Cai Y et al (2006) Rigidity sensing at the leading edge through avb3 integrins and RPTPa. Biophys J 90:1804–1809
Karin M, Hunter T (1995) Transcriptional control by protein phosphorylation: signal transmission from the cell surface to the nucleus. Curr Biol 5:747–757
Katsch K, De Jong SJ, Albrecht JC et al (2012) Actin-dependent activation of serum response factor in T cells by the viral oncoprotein tip. Cell Commun Signal 10:5
Katta SS, Smoyer CJ, Jaspersen SL (2014) Destination: inner nuclear membrane. Trends Cell Biol 24:221–229
Kaur I, Simons ER, Castro VA et al (2004) Changes in neutrophil functions in astronauts. Brain Behav Immun 18:443–450
Kaur I, Simons ER, Castro VA et al (2005) Changes in monocyte functions of astronauts. Brain Behav Immun 19:547–554
Kimzey SL (1977) Hematology and immunology studies. In: Johnson RS, Dietlein LF (eds) Biomedical results from Skylab, NASA SP-377. Scientific and Technical Information Office, National Aeronautics and Space Administration, Washington, DC, pp 249–282
Klopp E, Graff D, Struckmeier J et al (2002) The osteoblast mechano-receptor, microgravity perception and thermodynamics. J Gravit Physiol 9:269–270
Kole TP, Tseng Y, Huang L et al (2004) Rho kinase regulates the intracellular micromechanical response of adherent cells to rho activation. Mol Biol Cell 15:3475–3484
Kondrachuk AV, Sirenko SP (1996) The theoretical consideration of microgravity effects on a cell. Adv Space Res 17:165–168

Konstantinova IV, Antropova YN, Legenkov VI et al (1973) Study of reactivity of blood lymphoid cells in crew members of the Soyuz-6, Soyuz-7 and Soyuz-8 spaceships before and after flight. Space Biol Med 7:48–55
Kossmehl P, Shakibaei M, Cogoli A et al (2002) Simulated microgravity induces programmed cell death in human thyroid carcinoma cells. J Gravit Physiol 9:P295–P296
Langenbach KJ, Sottile J (1999) Identification of protein-disulfide isomerase activity in fibronectin. J Biol Chem 274:7032–7038
Lauffenburger DA, Horwitz AF (1996) Cell migration: a physically integrated molecular process. Cell 84:359–369
Le HQ, Ghatak S, Yeung CY et al (2016) Mechanical regulation of transcription controls Polycomb-mediated gene silencing during lineage commitment. Nat Cell Biol 18:864–875
Lebert M, Haeder DP (1996) How Euglena tells up from down. Nature 379:590
Lebert M, Richter P, Haeder DP (1997) Signal perception and transduction of gravitaxis in the flagellate Euglena gracilis. J Plant Physiol 150:685–690
Lebert M, Porst M, Richter P et al (1999) Physical characterization of gravitaxis in Euglena gracilis. J Plant Physiol 155:338–343
Lee JS, Gotlieb AI (2002) Microtubule-actin interactions may regulate endothelial integrity and repair. Cardiovasc Pathol 11:135–140
LeMasurier M, Gillespie PG (2005) Hair-cell mechanotransduction and cochlear amplification. Neuron 48:403–415
Lewis ML (2004) The cytoskeleton in spaceflown cells: an overview. Gravit Space Biol 17:1–12
Lewis ML, Reynolds JL, Cubano LA et al (1998) Spaceflight alters microtubules and increases apoptosis in human lymphocytes (Jurkat). FASEB J 12:1007–1018
Limouse M, Manié S, Konstantinova I et al (1991) Inhibition of phorbol ester-induced cell activation in microgravity. Exp Cell Res 197:82–86
Lombardi ML, Jaalouk DE, Shanahan CM et al (2011) The interaction between nesprins and sun proteins at the nuclear envelope is critical for force transmission between the nucleus and cytoskeleton. J Biol Chem 286:26743–26753
Machemer H, Machemer-Roehnisch S, Braeucker R et al (1991) Gravikinesis in Paramecium: theory and isolation of a physiological response to the natural gravity vector. J Comp Physiol A 168:1–12
Maiuri P, Knezevich A, De Marco A et al (2011) Fast transcription rates of RNA polymerase II in human cells. EMBO Rep 12:1280–1285
Makhija E, Jokhun D, Shivashankar G (2016) Nuclear deformability and telomere dynamics are regulated by cell geometric constraints. Proc Natl Acad Sci U S A 113:E32–E40
Mammoto A, Huang S, Moore K et al (2004) Role of RhoA, mDia, and ROCK in cell shape-dependent control of the Skp2-p27kip1 pathway and the G1/S transition. J Biol Chem 279:26323–26330
Mammoto A, Huang S, Ingber DE (2007) Filamin links cell shape and cytoskeletal structure to Rho regulation by controlling accumulation of p190RhoGAP in lipid rafts. J Cell Sci 120:456–467
Mangala LS, Zhang Y, He Z et al (2011) Effects of simulated microgravity on expression profile of microRNA in human lymphoblastoid cells. J Biol Chem 286:32483–32490
Maniotis AJ, Chen CS, Ingber DE (1997) Demonstration of mechanical connections between integrins, cytoskeletal filaments, and nucleoplasm that stabilize nuclear structure. Proc Natl Acad Sci U S A 94:849–854
Maroto R, Raso A, Wood TG et al (2005) TRPC1 forms the stretch-activated cation channel in vertebrate cells. Nat Cell Biol 7:179–185
Martinac B (2004) Mechanosensitive ion channels: molecules of mechanotransduction. J Cell Sci 117:2449–2460
Mazumder A, Roopa T, Basu A et al (2008) Dynamics of chromatin decondensation reveals the structural integrity of a mechanically prestressed nucleus. Biophys J 95:3028–3035
Mehta SK, Cohrs RJ, Forghani B et al (2004) Stress-induced subclinical reactivation of varicella zoster virus in astronauts. J Med Virol 72:174–179

Meissner K, Hanke W (2005) Action potential properties are gravity dependent. Microgravity Sci Technol 17:38–43
Meloni MA, Galleri G, Pippia P et al (2006) Cytoskeleton changes and impaired motility of monocytes at modelled low gravity. Protoplasma 229:243–249
Meloni MA, Galleri G, Pani G et al (2008) Effects of real microgravity aboard international space station on monocytes motility and interaction with T-lymphocytes. Paper presented at the 10th ESA Life Sciences Symposium/29th Annual ISGP Meeting/24th Annual ASGSB Meeting/ ELGRA Symposium "Life in Space for Life on Earth", 22–27 June 2008 Angers, France
Miralles F, Posern G, Zaromytidou AI et al (2003) Actin dynamics control SRF activity by regulation of its coactivator MAL. Cell 113:329–342
Miroshnikova YA, Nava MM, Wickström SA (2017) Emerging roles of mechanical forces in chromatin regulation. J Cell Sci 130:2243–2250
Morita T, Mayanagi T, Sobue K (2007) Reorganization of the actin cytoskeleton via transcriptional regulation of cytoskeletal/focal adhesion genes by myocardin-related transcription factors (MRTFs/MAL/MKLs). Exp Cell Res 313:3432–3445
Mossman KD, Campi G, Groves JT et al (2005) Altered TCR signaling from geometrically repatterned immunological synapses. Science 310:1191–1193
Najrana T, Sanchez-Esteban J (2016) Mechanotransduction as an adaptation to gravity. Front Pediatr 4:140
Nakamura H, Kumei Y, Morita S et al (2003) Antagonism between apoptotic (Bax/Bcl-2) and anti-apoptotic (IAP) signals in human osteoblastic cells under vector-averaged gravity condition. Ann N Y Acad Sci 1010:143–147
NASA (1967) Gemini Summary Conference. NASA-SP-138
NASA (1968) Gemini midprogram conference including experiment results. NASA-SP-121, JSC-CN-29009
Navarro AP, Collins MA, Folker ES (2016) The nucleus is a conserved mechanosensation and mechanoresponse organelle. Cytoskeleton 73:59–67
Niggli V (2003) Microtubule-disruption-induced and chemotactic-peptide-induced migration of human neutrophils: implications for differential sets of signalling pathways. J Cell Sci 116:813–822
Ohnishi T, Takahashi A, Wang X et al (1999) Accumulation of a tumor suppressor p53 protein in rat muscle during a space flight. Mutat Res 430:271–274
Ohta Y, Hartwig JH, Stossel TP (2006) FilGAP, a Rho- and ROCK-regulated GAP for Rac binds filamin A to control actin remodelling. Nat Cell Biol 8:803–814
Orr AW, Helmke BP, Blackman BR et al (2006) Mechanisms of mechanotransduction. Dev Cell 10:11–20
Otey CA, Carpen O (2004) Alpha-actinin revisited: a fresh look at an old player. Cell Motil Cytoskeleton 58:104–111
Papaseit C, Pochon N, Tabony J (2000) Microtubule self-organization is gravity-dependent. Proc Natl Acad Sci U S A 97:8364–8368
Paulsen K, Thiel C, Timm J et al (2010) Microgravity-induced alterations in signal transduction in cells of the immune system. Acta Astronaut 67(9–10):1116–1125
Paulsen K, Tauber S, Goelz N et al (2014) Severe disruption of the cytoskeleton and immunologically relevant surface molecules in a human macrophageal cell line in microgravity—results of an in vitro experiment on board of the Shenzhou-8 space mission. Acta Astronaut 94:277–292
Paulsen K, Tauber S, Dumrese C et al (2015) Regulation of ICAM-1 in cells of the monocyte/ macrophage system in microgravity. Biomed Res Int 2015:538786
Pellis NR, Goodwin TJ, Risin D et al (1997) Changes in gravity inhibit lymphocyte locomotion through type I collagen. In Vitro Cell Dev Biol Anim 33:398–405
Penard E (1917) Le genre Loxodes. Rev Suisse Zool 25:453–489
Planel H (2004) Space and life: an introduction to space biology and medicine. CRC Press, Boca Raton
Planel H, Richoilley G, Tixador R et al (1981) Space flight effects on Paramecium tetraurelia flown aboard Salyut 6 in the Cytos 1 and Cytos M experiment. Adv Space Res 1:95–101

Planel H, Tixador R, Nefedov Y et al (1982) Effect of space flight factors at the cellular level: results of the CYTOS experiment. Aviat Space Environ Med 53:370–374

Pletser V (2016) European aircraft parabolic flights for microgravity research, applications and exploration: a review. REACH-Rev Human Space Explor 1:11–19

Plett PA, Abonour R, Frankovitz SM et al (2004) Impact of modeled microgravity on migration, differentiation, and cell cycle control of primitive human hematopoietic progenitor cells. Exp Hematol 32:773–781

Pollard EC (1965) Theoretical studies on living systems in the absence of mechanical stress. J Theor Biol 8:113–123

Pourati J, Maniotis A, Spiegel D et al (1998) Is cytoskeletal tension a major determinant of cell deformability in adherent endothelial cells? Am J Phys 274:C1283–C1289

Ramdas NM, Shivashankar G (2015) Cytoskeletal control of nuclear morphology and chromatin organization. J Mol Biol 427:695–706

Rieder N (1977) Die Müllerschen Körperchen von Loxodes magnus (Ciliata, Holotricha): Ihr Bau und ihre mögliche Funktion als Schwererezeptor. In: Verhandlungen der Deutschen Zoologischen Gesellschaft, vol 70. Jahresversammlung, Erlangen, Gustav Fisher Verlag, Stuttgart, p 254

Rief M, Pascual J, Saraste M et al (1999) Single molecule force spectroscopy of spectrin repeats: low unfolding forces in helix bundles. J Mol Biol 286:553–561

Riveline D, Zamir E, Balaban NQ et al (2001) Focal contacts as mechanosensors externally applied local mechanical force induces growth of focal contacts by an mdia1-dependent and rock-independent mechanism. J Cell Biol 153:1175–1186

Roesner H, Wassermann T, Moeller W et al (2006) Effects of altered gravity on the actin and microtubule cytoskeleton of human SH-SY5Y neuroblastoma cells. Protoplasma 229: 225–234

Romanov YA, Buravkova LB, Rikova MP et al (2001) Expression of cell adhesion molecules and lymphocyte-endothelium interaction under simulated hypogravity in vitro. J Gravit Physiol 8:5–8

Sathe AR, Shivashankar G, Sheetz MP (2016) Nuclear transport of paxillin depends on focal adhesion dynamics and FAT domains. J Cell Sci 129:1981–1988

Schatten H, Lewis ML, Chakrabarti A (2001) Spaceflight and clinorotation cause cytoskeleton and mitochondria changes and increases in apoptosis in cultured cells. Acta Astronaut 49:399–418

Schmitt DA, Hatton JP, Emond C et al (1996) The distribution of protein kinase C in human leukocytes is altered in microgravity. FASEB J 10:1627–1634

Schnepel J, Tschesche H (2000) The proteolytic activity of the recombinant cryptic human fibronectin type IV collagenase from E. coli expression. J Protein Chem 19:685–692

Schwarzenberg M, Pippia P, Meloni MA et al (1999) Signal transduction in T lymphocytes–a comparison of the data from space, the free fall machine and the random positioning machine. Adv Space Res 24:793–800

Schwer CI, Lehane C, Guelzow T et al (2013) Thiopental inhibits global protein synthesis by repression of eukaryotic elongation factor 2 and protects from hypoxic neuronal cell death. PLoS One 8:e77258

Sciola L, Cogoli-Greuter M, Cogoli A et al (1999) Influence of microgravity on mitogen binding and cytoskeleton in Jurkat cells. Adv Space Res 24:801–805

Sheetz MP (2001) Cell control by membrane–cytoskeleton adhesion. Nat Rev Mol Cell Biol 2:392–396

Shevelyov YY, Nurminsky DI (2012) The nuclear lamina as a gene-silencing hub. Curr Issues Mol Biol 14:27

Shyy JY, Chien S (2002) Role of integrins in endothelial mechanosensing of shear stress. Circ Res 91:769–775

Singh KP, Kumari R, Dumond JW (2010) Simulated microgravity-induced epigenetic changes in human lymphocytes. J Cell Biochem 111(1):123–129

Solovei I, Wang AS, Thanisch K et al (2013) LBR and lamin A/C sequentially tether peripheral heterochromatin and inversely regulate differentiation. Cell 152:584–598

Spisni E, Toni M, Strillacci A et al (2006) Caveolae and caveolae constituents in mechanosensing: effect of modeled microgravity on cultured human endothelial cells. Cell Biochem Biophys 46:155–164

Stamenovic D, Mijailovich SM, Tolic-Norrelykke IM et al (2002) Cell prestress. II: Contribution of microtubules. Am J Physiol Cell Physiol 282:C617–C624

Stossel TP, Condeelis J, Cooley L et al (2001) Filamins as integrators of cell mechanics and signalling. Nat Rev Mol Cell Biol 2:138–145

Stowe RP, Sams CF, Mehta SK et al (1999) Leukocyte subsets and neutrophil function after short-term spaceflight. J Leukoc Biol 65:179–186

Streb C, Richter P, Ntefidou M et al (2002) Sensory transduction of gravitaxis in *Euglena gracilis*. J Plant Physiol 159:855–862

Strohman RC (1997) The coming Kuhnian revolution in biology. Nat Biotechnol 15:194–200

Studer M, Thiel C, Bradacs G et al (2010) Parabolic maneuvers of the Swiss Air Force fighter jet Northrop F5-E as a new platform to identify rapid gravi-responsive mechanisms in cultured mammalian cells. Paper presented at the 61st International Astronautical Congress, IAC-10. A1.7.9, 27 Sep–01 Oct 2010, Prague Czech Republic

Sundaresan A, Risin D, Pellis NR (2002) Loss of signal transduction and inhibition of lymphocyte locomotion in a ground-based model of microgravity. In Vitro Cell Dev Biol Anim 38:118–122

Tabony J, Rigotti N, Glade N et al (2007) Effect of weightlessness on colloidal particle transport and segregation in self-organisingmicrotubule preparations. Biophys Chem 127:172–180

Tahedl H, Richter P, Lebert M et al (1997) cAMP is involved in gravitxis signal transduction of Euglena gracilis. Microgravit Sci Technol 10:53–57

Tairbekov MG (1996) The role of signal systems in cell gravisensitivity. Adv Space Res 17:113–119

Tajik A, Zhang Y, Wei F et al (2016) Transcription upregulation via force-induced direct stretching of chromatin. Nat Mater 15:1287–1296

Tamada M, Sheetz MP, Sawada Y (2004) Activation of a signaling cascade by cytoskeleton stretch. Dev Cell 7:709–718

Tauber S, Hauschild S, Crescio C et al (2013) Signal transduction in primary human T lymphocytes in altered gravity–results of the MASER-12 suborbital space flight mission. Cell Commun Signal 11:32

Tauber S, Hauschild S, Paulsen K et al (2015) Signal transduction in primary human T lymphocytes in altered gravity during parabolic flight and clinostat experiments. Cell Physiol Biochem 35:1034–1051

Tauber S, Lauber B, Paulsen K et al (2017) Cytoskeletal stability and metabolic alterations in primary human macrophages in long-term microgravity. PLoS One 12:e0175599. https://doi.org/10.1371/journal.pone.0175599

Thiam HR, Vargas P, Carpi N et al (2016) Perinuclear Arp2/3-driven actin polymerization enables nuclear deformation to facilitate cell migration through complex environments. Nat Commun 7:10997

Thiel CS, Paulsen K, Bradacs G et al (2012) Rapid alterations of cell cycle control proteins in human T lymphocytes in microgravity. Cell Commun Signal 10:1

Thiel CS, Hauschild S, Tauber S et al (2015) Identification of reference genes in human myelomonocytic cells for gene expression studies in altered gravity. Biomed Res Int 2015:363575

Thiel CS, Lauber BA, Polzer J et al (2017a) Time course of cellular and molecular regulation in the immune system in altered gravity: progressive damage or adaptation ? REACH-Rev Human Space Explor 5:22–32

Thiel CS, de Zélicourt D, Tauber S et al (2017b) Rapid adaptation to microgravity in mammalian macrophage cells. Sci Rep 7:43

Thiel CS, Hauschild S, Huge A et al (2017c) Dynamic gene expression response to altered gravity in human T cells. Sci Rep 7:5204

Thiel CS, Huge A, Hauschild S et al (2017d) Stability of gene expression in human T cells in different gravity environments is clustered in chromosomal region 11p15. 4. NPJ Microgravity 3:22

Thiel CS, Tauber S, Christoffel S et al (2018) Rapid coupling between gravitational forces and the transcriptome in human myelomonocytic U937 cells. Sci Rep 8(1)

Thiel CS, Tauber S, Lauber B et al (2019) Rapid morphological and cytoskeletal response to microgravity in human primary macrophages. Int J Mol Sci 20(10):2402
Thorpe SD, Lee DA (2017) Dynamic regulation of nuclear architecture and mechanics—a rheostatic role for the nucleus in tailoring cellular mechanosensitivity. Nucleus 8(3):287–300
Tsang E, Giannetti AM, Shaw D et al (2008) Molecular mechanism of the Syk activation switch. J Biol Chem 283:32650–32659
Tzima E, Irani-Tehrani M, Kiosses WB et al (2005) A mechanosensory complex that mediates the endothelial cell response to fluid shear stress. Nature 437:426–431
Uhler C, Shivashankar GV (2016) Geometric control and modeling of genome reprogramming. BioArchitecture 6:76–84
Uhler C, Shivashankar GV (2018) Regulation of genome organization and gene expression by nuclear mechanotransduction. Nat Rev Mol Cell Biol 18(12):717–727
Uva BM, Masini MA, Sturla M et al (2002a) Microgravity-induced programmed cell death in astrocytes. J Gravit Physiol 9:P275–P276
Uva BM, Masini MA, Sturla M et al (2002b) Clinorotation-induced weightlessness influences the cytoskeleton of glial cells in culture. Brain Res 934:132–139
Uva BM, Strollo F, Ricci F et al (2005) Morpho-functional alterations in testicular and nervous cells submitted to modelled microgravity. J Endocrinol Investig 28:84–91
Vargas P, Barbier L, Sáez PJ et al (2017) Mechanisms for fast cell migration in complex environments. Curr Opin Cell Biol 48:72–78
Vartiainen MK, Guettler S, Larijani B et al (2007) Nuclear actin regulates dynamic subcellular localization and activity of the SRF cofactor MAL. Science 316:1749–1752
Versaevel M, Grevesse T, Gabriele S (2012) Spatial coordination between cell and nuclear shape within micropatterned endothelial cells. Nat Commun 3:671
Verschueren H, van der Taelen I, Dewit J et al (1995) Effects of *Clostridium botulinum* C2 toxin and cytochalasin D on in vitro invasiveness, motility and F-actin content of a murine T-lymphoma cell line. Eur J Cell Biol 66:335–341
Vogel V, Sheetz M (2006) Local force and geometry sensing regulate cell functions. Nat Rev Mol Cell Biol 7:265–275
Volkmann D, Baluska F (2006) Gravity: one of the driving forces for evolution. Protoplasma 229:143–148
Vorselen D, Roos WH, MacKintosh FC et al (2014) The role of the cytoskeleton in sensing changes in gravity by nonspecialized cells. FASEB J 28:536–547
Walther I, Pippia P, Meloni MA et al (1998) Simulated microgravity inhibits the genetic expression of interleukin 2 and its receptor in mitogen-activated T lymphocytes. FEBS Lett 436:115–118
Wang Y, Gilmore TD (2003) Zyxin and paxillin proteins: focal adhesion plaque LIM domain proteins go nuclear. Biochim Biophys Acta Mol Cell Res 1593:115–120
Wang N, Stamenovic D (2000) Contribution of intermediate filaments to cell stiffness, stiffening, and growth. Am J Physiol Cell Physiol 279:C188–C194
Wang N, Naruse K, Stamenović D et al (2001) Mechanical behavior in living cells consistent with the tensegrity model. Proc Natl Acad Sci U S A 98:7765–7770
Wang N, Tolić-Nørrelykke IM, Chen J et al (2002) Cell prestress. I. Stiffness and prestress are closely associated in adherent contractile cells. Am J Physiol Cell Physiol 282:C606–C616
Wang Y, Botvinick EL, Zhao Y et al (2005) Visualizing the mechanical activation of Src. Nature 434:1040–1045
Wang N, Tytell JD, Ingber DE (2009) Mechanotransduction at a distance: mechanically coupling the extracellular matrix with the nucleus. Nat Rev Mol Cell Biol 10:75
Wang Y, Nagarajan M, Uhler C et al (2017) Orientation and repositioning of chromosomes correlate with cell geometry–dependent gene expression. Mol Biol Cell 28(14):1997–2009
Wilson KL, Foisner R (2010) Lamin-binding proteins. Cold Spring Harb Perspect Biol 2:a000554
Zheng B, Han M, Bernier M et al (2009) Nuclear actin and actin-binding proteins in the regulation of transcription and gene expression. FEBS J 276:2669–2685

Microbial Stress: Spaceflight-Induced Alterations in Microbial Virulence and Infectious Disease Risks for the Crew

18

C. Mark Ott, Aurélie Crabbé, James W. Wilson, Jennifer Barrila, Sarah L. Castro-Wallace, and Cheryl A. Nickerson

Abbreviations

C4-HSL	*N*-Butanoyl-L-homoserine lactone
CF	Cystic fibrosis
EPEC	Enteropathogenic *E. coli*
ETEC	Enterotoxigenic *E. coli*
Fur	Ferric uptake regulator
HARV	High aspect ratio (or rotating) vessel
ISS	International Space Station
LD	Lethal dose
LPS	Lipopolysaccharide
LSMMG	Low-shear modeled microgravity
MEED	Microbial ecology evaluation device
MMG	Modeled microgravity
NASA	National Aeronautics and Space Administration

C. M. Ott (✉) · S. L. Castro-Wallace
Biomedical Research and Environmental Sciences Division, NASA/Johnson Space Center, Houston, TX, USA
e-mail: c.m.ott@nasa.gov; sarah.wallace@nasa.gov

A. Crabbé
Laboratory of Pharmaceutical Microbiology, Ghent University, Ghent, Belgium
e-mail: aurelie.crabbe@ugent.be

J. W. Wilson
Department of Biology, Villanova University, Villanova, PA, USA
e-mail: james.w.wilson@villanova.edu

J. Barrila · C. A. Nickerson
Center for Fundamental and Applied Microbiomics, The Biodesign Institute, Arizona State University, Tempe, AZ, USA
e-mail: jbarrila@asu.edu; cheryl.nickerson@asu.edu

© Springer Nature Switzerland AG 2020

A. Choukér (ed.), *Stress Challenges and Immunity in Space*,
https://doi.org/10.1007/978-3-030-16996-1_18

OES	Orbital Environmental Simulator
RT-PCR	Reverse transcriptase-polymerase chain reaction
RWV	Rotating wall vessel
SMG	Simulated microgravity
STLV	Slow turning lateral vessel

Although preventative measures to mitigate infectious disease risks to the crew are stringently enforced prior to the launch of spacecraft, pathogenic organisms are still carried by crewmembers, the spacecraft, and its cargo (Taylor 1974; Castro et al. 2004; Gueguinou et al. 2009) (see also Chap. 25). Of additional concern is spaceflight food, which is randomly monitored for microbial content prior to flight, yet remains a potential route of infection for food-borne pathogens, such as *Salmonella* sp. and *Staphylococcus aureus*. While the crewmembers are exceptionally healthy, dysfunction of their immune system has been repeatedly associated with spaceflight missions (Gueguinou et al. 2009), suggesting an increased susceptibility to infection. Other factors, such as the relatively crowded living conditions on flight vehicles, also increase the risk of infectious disease during spaceflight. Indeed, transfer of microbial flora between crewmembers has been demonstrated (Taylor 1974; Pierson et al. 1996). In addition, infections from the astronauts' own normal microbiological flora are still a risk, such as staphylococcal and streptococcal skin infections and urinary tract infections. Evaluations of the environmental microbiome aboard Mir and the International Space Station (ISS) indicated a predominance of common members of the environmental flora (Castro et al. 2004), although the appearance of medically significant organisms has been documented (Ott 2004). Moreover, increased antibiotic resistance for some bacteria during culture in spaceflight has been reported (Tixador et al. 1985; Kacena and Todd 1999), which could potentially compromise effective prophylactic treatment if a crew member were to acquire an in-flight infection from such an organism. Latent viruses also remain a risk to the astronauts (see Chap. 19) because of their ubiquity, the ineffective current preventive practices (e.g., quarantine), and their immunocompromised state (Pierson et al. 2007). Thus, the presence of opportunistic and obligate pathogens and the corresponding risk of infectious diseases cannot be completely prevented during spaceflight.

In order to fully understand the impact of spaceflight on infectious disease risks to the crew, it is critical to advance our knowledge of the effects of spaceflight on the human immune system in a synergistic approach with studies to characterize spaceflight-associated changes in microorganisms, alterations in the human and environmental microbiome, and the resulting impact on host–pathogen interactions. A wide variety of spaceflight experiments have been performed over the past 50 years demonstrating an extensive range of observed phenotypic and, recently, molecular genetic changes in microorganisms (Dickson 1991; Nickerson et al. 2004; Klaus and Howard 2006; Horneck et al. 2010); however, information elucidating the mechanism(s) behind these changes and how spaceflight affects microbial virulence has only recently begun to emerge.

18.1 Modeling Aspects of Spaceflight Culture on Earth

Our knowledge of spaceflight-induced alterations in microbial virulence has been enhanced by the use of ground-based spaceflight analog culture systems, such as the NASA-designed Rotating Wall Vessel (RWV) bioreactor (Fig. 18.1). The RWV is an optimized form of suspension culture in which cells are grown in cylindrical bioreactors, called high aspect ratio vessels (HARV) or slow turning lateral vessels (STLV) in physiologically relevant low fluid-shear conditions. The RWV consists of a hollow disk (HARV) or cylinder (STLV) that is completely filled with culture medium and rotates on an axis parallel to the ground (Klaus 2001; Nickerson et al.

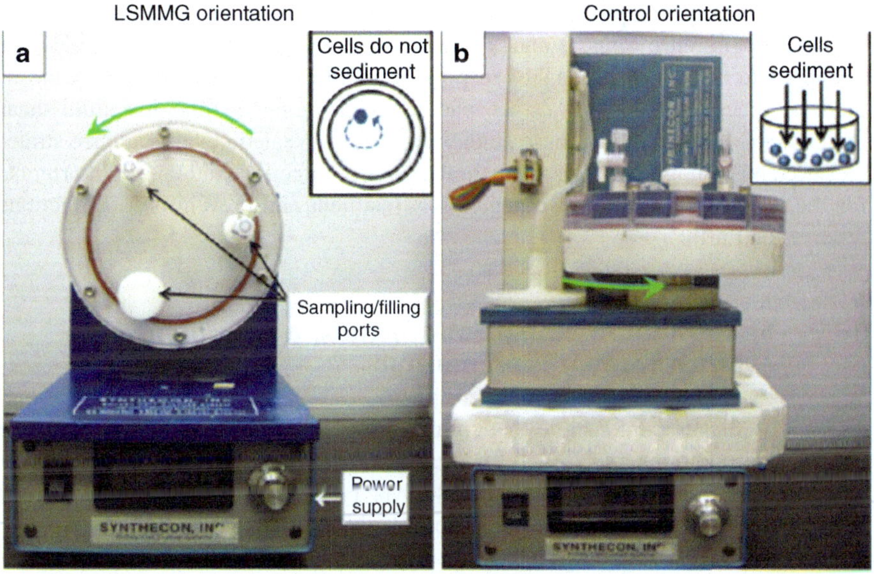

Fig. 18.1 The rotating wall vessel (RWV) bioreactor and power supply. The HARV RWV bioreactor is depicted in the (**a**) Low-shear modeled microgravity (LSMMG) and (**b**) control orientations. For both orientations, the cylindrical culture vessel is completely filled with culture medium through ports on the face of the vessel (indicated by black arrows in **a**) and operates by rotating around a central axis. Cultures are aerated through a hydrophobic membrane that covers the back of the reactor. In the LSMMG orientation (**a**), the axis of rotation of the RWV is perpendicular to the direction of the gravity force vector. In the control orientation (**b**), the axis of rotation is parallel with the gravity force vector. The direction of rotation is indicated by a green arrow in both orientations. The effect of RWV rotation on particle suspension is depicted for each orientation (insets). When the RWV is not rotating, or rotating in the control orientation, the force of gravity will cause particles in the apparatus to sediment and eventually settle on the bottom of the RWV (**b**, inset). When the RWV is rotating in the LSMMG position, particles are continually suspended in the media (**a**, inset). The result is a solid-body rotation of the medium and cells within the RWV, with the sedimentation of the particles/cells due to gravity being offset by the upward forces of rotation. The result is low fluid-shear aqueous suspension that is similar to what would occur in true microgravity

2004). The result is solid-body rotation of the medium and the cells within and a constant rotation perpendicular to the gravitational field that results in an environmental culture, which mimics aspects of the spaceflight environment (Nickerson et al. 2004). As a result, sedimentation of cells due to gravity is offset by the forces of the RWV rotation. The culture environment experienced by cells in the RWV is commonly referred to as low-shear modeled microgravity (LSMMG), modeled microgravity (MMG), or simulated microgravity (SMG). Interestingly, the low fluid-shear culture conditions in the RWV are relevant to those encountered by numerous microbial pathogens and commensals during their normal life cycles in the gastrointestinal, respiratory, and urogenital tracts (Nauman et al. 2007). A gas-permeable membrane on one side of the RWV (HARV) or a central core gas exchange membrane (STLV) allows constant air exchange during growth (Nickerson et al. 2004). Several studies culturing microbes in both the RWV and true spaceflight have focused on profiling molecular genetic (transcriptomic and proteomic), phenotypic (in vitro stress), and virulence responses (in vivo) to provide new insights into how microorganisms respond to culture in the microgravity environment of spaceflight. Notably, since pathogens encounter similar low fluid-shear regions in the human body, these studies have also revealed novel virulence strategies used by pathogens during the natural course of infection, and thus hold promise for the development of new strategies for treatment and prevention of infectious diseases on Earth.

18.2 Spaceflight Analog (LSMMG) Culture of *Salmonella enterica* Serovar Typhimurium

The first studies to demonstrate that culture of microbes in both LSMMG and true spaceflight conditions alters microbial virulence was performed using the obligate bacterial pathogen, *Salmonella enterica* serovar Typhimurium in a murine model of infection (Nickerson et al. 2000; Wilson et al. 2007). Indeed, *S.* Typhimurium remains the best characterized microorganism in response to spaceflight and spaceflight analog culture. As a common food-borne pathogen, *S.* Typhimurium was chosen as the model organism for these studies because (1) it has been extensively studied and well characterized, (2) it is a leading cause of intestinal and diarrheal disease in healthy individuals, and serious systemic illness in the immunocompromised, and (3) it is one of the five basic categories of organisms targeted by NASA for preflight monitoring of spaceflight food. Cultures of *S.* Typhimurium grown in the RWV environment (LSMMG) (Nickerson et al. 2004) displayed a significant increase in virulence as evidenced by a decreased time-to-death, decreased lethal dose 50 (LD_{50}), and increased tissue colonization (liver and spleen) in murine infections as compared to control cultures. The *S.* Typhimurium LSMMG cultures also displayed increased survival in cultured macrophages and increased resistance to acid stress, two key pathogenesis-related responses that are relevant to bacterial virulence. In addition, this study was the first to demonstrate that the LSMMG environment elicits a global molecular genetic response in bacteria using 2-D protein gel electrophoresis to show that *S.* Typhimurium protein levels changed during LSMMG culture as compared to controls (Nickerson et al. 2000). This study established the paradigm that LSMMG can alter bacterial virulence and serve as a master signal to globally reprogram bacterial gene expression. Moreover, this work provided the first evidence that fluid-shear levels relevant to those encountered by *Salmonella* between the brush border microvilli of intestinal epithelial cells within the infected host act as a novel environmental signal that regulates the virulence, stress resistance, and gene expression of this pathogen (Nickerson et al. 2000, 2004).

To identify the *Salmonella* genes that changed expression in response to LSMMG culture, whole genome microarray analysis was performed using RNA harvested from *S.* Typhimurium cultures grown in LSMMG and control conditions (Wilson et al. 2002a). The results demonstrated that 163 genes globally distributed across the *S.* Typhimurium genome are either upregulated (97 genes) or downregulated (68 genes) during growth in LSMMG. These genes belonged to a variety of functional groups including protein secretion systems, lipopolysaccharide (LPS) synthesis, ribosomal subunits, starvation/stress response, virulence factors, transcriptional regulation, iron-utilization enzymes, and several of unknown functions. Interestingly, none of the upregulated genes corresponded to known virulence factors, even though LSMMG enhanced *S.* Typhimurium virulence. This suggests that LSMMG may alter *Salmonella* virulence by a previously uncharacterized mechanism(s) that could involve novel virulence functions. Alternatively, the increase in *Salmonella* virulence due to LSMMG may be the result of contributions of multiple genes of different functions that are regulated as part of the global reprogramming of *Salmonella* under LSMMG conditions. Secondary assays including RT-PCR and LPS gels were used to confirm the hits obtained from the microarray analysis. In addition, since the authors noticed that ferric uptake regulator (Fur) protein-binding sites were associated with many of the genes found in the analysis, they tested the ability of an *S.* Typhimurium *fur* mutant strain to increase acid stress resistance under LSMMG conditions (as previously observed with the wild-type strain). The *fur* mutant did not display this phenotype, thus indicating that the *fur* gene may play a role in the response of *S.* Typhimurium to LSMMG.

Given the global alterations in molecular genetic and phenotypic responses of *S.* Typhimurium to LSMMG culture, it was hypothesized that the *rpoS* gene was a likely candidate for playing a role in LSMMG signal transmission, as it is a master regulator of the stress and virulence responses in many bacteria (Hengge-Aronis 2000; Dong and Schellhorn 2010). Specifically, an *S.* Typhimurium strain containing an *rpoS* mutation was extensively and systematically compared to an isogenic wild-type strain for responses to LSMMG culture (Wilson et al. 2002b). This study provided key information regarding the bacterial LSMMG response:

(1) the *rpoS* gene is not required for *S.* Typhimurium to display LSMMG-induced phenotypes (analyzed in exponential phase of growth), (2) LSMMG alters resistance to other stresses besides acidic and intracellular macrophage survival, including osmotic, thermal, and oxidative stresses, and (3) cells grown in LSMMG in minimal media show a shorter lag phase and doubling time compared to control cultures.

A follow-up study demonstrated a progressive relationship between the applied fluid-shear in the RWV bioreactor and pathogenesis-related molecular genetic and phenotypic responses of *S.* Typhimurium (Nauman et al. 2007). When exposed to progressively increasing fluid-shear levels in the RWV, planktonic cultures of *S.* Typhimurium displayed corresponding progressive changes in acid and thermal stress responses and targeted gene expression profiles, including *rtsA*, a regulatory protein implicated in *Salmonella* intestinal invasion. This was the first study to provide evidence that incremental changes in fluid-shear can cause corresponding changes in biological responses in *S.* Typhimurium during the infection process and may lead to discovery of new targets for antimicrobial therapeutic development against *Salmonella* and other pathogens.

The initial studies investigating the impact of LSMMG (and later the spaceflight environment) on microbial virulence focused on *S.* Typhimurium strain χ3339, which causes gastroenteritis in humans. Interestingly, a subsequent study investigated the effect of LSMMG culture on a different *S.* Typhimurium strain (D23580), which is a multidrug-resistant clinical isolate of ST313 causing life-threatening systemic infections (Yang et al. 2016). Unlike classic gastrointestinal *Salmonella* strains (e.g., χ3339), gastroenteritis is often absent during ST313 clinical infections and isolates are most commonly recovered from blood, rather than from stool in patients—suggesting the possibility that these isolates may be routinely exposed to the higher fluid-shear conditions found in the blood stream—which in turn may shape their responses to different fluid shear forces. This study showed that D23580 does indeed respond to fluid shear forces; however, it does so in a distinctly different manner relative to classic *S.* Typhimurium strains that cause gastroenteritis. Specifically, exposure of D23580 to high fluid shear (relevant to those encountered in areas of the bloodstream) increased its virulence potential and enhanced resistance to select environmental stressors.

18.3 Spaceflight Culture of *Salmonella enterica* Serovar Typhimurium

To determine whether the true microgravity environment of spaceflight alters bacterial virulence and gene expression in a similar manner to that of spaceflight-analog (LSMMG) culture, a flight experiment designated as MICROBE was flown aboard Space Shuttle mission STS-115 (Wilson et al. 2007). MICROBE was the first experiment to examine the effect of spaceflight on the virulence of a pathogen, and the first to obtain the entire molecular genetic response (transcriptomic and proteomic) of a bacterium to spaceflight. In this experiment, split samples of *S.*

Typhimurium were grown in otherwise identical environmental conditions aboard the Shuttle during spaceflight and on the ground in the Orbital Environmental Simulator (OES) room at the Kennedy Space Center. Growth of *S.* Typhimurium was initiated in both settings after the Shuttle was established in microgravity conditions of orbit. A portion of the *S.* Typhimurium spaceflight cultures were preserved with fixative on orbit to preserve samples for RNA/protein analysis to measure gene expression changes (via microarray and proteomic assays); while the

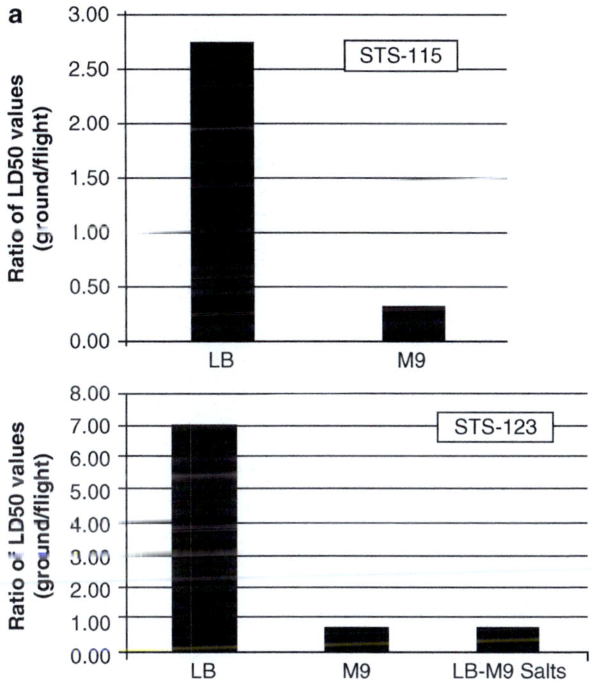

Fig. 18.2 *S.* Typhimurium virulence in LB, M9, and LB-M9 spaceflight cultures. (**a**) Ratio of LD_{50} values of *S.* Typhimurium spaceflight and ground cultures grown in LB (Lennox Broth), M9 (minimal carbon, high salt medium), or LB-M9 salts media (Lennox Broth supplemented with NaH_2PO_4, KH_2PO_4, NH_4Cl, NaCl, and $MgSO_4$) from STS-115 and STS-123 shuttle mission. Female Balb/c mice were perorally infected with a range of bacterial doses from either spaceflight or ground cultures and monitored over a 30-day period for survival. (**b**) Time-to-death curves of mice infected with spaceflight and ground cultures from STS-115 (infectious dosage: 10^7 bacteria for both media). (**c**) Time-to-death curves of mice infected with spaceflight and ground cultures from STS-123 (infectious dosage: 10^6 bacteria for LB and 10^7 bacteria for M9 and LB-M9 salts). Infectious dosages were selected such that the rates in time-to-death facilitated normalized comparisons across the different media. (**d**) SEM of spaceflight and synchronous ground control cultures of *S.* Typhimurium bacteria showing the formation of an extracellular matrix and associated cellular aggregation of spaceflight cells suggesting biofilm formation (magnification: ×3500) (Wilson et al. 2007)

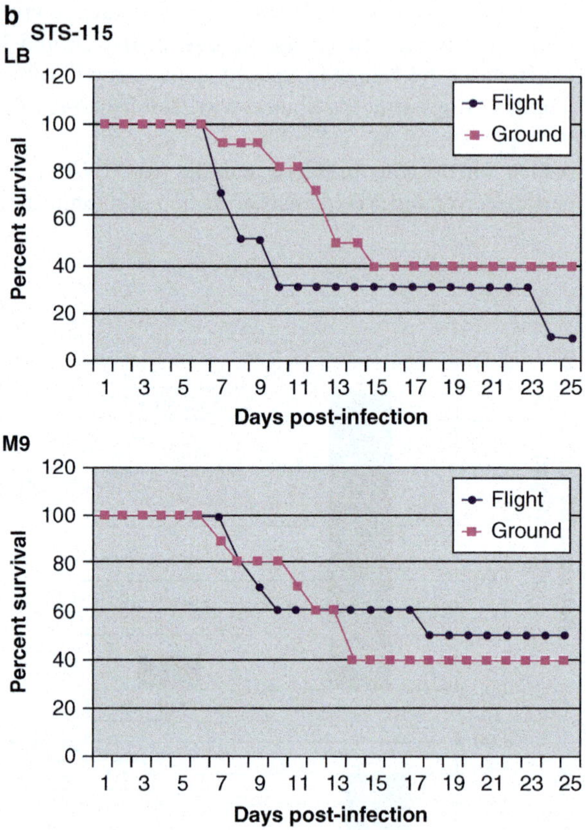

Fig. 18.2 (continued)

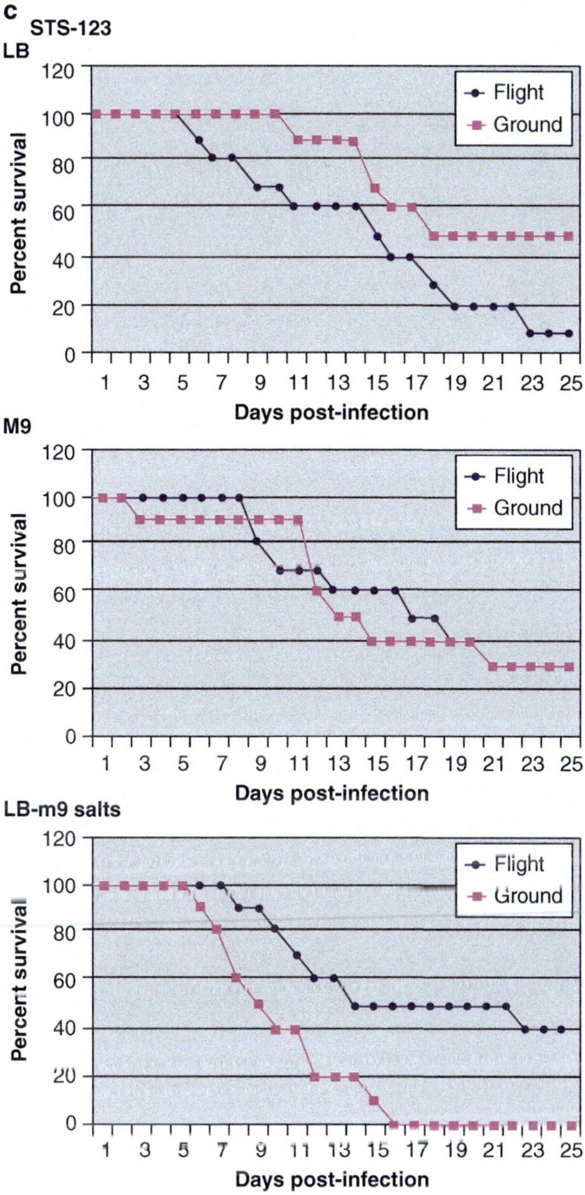

Fig. 18.2 (continued)

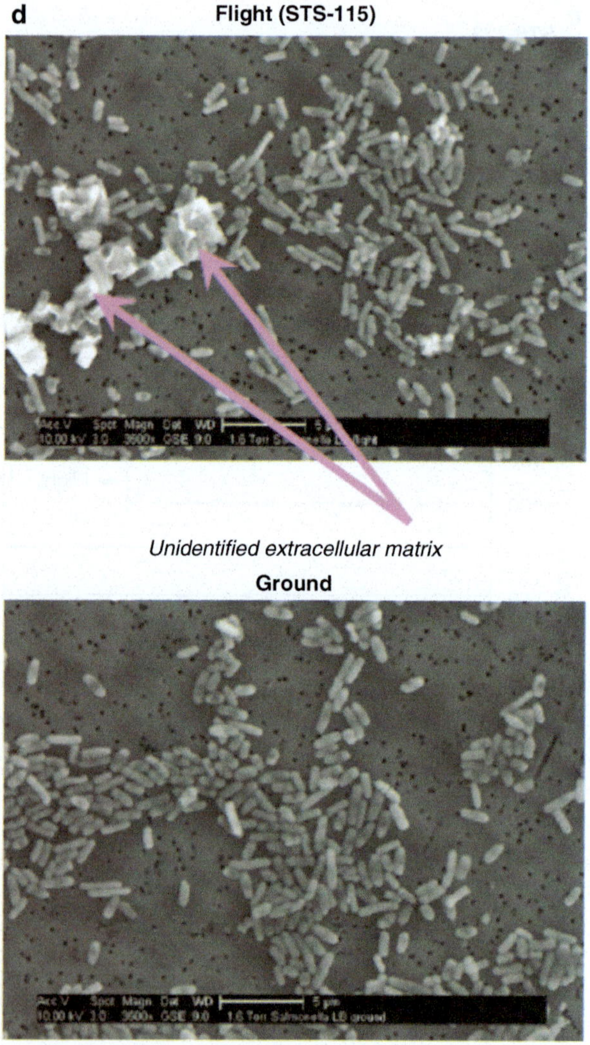

Fig. 18.2 (continued)

other portion of cultures were supplemented with fresh media and used (upon return to ground) for murine infections to measure virulence. Remarkably, the virulence assay results mimicked what was observed in LSMMG conditions in that the spaceflight cultures displayed increased virulence as measured by (1) decreased time to death, (2) decreased LD_{50}, and (3) increased percent mortality across multiple infectious dosages (given perorally) in murine infections as compared to ground controls (Fig. 18.2). In addition, 167 transcripts and 73 proteins were

Table 18.1 Genes of the Hfq-regulon of *S.* Typhimurium and *P. aeruginosa* differentially expressed in response to spaceflight culture

Salmonella spaceflight			*Pseudomonas* spaceflight		
Gene number	Gene name	Function	Gene number	Gene name	Function
Stress resistance/virulence					
STM0831	*dps*	Stress response protein	PA2300[a]	*chiC*	Chitinase
STM1070	*ompA*	Outer membrane porin	PA3479[a]	*rhlA*	Rhamnosyltransferase chain A
STM1572	*ompD*	Outer membrane porin	PA4944	*hfq*	RNA-binding protein Hfq
STM1652	*ynaF*	Putative universal stress protein			
STM2267	*ompC*	Outer membrane porin	*Microaerophilic/anaerobic metabolism*		
STM2638	*rseB*	Anti-αE factor	PA0518	*nirM*	Cytochrome c-551 precursor
STM2640	*rpoE*	SigmaE (α24) factor	PA0519	*nirS*	Nitrite reductase precursor
STM2884	*sipC*	Cell invasion protein	PA0524	*norB*	Nitric-oxide reductase subunit B
STM4361	*hfq*	RNA-binding protein Hfq	PA1557	PA1557	Probable cytochrome oxidase subunit (cbb3-type)
STM4561	*osmY*	Hyperosmotically inducible protein	PA5491	PA5491	Probable cytochrome
Plasmid transfer apparatus			*Hypothetical proteins*		
PSLT081	*traB*	Conjugative transfer	PA0200[a]	PA0200	Hypothetical protein
PSLT095	*traN*	Conjugative transfer	PA1123[a]	PA1123	Hypothetical protein
PSLT099	*trbB*	Conjugative transfer	PA2225	PA2225	Hypothetical protein
PSLT100	*traH*	Conjugative transfer	PA2453	PA2453	Hypothetical protein
PSLT101	*traG*	Conjugative transfer	PA2747	PA2747	Hypothetical protein
PSLT104	*traD*	Conjugative transfer	PA2753[a]	PA2753	Hypothetical protein
PSLT110	*traX*	Conjugative transfer	PA3369	PA3369	Hypothetical protein
SSL_T12	*traT*	Conjugative transfer	PA3520[a]	PA3520	Hypothetical protein
SSL_T20	*traK*	Conjugative transfer	PA3795	PA3795	Probable oxidoreductase
SSL_T24	*traF*	Conjugative transfer	PA4220	PA4220	Hypothetical protein
SSL_T3	*trbC*	Conjugative transfer	PA4351	PA4351	Probable acyltransferase
SSL_T5	*trbA*	Conjugative transfer	PA4352[a]	PA4352	Hypothetical protein

(continued)

Table 18.1 (continued)

Salmonella spaceflight			*Pseudomonas* spaceflight		
Gene number	Gene name	Function	Gene number	Gene name	Function
STM_PSLT085	*traR*	Conjugative transfer	PA4633	PA4633	Probable chemotaxis transducer
Small RNAs[b]			PA4739	PA4739	Hypothetical protein
STM_sRNA_α_RBS	*αRBS*	Small RNA			
STM_sRNA_CsrB	*csrB*	Small RNA			
STM4124	*oxyS*	Small RNA			
STM_sRNA-RFN	*RFN*	Small RNA			
STM_sRNA_RNaseP	*rnaseP*	Small RNA			
STM_sRNA_rne5	*rne5*	Small RNA			
STM_sRNA_tke1	*Tke1*	Small RNA			
Ribosomal proteins					
STM0469	*rpme2*	50 ribosomal protein L31	PA4433	*rplM*	50S ribosomal protein L13
STM3425	*rpsF*	30S ribosomal subunit protein S6			
STM3433	*rplP*	50S ribosomal subunit protein L16			
STM3436	*rpsS*	30S ribosomal subunit protein S19			
STM3439	*rplD*	50S ribosomal subunit protein L4			
STM3448	*rpsL*	30S ribosomal subunit protein S12			
STM4150	*rplA*	50S ribosomal subunit protein L1			
STM0469	*rpme2*	50S ribosomal protein L31			
Iron utilization/storage (Fur or Hfq-regulated)					
STM0596	*entE*	2,3-Dihydroxybenzoate-AMP ligase	PA3621	*fdxA*	Ferredoxin I
STM0831	*dps*	Stress response protein and ferritin	PA4880	PA4880	Probable bacterioferritin
STM1660[a]	*fnr*	Transcriptional regulator, Fe binding			
STM1732	*ompW*	Outer membrane protein W			
STM3443	*bfr*	Bacterioferritin, iron storage			

Biofilm formation					
STM1070	*ompA*	Outer membrane porin			
STM1172	*flgM*	Flagellar biosynthesis			
STM1916	*cheY*	Flagellar biosynthesis			
STM1925	*flhD*	Regulator of flagellar biosynthesis			
STM1962	*fliT*	Flagellar biosynthesis			
Various cellular functions					
–	*mysB*	Suppresses protein export mutants	PA1156	*nrdA*	Ribonucleotide-diphosphate reductase alpha subunit
STM0376	*sbmA*	ABC superfamily transporter	PA1183	*dctA*	C4-dicarboxylate transport protein
STM0536	*ppiB*	Peptidyl-prolyl isomerase B	PA1610	*fabA*	3-hydroxydecanoyl-ACP dehydratase
STM0665	*gltI*	ABC glutamate/aspartate transporter	PA1776	*sigX*	ECF sigma factor SigX
STM0833	*ompX*	Outer membrane protein	PA2003	*bdhA*	3-Hydroxybutyrate dehydrogenase
STM0943	*cspD*	Similar to CspA; not cold-induced	PA2247	*bkdA1*	2-Oxoisovalerate dehydrogenase (alpha subunit)
STM1066	*rmf*	Ribosome modulation factor	PA2248	*bkdA2*	2-Oxoisovalerate dehydrogenase (beta subunit)
STM1290	*gapA*	Glyceraldehyde dehydrogenase	PA2634	PA2634	Isocitrate lyase
STM1682	*tpx*	Thiol peroxidase	PA2851	*efp*	Elongation factor P
STM1749	*adhE*	Fe-dependent dehydrogenase	PA2966	*acpP*	Acyl carrier protein
STM1751	*hns*	DNA-binding protein	PA3686	*adk*	Adenylate kinase
STM1959	fliC	Flagellin, structural protein	PA4031	*ppa*	Inorganic pyrophosphatase
STM2282	*glpQ*	Glycerophosphodiesterase	PA4569	*ispB*	Octaprenyl-diphosphate synthase
STM2488	*nlpB*	Lipoprotein-34	PA5355	*glcD*	Glycolate oxidase subunit GlcD
STM2542	*nifU*	Fe-S cluster formation protein			
STM2665	*yfiA*	Ribosome-associated factor			
STM2801	*ygaC*	Putative cytoplasmic protein			
STM2802	*ygaM*	Putative inner membrane protein			

(continued)

Table 18.1 (continued)

Salmonella spaceflight			*Pseudomonas* spaceflight		
Gene number	Gene name	Function	Gene number	Gene name	Function
STM3060	*ygfE*	Putative cytoplasmic protein			
STM3285	*rbfA*	Ribosome-binding factor			
STM3648	*yiaG*	Putative transcriptional regulator			
STM3840	*rnpA*	RNase P, protein component			
STM3870	*atpE*	Membrane-bound ATP synthase			
STM3915	*trxA*	Thioredoxin 1, redox factor			
STM4231	*lamB*	Phage λ receptor protein			
STM4392	*priB*	Primosomal replication protein N			

Note: Both up- and downregulated genes with a differential expression value of more than twofold are included and are organized in major functional categories

[a]Also involved in microaerophilic/anaerobic metabolism

[b]Small RNAs were not included in the microarray analysis of *P. aeruginosa*. In addition, the RNA purification kit used for the *S*. Typhimurium microarray analysis had a 200 nucleotide cut-off, therefore additional smaller RNA species may have been inadvertently excluded in this work

identified to change expression in response to spaceflight, and these genes were globally distributed across the *S.* Typhimurium genome and belonged to a variety of functional groups. Of the genes identified in microarray analysis, a preponderance belonged to the Hfq regulon (including those encoding small regulatory RNAs, outer membrane proteins, ribosomal proteins, stress response proteins, plasmid transfer functions, iron metabolism, and ion transport) as well as the *hfq* gene itself (which was downregulated) (Table 18.1). Hfq is a highly conserved bacterial RNA chaperone protein that binds to small regulatory RNAs thereby facilitating their association with mRNAs, the result of which plays a diverse role in global regulation of prokaryotic gene expression, virulence, and physiology in response to stress (Gottesman 2004; Majdalani et al. 2005; Gottesman et al. 2006; Guisbert et al. 2007; Pfeiffer et al. 2007; Sittka et al. 2007, 2008). The spaceflight-induced Hfq regulon gene changes were up- or downregulated in correlation with a decrease in *hfq* gene expression. This finding corroborates previous microarray analysis during *S.* Typhimurium culture in LSMMG, where the *hfq* gene is also downregulated (Wilson et al. 2002a). Moreover, the number of downregulated genes (98) was larger than the number of upregulated genes (69) in response to spaceflight, another similarity to the LSMMG microarray results. Interestingly, Hfq also regulates expression of the Fur protein, which was found to play a role in the LSMMG-induced acid stress response in *S.* Typhimurium. Subsequent LSMMG ground-based studies using an isogenic *hfq* mutant strain of *S.* Typhimurium not only supported involvement of Hfq in the *S.* Typhimurium response to microgravity but also established the utility of using the RWV in the laboratory to confirm observations obtained from spaceflight experiments. Interestingly, electron microscopic evaluation of *S.* Typhimurium spaceflight samples revealed striking differences in cellular aggregation and clumping that was associated with the formation of an extracellular matrix reminiscent of biofilms as compared to the ground control cultures (Fig. 18.2d) (Wilson et al. 2007). This phenotypic observation was consistent with corresponding differences in the expression of genes associated with biofilm formation and may play a role in the enhanced virulence of the organisms grown in space.

Spaceflight studies of *Salmonella* demonstrate that microgravity culture impacts a wide range of microbial characteristics, including growth, morphology, survival, metabolism, and gene expression (Nickerson et al. 2004; Wilson et al. 2007, 2008). However, these experiments, as well as experiments with other microorganisms, have been done with pure cultures using relatively short-duration studies (typically ≤96 h). Long-term heritable changes, resulting from natural selection and microbial evolution, also need to be addressed particularly in the context of human exploration class missions (e.g., Mars mission) during which changes in the spacecraft and human microbiome would undoubtedly occur and could pose a risk to mission success. Indeed, the first experiment to look at the impact of long-duration spaceflight culture on microbial responses will launch to the ISS in late 2018. This study (entitled EVOLVES) is led by Principal Investigator Cheryl Nickerson from Arizona State University and will characterize the functional response of wildtype and mutant *Salmonella* strains to long-term multigenerational growth in the chronic stress of microgravity by examining a range of genotypic, molecular genetic and phenotypic responses (https://humanresearchroadmap.nasa.gov/Tasks/task.aspx?i=436). These studies will provide clear evidence as to whether microgravity creates selective mutations that could impact human exploration of deep space.

18.4 Role of Ion Composition on Spaceflight-Induced Virulence

The results of the MICROBE experiment aboard STS-115 led to a follow-up experiment, designated MDRV, aboard STS-123. The goal of this experiment was to (1) confirm the experimental results from STS-115, and (2) determine if altering the ion composition of the growth medium could decrease the spaceflight-induced increase in virulence (Wilson et al. 2008). The hypothesis that manipulation of ion concentrations could counteract or inhibit the spaceflight-associated increase in *Salmonella* virulence was based on initial results from the MICROBE experiment, which suggested that *S.* Typhimurium cultures grown in a minimal carbon, high-salt medium called M9 did not respond with the same spaceflight-induced increase in virulence observed with Lennox broth (LB). In a rare opportunity to replicate a spaceflight result, the data from STS-123 fully supported the results from STS-115 in that (1) *S.* Typhimurium spaceflight cultures grown in LB medium displayed increased virulence in murine infection compared to ground controls, and (2) the spaceflight cultures grown in M9 did not display increased virulence compared to controls (Fig. 18.2). Moreover, the addition of similar concentrations of five key inorganic salts found in M9 medium (NaH_2PO_4, KH_2PO_4, NH_4Cl, NaCl, and $MgSO_4$) to the LB medium reversed the increase in virulence of spaceflight cultures grown in LB medium alone. Interestingly, although different virulence responses were observed in spaceflight cultures grown in the LB and M9 media, significant similarities in gene and protein expression profiles indicated involvement of the Hfq regulon in either media. Subsequent ground-based investigations using the RWV reinforced the flight data indicating an inhibitory effect of high ion concentrations in the growth medium on pathogenesis-related responses. By systematically adding different combinations of the inorganic salts to the LB medium in the RWV, phosphate ion (PO_4) was isolated as the key component to repressing this microbial pathogenesis-related response (Wilson et al. 2008).

New discoveries using *S.* Typhimurium as a model organism are continuing as a follow-up flight experiment to build upon results obtained from MICROBE and MDRV, and were flown on STS-131 (April, 2010). This experiment, designated as STL-IMMUNE, was the first experiment to conduct an in-flight infection of human cells (intestinal) with a microbial pathogen (*S.* Typhimurium). The data from this spaceflight experiment will provide insight into alterations in host–pathogen interactions that occur during spaceflight and will unveil the cellular and molecular mechanisms behind those changes. This information has the potential to significantly change microbial risk assessment and operational requirements during a mission.

The discovery that spaceflight culture increased the virulence of *Salmonella*, yet genes known to be important for the virulence of this pathogen were not regulated as expected when this organism is grown on Earth, led to a follow-up experiment aboard Space Shuttle mission STS-135 in an effort to translate these research findings toward medical applications. Specifically, researchers investigated the impact of spaceflight culture on the protective immunogenicity and gene expression of live Recombinant Attenuated *Salmonella* Vaccine (RASV) strains, including those in clinical trials. These genetically engineered vaccine strains are used as carriers to infect the host and deliver protective antigens against different microbial pathogens to the immune system (Curtiss et al. 2009, 2010; Li et al. 2009; Shi et al. 2010a, b). The ultimate goal of this spaceflight vaccine initiative experiment was to accelerate the development of RASV strains carrying a protective antigen against *Streptococcus pneumonia* (or pneumococcus) by (1) enhancing their ability to safely induce a potent and protective immune response and (2) unveiling novel gene targets to develop new and improved existing vaccine strains. Pneumococcus causes life-threatening diseases, such as pneumonia, meningitis, and bacteremia, and kills over ten million people annually—and is particularly dangerous for newborns and the elderly, who are less responsive to current anti-pneumococcal vaccines. Experiments like these hold the potential to benefit astronauts on future exploration missions and the general public on Earth (Sarker et al. 2010).

18.5 The Response of *Pseudomonas aeruginosa* to Spaceflight and Spaceflight Analog Conditions: Similarities and Differences as Compared to *Salmonella*

As a versatile, ubiquitous bacterium that is occasionally part of the normal human flora, *P. aeruginosa* can also survive in extraterrestrial habitats, as evidenced by its isolation from the potable water system on the ISS and from Apollo crewmembers (Taylor 1974; Hawkins and Ziegelschmid 1975; Castro et al. 2004; Bruce et al. 2005). Astronaut cross-contamination with *P. aeruginosa* has been reported during short-term missions emphasizing the potential of this infectious agent to rapidly spread among crewmembers (Taylor 1974). Thus far, the presence of *P. aeruginosa* in the spaceflight vehicle has led to one reported incapacitating urinary tract infection in-flight (Taylor 1974). In addition to the importance for astronaut safety, studying the behavior of *P. aeruginosa* in the low fluid shear conditions of micro gravity provides insights into the role that low fluid-shear regions in the human body play in triggering virulence characteristics.

In a seminal spaceflight study, McLean et al., demonstrated that *P. aeruginosa* formed biofilms on polycarbonate membranes that were strongly resistant to mechanical disruption (McLean et al. 2001). More recent studies discovered that the microgravity environment of spaceflight increased the formation of biofilms by *P. aeruginosa*, and resulted in a unique biofilm architecture, referred to as column-and-canopy by the authors (Kim et al. 2013b). Specifically, biofilms grown in spaceflight on cellulose ester membrane discs generated column-shaped structures overlaid by canopies (resembling mushroom-shaped biofilms typically observed in flow cells), while biofilms formed under ground control conditions were flat (commonly observed under static conditions). Since an increased oxygen supply abolished the observed differences between biofilms grown under microgravity and control conditions, oxygen limitation in microgravity conditions was proposed to play a role in the observations (Kim et al. 2013a). The spaceflight environment has also been shown to result in higher *P. aeruginosa* densities following 72 h of culture in modified artificial urine medium (Kim et al. 2013a). The authors proposed that phosphate and oxygen limitations under spaceflight growth conditions were the causative factors for the observed increased bacterial densities (Kim et al. 2013a). In a separate study, the susceptibility of *P. aeruginosa* to antibiotics was examined with cultures grown in spaceflight and phenotypic analysis done on Earth using antibiotic disc tests on solid media. A decreased susceptibility to the polymyxin antibiotic colistin was observed, as well as an increased susceptibility to cephalothin, polymyxin B, and rifampixin (Juergensmeyer et al. 1999).

Future research into the development and impact of biofilms during spaceflight is critical to protect crew health, vehicle systems/integrity, and mission success. However, biofilm formation and architecture has only been studied in spaceflight experiments using single, pure cultures of microorganisms. To better understand the impact of microgravity on the formation, architecture, disinfection sensitivity, and corrosion potential of polymicrobial biofilms, a new spaceflight study led by Robert McLean at Texas State University will investigate the development of biofilms created by co-cultures of *P. aeruginosa* and *Escherichia coli*, the ability of silver solutions to disinfect these biofilms, and the corrosion caused by these biofilms on stainless steel. These studies will provide new evidence as to whether current biofilm control is adequate for spacecraft during human exploration of deep space.

The transcriptional and proteomic responses of *P. aeruginosa* to spaceflight conditions were profiled as a part of the MICROBE experiment (Crabbé et al. 2011). Intriguingly, Hfq and a significant part of the Hfq regulon were differentially regulated by *P. aeruginosa* in spaceflight. As described above, Hfq was initially identified as a key regulator in the LSMMG and spaceflight response of *S.* Typhimurium (Wilson et al. 2002a, 2007, 2008). Hence, Hfq is the first transcriptional regulator ever shown to be involved in the spaceflight response across bacterial species. Among the genes with the highest fold inductions in spaceflight-grown *P. aeruginosa* were those encoding the lectins, LecA and LecB. Lectins play a role in the bacterial adhesion process to eukaryotic cells, and have clinically important cytotoxic effects (Gilboa-Garber et al. 1977; Bajolet-Laudinat et al. 1994; Chemani et al. 2009). Another virulence gene that was induced by *P. aeruginosa* in response to spaceflight culture conditions was *rhlA*, which encodes rhamnosyltransferase I involved in rhamnolipid surfactant biosynthesis. Rhamnolipids are glycolipidic surface-active molecules with cytotoxic and immunomodulatory effects in eukaryotic cells (McClure and Schiller 1996; Davey et al. 2003; Pamp and Tolker-Nielsen 2007). Furthermore, spaceflight induced the expression of genes and proteins involved in the anaerobic growth of *P. aeruginosa*. Indeed, more limited oxygen availability could occur in spaceflight conditions due to low fluid-shear and thus, low mixing growth conditions. In a separate study, the gene expression profiles of *P. aeruginosa* and *S.* Typhimurium cultured in spaceflight were compared using a systems biology approach. Common pathways that were differentially regulated under spaceflight conditions in both organisms included pathways related to ribosome synthesis, RNA degradation, protein export, flagellar assembly, methane metabolism, toluene degradation, oxidative phosphorylation, TCA cycle, glycolysis, and purine and pyrimidine metabolism (Roy et al. 2016).

The cultivation of *P. aeruginosa* PAO1 in spaceflight analog conditions (LSMMG) in the RWV (28°C) in LB medium induced a transcriptomic and

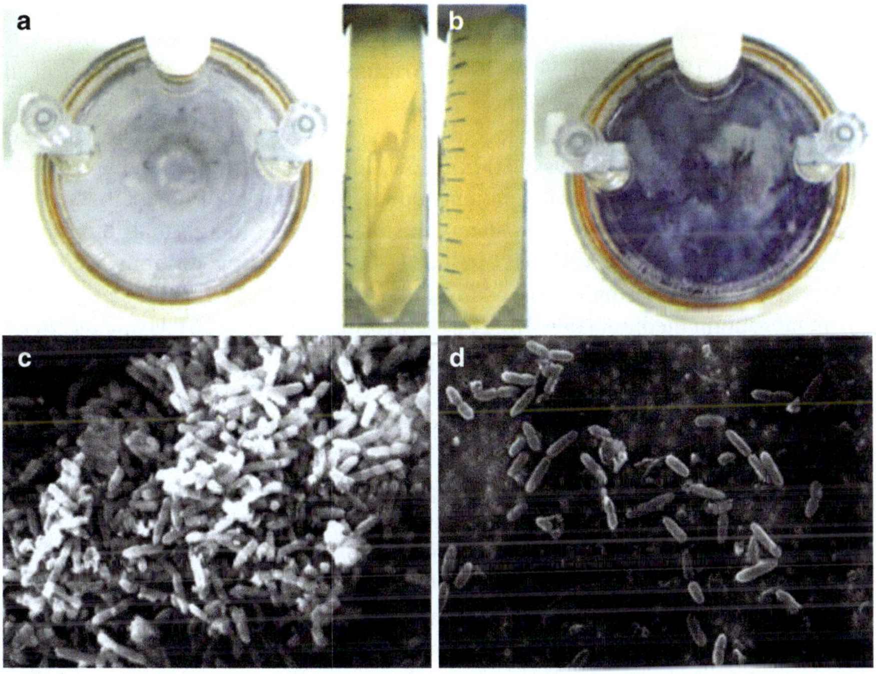

Fig. 18.3 Biofilm formation of *P. aeruginosa* in LSMMG (**a** and **c**) and higher shear (**b** and **d**) conditions using the RWV bioreactor. Panels **a** and **b** show *P. aeruginosa* PAO1 grown in LB medium and the gas-permeable membrane on the backside of the vessel was stained with Crystal Violet to detect biofilm formation. The decanted culture was collected in Falcon tubes (center). Panels **c** and **d** show SEM images of *P. aeruginosa* CF isolate PA39 grown in artificial sputum medium. Magnification = 4500× (Dingemans et al. 2016)

phenotypic profile related to virulence (Crabbé et al. 2010). More specifically, PAO1 produced higher amounts of the exopolysaccharide alginate when grown in LSMMG, as compared to the control. Alginate is an important virulence factor in *P. aeruginosa* since it restricts the diffusion of antimicrobial agents and confers resistance to immune defense mechanisms by avoiding phagocytic uptake, scavenging reactive oxygen species and suppressing leukocyte function (Learn et al. 1987; Pier et al. 2001). Accordingly, an increased oxidative stress resistance in LSMMG-grown *P. aeruginosa* was observed, as well as a higher transcription of the alternative sigma factor *algU*, essential for alginate production. As mentioned above, Hfq was also found to be an important regulator in the LSMMG response of *P. aeruginosa* (Table 18.1) (Wilson et al. 2002a, 2007, 2008; Crabbé et al. 2010, 2011). In addition, the *P. aeruginosa* LSMMG regulon (comprised of 134 genes) included genes involved in stress resistance, motility, and microaerophilic/anaerobic metabolism.

Another impact of low fluid-shear was demonstrated in a study where LSMMG-grown *P. aeruginosa* (37°C) formed dense self-aggregating biofilms in LB medium (Fig. 18.3a) (Crabbé et al. 2008) in contrast to membrane-attached biofilms formed in the higher fluid shear control orientation (Fig. 18.3b). Interestingly, the phenotypic and gene expression profiles grown in LSMMG showed similarities with those of *P. aeruginosa* found in lung secretions of cystic fibrosis (CF) patients (Lam 1980; Singh et al. 2000; Sriramulu et al. 2005; Bjarnsholt et al. 2009). In CF patients, the formation of drug-resistant and/or tolerant microcolonies by *P. aeruginosa* in the dense and viscous lung mucus is the major cause of mortality (Wagner and Iglewski 2008). Low fluid-shear zones are believed to be present in the lung mucus of CF patients due to the absence of mucociliary clearance, which represents the main shear-causing factor in the normal lung mucus (Blake 1973). Recently, LSMMG-induced self-aggregating biofilms were also reported for a highly adapted, transmissible *P. aeruginosa* CF isolate cultured in artificial sputum medium (Fig. 18.3c, d) (Dingemans et al. 2016). Using a *P. aeruginosa* strain and growth medium relevant for the CF lung environment resulted in the induction of additional pathways by LSMMG that are involved in the metabolism and virulence of this microorganism in the CF patient population (Dingemans et al. 2016).

18.6 How Universal Is the Microbial Response to Spaceflight and Spaceflight Analog Culture?

The studies described in the previous sections involving *S.* Typhimurium and *P. aeruginosa* provide key examples of two biomedically important human pathogens that exhibit a variety of similarities and differences in their responses to spaceflight and spaceflight-analog culture. While these organisms have received the most extensive degree of study, the response of other pathogens to these environments have also been investigated, and results from these studies suggest both potential common mechanisms as well as a myriad of different responses.

For example, *Staphylococcus aureus* is a Gram-positive bacterium of particular importance to crew health due to its prevalence and reported transmission aboard spacecraft (Pierson et al. 1996; Castro et al. 2004). RWV-cultured *S. aureus* displayed slower growth and generally repressed virulence characteristics, including decreased carotenoid production (Rosado et al. 2010; Castro et al. 2011), decreased capacity to lyse red blood cells (Rosado et al. 2010), increased susceptibility to oxidative stress (Castro et al. 2011), reduced survival in whole blood (Castro et al. 2011), and intriguingly, increased formation of a biofilm phenotype (Castro et al. 2011). Furthermore, molecular genetic expression analysis revealed the downregulation of the RNA chaperone protein Hfq, which parallels the response of *S.* Typhimurium to LSMMG culture. This common association with Hfq in both Gram-positive and Gram-negative organisms suggests an evolutionarily conserved response to fluid-shear among structurally diverse prokaryotes (Castro et al. 2011). However, unlike *S.* Typhimurium and *P. aeruginosa*, these results suggest *S. aureus* responds to the RWV environment by initiating a biofilm phenotype with diminished virulence characteristics that may enable the organism to establish a long term commensal relationship with the host. Collectively, these comparisons may provide unique insight into key factors influencing the delicate balance between infection and colonization by *S. aureus* during the initial host–pathogen interaction.

In addition, other Enterobacteriaceae have been profiled in response to LSMMG culture to determine the conserved nature of this response (Pacello et al. 2012; Soni et al. 2014). For example, a study by Soni et al. evaluated various Enterobacteriaceae from different genera in a systematic "side-by-side" manner. Evaluations of *S.* Typhimurium, *E. coli*, *Enterobacter cloacae*, *Citrobacter freundii*, and *Serratia marcescens* revealed essentially identical growth kinetics in both the LSMMG and control orientation for each organism. Each species was also profiled for LSMMG-induced stress resistance at stationary phase, including acid and oxidative stress. These studies confirmed that culture in LSMMG altered the acid stress response of most of these microorganisms, however some became more sensitive while others became more tolerant. Notably, *C. freundii* did not display any change in acid stress response. When evaluated for changes in oxidative stress response, all cultures grown in LSMMG became more sensitive to oxidative stress. In addition, qRT-PCR analysis demonstrated that the molecular genetic response of these species to LSMMG is conserved across Enterobactericeae (e.g., *hfq*, *trpD*, and *ydcI*), but the direction of gene expression changes (i.e. up or down) can vary depending on the genus.

The number and variety of microorganisms that have been studied in spaceflight and spaceflight analog culture is extensive, and beyond the scope of this chapter. However, several other findings from experiments investigating the responses of medically significant microorganisms during culture in these environments are presented in Table 18.2.

Table 18.2 Other pertinent findings from experiments investigating medically significant microorganisms cultured in spaceflight and spaceflight analog conditions

Microorganism	Environment	Finding
E. coli	LSMMG	Increased resistance to a variety of pathogenesis-related stresses, including low pH, osmotic, alcohol, and thermal stress (Gao et al. 2001; Lynch et al. 2004, 2006; Allen et al. 2008).
	LSMMG	Acid and osmotic stress resistance was shown to be RpoS-independent in the exponential phase of culture (similar to the findings for *S.* Typhimurium cultured to exponential phase in the RWV (Wilson et al. 2002b)), but RpoS-dependent in stationary phase (Lynch et al. 2004).
	LSMMG	Dense biofilms that exhibited increased resistance to some environmental stresses and antibiotics were observed when *E. coli* was cultured to stationary phase under LSMMG conditions, as compared to control conditions, in the presence of glass microcarrier beads (Lynch et al. 2006).
	LSMMG	Carvalho et al. (2005) infected a 3-D model of colonic epithelium with either LSMMG or control-cultured Enteropathogenic *E. coli* (EPEC) or Enterohaemorrhagic *E. coli* (EHEC), respectively, when both the host and pathogen were simultaneously cultured in the RWV bioreactor. Formation of attaching and effacing lesions similar to that observed during the normal course of infection in vivo, and increased intimin expression were observed during EHEC infection of 3-D colon cells in the RWV.
	LSMMG	Increased toxin production observed in Enterotoxigenic *E. coli* (ETEC) cultured under LSMMG conditions as compared to control cultures, which correlated with increased fluid accumulation in mice infected with LSMMG-cultured ETEC (Chopra et al. 2006). Increased TNF production was observed in macrophages infected with EPEC cultured under LSMMG as compared to control cultures.
	LSMMG	Increased adherence of adherent-invasive *E. coli* to Caco-2 colonic epithelial cells when the bacteria were cultured under LSMMG; an effect which became even more pronounced with an *rpoS* mutant (Allen et al. 2008).
	LSMMG	Multigenerational growth of *E. coli* in LSMMG induced genetic changes compared to an unadapted control suggesting the microgravity analog environment may selectively adapt microorganisms over time (Tirumalai et al. 2017).
	LSMMG	A comparison was performed evaluating *E. coli* strain MG1655 transcriptional responses when cultures were grown in rich and minimal medium under LSMMG and control conditions. While reproducible patterns were identified under each condition, no specific genes were identified that would suggest a single gene that was consistently differentially regulated under every LSMMG condition (Tucker et al. 2007).
	LSMMG	Evaluation of several strains of *E. coli* O157:H7 cultured under LSMMG provided evidence that cells were increasing in size in a media dependent fashion (Kim et al. 2014).
	LSMMG	Evaluation of several strains of *E. coli* O157:H7 cultured under LSMMG indicated a decreased resistance to thermal stress (55°C) and higher membrane fluidity based on fatty acid content (Kim and Rhee 2016).
	LSMMG	*E. coli* cultured in LSMMG displayed a decreased susceptibility to ciprofloxacin compared to control cultures. One possible explanation was increased expression of efflux pump genes *acrAB-tolC* (Xu et al. 2015).

Microorganism	Environment	Finding
Streptococcus pneumoniae	LSMMG	Differential transcriptional regulation of 101 genes, including those involved in the adherence and invasion (Allen et al. 2007).
Saccharomyces cerevisiae	Apollo 16	Increased phosphate uptake observed during spaceflight culture (Berry and Volz 1979).
	Apollo 16	Increased survival in tissues following multiple infection routes in a mouse model of infection (intraperitoneal, tail vein, and epidermal) as compared to ground-based controls (Hiebel and Volz 1977; Volz 1990).
	LSMMG	Random budding pattern and tendency to self-aggregate and clump in response to LSMMG culture in contrast to normal bipolar budding normally observed in controls. Corresponding changes in expression of genes important for polarity, budding, and cell separation (Purevdorj-Gage et al. 2006).
	LSMMG	Microarray analysis showed differential regulation of *S. cerevisiae* genes including those important for environmental stress responses (Johanson et al. 2002; Sheehan et al. 2007).
Candida albicans	LSMMG	Increased frequency of filamentous forms in LSMMG cultures accompanied by changes in expression of *hwp1* and *ywp1*, associated with yeast-hyphal transition. The morphogenic switch from round, budding yeast to filamentous form is associated with enhanced virulence (Altenburg et al. 2008).
	Space Shuttle STS-115	The first global transcriptional profiling and phenotypic characterization of the fungal pathogen, *C. albicans*, grown in spaceflight conditions. Enhanced random budding of spaceflight-cultured cells was observed as opposed to bipolar budding patterns for ground controls (Fig. 18.4). Spaceflight differentially regulated 452 genes compared to ground controls, including those involved in antifungal agent and stress resistance, e.g., ABC transporters, ergosterol, and oxidative stress resistance (Crabbé et al. 2013).
Streptococcus mutans	LSMMG	Increased sensitivity to oxidative stress. Transcriptomic analysis showed differential expression of 247 genes compared to control cultures, including those involved in carbohydrate metabolism, translation, and stress responses (Orsini et al. 2017).
Mycobacterium marinum	LSMMG	LSMMG enhanced sensitivity to oxidative stress as compared to control conditions. Transcriptomic analysis demonstrated differential gene expression of 562 genes, including those involved in metabolism, lipid degradation, and chaperone and mycobactin expression. Role for SigH identified in LSMMG response (Abshire et al. 2016).
Salmonella enterica (various strains)	LSMMG	*S. enterica* strains were challenged with hydrogen peroxide to determine conservation of their oxidative stress response. All strains displayed enhanced resistance. In addition, the deletion of the genes encoding for the catalases KatG and KatN removed the enhanced resistance induced by LSMMg culture. Interestingly, deletion of Hfq, RpoE, RpoS or OxyR from strains did not affect the enhanced resistance phenotype (Pacello et al. 2012).
Klebsiella pneumoniae	LSMMG	*K. pneumoniae* grown under LSMMG formed thicker biofilms and higher production of cellulose compared to control orientation cultures. RNA seq transcriptomic analysis showed 171 diferentially regulated genes between the two growth conditions belonging to 15 functional categories (Wang et al. 2016).

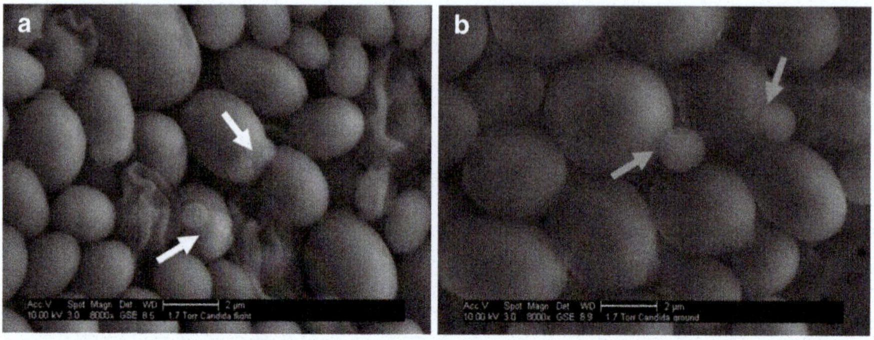

Fig. 18.4 Scanning electron microscopy image of *Candida albicans* cultured under spaceflight conditions aboard Space Shuttle STS-115 (**a**) or in ground control conditions (**b**). Random budding was only observed in cells cultured under spaceflight conditions (white arrows), while bipolar budding patterns were observed for ground controls (grey arrows). Magnification = 8000× (Crabbé et al. 2013)

18.7 Importance to Spaceflight and Life on Earth: The Future Has Started Now

As humans explore space, microorganisms will travel with them. Thus, understanding microbial responses to spaceflight will have a tremendous impact on how we design our spaceflight vehicles, build bioregenerative systems, grow food during a mission, and mitigate the risk of infectious disease to the crew. While several key studies have provided critical mechanistic insight into how microbial responses to the spaceflight environment may affect virulence, many questions still remain unanswered. Indeed, spaceflight-induced alterations in microbial virulence have just begun to be investigated. The impact of these changes on the host–pathogen interaction during spaceflight and corresponding clinical implications for a potentially susceptible crew is still unclear. Beyond prevention of exposure, antibiotics are the primary countermeasure to microbial infection during a spaceflight mission. Previous spaceflight experiments have identified increases in antibiotic resistance for organisms such as *E. coli* (kanamycin and colistin) and *S. aureus* (oxacillin, chloramphenicol, and erythromycin) in response to spaceflight culture (Tixador et al. 1985). The lack of conclusive information on changes in antibiotic resistance for a broad range of microorganisms and corresponding pharmacokinetics indicates a large knowledge gap in infectious disease control, although recent approaches may allow a better prediction (Sommer et al. 2017). Interestingly, the use of microgravity and space flight conditions and its distinct effects on microbial resistance may serve as an additional tool in this direction. Moreover, understanding the microbiota to which the crew will be exposed, and how spaceflight alters the microbial consortia and interactions with the crew, is one of the cornerstones of microbiological risk assessment during a mission. While microorganisms associated with the environment and food supply have been the focus of operational monitoring efforts, very little is known about the changes in crew and spacecraft microbiome during a mission while multiple new investigations are under way. Likewise, little is known about mutation rates and heritable changes in all microorganisms associated with the crew and their environment during a mission—and nothing is known of long-term spaceflight-induced changes to either the host or pathogen. Alone or in combination, these factors could dramatically affect the impact of microorganisms on spaceflight mission success since the resistancies and the specific antibiotic' availabilities are unforeseeable variables. As we look to the future and introduction of commercial spaceflights with greater civilian participation, including spaceflight tourism, we need a greater understanding of the unique microbial risks associated with human spaceflight. Moreover, lessons learned from spaceflight studies have profound implications for the general public, in terms of expanding our knowledge of (1) the mechanisms of microbial pathogenesis, which hold potential for development of novel strategies to a point of care diagnosis, allowing optimized treatments, and—most efficiently—to prevent infectious disease, and (2) the human microbiome and how stressful environments (see Chap. 34) alter the relationship between host and commensal that determine the transition between normal homeostasis and disease progression (http://commonfund.nih.gov/hmp/).

References

Abshire CF, Prasai K, Soto I, Shi R, Concha M, Baddoo M et al (2016) Exposure of *Mycobacterium marinum* to low-shear modeled microgravity: effect on growth, the transcriptome and survival under stress. NPJ Microgravity 2:16038

Allen CA, Galindo CL, Pandya U, Watson DA, Chopra AK et al (2007) Transcription profiles of *Streptococcus pneumoniae* grown under different conditions of normal gravitation. Acta Astronaut 60:133–144

Allen CA, Niesel DW, Torres AG (2008) The effects of low-shear stress on adherent-invasive *Escherichia coli*. Environ Microbiol 10:1512–1525

Altenburg SD, Nielsen-Preiss SM, Hyman LE (2008) Increased filamentous growth of *Candida albicans* in simulated microgravity. Genomics Proteomics Bioinformatics 6:42–50

Bajolet-Laudinat O, Girod-de Bentzmann S, Tournier JM, Madoulet C, Plotkowski MC, Chippaux C et al (1994) Cytotoxicity of *Pseudomonas aeruginosa* internal lectin PA-I to respiratory epithelial cells in primary culture. Infect Immun 62:4481–4487

Berry D, Volz PA (1979) Phosphate uptake in *Saccharomyces cerevisiae* Hansen wild type and phenotypes exposed to space flight irradiation. Appl Environ Microbiol 38:751–753

Bjarnsholt T, Jensen PO, Fiandaca MJ, Pedersen J, Hansen CR, Andersen CB et al (2009) *Pseudomonas aeruginosa* biofilms in the respiratory tract of cystic fibrosis patients. Pediatr Pulmonol 44:547–558

Blake J (1973) A note on mucus shear rates. Respir Physiol 17:394–399

Bruce RJ, Ott CM, Skuratov VM, Pierson DL (2005) Microbial surveillance of potable water sources of the International Space Station. SAE Trans 114:283–292

Carvalho HM, Teel LD, Goping G, O'Brien AD (2005) A three-dimensional tissue culture model for the study of attach and efface lesion formation by enteropathogenic and enterohaemorrhagic *Escherichia coli*. Cell Microbiol 7:1771–1781

Castro VA, Trasher AN, Healy M, Ott CM, Pierson DL (2004) Microbial characterization during the early habitation of the International Space Station. Microb Ecol 47:119–126

Castro SL, Nelman-Gonzalez M, Nickerson CA, Ott CM (2011) Low fluid shear culture of *Staphylococcus aureus* induces attachment-independent biofilm formation and represses *hfq* expression. Appl Environ Microbiol 77(18):6368–6378

Chemani C, Imberty A, de Bentzmann S, Pierre M, Wimmerova M, Guery BP et al (2009) Role of LecA and LecB lectins in *Pseudomonas aeruginosa*-induced lung injury and effect of carbohydrate ligands. Infect Immun 77:2065–2075

Chopra V, Fadl AA, Sha J, Chopra S, Galindo CL, Chopra AK (2006) Alterations in the virulence potential of enteric pathogens and bacterial-host cell interactions under simulated microgravity conditions. J Toxicol Environ Health A 69:1345–1370

Crabbé A, De Boever P, Van Houdt R, Moors H, Mergeay M, Cornelis P (2008) Use of the rotating wall vessel technology to study the effect of shear stress on growth behaviour of *Pseudomonas aeruginosa* PA01. Environ Microbiol 10:2098–2110

Crabbé A, Pycke B, Van Houdt R, Monsieurs P, Nickerson C, Leys N et al (2010) Response of *Pseudomonas aeruginosa* PAO1 to low shear modelled microgravity involves AlgU regulation. Environ Microbiol 12:1545–1564

Crabbé A, Schurr MJ, Monsieurs P, Morici L, Schurr J, Wilson JW et al (2011) Transcriptional and proteomic responses of *Pseudomonas aeruginosa* PAO1 to spaceflight conditions involve Hfq regulation and reveal a role for oxygen. Appl Environ Microbiol 77:1221–1230

Crabbé A, Nielsen-Preiss SM, Woolley CM, Barrila J, Buchanan K et al (2013) Spaceflight enhances cell aggregation and random budding in *Candida albicans*. PLoS One 8(12):e80677

Curtiss R III, Wanda SY, Gunn BM, Zhang X, Tinge SA et al (2009) *Salmonella enterica* serovar Typhimurium strains with regulated delayed attenuation in vivo. Infect Immun 77(3):1071–1082

Curtiss R III, Xin W, Li Y, Kong W, Wanda SY et al (2010) New technologies in using recombinant attenuated *Salmonella* vaccine vectors. Crit Rev Immunol 30(3):255–270

Davey ME, Caiazza NC, O'Toole GA (2003) Rhamnolipid surfactant production affects biofilm architecture in *Pseudomonas aeruginosa* PAO1. J Bacteriol 185:1027–1036

Dickson KJ (1991) Summary of biological spaceflight experiments with cells. ASGSB Bull 4:151–260

Dingemans J, Monsieurs P, Yu SH, Crabbé A, Förstner KU et al (2016) Effect of shear stress on *Pseudomonas aeruginosa* isolated from the cystic fibrosis lung. MBio 7(4). https://doi.org/10.1128/mBio.00813-16

Dong T, Schellhorn HE (2010) Role of RpoS in virulence of pathogens. Infect Immun 78:887–897

Gao Q, Fang A, Pierson DL, Mishra SK, Demain AL (2001) Shear stress enhances microcin B17 production in a rotating wall bioreactor, but ethanol stress does not. Appl Microbiol Biotechnol 56:384–387

Gilboa-Garber N, Mizrahi L, Garber N (1977) Mannose-binding hemagglutinins in extracts of *Pseudomonas aeruginosa*. Can J Biochem 55:975–981

Gottesman S (2004) The small RNA regulators of *Escherichia coli*: roles and mechanisms. Annu Rev Microbiol 58:303–328

Gottesman S, McCullen CA, Guillier M, Vanderpool CK, Majdalani N, Benhammou J et al (2006) Small RNA regulators and the bacterial response to stress. Cold Spring Harb Symp Quant Biol 71:1–11

Gueguinou N, Huin-Schohn C, Bascove M, Bueb JL, Tschirhart E, Legrand-Frossi C et al (2009) Could spaceflight-associated immune system weakening preclude the expansion of human presence beyond Earth's orbit? J Leukoc Biol 86:1027–1038

Guisbert E, Rhodius VA, Ahuja N, Witkin E, Gross CA (2007) Hfq modulates the sigmaE-mediated envelope stress response and the sigma32-mediated cytoplasmic stress response in *Escherichia coli*. J Bacteriol 189:1963–1973

Hawkins WR, Ziegelschmid JF (1975) Clinical aspects of crew health. In: Johnson RS, Dietlein LF, Berry CA (eds) Biomedical results of Apollo, SP-368. NASA Spec. Rep, Washington, DC, pp 43–81

Hengge-Aronis R (2000) The general stress response in *Escherichia coli*. In: Storz G, Hengge-Aronis R (eds) Bacterial stress responses. ASM Press, Washington, DC

Hiebel TL, Volz PA (1977) Foreign body reactions induced by fungi irradiated in space. Phytologia 35:365–372

Horneck G, Klaus DM, Mancinelli RL (2010) Space microbiology. Microbial Mol Biol Rev 74(1):121–156

Johanson K, Allen PL, Lewis F, Cubano LA, Hyman LE, Hammond TG (2002) *Saccharomyces cerevisiae* gene expression changes during rotating wall vessel suspension culture. J Appl Physiol 93:2171–2180

Juergensmeyer MA, Juergensmeyer EA, Guikema JA (1999) Long-term exposure to spaceflight conditions affects bacterial response to antibiotics. Microgravity Sci Technol 12(1):41–47

Kacena MA, Todd P (1999) Gentamicin: effect on *E. coli* in space. Microgravity Sci Technol 12:135–137

Kim HW, Rhee MS (2016) Influence of low-shear modeled microgravity on heat resistance, membrane fatty acid composition, and heat stress-related gene expression in *Escherichia coli* O157:H7 ATCC 35150, ATCC 43889, ATCC 43890, and ATCC 43895. Appl Environ Microbiol 82(10):2893–2901

Kim W, Tengra FK, Shong J, Marchand N, Chan HK et al (2013a) Effect of spaceflight on *Pseudomonas aeruginosa* final cell density is modulated by nutrient and oxygen availability. BMC Microbiol 13:241

Kim W, Tengra FK, Young Z, Shong J, Marchand N et al (2013b) Spaceflight promotes biofilm formation by *Pseudomonas aeruginosa*. PLoS One 8(4):e62437

Kim HW, Matin A, Rhee MS (2014) Microgravity alters the physiological characteristics of *Escherichia coli* O157:H7 ATCC 35150, ATCC 43889, and ATCC 43895 under different nutrient conditions. Appl Environ Microbiol 80(7):2270–2278

Klaus DM (2001) Clinostats and bioreactors. Gravit Space Biol Bull 14:55–64

Klaus DM, Howard HN (2006) Antibiotic efficacy and microbial virulence during space flight. Trends Biotechnol 24:131–136

Lam J (1980) Production of mucoid microcolonies by *Pseudomonas aeruginosa* within the infected lungs in cystic fibrosis. Infect Immun 28:546–556

Learn DB, Brestel EP, Seetharama S (1987) Hypochlorite scavenging by *Pseudomonas aeruginosa* alginate. Infect Immun 55:1813–1818

Li Y, Wang S, Scarpellini G, Gunn B, Xin W et al (2009) Evaluation of new generation *Salmonella enterica* serovar Typhimurium vaccines with regulated delayed attenuation to induce immune responses against PspA. Proc Natl Acad Sci U S A 106(2):593–598

Lynch SV, Brodie EL, Matin A (2004) Role and regulation of sigma S in general resistance conferred by low-shear simulated microgravity in *Escherichia coli*. J Bacteriol 186:8207–8212

Lynch SV, Mukundakrishnan K, Benoit MR, Ayyaswamy PS, Matin A (2006) *Escherichia coli* biofilms formed under low-shear modeled microgravity in a ground-based system. Appl Environ Microbiol 72:7701–7710

Majdalani N, Vanderpool CK, Gottesman S (2005) Bacterial small RNA regulators. Crit Rev Biochem Mol Biol 40:93–113

McClure CD, Schiller NL (1996) Inhibition of macrophage phagocytosis by *Pseudomonas aeruginosa* rhamnolipids *in vitro* and *in vivo*. Curr Microbiol 33:109–117

McLean RJ, Cassanto JM, Barnes MB, Koo JH (2001) Bacterial biofilm formation under microgravity conditions. FEMS Microbiol Lett 195(2):115–119

Nauman EA, Ott CM, Sander E, Tucker DL, Pierson D, Wilson JW et al (2007) Novel quantitative biosystem for modeling physiological fluid shear stress on cells. Appl Environ Microbiol 73:699–705

Nickerson CA, Ott CM, Mister SJ, Morrow BJ, Burns-Keliher L, Pierson DL (2000) Microgravity as a novel environmental signal affecting *Salmonella enterica* serovar Typhimurium virulence. Infect Immun 68:3147–3152

Nickerson CA, Ott CM, Wilson JW, Ramamurthy R, Pierson DL (2004) Microbial responses to microgravity and other low-shear environments. Microbiol Mol Biol Rev 68:345–361

Orsini SS, Lewis AM, Rice KC (2017) Investigation of simulated microgravity effects on *Streptococcus mutans* physiology and global gene expression. NPJ Microgravity 3:4

Ott CM (2004) Human immune function and microbial pathogenesis in human spaceflight. Paper presented at the 10th International Symposium on Microbial Ecology, Cancun, Mexico

Pacello F, Rotilio G, Battistoni A (2012) Low-shear modeled microgravity enhances *Salmonella* enterica resistance to hydrogen peroxide through a mechanism involving KatG and KatN. Open Microbiol J 6:53–64

Pamp SJ, Tolker-Nielsen T (2007) Multiple roles of biosurfactants in structural biofilm development by *Pseudomonas aeruginosa*. J Bacteriol 189:2531–2539

Pfeiffer V, Sittka A, Tomer R, Tedin K, Brinkmann V, Vogel J (2007) A small non-coding RNA of the invasion gene island (SPI-1) represses outer membrane protein synthesis from the *Salmonella* core genome. Mol Microbiol 66:1174–1191

Pier GB, Coleman F, Grout M, Franklin M, Ohman DE (2001) Role of alginate O acetylation in resistance of mucoid *Pseudomonas aeruginosa* to opsonic phagocytosis. Infect Immun 69:1895–1901

Pierson DL, Chidambaram M, Heath JD, Mallary L, Mishra SK, Sharma B et al (1996) Epidemiology of *Staphylococcus aureus* during space flight. FEMS Immunol Med Microbiol 16:273–281

Pierson DL, Mehta SK, Stowe RP (2007) Reactivation of latent herpes viruses in astronauts. In: Ader R (ed) Psychoneuroimmunology. Academic, San Diego, pp 851–868

Purevdorj-Gage B, Sheehan KB, Hyman LE (2006) Effects of low-shear modeled microgravity on cell function, gene expression, and phenotype in *Saccharomyces cerevisiae*. Appl Environ Microbiol 72:4569–4575

Rosado H, Doyle M, Hinds J, Taylor PW (2010) Low-shear modelled microgravity alters expression of virulence determinants of *Staphylococcus aureus*. Acta Astronaut 66:408–413

Roy R, Shilpa PP, Bagh S (2016) A systems biology analysis unfolds the molecular pathways and networks of two proteobacteria in spaceflight and simulated microgravity conditions. Astrobiology 16(9):677–689

Sarker S, Ott CM, Barrila J, Nickerson CA (2010) Discovery of spaceflight regulated virulence mechanisms in *Salmonella* and other microbial pathogens: novel approaches to commercial vaccine development. Gravit Space Biol 23(2):75–78

Sheehan KB, McInnerney K, Purevdorj-Gage B, Altenburg SD, Hyman LE (2007) Yeast genomic expression patterns in response to low-shear modeled microgravity. BMC Genomics 8:3

Shi H, Santander J, Brenneman KE, Wanda SY, Wang S et al (2010a) Live recombinant *Salmonella* Typhi vaccines constructed to investigate the role of *rpoS* in eliciting immunity to a heterologous antigen. PLoS One 5(6):e11142

Shi H, Wang S, Roland KL, Gunn BM, Curtiss R III (2010b) Immunogenicity of a live recombinant *Salmonella enterica* serovar typhimurium vaccine expressing pspA in neonates and infant mice born from naive and immunized mothers. Clin Vaccine Immunol 17(3):363–371

Singh PK, Schaefer AL, Parsek MR, Moninger TO, Welsh MJ, Greenberg EP (2000) Quorum-sensing signals indicate that cystic fibrosis lungs are infected with bacterial biofilms. Nature 407:762–764

Sittka A, Pfeiffer V, Tedin K, Vogel J (2007) The RNA chaperone Hfq is essential for the virulence of *Salmonella* Typhimurium. Mol Microbiol 63:193–217

Sittka A, Lucchini S, Papenfort K, Sharma CM, Rolle K, Binnewies TT et al (2008) Deep sequencing analysis of small noncoding RNA and mRNA targets of the global post-transcriptional regulator, Hfq. PLoS Genet 4:e1000163

Sommer MOA, Munck C, Toft-Kehler RV, Andersson DI (2017) Prediction of antibiotic resistance: time for a new preclinical paradigm? Nat Rev Microbiol 15(11):689–696. https://doi.org/10.1038/nrmicro.2017

Soni A, O'Sullivan L, Quick LN, Ott CM, Nickerson CA, Wilson JW (2014) Conservation of the low-shear modeled microgravity response in Enterobacteriaceae and analysis of the *trp* genes in this response. Open Microbiol J 13:51–58

Sriramulu DD, Lunsdorf H, Lam JS, Römling U (2005) Microcolony formation: a novel biofilm model of *Pseudomonas aeruginosa* for the cystic fibrosis lung. J Med Microbiol 54:667–676

Taylor GR (1974) Recovery of medically important microorganisms from Apollo astronauts. Aerosp Med 45:824–828

Tirumalai MR, Karouia F, Tran Q, Stepanov VG, Bruce RJ et al (2017) The adaptation of *Escherichia coli* cells grown in simulated microgravity for an extended period is both phenotypic and genomic. NPJ Microgravity 3:15

Tixador R, Richoilley G, Gasset G, Templier J, Bes JC, Moatti N et al (1985) Study of minimal inhibitory concentration of antibiotics on bacteria cultivated *in vitro* in space (Cytos 2 experiment). Aviat Space Environ Med 56:748–751

Tucker DL, Ott CM, Huff S, Fofanov Y, Pierson DL et al (2007) Characterization of *Escherichia coli* MG1655 grown in a low-shear modeled microgravity environment. BMC Microbiol 7:15

Volz PA (1990) Mycology studies in space. Mycopathologia 109:89–98

Wagner VE, Iglewski BH (2008) *P. aeruginosa* biofilms in CF infection. Clin Rev Allergy Immunol 35:124–134

Wang H, Yan Y, Rong D, Wang J, Wang H et al (2016) Increased biofilm formation ability in *Klebsiella pneumoniae* after short-term exposure to a simulated microgravity environment. Microbiology Open 5(5):793–801

Wilson JW, Ramamurthy R, Porwollik S, McClelland M, Hammond T, Allen P et al (2002a) Microarray analysis identifies *Salmonella* genes belonging to the low-shear modeled microgravity regulon. Proc Natl Acad Sci U S A 99:13807–13812

Wilson JW, Ott CM, Ramamurthy R, Porwollik S, McClelland M, Pierson DL et al (2002b) Low-Shear modeled microgravity alters the *Salmonella enterica* serovar Typhimurium stress response in an RpoS-independent manner. Appl Environ Microbiol 68:5408–5416

Wilson JW, Ott CM, Zu Bentrup KH, Ramamurthy R, Quick L, Porwollik S et al (2007) Space flight alters bacterial gene expression and virulence and reveals a role for global regulator Hfq. Proc Natl Acad Sci U S A 104:16299–16304

Wilson JW, Ott CM, Quick L, Davis R, zu Bentrup KH, Crabbé A et al (2008) Media ion composition controls regulatory and virulence response of *Salmonella* in spaceflight. PLoS One 3:e3923

Xu B, Li C, Zheng Y, Si S, Shi Y (2015) Simulated microgravity affects ciprofloxacin susceptibility and expression of acrAB-tolC genes in *E. coli* ATCC25922. J Clin Exp Pathol 8(7):7945–7952

Yang J, Barrila J, Roland KL, Ott CM, Nickerson CA (2016) Physiological fluid shear alters the virulence potential of invasive multidrug-resistant non-typhoidal *Salmonella typhimurium* D23580. NPJ Microgravity 2:16021

Stress, Spaceflight, and Latent Herpes Virus Reactivation

19

Raymond P. Stowe, Duane L. Pierson, and Satish K. Mehta

19.1 Stress Responses Associated with Spaceflight

Stress responses are controlled through a complex network of neural, humoral, and metabolic factors among which the hypothalamic–pituitary–adrenocortical (HPA) axis and the sympathetic–adrenal–medullary (SAM) axis play a crucial effector role (see also Chaps. 6–8). In brief, while activation of the HPA axis is thought to occur during events perceived as more overwhelming and less readily coped with, activation of the SAM axis is associated with stressful emotions such as anger and fear. The neuroendocrine response is extremely complex and involves numerous feedback mechanisms. Individuals may manifest different levels of stress hormones due to variables such as prior exposure to the stressor, behavior, and coping responses. Due to sampling and stowage constraints during spaceflight, most of the data on stress hormone changes have been derived from samples taken after landing and comparing them to samples taken before launch.

Even before spaceflight, neuroendocrine changes can occur. An increase in serum cortisol was found in two cosmonauts 15 days prior to launch as compared to 2 months before (Caillot-Augusseau et al. 1998). Plasma cortisol was also elevated in several Shuttle astronauts 10 days prior to launch as compared to their annual medical exams, a period well removed from launch and thought to be a period of

R. P. Stowe (✉)
Microgen Laboratories, La Marque, TX, USA
e-mail: rpstowe@microgenlabs.com

D. L. Pierson
NASA-Johnson Space Centre, Houston, TX, USA
e-mail: duane.l.pierson@nasa.gov

S. K. Mehta
NASA-Johnson Space Centre, Houston, TX, USA

JesTech., Houston, TX, USA
e-mail: satish.k.mehta@nasa.gov

© Springer Nature Switzerland AG 2020

A. Choukér (ed.), *Stress Challenges and Immunity in Space*,
https://doi.org/10.1007/978-3-030-16996-1_19

low stress (Stowe et al. 2000). More recent studies have confirmed that astronauts experience considerable stress just prior to launch (Mehta et al. 2000, 2001; Pierson et al. 2005). Thus, timing of biological sample collection relative to launch must be considered during the experimental design phase to account for preflight stress.

Although in-flight data are very limited, a pattern has emerged during the early phase of flight. A significant elevation in urinary cortisol occurred immediately after launch in one mission specialist (Leach 1987). In the Spacelab Life Sciences (SLS) missions, urinary cortisol was also significantly increased immediately after launch in both crews (Leach et al. 1996). Cortisol returned to preflight levels on flight days 2–9 of SLS-1, whereas in SLS-2 urinary cortisol was significantly elevated throughout most of the flight. In the STS-95 mission, plasma cortisol was significantly elevated after launch in one crewmember and remained elevated for the first 4 flight days (Stowe et al. 2001a). Catecholamines generally are not elevated during Shuttle spaceflights indicating little or no sympathetic adrenal activation (Smith et al. 1997), but as previously stated landing stresses result in significant elevations in plasma and urinary catecholamines.

In the later, "adaptive" phase of spaceflight, cortisol has been reported to either not change or increase. During Skylab, plasma and urinary cortisol were generally increased throughout the flight (Leach and Rambaut 1977). Plasma cortisol was increased on days 216–219 of the 237-day Salyut flight; no change was observed in urinary cortisol (Gazenko et al. 1988). Urinary cortisol was reduced on flight days 43–45 but increased on flight day 88 during the Salyut-7 (Vorobyev et al. 1986). Plasma cortisol was unchanged from preflight levels toward the end of a 241-day Mir mission (Grigoriev et al. 1990). During Mir, samples obtained from two subjects on flight days 88–186 showed an increase in urinary cortisol while four others showed a decrease (Stein et al. 1999). It has been proposed that the variability in cortisol levels during long-duration missions may be due to mission-specific stress (Stein 1999), changes in the steroidogenesis pathway (Vorobyev et al. 1986), disruption of circadian rhythm (Leach-Huntoon and Cintron 1996), and negative feedback loops (Leach et al. 1996).

In summary, multiple factors associated with spaceflight (e.g., hypergravity, microgravity, confinement, separation from family, sleep deprivation) elevate stress hormones and reflect activation of the HPA axis. There is increasing evidence that spaceflight and associated stressors result in a shift toward Th2 cytokine profile (see Chap. 12). Glucocorticoids are primarily anti-inflammatory in that they inhibit Interleukin (IL)-12 production and increase IL-10 production by monocytes which drives the immune response toward a Th2 cytokine profile (IL-4, IL-5, IL-10) and away from a Th1 profile (IL-2, IL-12, IFN-γ) (Elenkov et al. 1996). Decreased production of Th1 cytokines has been documented repeatedly in astronauts (reviewed in Taylor et al. 1997); these findings are underscored by the inhibited delayed-type hypersensitivity response observed during short-term (Taylor 1993) and long-term missions (Konstantinova et al. 1993). The hypothesis is also supported by data from Stein and Schluter (1997), who showed that IL-10 levels spike immediately after launch and correlate with increased levels of cortisol. Moreover, glucocorticoids enhance IL-4 and IL-10 synthesis resulting in increased circulating IgE levels in vivo (Zieg et al. 1994; Ramierz et al. 1996), and we have recently noted increased IgE levels after short-term spaceflights (Stowe et al. 2001a, 2003). Accordingly, it has been proposed that a Th1-to-Th2 shift may increase susceptibility to opportunistic infections such as viral pathogens (Elenkov et al. 1996). The limitation of NK functions in space (see also Chap. 13) as well as enhanced T regulatory cell function are now under investigation as a key parameters affected and the conditions of space flight and viral replication and infection. To which degree these cell effects are challenged by microgravity or cellular stress per se (Chaps. 5, 17 and 18) or by the work and life conditions (De Lorenzo et al. 2015; also see Chaps. 6 and 9) remains an open issue and likely the combined effects perpetuate risk of viral reactivation.

19.2 Reactivation of Latent Herpes Viruses in Astronauts

Although infectious disease risks associated with most pathogens (i.e., those that cause acute infections) may be reduced by a quarantine period before spaceflight, latent viruses are not mitigated by a quarantine period and are subject to intermittent reactivation. Herpesviruses, the best known latent viruses, commonly establish infections in humans. The family of herpesviridae, which includes herpes simplex type-1 and type-2 (HSV-1, -2), cytomegalovirus (CMV), Epstein-Barr virus (EBV), varicella-zoster virus (VZV), human herpesvirus-6, -7, and -8 (HHV-6, -7, -8) are all double-stranded DNA viruses that may establish a lifelong latent infection in the host. These infections are characterized by an acute phase usually associated with minor morbidity and mortality: productive viral replication occurs in a sequential manner consisting of immediate-early gene expression, transcriptional transactivation, early gene expression, viral genome replication, late gene expression, nucleocapsid assembly, viral genome packaging, and release of enveloped virions from the cell. Primary infection is followed by a chronic latent phase reflecting a balance between viral replication and the host immune response.

Herpes viruses are medically important viruses; HSV-1 infects 70–80% of all adults and is classically associated with oropharyngeal lesions such as cold sores, pharyngitis, and tonsillitis. EBV infects over 85% of the adult population and is the causative agent of infectious mononucleosis, Burkitt's lymphoma, undifferentiated nasopharyngeal carcinoma, and diffuse polyclonal B-cell lymphoma. Most CMV infections in adults are asymptomatic but may result in an infectious mononucleosis-like syndrome, central nervous system infections, and febrile illnesses. Notably, CMV infections can be severe in immunocompromised individuals such as AIDS and posttransplant patients (Komanduri et al. 2001). VZV causes chicken pox on primary infection and remains latent thereafter; VZV may reactivate resulting in episodes of zoster or "shingles" (Arvin 1996).

Herpes virus reactivation appears to be triggered by stress (Fig. 19.1). Prior ground-based studies of chronic stress have demonstrated that reduction or loss of cellular immune function results in latent herpes viruses. Glaser and colleagues have demonstrated decreased cellular immunity and increased antibodies to EBV

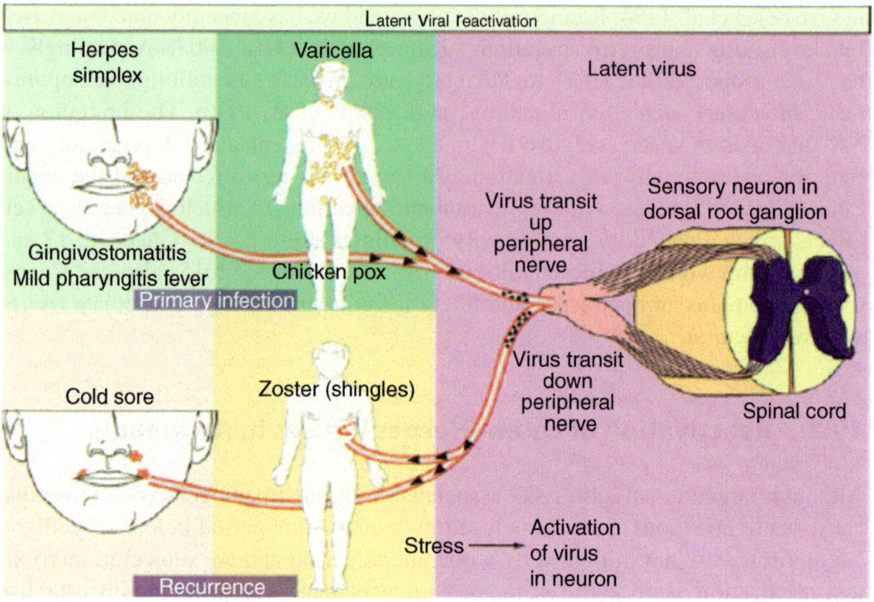

Fig. 19.1 Primary infection and reactivation of herpesvirus

in chronically stressed individuals (Glaser et al. 1985, 1991, 1993). Although virus reactivation in these individuals was not associated with clinical disease, other studies have linked psychological stress with onset and severity of infectious mononucleosis (Kasl et al. 1979). Moreover, significant immunosuppression (i.e., organ transplant recipients) correlates directly with the development of EBV-related lymphoproliferative diseases (Preiksaitis et al. 1992; Rea et al. 1994; Haque and Crawford 1997). Given the alterations in the immune system during spaceflight, a major concern is the development of an EBV-associated disease or lymphoma during long-term spaceflight. In addition, reinfection or transmission to a previously uninfected individual (resulting in primary infection) may be another concern.

We have been investigating reactivation of latent herpes viruses in astronauts by measuring antiviral antibodies and shedding of viral DNA, most notably to EBV, CMV, and VZV. The results of our studies and interpretation of the results are discussed.

19.2.1 Epstein-Barr Virus

EBV infects approximately 90% of the adult population and is closely associated with malignancies such as Burkitt's lymphoma and nasopharyngeal carcinoma (Cohen 2000). Following primary infection, EBV remains latent in healthy individuals and undergoes occasional reactivation. In previously infected (seropositive) individuals, antiviral capsid antigen (VCA) and anti-EBV nuclear antigen (EBNA)

IgG antibodies are always present. A dysfunction of the immune surveillance system against EBV infection permits productive cycles of viral replication, which lead to increased production of early antigen (EA) and VCA IgG antibodies. Increased viral shedding in saliva can also be detected.

In our initial study on latent virus reactivation, saliva samples were collected from 11 EBV-seropositive astronauts before, during, and after spaceflight (Payne et al. 1999). EBV DNA was detected more frequently before space shuttle missions than during ($P < 0.001$) or after ($P < 0.01$) missions; no significant difference in frequency was detected between the in-flight and postflight periods (Payne et al. 1999). In addition to viral DNA, the titers of antibodies to EBV antigens VCA and EA were measured in blood samples taken from astronauts at their annual medical exams (baseline), 10 days before launch (L − 10), a few hours after landing (R + 0), and 3 days after landing (R + 3). The titer of anti-VCA antibodies was greater at L − 10 than the baseline value (Stowe et al. 2000, 2001a, b; Pierson et al. 2005; Mehta et al. 2014) indicating increased viral replication. Notably, the titer of anti-EBNA antibodies (which positively correlates with T-cell function) was less than its titer at the baseline reflecting decreased cellular immunity (Stowe et al. 2001b). In a subsequent study that used quantitative real-time PCR, the mean EBV DNA copy number was significantly greater during flight than either preflight or postflight (Pierson et al. 2005). We also found that the amount of EBV DNA shed in the saliva of astronauts during spaceflight increased as the number of days in space increased. Similar data from two cosmonauts aboard the Russian space station Mir showed that EBV shedding in saliva occurred throughout the nearly 3-month mission, a much longer time than the relatively short (<14 days) duration of the shuttle missions. However, the number of EBV copies shed by cosmonauts aboard Mir did not increase as a function of days in flight, as observed with astronauts on the space shuttle. The maximum number of EBV copies shed by cosmonauts on Mir was 1130/mL of saliva, and the maximum number of copies shed by astronauts on the shorter shuttle flights was 738/mL of saliva.

Because the above results suggested that EBV was productively replicating in infected cells, we analyzed EBV gene expression in peripheral blood B lymphocytes before and immediately after spaceflight (Stowe et al. 2011). In samples from Shuttle astronauts ($n = 6$), EBV-encoded RNA (EBER)1 and β-actin (all astronaut samples; data not shown) were detected in all samples. EBV promotor EBNA1-Qp was expressed in 25% of the samples, a rate just slightly higher than in the controls ($n = 24$). Low-level latent membrane protein (LMP-1) and transcription of the transactivator protein BZLF1 was somewhat more frequent than in the controls. Unlike the control group, anti-apoptotic protein BHRF1 was frequently transcribed, sometimes at high levels. Notably, transcripts of LMP1, BZLF1, and BHRF1 were expressed both before (L − 10) and after (R + 0) spaceflight. Transcription of LMP2A, EBNA1-Cp/Wp, EBNA2, EBNA-1-Fp, SM, BALF5 (the EBV DNA polymerase), and gp220 (a polymerase-dependent virion envelope glycoprotein) was absent from all samples. Altogether, there was a significant increase in the number of immediate early and early gene transcripts in Shuttle astronaut samples at L − 10 and R + 0 as compared to healthy control samples.

EBER1 was detected in all samples from astronauts ($n = 6$) from the International Space Station (ISS). Fifty percent of samples expressed EBNA1-Qp, which was greater than the controls (17%) and the Shuttle astronauts (25%). Transcription of the latency III genes (EBNA1-Cp/Wp and EBNA2) was detected in samples from two ISS astronauts. Among the replicative EBV genes, early BHRF1 was transcribed in 9 of 12 ISS astronaut samples (4 samples were positive at L − 10). Transcription of early replicative SM was detected in four of six subjects, including two of them prior to flight. Early replicative BALF5 and late replicative gp220 transcripts were evident for the first time in samples from four of the six ISS astronauts, and they were found only in samples obtained immediately after spaceflight. The co-expression of BALF5 with gp220 in R + 0 samples from three ISS astronauts is strong evidence that complete productive EBV replication is occurring in the peripheral blood B-lymphocytes of these astronauts. Overall, there was a significant increase in the number of immediate early and early gene transcripts in ISS astronaut samples as compared to healthy control samples (L − 10 and R + 0) and Shuttle astronauts (R + 0 only). In addition, the number of latent and late lytic gene transcripts was also significantly increased at R + 0 as compared to healthy control samples.

Although the clinical significance of these findings remains to be determined, they are potentially significant for the health of astronauts who will spend long periods in space. Chronically high levels of latent EBV gene expression and productive EBV replication are etiologically linked to the development of a variety of EBV-associated diseases. Latent herpesvirus reactivation, in light of immune dysregulation in astronauts, could pose a health risk for crewmembers following prolonged interplanetary space travel, in particular due to the lack of specialized medical facilities in case of injury or illness.

19.2.2 Cytomegalovirus

Another latent herpesvirus, human CMV, may pose a similar kind of risk to astronauts' health during spaceflight. CMV infection is typically acquired asymptomatically during childhood. However, in individuals whose immune system is either immature or immunocompromised such as HIV infection (Dittmer et al. 2017), CMV can cause multiple diseases such as encephalitis, gastroenteritis, pneumonia, and chorioretinitis (Fiala et al. 1975). Moreover, several studies have suggested that CMV infection may contribute to preexisting immunosuppression by directly infecting leukocytes as well as hematopoietic cells (Carney and Hirsch 1981; De Pelsmaeker et al. 2018; Rice et al. 1984; Simmons et al. 1990). We therefore examined the effects of spaceflight on CMV reactivation and shedding.

In our first study of CMV in astronauts, CMV reactivation and shedding was examined in urine samples collected from 71 crew members from short-term space missions (Mehta et al. 2000). The frequency of CMV DNA shedding in either pre- or postflight urine samples from astronauts was significantly higher than that of the control population ($P < 0.05$). CMV DNA was detected in 27% of the crew members studied during the mission monitoring period. By contrast, only 1 of the 61 control subjects shed CMV during one sampling period. We had an opportunity to sample both blood and urine from two astronauts during a subsequent flight. Consistent with the earlier study (Mehta et al. 2000), CMV was shed in urine of one crew member before, during, and after flight, and in urine of the other crew member only during flight (on 2 days) (Stowe et al. 2001a).

Of the 71 crew members, 55 (77%) were seropositive. As a group, no significant changes in anti-CMV antibodies were found postflight. However, upon dividing these subjects between 40 nonshedders and 15 CMV shedders, an interesting difference was found. No significant change in CMV IgG antibody titer of nonshedders was found at any time point compared to the baseline values. In contrast, the anti-CMV antibody titer of the 15 shedders was significantly increased at all time points compared to their baseline values. Furthermore, anti-CMV antibodies was significantly increased at R + 3 compared to L − 10. The anti-CMV antibody titers of the control subjects did not differ significantly from the baseline levels of astronauts, and no changes in anti-CMV IgG antibody titer from 11 CMV-seropositive control subjects were found across three sampling times (Mehta et al. 2000). Overall the results of our CMV studies demonstrated that CMV reactivation occurred in astronauts and may pose another health problem during longer-duration missions.

19.2.3 Varicella Zoster Virus

VZV is a highly successful human pathogen. Primary VZV infection typically results in childhood varicella (chickenpox). Varicella is characterized by malaise, fever, and an extensive vesicular rash. Varicella is normally self-limiting and resolves with the development of humoral and cell-mediated immunity. VZV is transmitted by direct contact with vesicle fluid or aerosolized respiratory tract secretions (Kavaliotis et al. 1998). VZV DNA is present in the air in hospital rooms of varicella or zoster patients (Sawyer et al. 1994). While the air and even remote surfaces around zoster patients can contain VZV DNA (Yoshikawa et al. 2001), only recently has infectious virus been recovered in saliva from a zoster patient (Mehta et al. 2008). VZV is highly contagious with a secondary attack rate approaching 100% (Asano et al. 1977) that can lead to epidemics in places of close confinement.

While widespread aggressive vaccination has greatly lessened the morbidity and mortality associated with varicella (Arvin and Gershon 1996), in countries where vaccination is not mandatory, deaths still occur (Rawson et al. 2001).VZV reactivation, predominately in the elderly, most often results in zoster (shingles), but virus may spread to the spinal cord to cause myelitis or to blood vessels of the brain to cause vasculopathy. VZV reactivation often produces prolonged pain after zoster (postherpetic neuralgia). While the clinical features of VZV reactivation are well recognized, subclinical VZV reactivation and shedding has recently been reported in astronauts. Physical and physiological stressors associated with spaceflight (Taylor and Janney 1992; White and Averner 2001; Williams 2003) appear to induce virus reactivation and subsequent shedding of VZV in saliva (Mehta et al. 2004;

Cohrs et al. 2008). Herein, asymptomatic VZV reactivation during spaceflight, in ground-based spaceflight analogs, and in the astronauts before, during, and after short-duration spaceflight is reviewed.

19.2.3.1 Asymptomatic VZV Reactivation and Virus Transmission

Prolonged maintenance of anti-VZV antibodies in small, isolated populations supports the notion of subclinical reactivation; however, this finding equally supports periodic boost in anti-VZV titer due to subclinical reactivation of latent VZV (Hope-Simpson 1965; Black et al. 1974). The eyes are a likely site to demonstrate subclinical VZV reactivation since, both HSV-1 and VZV become latent in trigeminal ganglia (Cohrs et al. 2005) and even within the same neuron (Theil et al. 2003), both viruses cause keratitis following reactivation (Reijo et al. 1983; Cook et al. 1986; Kaye et al. 2000). However, using equally sensitive PCR technology, VZV DNA was not detected in tear film from 35 normal subjects (Yamamoto et al. 1994; Willoughby et al. 2002) or conjunctival scrapings from 30 individuals with eye trauma or unrelated disease (Lee-Wing et al. 1999). Thus, unlike HSV-1, where asymptomatic virus reactivates often and can be detected frequently in tear film, asymptomatic VZV reactivation is either a rare event or does not progress to yield virus in the eyes. The most obvious signs of VZV reactivation are the vesicular rash and the pain associated with zoster; however, even in the absence of rash, the virus is active and can spread to the retina causing blindness, to the spinal cord causing paralysis and incontinence, and to cerebral arteries resulting in stroke (Kleinschmidt-DeMasters and Gilden 2001; Orme et al. 2007). Associating VZV with a disease asymptomatically can be challenging. For example, when stroke occurs in the elderly, especially many months following zoster, the association with VZV reactivation requires cerebral spinal fluid (CSF) analysis for VZV antibodies (Nagel et al. 2007). Likewise, detection of asymptomatic VZV reactivation, which often is only seen as an increase in antibody titer against VZV (but may also result in virus transmission), is difficult to detect. In such instances, virological verification of VZV disease has relied on the detection of VZV DNA or anti-VZV IgG antibodies in CSF or, less often, the presence of VZV DNA in blood mononuclear cells or anti-VZV IgM antibodies in serum.

In an effort to understand the prevalence of virus reactivation in the unique environment of space, DNA extracted from 312 saliva samples were collected from eight astronauts (two short-term shuttle missions). VZV DNA was detected in only 1 of 112 saliva samples taken 234–265 days before flight, whereas during and shortly after spaceflight 61 of 200 saliva samples were positive for the virus DNA (Mehta et al. 2004). No VZV DNA was detected in saliva from 10 age-matched control subjects (88 samples). Subsequently, VZV DNA was again detected in saliva samples from two of three space shuttle astronauts during and within days following spaceflight, but not before launch (Cohrs et al. 2008). Importantly, infectious VZV was isolated in saliva samples following landing from the two astronauts whose saliva contained VZV DNA. These results suggest asymptomatic VZV reactivation does occur, and saliva is a convenient source material for its detection. While it is impossible to determine if the virus in both astronauts resulted from simultaneous reactivation, or if a single reactivation and shed in one subject resulted in asymptomatic infection of the second subject, the presence of a rare restriction endonuclease recognition site in the VZV DNA isolated from both subjects argues for the latter case.

19.2.3.2 Clinical Importance of Asymptomatic VZV Shedding

The most obvious sign of VZV reactivation is vesicular rash and pain associated with zoster (Fig. 19.2); however, in the absence of rash, the virus can also spread to the retina causing blindness, to the spinal cord causing paralysis and incontinence, and to cerebral arteries resulting in stroke (Kleinschmidt-DeMasters and Gilden 2001; Orme et al. 2007). While associating VZV with disease is straightforward when rash and vesicles are present, correlating VZV with disease is difficult when the outward signs are absent. For example, when stroke occurs in the elderly, especially many months following zoster, the association with VZV reactivation requires cerebral spinal fluid analysis for VZV antibodies (Nagel et al. 2007). Likewise, detection of asymptomatic VZV reactivation which often is only seen as an increase in antibody titer against VZV (but may also result in virus transmission) is difficult to detect.

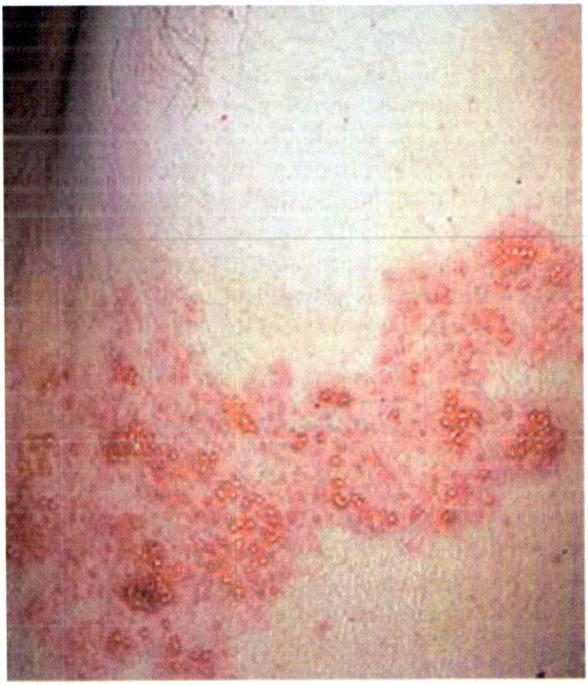

Fig. 19.2 The characteristic painful rash of shingles

19.2.4 Viral Reactivation in Long Duration Spaceflight

Recently, reactivation of latent herpes viruses, EBV, VZV, and CMV was reported in 23 astronauts (18 male and 5 female) before, during, and after long-duration (up to 180 days) spaceflight onboard the ISS (Crucian et al. 2016; Mehta et al. 2017). These viruses reactivated independently of each other. Reactivation of EBV, VZV, and CMV increased in frequency, duration, and amplitude (viral copy numbers) when compared to short duration (10–16 days) Space Shuttle missions. No evidence of reactivation of HSV-1, HSV-2, or human herpes virus 6 was found. The mean diurnal trajectory of salivary cortisol changed significantly during flight as compared to before flight ($P = 0.010$). There was no statistically significant difference in levels of plasma cortisol or DHEA concentrations among time points before, during, and after flight for these ISS crew members, although observed cortisol levels were lower at the mid- and late-flight time points. The data confirm that astronauts undertaking long-duration spaceflight experience both increased latent viral reactivation and changes in diurnal trajectory of salivary cortisol concentrations. Moreover cumulative case reports indicate also that other immunotropic stress response systems are activated in space and might have aggravated such steroid hormone compound related changes. Here it was observed that endogenous peripheral endocannabinoids (see Chap. 10) are increasingly produced during long-duration mission on the ISS (Strewe et al. 2012).

Spaceflight Analog Studies: In a recent collaborative study with Alexander Choukér (physician researcher and PI in ESA lead projects), 19 subjects at Concordia Station in Antarctica were examined during overwintering for their latent viral reactivation and shedding patterns in their saliva samples collected each month for before, during, and after winter-over period. VZV DNA was found by real time polymerase assay in 10 out of 19 subjects (52%) (Fig. 19.3) with most of the shedding occurring during the study than before or afterwards. This is about the same rate of VZV shedding as we have found in astronauts during short and long duration spaceflights (50–65%) (Mehta et al. 2004). These data suggest that "wintering over" in Antarctica is an excellent analog to spaceflight for studying the efficacy of a countermeasure against viral shedding.

19.2.5 Clinical Application of NASA-Developed Technology

The PCR assay is specific, sensitive, and rapid (same day results), but a skilled operator and an expensive equipment are required. We developed a new kit for physicians that can detect the presence of VZV DNA in saliva (or urine) in ~15–20 min with no additional equipment needed (Provisional Patent Application). This technology is especially useful for an individual with dermatomal pain of unknown origin. The primary application for this technology is for early diagnosis of shingles patients. Typically, individuals destined to develop shingles experience pain in a dermatomal region with no apparent cause. If the pain is due to the onset of shingles, early diagnosis with this technology allows early intervention with prevention

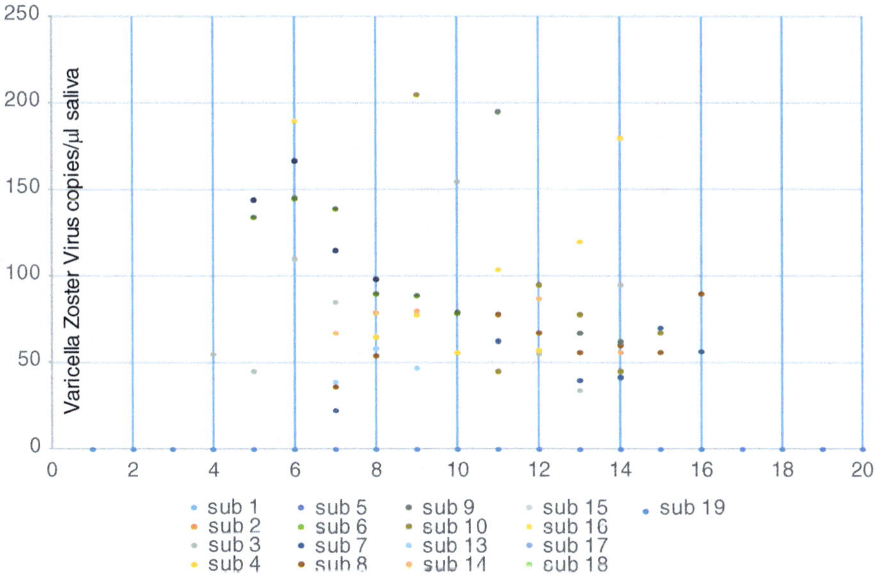

Fig. 19.3 Shedding of Varicella Zoster Virus in the saliva of 19 expeditioners in a Concordia study

or blunting of the zoster rash and less nerve damage. Early treatment will likely reduce the number of postherpetic neuralgia patients.

The PCR protocol developed from spaceflight flights at NASA was used for detection of VZV in saliva to diagnose a case of varicella (chicken pox) in an older adult (Mehta et al. 2012). This patient presented with a single vesicle on her back. The next day, the lesions had spread, and both vesicles and saliva were analyzed for VZV DNA. Although a clinical diagnosis was uncertain on the third day, PCR revealed a high copy number of VZV DNA in both vesicles and saliva. Importantly, immediate antiviral treatment was followed by a quick recovery without the complicating prolonged fatigue and weakness that is characteristically seen in adults with varicella. This is the first case of varicella in an adult in which PCR was used to diagnose disease and in which genotyping of VZV DNA recovered from a patient and her spouse, who had zoster 1 week earlier, confirmed spouse-to-patient transmission. Similarly, this assay was used to diagnose a case of zoster in a 21-year-old female patient before the full blown symptoms appeared (Mehta et al. 2008). This study demonstrated the practical application of NASA-developed technology to everyday life.

VZV DNA has been detected in the saliva samples from patients with acute zoster (Mehta et al. 2008), zoster sine herpete, chickenpox (Mehta et al. 2013) and PHN even in the absence of clinical symptoms, which now makes diagnosis less invasive and less time consuming. In fact, a rapid and sensitive virus detection method has been developed and used to detect virus in saliva samples taken from asymptomatic patients with neurologic and other VZV related disease (Mehta et al.

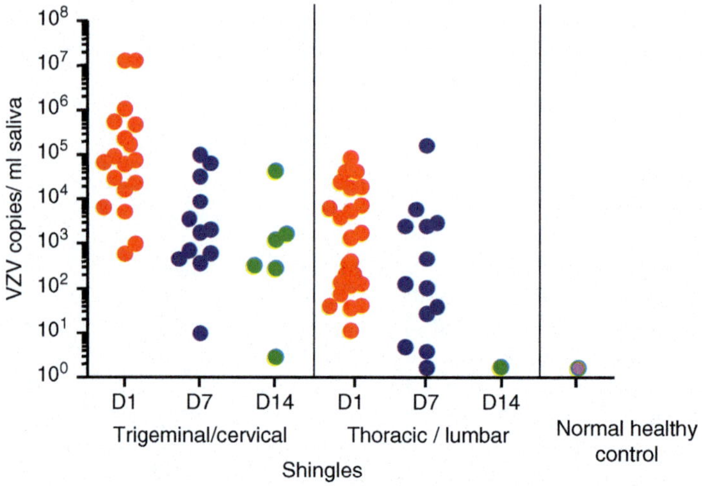

Fig. 19.4 VZV copies in shingles patients and in normal healthy controls

2013). Figure 19.4 illustrates VZV copy numbers in saliva from shingles patients before antiviral treatment. For this method, the saliva is collected by passive drool or by way of a synthetic swab and then processed for DNA within an hour from sample collection. The results from a few studies using this technique have shown that VZV DNA is present in 100% of patients tested before antiviral treatment and is exclusively in the cell pelleted fraction of saliva. These studies further showed that VZV, isolated from zoster patient saliva, was primarily associated with the epithelial cell membrane but could also be inside the cell. Epithelial cells with VZV continued to be present in the saliva of a single zoster patient up to 10 months after recovery. These kinds of studies are ongoing and our spaceflight-developed technology for rapid viral detection continues to be used locally and around the world for patients with zoster (Mehta et al. 2013), chicken pox (Mehta et al. 2008), PHN, multiple sclerosis (Ricklin et al. 2013), and various other neurological disorders (Gilden et al. 2010).

19.3 Summary

Spaceflight represents a unique environment that is comprised of multiple stresses before, during, and after flight and result in activation of the HPA and SAM axes and subsequent increased circulating concentrations of cortisol, catecholamines, and neuropeptides. Many immune functions are consistently diminished in astronauts immediately after spaceflight, and a spaceflight-induced shift to a Type 2 cytokine pattern is becoming evident. Alterations in immunity and stress hormone levels result in reactivation of latent herpes viruses. The medical significance of altered immunity and increased, but asymptomatic viral reactivation in astronauts is unknown, but it is possible that spaceflight-induced changes in immunity may increase during the exploration missions, some of which will have much longer durations than current ISS missions. Further decreases in immunity may result in greater risks of active infection by opportunistic pathogens, including latent herpes viruses. These studies of asymptomatic reactivation of latent herpes viruses in an extremely healthy group of subjects may provide new insight into stress, immunity, and viral disease in the general population.

References

Arvin AM (1996) Varicella-zoster virus. Clin Microbiol Rev 9(3):361–381

Arvin AM, Gershon AA (1996) Live attenuated varicella vaccine. Annu Rev Microbiol 50:59–100

Asano Y, Nakayama H et al (1977) Protection against varicella in family contacts by immediate inoculation with live varicella vaccine. Pediatrics 59(1):3–7

Black FL, Hierholzer WJ et al (1974) Evidence for persistence of infectious agents in isolated human populations. Am J Epidemiol 100(3):230–250

Caillot-Augusseau A, Lafage-Proust MH et al (1998) Bone formation and resorption biological markers in cosmonauts during and after a 180-day space flight (Euromir 95). Clin Chem 44(3):578–585

Carney WP, Hirsch MS (1981) Mechanisms of immunosuppression in cytomegalovirus mononucleosis. II. Virus-monocyte interactions. J Infect Dis 144(1):47–54

Cohen JI (2000) Epstein-Barr virus infection. N Engl J Med 343(7):481–492

Cohrs RJ, Laguardia JJ et al (2005) Distribution of latent herpes simplex virus type-1 and varicella zoster virus DNA in human trigeminal Ganglia. Virus Genes 31(2):223–227

Cohrs RJ, Mehta SK et al (2008) Asymptomatic reactivation and shed of infectious varicella zoster virus in astronauts. J Med Virol 80(6):1116–1122

Cook SD, Aitken DA et al (1986) Herpes simplex virus in the cornea; an ultrastructural study on viral reactivation. Trans Ophthalmol Soc U K 105(Pt 6):634–641

Crucian B, Johnston S et al (2016) A case of persistent skin rash and rhinitis with immune system dysregulation onboard the International Space Station. J Allergy Clin Immunol Pract 4(4):759–762

De Lorenzo BH, de Oliveira Marchioro L et al (2015) Sleep-deprivation reduces NK cell number and function mediated by β-adrenergic signalling. Psychoneuroendocrinology 57:134–143

De Pelsmaeker S, Romero N et al (2018) Herpesvirus evasion of natural killer cells. J Virol 92:e02105–e02117

Dittmer DP, Tamburro K et al (2017) Oral shedding of herpesviruses in HIV-infected patients with varying degrees of immune status. AIDS 31(15):2077–2084

Elenkov IJ, Papanicolaou DA et al (1996) Modulatory effects of glucocorticoids and catecholamines on human interleukin-12 and interleukin-10 production: clinical implications. Proc Assoc Am Physicians 108(5):374–381

Fiala M, Payne JE et al (1975) Epidemiology of cytomegalovirus infection after transplantation and immunosuppression. J Infect Dis 132(4):421–433

Gazenko OG, Schulzhenko EB et al (1988) Review of basic medical results of the Salyut-7–Soyuz-T 8-month manned flight. Acta Astronaut 17(2):155–160

Gilden D, Cohrs RJ et al (2010) Neurological disease produced by varicella zoster virus reactivation without rash. Curr Top Microbiol Immunol 342:243–253

Glaser R, Kiecolt-Glaser JK et al (1985) Stress, loneliness, and changes in herpesvirus latency. J Behav Med 8(3):249–260

Glaser R, Pearson GR et al (1991) Stress-related activation of Epstein-Barr virus. Brain Behav Immun 5(2):219–232

Glaser R, Pearson GR et al (1993) Stress and the memory T-cell response to the Epstein-Barr virus in healthy medical students. Health Psychol 12(6):435–442
Grigoriev AJ et al (1990) Medical results of the fourth prime expedition on the orbital station Mir. In: Proceedings of the 4th European Symposium on Life Sciences Research in Space, Trieste, Italy, 28 May–1 June 1990
Haque T, Crawford DH (1997) PCR amplification is more sensitive than tissue culture methods for Epstein-Barr virus detection in clinical material. J Gen Virol 78(Pt 12):3357–3360
Hope-Simpson RE (1965) The nature of Herpes Zoster: a long-term study and a new hypothesis. Proc R Soc Med 58:9–20
Kasl SV, Evans AS et al (1979) Psychosocial risk factors in the developmental of infectious mononucleosis. Psychosom Med 41(6):445–466
Kavaliotis J, Loukou I et al (1998) Outbreak of varicella in a pediatric oncology unit. Med Pediatr Oncol 31(3):166–169
Kaye SB, Baker K et al (2000) Human herpesviruses in the cornea. Br J Ophthalmol 84(6):563–571
Kleinschmidt-DeMasters BK, Gilden DH (2001) Varicella-Zoster virus infections of the nervous system: clinical and pathologic correlates. Arch Pathol Lab Med 125(6):770–780
Komanduri KV, Feinberg J et al (2001) Loss of cytomegalovirus-specific CD4+ T cell responses in human immunodeficiency virus type 1-infected patients with high CD4+ T cell counts and recurrent retinitis. J Infect Dis 183(8):1285–1289
Konstantinova IV, Rykova MP et al (1993) Immune changes during long-duration missions. J Leukoc Biol 54(3):189–201
Leach CS (1987) Fluid control mechanisms in weightlessness. Aviat Space Environ Med 58(9 Pt 2):A74–A79
Leach CS, Rambaut PC (1977) Biochemical responses of the Skylab crewmen: an overview. In: Johnston RS, Dietlein LF (eds) Biomedical results from Skylab (NASA SP-377). National Aeronautics and Space Administration, Washington, DC, pp 204–216
Leach CS, Alfrey CP et al (1996) Regulation of body fluid compartments during short-term spaceflight. J Appl Physiol 81(1):105–116
Leach-Huntoon CS, Cintron NM (1996) Endocrine system and fluid and electrolyte balance. In: Leach Huntoon CS, Antipov VV, Grigoriev AI (eds) Humans in spaceflight, vol III, 1st edn. American Institute of Aeronautics and Astronautics, Reston, pp 89–104
Lee-Wing MW, Hodge WG et al (1999) The prevalence of herpes family virus DNA in the conjunctiva of patients positive and negative for human immunodeficiency virus using the polymerase chain reaction. Ophthalmology 106(2):350–354
Mehta SK, Stowe RP et al (2000) Reactivation and shedding of cytomegalovirus in astronauts during spaceflight. J Infect Dis 182(6):1761–1764
Mehta SK, Kaur I et al (2001) Decreased non-MHC-restricted (CD56+) killer cell cytotoxicity after spaceflight. J Appl Physiol 91(4):1814–1818
Mehta SK, Cohrs RJ et al (2004) Stress-induced subclinical reactivation of varicella zoster virus in astronauts. J Med Virol 72(1):174–179
Mehta SK, Tyring SK et al (2008) Varicella-zoster virus in the saliva of patients with herpes zoster. J Infect Dis 197(5):654–657
Mehta SK, Gilden D et al (2012) A case report: PCR-assisted diagnosis of varicella in an adult. Open J Med Microbiol 2:3
Mehta SK, Tyring SK et al (2013) Rapid and sensitive detection of varicella zoster virus in saliva of patients with herpes zoster. J Virol Methods 193:128–130
Mehta SK, Laudenslager ML et al (2014) Multiple latent viruses reactivate in astronauts during Space Shuttle missions. Brain Behav Immun 41:210–217
Mehta SK, Laudenslager ML et al (2017) Latent virus reactivation in astronauts on the International Space Station. NPJ Microgravity 3:11
Nagel MA, Forghani B et al (2007) The value of detecting anti-VZV IgG antibody in CSF to diagnose VZV vasculopathy. Neurology 68(13):1069–1073
Orme HT, Smith AG et al (2007) VZV spinal cord infarction identified by diffusion-weighted MRI (DWI). Neurology 69(4):398–400

Payne DA, Mehta SK et al (1999) Incidence of Epstein-Barr virus in astronaut saliva during spaceflight. Aviat Space Environ Med 70(12):1211–1213
Pierson DL, Stowe RP et al (2005) Epstein-Barr virus shedding by astronauts during space flight. Brain Behav Immun 19(3):235–242
Preiksaitis JK, Diaz-Mitoma F et al (1992) Quantitative oropharyngeal Epstein-Barr virus shedding in renal and cardiac transplant recipients: relationship to immunosuppressive therapy, serologic responses, and the risk of posttransplant lymphoproliferative disorder. J Infect Dis 166(5):986–994
Ramierz F, Fowell DJ et al (1996) Glucocorticoids promote a TH2 cytokine response by CD4+ T cells in vitro. J Immunol 156(7):2406–2412
Rawson H, Crampin A et al (2001) Deaths from chickenpox in England and Wales 1995–7: analysis of routine mortality data. BMJ 323(7321):1091–1093
Rea D, Delecluse HJ et al (1994) Epstein-Barr virus latent and replicative gene expression in posttransplant lymphoproliferative disorders and AIDS-related non-Hodgkin's lymphomas. French Study Group of Pathology for HIV-associated Tumors. Ann Oncol 5(Suppl 1):113–116
Reijo A, Antti V et al (1983) Endothelial cell loss in herpes zoster keratouveitis. Br J Ophthalmol 67(11):751–754
Rice GP, Schrier RD et al (1984) Cytomegalovirus infects human lymphocytes and monocytes: virus expression is restricted to immediate-early gene products. Proc Natl Acad Sci U S A 81(19):6134–6138
Ricklin ME, Lorscheider J et al (2013) T-cell response against varicella-zoster virus in fingolimod-treated MS patients. Neurology 81:174–181
Sawyer MH, Chamberlin CJ et al (1994) Detection of varicella-zoster virus DNA in air samples from hospital rooms. J Infect Dis 169(1):91–94
Simmons P, Kaushansky K et al (1990) Mechanisms of cytomegalovirus-mediated myelosuppression: perturbation of stromal cell function versus direct infection of myeloid cells. Proc Natl Acad Sci U S A 87(4):1386–1390
Smith SM, Krauhs JM et al (1997) Regulation of body fluid volume and electrolyte concentrations in spaceflight. Adv Space Biol Med 6:123–165
Stein TP (1999) Nutrition and muscle loss in humans during spaceflight. Adv Space Biol Med 7:49–97
Stein TP, Schluter MD (1997) Human skeletal muscle protein breakdown during spaceflight. Am J Phys 272(4 Pt 1):E688–E695
Stein TP, Leskiw MJ et al (1999) Protein kinetics during and after long duration spaceflight on MIR. Am J Phys 276(6 Pt 1):E1014–E1021
Stowe RP, Pierson DL et al (2000) Stress-induced reactivation of Epstein-Barr virus in astronauts. Neuroimmunomodulation 8(2):51–58
Stowe RP, Mehta SK et al (2001a) Immune responses and latent herpesvirus reactivation in spaceflight. Aviat Space Environ Med 72(10):884–891
Stowe RP, Pierson DL et al (2001b) Elevated stress hormone levels relate to Epstein-Barr virus reactivation in astronauts. Psychosom Med 63(6):891–895
Stowe RP, Sams CF et al (2003) Effects of mission duration on neuroimmune responses in astronauts. Aviat Space Environ Med 74(12):1281–1284
Stowe RP, Kozlova EV et al (2011) Unrestricted latent and lytic Epstein-Barr virus gene expression in the peripheral blood of astronauts. J Med Virol 83(6):1071–1077
Strewe C, Feuerecker M et al (2012) Effects of parabolic flight and spaceflight on the endocannabinoid system in humans. Rev Neurosci 23(5–6):673–680
Taylor GR (1993) Immune changes during short-duration missions. J Leukoc Biol 54(3):202–208
Taylor GR, Janney RP (1992) In vivo testing confirms a blunting of the human cell-mediated immune mechanism during space flight. J Leukoc Biol 51(2):129–132
Taylor GR, Konstantinova I et al (1997) Changes in the immune system during and after spaceflight. Adv Space Biol Med 6:1–32
Theil D, Paripovic I et al (2003) Dually infected (HSV-1/VZV) single neurons in human trigeminal ganglia. Ann Neurol 54(5):678–682

Vorobyev YI, Gazenko OG et al (1986) Preliminary results of medical investigations during 5-month spaceflight aboard Salyut-7–Soyuz-T orbital complex. Kosm Biol Aviakosm Med 20:27–34

White RJ, Averner M (2001) Humans in space. Nature 409(6823):1115–1118

Williams DR (2003) The biomedical challenges of space flight. Annu Rev Med 54:245–256

Willoughby CE, Baker K et al (2002) Epstein-Barr virus (types 1 and 2) in the tear film in Sjogren's syndrome and HIV infection. J Med Virol 68(3):378–383

Yamamoto S, Shimomura Y et al (1994) Detection of herpes simplex virus DNA in human tear film by the polymerase chain reaction. Am J Ophthalmol 117(2):160–163

Yoshikawa T, Ihira M et al (2001) Rapid contamination of the environments with varicella-zoster virus DNA from a patient with herpes zoster. J Med Virol 63(1):64–66

Zieg G, Lack G et al (1994) In vivo effects of glucocorticoids on IgE production. J Allergy Clin Immunol 94(2 Pt 1):222–230

Stress and Radiation Responsiveness 20

Marjan Moreels, Bjorn Baselet, Olivier Van Hoey,
Filip Vanhavere, and Sarah Baatout

20.1 The Space Radiation Environment

20.1.1 Introduction

Ionizing radiation is radiation having enough energy to induce cell damage through ionization. This can be in the form of photons such as gammas or X-rays or subatomic particles such as electrons, muons, neutrons, protons, and heavier nuclei. A first important characteristic of radiation is its energy, which is typically in mega-electronvolt (MeV). One electronvolt is the energy an electron gains during acceleration by a one volt potential difference. A second important characteristic of radiation is the linear energy transfer (LET), typically in keV/μm. The LET specifies the amount of energy deposited per unit of length. The higher the LET of the radiation, the more complex the cell damage it creates and the more harmful the radiation is. The LET increases with increasing mass and increasing charge. Therefore, heavy nuclei have high LET and are a very harmful radiation type (Fig. 20.1). A third important characteristic of radiation is its intensity. For radiation protection purposes the radiation intensity is expressed in terms of the effective dose rate, typically in microsievert per hour (μSv/h) (ICRP 2007). The effective dose takes into account the amount of energy the radiation deposits in the different tissues, the harmfulness of the radiation type and the sensitivity of the exposed tissues and is proportional to the chance for developing radiation-induced cancer. Based on epidemiological studies this chance is estimated at about 5% per sievert.

M. Moreels (✉) · B. Baselet · S. Baatout
Radiobiology Unit, Belgian Nuclear Research Centre, SCK•CEN, Mol, Belgium
e-mail: marjan.moreels@sckcen.be

O. Van Hoey · F. Vanhavere
Radiation Protection, Dosimetry and Calibration, Belgian Nuclear Research Centre, SCK•CEN, Mol, Belgium

© The Author(s) 2020

A. Choukér (ed.), *Stress Challenges and Immunity in Space*,
https://doi.org/10.1007/978-3-030-16996-1_20

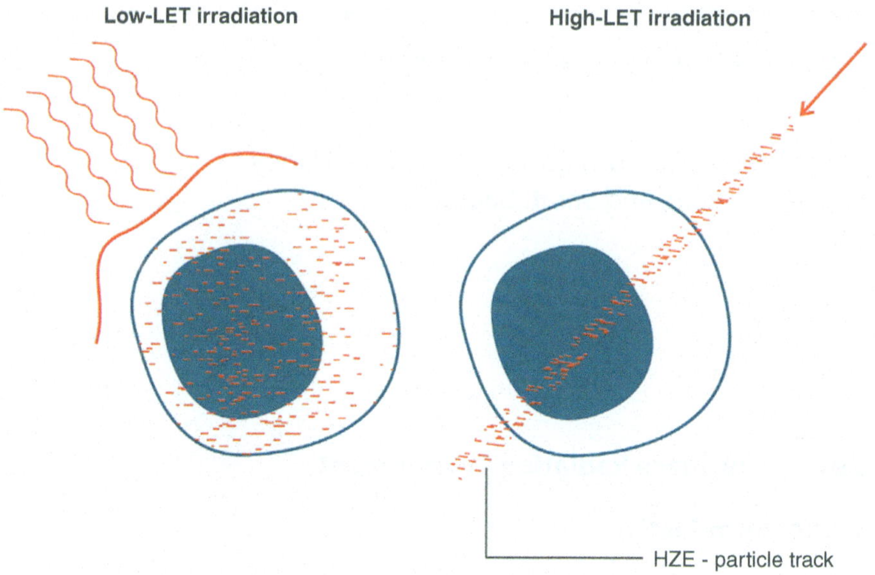

Fig. 20.1 Comparative diagram on DNA damage induced by low- and high-LET radiation. HZE particles, also called "densely ionizing radiation" typically deposit a large amount of their energy along linear tracks referred to as cores, while the remaining energy is deposited radially and uniformly by secondary electrons (i.e. Delta-rays). In contrast, low-LET radiations deposit their energy uniformly and are often referred as "sparsely ionizing radiation" (Cortese et al. 2018)

Ionizing radiation is omnipresent. Even on Earth we are continuously exposed to gamma radiation from natural radionuclides in the soil and to neutrons and muons created in the atmosphere by cosmic radiation from the sun and supernova remnants. The typical dose rate on Earth is about 0.1 μSv/h. This gives rise to a yearly dose due to natural background radiation of about 1 mSv/year. Radiation workers in hospitals and nuclear power plants are exposed to artificial radiation sources with dose rates up to a few μSv/h. However, their yearly dose should remain below the legal effective dose limit of 20 mSv.

In space the dose rate is much higher than on Earth because the protection by the Earth's atmosphere and the geomagnetic field against cosmic radiation is very limited in Low Earth Orbit (LEO) or even absent in deep space (Hassler et al. 2014; Kleiman 2012). In the International Space Station (ISS) the dose rate is typically about 20 μSv/h or 200 times higher than on Earth. For a typical stay of 6 months this leads to a dose of about 100 mSv, which is five times higher than the yearly legal dose limit for radiation workers on Earth. The dose rate on the surface of Mars is slightly higher, about 25 μSv/h or 250 times higher than on Earth. In deep space the dose rate is about 75 μSv/h or 750 times higher than on Earth. A typical manned Mars mission scenario with 180 days transit to Mars, 500 days on the Mars surface and 180 days transit back to Earth would lead to a total dose of about 1 Sv and thus about 5% risk of radiation-induced cancer (Hassler et al. 2014; Cucinotta et al. 2017). During strong solar storms the dose rate in space can temporarily increase a

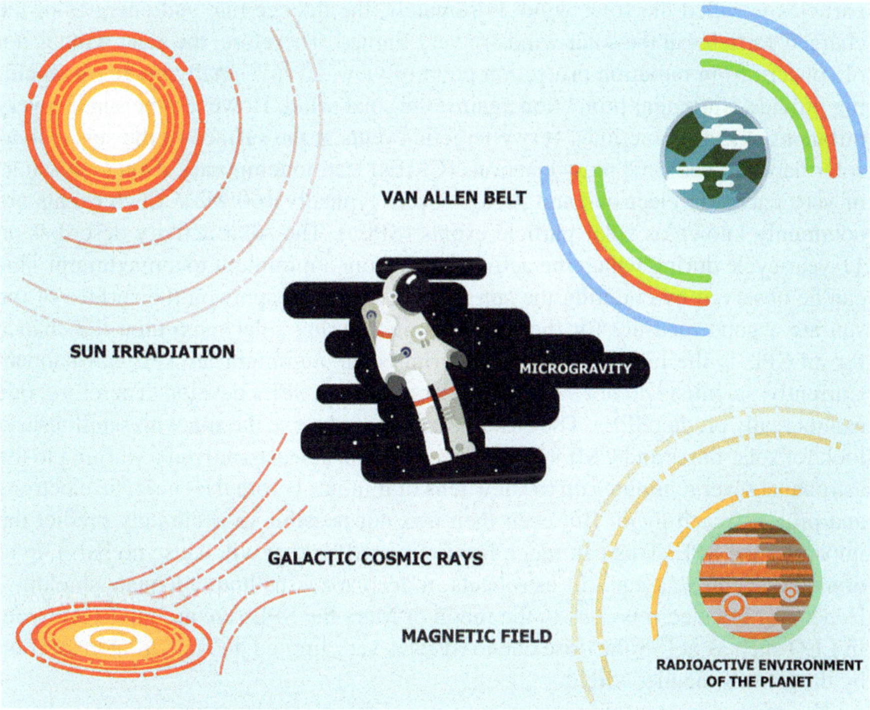

Fig. 20.2 Major sources of space radiation. The space radiation comes from three major sources including galactic cosmic rays, sun radiation, and Van Allen radiation belts of the Earth (Cortese et al. 2018)

few orders of magnitude. Without adequate shielding astronauts on the surface of the moon or Mars or in transit could be exposed to doses leading to serious health effects or even death (Benton and Benton 2001).

The radiation environment in space is also very different from that on Earth (Fig. 20.2) (Benton and Benton 2001). It consists of electrons, muons, neutrons, protons, and heavier nuclei up to extremely high energies. Furthermore, the radiation environment depends strongly on the solar activity, local shielding, and the location in space. This makes it very challenging to predict and monitor radiation doses received by astronauts. There is also much more uncertainty on the radiation-induced health effects because the epidemiological data for exposure to this type of radiation is very scarce.

20.1.2 Primary Cosmic Radiation

There are two primary sources of cosmic radiation: the sun and supernova remnants (Benton and Benton 2001). The sun is continuously emitting energetic electrons, protons, and a limited amount of heavier nuclei. This continuous flux of charged particles is called the solar wind. Fortunately, the fluence rate and energies of the charged particles in the solar wind are very limited. Therefore, the solar wind is not of concern from radiation protection point of view. Even a small amount of shielding provides sufficient protection against the solar wind. However, the sun is a very turbulent object. Sometimes very energetic events at the surface of the sun such as solar flares and coronal mass ejections (CMEs) lead to temporary strong emissions of very energetic electrons and protons up to typically 100 MeV. Such events are commonly known as solar particle events (SPEs). The solar activity describes an 11-year cycle during which the activity goes from a minimum to a maximum. This can be observed by counting the sun spots. These dark spots on the surface of the sun are a good measure for the solar activity. During solar maximum the chance for an SPE is the highest. But even during solar minimum an SPE can happen. Currently, scientists do not yet fully understand how SPEs develop. Therefore, one cannot really predict SPEs. The best one can do is observe the sun with satellites and look for solar flares and CMEs. In that way one can at least send out a warning to the astronauts several minutes up to a few tens of minutes before the energetic electrons and protons reach them. But even then it is not possible to accurately predict the impact of the SPE. This can range from minor SPEs that will cause no harm up to dramatic events that can kill astronauts in deep space without adequate shielding. Hence, for manned missions to the moon or Mars the SPEs are an important issue. In LEO such as at ISS the dose due to SPEs is very limited due to partial protection by the Earth's magnetic field.

The second source of primary cosmic radiation is supernova remnants. Some stars undergo a very energetic explosion by the end of their life. After such an explosion or supernova very strong electromagnetic fields remain. In these electromagnetic fields charged particles can be accelerated up to extremely high energies. This is probably the most important source of the galactic cosmic radiation (GCR) coming from outside of our solar system. The GCR is a continuous flux of energetic charged particles coming isotropically from all directions. It is composed mainly of protons (85%) and ^{4}He nuclei (12%) and smaller amounts of heavier nuclei (1%) and electron and positrons (2%). The heavier nuclei are also called high-atomic number (Z), high-energy, or HZE particles. Although they are not so numerous, they are important from radiation protection point of view because of their high LET. The GCR energies are extremely high up to 10^{12} MeV with a peak around 1000 MeV. These high energies make it extremely challenging to shield from GCR. The most energetic particles can even penetrate the geomagnetic field. Proper shielding requires a material shield of at least a few meters thick. On Earth our atmosphere provides this protection. In spacecraft it is impossible to shield from GCR because of the weight limits. At ISS about 75% of the radiation dose is coming from GCR. In possible future habitats on the surface of the moon or Mars adequate protection could be provided by using local soil material. The GCR contribution to the dose is easy to predict. It is fairly constant in time. There is only a slight modulation of the GCR due to shielding by the magnetic field carried by the solar wind. The GCR dose rate is highest during solar minimum and lowest during solar maximum.

20.1.3 The Geomagnetic Field and the Van Allen Radiation Belts

Convection currents of molten iron in the Earth's outer core lead to electric currents and the generation of the geomagnetic field. This field is almost a magnetic dipole. The dipole axis is tilted about 11° with respect to the Earth's rotational axis. Also the dipole centre is slightly displaced with respect to the Earth's gravitational centre. The geomagnetic field shields the Earth and its close environment efficiently from cosmic radiation. Charged particles follow the magnetic field lines while gyrating around them with a certain radius. The more energetic the particle, the larger this gyration radius. Only the most energetic particles manage to reach the Earth's atmosphere. Less-energetic particles are captured by the magnetic field lines and diverted towards the poles. The equatorial region is best protected because the magnetic field lines there are parallel to the Earth's surface, while the polar regions are least protected because the magnetic field lines there intersect the Earth's surface. Therefore, the GCR dose rate is higher around the poles and the dose received by spacecraft in LEO increases with the inclination of the orbit.

A second important effect of the geomagnetic field is the creation of the Van Allen radiation belts. Some of the energetic charged particles of the SPEs and the GCR are trapped in the geomagnetic field. These trapped particles form two belts around the Earth that are called the Van Allen belts. The inner belt has its centre around 3000 km above the Earth's surface and contains electrons with energies up to 5 MeV and protons with energies up to 700 MeV. This belt is mainly filled by GCR. Therefore, its size and dose rate are inversely proportional to the solar activity. When spacecraft pass through this belt they are exposed to relatively high dose rates mainly due to the energetic protons. In LEO spacecraft are normally below the inner belt. Only above the coast of Brazil there is a region that is called the South Atlantic Anomaly (SAA) where the inner belt reaches down to 200 km above the Earth's surface. This is caused by the fact that the magnetic axis is not coincident with the rotational axis and does not go through the gravitational centre of the Earth. The ISS has an orbit with a typical altitude of 400 km and thus crosses the inner belt significantly in the SAA. These SAA crossings lead to about 25% of the radiation dose received by the astronauts in the ISS. Therefore, increasing the altitude of the ISS leads to increase of the radiation dose. The total radiation dose received in the ISS is also increasing for decreasing solar activity because both the GCR and SAA contributions are inversely proportional to the solar activity, while the SPEs do not contribute significantly to the radiation dose. Figure 20.3 shows a map of the absorbed dose rate measured in the Columbus module onboard the ISS with the DOSTEL detector in the framework of the DOSIS experiment during solar minimum in 2009 (Berger et al. 2017). This map clearly illustrates the increased dose rate in the SAA and the polar areas. The outer radiation belt has its centre around 22000 km above the Earth's surface and contains electrons with energies up to 7 MeV. This belt is mainly filled by SPEs. Therefore, its size and dose rate are proportional to the solar activity. This belt is only crossed for missions to the moon and Mars and is less of an issue because it only contains relatively low-energy electrons.

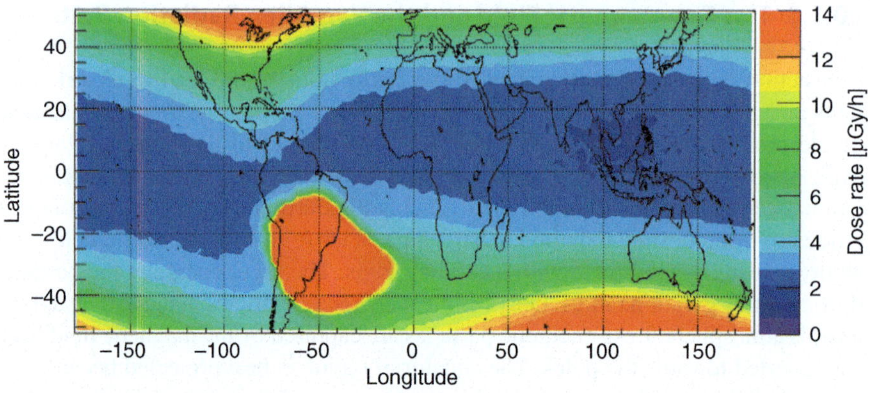

Fig. 20.3 Map of the absorbed dose rate measured in the Columbus module onboard the International Space Station with the DOSTEL detector in the framework of the DOSIS experiment during solar minimum in 2009 (Berger et al. 2017)

20.1.4 Shielding and Secondary Cosmic Radiation

Shielding is one of the primary methods to decrease radiation doses. However, interaction of cosmic radiation with shielding materials does not only attenuate the incoming radiation. Collisions of the very energetic GCR charged particles with the atoms of the shielding also create secondary cosmic radiation by nuclear reactions. This secondary cosmic radiation consists of neutrons, muons, pions, gammas, electrons, protons, and heavier nuclei and thus makes the radiation environment in space even more complex. Especially the secondary heavy nuclei and neutrons typically represent a substantial contribution to the radiation dose of the astronauts. The creation of the secondary cosmic radiation depends strongly on the shielding material composition. Ideally one should use shielding materials containing atoms with low atomic number such as hydrogen. Such materials give the best shielding per unit of mass and create the least secondary cosmic radiation. Constructive materials such as aluminium are thus not optimal with respect to radiation shielding. Using only a few millimetre of aluminium shielding can even increase the radiation dose. So, shielding could be optimized by using also low-atomic number materials. For habitats on the surface of the moon or Mars one could use several meters of soil material to get proper shielding.

20.2 Space Radiation Dosimetry

20.2.1 General Methodology

Radiation dosimetry for astronauts is very different from radiation dosimetry for radiation workers on Earth. Astronauts are exposed to much higher radiation doses and the type of radiation is also very different from the radiation typically encountered by radiation workers on Earth. Therefore, the very approximate concept of the effective dose is not applicable for astronaut risk assessment. A more precise and personalized risk assessment has to be performed (Dietze et al. 2013).

A detailed risk assessment should be performed before each mission. This risk assessment is based on the expected radiation energy deposition in the different organ in terms of the organ absorbed dose in joule per kilogram (J/kg) or gray (Gy). These organ absorbed doses should be assessed separately for different radiation particles and energies and multiplied with the appropriate quality factors taking into account the radiation harmfulness, the organ sensitivity and the age and sex of the astronaut. This product of the organ absorbed dose with the quality factor is called the organ dose equivalent. For assessing the organ absorbed doses one starts from models of the GCR, Van Allen belts and worst case SPEs. These models are used as input for radiation transport simulations with Monte Carlo codes such as MCNP (MCNP website), FLUKA (FLUKA website), GEANT (GEANT website) and PHITS (PHITS website). These codes simulate how the primary cosmic radiation interacts with the spacecraft structure and the human body and eventually how much energy is deposited in each of the organs by the different radiation types.

However, these calculations should be accompanied by radiation measurements. There are still significant uncertainties in the cosmic radiation models and the interaction cross sections used in the Monte Carlo codes. So, measurements are required to validate and improve the models and the Monte Carlo codes. This can be done by placing radiation detectors in manned and unmanned spacecraft, satellites, and rovers on the moon and Mars. Furthermore, SPEs still cannot be predicted and the radiation dose also depends strongly on the local shielding inside spacecraft. Therefore, ambient radiation detectors and personal dosimeters for astronauts are required to alert in case of abnormally high dose rates due to SPEs and to accurately monitor the actually received dose.

20.2.2 Space-Related Constraints for Radiation Detectors

Radiation detectors used for radiation dosimetry in space are bound by several constraints (Benton and Benton 2001). Because of the high cost of launching equipment into orbit, radiation detectors must be small and of low mass. Furthermore, they should be of a robust design and able to withstand a long period of use without failing. Finally, they need to consume as little power as possible. The types of materials that can be used are also bound by constraints on crew safety, such as the possible outgassing of certain polymers and the limited bandwidth available for the transmission of data to Earth. From a technological point of view it is very challenging to measure the large dynamic range of energies, fluence rates and particle types. During SPEs the fluence rates can increase by several orders of magnitude. Ideally, the detector should be able to distinguish different particles and energies. Currently there is no detector that can fulfil all these requirements. Therefore, the results from different detectors need to be combined with simulations. Measurements are necessary to validate simulations and to allow personalized dose assessment. But also the other way around, simulations are necessary to interpret radiation measurements. The design and development of space radiation detectors is also assisted by simulations. Furthermore, space radiation detectors are tested and characterized extensively at reference ion beam facilities on Earth such as the Heavy Ion Medical Accelerator in Chiba (HIMAC) in Japan before sending them into space.

20.2.3 Radiation Detectors Used in Space

Radiation cannot be observed directly. Therefore, radiation detectors rely on ionization and excitation effects induced by radiation in the detector material (Knoll 2010). Most radiation detectors are based on collection of radiation-induced charges such as gas ionization chambers and semiconductor detectors or on collection of visible light emitted by radioluminescent materials such as scintillators and optically or thermally stimulated luminescence detectors. There are also detectors in which radiation creates visible changes such as film, plastic nuclear track, and superheated emulsion or bubble detectors.

Different types of radiation detectors are used in space, depending on their purpose. There are active detectors providing real-time dose information and passive detectors that accumulate the dose until they are read out at a certain moment. Some detectors are mounted as ambient monitors somewhere fixed inside or outside a spacecraft or rover to monitor the radiation environment, while other detectors are carried by astronauts as personal dosimeter to monitor their personal dose.

Active detectors typically give the most detailed information because the data are time resolved and typically also give information on the radiation type. However, active detectors require power supply, are typically relatively bulky and expensive and require complex analysis. The oldest active detectors are based on simple gas ionization chambers. The R-16 (Tverskaya et al. 2004) is a combination of two such gas ionization chambers with different shielding. It was used already on board the Mir space station and is still used inside the Russian segment of the ISS. It cannot provide information on the radiation type. A more advanced type of gas ionization chamber used in space is the Tissue Equivalent Proportional Counter (TEPC). It is composed of a chamber of tissue equivalent wall material containing a tissue equivalent gas. The TEPC simulates the radiation energy deposition in a typically 1 or 2 μm diameter sphere of tissue. After appropriate analysis an approximate LET spectrum of the radiation can be derived. With the ISS-TEPC (Badhwar et al. 1996) and the updated version IV-TEPC, NASA currently has two TEPCs operational in the ISS. The newer active detectors are based on telescopes containing several semiconductor silicon diodes. Examples are the MSL-RAD (Mars Science Laboratory—Radiation Assessment Detector) (Hassler et al. 2014) on board the Curiosity rover on the surface of Mars and the DOSTEL (Berger et al. 2017), CPDS (Lee et al. 2007), ISS-RAD (similar to MSL-RAD) and Liulin (Dachev et al. 2015) detectors on board ISS. Such detectors provide more precise LET spectra and give also directional information. Both the gas ionization chambers and silicon telescopes are relatively bulky. Currently, there is a transition towards smaller active detectors. NASA has the ISS-REM monitoring network on board ISS and is working on the BIRD and HERA detectors for the Orion missions (Kroupa et al. 2015). These detectors are all based on the Timepix technology which consists of a compact pixelated silicon detector that requires limited power and is able to provide even more information on radiation type and angle. Both ESA and NASA are also developing very compact active personal dosimeters to replace the current passive personal dosimeters of the ISS astronauts. These are based on silicon diodes and Direct Ion Storage (DIS) detectors, which are miniature ionization chambers storing the liberated charges on a nonvolatile semiconductor memory cell. However, these active personal dosimeters do not provide detailed information on the radiation type. So, they still need to be complemented with more sophisticated active ambient monitors.

Passive detectors typically give less information because they provide a measurement integrated over time and radiation type. However, they are very compact, cheap and don't require any power. This makes them very complementary to active detectors. They are used for instance to perform detailed mapping of the dose rate such as inside the Columbus module of the ISS in the DOSIS 3D experiment (Berger et al. 2016), to measure inside anthropomorphic phantoms to experimentally assess organ absorbed doses such as in the MATROSHKA experiment (Berger et al. 2013), to monitor radiation doses for biological experiments in space (Vanhavere et al. 2008) and as personal dosimeter for the ISS astronauts. There are three main types of passive radiation detectors used in space. The first type are the optically stimulated luminescence detectors or OSLDs (e.g. Al_2O_3:C) and the thermoluminescent detectors or TLDs (e.g. LiF:Mg,Ti, LiF:Mg,Cu,P, $CaSO_4$:Dy). These detectors absorb the radiation energy and emit it as visible light when stimulated with heat or visible light during read out. The amount of light emitted is proportional with the received dose. There is also the Pille system (Szanto et al. 2015) which allows read out of $CaSO_4$:Dy detectors inside the ISS. It is used by the Russians for dose mapping and personal dose monitoring during Extra Vehicular Activities (EVAs). The disadvantage of the stimulated luminescence detectors is that their sensitivity drops rapidly for radiation with LET above 10 keV/μm. But limited information on the high LET part of the radiation can be obtained by combining different OSLDs and TLDs and by looking at the high temperature signal of the TLDs (Parisi et al. 2017). The second type of passive detectors are Plastic Nuclear Track Detectors (PNTDs). These detectors are based on a polymer, typically polyallyl diglycol carbonate or CR-39. Radiation with high LET creates tracks in this detector that can be visualized under a microscope after chemical etching. The number of tracks, their size, and shape can be used to calculate the dose and even an LET spectrum. The PNTDs are only sensitive for radiation with LET above 10 keV/μm. Therefore, the OSLDs and TLDs are typically combined with the PNTDs such as in the passive dosimeter of the ISS astronauts and the DOSIS 3D experiment. Finally, there are also the superheated emulsion or bubble detectors. These detectors contained small droplets of superheated liquid dispersed in a polymer. Radiation with high LET can deposit enough energy in these droplets to evaporate them into gas bubbles. The number of bubbles is a measure of the dose. The bubble detectors are typically used for neutron measurements (Smith et al. 2016).

20.3 Radiobiology

20.3.1 Introduction

All biological consequences of ionizing radiation on living tissues are a result of the interaction with atoms in a process called ionization. Ionizing radiation has sufficient energy to remove one orbital electron from an atom, thereby creating an ion pair. There are two mechanisms by which radiation ultimately affects cells: either by direct ionization of the target molecule, or indirectly by the production of free radicals which may ultimately affect the target molecule (Fig. 20.4). During direct ionization, radiation transfers energy directly to the atoms of cellular components such as deoxyribonucleic acid (DNA), proteins, and lipids. If a cell is exposed to ionizing radiation, the probability of directly interacting with DNA is low, as it makes up only a small part (1%) of the cell. The main constituent of all cells of the human body is water (up to 80%). Therefore, most of the energy produced by ionizing radiation (particularly photon irradiation) leads to water radiolysis, which results in free radical production that ultimately can lead to indirect DNA damage. Free radicals are highly reactive compounds, such as hydroxyl radicals and superoxide anions, which are characterized by an unpaired electron. In addition, other reactive compounds including hydrogen peroxide and hydroxyl ions can also be produced.

Radiation induces a large spectrum of DNA lesions, including single-strand breaks (SSB), double-strand breaks (DSB), base loss, base changes, and cross-links. The DSB constitute the principle cytotoxic lesion in response to both photon and particle radiation, and is considered to be the critical primary lesion in the formation of chromosomal aberrations. The quantification of DSBs as well as chromosomal aberrations after radiation exposure is frequently used as a biological dosimeter or to evaluate radiosensitivity of individuals (see Chap. 28). DSBs are detected in the

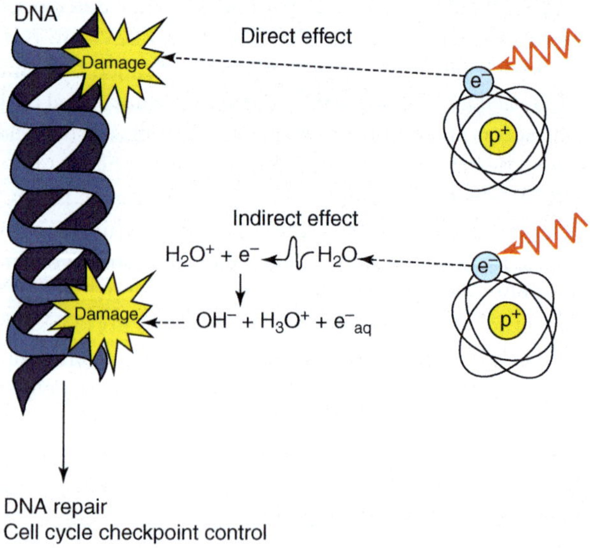

Fig. 20.4 Ionizing radiation induces direct DNA damage and indirect damage through the radiolysis of water (figure is *adapted* from a figure in (Morgan and Sowa 2007))

cell by sensing molecules which activate a signaling cascade by phosphorylating the histone H2AX (γ-H2AX) (Kinner et al. 2008; Rogakou et al. 1998). Repair enzymes will be attracted to the damaged site and the cell will go into cell cycle arrest to allow time for repair. It is well known that the number of γ-H2AX foci is proportional to the amount of DSBs. By immunofluorescent staining of the γ-H2AX foci, quantitative and qualitative evaluation of the amount of DSBs as well as subsequent DNA repair kinetics can be performed (Fig. 20.5) (Ghardi et al. 2012). Substantial evidence indicates that particle radiation such as protons or heavy ions induces DNA damage that is quantitatively and qualitatively different from that caused by photons (Fakir et al. 2006). Particle radiation (especially high-LET) produces dense ionization tracks, thereby inducing a greater number of, and more complex, "clustered" DNA lesions than photon radiation (Hada and Georgakilas 2008; Terato et al. 2008). These clusters contain various types of DNA damage (e.g., SSB, DSB) within a localized region of the DNA molecule and are associated with the increased relative biological effectiveness (RBE) of particle radiation beams (Fakir et al. 2006; Hada and Georgakilas 2008; Cortese et al. 2018).

The global response of a cell to DNA damage triggers multiple pathways involved in sensing DNA damage, activating cell cycle checkpoints, and inducing DNA repair (Su 2006).

However, when the damage is severe, cellular apoptosis can be induced. Failure of DSB repair or misrepair can initiate genomic instability, causing chromosome aberrations and genetic mutations, and may eventually lead to cancer. Besides damaging the DNA molecule, ionizing radiation can cause a number of lesions in other macromolecules as well (e.g., lipid peroxidation, reactive oxygen species). These non-DNA lesions trigger multiple signaling pathways including Protein Kinase C (PKC), Mitogen-activated protein Kinase (MAPK), and c-Jun NH_2-Terminal Kinase (JNK), which are involved in cell cycle control, DNA repair, and apoptosis. In addition, other radiation-induced phenomena have been described. These include nontargeted and delayed effects such as bystander effects, genomic instability, and adaptive response (Averbeck 2010). Bystander effects occur in cells that are not hit directly, but which are affected by signals derived from neighboring irradiated cells. Genomic instability is characterized by the increased rate of acquisition of genomic alterations (e.g., chromosomal aberrations, mutations) of the progeny of an originally irradiated cell appearing several generations after irradiation. On the other hand, the adaptive response model postulates that certain doses of low-dose radiation may be beneficial, and renders cells less susceptible to the damaging effects of radiation.

20.3.2 Biological Effects of Radiation

20.3.2.1 Tissue Reactions and Stochastic Effects of Ionizing Radiation

Radiation risk assessment by advisory bodies such as ICRP (International Commission on Radiological Protection) and NCRP (National Council on

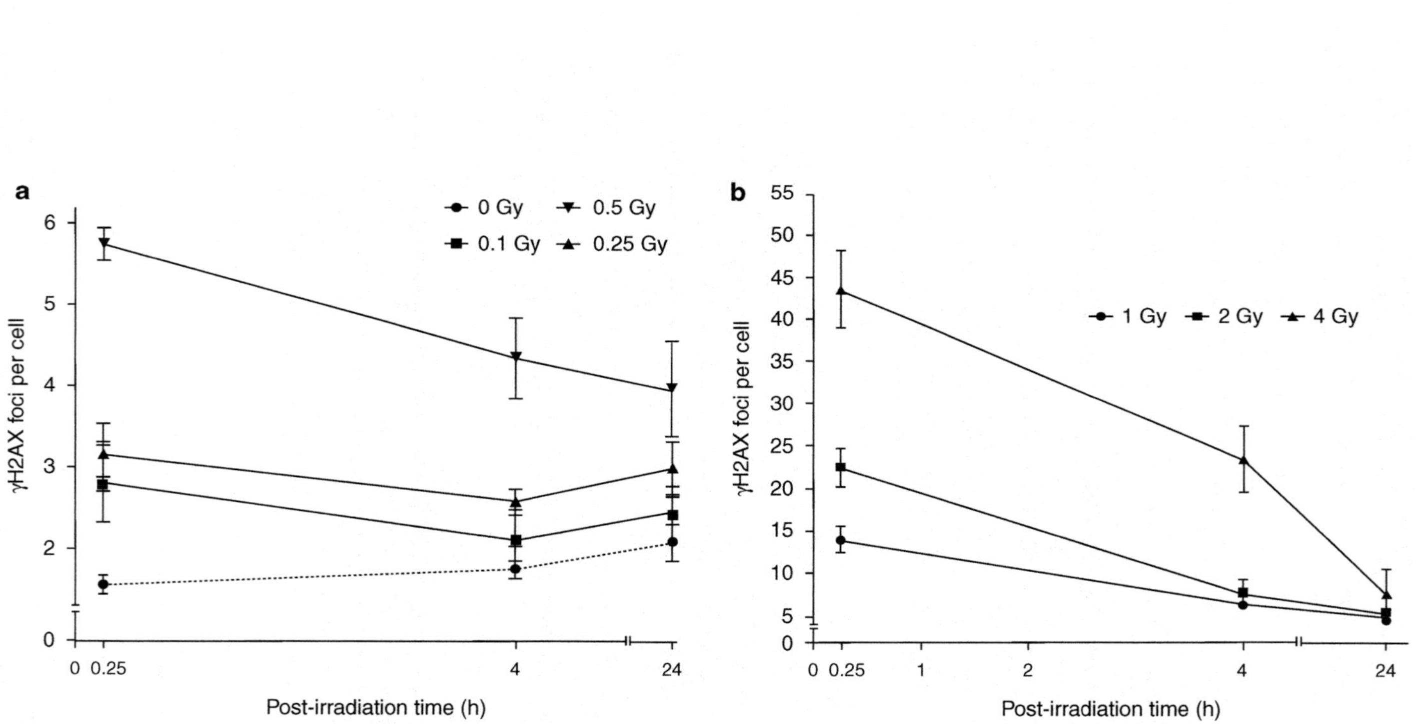

Fig. 20.5 Kinetics of γH2AX foci formation in human lymphocytes analyzed by fluorescence microscopy. Time response curves represent the relative mean of γH2AX foci at different time points after irradiation. (**a**) Low and moderate doses (0.1, 0.25 and 0.5 Gy) or (**b**) high doses (1, 2 and 4 Gy) of X-irradiation. Error bars represent ± SEM ($n = 3$). (Figure from (Ghardi et al. 2012))

Radiation Protection & Measurements) have classified biological effects of radiation as either tissue reactions or stochastic (ICRP 2012). *Tissue reactions*, previously called deterministic effects, are associated with high doses of radiation exposure over a short period of time. The severity of the effects in affected individuals depends on the dose and increases with the magnitude of the radiation. There is a threshold radiation dose, below which the tissue reactions has, so far, not been detected clinically (Fig. 20.6a). *Stochastic* effects are usually associated with exposure to low doses of radiation over a longer period of time, typically like those encountered by astronauts on board the ISS. The probability of inducing the effect, but not the severity of the effect, is dose-related (Fig. 20.6b). Dose-effect curves for these changes are considered to be nonthreshold in type. It is assumed that there is always a small probability of an effect even at very low doses. Low doses can be defined as a dose, and dose rate, at which, on the average, only a fraction of all targets (cell nuclei) is affected by an energy deposition event. In this dose range, the risk for one cell to be transformed is very low. However, the risk to the organism of having one transformed cell depends on the number of cells being hit. The upper limit according to this criterion is 0.020 Gy (ICRP). Increased incidence of cancer after low-dose radiation is an example of a stochastic effect. For the induction of tumors, doses below 0.1 Gy are considered as low doses. The linear nonthreshold model presupposes that the damage caused by ionizing radiation linearly increases in response to the dose. Furthermore, nontargeted as well delayed effects including bystander effects, genomic instability, and adaptive response might be involved in the response to low doses.

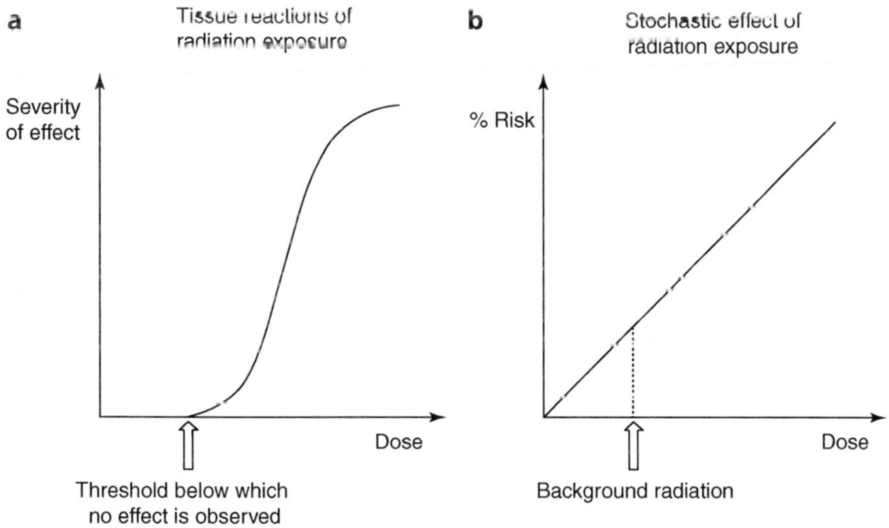

Fig. 20.6 Biological effects of ionizing radiation: (**a**) tissue reactions and (**b**) stochastic effect

20.3.2.2 Early and Late Effects

Radiation effects can appear rapidly, but also after a delay. *Early* or *acute effects* are those occurring within hours, days, or a few weeks following high-dose exposure (>1 Gy) in a short period of time. These effects appear to be threshold phenomena and are classified as *tissue reactions*. Exposure to high doses of ionizing radiation can cause a rapid whole-body response, often referred to as acute radiation syndrome (ARS) or radiation sickness. These high doses tend to kill cells in such a way that tissues and organs become damaged and their functioning impaired. The principal sites of biological action are rapidly proliferating cells from the bone marrow, gastrointestinal cells, skin, and testes. These tissues and organs are classified as radiosensitive organs. ARS is characterized by a sequence of phased symptoms in which the clinical effects develop proportionally to the dose amount (Xiao and Whitnall 2009). Moderate doses (1–7 Gy in humans) induce depression of bone marrow function, known as hematopoietic syndrome. This syndrome leads to decreased resistance to infections and hemorrhage. Higher single doses (about 8 Gy or more) will result in gastrointestinal syndrome leading to small intestinal cell killing. Very severe whole-body exposure (20–40 Gy) is characterized by a deteriorating state of consciousness with eventual coma and death (neurovascular syndrome).

Late or *long-term effects* usually occur after a number of months or years following radiation exposure. For radiation protection purposes, these late effects are classified as being stochastic or tissue reactions in nature. Epidemiological studies have clearly identified an increased risk of several types of cancers induced by ionizing radiation (Preston et al. 2007), rendering *carcinogenesis* as the main somatic *stochastic late effect*. It is generally agreed that DNA damage, including DSB, mutations, and chromosomal rearrangements are the initiating event in the multistep process leading to malignant transformation. In this context, radiation-induced chromosomal aberrations in peripheral blood lymphocytes are considered as a validated biomarker of cancer risk estimation (Durante 2005; Norppa et al. 2006; Boffetta et al. 2007). In addition to an increased risk of carcinogenesis, long-term immune dysfunction should also not be underestimated (Kusunoki and Hayashi 2008). A classic example of a *late tissue reaction* induced by ionizing radiation exposure is cataract formation (reviewed in Kleiman 2012). The eye, and more specifically the crystalline lens, is one of the most radiosensitive organs of the human body. Ocular ionizing radiation exposure causes dose-related, progressive changes in the lens, finally leading to cataract. Since 2012 and 2016, respectively the ICRP and NCRP have reported threshold values for visually disabling cataracts of 0.5 Gy. Furthermore, apart from lens opacity, it has long been realized that high-radiation doses also have the potential to cause effects such as cardiovascular diseases as well as cognitive impairment and that such non-cancer effects at low doses cannot be readily explained by the extrapolation of cancer risk from high to low doses (LNT model), pointing out the need for more experimental (mechanistic) and epidemiological studies to address this particular extrapolation issue to radiation-induced non-cancer effects.

20.4 Biological Effects of Cosmic Radiation

Compared to individuals on Earth, astronauts receive much higher doses of ionizing radiation during spaceflight. Dose estimates for interplanetary space travel indicate that astronauts receive 0.5–1.4 Gy/year. Recently, estimates for a round-trip to Mars show that total cumulative doses of up to 0.662 ± 0.108 Sv are likely (Zeitlin et al. 2013). However, these doses are not expected to result in significant acute radiation effects (thereby assuming no significant SPEs during the mission), but they may increase the long-term health risks that are associated with radiation exposure. Besides this, the type of radiation in space differs from the typical terrestrial radiation (e.g., X-rays, γ-rays), and consists mainly of protons and HZE particles (e.g., iron, oxygen, carbon, and silicon ions). Although protons account for >90% of deep space radiation, charged particles are predicted to account for most of the biological consequences of cosmic radiation exposure, due to their high-LET and their linear track structure (Swenberg et al. 1991; Green et al. 2001). Unfortunately, shielding against this type of ionizing radiation is not feasible. Exposure to high-LET radiation is therefore considered as a major health risk for astronauts and is of major concern during long-term spaceflights.

Cataract formation was the first proven late tissue reaction of space radiation on astronauts. Previous studies showed that an increased risk of cataract with lens doses greater than 8 mGy was observed in astronauts (Cucinotta et al. 2001; Rastegar et al. 2002; Chylack et al. 2009). In addition, a significant association was found between the space radiation doses received by US astronauts, and the progression rate of cortical cataracts (Chylack et al. 2012). The induction of late stochastic effects such as cancer risk after cosmic radiation exposure is so far unknown (Cucinotta and Durante 2006; Durante and Cucinotta 2008). Estimations of cancer risk for space missions are mainly based on epidemiological settings from data obtained from atomic bomb survivors and patients exposed to radiation therapy (Little 2009). However, most of these studies are related to photon radiation types (e.g., X-rays, γ-rays), and do not take into account the important biological differences that exist between photons and particles. As a result, extrapolation of photon data for radiation risk assessment on astronauts can lead to wrong conclusions (Cucinotta and Durante 2006). Therefore, at present, prediction of cancer risk for humans exposed to heavy ions during deep space mission has very large uncertainties since there are no data available that address the risk from extended exposure to complex radiation fields. For this reason, NCRP recommended age and gender-dependent career dose limits based on a 3% excess cancer fatality risk. In order to improve the risk modeling during deep space missions, calculations should be based on biological and mechanistic studies on the effects of different radiation qualities on carcinogenesis (Barcellos-Hoff et al. 2015). Estimation of cancer risk following occupational deep space ionizing radiation exposure could be based on studies performed on cancer patients treated with charged particles including protons and carbon ions (hadron therapy). However, recent studies on patients treated with particle irradiation only reported a few cases of secondary tumors and suffer from a limited follow-up time and inadequate statistics (Kamran et al. 2016; Eaton et al. 2015). Nevertheless, the increased use of charged particles (protons and carbon ions) for radiotherapy purposes can increase our general knowledge about the biological effects of particles in general. Recently, ground-based studies on animals showed that exposure to particle radiation has a higher carcinogenic effect compared to photon exposure (Weil et al. 2014; Suman et al. 2015; Trani et al. 2010). In addition, our lab performed many in vitro experiments with cancer cells exposed to carbon ions (Suetens et al. 2016, 2014, 2015). Further studies in this field will definitely contribute in reducing the present uncertainties associated with cancer risk estimates on astronauts.

Chromosomal abberations have been used as surrogate endpoints to investigate radiation quality effects related to cancer risk estimation. Caution should be given, however, when extrapolating the number of chromosomal abberations to a cancer risk as it has been observed that the dose response for chromosomal aberrations following particle exposure is nonlinear at lower (<0.1 Gy) doses (Hada et al. 2014). As mentioned before, measurement of chromosomal aberration frequencies in lymphocytes after radiation exposure have been proposed as an indicator of cancer risk (Durante 2005). So far, several papers have described radiation-induced chromosomal damage in astronauts' lymphocytes after missions in low-Earth orbit (Testard et al. 1996; Obe et al. 1997; Yang et al. 1997; George et al. 2001; Durante et al. 2003; Greco et al. 2003; George et al. 2005). Short-duration missions did not result in significant detectable differences in chromosome abnormalities measured before and after flight. In contrast, an increase in post-flight aberrations was found in the case of long-term flights. However, in another study no correlation was found between aberration frequencies and total mission duration or in the cumulative dose equivalent in space for 13 crew members involved in multiple spaceflights (Durante et al. 2003). In addition, it appeared that the post-flight decline in time of chromosomal aberrations was faster than expected and significant heterogeneity was observed among individuals. Besides differences in individual radiosensitivity, the adaptive response mechanism to space radiation might explain these observations. This adaptation might take place after the first exposure, leading to an increased radioresistance in exposed individuals. Recently, the effect of repeated space flight on the yield of translocations in individual crewmembers was analyzed (George et al. 2013). All crewmembers showed a consistent increase in total exchanges and translocations after both the first and second flight. These results support the assumption of addividity of biological doses for ISS crew exposures. The long-term consequences of these persistent chromosomal rearrangements are not well understood so far, and further investigation in this field is definitely needed.

20.4.1 Immune Dysfunction in Space: Impact of Cosmic Radiation?

Immune dysfunction during spaceflight is of paramount concern and can lead to serious health problems. The first study about immune abnormalities in space dates back to the 1970s, where reduced reactivity of peripheral blood lymphocytes in crew members was observed (Konstantinova et al. 1973). Currently, numerous immune aberrations have been reported in astronauts, cosmonauts, and animals flown in space (reviewed in Gueguinou et al. 2009). Some of these observed alterations include variations in peripheral blood leukocyte populations, changes in function of cells involved in innate immunity, decreased cytokine expression, depression of T cell activation and proliferation, and many others. So far, the precise nature of immune dysregulation related to spaceflight is unknown (see Chaps. 11–19) and multiple underlying causes might be involved. In this context, the impact of cosmic radiation, which is always present during spaceflight, should not be underestimated.

To gain more insight into the specific influence of radiation on the immune system, Earth-based experiments are increasingly used and represent a more reproducible alternative to in-flight experiments. To date, several ground-based studies in animals demonstrated that ionizing radiation influences many aspects of the immune system and can cause immune dysfunction. Until recently, the majority of radiobiology studies were related to photon radiation such as X- and γ-rays (Harrington et al. 1997; Shankar et al. 1999; Gridley et al. 2001; Pecaut et al. 2001; Shearer et al. 2005). However, due to the recent worldwide increase in particle accelerator facilities, several studies investigated the effect of whole-body exposure to particle radiation including protons and heavy ions. In this way, it is possible to expose cells and animals to types of radiation encountered by astronauts during interplanetary space travel. Laboratory studies have shown that whole-body radiation of rodents can result in significant acute and long-term effects on the immune system. Table 20.1 gives a summary of the main papers describing ground-based experiments that investigate acute and chronic immune changes after exposure to photon and particle radiation.

20.4.1.1 Acute Effects

In general, when mice are exposed to a single exposure of up to 3 Gy of protons, photons, or high-LET particles, a clear effect on different immune cell populations is observed. There is a large dose response on day 4 followed by general recovery over the following 2–3 weeks. Independent of the radiation type, relative radiosensitivities of the various lymphocyte populations (B cells > T cells > NK) and T cell subsets ($CD8^+$ > $CD4^+$) are observed (Kajioka et al. 1999, 2000; Gridley and Pecaut 2006; Pecaut et al. 2006; Gridley et al. 2008; Pecaut and Gridley 2008). Lower doses of high-LET particles (<2 Gy) were shown to induce acute damage to the hematopoietic stem cells and their progenitor cells (Chang et al. 2016, 2017a, b). To better mimic the space radiation environment, experiments with simulated solar particle event (sSPE) were performed (Gridley et al. 2008). For this purpose, mice were exposed to a high dose of sSPE protons over a period of 36 h. The results obtained were compared to γ-ray and proton exposure with the same dose. Acute effects including general immune depression and leukocyte abnormalities were present; however, the damaging effects of sSPE on leukocytes were generally less pronounced compared to the acute photon and proton radiation. Furthermore, it has been demonstrated that inhomogeneity of the proton dose distribution (30–74 MeV) does not affect white blood cell counts (Sanzari et al. 2014). Besides the risk of

Table 20.1 Ground-based studies—impact of photon and particle radiation on the immune system of rodents

Reference	Irradiation source	Animal model	Time point of analysis	Effect
Harrington et al. (1997)	γ-rays: 1–7 Gy	C57Bl/6 mice	Acute (1–4–7 days)	Immune suppression
Kajioka et al. (1999, 2000)	Proton: 3 Gy (0.4 Gy/min)	C57Bl/6 mice	Acute (4–10–15–29 days)	Decrease in lymphocyte populations
	γ-rays: 3 Gy			→ B sens > CD4 > CD8 > NK resist
				Decrease in acute response to antigen
				No significant differences between proton and γ-ray exposure
Pecaut et al. (2001)	γ-rays: 0–0.5–1.5–3 Gy	C57Bl/6 mice	Acute (4 days)	Decrease in lymphocytes populations
	LDR: 1 cGy/min			B sens > CD4 > CD8 > NK resist
	HDR: 80 cGy/min			Immune changes depend more on dose than on dose rate
Gridley et al. (2001)	γ-rays: 0–0.5–1.5–3 Gy	C57Bl/6 mice	Acute (4 days)	Decrease in number of blood cells
	LDR: 1 cGy/min			Decrease in IL-2 secretion by activated spleen cells
	HDR: 80 cGy/min			Changes depend more on the dose, than on the dose rate
Gridley et al. (2002a)	Proton: 0–0.5–1.5–3 Gy (entry region Bragg peak)	C57Bl/6 mice	Acute (4 days)	Decrease in number of lymphocytes
	LDR: 1 cGy/min			NK are more radioresistant
	HDR: 80 cGy/min			Significant dose and dose rate effects
	γ-rays: 3 Gy			Effect of proton irradiation (3 Gy) is larger than γ-ray exposure (3 Gy)
Pecaut et al. (2002)	Proton: 0–0.5–1.5–3 Gy (entry region Bragg peak)	C57Bl/6 mice	Acute (4 days)	Decreased splenocyte response
	LDR: 1 cGy/min			Increased blastogenesis
	HDR: 80 cGy/min			Effects depend on dose (and not dose rate)
	γ-rays: 3 Gy			Effects are more pronounced with proton exposure compared to γ-rays

Gridley et al. (2002b)	Fe ion: 0.1–0.5–2.0 Gy (LET = 148.2 keV/μm)	C57Bl/6 mice	Acute (Fe ion: 4 days)	Acute effects: decrease in lymphocytes (B sens > CD8 > CD4 > NK resist)
	Si ion: 2.0 Gy (LET = 42.1 keV/μm)		Chronic (Fe ion, Si ion: 113 days)	Chronic effects after Fe exposure: high number of B cells, low number of CD8 cells
				Chronic effects after Si exposure: low number of NK
				Immune aberrations persist long after exposure
				Effects depend on radiation type
Pecaut et al. (2003)	Proton: 3–4 Gy (± shielding)	C57Bl/6 mice	Chronic (122 days)	Dose-dependent decrease in lymphocyte populations
Shearer et al. (2005)	γ-rays: 0.3 Gy	Balb/c mice		Decrease in number of immune cells
Pecaut et al. (2006)	Fe ion: 0–0.5–2–3 Gy (LET = 148.2 keV)	C57Bl/6 mice	Acute (4 days)	Decrease in number of lymphocyte populations
				B sens > CD8 > CD4 > NK resist
Gridley et al. (2006)	Fe ion: 0–0.5–2–3 Gy (LET = 148.2 keV)	C57Bl/6 mice	Acute (4 days)	Alterations in leukocyte response and function
Gridley et al. (2006)	γ-rays: 2 Gy	C57Bl/6 mice	Chronic (110 days)	Significant aberrations in immune parameters observed 4 months after exposure
	Proton: 2 Gy			
	C ion: 2 Gy (LET = 12.9 keV/μm)			
	Fe ion: 2 Gy (151.5 keV/μm)			
Gridley et al. (2008)	Fe ion: 0–1–2–4 Gy (LET = 148.2 keV/μm)	SD Rat	Chronic (9 months)	Decrease in number of lymphocytes
Gridley et al. (2008)	SPE protons: 2 Gy (chronic)	C57Bl/6 mice	Acute (4–21 days)	Effects on immune system with SPE protons less pronounced compared to other types of radiation (acute γ-rays or acute protons)
	Protons: 2 Gy (acute)			
	γ-rays: 2 Gy (acute)			
Gridley et al. (2009)	Proton: 0.01–0.05–0.1 Gy	C57Bl/6 mice	Acute (4–21 days)	Changes in CD4 T cell gene expression after low-dose proton irradiation
	LDR: 0.1 cGy/h (delivered over a 2 week period)			LDR enhances CD4 T cell responsiveness

(continued)

Table 20.1 (continued)

Reference	Irradiation source	Animal model	Time point of analysis	Effect
Pecaut and Gridley (2010)	Fe ion: 0.5–2–3 Gy	C57Bl/6 mice	Acute (4–30 days)	Mouse strain influences Fe radio-immune response
		CBA/Ca		
Gridley et al. (2010)	± SPE protons: 1.7 Gy (chronic) ± pre exposure to γ-rays: 0.01 Gy (LDR: 0.179 mGy/h)	C57Bl/6 mice	Acute (4–21 days)	Preexposure to LDR photons does not protect against adverse effects of radiation mimicking a SPE on the immune system
Rizvi et al. (2011)	± SPE protons: 1.7 Gy (chronic) ± pre exposure to γ-rays: 0.01 Gy (LDR: 0.179 mGy/h)	C57Bl/6 mice	Acute (4–21 days)	Protracted exposure to LDR γ-rays modifies effect of SPE protons on lymphocyte signaling proteins and secretion of cytokines
Luo-Owen et al. (2012)	Chronic preexposure to γ-rays: 0.05 Gy (LDR: 0.025 cGy/h) + acute proton irradiation (2 or 3 Gy)	C57Bl/6 mice	Acute (4–17 days)	Preexposure modulates the response to acute proton exposure
Gridley et al. (2013)	2 Gy proton (1 Gy/min) or γ-ray (0.9 Gy/min) + preexposure to γ-rays: 0.01 Gy (LDR: 0.03 cGy/h)	C57Bl/6 mice	Acute/chronic (21–56 post-exposure)	Some immune responses to acute 2 Gy radiation are dependent on radiation quality, time of assessment, and preexposure to LDR γ-rays
Sanzari et al. (2014)	SPE proton radiation (homogenous and inhomogenous dose distribution)	ICR mice	Acute (4–24 h)	Reduced number of blood cells, comparable for both dose distributions
Chang et al. (2015)	1 Gy whole-body proton irradiation (150 MeV/n)	C57BL/6J mice	Chronic (22 weeks)	Persistent reduction of bone marrow hematopoietic stem cells linked to increased oxidative stress, reduced quiescence, and increased DNA damage
Chang et al. (2016)	Low dose of ^{16}O exposure (0.1, 0.25 and 1.0 Gy)	C57Bl/6 mice	Acute (14 days)	Acute damage to hematopoetic progenitor and stem cells

Gridley and Pecaut (2016)	^{56}Fe ion: 0–1–2–3 Gy	C57BL/6 mice	Chronic (40 days)	Aberrations in a variety of immune parameters
Chang et al. (2017b)	0.5 and 1 Gy whole-body proton irradiation (150 MeV/n)	C57BL/6 mice	Acute (14 days)	Decreased numbers of common myeloid progenitor and $Lin^{-}SCA1^{+}c\text{-}KIT^{+}$ bone marrow stem cells
Chang et al. (2017a)	Low dose of ^{28}Si ions (0.3, 0.6 and 0.9 Gy)	C57BL/6 mice	Acute (28 days)	Reduced number of hematopoietic stem cells
Wang et al. (2017)	0.05, 0.1, 0.25 and 1.0 Gy whole body ^{16}O (600 MeV/n)	C57BL/6J mice	Chronic (3 months)	Long-term decrease in number of hematopoietic stem cells, primarily via increased intracellular ROS production

LDR low dose rate, *HDR* high dose rate

relatively high-dose exposure (e.g., during an SPE), low dose/low-dose-rate (LDR) radiation must be taken into account when performing research in the context of radiation risks for astronauts. Rizvi et al. (2011) and Luo-Owen et al. (2012) demonstrated that total body LDR γ-radiation can modify the response of leukocytes exposed to simulated SPE protons, thereby increasing cellular tolerance.

In the context of individual radiosensitivity, the potential impact of mouse strains with a genetic background on various immune parameters after acute iron ion exposure was compared. These results showed that the impact of the genetic background on radiation-induced immune aberrations appeared to be minimal, and only included changes in circulating phagocytic populations, erythrocytes, and liver mass (Pecaut and Gridley 2010).

20.4.1.2 Chronic Effects

So far, knowledge about the long-term effects on the immune system after exposure to different types of radiation is limited (Gridley et al. 2002b, 2008, 2013; Pecaut et al. 2003; Gridley and Pecaut 2006, 2016). Summarized in Table 20.1, these studies investigated the potential chronic effects on lymphoid cells 4 months after exposure to a single high dose of photons, protons, iron, silicon, or carbon ion whole-body irradiation (Gridley et al. 2002a, b; Pecaut et al. 2003; Gridley and Pecaut 2006). The first study in this field investigated long-term effects (3 months) after iron and silicon irradiation (Gridley et al. 2002b). In response to Fe ion irradiation animals had significantly increased total lymphocyte and B cell numbers, whereas $CD8^+$ T cell proportions were low, compared with nonirradiated controls. However, whether these changes result in abnormal/compromised immune responses is not clear. Interestingly, these changes could not be observed after exposure to Si ion beams. Long-term changes in mice exposed to Si beams resulted in a lower number of NK. After whole-body exposure to proton irradiation at doses of the order of large SPE, dose-dependent decreases in $CD8^+$ and NK were observed (= depression of peripheral white blood cell count) (Pecaut et al. 2003). In contrast, B and T helper cell numbers in the spleen were significantly elevated following total body irradiation with iron ions (Gridley and Pecaut 2016). Another study showed increases in the number of T cells and a decrease in NK in response to proton and carbon ions (Gridley and Pecaut 2006). Gridley et al. (2008) focused on the impact of a single iron ion 9 months after whole-body irradiation in rats and showed lower numbers of circulating lymphocytes and monocytes, indicating that the intrinsic quality of a particle beam is of importance as well, and can evoke different long-term effects on the immune system. When investigating the priming effect of low-dose radiation on the sensitivity to a subsequent high proton dose, expression of apoptosis and inflammation related genes was still affected on day 56 post-exposure (Gridley et al. 2013). Furthermore, high LET irradiation has been demonstrated to induce a persistent reduction in murine bone marrow hematopoietic stem cells via mechanisms related to oxidative stress, DNA damage and stem cell quiescence (Chang et al. 2015, 2016; Wang et al. 2017).

In summary, several ground-based studies clearly demonstrated acute and long-term changes in the immune system status of whole-body irradiated animals exposed to photon and/or particle radiation. Some of these observed alterations include changes in the numbers of T- and NK which are important cellular components to suppress infections and kill virus-infected or neoplastic cells. Prolonged deficiency in any of these lymphocyte populations can have serious consequences for astronaut health. In addition this immune cell deficiency can even be exaggerated as a persistent reduction in bone marrow hematopoietic stem cell after radiation exposure has also been observed. However, it still needs to be determined whether these observed immunological aberrations will actually result in impaired immune function.

Another important observation in this field is that results obtained with high-LET are not always similar to those obtained after photon radiation such as X-rays and γ-rays. This clearly demonstrates that caution is important when extrapolating to photon results. In addition, long-term immune changes differ significantly between the various high-LET particles, thereby indicating that the intrinsic quality of the particle beam may be important as well.

Ground-based experiments are increasingly being used and represent a more reproducible alternative to inflight experiments. Although these experiments are a good model to investigate the impact of radiation on the immune system, they most often evaluate the effect of exposure from a single source of radiation. In this context, simultaneous exposure to radiation of different types, thereby better simulating the radiation spectrum to which astronauts are exposed to, might be interesting to decipher whether this might affect a broader range of immunological parameters.

20.4.1.3 Spaceflight-Associated Immune Changes: Examples of Tentative Interaction Between Radiation and Other Space Flight Stressors

To date, it is beyond doubt that spaceflight can induce changes in the immune system. Most, if not all, of these immune alterations have been attributed to both psychological stress and the microgravity (μg) environment. In the context of cosmic ray exposure, it is only after long-term missions that increased levels of cosmic radiation may play a more significant role in this immune dysfunction. However, it is likely that several of the space stressors can interact with one another. These interactions may be additive or synergistic, but can be antagonistic as well, thereby resulting in a final common effect on the immune system that might compromise astronaut resistance to infections and other diseases (see Chap. 3). In the next paragraph an example of a tentative interaction between two spaceflight specific stressors is given.

20.4.1.4 The Combined Effects of μG and Radiation

During spaceflight, μg induces numerous systemic effects including alterations in the musculoskeletal system, cardiovascular system, sensory-motor system, and immune system. With regard to the latter, changes such as decreased number and responsiveness of T lymphocytes, reduced cytotoxic activity of NK, and alterations in cytokine and chemokine activity have been reported (reviewed by Frippiat et al. 2016). However, one might ask whether this reduced gravity can alter the cellular response to ionizing radiation. Experiments performed on living embryonic systems in space showed a synergistic interaction between both space stressors, thereby decreasing cell survival and inducing chromosomal aberrations (Reitz et al. 1989; Horneck 1999).

However, further studies demonstrated that the interplay between both could not be explained by a decreased capacity to repair damaged DNA (Kiefer and Pross 1999; Pross et al. 2000). Therefore, other mechanisms have to be postulated for this synergism. One may hypothesize that immune cells respond to decreased gravity as well as to radiation challenges by activating similar cell signaling pathways (see Chap. 28). In the context of μg, "gravi-sensitive" signal transduction components are present in different cell compartments of immune cells such as on the cell surface (e.g., IL-2 receptor, which can result in a diminished proliferative response of T cells), in the cell cytoplasm (e.g., intracellular signaling pathways), and in the nucleus (e.g., expression of the genes regulating a number of cellular processes including differentiation and proliferation) (Ullrich et al. 2008; Tauber et al. 2015). With regard to intracellular signaling pathways, various kinases such as tyrosine, PKC, and MAPK play important roles in response to μg. Besides μg, it has been shown that radiation also induces changes in the activation of different kinases in immune cells (Varadkar et al. 2003; Varadkar and Krishna 2004; Mitra et al. 2007). Tyrosine kinase, PKC, and MAPK activity increase with increasing dose after irradiation in lymphocytes in vivo. However, in contrast, MAPK activity decreased with an increasing dose in ex vivo irradiated lymphocytes. The effect might become even more complicated when comparing photon data with results obtained after high-LET radiation (Narang et al. 2009). In this context, the hindlimb unloading mice model was used to investigate the combined impact of microgravity and proton radiation on several immune parameters (Sanzari et al. 2013a). Results demonstrated that exposure to combined stressors decreased leukocyte numbers and function. In addition, whole-body proton or γ-ray radiation in a ferret model resulted in a significant reduction in circulating white blood cells (Sanzari et al. 2013b). The importance of the radiation-counterpart in cellular changes in response to physical space stressors such as μg is still not clear, and more research is definitely needed to gain additional insight into this complex matter. Moreover, the differences between the response of a single cell and the one of the whole animal must be considered as well. Once again, several studies underscore that in vitro data cannot be extrapolated indiscriminately to in vivo conditions.

Besides similar cell signaling pathways, μg and radiation can indirectly affect inflammation by influencing components that are essential in mediating the inflammatory response. A crucial step in inflammation is the trafficking of leukocytes from the blood stream into the tissue. This leukocyte–endothelial adhesion involves dynamic interactions between leukocytes and endothelial cells, and is mediated by several families of cell adhesion molecules (CAMs). CAMs that are expressed on the surface of vascular endothelial cells include the selectin family (E-selectin and P-selectin) and the Ig superfamily (e.g., ICAM-1). These CAMs interact with leukocytes to initiate cell extravasation and migration. The impact of both radiation and μg on the induction of cell adhesion molecules on endothelial cells has been studied (Zhang et al. 2008; Hallahan et al. 1996; Romanov et al. 2001). Both E-selectin and ICAM-1 are increased after X-irradiation, whereas VCAM and ICAM-1 also increase under hypogravity conditions. These experiments indicate that leukocyte adhesion (and consequently inflammation) might be promoted both by radiation and μg. Additional experiments are needed to gain more insight into the combined effects of both physical spaceflight stressors.

To gain more fundamental insight into the potential interplay between these two physical space stressors in immune cells, Earth-based experiments that simulate space conditions can be useful. In this light, µg-simulating devices such as rotating-wall bioreactors (clinostat) or the Random Positioning Machine (RPM) are currently used to perform in vitro experiments under hypogravity conditions. Ideally, cells should be exposed to µg and different radiation qualities at the same time. However, only a limited number of studies have yet been performed in which cells are simultaneously exposed to both these physical space stressors (Beck et al. 2014; Pani et al. 2016; Fernandez-Gonzalo et al. 2017). In a recent study simulated µg was found to increase particle radiation-induced apoptosis in B lymphoblast cells (Dang et al. 2014). The combined impact of space radiation and microgravity on DNA integrity is also of a major concern for their impact on astronaut health. Unfortunately, results from either ground-based or inflight studies are ambiguous and require validation in the true space environment (reviewed in Moreno-Villanueva et al. 2017).

20.5 Conclusion

It has become clear that several aspects of the spaceflight environment lead to acute and long-term changes in the immune system. Therefore, to reduce the health risks for astronauts, it is important to better understand the mechanisms responsible for the changes observed in the immune parameters (Dang et al. 2014). Ideally, to discriminate between different factors and their impact on the immune system, data should be obtained e.g., from the same cohort of animals that are exposed simultaneously to different spaceflight stressors. However, various experimental setups (space vs. Earth-based models) are concomitant with different uncontrolled stressors e.g., shipping animals, housing. These conditions may elicit immune system alterations that always make direct comparison problematic. In addition, to study the effect of space radiation on ground-based models, radiation should be delivered at a very low dose rate for extended periods in order to be as relevant as possible to spaceflight. Unfortunately, these conditions are not easy to achieve on Earth due to facility limitations and high demand for beam time. Currently, researchers are starting to elucidate the different effects of cosmic radiation on the immune system. Nevertheless, several important questions remain: Which pathways are responsible for repairing radiation-induced changes? How capable is the irradiated immune system of responding to an immune challenge? How are other metabolic cofactors (see Chaps. 5, 16, 28, 32 and 33), hormones and transmitters affecting this process?

In conclusion, gaining more insight into changes in immune responses after radiation exposure is needed to more accurately predict health risks associated with long-duration spaceflight. This knowledge might not only be relevant to extended ISS missions (more than 1 year) and to future exploration and colonization missions in space but also of significant importance to life on Earth and to patient care, e.g., during cancer radiation therapy, when metabolic effects, the immune system, and the radiation effects are interacting closely on the therapeutic goals of curing and limiting cancer growth, respectively.

References

Averbeck D (2010) Non-targeted effects as a paradigm breaking evidence. Mutat Res 687:7–12

Badhwar GD, Golightly MJ, Konradi A, Atwell W, Kern JW, Cash B, Benton EV, Frank AL, Sanner D, Keegan RP, Frigo LA, Petrov VM, Tchernykh IV, Akatov Yu A, Shurshakov VA, Arkhangelsky VV, Kushin VV, Klyachin NA, Vana N, Schoner W (1996) In-flight radiation measurements on STS-60. Radiat Meas 26:17–34

Barcellos-Hoff MH, Blakely EA, Burma S, Fornace AJ Jr, Gerson S, Hlatky L, Kirsch DG, Luderer U, Shay J, Wang Y, Weil MM (2015) Concepts and challenges in cancer risk prediction for the space radiation environment. Life Sci Space Res (Amst) 6:92–103

Beck M, Moreels M, Quintens R, Abou-El-Ardat K, El-Saghire H, Tabury K, Michaux A, Janssen A, Neefs M, Van Oostveldt P, De Vos WH, Baatout S (2014) Chronic exposure to simulated space conditions predominantly affects cytoskeleton remodeling and oxidative stress response in mouse fetal fibroblasts. Int J Mol Med 34:606–615

Benton ER, Benton EV (2001) Space radiation dosimetry in low-earth orbit and beyond. Nucl Instrum Methods Phys Res B 184:255–294

Berger T, Bilski P, Hajek M, Puchalska M, Reitz G (2013) The MATROSHKA experiment: results and comparison from extravehicular activity (MTR-1) and intravehicular activity (MTR-2A/2B) exposure. Radiat Res 180:622–637

Berger T, Przybyla B, Matthia D, Reitz G, Burmeister S, Labrenz J, Bilski P, Horwacik T, Twardak A, Hajek M, Fugger M, Hofstatter C, Sihver L, Palfalvi JK, Szabo J, Stradi A, Ambrozova I, Kubancak J, Brabcova KP, Vanhavere F, Cauwels V, Van Hoey O, Schoonjans W, Parisi A, Gaza R, Semones E, Yukihara EG, Benton ER, Doull BA, Uchihori Y, Kodaira S, Kitamura H, Boehme M (2016) DOSIS & DOSIS 3D: long-term dose monitoring onboard the Columbus Laboratory of the International Space Station (ISS). J Space Weather Space Clim 6:A39

Berger T, Burmeister S, Matthiae D, Przybyla B, Reitz G, Bilski P, Hajek M, Sihver L, Szabo J, Ambrozova I, Vanhavere F, Gaza R, Semones E, Yukihara EG, Benton ER, Uchihori Y, Kodaira S, Kitamura H, Boehme M (2017) DOSIS & DOSIS 3D: radiation measurements with the DOSTEL instruments onboard the Columbus Laboratory of the ISS in the years 2009–2016. J Space Weather Space Clim 7. https://doi.org/10.1051/swsc/2017005

Boffetta P, van der Hel O, Norppa H, Fabianova E, Fucic A, Gundy S, Lazutka J, Cebulska-Wasilewska A, Puskailerova D, Znaor A, Kelecsenyi Z, Kurtinaitis J, Rachtan J, Forni A, Vermeulen R, Bonassi S (2007) Chromosomal aberrations and cancer risk: results of a cohort study from Central Europe. Am J Epidemiol 165:36–43

Chang J, Feng W, Wang Y, Luo Y, Allen AR, Koturbash I, Turner J, Stewart B, Raber J, Hauer-Jensen M, Zhou D, Shao L (2015) Whole-body proton irradiation causes long-term damage to hematopoietic stem cells in mice. Radiat Res 183:240–248

Chang J, Luo Y, Wang Y, Pathak R, Sridharan V, Jones T, Mao XW, Nelson G, Boerma M, Hauer-Jensen M, Zhou D, Shao L (2016) Low doses of oxygen ion irradiation cause acute damage to hematopoietic cells in mice. PLoS One 11:e0158097

Chang J, Feng W, Wang Y, Allen AR, Turner J, Stewart B, Raber J, Hauer-Jensen M, Zhou D, Shao L (2017a) 28Si total body irradiation injures bone marrow hematopoietic stem cells via induction of cellular apoptosis. Life Sci Space Res 13:39–44

Chang J, Wang Y, Pathak R, Sridharan V, Jones T, Mao XW, Nelson G, Boerma M, Hauer-Jensen M, Zhou D, Shao L (2017b) Whole body proton irradiation causes acute damage to bone marrow hematopoietic progenitor and stem cells in mice. Int J Radiat Biol 93:1312–1320

Chylack LT Jr, Feiveson AH, Peterson LE, Tung WH, Wear ML, Marak LJ, Hardy DS, Chappell LJ, Cucinotta FA (2012) NASCA report 2: Longitudinal study of relationship of exposure to space radiation and risk of lens opacity. Radiat Res 178:25–32

Chylack LT Jr, Peterson LE, Feiveson AH, Wear ML, Manuel FK, Tung WH, Hardy DS, Marak LJ, Cucinotta FA (2009) NASA study of cataract in astronauts (NASCA). Report 1: Cross-sectional study of the relationship of exposure to space radiation and risk of lens opacity. Radiat Res 172:10–20

Cortese F, Klokov D, Osipov A, Stefaniak J, Moskalev A, Schastnaya J, Cantor C, Aliper A, Mamoshina P, Ushakov I, Sapetsky A, Vanhaelen Q, Alchinova I, Karganov M, Kovalchuk O,

Wilkins R, Shtemberg A, Moreels M, Baatout S, Izumchenko E, de Magalhaes JP, Artemov AV, Costes SV, Beheshti A, Mao XW, Pecaut MJ, Kaminskiy D, Ozerov IV, Scheibye-Knudsen M, Zhavoronkov A (2018) Vive la radioresistance!: converging research in radiobiology and biogerontology to enhance human radioresistance for deep space exploration and colonization. Oncotarget 9:14692–14722

Cucinotta FA, Durante M (2006) Cancer risk from exposure to galactic cosmic rays: implications for space exploration by human beings. Lancet Oncol 7:431–435

Cucinotta FA, Schimmerling W, Wilson JW, Peterson LE, Badhwar GD, Saganti PB, Dicello JF (2001) Space radiation cancer risks and uncertainties for Mars missions. Radiat Res 156:682–688

Cucinotta FA, To K, Cacao E (2017) Predictions of space radiation fatality risk for exploration missions. Life Sci Space Res (Amst) 13:1–11

Dachev TP, Semkova JV, Tomov BT, Matviichuk YN, Dimitrov PG, Koleva RT, Malchev S, Bankov NG, Shurshakov VA, Benghin VV, Yarmanova EN, Ivanova OA, Hader DP, Lebert M, Schuster MT, Reitz G, Horneck G, Uchihori Y, Kitamura H, Ploc O, Cubancak J, Nikolaev I (2015) Overview of the Liulin type instruments for space radiation measurement and their scientific results. Life Sci Space Res (Amst) 4:92–114

Dang B, Yang Y, Zhang E, Li W, Mi X, Meng Y, Yan S, Wang Z, Wei W, Shao C, Xing R, Lin C (2014) Simulated microgravity increases heavy ion radiation-induced apoptosis in human B lymphoblasts. Life Sci 97:123–128

Dietze G, Bartlett DT, Cool DA, Cucinotta FA, Jia X, McAulay IR, Pelliccioni M, Petrov V, Reitz G, Sato T (2013) ICRP, 123. Assessment of radiation exposure of astronauts in space. Ann ICRP 42(4):1–339

Durante M (2005) Biomarkers of space radiation risk. Radiat Res 164:467–473

Durante M, Cucinotta FA (2008) Heavy ion carcinogenesis and human space exploration. Nat Rev Cancer 8:465–472

Durante M, Snigiryova G, Akaeva E, Bogomazova A, Druzhinin S, Fedorenko B, Greco O, Novitskaya N, Rubanovich A, Shevchenko V, Von Recklinghausen U, Obe G (2003) Chromosome aberration dosimetry in cosmonauts after single or multiple space flights. Cytogenet Genome Res 103:40–46

Eaton BR, MacDonald SM, Yock TI, Tarbell NJ (2015) Secondary malignancy risk following proton radiation therapy. Front Oncol 5:261

Fakir H, Sachs RK, Stenerlow B, Hofmann W (2006) Clusters of DNA double-strand breaks induced by different doses of nitrogen ions for various LETs: experimental measurements and theoretical analyses. Radiat Res 166:917–927

Fernandez-Gonzalo R, Baatout S, Moreels M (2017) Impact of particle irradiation on the immune system: from the clinic to Mars. Front Immunol 8:177

Frippiat JP, Crucian BE, de Quervain DJ, Grimm D, Montano N, Praun S, Roozendaal B, Schelling G, Thiel M, Ullrich O, Choukér A (2016) Towards human exploration of space: the THESEUS review series on immunology research priorities. NPJ Microgravity 2:16040

George K, Durante M, Wu H, Willingham V, Badhwar G, Cucinotta FA (2001) Chromosome aberrations in the blood lymphocytes of astronauts after space flight. Radiat Res 156: 731–738

George K, Willingham V, Cucinotta FA (2005) Stability of chromosome aberrations in the blood lymphocytes of astronauts measured after space flight by FISH chromosome painting. Radiat Res 164:474–480

George K, Rhone J, Beitman A, Cucinotta FA (2013) Cytogenetic damage in the blood lymphocytes of astronauts: effects of repeat long-duration space missions. Mutat Res 756:165–169

Ghardi M, Moreels M, Chatelain B, Chatelain C, Baatout S (2012) Radiation-induced double strand breaks and subsequent apoptotic DNA fragmentation in human peripheral blood mononuclear cells. Int J Mol Med 29:769–780

Greco O, Durante M, Gialanella G, Grossi G, Pugliese M, Scampoli P, Snigiryova G, Obe G (2003) Biological dosimetry in Russian and Italian astronauts. Adv Space Res 31:1495–1503

Green LM, Murray DK, Bant AM, Kazarians G, Moyers MF, Nelson GA, Tran DT (2001) Response of thyroid follicular cells to gamma irradiation compared to proton irradiation.

I. Initial characterization of DNA damage, micronucleus formation, apoptosis, cell survival, and cell cycle phase redistribution. Radiat Res 155:32–42
Gridley DS, Dutta-Roy R, Andres ML, Nelson GA, Pecaut MJ (2006) Acute effects of iron-particle radiation on immunity. Part II: Leukocyte activation, cytokines and adhesion. Radiat Res 165:78–87
Gridley DS, Luo-Owen X, Rizvi A, Makinde A, Pecaut M, Mao XW, Slater JM (2010) Low-dose photon and simulated solar particle event proton effects on Foxp3+ T regulatory cells and other leukocytes. Technol Cancer Res Treat 9:637–649
Gridley DS, Pecaut MJ (2006) Whole-body irradiation and long-term modification of bone marrow-derived cell populations by low- and high-LET radiation. In Vivo 20:781–789
Gridley DS, Pecaut MJ (2016) Changes in the distribution and function of leukocytes after whole-body iron ion irradiation. J Radiat Res 57:477–491
Gridley DS, Pecaut MJ, Miller GM, Moyers MF, Nelson GA (2001) Dose and dose rate effects of whole-body gamma-irradiation: II. Hematological variables and cytokines. In Vivo 15:209–216
Gridley DS, Pecaut MJ, Dutta-Roy R, Nelson GA (2002a) Dose and dose rate effects of whole-body proton irradiation on leukocyte populations and lymphoid organs: Part I. Immunol Lett 80:55–66
Gridley DS, Pecaut MJ, Nelson GA (2002b) Total-body irradiation with high-LET particles: acute and chronic effects on the immune system. Am J Physiol Regul Integr Comp Physiol 282:R677–R688
Gridley DS, Rizvi A, Luo-Owen X, Makinde AY, Coutrakon GB, Koss P, Slater JM, Pecaut MJ (2008) Variable hematopoietic responses to acute photons, protons and simulated solar particle event protons. In Vivo 22:159–169
Gridley DS, Rizvi A, Luo-Owen X, Makinde AY, Pecaut MJ (2009) Low dose, low dose rate photon radiation modifies leukocyte distribution and gene expression in CD4(+) T cells. J Radiat Res 50:139–150
Gridley DS, Rizvi A, Makinde AY, Luo-Owen X, Mao XW, Tian J, Slater JM, Pecaut MJ (2013) Space-relevant radiation modifies cytokine profiles, signaling proteins and Foxp3+ T cells. Int J Radiat Biol 89:26–35
Gueguinou N, Huin-Schohn C, Bascove M, Bueb JL, Tschirhart E, Legrand-Frossi C, Frippiat JP (2009) Could spaceflight-associated immune system weakening preclude the expansion of human presence beyond Earth's orbit? J Leukoc Biol 86:1027–1038
Hada M, Georgakilas AG (2008) Formation of clustered DNA damage after high-LET irradiation: a review. J Radiat Res 49:203–210
Hada M, Chappell LJ, Wang M, George KA, Cucinotta FA (2014) Induction of chromosomal aberrations at fluences of less than one HZE particle per cell nucleus. Radiat Res 182:368–379
Hallahan D, Kuchibhotla J, Wyble C (1996) Cell adhesion molecules mediate radiation-induced leukocyte adhesion to the vascular endothelium. Cancer Res 56:5150–5155
Harrington NP, Chambers KA, Ross WM, Filion LG (1997) Radiation damage and immune suppression in splenic mononuclear cell populations. Clin Exp Immunol 107:417–424
Hassler DM, Zeitlin C, Wimmer-Schweingruber RF, Ehresmann B, Rafkin S, Eigenbrode JL, Brinza DE, Weigle G, Bottcher S, Bohm E, Burmeister S, Guo J, Kohler J, Martin C, Reitz G, Cucinotta FA, Kim MH, Grinspoon D, Bullock MA, Posner A, Gomez-Elvira J, Vasavada A, Grotzinger JP (2014) Mars' surface radiation environment measured with the Mars Science Laboratory's Curiosity rover. Science 343:1244797
Horneck G (1999) Impact of microgravity on radiobiological processes and efficiency of DNA repair. Mutat Res 430:221–228
ICRP (2007) The 2007 recommendations of the International Commission on Radiological Protection. ICRP publication 103. Ann ICRP 37:1–332
ICRP (2012) ICRP statement on tissue reactions and early and late effects of radiation in normal tissues and organs – threshold doses for tissue reactions in a radiation protection context. ICRP publication 118. Ann ICRP 41:1–322
Kajioka EH, Gheorghe C, Andres ML, Abell GA, Folz-Holbeck J, Slater JM, Nelson GA, Gridley DS (1999) Effects of proton and gamma radiation on lymphocyte populations and acute response to antigen. In Vivo 13:525–533

Kajioka EH, Andres ML, Li J, Mao XW, Moyers MF, Nelson GA, Slater JM, Gridley DS (2000) Acute effects of whole-body proton irradiation on the immune system of the mouse. Radiat Res 153:587–594

Kamran SC, Berrington de Gonzalez A, Ng A, Haas-Kogan D, Viswanathan AN (2016) Therapeutic radiation and the potential risk of second malignancies. Cancer 122:1809–1821

Kiefer J, Pross HD (1999) Space radiation effects and microgravity. Mutat Res 430:299–305

Kinner A, Wu W, Staudt C, Iliakis G (2008) Gamma-H2AX in recognition and signaling of DNA double-strand breaks in the context of chromatin. Nucleic Acids Res 36:5678–5694

Kleiman NJ (2012) Radiation cataract. Ann ICRP 41:80–97

Knoll GF (2010) Radiation detection and measurement. Wiley, Chichester

Konstantinova IV, Antropova EN, Legen'kov VI, Zazhirei VD (1973) Reactivity of lymphoid blood cells in the crew of "Soiuz-6", "Soiuz-7" and "Soiuz-8" spacecraft before and after flight. Kosm Biol Med 7:35–40

Kroupa M, Bahadori A, Campbell-Ricketts T, Empl A, Hoang SM, Idarraga-Munoz J, Rios R, Semones E, Stoffle N, Tlustos L, Turecek D, Pinsky L (2015) A semiconductor radiation imaging pixel detector for space radiation dosimetry. Life Sci Space Res (Amst) 6:69–78

Kusunoki Y, Hayashi T (2008) Long-lasting alterations of the immune system by ionizing radiation exposure: implications for disease development among atomic bomb survivors. Int J Radiat Biol 84:1–14

Lee K, Flanders J, Semones E, Shelfer T, Riman F (2007) Simultaneous observation of the radiation environment inside and outside the ISS. Adv Space Res 40:1558–1561

Little MP (2009) Cancer and non-cancer effects in Japanese atomic bomb survivors. J Radiol Prot 29:A43–A59

Luo-Owen X, Pecaut MJ, Rizvi A, Gridley DS (2012) Low-dose total-body gamma irradiation modulates immune response to acute proton radiation. Radiat Res 177:251–264

Mitra AK, Singh RK, Krishna M (2007) MAP kinases: differential activation following in vivo and ex vivo irradiation. Mol Cell Biochem 294:65–72

Moreno-Villanueva M, Wong M, Lu T, Zhang Y, Wu H (2017) Interplay of space radiation and microgravity in DNA damage and DNA damage response. NPJ Microgravity 3:14

Morgan WF, Sowa MB (2007) Non-targeted bystander effects induced by ionizing radiation. Mutat Res 616:159–164

Narang H, Bhat N, Gupta SK, Santra S, Choudhary RK, Kailash S, Krishna M (2009) Differential activation of mitogen-activated protein kinases following high and low LET radiation in murine macrophage cell line. Mol Cell Biochem 324:85–91

Norppa H, Bonassi S, Hansteen IL, Hagmar L, Stromberg U, Rossner P, Boffetta P, Lindholm C, Gundy S, Lazutka J, Cebulska-Wasilewska A, Fabianova E, Sram RJ, Knudsen LE, Barale R, Fucic A (2006) Chromosomal aberrations and SCEs as biomarkers of cancer risk. Mutat Res 600: 37–45

Obe G, Johannes I, Johannes C, Hallman K, Reitz G, Facius R (1997) Chromosomal aberrations in blood lymphocytes of astronauts after long-term space flights. Int J Radiat Biol 72:727–734

Pani G, Verslegers M, Quintens R, Samari N, de Saint-Georges L, van Oostveldt P, Baatout S, Benotmane MA (2016) Combined exposure to simulated microgravity and acute or chronic radiation reduces neuronal network integrity and survival. PLoS One 11:e0155260

Parisi A, Van Hoey O, Megret P, Vanhavere F (2017) Deconvolution study on the glow curve structure of LiF:Mg,Ti and LiF:Mg,Cu,P thermoluminescent detectors exposed to H-1, He-4 and C-12 ion beams. Nucl Instrum Methods Phys Res B Beam Interact Mater Atoms 407:222–229

Pecaut MJ, Gridley DS (2008) Radiation and secondary immune response to lipopolysaccharide. In Vivo 22:423–434

Pecaut MJ, Gridley DS (2010) The impact of mouse strain on iron ion radio-immune response of leukocyte populations. Int J Radiat Biol 86:409–419

Pecaut MJ, Haerich P, Zuccarelli CN, Smith AL, Zendejas ED, Nelson GA (2002) Behavioral consequences of radiation exposure to simulated space radiation in the C57BL/6 mouse: open field, rotorod, and acoustic startle. Cogn Affect Behav Neurosci 2:329–340

Pecaut MJ, Nelson GA, Gridley DS (2001) Dose and dose rate effects of whole-body gamma-irradiation: I. Lymphocytes and lymphoid organs. In Vivo 15:195–208

Pecaut MJ, Gridley DS, Nelson GA (2003) Long-term effects of low-dose proton radiation on immunity in mice: shielded vs. unshielded. Aviat Space Environ Med 74:115–124
Pecaut MJ, Dutta-Roy R, Smith AL, Jones TA, Nelson GA, Gridley DS (2006) Acute effects of iron-particle radiation on immunity. Part I: Population distributions. Radiat Res 165:68–77
Preston DL, Ron E, Tokuoka S, Funamoto S, Nishi N, Soda M, Mabuchi K, Kodama K (2007) Solid cancer incidence in atomic bomb survivors: 1958–1998. Radiat Res 168:1–64
Pross HD, Casares A, Kiefer J (2000) Induction and repair of DNA double-strand breaks under irradiation and microgravity. Radiat Res 153:521–525
Rastegar N, Eckart P, Mertz M (2002) Radiation-induced cataract in astronauts and cosmonauts. Graefes Arch Clin Exp Ophthalmol 240:543–547
Reitz G, Bucker H, Facius R, Horneck G, Graul EH, Berger H, Ruther W, Heinrich W, Beaujean R, Enge W, Alpatov AM, Ushakov IA, Zachvatkin Yu A, Mesland DA (1989) Influence of cosmic radiation and/or microgravity on development of Carausius morosus. Adv Space Res 9:161–173
Rizvi A, Pecaut MJ, Slater JM, Subramaniam S, Gridley DS (2011) Low-dose gamma-rays modify CD4(+) T cell signalling response to simulated solar particle event protons in a mouse model. Int J Radiat Biol 87:24–35
Rogakou EP, Pilch DR, Orr AH, Ivanova VS, Bonner WM (1998) DNA double-stranded breaks induce histone H2AX phosphorylation on serine 139. J Biol Chem 273:5858–5868
Romanov SR, Kozakiewicz BK, Holst CR, Stampfer MR, Haupt LM, Tlsty TD (2001) Normal human mammary epithelial cells spontaneously escape senescence and acquire genomic changes. Nature 409:633–637
Sanzari JK, Romero-Weaver AL, James G, Krigsfeld G, Lin L, Diffenderfer ES, Kennedy AR (2013a) Leukocyte activity is altered in a ground based murine model of microgravity and proton radiation exposure. PLoS One 8:e71757
Sanzari JK, Wan XS, Krigsfeld GS, Wroe AJ, Gridley DS, Kennedy AR (2013b) The effects of gamma and proton radiation exposure on hematopoietic cell counts in the ferret model. Gravit Space Res 1:79–94
Sanzari JK, Cengel KA, Wan XS, Rusek A, Kennedy AR (2014) Acute hematological effects in mice exposed to the expected doses, dose-rates, and energies of solar particle event-like proton radiation. Life Sci Space Res (Amst) 2:86–91
Shankar G, Scott Bryson J, Darrell Jennings C, Kaplan AM, Cohen DA (1999) Idiopathic pneumonia syndrome after allogeneic bone marrow transplantation in mice. Role of pretransplant radiation conditioning. Am J Respir Cell Mol Biol 20:1116–1124
Shearer WT, Zhang S, Reuben JM, Lee BN, Butel JS (2005) Effects of radiation and latent virus on immune responses in a space flight model. J Allergy Clin Immunol 115:1297–1303
Smith MB, Khulapko S, Andrews HR, Arkhangelsky V, Ing H, Koslowksy MR, Lewis BJ, Machrafi R, Nikolaev I, Shurshakov V (2016) Bubble-detector measurements of neutron radiation in the International Space Station: ISS-34 to ISS-37. Radiat Prot Dosim 168:154–166
Su TT (2006) Cellular responses to DNA damage: one signal, multiple choices. Annu Rev Genet 40:187–208
Suetens A, Moreels M, Quintens R, Chiriotti S, Tabury K, Michaux A, Gregoire V, Baatout S (2014) Carbon ion irradiation of the human prostate cancer cell line PC3: a whole genome microarray study. Int J Oncol 44:1056–1072
Suetens A, Moreels M, Quintens R, Soors E, Buset J, Chiriotti S, Tabury K, Gregoire V, Baatout S (2015) Dose- and time-dependent gene expression alterations in prostate and colon cancer cells after in vitro exposure to carbon ion and X-irradiation. J Radiat Res 56:11–21
Suetens A, Konings K, Moreels M, Quintens R, Verslegers M, Soors E, Tabury K, Gregoire V, Baatout S (2016) Higher initial DNA damage and persistent cell cycle arrest after carbon ion irradiation compared to X-irradiation in prostate and colon cancer cells. Front Oncol 6:87
Suman S, Kallakury BV, Fornace AJ Jr, Datta K (2015) Protracted upregulation of leptin and IGF1 is associated with activation of PI3K/Akt and JAK2 pathway in mouse intestine after ionizing radiation exposure. Int J Biol Sci 11:274–283

Swenberg CE, Birke S, Geacintov NE (1991) Characterization of the interaction of the radioprotector 1-methyl-2-[2-(methylthio)-2-piperidinovinyl]quinolinium iodide with supercoiled DNA. Radiat Res 127:138–145

Szanto P, Apathy I, Deme S, Hirn A, Nikolaev IV, Pazmandi T, Shurshakov VA, Tolochek RV, Yarmanova EN (2015) Onboard cross-calibration of the Pille-ISS Detector System and measurement of radiation shielding effect of the water filled protective curtain in the ISS crew cabin. Radiat Meas 82:59–63

Tauber S, Hauschild S, Paulsen K, Gutewort A, Raig C, Hurlimann E, Biskup J, Philpot C, Lier H, Engelmann F, Pantaleo A, Cogoli A, Pippia P, Layer LE, Thiel CS, Ullrich O (2015) Signal transduction in primary human T lymphocytes in altered gravity during parabolic flight and clinostat experiments. Cell Physiol Biochem 35:1034–1051

Terato H, Tanaka R, Nakaarai Y, Nohara T, Doi Y, Iwai S, Hirayama R, Furusawa Y, Ide H (2008) Quantitative analysis of isolated and clustered DNA damage induced by gamma-rays, carbon ion beams, and iron ion beams. J Radiat Res 49:133–146

Testard I, Ricoul M, Hoffschir F, Flury-Herard A, Dutrillaux B, Fedorenko B, Gerasimenko V, Sabatier L (1996) Radiation-induced chromosome damage in astronauts' lymphocytes. Int J Radiat Biol 70:403–411

Trani D, Datta K, Doiron K, Kallakury B, Fornace AJ Jr (2010) Enhanced intestinal tumor multiplicity and grade in vivo after HZE exposure: mouse models for space radiation risk estimates. Radiat Environ Biophys 49:389–396

Tverskaya LV, Panasyuk MI, Reizman SY, Sosnovets EN, Teltsov MV, Tsetlin VV (2004) The features of radiation dose variations onboard ISS and Mir space station: comparative study. Adv Space Res 34:1424–1428

Ullrich O, Huber K, Lang K (2008) Signal transduction in cells of the immune system in microgravity. Cell Commun Signal 6:9

Vanhavere F, Genicot JL, O'Sullivan D, Zhou D, Spurny F, Jadrnickova I, Sawakuchi GO, Yukihara EG (2008) DOsimetry of BIological EXperiments in SPace (DOBIES) with luminescence (OSL and TL) and track etch detectors. Radiat Meas 43:694–697

Varadkar PA, Krishna M (2004) Differential activation of kinases in ex vivo and in vivo irradiated mice lymphocytes. J Radiat Res 45:127–131

Varadkar PA, Krishna M, Verma NC (2003) Dose-dependent differential expression of protein kinase C isozymes in mouse lymphocytes after gamma irradiation in vivo and ex vivo. Radiat Res 159:453–457

Wang Y, Chang J, Li X, Pathak R, Sridharan V, Jones T, Mao XW, Nelson G, Boerma M, Hauer-Jensen M, Zhou D, Shao L (2017) Low doses of oxygen ion irradiation cause long-term damage to bone marrow hematopoietic progenitor and stem cells in mice. PLoS One 12:e0189466

Weil MM, Ray FA, Genik PC, Yu Y, McCarthy M, Fallgren CM, Ullrich RL (2014) Effects of 28Si ions, 56Fe ions, and protons on the induction of murine acute myeloid leukemia and hepatocellular carcinoma. PLoS One 9:e104819

Xiao M, Whitnall MH (2009) Pharmacological countermeasures for the acute radiation syndrome. Curr Mol Pharmacol 2:122–133

Yang TC, George K, Johnson AS, Tavakoli A, Durante M, Fedorenko BS (1997) Cytogenetic effects of space radiation in lymphocytes of MIR-18 crews. Aviakosm Ekolog Med 31:8–14

Zeitlin C, Hassler DM, Cucinotta FA, Ehresmann B, Wimmer-Schweingruber RF, Brinza DE, Kang S, Weigle G, Bottcher S, Bohm E, Burmeister S, Guo J, Kohler J, Martin C, Posner A, Rafkin S, Reitz G (2013) Measurements of energetic particle radiation in transit to Mars on the Mars Science Laboratory. Science 340:1080–1084

Zhang R, Jia G, Bao J, Zhang Y, Bai Y, Lin L, Tang H, Ma J (2008) Increased vascular cell adhesion molecule-1 was associated with impaired endothelium-dependent relaxation of cerebral and carotid arteries in simulated microgravity rats. J Physiol Sci 58:67–73

Open Access This chapter is licensed under the terms of the Creative Commons Attribution 4.0 International License (http://creativecommons.org/licenses/by/4.0/), which permits use, sharing, adaptation, distribution and reproduction in any medium or format, as long as you give appropriate credit to the original author(s) and the source, provide a link to the Creative Commons license and indicate if changes were made.

The images or other third party material in this chapter are included in the chapter's Creative Commons license, unless indicated otherwise in a credit line to the material. If material is not included in the chapter's Creative Commons license and your intended use is not permitted by statutory regulation or exceeds the permitted use, you will need to obtain permission directly from the copyright holder.

Part IV

Preventive and Diagnostic Tool and Strategies

Considerations for Development and Application of Health Monitoring Tools in Space

21

Ines Kaufmann, Teodor Veres, Judith-Irina Buchheim, and Alexander Choukér

21.1 Introduction

Maintaining astronauts' health is one of the most important strategic objectives during manned space mission which is necessary to attain versatile mission goals. For astronauts, the next physician is far away and problems in radiocircuits with delay or loss of communication will likely occur. Therefore, it is of great interest to develop innovative health care screening tools for astronauts affording to appreciate the certainty of the presence or absence of diseases and to judge the demand for therapeutic strategies. In space, diseases like acute or chronic infections of bacterial, fungal, and viral nature as well as cancer development are likely to occur and are strongly linked to the adequate performance of the immune system. Moreover, the latter is eminently modified and compromised as a consequence of stressors in space. Accordingly, monitoring of health in space is also a monitoring of stress and changes of organ functions like immunity.

The pace of development of new diagnostic techniques and lab-on-a-chip microtechnologies has greatly accelerated over the past decades (Sackmann et al.

I. Kaufmann (✉)
Munich Clinic Neuperlach, Munich, Germany
e-mail: ines.kaufmann@klinikum-muenchen.de

T. Veres
Life Sciences Division, National Research Council of Canada, Ottawa, ON, Canada

Biomedical Engineering Department, McGill University, Montreal, QC, Canada

Bioengineering Department, McGill University, Montreal, QC, Canada

J.-I. Buchheim · A. Choukér
Laboratory of Translational Research Stress and Immunity, Department of Anaesthesiology, Hospital of the University of Munich (LMU), Munich, Germany

© Springer Nature Switzerland AG 2020
A. Choukér (ed.), *Stress Challenges and Immunity in Space*,
https://doi.org/10.1007/978-3-030-16996-1_21

2014). Since quality of the diagnostic procedure largely determines quality of care, overcoming deficiencies in standards and methodology deserves high priority. It is self-understanding that it is important to test and to further evaluate the benefits of diagnostic tools on Earth — under laboratory and field conditions — before considering its use in space. This is critical not only because of the limited number of astronauts but also of the high costs of developing the specifically certified hardware for space crafts. The most important challenges result from the fact that after detecting changes in a promising parameter/variable, the description and definition of the latter as a biomarker (Baumgartner et al. 2010) requires an adequate sample size for either the "sick" or their controls in order to provide the desired information and certainty. However, there is a big ambiguousness and uncertainty regarding biomarker-guided adaptive designs of clinical trials which should be taken into consideration because they can get an appealing character. The main reason is their application to real clinical practice *and* their ability to evaluate both multiple experimental treatments and biomarkers simultaneously (Antoniou et al. 2016). The number of patients in a trial can be adjusted to the results from an interim analysis and the duration of the study is allowed to become minimized. However, both procedures require a precise detailed data analysis with the right conclusions under careful attention of the study issue and the key questions which have to be answered. Hence it is absolutely necessary to evaluate a diagnostic procedure accurately and extensively with respect to the Good Clinical Practice Guidelines (GCP Guidelines) on Earth before taking into account the use in astronauts during space missions.

21.2 The Ideal Diagnostic Tools for Space

21.2.1 Definitions

Diagnostic investigations collect information to clarify the individual health status. The "ideal" diagnostic tool is able to elevate certainty of the presence or absence of disease—meaning to have high sensitivity and specificity—to support treatment of pathophysiological conditions, to assess prognosis, and to monitor the clinical course (Kottnerus et al. 2002).

Diagnostic procedures to monitor stress effects and health-relevant changes must be evaluated not only in accordance with their intended objectives but also taking into account possible discomfort and complications. The guiding principle for application should be: "Minimize astronauts' risk. Maximize astronauts' care." With regard to the risks and invasiveness of medical procedures, there are three main categories (Table 21.1):

(a) Noninvasive procedures,
(b) Minimally invasive procedures,
(c) Invasive techniques.

Table 21.1 Examples for noninvasive, minimally invasive, and invasive procedures

Noninvasive procedures
– Cognitive tests
– Monitoring group interaction (see Chap. 22)
– Electrocardiography (ECG, see Chap. 23)
– Electroencephalography (EEG)
– Temperature (double sensor, see Chap. 26)
– Breath gas analysis (e.g., E-NOSE, see Chap. 24)
– Saliva and urine analyses
– Hair analyses for stress hormones (see Chap. 29)
– Wearables with flexible sensors detecting several vital data, e.g., the blood pulse wave, electrocardiogram, blood pressure, breathing rate/volume, skin temperature, oxygen saturation, and other physical activity levels
– Sleep monitoring by new devices (e.g., SomnoART)
Minimally invasive procedures
– Lab-on-a-chip technologies
– Capillary blood analyses (e.g., analyses by flow cytometry, see Chap. 27)
Invasive procedures
– Invasive catheterization
– Surgical procedures

21.2.2 Examples

Medical techniques which do not penetrate the mucosa, skin, or body cavity beyond a natural or artificial body orifice are called *noninvasive*. For centuries, physicians have used many ordinary noninvasive methods rested upon physical parameters. For example, they auscultated heart and lung sounds, percutated or palpated in order to assess organ functions. The discovery of the first modern noninvasive procedures dates back to the end of the nineteenth century: In 1895, the German Physicist Conrad Röntgen discovered a new form of radiation; he called this radiation "X-rays" to denote its unknown structure. X-radiation had the unique property to penetrate through many materials that absorb visible light and to knock electrons loose from atoms. For first time, it was possible to depict anatomical structures and to make a diagnosis in ill patients without exploratory surgery. This was the beginning of a new era. Since then, noninvasive methods have continuously enlarged and have been tested to provide important information on the health of patients in remote areas using real-time telemedicine devices, e.g., an electrocardiogram and ultrasound (Kirkpatrick et al. 2013; Nicogossian et al. 2001; Otto et al. 2009, 2010). Noninvasive methods permit accomplishment of diagnostic procedures without extravagant expenses and with a minimal risk for the individual. They are of outstanding impact for observation of strain and its consequences in individuals under extreme living conditions by considering stress-associated neurocognitive changes (see for details Chap. 7), by using strategies for psychological (Chap. 22), vegetative (Chaps. 6 and 23) and circadian rhythm (Chaps. 9 and 26) and sleep changes, respectively. Easily collectable specimens allow the analyses of particular hormones, pharmaceuticals, or contaminants in body fluids (saliva, urine). Fibrous keratin structures (e.g., hair, see Chap. 29) can also serve as appropriate material to monitor the stress effects (Ballantyne 2007; Gidlow et al. 2016; Stalder et al. 2017). Moreover, the analyses of volatile organic and inorganic compounds in exhaled air may become an emerging tool (Chap. 24) for maintenance of "crew health," their "system performance" as well as "environmental integrity in space" (Nicogossian et al. 2001), which involves the regular assessment of the microbial environment (Chap. 25) as well. A new wearable device consisting of a shirt and dedicated tablet application is able to measure continuously physiological data including pulse wave, electrocardiogram, blood pressure, breathing parameters, skin temperature, oxygen saturation, and other physical activity levels thereby simplifying the process by combining numerous devices into one wireless article of clothing. This easy-to-use garment has been launched into space in December 2018 (http://www.asc-csa.gc.ca/eng/sciences/bio-monitor.asp). Sleep recording can be made by a novel approach which is placed within a few seconds on one forearm and the person can be self-recorded in this environment. It is based on instantaneous heart rate, body movement recordings and also combine with more than 40 sleep experts' rules to assess all sleep parameters instead of polysomnography (Muzet et al. 2016). In comparison to minimally or more invasive procedures noninvasive techniques will be the "winners" if they are based on clear data in terms of accuracy, safety, and reproducibility.

Minimally invasive procedures are characterized by minimizing potentially detrimental diagnostic and/or surgical trauma to the individual. This includes, as an example, peripheral venous or capillary blood withdrawal from the finger tip. Weighing the risks against the benefits of blood draws the benefits will significantly rise, when the number, reliability, and the impact of the readout parameters can support the definition of the individual health status and may direct to appropriate measures for counterbalancing of unfavorable changes (Chaps. 30–35). Analytical tools can involve flow cytometric measurements (Chap. 27).

An *invasive* procedure penetrates or breaks through the skin barrier or enters a body cavity, e.g., the mouth, and includes techniques that comprise relevant perforation, an incision, a transection, a catheterization, or another entry into the human body. Invasive techniques are mostly well validated. Their disadvantages are the time-consuming and cost-intensive application requiring the presence of an on-site expert which, however, can be assisted by telemedicine and robotics in the future as well (Ushakov et al. 2015).

21.2.3 The Next Stage of Monitoring: Bio-microdevices

21.2.3.1 The Need for Molecular Analytics

Fueled by their potential for diagnostics and personalized therapy methods for isolation and on-site analysis of blood components, such as plasma, red blood cells, white blood cells, extracellular vesicles (EV, e.g., exosomes, microvesicles), and circulating nucleic acids (cfDNA, miRNA) are developing rapidly (Andaloussi et al. 2013; Nagrath et al. 2007; Ratajczak and Ratajczak 2016; Tan et al. 2009; Tkach and Thery 2016). They will be used to characterize immune system state, inflammation, bone loss, radiation effects; and are becoming the key to personalized precision medicine approaches. Examples include the need for rapid detection and time evolution of the expression level for a number of soluble proteins currently used in clinical diagnostics such as the immune system panels (IL2, IL4, IL5, IL6, and IFNγ). This is particularly relevant for remote applications, such as exposure to radiation, stress, or low gravity exposures where body fluids' analysis is required. Today these analyses are almost exclusively performed on Earth using samples collected and stored on the International Space Station. In practice, this mode of operation is costly and detrimental to timely assessment since biological samples collected from space crew must be preserved (often over extended periods of time) and returned to Earth before they are transferred to a dedicated laboratory for investigation. These bio-analytical challenges compounded with the need to process a diversity of samples large, makes the sample preparation and automated molecular analytics a critical barrier particularly relevant for analysis in remote environments and applications, such exposure to radiation, stress, or low gravity exposures.

In addition, the life science space research community requires genomic/transcriptomic information for the monitoring of astronauts' health throughout long-duration missions in order to assess extent of adaptation or deconditioning arising during space exploration. A growing body of literature suggests that epigenetic regulation of immune genes plays an important part in tailoring the response to therapy and defining outcomes. Therefore, a capacity of rapid and automated on-site analysis from small volumes of body fluids of biomarkers, including DNA and RNA at very low concentrations, is crucial to a timely and efficient decision-making process together with (immune-)cell functional analyses.

21.2.3.2 Bio-devices for Remote Diagnostics—Need and Opportunity for Space

Given the complexity of performing molecular, cell-based analysis and the constraints related to performing experimental work aboard both ISS and beyond, there is a request of new approaches and medical devices allowing to harvest and process information from a broad range of sample types with limited hands-on crew engagement during operations. Particularly, low-form factor molecular-based bio-devices that are portable, wearable, or implantable, and are IT-enabled to provide continuous monitoring are required to address the need of remote medical support.

Bio-microdevices technologies have been developed significantly over the last decade particularly fueled by microfluidic-powered lab-on-chip (LOC) technologies that facilitate integration of multiple and complex analytical protocols, most of which are typically done using multiple bulky instruments. LOC technologies allow such assays to be conducted in a rapid and inexpensive manner. They also enable minimally trained personnel to perform analytical procedures outside of laboratory settings (Cui et al. 2015; Erickson et al. 2014; Mielczarek et al. 2016; Patabadige et al. 2016; Sackmann et al. 2014). LOC devices have a wide range of applications in fields such as point-of-care diagnostics, high-throughput screening, DNA analysis, protein analysis, cell-based assays, or companion diagnostic devices. These needs and the subsequent developments include advances in hardware miniaturization, integration of the multiple steps in modular and reconfigurable devices, producing clean samples required for analysis and integration of full sample-to-answer assay in automated systems. Microfluidics is one of the key technologies enabling lab-on-chip devices. Several microfluidic approaches for manipulation liquids have been developed including external on-chip pressure-mediated pumping, electrowetting, centrifugal, and capillary forces. To be suitable for handling of complex, very different sample natures and in remote operations, the methods used to control the liquid displacements inside the microfluidics cartridges must have large range of liquid–solid contact angles. Among the various arising microfluidic technologies, centrifugal microfluidics stands out as a promising platform for sample preparation and integration of sample-to-answer assays able to handle samples ranging from μL to mL with almost zero dead volume. Centrifugal microfluidics uses the centrifugal forces from high-speed rotation to drive the liquids radially from the center to the periphery of the devices. For space applications, this type of microfluidic platform is particularly appealing as the presence of a strong on-chip centrifugal acceleration not only offers unique sample preparation capabilities (sedimentation of debris, density fractionation, etc.) but also makes the entire microfluidic process independent of normal gravitation. Also, the centrifugal force helps providing predictable and repeatable liquid positioning despite the intrinsic variability associated with device filling and facilitates working with complex samples such as blood or saliva that can contain bubbles or show complex interactions with the sidewalls of the microfluidic devices.

Among the very promising microfluidic approaches currently available is the "*Power-Blade*" microfluidic platform developed by the National Research Council of Canada. "*Power-Blade*" combines regulated pressure control in a centrifugal microfluidics platform as a means to facilitate integration of complex assays in centrifugal microfluidics and improves reliability of fluidic manipulations. In this method (Fig. 21.1), a programmable air pump and multiple miniature electromechanical valves are placed on a rotating stage and connected to microfluidic devices. The electronic system is designed to ensure that the valves, the pump, and the other active elements can be computer-controlled in real time while the platform is rotating at high speed. Centrifugal fluidic actuation is independent of liquid–solid contact angles, the nature of the samples, and the chip fabrication materials, making this method very suitable for space applications (Clime et al. 2014, 2015; Brassard et al. 2015, 2016, 2018a, b).

The Power-Blade platform's exceptional capabilities to implement automation of complex biological samples preparation and integration of biomolecular assays has been demonstrated in a number of very diverse applications.

One example is the extraction of the genomic DNA from cell culture lysates (Clime et al. 2015) and rapid (1 h) VTEC *E.coli* bacteria's molecular subtyping workflows including bacterial lysis, on-chip DNA amplification implementing an molecular assay has been performed (Fig. 21.1c).

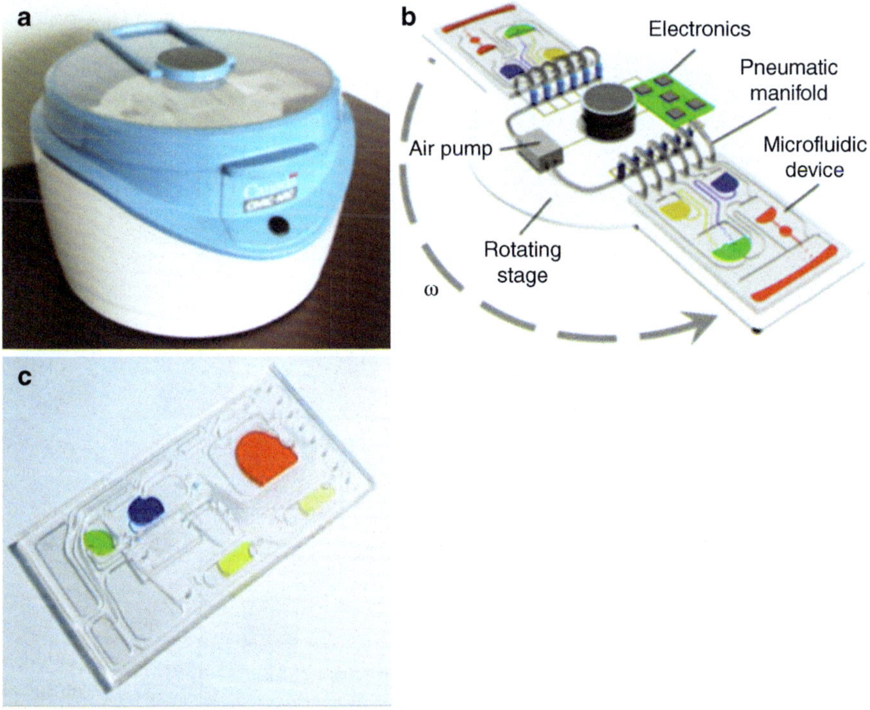

Fig. 21.1 (**a**) Picture showing the current laboratory earth prototype of the Power-Blade platform. (**b**) 3D schematics showing the various elements of the Power-Blade technology. (**c**) Power-Blade-operated polymer microfluidic device (2 × 4 in.) performing bacterial lysis, on-chip DNA amplification implementing an molecular assay for VTEC *E. coli* molecular subtyping

The P-Blade technology has been validated in collaboration with Canadian Space Agency (Brassard et al. 2017, 2018a, b); for low-gravity operation and for performing liquid-biopsy-like sample preparation from body fluids. P-Blade has demonstrated protein and DNA extraction from whole blood in less than 60 min using dedicated microfluidic cartridges that undergo automated centrifugal and pneumatic liquid manipulation. One example is the automated extraction and purification of multiple proteins and total genomic DNA from large volumes of whole blood samples (Brassard et al. 2018a, b). In Fig. 21.2 it is demonstrated that very high efficiencies for protein extraction from blood samples can be achieved across the entire physiological range of interest. Three target protein biomarkers, TNF, PTH, and ALP, encompass the whole physiological concentration range of interest for proteins biomarkers (TNF, low 0.3–1.3 pg/mL; PTH, average 10–70 pg/mL; ALP, high 10–70 ng/mL).

Power-Blade Microfluidic Platform is modular by design. It is a dynamic bioanalysis hardware platform that accepts application-specific cartridges or consumables to enable a wide variety of functions and assays, each with custom operational protocols.

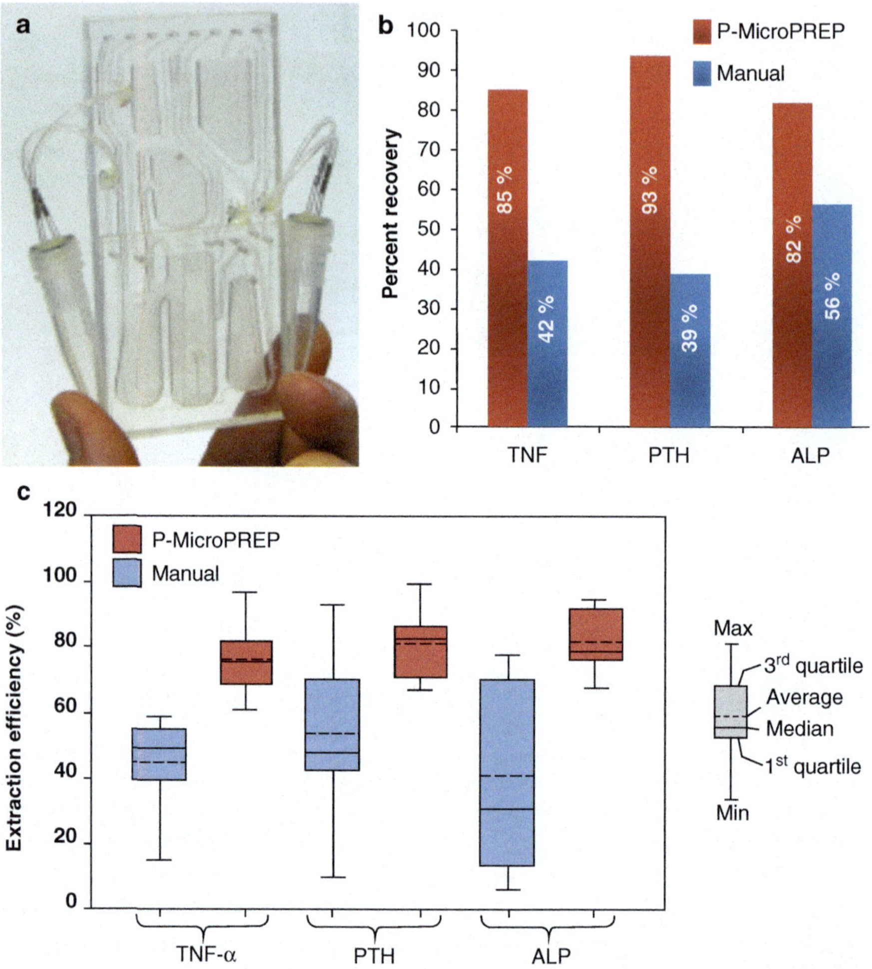

Fig. 21.2 (**a**) 2 in. × 4 in. three layers thermoplastic polymer device operated on the Power-Blade used for automated (45 min) extraction from 1 ml whole blood of multiple proteins; (**b**) Comparison of protein extraction efficiency for tests performed with the automated microfluidic cartridge and test performed manually; and (**c**) Boxplot showing a summary of the protein extraction efficiency measured for all experiments performed with a manual procedure and with the automated microfluidic cartridge (Brassard et al. 2018a, b)

21.2.3.3 Intelligent Assistance

Psychological impairment due to the prolonged perception of stress is a major risk factor for manned deep space missions (Kraft et al. 2003). Peer pressure or the so-called groupthink effect is well known to impair opinion and objective reasoning (Turner and Pratkanis 1998). This effect was observed in autonomous groups that worked under stress and caused a major degradation of the communication between crew and ground personnel. The characteristics can be overestimation of one's own capabilities, high group pressure towards uniformity, not questioning the leader's decisions and an overcritical view on outsiders, meaning people outside the own group. A successful interplanetary mission will critically depend on individual adaptation and performance as well as successful joint crew interaction and appropriate psychological prevention or countermeasures (Manzey 2004). Because of that, innovations are needed, which autonomously prevent stress and its negative attributes by effectively supporting crew in a pleasant way. Apart from training crew and crew surgeons, artificial intelligence can provide another form of support for long-duration space missions (Fig. 21.3).

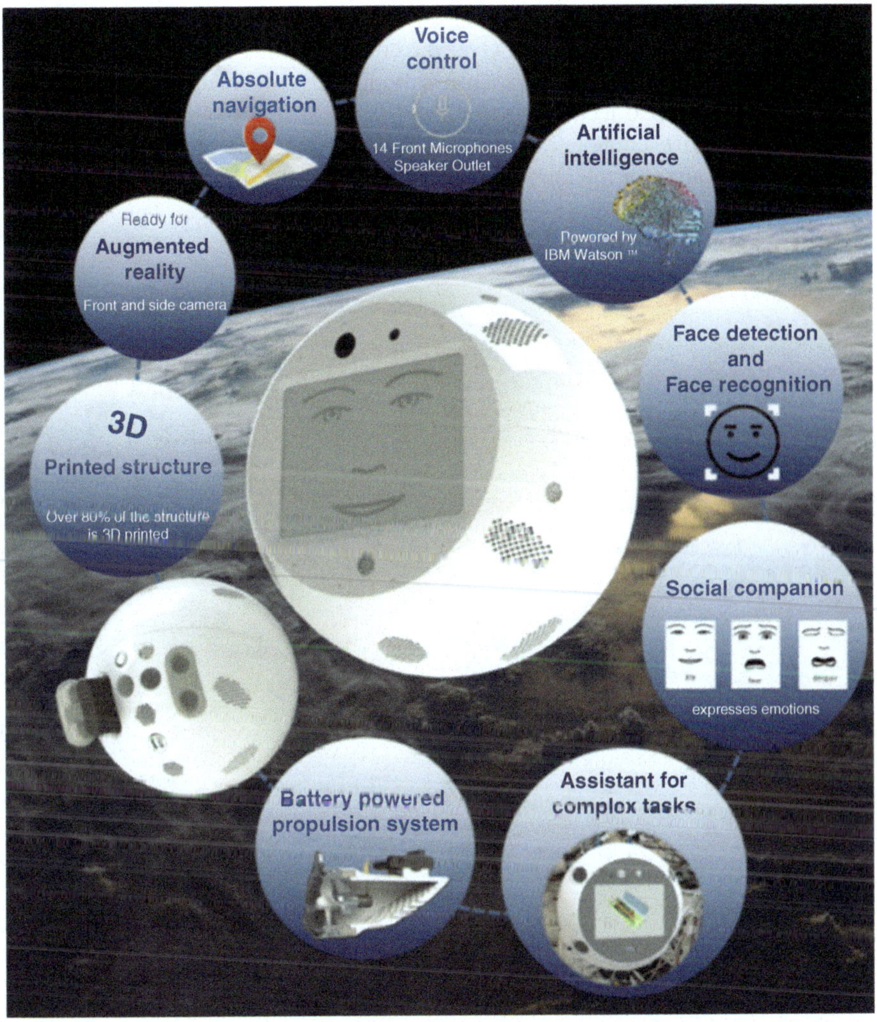

Fig. 21.3 Design concept of CIMON and technical features: The CIMON systems consists of a 3D printed hardware with a battery-powered propulsion system and a variety of intelligent functions such as navigation guidance and control, voice control, and face detection and recognition. Over the time-course of development more features and functions will become available (Source: DLR)

On behalf of the German Aerospace Center (DLR) the company Airbus Defence and Space developed and designed an autonomous robotic sphere equipped with the artificial intelligence (AI) "Watson" by IBM. This crew interactive mobile companion (CIMON) has a size of 32 cm in diameter and 5 kg weight. CIMON was launched to the International Space Station (ISS) in June 2018. After successful commissioning and successful proof of safety in a technical demonstration by ESA Astronaut Alexander Gerst (Fig. 21.4), CIMON will be deployed for human interaction in form of preliminary basic experiments. It/he shall assist during simple procedures, display videos or read out loud instructions for crew activities. In his actual form, he is capable of communicating through verbal commands and identifying crew members through face recognition. Access to the AI can currently occur only when signal to ground is possible. Over the course of time, CIMON is expected to evolve, not only regarding his personality through crew interaction but also regarding tasks assigned and technology equipped with. The employment of AI carries also some negative connotations through movies and other media leading to a potential reservation in interaction with CIMON. It is clear, however, that further

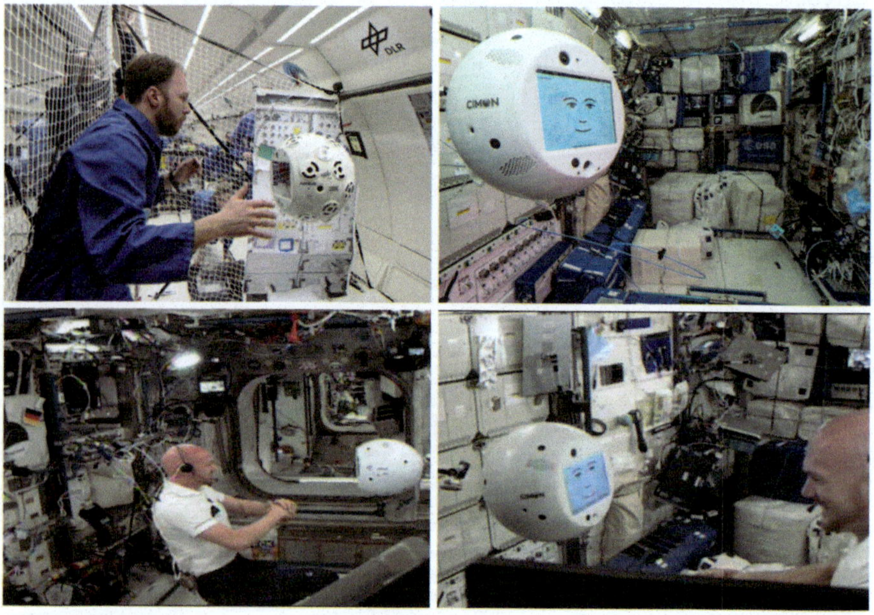

Fig. 21.4 Feasibility study of CIMON: Upper left panel: Testing of the navigation control system during parabolic flight: During 22 s of microgravity, CIMON showed stable facial orientation towards a study participant supporting implementation aboard the ISS (Source: Novespace/DLR). Upper right panel: First power-on of CIMON in his new working environment inside the Columbus module (Source: ESA/NASA/DLR). Lower panels left and right: commissioning aboard the ISS with support of ESA-Astronaut Alexander Gerst on November 15, 2018. CIMON proved stable navigation control as well as fully functional verbal interaction, face detection and features of the artificial intelligence (Source: ESA/NASA/DLR)

exploration of his potential capabilities cannot proceed without the support of the crew. Therefore, he will not only have to prove himself as technically reliable and safe in this working environment but also need to encourage sympathy and acceptance. The achievement of these highly estimated values will be decisive for the further development of AI as crew support in the future.

21.2.4 Ready for Space?

This introductory chapter presented herein summarized a number of analytical capabilities that can be of strong advantage to health care for extended space missions. The noninvasive or minimally invasive monitoring tools, the versatile lab-on-chip platforms such as the ones outlined above, and artificial intelligence can altogether pave the way to develop exceptional capacities. We envision that artificial intelligence-based tools will be capable also to remotely monitor the micro-devices, hardware operation, assist with experiments analysis, benchmark and make decisions on experiments workflows. Those opportunities are particularly interesting for their potentially disruptive effects in within 5–10 years for both space exploration and terrestrial applications. While responding directly to the particular context of space-based applications, we readily assert that these investments will also address key needs for (1) Earth-based, distributed and automated remote medicine and (2) the next generation of medical technology that can offer personalized and precise medical care over the next decades.

In space medicine, the regnant trend is to attach importance to the molecular biology and genetic testing which will become more easy as described. However, in order to translate these new findings into any improvement of living conditions or relevant impact on health preservation during space missions, it is necessary (1) to clearly determine space-relevant problems in humans and (2) to differentially monitor and to hereby link the changes observed in one organ to other organ systems. However, especially in space, the implementation of health-monitoring tools will depend on the following factors:

1. Technical feasibility, upload weight restrictions, and onboard energy consumption,
2. Relevance in respect of health problems occurring in astronauts during missions,
3. Its value for identifying a disease and then monitoring its response to treatment,
4. Manual feasibility by astronauts as well as.
5. Safe use and convenience application.

The diversity of diagnostic methods requires a close collaboration of experts in the several fields of medicine and engineering. For the single expert this prevailing trend is a challenge. He has to acquire as much as possible theoretical and practical expertise in the various diagnostic techniques. Hence the goal of education of physicians in space organizations should be to train medical experts who have the competence, the occupational skills, and the experience to take the responsibility for engineering, supervising, and referring many diagnostic tests in space. In future, the results of noninvasive or minimally diagnostic procedures could be generated on-site in space and the space crew must be able to autonomously interpret the data and to draw the correct conclusions. This will be only possible if astronauts are trained extensively by medical experts. Alternatively, telemedicine will be established and wireless tools will be provided for astronaut care ("e-health"). However, it must be taken in consideration that the extended communication delays in exploration missions will eliminate the option of real-time guidance, thus requiring autonomous stuff operation and training of astronauts before space missions cannot be eliminated (Hurst et al. 2015).

21.3 Summary

"Long-duration missions necessitate further technological breakthroughs in tele-operations and autonomous technology." This statement made by Nicogoassian and coworkers in 2001 confirms the importance of onboard health care and extends to the critical need, respectively, to make noninvasive or minimally invasive stress and health-monitoring tools available. These tools should deliver comprehensive data on the degree of stress and health-relevant immune changes of astronauts, their related risks as well to be able to prevent from health threats. To realize the latter in space, existing technical approaches has to be improved and new innovative methods are under way to be developed and tested, respectively. The (1) integration of environmental monitoring and health monitoring devices with (2) "on-the-spot" sample analyses, data processing, and interpretation as (3) potentially supported by artificial intelligence will all help and be of advantage for the space crew on their mission to outer space, and also equally important for the benefit for patient care on Earth. Especially, they will have the potential to help bedridden or housebound patients, people who live in rural communities, or workers in dangerous environments with limited access to medical support.

References

Andaloussi SE, Mager I, Breakefield XO, Wood MJ (2013) Extracellular vesicles: biology and emerging therapeutic opportunities. Nat Rev Drug Discov 12:347–357

Antoniou M, Jorgensen AL, Kolamunnage-Dona R (2016) Biomarker-guided adaptive trial designs in phase II and phase III: a methodological review. PLoS One 11(2):e0149803

Ballantyne C (2007) Fact or fiction? Stress causes gray hair. Sci Am. http://www.scientificamerican.com/article.cfm?id=fact-or-fiction-stress-causes-gray-hair

Baumgartner C, Lewis GD, Netzer M, Pfeifer B, Gerszten RE (2010) A new data mining approach for profiling and categorizing kinetic patterns of metabolic biomarkers after myocardial injury. Bioinformatics 26(14):1745–1751

Brassard D, Clime L, Geissler M, Veres T (2015) Combining active pneumatic pumping and centrifugal forces : a new paradigm for the integration of bioanalytical assays. Proceedings of the 19th international conference on miniturized systems for chemistry and life sciences, MicroTAS, p 191

Brassard D, Clime L, Mounier M, Veres T (2016) Programmable aliquots in passive microfluidic devices using a centrifugal platform with active pneumatic pumping. Proceedings of 20th international conference on miniaturized systems for chemistry and life sciences, MicroTAS, p 857
Brassard D et al (2017) Advanced centrifugal microfluidic platform for the automation of clinical assays. American Association for Clinical Chemistry (AACC) annual meeting
Brassard D et al (2018a) High-yield automated extraction of nucleic acids from whole blood using a centrifugal microfluidic platform with active pneumatic pumping. Proceedings of the 22th international conference on miniaturized systems for chemistry and life sciences, MicroTAS, p 141
Brassard D et al (2018b) Microfluidic-based platform for universal sample preparation and biological assays automation for life-sciences research and remote medical applications. Deep space gateway science workshop (2018) LPI Contrib. No. 2063
Clime L, Brassard D, Veres T (2014) Centrifugal microfluidic chip platform. PCT/IB2015/051591-US20170036208A1
Clime L, Brassard D, Geissler M, Veres T (2015) Active pneumatic control of centrifugal microfluidic flows for lab-on-a-chip applications. Lab Chip 15:2400–2411
Cui F, Rhee M, Singh A, Tripathi A (2015) Microfluidic sample preparation for medical diagnostics. Annu Rev Biomed Eng 17:11.1–11.20
Erickson D, O'Dell L, Jiang L, Oncescu V, Gumus A, Lee S, Mancuso M, Mehta S (2014) Smartphone technology can be transformative to the deployment of lab-on-chip diagnostics. Lab Chip 14:3159–3167
Gidlow CJ, Randall J, Gillman J, Silk S, Jones MV (2016) Hair cortisol and self-reported stress in healthy, working adults. Psychoneuroendocrinology 63:163–169
Hurst VW, Peterson S, Garcia K, Ebert D, Ham D, Amponsah D, Dulchavsky S (2015) Concept of operations evaluation for using remote-guidance ultrasound for exploration spaceflight. Aerosp Med Hum Perform 86(12):1034–1038
Kirkpatrick AW, Blaivas M, Sargsyan AE, McBeth PB, Patel C, Xiao Z, Pian L, Panebianco N, Hamilton DR, Ball CG, Dulchavsky SA (2013) Enabling the mission through trans-atlantic remote mentored musculoskeletal ultrasound: case report of a portable hand-carried tele-ultrasound system for medical relief missions. Telemed J E Health 19(7):530–534
Kottnerus JA, van Weel C, Muris JWM (2002) Evaluation of diagnostic procedures. BMJ 324:477–480
Kraft NO, Lyons TJ, Binder H (2003) Intercultural crew issues in long-duration spaceflight. Aviat Space Environ Med 74:575–578
Manzey D (2004) Human missions to Mars: new psychological challenges and research issues. Acta Astronaut 55:781–790
Mielczarek WS, Obaje EA, Bachmann TT, Kersaudy-Kerhoas M (2016) Microfluidic blood plasma separation for medical diagnostics: is it worth it? Lab Chip 16:3441–3448
Muzet A, Werner S, Fuchs G, Roth T, Saoud JB, Viola AU, Schaffhauser JY, Luthringer R (2016) Assessing sleep architecture and continuity measures through the analysis of heart rate and wrist movement recordings in healthy subjects: comparison with results based on polysomnography. Sleep Med 21:47–56
Nagrath S, Sequist L, Maheswaran S, Bell D, Irimia D, Ulkus L, Smith M, Kwak E, Digumarthy S, Muzikansky A, Ryan P, Balis U, Tompkins R, Haber D, Toner M (2007) Isolation of rare circulating tumour cells in cancer patients by microchip technology. Nature 450:1235–1239
Nicogossian AE, Pober DF, Roy SA (2001) Evolution of telemedicine in the space program and earth applications. Telemed J E Health 7(1):1–15
Otto C, Hamilton DR, Levine BD, Hare C, Sargsyan AE, Altshuler P, Dulchavsky SA (2009) Into thin air: extreme ultrasound on Mt Everest. Wilderness Environ Med 20(3):283–289
Otto C, Comtois JM, Sargsyan A, Dulchavsky A, Rubinfeld I, Dulchavsky S (2010) The Martian chronicles: remotely guided diagnosis and treatment in the Arctic Circle. Surg Endosc 24(9):2170–2177
Patabadige DEW, Jia S, Sibbitts J, Sadeghi J, Sellens K, Culbertson CT (2016) Micro total analysis systems: Fundamental advances and applications. Anal Chem 88:320–338

Ratajczak MZ, Ratajczak J (2016) Horizontal transfer of RNA and proteins between cells by extracellular microvesicles: 14 years later. Clin Transl Med 5:7

Sackmann EK, Fulton AL, Beebe DJ (2014) The present and future role of microfluidics in biomedical research. Nature 507:181–189

Stalder T, Steudte-Schmiedgen S, Alexander N, Klucken T, Vater A, Wichmann S, Kirschbaum C, Miller R (2017) Stress-related and basic determinants of hair cortisol in humans: A meta-analysis. Psychoneuroendocrinology 77:261–274

Tan SJ, Yobas L, Lee GY, Ong CN, Lim CT (2009) Microdevice for the isolation and enumeration of cancer cells from blood. Biomed Microdevices 11:883–892

Tkach M, Thery C (2016) Communication by extracellular vesicles: where we are and where we need to go. Cell 164:1226–1232

Turner ME, Pratkanis AR (1998) Twenty-five years of groupthink theory and research: lessons from the evaluation of a theory. Organ Behav Hum Decis Process 73:105–115

Ushakov IB, Karpov AA, Kryuchkov BI, Poliakov AV, Usov VM (2015) Promising options for medical robotics application in support of crew life activities and mitigation of medical risks during space flight. Aviakosm Ekolog Med 49(6):76–78

Psychological Monitoring 22

Bernd Johannes and Berna van Baarsen

22.1 Introduction

Irrespective if and how psychological monitoring is to be fully understood during extend missions to (outer) space, it must first be viewed as an essential support tool for the crew, parallel to the technical monitoring systems e.g., of fuel and water quality. Rigorous selection of the crew favors those who are tough, enduring, well-trained, performance-oriented, and unaccustomed to "needing" the help of psychologists. Although these participants are honest in subjective self-evaluation, they tend to exhibit repression. This requires rethinking, and reevaluation of psychological methodology. This chapter of this books second edition addresses and reviews some general principles of psychological monitoring, summarizes the definition of the terms but also links the results from recent applications to these methods, such as psychological monitoring, team talk sessions and crewmember support. The methods used should be objective, nonobtrusive, nondisrupting, mission related, useful for the crew itself (immediate expert-system-based feedback) as well as that they should be neither tedious nor artificial or abstract, so that they will not be avoided but accepted by the crew. There is an increasing need for new measurements and methods which are objective, reliable, computerized, continuously, or at least repeatedly, applicable in order to achieve—in a perfect setting—monitoring of nearly all behavior and performances during a mission. Numerous (newly developed) methods that can be helpful to achieve this goal are based on video observation, event counting, performance measurement, time measurement, and

B. Johannes (✉)
German Aerospace Center (DLR), Institute of Aerospace Medicine, Cologne, Germany
e-mail: bernd.johannes@dlr.de

B. van Baarsen
VU Medical Centre Amsterdam, Amsterdam, The Netherlands

© Springer Nature Switzerland AG 2020
A. Choukér (ed.), *Stress Challenges and Immunity in Space*,
https://doi.org/10.1007/978-3-030-16996-1_22

psychophysiological measurement, and also self-reports. To which degree they are helpful in assessing stress and stress-responses and health relevant issues is described and discussed.

22.2 Monitoring and Preparation Prior to the Mission

Preparation and training procedures will become more effective when, in addition to technical information and job-related factors, also relevant psychological, existential, and social aspects are taken into account (Kanas and Manzey 2003; Van Baarsen 2011, 2013; De la Torre et al. 2012). Attention should be given to informed consent procedures in which preflight information and (job) expectations should adequately match. Other relevant training aspects are personal autonomy among crewmembers in relation to crew motivation and perseverance mission behaviour (Van Baarsen 2013). The training of astronauts and the preparation of a space mission take several years; this is sufficient time to analyze and verify the behavior and performance of individuals and crews.

On the individual level, crewmember support should be provided early on in the training period (Baranov et al. 2001). Coping with unknown and perhaps dangerous situations is a unique personal event, for which one must be adequately prepared. On the team level, the preparation period is highly important for the crew development, the intercultural adaptation, and the establishment of mission goals and motivation. Teams perform better when goals originate from and are relevant to them and are evaluated and restructured periodically, also in relation to mission and private expectations (Van Baarsen et al. 2012a). Encouragement of reasoning and deliberation among team members may increase group cohesion and decrease the risk of spreading blame and finding scapegoats. Prior companionship and team-talk sessions are of great importance for the development of common goals (Sandal 2001; Tomi 2001). The management of emotions and the building of supportive and constructive relations between the crew (Kass and Kass 2001) have to be established. Possible effects of social-related personality characteristics can lessening or enhance feelings of loneliness, isolation, and boredom, for example, the ability to enjoy one's own company (Van Baarsen et al. 2009), the desire to be on one's own, or social anxiety, may influence selection, monitoring, and support programs. These effects stress the necessity to compare different countermeasures and determine which ones are most efficient in preventing boredom, distraction, and loneliness (Van Baarsen 2011).

The preparation time provides a good opportunity to support the acceptance of monitoring methods and psychological support tools. Regarding the mutual aspects of communication, reflection, and support, crew and mission control should both participate in (parts of) the training (Kanas 2009; Van Baarsen et al. 2012a). The final mission selection is a crew composition with possible individual out-selections and crew rebuilding. One specific crew problem is the late mission assignment of load specialists, for example, from foreign space agencies.

22.3 Monitoring and Support During the Mission

Psychological monitoring and support is most difficult during the mission. However, there are more publications about that mission period than about others (Grigoriev et al. 1985; Kanas 1991; Manzey et al. 1995; Sandal et al. 1996). There are well-documented physical influences of long-term weightlessness on the human body, but health and fitness are not identical with well-being. Well-being depends how the individual copes with stress due to physical changes caused by weightlessness, fluctuating O_2 and CO_2 concentrations, radiation exposure, muscle and bone loss, and cardiovascular deconditioning, as well as the adaptation to the psychological effects of long-term isolation. These effects can include asthenia, exhaustion, motivation loss, passivity, feelings of isolation, symptoms of depression, excitability, stimulus deprivation, and movement or activity deficits (Table 22.1).

Table 22.1 Definitions

Psychological terms used in long-term isolation research	
Asthenia	Weakness, or a loss of strength or energy, that can be psychological and/or physical, and which is marked, e.g., by fatigue, muscular pain, lack of motivation
Boredom	Experiences of boredom differ between persons and depend, e.g., on the levels of excitability and action preferences, and the number and intensity of inputs from the environment
Depression	A persistent feeling of sadness and worthlessness and a lack of interest or desire to search for and/or engage in joyful activities. Depression is related to symptoms such as motivation loss, exhaustion, passivity, sleeping problems, disturbed appetite, and adaptation and eating disorders
Distraction	An interruption to attention or anything that draws attention away from the desired task thereby blocking or diminishing the reception of primary information
Excitability	The extent to which people are sensitive or (neurologically) aroused to emotional responses, which affect the ways in which they experience the environment
Feelings of isolation	Feelings due to a perceived, internalized isolation. Feelings of isolation may be positive or negative depending on person, social and health characteristics such as social embeddedness, personal vulnerability, support desires, preferences (solitude) and expectations, and coping style
Hygienic discipline	The ability to perform practices in order to preserve health, on a consistent, repetitive basis
Loneliness	In general, loneliness is expected to be experienced due to a discrepancy between actual and desired personal relationships. Loneliness can be socially (the feeling you can count on others for support) or emotionally (the feeling of lacking reliable emotional attachments to others) defined. Existential loneliness is expected to be inherent to human existence and meaning of life
Mood	An emotional state or a general feeling that does not result from a particular situation. Awareness of a mood often results following reflection on a response to a situation
Stimulus deprivation	Lack of environmental input or deprivation or reduction or removal of stimuli from one or more of the senses

Maintaining health depends much on physical training discipline and hygienic discipline. The sleep–wake cycle within a 24-h day and night cycle as well as the work–rest schedule are disturbed (see also Chap. 9). It is very informative to observe how these daily duties are permanently solved. If there are decreases or changes in spare time use, such as skipping usual hobbies and so forth, this could be an early indication of mood change or motivation loss.

When small teams have to work in highly autonomous conditions, where team members are fully dependent on each other in managing technology, carrying out tasks, and with additional communication delays as well as daily (inter)personal and occupational hazards, successful adaptation and adherence to the primary jobs are difficult. Especially when questions of personal goal setting and existential events arise, it is crucial to analyze matters of challenge, dedication, and inspiration on the one hand, which can mean fulfillment (Van Baarsen 2011). On the other hand, there are loneliness, (herein defined as perceived, internalized isolation, Cacioppo et al. 2010) demotivation, and boredom, which can manifest itself as frustration (Van Baarsen 2001, 2011). While positive feelings of personal achievement can increase team performance, coherence, and well-being, negative impressions of attainment, failure, and boredom may enhance difficulties with interpersonal and intercultural contacts and communication, supportive behavior, and companionship. The availability of intimate contact as well as emotional and instrumental support is an influential social resource which may attenuate feelings of distress and loneliness (De Jong Gierveld 1998; Van Baarsen 2002). During powerful isolating events such as a long-duration spaceflight, loneliness can be experienced due to changes in the size of the personal network and its interpersonal dynamics as well as the changes in perceived intimacy and support. Particularly members of small intercultural teams are prone to experience loneliness: Increased dependency, different cultural backgrounds and communication styles (Gushin et al. 2001; Kanas et al. 2000; Palinkas et al. 2004) as well as the limited number of contacts without the possibility of altering the crew composition can result in decreased satisfaction and/or increased need for intimacy and support. In fact, smaller networks discourage social seeking behavior and decrease the chance that emotions can be shared (Van Baarsen and Broese van Groenou 2001). High-fidelity studies in space analog environments have been seen to be very helpful for those investigations bridging environmental stressors to psychosocial challenges (see Chaps. 36 and 37). In each six crewmembers participating in the Mars105 and Mars500 simulation programs (Van Baarsen et al. 2009, 2012b) an increase in loneliness, particularly shortly after confinement, with considerable variation in how crewmembers adjusted to the confinement in terms of loneliness was observed over time. Loneliness frequently results in a decrease of well-being which is measured for instance, as depression, sleeping problems, disturbed appetite, and adaptation disorders (De Jong Gierveld 1998). Loneliness and poor health, interrelated through a complex process, may result in feelings of alienation and estrangement (Van Baarsen 2002).

Presently, psychological monitoring focuses usually only on performance in special and common tasks, for example, docking training (Hockey 1986; Manzey et al. 1998; Salnitski et al. 2001; Sandal et al. 1996), or have a scientific background. The long-term effect of limited resources in insignificant daily tasks, missing personal contact, and the technocratic environment, however, can drastically influence the daily attention to house-keeping and the maintenance of equipment, resources, animals, and plants. Decreased personal or team motivation can lead to less focused and misplaced goal orientation. It is important to mention changes in the individual coping with the reduced privacy, increases or decreases of family contact seeking, sensation seeking, other mood changes resulting from maladaptation to confinement and isolation or the reaction to periods of high work load alternating with steep boredom. A reduced variability in self-reports and work statements are clues for the evaluation of the psychological state of crewmembers.

We still only have limited methods to objectively monitor all the indicators mentioned. Sensitive observation and private conversation based on a trusting personal relation between ground and crew remain the most important. Likewise, the monitoring of group cohesion and group conflict should focus on the group dynamics, cultural life, events, group activities, and sub-group separation (Sandal et al. 1995; Vessey and Blackwell Landon 2017). Changes occurring in the group roles and functions must be noted: The possibility to develop informal leadership can disturb or support group cohesion. All interpersonal or even intercultural differences can enhance as well as hinder the group functioning. Limited resources and/or impaired competence can provoke feelings of failure or frustration, specifically anger, sense of loss, fallibility, blame, limited control and boredom (Van Baarsen 2011), ultimately leading to depression, self-derogation, and risky behavior (Frankl 1959; Molasso 2006; Van Baarsen 2002). When unexpected competence deficits arise in individual crewmembers, for example during a sudden occurrence of a life-threatening event, efficient group performance may be in danger.

For long-term missions it is nearly unavoidable that conflicts appear. The questions about why they occur and how the crew should cope with them must be answered. A rising problem is the crew autonomy of long-term missions. Long durations and long distances will heavily influence the relationship between the crew and the ground staff. Due to the distance between the Earth and the space craft, a delay in communication occurs of up to 20 min into one direction. This hinders, for example, a "usual" talk or an online support in case of trouble. It is unknown how the crew may react if they can't see their home planet anymore ("Earth out of view" effect, Kanas and Manzey 2003, p. 186) or realize that they were traveling behind the "point of no return." Even Christopher Columbus could have had returned each moment with his sailer "Santa Maria" if he would have had wanted to. There is a growing risk of misunderstanding, loss of belief, and conflicts (Kanas et al. 2007) between the crew and ground. The crew is no longer controlled by ground control, but guided, informed, and supported in decision-making. Crew autonomy and time delays in communicating with outside monitors in "mission control" may affect crew motivation (Van Baarsen 2013) as well as crew emotions and performance, and influence the work activities of mission control personnel (Kanas et al. 2010). Tensions in interpersonal relations affect the ability to interact with each other and increase towards the end of the mission (Cazes et al. 1996). For an autonomous decision-making of the crew, an independent computerized support system on board could be a great assistance not only for technical problems (Hoermann et al. 2008).

22.4 Relevance of Isolation Stress for the Cardiovascular and Immune System

We can herein only shortly connect the effects of stress, isolation, and confinement on a human's health. These effects are to be seen very holistically affecting central and peripheral regulation loops and are addressed and discussed in more detail (see Chaps. 3, 6 and 7). It is know from a steady stream of recent papers about terrestrial research that loneliness in the human society (Holt-Lunstad et al. 2010; House et al. 1988; Van Baarsen 2001) is related to several potentially unhealthy changes in the cardiovascular, immune, and nervous systems (Cacioppo et al. 2002, 2010; Miller 2011). This is even more relevant for small crews with limited social alternatives. A lack of group cohesion and group conflicts changes levels and kinetics of stress hormones (Choukér et al. 2002; Shimamaiya et al. 2004). The confinement-related limitations of activity hinder the positive effects of physical activity on immune system and mood and alter brain functions (Schneider et al. 2009, 2010). There are several findings supporting the linkage of behavioral stress and cardiovascular diseases via the immune system (Applegate et al. 1999; Cacioppo et al. 2002). That underlines the necessity of a careful monitoring of psychological changes, e.g., the occurrence of symptoms of loneliness or asthenia. Group cohesion and group functioning is one of the most important factors, compensating or enhancing the deficits of life in isolation, promoting team-spirit ("Teamgeist") and mission motivation or evoking loneliness. Understanding these complex interactions will help to support appropriate measures to counteract unfavorable developments during a long-duration spaceflight or in other conditions in which a crew had to cope with extreme environmental conditions (see Chap. 31).

22.5 New Method Examples

As mentioned above, psychological monitoring needs new methods and technologies which are applicable, not impeding, and accepted by the crew. There are first successful attempts in *video face emotion* analysis (Dinges et al. 2007) even tested already on the ISS in space. Actually there were different software approaches developed for an automated face emotion recognition (FaceReader, NOLDUS, AFFDEX SDK, McDuff et al. 2017), used already in clinical investigations (Bandini et al. 2017) but also promising for space research.

Voice frequency analysis is a very promising approach for space psychology (Johannes et al. 2000, 2007). Figure 22.1 illustrates the complex structure of voice frequencies differing between the used two Russian word commands "ok" and "error." For a subject's state analysis the changes in frequency ranges are more important than the energy distribution. Insofar loud-spoken and soft-spoken commands can be compared. The voice pitch and higher formants are slightly increased

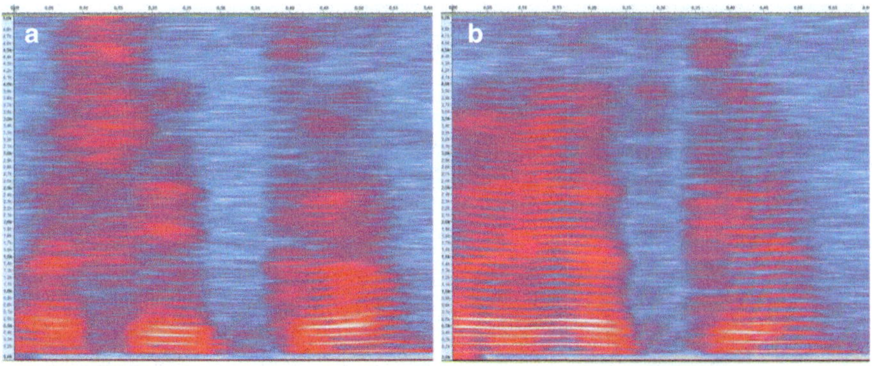

Fig. 22.1 Sonograms of voice samples during the inflight space experiment "Pilot" (docking training), (**a**) "error" (ru: "oshipka"), (**b**) "ok" (ru: "vporyadke"). They illustrate different frequency intensities by diffent colors (from blue to yellow). The ordinate is given in hertz (Hz), the time abscissa in milliseconds. The lowest frequency is called fundamental frequency (voice pitch) and is related to the psychological state of the speaker. The higher frequency bands are reflections of the articulary tract and form several formants, mostly three. Formants provide relevant information for word and speaker recognition

in space. Voice analysis was first applied in space in 1965, e.g., during the very first EVA by Leonov (Simonov and Frolov 1977), and continued in later Mir missions (Sulc 1979; Vaic et al. 1981). Voice pitch (lowest, fundamental frequency) analysis was also verified in a long-term isolation study (SFINCSS) to indicate the actual speaker's excitation during the phone communication with the ground staff (Johannes et al. 2002). The in parallel assessed subjective self-ratings became constant after some weeks and were not anymore informative.

The voice pitch can be registered also by means of small mobile systems and can be put together for communication analysis as illustrated for three subjects in Fig. 22.2. Such kind of voice cross analysis could indicate the common and individual amounts of communication but also the emotional states and reactions of the speakers.

A *wireless group structure monitoring* could provide information about the dynamics of the crew structure during the mission (Sandal et al. 1995). The first results of the 105-day phase of the Mars500 project promised correlation between the objectively measured time by means of a wireless system, crewmembers spend together (Johannes et al. 2015) and the classical sociogram assessment by questionnaires (Salnitski et al. 2009). Figure 22.3 illustrates the resulting group structure and its changes between different days. The average time as well as the equal distribution between crewmembers provides information about the crew cohesion. It will become visible if one crewmember runs out of the crew and becomes isolated within isolation, which is a strong indication for psychological disturbances such as depression. Meanwhile some different commercial systems (batches, Actiwatches with proximity assessment) are available for this approach.

Work sample analysis (Salnitski et al. 1999, 2001, 2004; Johannes et al. 2016b) represents an already well established method and is used as a routine monitoring. Real mission relevant skills (e.g., the hand controlled docking of a space craft on a

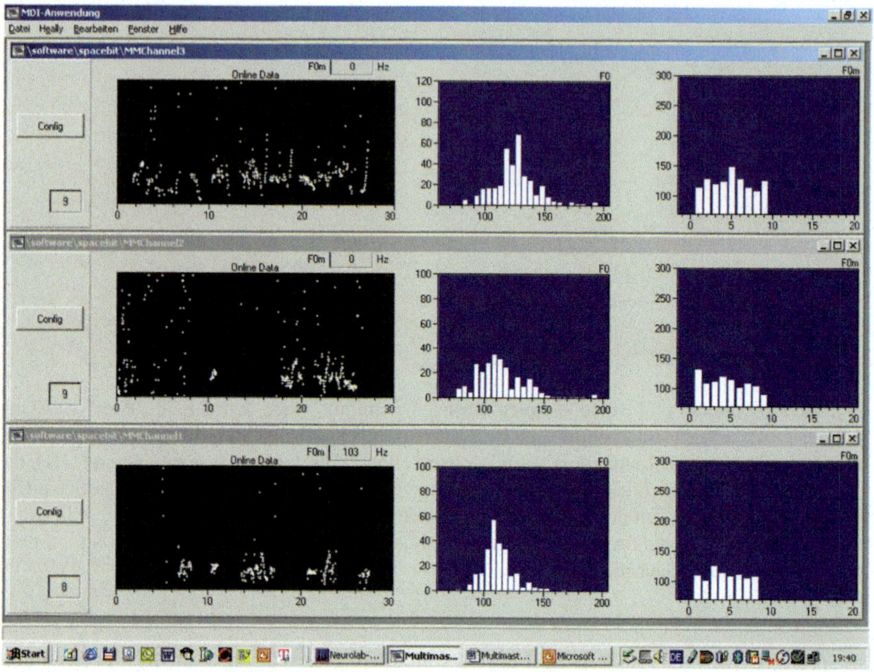

Fig. 22.2 During the (simulated) communication of three subjects their single voice pitch values are registered in 30-s intervals (black panel), put into histograms (left blue panel), and the mode of this histogram (F0 m) sampled as intonation indicator and monitored over time (right blue panel)

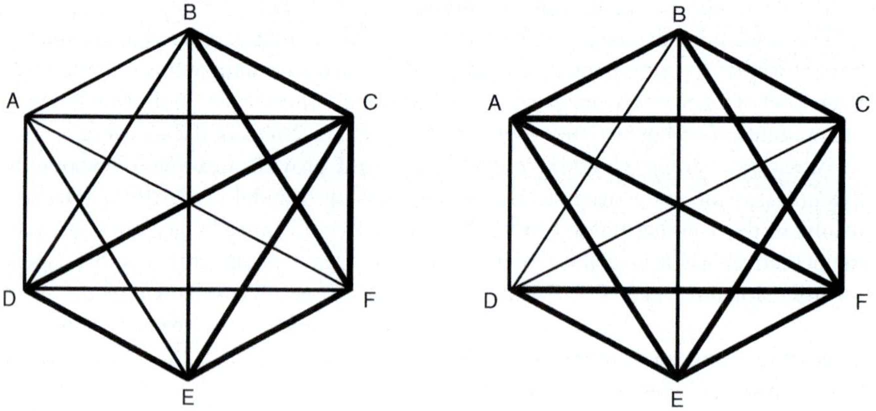

Fig. 22.3 Two examples of daily sociograms, assessed by a wireless system of small satellites, brought by each crewmember during the day, measuring the time spent together. The thickness of the lines represent the relative times spend together as a measure of personal closeness

space station or catching a free flying object by means of a hand controlled robot arm) are refreshed and trained during the mission, thus, providing information about the proficiency of the operators and about their actual motivational state (Johannes et al. 2017). To evaluate, if this professional monitoring is associated with a physiological strain assessment, an indirect monitoring of the psychophysiological state is provided (Johannes et al. 2001, 2003; Johannes and Gaillard 2014). It seems possible to implement features of embedded testing of fundamental cognitive functions into these refreshment trainers. This may be supported by the assessment of *least invasive neurophysiological parameters* like the P300, an EEG reaction to a relevant stimulus (e.g., acoustic) with a latency time of 300 ms (Johannes et al. 2016a). However, areas such as automated daily duty analysis, analysis of the cooperation quality during shared duties, analysis of readiness and progress in coping with expert-system feedback or learning new things (e.g., the language of crew mates or new skills) as well as a spare time use analysis are currently only at the proposal level.

Some new approaches aim to provide support to the astronauts using expert systems (AI). The DLR proposed a "Digital Friend" system (Hoermann et al. 2008). Astrium (now Airbus) suggested an Astronaut supporting system "ASTRID." In late 2018 the prototype CIMON (Crew Interactive Mobile Companion) was brought up and tested on the ISS. This could be the beginning and an example of how an artificial supporting system could look and work like (see also Chap. 21).

22.6 Summary

Because the psychological consequences of living in a hostile and potentially life-threatening environment are closely interlinked with biological responses as well, monitoring with psychological tools is an important tool to monitor stress-related neural and hormonal changes which are potentially affecting immunity and health during long-duration mission. Actually we have to state still an enormous deficit of new objective methods which have to be developed and tested, especially because they might be of huge advantage in the field of neurosciences, tele-medicine and for a crew on a long space mission. However, psychological monitoring is a sensitive case. It will work under enhanced autonomy only if it provides primary support to the crew itself and is only secondary meant for information for the ground. If agencies on Earth will have once selected a Mars-crew, they have to believe in it and support it on the crew's demand.

References

Applegate KL, Hay J, Cacioppo JT, Kiecolt-Glaser JK, Glaser R (1999) Effects of stress on the immune system: implications for reactivation of latent herpesviruses. In: Schedlowski M, Tewes U (eds) Psychoneuroimmunology: an interdisciplinary introduction. Plenum, New York, pp 517–524

Bandini A, Orlandi S, Escalante HJ, Giovannelli F, Cincotta M, Reyes-Garcia CA, Vanni P, Zaccara G, Manfredi C (2017) Analysis of facial expressions in parkinson's disease through video-based automatic methods. J Neurosci Methods 281:7–20. https://doi.org/10.1016/j.jneumeth.2017.02.006. Epub 2017 Feb 20. PubMed PMID: 28223023

Baranov V, Demin E, Gushin V, Belakovsky M, Sekigushi C, Inoue N, Mizuno K, Vachon M, Tomi L (2001) SFINCSS-99 (simulation of flight on international crew on Space Station): experience and lessons learned. In: Baranov VM (ed) Simulation of extended isolation: advances and problems. Firm SLOVO, Moscow, pp 524–530

Cacioppo JT, Hawkley LC, Crawford LE, Ernst JM, Burleson MH, Kowalewski RB, Malarkey WB, Van Cauter E, Berntson GG (2002) Loneliness and health: potential mechanisms. Psychosom Med 64:407–417

Cacioppo JT, Hawkley LC, Thisted RA (2010) Perceived social isolation makes me sad: five year cross-lagged analyses of loneliness and depressive symptomatology in the Chicago health, aging, and social relations study. Psychol Aging 25:453–463

Cazes C, Rosnet E, Bachelard C, Le Scanff C, Rivolier J (1996) Group dynamics during the EXEMSI isolation study. In: Bonting SL (ed) Advances in space biology and medicine, vol 5. JAI Press, Greenwich, pp 245–262

Choukér A, Smith L, Christ F, Larina I, Nichiporuk I, Baranov V, Bobrovnik E, Pastushkova L, Messmer K, Peter K, Thiel M (2002) Effects of confinement (110 and 240 days) on neuroendocrine stress response and changes of immune cells in men. J Appl Physiol 92:1619–1627

De Jong Gierveld J (1998) A review of loneliness: concept and definitions, determinants and consequences. Rev Clin Gerontol 8:73–80

De la Torre GG, Van Baarsen B, Ferlazzo F, Kanas N, Weiss K, Schneider S, Whiteley I (2012) Future perspectives on space psychology: recommendations on psychosocial and neurobehavioural aspects of human spaceflight. Acta Astronaut 81:587–599

Dinges DF, Venkataram S, McGlinchey EL, Metaxas DN (2007) Monitoring of facial stress during space flight: optical computer recognition combining discriminative and generative methods. Acta Astronaut 60:341–350

Frankl VE (1959) The doctor and the soul. Vintage, New York

Grigoriev AI, Kozerenko OP, Myasnikov VI (1985) Selected problems of psychological support of prolonged spaceflight. Paper presented to 36th IAF congress, Stockholm

Gushin VI, Zaprisa NS, Pustinnikova JM, Smirnova TM, Popova II (2001) Characteristics of Russian and non-Russian crewmembers' communication with external parties under prolonged isolation. In: Baranov VM (ed) Simulation of extended isolation: advances and problems. Firm SLOVO, Moscow, pp 85–100

Hockey GRJ (1986) Changes in operator efficiency as a function of environmental stress, fatigue, and circadian rhythms. In: Boff KR, Kaufman L, Thomas JP (eds) Handbook of perception and human performance, vol II. Wiley, New York, pp 44-1–44-49

Hoermann HJ, Johannes B, Salnitski VP, Pecena Y (2008) The "digital friend": design features of a psychological support system for space-crews. Acta Astronaut 63:848–854

Holt-Lunstad J, Smith TB, Layton JB (2010) Social relationship and mortality risk: a meta-analytic review. PLoS Med 7(7):e1000316

House JS, Landis KR, Umberson D (1988) Social relationships and health. Science 214:540–545

Johannes B, Gaillard AWK (2014) A methodology to compensate for individual differences in psychophysiological assessment. Biol Psychol 96:77–85

Johannes B, Salnitski VP, Gunga HC, Kirsch KA (2000) Voice stress monitoring in space – possibilities and limits. Aviat Space Environ Med 71:A58–A65

Johannes B, Salnitski VP, Lukjanuk VV, Gunga HC, Kirsch K (2001) Monitoring of autonomic outlet types during long-term confinement. In: Baranov VM (ed) Simulation of extended isolation: advances and problems. Slovo, Moscow, pp 51–60

Johannes B, Salnitski VP, Petsch J , Karashtin VV, Kirsch K (2002) Continuous voice-frequency monitoring of vocal outgoing communication during long-term confinement. Abstract book, World Space Congress, Houston, TX, USA, 9.-19.10

Johannes B, Salnitski VP, Polyakov VV, Kirsch K (2003) Changes in the autonomic reactivity pattern to psychological load under long-term microgravity - twelve men during 6-month spaceflights. Aviakosm i Ecolog Med 37:6–16

Johannes B, Wittels P, Enne R, Eisinger G, Castro C, Thomas J, Adler A, Gerzer R (2007) Non-linear function model of voice pitch dependency on physical and mental load. Eur J Appl Physiol 101:267–276

Johannes B, Sitev AS, Vinokhodova AG, Salnitski VP, Savchenko EG, Artyukhova AE, Bubeev YA, Morukov B, Tafforin C, Basner M, Dinges D, Rittweger J (2015) Wireless monitoring of changes in crew relations during long-duration mission simulation. PLoS One 10(8):e0134814. https://doi.org/10.1371/journal.pone.0134814

Johannes B, Gaillard AWK, Bronnikov S, Bubeev Y, Kotrovskaya T, Rittweger J (2016a) Extended psychophysiological assessment during a simulated spacecraft docking experiment. ASMA 87th annual scientific meeting "Human performance and the year of the aerospace medicine professional". Atlantic City, NY, 24–28 April 2016, Abstract book, p 275

Johannes B, Salnitski VP, Dudukin AV, Shevchenko LG, Shebuchev AE, Bronnikov SV (2016b) Performance assessment in the experiment PILOT on-board space stations MIR and ISS. Aerosp Med Hum Perform 87(6):534–544

Johannes B, Bronnikov S, Bubeev YA, Dudukin AV, Hoermann HJ, Frett T, Rittweger J, Gaillard AKW (2017) A tool to facilitate learning in a complex manual control task. Int J Appl Psychol 7(4):79–85

Kanas N (1991) Psychosocial support for cosmonauts. Aviat Space Environ Med 62:353–355

Kanas N (2009) From Earth's orbit to the outer planets and beyond: Psychological issues in space. The sixth IAA symposium on realistic near term advanced scientific space missions, 6–9 July 2009, Aosta, Italy

Kanas N, Manzey D (2003) Space psychology and psychiatry. Microcosm Press, Kluwer Academic Press, El Segundo, London

Kanas N, Salnitskiy V, Grund EM, Gushin V, Weiss DS, Kozerenco O, Sled A, Marmar CR (2000) Social and cultural issues during shuttle/MIR space missions. Acta Astronaut 47:647–655

Kanas N, Salnitski VP, Ritsher JB, Gushin VI, Weiss DS, Saylor SA, Kozerenko OP, Marmar CR (2007) Crewmember and mission control personnel interactions during international Space Station missions. Aviat Space Environ Med 78:601–607

Kanas N, Saylor S, Harris M, Neylan T, Boyd J, Weiss DS, Baskin P, Cook C, Marmar C (2010) High vs. low crewmember autonomy in space simulation environments. Acta Astronaut 67:731–738

Kass R, Kass J (2001) Team-work during long-term isolation: Sfincss experiment GP-006. In: Baranov VM (ed) Simulation of extended isolation: advances and problems. Firm SLOVO, Moscow, pp 124–147

Manzey D, Schiewe A, Fassbender C (1995) Psychological countermeasures for prolonged manned spaceflight. Acta Astronaut 35:339–361

Manzey D, Lorenz B, Polyakov V (1998) Mental performance in extreme environments: results from a performance monitoring study during a 438-day space mission. Ergonomics 41:537–559

McDuff D, Kodra E, Kaliouby RE, LaFrance M (2017) A large-scale analysis of sex differences in facial expressions. PLoS One 12(4):e0173942. https://doi.org/10.1371/journal.pone.0173942. eCollection 2017. PubMed PMID: 28422963; PubMed Central PMCID: PMC5396880

Miller G (2011) Why loneliness is hazardous to your health. Science 6014:138–140

Molasso WR (2006) Exploring Frankl's purpose in life with college students. J Coll Char 7:1–10

Palinkas LA, Johnson JC, Boster JS, Rakusa-Suszczewski S, Klopov VP, Fu XQ, Sachdeva U (2004) Cross-cultural differences in psychosocial adaptation to isolated and confined environments. Aviat Space Environ Med 75:973–980

Salnitski VP, Myasnikov VI, Bobrov AF, Shevchenko LG (1999) Integralnaya otsenka I prognoz professionalnoij nadezhnosti kosmonavtov v polote (integrated evaluation and prognosis of cosmonaut's professional reliability during space flight) (russ). Aviakosm i Ecolog Med 33:16–22

Salnitski VP, Dudukin AV, Johannes B (2001) Evaluation of operator's reliability in long-term isolation (the "pilot"-test). In: Baranov VM (ed) Simulation of extended isolation: advances and problems (30–50). Slovo, Moscow

Salnitski VP, Bobrov AF, Dudukin AV, Johannes B (2004) Reanalysis of operators reliability in professional skills under simulated and real space flight conditions. In: Proceedings of the 55th IAC congress, Vancouver, Canada, CD

Salnitski VP, Savchenko EG, Arthyukhova A, Johannes B (2009) Drahtloses monitoring von Gruppenstrukturen im 'Mars500'-Projekt. DGLRM-Jahrestagung, 18–19 Sept. 2009, Fürstenfeldbruck, Vortrag, Abstraktband, p. 25

Sandal G (2001) Psychosocial issues in space: future challenges. Gravit Space Biol Bull 14:47–54

Sandal G, Vaernes R, Ursin H (1995) Interpersonal relations during simulated space missions. Aviat Space Environ Med 66:617–624

Sandal G, Vaernes R, Bergan T, Warncke M, Ursin H (1996) Psychological reactions during polar expeditions and isolation in hyperbaric chambers. Aviat Space Environ Med 67:227–234

Schneider S, Mierau A, Diehl J, Askew CD, Strüder HK (2009) EEG activity and mood in health oriented runners after different exercise intensities. Physiol Behav 96:706–716

Schneider S, Brümmer V, Carnahan H, Kleinert J, Piacenti MF, Meeusen RM, Strüder HK (2010) Exercise as a countermeasure to psycho-physiological deconditioning during long-term confinement. Behav Brain Res 25:208–214

Shimamaiya T, Terada N, Hiejima Y, Wakabayashi S, Kasai H, Mohri M (2004) Effects of 10-day confinement on the immune system and psychological aspects in humans. J Appl Physiol 97:920–924

Simonov PV, Frolov MV (1977) Analysis of the human voice as a method of controlling emotional state: achievements and goals. Aviat Space Environ Med 48(1):23–25

Sulc J (1979) Analysis of verbal behavior of the first czechoslovak cosmonaut during space flight. Ceskoslovenska Psychol XXIII:42–49

Tomi L (2001) The role of cross-cultural factors in long-duration international space missions: lessons from the Sfincss study. In: Baranov M (ed) Simulation of extended isolation: advances and problems. Firm SLOVO, Moscow, pp 117–124

Vaic H, Friedrich J, Kolinchenko TB (1981) Analyse des Flugfunkverkehrs als Beitrag zur Beurteilung der Arbeitsfähigkeit von Kosmonauten. Z Militärmed 2:73–76

Van Baarsen B (2001) How's life? Adaptation to widowhood in later life and the consequences of partner death on the experienced emotional and social loneliness. Dissertation, Vrije University Amsterdam, Amsterdam

Van Baarsen B (2002) Eenzaamheid en zingeving: Signaleren en meten van eenzaamheid in de praktijk. [loneliness and meaning of life: observing and measuring loneliness in daily life.]. Geron 4:60–68

Van Baarsen B (2011) Humans in outer space: existential Fulfilment or frustration? Psychological, existential, social and ethical issues for crew on a long term space mission beyond earth orbit. In: Landfester U, Worms JC, Schrogl K-U (eds) Humans in outer space – interdisciplinary perspectives. Springer, Wien, pp 222–238

Van Baarsen B (2013) Person autonomy and voluntariness as important factors in motivation, decision making, and astronaut safety: first results from the Mars500 LODGEAD study. Acta Astronaut 87:139–146. https://doi.org/10.1016/j.actaastro.2013.02.006

Van Baarsen B, Broese van Groenou MI (2001) Partner loss in later life: gender differences in coping shortly after bereavement. J Loss Trauma 6:243–262

Van Baarsen B, Ferlazzo F, Ferravante D,Di Nocera F, Jörgensen J, Smit J, Van Duijn M, Giannini A-M, Kuipers A, Van der Pligt J (2009) Digging into space psychology and isolation: the Mars520 LODGEAD study. Preliminary results of the Mars105 pilot study. The 60th International Astronautical Congress, 12–16 Oct 2009. Daejeon, Republic of Korea

Van Baarsen B, Vinokhodova AG, & Ferlazzo F (2012a) Person autonomy of crew members in extreme confinement as seen from the viewpoint of mission ground control: Implications for communication and decision making. The 63th International Astronautical Congress, 3–7 Oct 2012. Naples, Italy

Van Baarsen B, Ferlazzo F, Ferravante D, Smit JH, Van der Pligt J, & Van Duijn MAJ (2012b) Emotional and cognitive adaptation during 520 days of isolation: results from the LODGEAD Mars500 study. The 63th International Astronautical Congress, 3–7 Oct 2012. Naples, Italy

Vessey WB, Blackwell Landon L (2017) Team performance in extreme environments. In: Salas E, Rico R, Passmore J (eds) The Wiley Blackwell handbook of the psychology of team working and collaborative processes. Wiley & Sons, Hoboken, pp 531–553

Monitoring of Autonomic Activity by Cardiovascular Variability: How to Measure?

23

André E. Aubert and Bart Verheyden

23.1 Introduction: *Why* Can Heart Rate Variability Be a Suitable Tool to Monitor Autonomic Activity?

The study of cardiovascular variability in heart rate (heart rate variability: HRV) and blood pressure (blood pressure variability: BPV) allows insight into the autonomic modulation of cardiovascular function (Malliani 2000). Most of cardiovascular variables such as heart rate, blood pressure, and stroke volume fluctuate on a beat-to-beat basis, thus creating a complex system with feedback (Aubert and Ramaekers 1999), with both linear and nonlinear fluctuations. Such a control system, well known in engineering, is more adequate to maintain stability and adapt to functional needs (Guyton et al. 1981). Therefore, cardiovascular control, as expressed by the time-dependence of hemodynamic variables, is a direct reflection of autonomic activity (Malliani 2000). An alteration of variability has been reported in several cardiovascular and non-cardiological diseases

(Lees et al. 2018; da Silva et al. 2018; McIntosh 2016). Moreover HRV has a prognostic value and is therefore important in modeling risk stratification (Malliani 2000), such as a predictor after acute myocardial infarction (Kleiger et al. 1987) and as an early warning sign of diabetic neuropathy (Benichou et al. 2018). Therefore, decreased cardiovascular variability is a sensitive indicator of altered autonomic modulation and can identify high-risk patients. Thus a continuous measurement of heart rate, blood pressure, and their relationship, arterial baroreflex mechanism, allows an indirect insight into the functioning of the cardiovascular control mechanism (Aubert and Ramaekers 1999). An additional advantage is that these measurements are noninvasive and thus well adapted to a clinical setting or experimental use with astronauts in space.

A. E. Aubert (✉) · B. Verheyden
Laboratory of Experimental Cardiology, Univ. Hospital Gasthuisberg, Leuven, Belgium
e-mail: andre.aubert@kuleuven.be

© Springer Nature Switzerland AG 2020

A. Choukér (ed.), *Stress Challenges and Immunity in Space*,
https://doi.org/10.1007/978-3-030-16996-1_23

The normal heartbeat varies secondary to respiration (respiratory sinus arrhythmia) (Toska and Eriksen 1993), in response to physical (Aubert et al. 2003) and mental stress (Aubert et al. 2010) and multiple other factors such as aging and gender (Ramaekers et al. 1998); finally it is characterized by a circadian variation (Umetani et al. 1998). Both the basic heart rate and its modulation are primarily determined by alterations in autonomic activity (Cohen and Taylor 2002). Increased parasympathetic or vagal activity slows heart rate and increased sympathetic activity increases heart rate (also related to biochemical processes, see Chaps. 7 and 8). Therefore, there is a continuous balance between the activities of both antagonist systems that regulates cardiovascular variables around a mean value. In a healthy individual, the role of the autonomic nervous system is essential to adequate cardiovascular functioning (Verheyden 2007).

However in research and hospital settings, it remained a problem how to evaluate and relate these oscillations to sympathetic and parasympathetic activities. This became only possible after the advent of computer techniques in the seventies for automatic detection of fiducial points on the signals and further digital processing (Hyndman et al. 1971; Hyndman 1974; Sayers 1973). Cardiovascular changes (HRV and BPV) may be measured by a number of techniques. After peak detection on the basic parameters such as ECG and blood pressure, time series of RR intervals (tachogram) and of systolic and diastolic blood pressure values (systogram and diastogram) are obtained (Aubert et al. 1999). These signals consist of discrete time series that can be further processed either in time- or in frequency domain (Malik and Camm 1995) or using nonlinear processing methods (Beckers et al. 2006a, b).

23.2 Data Sampling: *What* Do You Need to Determine Cardiovascular Variability?

In general, three steps are needed for data recording and analysis:

The first step is the recording of a high-quality ECG and blood pressure tracing. The former is obtained with electrocardiography. The latter is most often obtained from noninvasive recordings with a finger pulse pressure method (Imholz et al. 1998), based on the volume-clamp method of Penaz from a finger cuff. Figure 23.1 shows the elements of the equipment needed to determine HRV and BPV and Fig. 23.2 how this is implemented in the ISS. The model shown is the one used for training purpose of the astronauts before spaceflight in Star City (Moscow, Russia).

Although the finger pressure cuff works perfectly almost all the time, under certain circumstances malfunction can occur: it is the so-called Cold finger syndrome." This is similar to Raynaud's disease, which is a type of disorder that affects the small blood vessels. It causes a problem with the vessels in the extremities, and it results in reduced blood flow, thus leading to a loss of signals.

During a spaceflight pre-flight base-line measurements of the astronaut were satisfactory. However, signals obtained during a first session in-flight in the ISS were unusable (Fig. 23.3, upper part). The same equipment had been used 2 months earlier on another astronaut with total satisfaction. As hardware was suspected of malfunctioning,

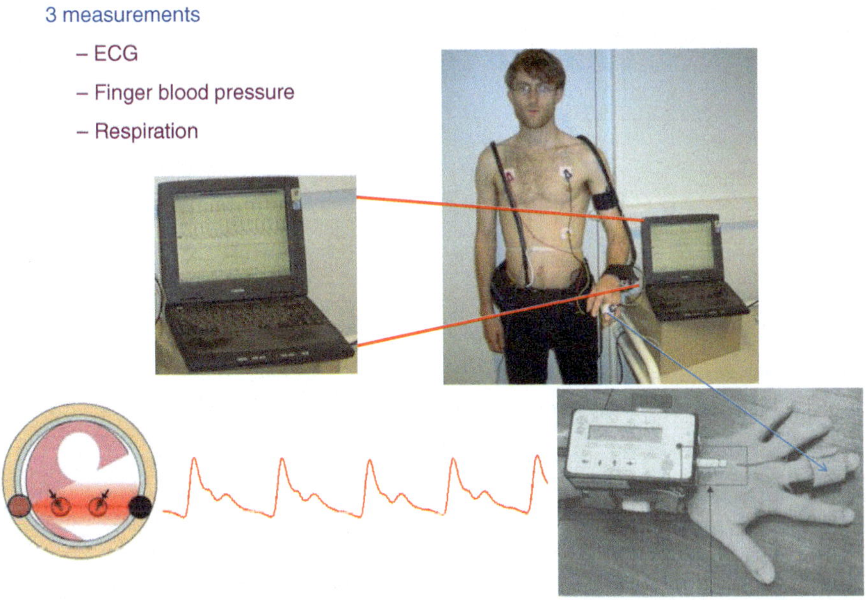

Fig. 23.1 Upper right: subject with ECG, finger blood pressure cuff, and respiration detector. Lower left: illustration of volume-clamp method and resulting blood pressure signal. Lower right: finger cuff

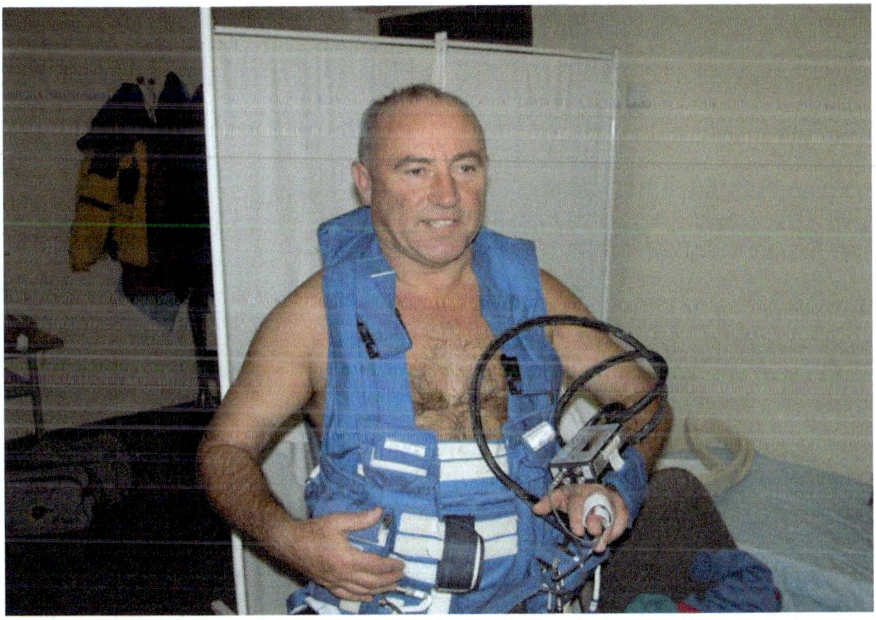

Fig. 23.2 Implementation of measurement device (ECG, finger blood pressure cuff, and respiration detector) and used in the ISS in the period 2000–2011. The system supplied several hundred hours of laboratory-quality recordings on successive astronauts and cosmonauts

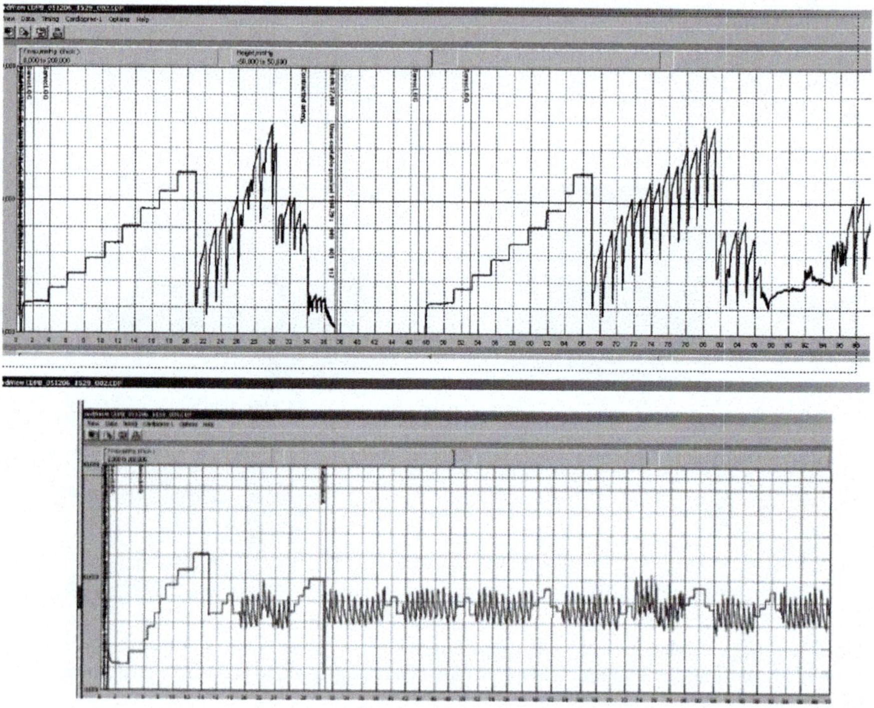

Fig. 23.3 Upper part: due to "Cold finger syndrome" the pressure device keeps on trying to calibrate the pressure signal (step signal), but to no avail: no good pressure signals. Lower part: after warming the finger the problem disappeared and normal signals were obtained

new finger cuffs were shipped to the station by a Proton supply vessel. Despite new cuffs, the blood pressure waveforms remained unacceptable. The remaining part of the hardware seemed to function normally, as no error messages were generated. It was found the astronaut had "Cold finger syndrome" and after heating his finger with a warm cloth, the signal became normal again (Fig. 23.3, lower part).

Duration of recordings can extend from a minimum of 10 min (Task force 1996) to 24 h in Holter recordings. These signals are analog/digital converted for computer processing. In order to have a good time resolution and event definition, a sampling rate of at least 250 Hz and preferentially of 1000 Hz (giving a time resolution of 1 ms) is recommended (Pinna et al. 1994). A typical ECG signal is shown in Fig. 23.1a. For analysis in the frequency domain, the ECG signal must meet several conditions: it must be sufficiently long (Task force 1996), random, stationary, and free of arrhythmias and noise. Random means that sequences cannot be determined by a mathematical expression or rule (Bendat and Piersol 1971). Stationary means that the probability function of the ECG signal does not change over time (Aubert et al. 1999). For frequency domain measurements, at least two times the wavelength of the lowest frequency component is needed for data recording. Consequently, the lowest limit for the analysis of the high frequency (HF) component (0.15 Hz) would be 13.3 s and for the low frequency (LF) component (0.04 Hz) would be 50 s.

The second step for HRV analysis consists of peak detection of the QRS complex (representing the ventricular depolarization) (Fig. 23.4a, lower part) or of systolic blood pressure on the blood pressure tracing. The result is a discrete, unevenly spaced time event series (Fig. 23.4b) the tachogram, a normal-to-normal interval (R-R interval, RRI). For BPV analysis, both blood pressure amplitude and the timing of its occurrence have to be synchronously recorded. The variations in systolic blood pressure result in the systogram and the variations in diastolic pressure lead to the diastogram. Before processing, these point series have to be corrected for ectopic and missed beats. This correction is obtained by filtering and interpolation algorithms (Aubert et al. 1999).

Finally, as a third step, for most spectrum analysis methods, an evenly spaced point series is needed. Therefore an evenly spaced point series is created by interpolation and a series consisting of equidistant points (0.5 s) is created. Although newer spectrum transform techniques (Lomb method) allow to use non equidistant time series (Holland and Aboy 2009).

All the above steps are illustrated in Fig. 23.4:

(a) ECG recording of a young healthy subject and peak detection.
(b) Unfiltered RR-intervals depicted in a tachogram, as obtained from panel a with a time length of 600 s and insert and (b1) blow-up of first 20 points of the tachogram.

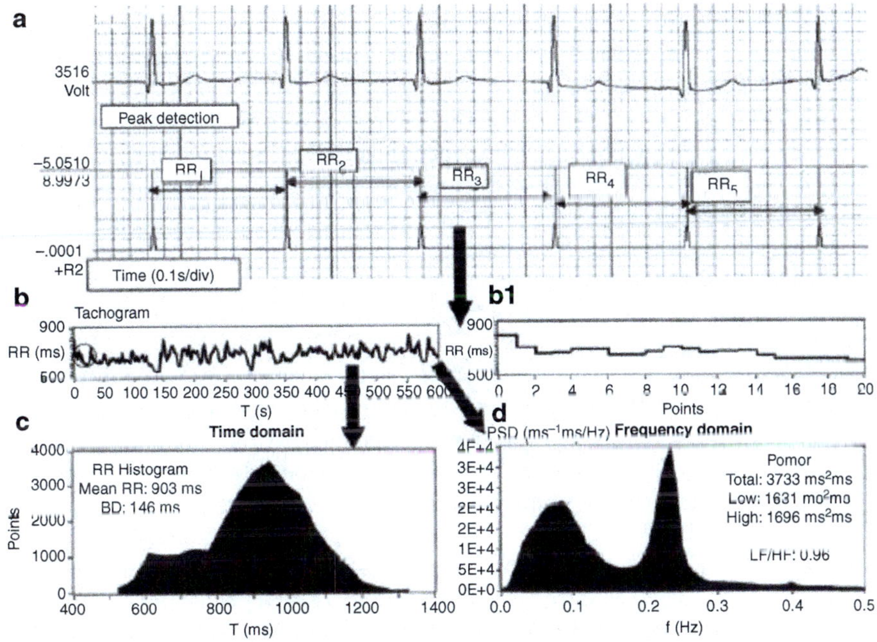

Fig. 23.4 Schematic representation of the different steps involved in HRV analysis, with courtesy from Acta Cardiologica 199:54:107107–120. Upper tracing: ECG signal, second lower: peak detection on ECG (**a**), middle left: tachogram (**b**), lowest: histogram (**c**); right middle: tachogram, first 20 points of tracing on the left (within circle) (**b1**), lowest right: power spectrum plot (**d**)

(c) HRV computed in the time domain and depicted as an histogram giving the distribution of RR intervals. This histogram was obtained from a 24 h Holter recording.
(d) Analysis in the frequency domain: power spectral density (PSD), obtained from the same recording. Low-frequency band: 0.04–0.15 Hz (LF); High-frequency band: 0.15–0.4 Hz (HF). Below LF there is a very low-frequency band: DC-0.04 Hz (VLF).

23.2.1 Time Domain Analysis

Parameters in the time domain are easily computed with simple statistical methods, even from short-duration time windows. Their main limitation is the lack of discrimination between activities of both branches of the autonomic system (Pinna et al. 1996).

Recommendations for a standardization for different parameters have been published (Task force 1996). The most frequently used time domain parameters (Malik and Camm 1995) include SD and SDANN, both representing global variability and rMSSD and pNN50, both highly correlated to high-frequency power in the frequency domain (Ramaekers et al. 1998) and as such are also markers of vagal activity.

- *SDNN (or SDRR)* (ms): is the standard deviation of the normal to normal (NN) interval over the recorded time interval (Copie et al. 1996; Kochiadakis et al. 1997). Theoretically, the heart rate variance which is equal to $(SDNN)^2$ is identical to the total power (ms^2) as determined from power spectral analysis (surface under the curve of PSD). It represents global HRV. SDNN depends upon the length of the recording. A longer recording of the same subject gives a higher SDNN value. Therefore it is inappropriate to compare SDNN values obtained from recordings of different durations.
- *SDANN* (ms): is the standard deviation of the 5 min mean NN interval over the entire recording. As such this estimates changes caused by cycles longer than 5 min.
- *rMSSD* (ms): is the square root of the mean squared successive differences between adjacent RR intervals over the entire recording. It correlates with the parasympathetic (vagal) activity as measured from HF.
- *pNN50 (%)*: is the percentage of successive interval differences greater than 50 ms computed over the entire recording.

The latter two are highly and positively correlated with each other and therefore they can be considered as surrogates for each other (Kleiger et al. 1991).

Geometric methods can be used as well to process RR-intervals. The simplest one is the sample density histogram representation (Fig. 23.4c). Another possibility consists in plotting the duration of each RR interval against the duration of the immediately preceding RR interval: Poincare maps. Assessment of HRV is based on quantification of dimension of shape of pattern.

An advantage of geometric methods is that they are very much insensitive to errors caused by ectopic beats or even arrhythmias. A disadvantage is that they are not as precise as pure time or frequency domain parameters. Also quite long-duration recordings are needed.

23.2.2 Frequency Domain Analysis

By definition, spectral analysis decomposes any steady, stationary, fluctuating signal into its sinus components. It allows plotting the power of each component as a function of its frequency (Fig. 23.4d) and the computation of the power in defined frequency regions. Power spectral analysis can be performed by nonparametric Fast Fourier Transform (FFT) (Akselrod et al. 1981) and by parametric autoregressive modeling (AR) (Baselli et al. 1985).

FFT: This approach is rather simple to apply and is computationally efficient. However its frequency resolution is limited and directly related to the duration of the recording period: the lowest frequency is the inverse of the recording length. The upper frequency limit is imposed by the Nyquist criterion: half the sampling frequency. Power spectral density can be computed in defined frequency bands: VLF, LF, and HF. These methods are based on the linear characteristics of the signal. However, due to windowing (the choice of a finite time segment), these methods suffer from spectral leakage, leading to masking of weak signals that are present in the data (Bendat and Piersol 1971). Using of parametric power spectrum estimation methods avoids this problem of leakage.

AR: In this approach the time series of RR-intervals is described by an autoregressive model: the signal at every time step is expressed as a linear function of its values at J previous steps. Therefore this method requires an a priori choice of the value of J; the order of the parametric model to provide the best fit to the data. The model order J, selected by information theory criteria, determines both centre frequency and the magnitude of the spectral components. The advantages of using parametric methods are: (1) Smoother spectral components which can be distinguished independently of preselected frequency bands, (2) Easy post-processing of the spectrum with an automatic calculation of low and high frequency power components and easy identification of the central frequency of each component, and (3) an accurate estimation of power spectral density even on a small number of samples on which the signal is supposed to maintain stationarity. The basic disadvantage of parametric methods is the need to verify the suitability of the chosen model and is complexity (the order of the model).

Spectral components and physiology (see also more extensively in Chap. 8): All methods for quantifying heart rate and blood pressure fluctuations are used for interpreting the link between mathematics to physiology. However there are still many contradictions and controversies.

In a typical cardiovascular power spectrum different frequency bands can be distinguished (Fig. 23.4d): high frequency (HF: 0.15–0.4 Hz), low frequency (LF: 0.04–0.15 Hz), and very low frequency (VLF: 0–0.04 Hz) (Task force 1996). The physiological explanation of VLF has not been well defined until yet. The HF component is generally considered to be linked to respiration (respiratory sinus arrhythmia) and thought to be mediated almost exclusively by fluctuations of efferent parasympathetic activity (Malliani et al. 1991). In healthy subjects heart rate and the entire arterial pressure waveform rise and fall at the frequency of respiration. These fluctuations represent both autonomic neural fluctuations and mechanically induced central blood volume changes in synchrony with respiration. During inspiration, the RRI on an ECG is shortened, whereas during expiration RRI is prolonged (Cohen and Taylor 2002).

HF oscillations are almost completely abolished by large dose of atropine (Akselrod et al. 1981; Ramaekers et al. 2002). However care has to be taken about respiration frequency: if that becomes lower than 0.15 Hz, it shifts to the LF component (Novak et al. 1993). The physiological background of the LF component is still rather controversial. It is believed to consist of a mixture of sympathetic and parasympathetic modulation (Eckberg 1997).

LF and HF components can be expressed in absolute units (ms^2 or $mmHg^2$), or in normalized units: LFnu and HFnu. This can be obtained by dividing the power of a given component by the total power (TP) (from which VLF has been subtracted) and multiplying by 100. As such LF and HF components are independent of total power:

$$\mathrm{LFnu} = \mathrm{LF}/\left(\mathrm{TP} - \mathrm{VLF}\right) \times 100 \qquad \mathrm{HFnu} = \mathrm{HF}/\left(\mathrm{TP} - \mathrm{VLF}\right) \times 100$$

Joint time-frequency analysis: Power spectrum analysis gives a global picture of oscillations present in a signal within a certain time window. Because of the problem of stationarity, as discussed before, frequency domain HRV parameters are not reliable in case of quick changes in heart rate or its autonomic modulation. The spectrum essentially shows which frequencies are contained in the signal as well as their corresponding amplitudes and phases, but does not show at which time these frequencies occur.

Joint time-frequency analysis on the other hand can visualize dynamic or transient changes over time (Chan et al. 2001) as it combines simultaneously time and frequency information. For example, fast changes in heart rate can occur during drug infusion, tilt testing, orthostatic stress, or any time-related event. Several methods have been proposed such as: short time Fourier Transform, Choi-Williams distribution, smoothed pseudo Wigner-Ville distribution, discrete wavelet transform (Shie and Dapang 1996), and the more advanced continuous wavelet transform (Verlinde et al. 2001). Several types of wavelet functions exist such as the Morlet wavelet, Haar wavelet, Daubechies wavelet, Gaussian wavelet and Mexican hat wavelet. The idea of continuous wavelet transform is to project a signal on a family of zero-mean functions (the wavelets) deduced from an elementary function by translations and dilatations. All these above described methods allow to determine instantaneous frequency and power or the evolution over time of the LF and HF powers and LF and HF frequencies separately (Shie and Dapang 1996).

23.2.3 Nonlinear Dynamics

Nonlinear control of heart rate and blood pressure poses potential physiological advantages in the possibility to adapt quickly and more subtly to changes in physiological needs. The basic idea behind nonlinear control mechanisms is that with a rather small amount of energy, a different end-point can be reached. Analysis methods derived from nonlinear system dynamics have opened up a new approach for studying and understanding the characteristics of cardiovascular dynamics. Nonlinear analysis methods differ from the conventional methods because they are not designed to assess the magnitude of variability but, rather, the quality, scaling, and correlation properties of the signals (Aubert et al. 2009). A nonlinear system is mathematically defined as a second- or higher order power system, meaning that the independent variable in the mathematical equation contains an exponent. A linear system can be decomposed into its component parts, whereas in a nonlinear system the parts interfere, cooperate, or compete with each other. Therefore, a small change can dramatically alter the nonlinear system because the initial condition of all variables along with the input stimulus influences the output response.

Chaos theory is a specialized subtheory of nonlinear system dynamics that describes systems that are low-dimensional (3–5 variables), have defined boundaries, and exhibit sensitive dependence on initial conditions.

However their physiological background or relationship to autonomic modulation is still a matter of debate (Beckers et al. 2006d; Dabire et al. 1998) and no up-to-date standards for nonlinear indices have been published (Task force 1996). Nonlinear indices seem rather to reflect overall, integrated control of heart rate (Lombardi 2000). Nevertheless the importance of nonlinear behavior of cardiovascular control was underlined by Yamamoto and Hughson (1991). These authors have shown that for persons in the supine (awake) position, the contribution of nonlinear fluctuations to total was 85.5 ± 4.4%.

These nonlinear methods of analyzing HRV aim to assess qualitative properties rather than the magnitude of the signal. The physiological background of these novel methods of analyzing heart rate dynamics is much more poorly understood compared to traditional linear methods.

Nonlinear analysis methods can be divided into two categories: indices that describe the scaling behavior of the nonlinear system (FD, 1/f, DFAα1 and DFAα2); and indices that describe the complexity of the system (CD, LE, ApEn, NL).

Scaling: fractal dimension. A fractal object is self-similar or shows scale invariance: the details of the structure are similar when zooming in at different resolutions. The FD of a waveform represents a powerful tool for transient detection. This feature has been used in the analysis of ECG to identify and distinguish specific states of physiologic function (Beckers et al. 2006b).

Scaling. 1/f slope. The 1/f slope of the log(power)—log(frequency) plot is obtained from the linear regression of the power spectral density from 10^{-4} to 10^{-2} Hz. A slope of −1 is an indication of scaling behavior (Beckers et al. 2006d). The slope of the power-law relationship analysis has been observed to decrease with advancing age (Beckers et al. 2006a).

Scaling: detrended fluctuation dynamics (DFA). Detrended fluctuation analysis quantifies fractal-like correlation properties of the time series and uncovers short-range and long-range correlations. The root mean square fluctuation of the integrated and detrended data is measured within observation windows of various sizes and then plotted against window size on a log–log scale. Both the short-term (4–11 beats) DFAα1 and the long-term (>11 beats) DFAα2 scaling exponents can be calculated. Values of alpha around 1 are an indication of scaling behavior of 1/f fluctuations. Fractal scaling seems to be age dependent (Beckers et al. 2006a).

Complexity: Correlation dimension (CD). The complexity of a system is described in phase space. The latter is a space in which all possible states of the system are represented, with each possible state of the system corresponding to one unique point in the phase space. The correlation dimension determines an order of the system i.e. the number of dimensions needed to model the dynamics of the system under consideration (Bogaert et al. 2001; Grassberger and Procaccia 1983; Raab et al. 2006). In the presence of nonlinear complex behavior, an attractor in phase space characterizes the dynamics of the system and its complexity can be quantified in terms of the properties of the attractor.

Complexity: Lyapunov exponent. The trajectories of chaotic signals in phase space follow typical patterns. Closely spaced trajectories converge and diverge exponentially, relative to each other. Lyapunov exponents measure the average rate of convergence/divergence of these neighboring trajectories. For dynamical systems, sensitivity to initial conditions is quantified by the Lyapunov exponents.

Complexity: Approximate entropy (ApEn or SampEn). Entropy refers to system randomness, regularity, and predictability and allows systems to be classified by rate of information loss or generation.

Complexity: Numerical noise titration. The method of numerical noise titration is an analytical technique that provides a sufficient and robust numerical test to detect chaos and it gives a relative measure of chaotic intensity, even in the presence of significant noise contamination.

Nonlinear dynamics methods are a powerful addition in the tools for studying cardiovascular oscillations (Huikuri et al. 2009). Although relationship with the branches of the ANS have been shown (Mansier et al. 1996; Beckers 2002; Beckers et al. 2006a, b, d), these methods are still not widely used. Major reasons are: (1) their lack of visual representation compared to the linear methods of frequency analysis; (2) many methods are still under development and need to be validated; (3) often conflicting results are presented. Some of the conflicting findings in the literature may be due to the use of selective or nonselective blocking agents, to the use of anesthetized animals, to specific interventions (hypoxia, hemorrhage), or specific mathematical techniques. Therefore there is still a need for new well-controlled experiments, specifically aimed at the different types of receptors. In general, data on the physiological counterparts of nonlinear methods of HRV are limited (Huikuri et al. 2009).

23.3 Application for Health and Disease: *When* and *Where* to Use in Weightlessness (Space) Conditions

Cardiovascular variability has been successfully applied in innumerable domains. The purpose of this chapter is not to give a complete overview, but just a nonlimitative list of applications: circadian variations, gender, aging, anxiety, hostility, depression, physical training, smoking, alcohol, caffeine, air pollution, recreational drugs, development biology, influence of gravity, myocardial infarction, sudden cardiac death, hypertension, diabetes mellitus, heart failure, and heart transplantation.

Noninvasive methods are especially adapted for studies of human spaceflight research. Indeed, availability and expertise of astronauts are limited and therefore experiments have to be made as simple as possible (Beckers et al. 2004). As cardiovascular variability methods fulfill these conditions, they since long enjoyed a wide popularity, as they allow, with relative simple methods, a window toward autonomic modulation of the cardiovascular system. Gravity, or its absence, causes a large adaptation of the cardiovascular system in astronauts during spaceflight and extended periods of weightlessness and after return to Earth's gravity (Verheyden et al. 2007, 2009, 2010).

Cardiovascular variability methods have been successfully applied in various simulation studies of weightlessness such as parabolic flight (Beckers et al. 2003b; Verheyden et al. 2005), head-out-of-water (Miwa et al. 1997), and head down bedrest (Pavy-Le et al. 2007; Liu et al. 2009). During parabolic flights short periods of about 20 s of weightlessness are obtained, between periods of 20 s of 1.8 g during acceleration and deceleration of the plane. Each maneuver can be split up into five phases of about 20 s according to the gravitational force: Phase I of normogravity (1 g) before each parabola when the plane is flying at an altitude of 6000 m. Phase II or hypergravity (1.8 g) when the plane is rapidly accelerating from 6000 m to an altitude of 10,000 m at the ascending leg of the parabola. Phase III or microgravity (weightlessness at 0 g) at the top of the parabola. Phase IV or hypergravity (1.8 g) at the descending leg of the parabola: the plane lowers from 10,000 m to 6000 m and phase V or normogravity after the parabola, again a flight path at 6000 m (Liu et al. 2012). Instantaneous gravity is continuously recorded with the aircraft gravity vector-accelerometer.

Simulation of weightlessness during head-out-of-water and head down bedrest are both based on altering body fluid redistribution, similar to real spaceflight: In case of head-out-of-water by the hydrostatic pressure on lower limbs and for head down bedrest by tilting the bed on which the subject is lying, under an angle of 6° with the head at the lower end (Liu et al. 2009).

Long-term human space missions as required for travel to deep space and other planets, make feel the need for long-duration analog studies on Earth such as long-term confinement (during many months to year-long) and isolation studies. Because of the long-lasting isolation in a small habitat, such as the Mars500 project (confinement for 520 days) at the IMBP Institute in Moscow (with an important contribution from ESA) (see also Chaps. 36 and 37), the confinement may be considered similar to a simulated Mars mission, except for the effects of microgravity and radiation (Aubert et al. 2005). Therefore, isolation studies represent a unique opportunity to evaluate the net effect on physiological variables, avoiding the net effect of confounding factors such as real microgravity and radiation hazards in space (Arbeille et al. 2014). The crew consists of a highly motivated and severely selected group of six volunteers, very much comparable to astronauts (Vigo et al. 2013).

Ground-based research is needed for preparatory research for human deep space missions to Mars and it can make an important contribution to enhance our knowledge about psychological and physiological issues of long-duration space missions (Manzey 2004).

Also during space missions to the International Space Station and the Chinese Space station, the experience with cardiovascular variability methods is increasing and results from studies performed from the early days on and until now, especially in the IMBP Institute in Moscow by the group around Baevsky (Baevsky et al. 1997, 1998, 2007) have given new insights. From their results and also from our group it could be concluded that general cardiovascular control remains stable in space (Beckers et al. 2003a, b, 2006c, 2009; Verheyden 2007; Verheyden et al. 2007, 2009, 2010), but can be altered at the transitions between Earth to space and back. In one occasion HRV methods were used to test the hypothesis that microgravity alters cardiovascular neural response to standardized cognitive load stimuli, induced by mathematical mental stress (Aubert et al. 2010). It was concluded that a mental arithmetic task in astronauts elicits sympathovagal shifts toward enhanced sympathetic modulation and reduced vagal modulation, in this cohort however, responsiveness was not different during microgravity. In another study a difference in HRV parameters was found between European (ESA and Roscosmos) and Chinese astronauts (Liu et al. 2015). Further studies are warranted to more often monitor and for longer periods of time and in conjunction with other biological read-out parameters.

After 50 years of spaceflight, we are beginning to understand some mechanisms of physiological adaptations; however, new questions develop all the time because of sometimes unpredicted results. There is still a long way to go before we can safely send humans to Mars and bring them back safely.

Experience over the past 50 years have shown no major influence on lung function during weightlessness. On the other hand a major problem occurs with lunar (and Mars) dust. Moon dust was so pervasive that no lunar rock boxes from any of the six Apollo missions to the moon ever maintained their lunar vacuum—they all leaked.

Lunar and Martian dust may be a toxic challenge to astronauts (Aubert et al. 2016). This represents not only a problem for spacesuits and equipment, but also to the cardiovascular system (Donaldson et al. 2013), as the particles will inevitably be transported into the habitats as it clings to the spacesuit as was shown from the Apollo missions (Fig. 23.5). Study of this problem has also Earth-related consequences as air pollution has a great impact on human health.

The importance of all these physiological adaptations also has to be assessed in relation to the duration of the space mission. Exploration missions into deep space, such as a journey to Mars or to asteroids, may raise a series of new questions about

Fig. 23.5 Lunar dust stuck to the astronauts spacesuit will inevitably be transferred to the space habitat. (Credit NASA). Lunar dust is covered in a glassy coating that can either be smooth or jagged (size of the particles around 20 μm)

the health and especially the immune-health of the human participants. Also psychological and mental health issues (see Chaps. 22 and 31) will grow increasingly important during long duration missions and are linked also to organ changes. The high noise levels, less than optimal light conditions, and confinement to small living quarters will be contributors to high psychological and social stress levels that are also nonnegligible factors, certainly in long-duration spaceflights.

To gain support of the general public for manned spaceflight, it is also important that it receives information about the Earth-bound applications of space research. The development of integrative diagnostic tools and of therapies for osteoporosis, syncope, heart failure, cardiovascular deconditioning and muscle deconditioning will enhance public support and political will to continue and increase the intensity of human space exploration (Aubert et al. 2016).

23.4 Summary

Medicine is still considered as an art by many. However, objective methods that are soundly founded and physiologically tested may turn it into a science. Therefore further studies are warranted to more often monitor and for longer periods of time, HRV and in conjunction with other biological read-out parameters. To use the HRV as one—and because of its noninvasiveness very attractive—stress monitoring tool is intriguing, on Earth and in space. This will require very reliable hardware (ECG), data transfer and analyses, and a solid database to allow for conclusive interpretation hereby providing adequate recommendations for respective countermeasures or further diagnosis of dysfunctional organs which are affected by imbalanced autonomic regulation, like the immune system.

References

Akselrod S, Gordon D, Ubel FA, Shannon DC, Barger AC, Cohen RJ (1981) Power spectrum analysis of heart-rate fluctuation–a quantitative probe of beat-to-beat cardiovascular control. Science 213:220–222

Arbeille P, Provost R, Vincent N, Aubert AE (2014) Adaptation of the main peripheral artery and vein to long term confinement (Mars 500). PLoS One 9(1):e83063. https://doi.org/10.1371/journal.pone

Aubert AE, Ramaekers D (1999) Neurocardiology: the benefits of irregularity - the basics of methodology, physiology and current clinical applications. Acta Cardiol 54:107–120

Aubert AE, Ramaekers D, Beckers F, Breem R, Denef C, Van de Werf F, Ector H (1999) The analysis of heart rate variability in unrestrained rats. Validation of method and results. Comput Methods Prog Biomed 60:197–213

Aubert AE, Seps B, Beckers F (2003) Heart rate variability in athletes. Sports Med 33:889–919

Aubert AE, Beckers F, Verheyden B (2005) Cardiovascular function and basics of physiology in microgravity. Acta Cardiol 60(2):129–151

Aubert AE, Vandeput S, Beckers F, Liu J, Verheyden B, Van Huffel S (2009) Complexity of cardiovascular regulation in small animals. Philos Transact A Math Phys Eng Sci 367: 1239–1250

Aubert AE, Verheyden B, D'Ydewalle C, Beckers F, Van den Bergh O (2010) Effects of mental stress on autonomic cardiac modulation during weightlessness. Am J Physiol Heart Circ Physiol 298:H202–H209

Aubert AE, Larina I, Momken I, Blanc S, White O, Prisk K, Linnarsson D (2016) Towards human exploration of space: the theseaus review on cardiovascular, respiratory and renal research. NPJ Microgravity 2:16031. https://doi.org/10.1038/npjmgrav.2016.31

Baevsky RM, Petrov VM, Cornelissen G, Halberg F, Orth-Gomer K, Akerstedt T, Otsuka K, Breus T, Siegelova J, Dusek J, Fiser B (1997) Meta-analyzed heart rate variability, exposure to geomagnetic storms, and the risk of ischemic heart disease. Scr Med (Brno) 70:201–206

Baevsky RM, Petrov VM, Chernikova AG (1998) Regulation of autonomic nervous system in space and magnetic storms. Adv Space Res 22:227–234

Baevsky RM, Baranov VM, Funtova II, Diedrich A, Pashenko AV, Chernikova AG, Drescher J, Jordan J, Tank J (2007) Autonomic cardiovascular and respiratory control during prolonged spaceflights aboard the international Space Station. J Appl Physiol 103:156–161

Baselli G, Bolis D, Cerutti S, Freschi C (1985) Autoregressive modeling and power spectral estimate of R-R interval time series in arrhythmic patients. Comput Biomed Res 18:510–530

Beckers F (2002) Linear and nonlinear analysis of cardiovascular variability. In: Validation and clinical applications (Acta Biomedica Lovaniensis). Leuven University Press, Leuven, pp 1–133. ISBN 90 5867 249 2

Beckers F, Verheyden B, Aubert AE (2003a) Evolution of heart rate varibility before, during and after spaceflight. J Gravit Physiol 10:107–108

Beckers F, Seps B, Ramaekers D, Verheyden B, Aubert AE (2003b) Parasympathetic heart rate modulation during parabolic flights. Eur J Appl Physiol 90:83–91

Beckers F, Verheyden B, De Winne F, Duque P, Chaput D, Aubert AE (2004) HICOPS: human interface computer program in space. J Clin Monit Comput 18(2):131–136

Beckers F, Verheyden B, Aubert AE (2006a) Aging and nonlinear heart rate control in a healthy population. Am J Physiol Heart Circ Physiol 290:H2560–H2570

Beckers F, Verheyden B, Couckuyt K, Aubert AE (2006b) Fractal dimension in health and heart failure. Biomed Tech 51:194–197

Beckers F, Verheyden B, Couckuyt K, Liu J, Aubert AE (2006c) Autonomic cardiovascular modulation after spaceflight during orthostatic stress. Eur Heart J 27:190–190

Beckers F, Verheyden B, Ramaekers D, Swynghedauw B, Aubert AE (2006d) Effects of autonomic blockade on non-linear cardiovascular variability indices in rats. Clin Exp Pharmacol Physiol 33:431–439

Beckers F, Verheyden B, Liu J, Aubert AE (2009) Cardiovascular autonomic control after short-duration spceflights. Acta Astronaut 65:804–812

Bendat JS, Piersol AG (1971) Random data: analysis and measurement procedures. Wiley Interscience, New York. ISBN 0 471 0470 X

Benichou T, Pereira B, Mermillod M, Tauveron I, Pfabigan D, Maqdasy S, Dutheil F (2018) Heart rate variability in type 2 diabetes mellitus: A systematic review and meta-analysis. PLoS One 13:4

Bogaert C, Beckers F, Ramaekers D, Aubert AE (2001) Analysis of heart rate variability with correlation dimension method in a normal population and in heart transplant patients. Auton Neurosci 90:142–147

Chan HL, Huang HH, Lin JL (2001) Time-frequency analysis of heart rate variability during transient segments. Ann Biomed Eng 29:983–996

Cohen MA, Taylor JA (2002) Short-term cardiovascular oscillations in man: measuring and modelling the physiologies. J Physiol 542:669–683

Copie X, Le Heuzey JY, Iliou MC, Khouri R, Lavergne T, Pousset F, Guize L (1996) Correlation between time-domain measures of heart rate variability and scatterplots in postinfarction patients. Pacing Clin Electrophysiol 19:342–347

Dabire H, Mestivier D, Jarnet J, Safar ME, Chau NP (1998) Quantification of sympathetic and parasympathetic tones by nonlinear indexes in normotensive rats. Am J Physiol Heart Circ Physiol 44:H1290–H1297

Donaldson K, Duffin R, Langrish JP, Miller MR, Mills NL, Poland CA, Raftis J, Shah A, Shaw CA, Newby DE (2013) Nanoparticles and the cardiovascular system: a critical review. Nanomedicine (Lond) 8(3):403–423. https://doi.org/10.2217/nnm.13.16. Review

Eckberg DL (1997) Sympathovagal balance - a critical appraisal. Circulation 96:3224–3232

Grassberger P, Procaccia I (1983) Measuring the strangeness of strange attractors. Physica D 9:189–208

Guyton AC, Hall JE, Lohmeier TE, Jackson TE, Kastner PR (1981) Blood pressure regulation: basic concepts. Fed Proc 40:2252–2256

Holland A, Aboy M (2009) A novel recursive Fourier transform for nonuniform sampled signals: application to heart rate variability spectrum estimation. Med Biol Eng Comput 47(7):697–707

Huikuri HV, Perkiömäki JS, Maestri R, Pinna GD (2009) Clinical impact of evaluation of cardiovascular control by novel methods of heart rate dynamics. Philos Trans A Math Phys Eng Sci 367:1223–1238

Hyndman BW (1974) The role of rhythms in homeostasis. Kybernetik 15:227–236

Hyndman BW, Kitney RI, Sayers BM (1971) Spontaneous rhythms in physiological control systems. Nature 233:339–341

Imholz BPM, Wieling W, van Montfrans GA, Wesseling KH (1998) Fifteen years experience with finger arterial pressure monitoring: assessment of the technology. Cardiovasc Res 38:605–616

Kleiger R, Miller J, Bigger JT Jr, Moss A (1987) Decreased heart rate variability and its association with increased mortality after acute myocardial infarction. Am J Cardiol 59:266–262

Kleiger RE, Bigger JT, Bosner MS, Chung MK, Cook JR, Rolnitzky LM, Steinman R, Fleiss JL (1991) Stability over time of variables measuring heart rate variability in normal subjects. Am J Cardiol 68:626–630

Kochiadakis GE, Orfanakis AE, Rombola AT, Chrysostomakis SI, Chlouverakis GI, Vardas PE (1997) Reproducibility of time-domain indexes of heart rate variability in patients with vasovagal syncope. Am J Cardiol 79:160–165

Lees T, Shad-Kaneez F, Simpson AM, Nassif NT, Lin Y, Lal S (2018) Heart rate variability as a biomarker for predicting stroke, post-stroke complications and functionality. Biomark Insights 13:1177271918786931
Liu J, Li Y, Verheyden B, Liu X, Chen Z, Chen S, Aubert AE (2009) Cardiovascular control during 60 days head-down bed rest: Chinese herbal medicine as a countermeasure. Space Med Med Eng 22:391–398
Liu J, Verheyden B, Beckers F, Aubert AE (2012) Hemodynamic adaptation during sudden gravity transitions. Eur J Appl Physiol 117:79–89
Liu J, Li Y, Verheyden B, Chen S, Chen Z, Gai Y, Liu J, Gao J, Xi Q, Yuan M, Lin Q, Aubert AE (2015) Is autonomic modulation different between European and Chinese astronauts? PLoS One 10(3):e0120920. DOI:10.13
Lombardi F (2000) Chaos theory, heart rate variability, and arrhythmic mortality. Circulation 101:8–10
Malik M, Camm AJ (1995) Heart rate variability. Futura Publ Co, Armonk, pp 1–543. ISBN 0 87993 607 X
Malliani A (2000) Principles for cardiovascular neural regulation in health and disease. Kluwer Academic Publishers, Boston, pp 1–222. ISBN 0-7923-7775-3
Malliani A, Pagani M, Lombardi F, Cerutti S (1991) Cardiovascular neural regulation explored in the frequency-domain. Circulation 84:482–492
Mansier P, Clairambault J, Charlotte N, Medigue C, Vermeiren C, LePape G, Carre F, Gounaropoulou A, Swynghedauw B (1996) Linear and non-linear analyses of heart rate variability: a mini review. Cardiovasc Res 31:371–379
Manzey D (2004) Human missions to Mars: new psychological challenges and research issues. Acta Astronaut 55(3–9):781–790
McIntosh RC (2016) A meta-analysis of HIV and heart rate variability in the era of antiretroviral therapy. Clin Auton Res 26(4):287–294
Miwa C, Sugiyama Y, Mano T, Iwase S, Matsukawa T (1997) Sympatho-vagal responses in humans to thermoneutral head-out water immersion. Aviat Space Environ Med 68:1109–1114
Novak V, Novak P, Dechamplain J, Leblanc AR, Martin R, Nadeau R (1993) Influence of respiration on heart-rate and blood-pressure fluctuations. J Appl Physiol 74:617–626
Pavy-Le TA, Heer M, Narici MV, Rittweger J, Vernikos J (2007) From space to earth: advances in human physiology from 20 years of bed rest studies (1986–2006). Eur J Appl Physiol 101:143–194
Pinna GD, Maestri R, Di CA, Colombo R, Minuco G (1994) The accuracy of power-spectrum analysis of heart-rate variability from annotated RR lists generated by Holter systems. Physiol Meas 15:163–179
Pinna GD, Maestri R, Di CA (1996) Application of time series spectral analysis theory: analysis of cardiovascular variability signals. Med Biol Eng Comput 34:142–148
Raab C, Wessel N, Schirdewan A, Kurths J (2006) Large-scale dimension densities for heart rate variability analysis. Phys Rev E Stat Nonlinear Soft Matter Phys 73(4 Pt 1):041907. Epub 2006 Apr 10
Ramaekers D, Ector H, Aubert AE, Rubens A, Van de Werf F (1998) Heart rate variability and heart rate in healthy volunteers - is the female autonomic nervous system cardioprotective? Eur Heart J 19:1334–1341
Ramaekers D, Beckers F, Demeulemeester H, Aubert AE (2002) Cardiovascular autonomic function in conscious rats: a novel approach to facilitate stationary conditions. Ann Noninvasive Electrocardiol 7:307–318
Sayers BM (1973) Analysis of heart rate variability. Ergonomics 16:17–32
Shie Q, Dapang C (1996) Joint time-frequency analysis. Prentice Hall, Upper Saddle River. ISBN 0 13 254384 2
da Silva TD, Massetti T, Crocetta TB, de Mello Monteiro CB, Carll A, Vanderlei LCM, Arbaugh C, Oliveira FR, de Abreu LC, Ferreira Filho C, Godleski J, Ferreira C (2018) Heart rate variability and cardiopulmonary dysfunction in patients with duchenne muscular dystrophy: a systematic review. Pediatr Cardiol 39(5):869–883

Task force (1996) Heart rate variability. Standards of measurement, physiological interpretation, and clinical use. Eur Heart J 17:354–381

Toska K, Eriksen M (1993) Respiration-synchronous fluctuations in stroke volume, heart rate and arterial pressure in humans. J Physiol 472:501–512

Umetani K, Singer DH, McCraty R, Atkinson M (1998) Twenty-four hour time domain heart rate variability and heart rate: relations to age and gender over nine decades. J Am Coll Cardiol 31:593–601

Verheyden B (2007) Cardiovascular control in space and on earth: the challenge of gravity (Acta Biomedica Lovaniensia). Leuven University Press, Leuven, pp 1–152. ISBN 978 90 5867 618 4

Verheyden B, Beckers F, Aubert AE (2005) Spectral characteristics of heart rate fluctuations during parabolic flight. Eur J Appl Physiol 95:557–568

Verheyden B, Beckers F, Couckuyt K, Liu J, Aubert AE (2007) Respiratory modulation of cardiovascular rhythms before and after short-duration human spaceflight. Acta Physiol (Oxf) 191:297–308

Verheyden B, Liu J, Beckers F, Aubert AE (2009) Adaptation of heart rate and blood pressure to short and long duration space missions. Respir Physiol Neurobiol 169S:S13–S19

Verheyden B, Liu J, Beckers F, Aubert AE (2010) Operational point of neural cardiovascular regulation in humans up to 6 months in space. J Appl Physiol 108:646–654

Verlinde D, Beckers F, Ramaekers D, Aubert AE (2001) Wavelet decomposition analysis of heart rate variability in aerobic athletes. Auton Neurosci 90(1–2):138–141

Vigo D, Tuerlinckx F, Ogrinz B, Li W, Simonelli G, Bersenev E, Van den Bergh O, Aubert AE (2013) Circadian rhythm of autonomic cardiovascular control during Mars500 simulated mission to Mars. Aviat Space Environ Med 84(10):1023–1028

Yamamoto Y, Hughson RL (1991) Coarse-graining spectral-analysis – new method for studying heart-rate-variability. J Appl Physiol 71:1143–1150

Breath Gas Analysis

24

Michael Dolch, Siegfried Praun, Johannes Villiger,
Alexander Choukér, and Gustav Schelling

24.1 Introduction

Since ancient times human breath has been used in assisting to diagnose diseases like diabetic coma, liver diseases, and renal failure. Similar to the exchange of carbon dioxide and oxygen by the lungs numerous volatile substances either endogenously produced or exogenously absorbed are exchanged via the lungs. However, despite the knowledge on the diagnostic properties of exhaled breath, the smell underlying substances in nature remained undiscovered until recently. A first step toward a better understanding was done in 1971 when Pauling detected about 250 substances in the exhaled breath of humans (Pauling et al. 1971). Thereafter, the implementation of increasingly sensitive diagnostic platforms since the 1980s has led to the identification of several chemical compounds within human exhaled breath. At present, around 1840 volatile organic and inorganic compounds (VOICs) are known to be released by the human body. So far, this collection of VOICs represents the so-called human volatilome. About 780 of these VOICs appear in breath, whereas the remaining portion appears in other gaseous or liquid components as saliva, blood, milk, skin secretions, urine, and feces of healthy individuals (de Lacy Costello et al. 2014). Examples of so-far identified VOICs in human exhaled breath together with a possible clinical application are given in Table 24.1.

M. Dolch (✉) · A. Choukér · G. Schelling
Department of Anaesthesiology, Hospital of the University of Munich (LMU), Munich, Germany
e-mail: michael.dolch@med.uni-muenchen.de, m.dolch@krk-aoe.de

S. Praun · J. Villiger
V&F Analyse- und Messtechnik GmbH, Absam, Austria
e-mail: siegfried.praun@vandf.com

© Springer Nature Switzerland AG 2020

A. Choukér (ed.), *Stress Challenges and Immunity in Space*,
https://doi.org/10.1007/978-3-030-16996-1_24

Table 24.1 Examples of exhaled breath compounds, potential origin, and potential clinical application

Compound	Source	Potential clinical application
Acetaldehyde (Moeskops et al. 2006; Smith et al. 2002)	Ethanol metabolism oxidative stress	Monitoring of ethanol metabolism and oxidative stress
Acetone (Turner et al. 2006a)	Decarboxylation of acetoacetate, dehydrogenation of iso-propanol	Monitoring of diabetes
Acetonitrile (Buszewski et al. 2009; Pinggera et al. 2005)	Uptake from cigarette smoke	Monitoring of smoking behavior
Alkanes and monomethylalkanes (Phillips et al. 2004)	Lipid peroxidation	Heart allograft rejection
Ammonia (Davies et al. 1997; Endre et al. 2011)	Protein metabolism	End-stage renal disease
Dimethyl sulfide (Van den Velde et al. 2008)	Liver failure	Monitoring of liver damage
Ethane (Kazui et al. 1994; Risby and Sehnert 1999; Stenseth et al. 2007)	Lipid peroxidation product	Oxidative stress monitoring
Ethanol (Kramer et al. 2007; Smith et al. 2002)	Gut bacteria, external uptake	Alcohol intoxication
Ethylene (Risby 2005)	Lipid peroxidaton product	Oxidative stress monitoring
$^{13}CO_2$ (Modak 2007; Graham et al. 1987)	Metabolism	*Helicobacter pylori* breath test, gastric emptying, gut disease, liver function, pharmacokinetic monitoring
Hydrogen (Christl et al. 1992; Bond and Levitt 1976)	Gut bacteria	Lactose malabsorption breath test
Isoprene (Karl et al. 2001; Turner et al. 2006b)	Cholesterol biosynthesis	Statine therapy monitoring
Malondialdehyde (Scholpp et al. 2002)	Lipid peroxidation product	Oxidative stress monitoring
Methane (Roccarina et al. 2010)	Gut bacteria	Gastrointestinal diseases
Nitric oxygen (Kövesi et al. 2003; Pedrosa et al. 2010; Turner 2007; Karlsson et al. 2009)	Nitric oxide synthetase	Monitoring of airway inflammation and ischemia-reperfusion injury
Propionaldehyde (Moeskops et al. 2006; Steeghs et al. 2006)	Lipid peroxidation product, propanol metabolism	Oxidative stress monitoring
Propionaldehyde (Dolch et al. 2015)	Lipid peroxidation	Oxidative stress monitoring
Propofol (Hornuss et al. 2007)	Intravenous anesthetic	Monitoring of pharmacon during anesthesia
Trimethylamine (Endre et al. 2011)	Uptake of trimethylamine or precursor	Monitoring of hemodialysis efficacy
Pentane (Li et al. 2009; Schubert et al. 1998, 2005)	Lipid peroxidation product	Monitoring of oxidative stress

24.2 Technical Approaches for Breath Analyses

As compared to most other diagnostic methods used in medicine, the analysis of exhaled breath is truly noninvasive and allows an unlimited number of repetitions. Furthermore, "breath gas analysis" is not only restricted to the analysis of breath. The same analytic method could be applied to gaseous headspace probes over any liquid originating from the human body (Pinggera et al. 2005). Currently operated diagnostic platforms for the analysis of exhaled breath and headspace VOIC composition include gas chromatographic (GC) and different direct mass spectrometric methods. Depending on the analytical system and the respectively used setups, sensitivities down to the parts per trillion per volume level are reported for both methods (Amann et al. 2007).

Recent developments lead to an increase in sensitivity and decreases in time requirements in GC methods. Furthermore, coupling of GC with mass spectrometric methods (GC-MS) improved rapid compound identification. However, the disadvantages associated with the use of GC methods are the ongoing need for preconcentration, substance calibration, and limited capability for online measurements. Nevertheless, if definite compound identification is warranted GC still represents the reference method. Among direct mass spectrometry methods the following systems

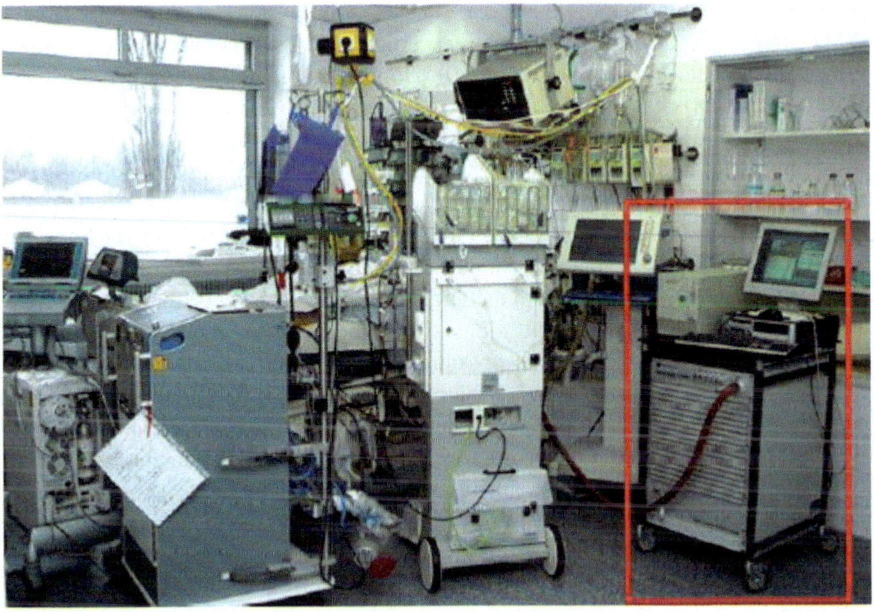

Fig. 24.1 Ion molecule reaction mass spectrometer (IMR-MS, *red box*) during online monitoring of exhaled breath in a patient with acute respiratory distress syndrome (ARDS). A directly to the patients endotracheal tube connected steel T-piece allows gas transfer from the patient to the IMR-MS (*red rubber tubing*). The vacuum pumps define the size of the hardware. In space, technical solutions might use the external vacuum instead

have been used recently for breath gas analysis: (a) ion-molecule reaction mass spectrometry (IMR-MS); (b) ion mobility spectroscopy (IMS); (c) proton transfer reaction mass spectrometry (PTR-MS); (d) and selected ion flow tube mass spectrometry (SIFT-MS). The major advantage of all these instruments is that preconcentration procedures are unnecessary, which allows their possible use for breath-by-breath online measurements of the exhaled breath VOIC composition. For IMR-MS even the online measurement of critically ill patients (Fig. 24.1) and in patients undergoing cranial surgery has been shown (Lindinger et al. 1998; Smith et al. 2002; Dolch et al. 2008; Hornuss et al. 2007). However, although these instruments allow for detailed in-depth analyses of exhaled breath VOIC composition the physical mass, energy consumption, demand for consumables, and maintenance expenses are factors limiting their use for breath gas analysis in space. Here, the rapidly evolving technical progress in the field of metal oxide sensor-based diagnostic devices possibly offers a promising alternative. Metal oxide sensor instruments are typically of low mass and size, low energy consumption, and low maintenance expenses. However, the disadvantage of these sensors is their so-far limited sensitivity and specifity.

24.3 Applications and Further Directions for Earth and Space

Clinical breath gas analysis is an emerging tool offering unique diagnostic capabilities for a large number of disorders, which include bacterial infection (Braden 2009), metabolic disorders (Romagnuolo et al. 2002), pulmonary disease (Turner 2007), and cancer (Szulejko et al. 2010). However, potential biomarkers for these disorders are part of a complex matrix of different VOICs present in human breath and the understanding of their pathophysiological role is still in its infancy. This is also due to (1) the high number of 780 VOICs reported in human breath (de Lacy Costello et al. 2014), (2) the instrumentation and techniques used for breath analyses (Buszewski et al. 2007) and (3) on the choice of physical methods for capturing volatile compounds in exhaled breath and sample preparation and preconcentration (Szulejko et al. 2010). Apart from technical and analytic aspects, the large variability in the number and concentrations of exhaled VOICs reported by different researchers has a number of other possible causes which include (a) subject demographics, (b) oral hygiene, (c) diet, (d) ambient air contaminants, and (e) the portion of expiratory air which was sampled. Therefore, further research is urgently needed to evaluate the impact of these factors on exhaled breath measurement results.

The impressive progress in breath gas analysis during the last decades has already resulted in some realized applications in medicine. So far the measurement of ethanol, carbon dioxide (CO_2, ventilation), $^{13}CO_2$ (*Helicobacter pylori* breath test, pharmacokinetic monitoring), hydrogen (H_2, lactose malabsorption breath test), nitrous oxide (N_2O, anesthetic agent), nitric oxygen (NO, asthma, airway inflammation), and volatile anesthetics in exhaled breath. Furthermore the pharmacokinetic monitoring of Propofol (Fig. 24.2) (Hornuss et al. 2007), a widely used intravenous anesthetic, in exhaled breath has only recently become available for clinical use (Exhaled Drug Monitor Edmon).

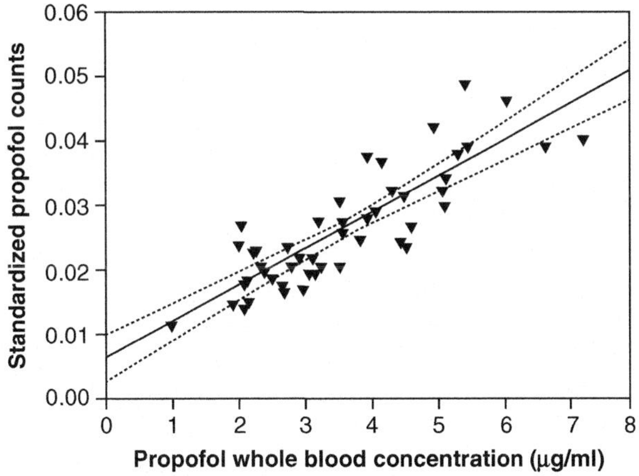

Fig. 24.2 Correlation of human propofol whole blood concentration and expiratory air propofol measured by ion molecule reaction mass spectrometry during surgery (From Hornuss et al. 2007)

As the result of ongoing research activity the number of VOICs with a probable stress and disease association identified is constantly increasing. Other metabolic markers, like acetone, as one of the most abundant VOICs in human exhaled breath, are closely linked to dextrose metabolism and lipolysis. Increased concentrations of ketone bodies in blood are present in patients with uncontrolled diabetes or during starvation as the result of acetyl-CoA decarboxylation (Miekisch et al. 2004; Schmoelz et al. 2007). Isoprene, one of the main hydrocarbons endogenously produced in mammals (Gelmont et al. 1981) is closely linked to cholesterol biosynthesis. Typically concentrations around 100 parts per billion of volume (ppbv) are present in the exhaled breath of adults (Karl et al. 2001; Turner et al. 2006b) and statin therapy caused a decrease in exhaled breath isoprene (Karl et al. 2001). Acetonitrile, a saturated aliphatic nitrile, for example, is absorbed from cigarette smoke and present within exhaled breath and urine headspace samples solely in smokers (Pinggera et al. 2005; Buszewski et al. 2009). Ammonia and trimethylamine were found to be increased in patients with end-stage renal failure and are possibly useful for the monitoring of dialysis efficacy (Davies et al. 1997; Endre et al. 2011). Dimethysulfide was found to be increased in patients with liver failure and accounts to the sweet and musty smell (Van den Velde et al. 2008). Several compounds present in human breath have been identified as end products of oxidative stress-mediated lipid peroxidation. Acetone, ethane, malondialdehyde, pentane, and propionaldehyde are increased in inflammatory processes as ischemia reperfusion injury (Miekisch et al. 2004; Li et al. 2009; Risby and Sehnert 1999) or "sterile" gravitational stress-dependent immune activation during parabolic flight (Choukèr et al. 2007), acute respiratory stress syndrome (Scholpp et al. 2002), and ultraviolet light-induced lipid peroxidation (Steeghs et al. 2006). Furthermore, NO, an important cell-signaling molecule that can be released in pathological amounts by the NO synthase during inflammation, asthma exacerbation (Pedrosa et al. 2010; Turner 2007), and lung transplant reperfusion injury (Kövesi et al. 2003). The VOICs importance and value to monitor tissue damage, oxidative and inflammatory states noninvasively has been investigated in patients with different degrees of respiratory infections (Lewis et al. 2017; Dolch et al. 2015) and warrants further trials to assess their sensitivity and specificity, as well as the interaction between diseases stated, exhaled metabolites and the gut microbiome (Smolinska et al. 2018) (see also Chap. 34).

Moreover, the future benefits for manned deep-space exploration resulting from this gain of knowledge and technical advancement for monitoring metobolic states and infectious diseases, are further complemented by recent studies starting to solidify the benefits of the use of single or mostly combined volatile organic compounds in the diagnosis and monitoring of disease states of cancer. In the light of high radiation exposure during interplanetary missions (see Chap. 20) and yet unknown environmental factors of spaceflight potentially leading to cancer induction and progression, e.g., due to dust, chemicals, and together with immune dysfunctional states, early diagnosis of cancer is needed. On Earth, in the last 5 years many and also well-designed clinical trials or observational studies reported on the advantage of exhaled volatile compunds to diagnose pancreatic cancer (Markar et al. 2018), breast cancer (Phillips et al. 2018; Wang et al. 2014), and lung cancer screening (Phillips et al. 2015), thereby distinguishing patients with cancer from those without cancer. Also first reports have been published on diagnosis for thyroid cancer and the authors concluded that "Breath analysis may provide a new, noninvasive, and directly qualitative method for the clinical diagnosis of thyroid disease" (Guo et al. 2015). To which extent these results will be applicable also to monitor tumor regression under therapeutic measures remains to be seen.

24.4 Current Space and Space Analogs Investigations

Manned spaceflights represent an extreme example of exposure to stressful environmental challenges and monitoring of crew health status represents unique technical and medical challenges. Especially in the light of future long-duration manned lunar and Martian exploration or outer space missions, the monitoring of health is a prerequisite for mission accomplishment. Here, the analysis of biomarkers in expiratory air offers promising noninvasive diagnostic options to monitor human health status because blood sample return to Earth will not be possible and on-site monitoring can be limited. Moreover, this technology represents a very promising technique for dual analyses of volatile masses in the exhaled air as well as in the spacecraft or habitat environment, respectively. As described above, some important biochemical pathways have been identified as the source of VOICs that are released from tissue into the gas phase indicating metabolic changes, immune activity, oxidative stress, infections, and solid cancer. Moreover, this technology can be also of important value for an individual pharmacokinetic monitoring when drugs are used and can be of high value to assess the compliance of taking the prescribed drugs (Hornuss et al. 2007). To better and further understand the readout parameter of exhaled air analyses also in the light of space applications, this technique has been implemented now in multiple scientific investigations, to

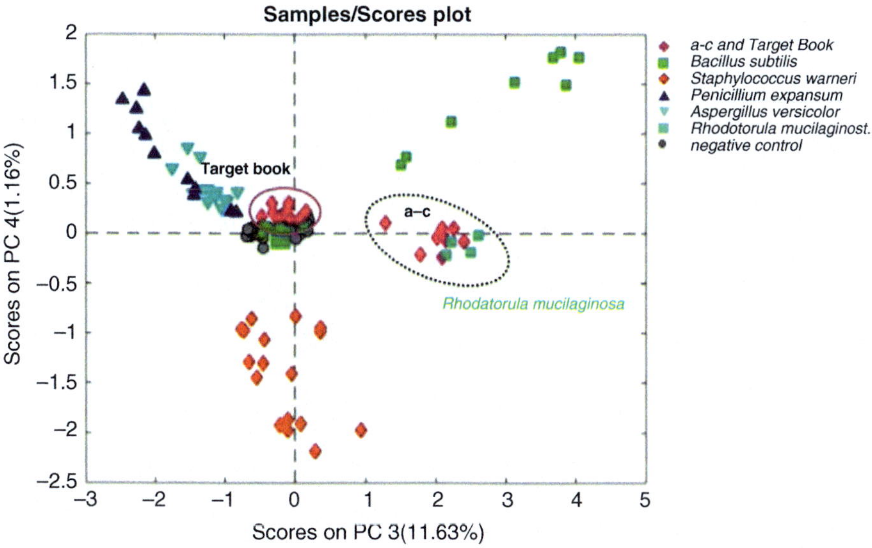

Fig. 24.3 Score plot of principal components 3/4 of E-NOSE signals derived from negative controls, trained microbial strains (*Bacillus subtilis*, *Staphylococcus warneri*, *Penicillium expansum*, and *Aspergillus versicolor*) and *Rhodotorula mucilaginosa* measurements gained in space and on Earth. *R. mucilaginosa* signals coincide sustainable (From Reidt et al. 2017)

understand the influence and the impact of stressor in space, like weightlessness (see Fig. 24.4), confinement, nutritional changes, hypoxia, and radiation exposure and to compare those to standardized baseline values gathered under full environmental control. The following studies have been conducted with our partner institutions to monitor human adaption to space-relevant conditions of life, accordingly: "CHOICE" study (Antarctica, hypoxia, and isolation, see also Chaps. 16 and 38), the "Mars500" project (long-term isolation, control for multiple stressors, nutritional modification), short- and mid-term bedrest (immobilization, test for artificial gravity or nutritional countermeasures), or the NEEMO (hyperbaric/−oxic stress and simulation of extravehicular activities (EVA, "spacewalks")) in the NASA aqueous habitat. In space, first investigations on exhaled volatile compound were successful to monitor exhaled breath NO in crew members of the International Space Station (Karlsson et al. 2009) to track lung inflammation. These studies are ongoing on the ISS (Airway Monitoring experiment, L. Karlsson/Karolinska Institutet, Sweden). Since health monitoring is strongly linked to microbial monitoring in the habitat, the recently published report by Reidt et al. shows the successful application of a metal oxide gas sensor array for the detection of surface contamination aboard the ISS (see Fig. 24.3, Reidt et al. 2017). The odor of selected surfaces analyzed with the so-called E-NOSE contained sufficient information for subsequent identification of the yeast *Rhodotorula mucilaginosa* as the odor underlying microorganism complementing and expanding the standard environmental monitoring of the habitat (see also Chap. 25).

Fig. 24.4 Breath gas analysis during a parabolic flight mission. The volunteers exhaled breath is collected in a specifically developed expiratory air buffer device from which expiratory air is continuously transferred to a metal oxide gas sensor array that was modified for its use under the condition of microgravity. With permission

This integrated approach will be further extended by VOIC values suitable for monitoring health alterations as well as general homeostasis, allostasis, or allostatic load (see Chap. 4) from expiratory air collected from the crew of the ISS. A first step towards this approach is the currently carried out development of the E-NOSE extending the instruments scope for breath gas analysis aboard the ISS (Dolch et al. 2017, Fig. 24.4).

24.5 Summary

Breath gas analysis is an emerging technology that has proven in the last decade to open the opportunity to intensify longitudinal health monitoring. Especially during long-duration space exploration it can provide a noninvasive research tool for the determination of organ functional changes when subjected to stressors in space, which includes also the effects of countermeasure (e.g., nutritional, exercise, pharmacologic) hereto. Here, especially the individual and repeated analyses offer data to help assessing the crews health and diseases as a function of time, and will allow also to monitor the drugs and metabolites individually in the exhaled air. This can be an important step for a successful and personalized therapy in space. Also, the dual use property can allow for the assessment of environmental changes in the habitat with respect to air composition and pollution. Results of either application can be evaluated on-site or transferred electronically to Earth for further interpretation. However, further in-depth investigations are warranted to better understand the results provided "breath-by-breath." Thereby, research for applications in patients as well as for space crews are as described fully complementary and cofertilizing and both will benefit from this technology in the future.

Acknowledgments Supported in part by the Department of Anaesthesiology of the University of Munich, the European Space Agency, the National Aeronautics and Space Administration (NASA), the Institute for Biomedical Problems (IBMP), the French and Italian Polar institutes (IPEV, PNRA), the German Space and Aeronautics Centre (DLR) & and the German Federal Ministry of Economics and Technology (50WB0523, 50WB0719, 50WB0919, 50WB1317), and the Kompetenzzentrum Medizin in Tirol, project 09A.

References

Amann A, Spanel P, Smith D (2007) Breath analysis: the approach towards clinical applications. Mini Rev Med Chem 7:115–129

Bond JH, Levitt MD (1976) Quantitative measurement of lactose absorption. Gastroenterology 70:1058–1062

Braden B (2009) Methods and functions: breath tests. Best Pract Res Clin Gastroenterol 23:337–352

Buszewski B, Kesy M, Ligor T et al (2007) Human exhaled air analytics: biomarkers of diseases. Biomed Chromatogr 21:553–566

Buszewski B, Ulanowska A, Ligor T et al (2009) Analysis of exhaled breath from smokers, passive smokers and non-smokers by solid-phase microextraction gas chromatography/mass spectrometry. Biomed Chromatogr 23:551–556

Choukèr A, Kaufmann I, Dolch M et al (2007) Stress responses during parabolic flight maneuvers are reflected by changes in expiratory air. In: (Jahr) SMOTIaOBR (ed) Breath analysis summit 2007: clinical applications of breath testings, Cleveland, 1–3 Nov 2007

Christl SU, Murgatroyd PR, Gibson GR et al (1992) Production, metabolism, and excretion of hydrogen in the large intestine. Gastroenterology 102:1269–1277

Davies S, Spanel P, Smith D (1997) Quantitative analysis of ammonia on the breath of patients in end-stage renal failure. Kidney Int 52:223–228

De Lacy Costello B, Amann A, Al-Kateb H et al (2014) A review of the volatiles from the healthy human body. J Breath Res 8(1):014001

Dolch M, Frey L, Hornuss C et al (2008) Molecular breath-gas analysis by online mass spectrometry in mechanically ventilated patients: a new software-based method of CO_2-controlled alveolar gas monitoring. J Breath Res 2(3):037010

Dolch ME, Choukèr A, Hornuss C et al (2015) Quantification ofpropionaldehydeinbreathofpatientsafter lung transplantation. Free Radic Biol Med 85:157–164

Dolch ME, Hummel T, Fetter V et al (2017) Electronic nose functionality for breath gas analysis during parabolic flight. Microgravity Sci Technol 29:201–207

Endre ZH, Pickering JW, Storer MK et al (2011) Breath ammonia and trimethylamine allow real-time monitoring of haemodialysis efficacy. Physiol Meas 32:115–130

Gelmont D, Stein RA, Mead JF (1981) Isoprene-the main hydrocarbon in human breath. Biochem Biophys Res Commun 99:1456–1460

Graham DY, Klein PD, Evans DJ Jr et al (1987) *Campylobacter pylori* detected noninvasively by the 13C-urea breath test. Lancet 1:1174–1177

Guo L, Wang C, Chi C et al (2015) Exhaled breath volatile biomarker analysis for thyroid cancer. Transl Res 166:188–195
Hornuss C, Praun S, Villinger J et al (2007) Real-time monitoring of propofol in expired air in humans undergoing total intravenous anesthesia. Anesthesiology 106:665–674
Karl T, Prazeller P, Mayr D et al (2001) Human breath isoprene and its relation to blood cholesterol levels: new measurements and modeling. J Appl Physiol 91:762–770
Karlsson LL, Kerckx Y, Gustafsson LE et al (2009) Microgravity decreases and hypergravity increases exhaled nitric oxide. J Appl Physiol 107:1431–1437
Kazui M, Andreoni KA, Williams GM et al (1994) Visceral lipid peroxidation occurs at reperfusion after supraceliac aortic cross-clamping. J Vasc Surg 19:473–477
Kövesi TS, Royston D, Yacoub MH et al (2003) Exhaled nitric oxide in human lung ischemia-reperfusion. In: Marczin N, Kharitonov SA, Yacoub MH et al (eds) Disease markers in exhaled breath. Marcel Dekker, Inc, New York/Basel
Kramer A, Below H, Bieber N et al (2007) Quantity of ethanol absorption after excessive hand disinfection using three commercially available hand rubs is minimal and below toxic levels for humans. BMC Infect Dis 7:117
Lewis JM, Savage RS, Beeching NJ et al (2017) Identifying volatile metabolite signatures for the diagnosis of bacterial respiratory tract infection using electronic nose technology: a pilot study. PLoS One 12:e0188879
Li P, Xu G, Wang C et al (2009) Breath pentane: an indicator for early and continuous monitoring of lipid peroxidation in hepatic ischaemia-reperfusion injury. Eur J Anaesthesiol 26:513–519
Lindinger W, Hansel A, Jordan A (1998) On-line monitoring of volatile organic compounds at pptv levels by means of proton-transfer-reaction mass spectrometry (PTR-MS) medical applications, food control and environmental research. Int J Mass Spectrom Ion Process 173:191–241
Markar SR, Brodie B, Chin ST et al (2018) Profile of exhaled-breath volatile organic compounds to diagnose pancreatic cancer. Br J Surg 105:1493–1500
Miekisch W, Schubert JK, Noeldge-Schomburg GFE (2004) Diagnostic potential of breath analysis - focus on volatile organic compounds. Clin Chim Acta 347:25–39
Modak AS (2007) Stable isotope breath tests in clinical medicine: a review. J Breath Res 1:014003
Moeskops BW, Steeghs MM, Van Swam K et al (2006) Real-time trace gas sensing of ethylene, propanal and acetaldehyde from human skin in vivo. Physiol Meas 27:1187–1196
Pauling L, Robinson AB, Teranish R et al (1971) Quantitative analysis of urine vapor and breath by gas-liquid partition chromatography. Proc Natl Acad Sci U S A 68:2374–2376
Pedrosa M, Cancelliere N, Barranco P et al (2010) Usefulness of exhaled nitric oxide for diagnosing asthma. J Asthma 47:817–821
Phillips M, Boehmer JP, Cataneo RN et al (2004) Heart allograft rejection: detection with breath alkanes in low levels (the HARDBALL study). J Heart Lung Transplant 23:701–708
Phillips M, Bauer TL, Cataneo RN et al (2015) Blinded validation of breath biomarkers of lung cancer, a potential ancillary to chest CT screening. PLoS One 10:e0142484
Phillips M, Cataneo RN, Cruz-Ramos JA et al (2018) Prediction of breast cancer risk with volatile biomarkers in breath. Breast Cancer Res Treat 170:343–350
Pinggera GM, Lirk P, Bodogri F et al (2005) Urinary acetonitrile concentrations correlate with recent smoking behaviour. BJU Int 95:306–309
Reidt U, Helwig A, Müller G et al (2017) Detection of microorganisms onboard the international Space Station using an electronic nose. Gravit Space Res 5:89–111
Risby TH (2005) Current status of clinical breath analysis. In: Amann A, Smith D (eds) Breath analysis for clinical diagnosis and therapeutic monitoring. World Scientific Publishing, Singapore
Risby TH, Sehnert SS (1999) Clinical application of breath biomarkers of oxidative stress status. Free Radic Biol Med 27:1182–1192
Roccarina D, Lauritano EC, Gabrielli M et al (2010) The role of methane in intestinal diseases. Am J Gastroenterol 105:1250–1256

Romagnuolo J, Schiller D, Bailey RJ (2002) Using breath tests wisely in a gastroenterology practice: an evidence-based review of indications and pitfalls in interpretation. Am J Gastroenterol 97:1113–1126

Schmoelz M, Praun S, Schmoeckel M et al (2007) Online measurement of acetone in expiratory air by mass spectrometry during cardiac surgery. In: Annual meeting of the American Society of Anesthesiologists, San Francisco, 16 Oct, A1426

Scholpp J, Schubert JK, Miekisch W et al (2002) Breath markers and soluble lipid peroxidation markers in critically ill patients. Clin Chem Lab Med 40:587–594

Schubert JK, Müller WPE, Benzing A et al (1998) Application of a new method for analysis of exhaled gas in critically ill patients. Intensive Care Med 24:415–421

Schubert JK, Miekisch W, Birken T et al (2005) Impact of inspired substance concentrations on the results of breath analysis in mechanically ventilated patients. Biomarkers 10:138–152

Smith D, Wang T, Spanel P (2002) On-line, simultaneous quantification of ethanol, some metabolites and water vapour in breath following the ingestion of alcohol. Physiol Meas 23:477–489

Smolinska A, Tedjo DI, Blanchet L et al (2018) Volatile metabolites in breath strongly correlate with gut microbiome in CD patients. Anal Chim Acta 1025:1–11

Steeghs MM, Moeskops BW, Van Swam K et al (2006) On-line monitoring of UV-induced lipid peroxidation products, from human skin in vivo using proton-transfer reaction mass spectrometry. Int J Mass Spectrom 253:58–64

Stenseth R, Nilsen T, Haaverstad R et al (2007) Frequent sampling allows detection of short and rapid surges of exhaled ethane during cardiac surgery. Perfusion 22:391–396

Szulejko JE, Mcculloch M, Jackson J et al (2010) Evidence for cancer biomarkers in exhaled breath. IEEE Sensors J 10:25

Turner S (2007) The role of exhaled nitric oxide in the diagnosis, management and treatment of asthma. Mini Rev Med Chem 7:539–542

Turner C, Spanel P, Smith D (2006a) A longitudinal study of ammonia, acetone and propanol in the exhaled breath of 30 subjects using selected ion flow tube mass spectrometry, SIFT-MS. Physiol Meas 27:321–337

Turner C, Spanel P, Smith D (2006b) A longitudinal study of breath isoprene in healthy volunteers using selected ion flow tube mass spectrometry (SIFT-MS). Physiol Meas 27:13–22

Van Den Velde S, Nevens F, Van Hee P et al (2008) GC-MS analysis of breath odor compounds in liver patients. J Chromatogr B Analyt Technol Biomed Life Sci 875:344–348

Wang C, Sun B, Guo L et al (2014) Volatile organic metabolites identify patients with breast cancer, cyclomastopathy, and mammary gland fibroma. Sci Rep 4:5383

25 Monitoring the Microbial Burden in Manned Space Stations

Rob Van Houdt and Natalie Leys

25.1 Introduction

Astronauts on mission aboard the International Space Station (ISS) experience the uniqueness of this space station in all its aspects. This high-tech small habitat confines alternating crews, ranging from 3 to 10 members, protects them from the extreme space environment (pressure, temperature, radiation) and provides them with the necessary supplies (air, water, food). While living in space and enclosed inside sophisticated space vehicles, astronauts face unique stressors that they did not encounter before on Earth and its natural environment. One of the stress challenges for astronauts in space is the special indoor microbial environment.

The ubiquity and resilience of microorganisms makes them unavoidable in most environments, including space habitats. Most microorganisms are not harmful but highly valuable to humans. Nevertheless, microbial contamination can be a stress factor and of health relevance in the condition of an impaired immunity. Furthermore, immune alterations are elicited and amplified by microgravity and/or other stress conditions in space (see Chaps. 18 and 19), thereby increasing the vulnerability to infections (Aviles et al. 2003; Mehta et al. 2000).

Not surprisingly, a man-made and occupied confined environment, such as the ISS, generates its own unique microbial population, which mainly originates from the crew (skin, upper respiratory tract, mouth, and gastrointestinal tract, see also Chap. 34) but also includes environmental microorganisms. There is an obvious requirement to control the total load and diversity of environmental microorganisms in manned spacecraft and to continuously monitor the specific parameters influencing their persistence, survival, and growth for guaranteeing adequate living quality and reducing the risks of harmful effects on the crew. Particularly, (opportunistic)

R. Van Houdt (✉) · N. Leys
Microbiology Unit, Interdisciplinary Biosciences, Belgian Nuclear Research Centre (SCK•CEN), Mol, Belgium
e-mail: rvhoudto@sckcen.be; nleys@sckcen.be

© Springer Nature Switzerland AG 2020
A. Choukér (ed.), *Stress Challenges and Immunity in Space*,
https://doi.org/10.1007/978-3-030-16996-1_25

pathogens—capable of causing infections and disease—are undesired and their manifestation, both in numbers and diversity, needs to be prevented.

Therefore, in an effort to reduce the health hazards posed by microbial contamination, international quality standards for air, surfaces, and water have been defined and prevention, monitoring, and mitigation measures have been implemented by the space agencies.

25.2 Biocontamination Control, Monitoring, and Mitigation Onboard ISS

To ensure the indoor microbial quality in space stations, the maximal concentration of microorganisms, in particular bacteria and fungi, allowed in the air and on the surfaces of the ISS were internationally defined and are described in the ISS Medical Operations Requirements Document (ISS MORD 2006; Duncan et al. 2008). The threshold levels are a trade-off between acceptable risk and realizable levels with the current prevention and monitoring technologies available and applicable for space (Table 25.1). Threshold levels are evaluated by scoring the total aerobic viable count of bacteria or fungi on a rich agar medium (ISS MORD 2006; Pierson et al. 2012; Van Houdt et al. 2012).

25.2.1 Air

To limit the introduction of contaminants in the ISS, stringent pre-flight microbial air limits, 300 CFU/m^3 for bacteria and 50 CFU/m^3 for fungi, have been established for cargo and vehicles traveling to the ISS (Pierson et al. 2012; ISS MORD 2006). In-flight ISS, the internationally defined limit for airborne bacteria and fungi are 1000 CFU/m^3 and 100 CFU/m^3, respectively (ISS MORD 2006) (Table 25.1). These are comparable to the threshold limits frequently used for healthy offices (Sessa et al. 2002; Dacarro et al. 2003) or more in general to assess indoor air quality (Brief and Bernath 1988). These ISS thresholds can roughly be placed in the category "intermediate" for bacteria and "low" for fungi based on the classification levels of

Table 25.1 Environmental microbial quality standards for air, surfaces, and water in ISS

		Maximum for bacteria	Maximum for fungi
Pre-flight	Air	300 CFU/m^3	50 CFU/m^3
	Internal surfaces	500 CFU/100 cm^2	10 CFU/100 cm^2
	Potable water[a]	50 CFU/mL	ND
In-flight	Air	1000 CFU/m^3	100 CFU/m^3
	Internal surfaces	10,000 CFU/100 cm^2	100 CFU/100 cm^2
	Potable water[a]	50 CFU/mL	ND

ND not determined, *CFU* colony forming unit
[a]Coliforms may not be detected

bacteria and fungi in homes and offices by the European Collaborative Action "Indoor Air Quality and its Impact on Man" (Verhoeff 1993).

To prevent microbial distribution and to assure that the microbial burden remains below the thresholds at all times, the air in ISS is filtered. While the air in the U.S. Segment of the ISS (including US Nodes 1, 2, and 3, US Lab, US Airlock), Japanese Experiment module JEM-PM, and European Laboratory Columbus is filtered through High Efficiency Particulate Air (HEPA) filters, the Russian Segment (Service Module, Functional Cargo Block, and Mini-Research Modules 1 & 2) uses pleated woven filters (Van Houdt et al. 2012). Both filter fine aerosols (particles 0.3 μm in diameter or larger) and constantly reduce the levels of dust particles and associated microorganisms in the air (99.97% efficiency for the HEPA filters) (NASA 2005a). Although filtration continuously scavenges the microbial cells, it does not inactivate them and they need to be replaced when saturated. In contrast, the POTOK 150MK Russian air filtration and disinfection systems onboard ISS do provide microbe inactivation by using electrostatic pulses and charged ions followed by filtration with an efficiency up to 99% for particle sizes ranging from 0.01 μm to 10 μm (Volodina et al. 2003). Two POTOK systems are currently implemented in the ISS, one in the Service Module and a second one in the Functional Cargo Block.

To monitor the level and composition of the airborne microbial population, air samples from different sites of the ISS are regularly (once every 90 days) collected within the framework of the standard onboard procedure (ISS MORD 2006). Although ISS is one interconnected habitat and standards have internationally been defined, different monitoring procedures apply for the different segments. In each module of the American segment of ISS, air samples (84.9 L) are collected with the American Microbial Air Sampler (MAS) Kit. The MAS is a modified portable impaction sampler (Burkard Manufacturing Co. Ltd., Hertfordshire, UK) and collects 84.9 L of air onto a standard Petri dish (Yamaguchi et al. 2014). One sample is collected on a Tryptone Soy agar (TSA) plate for bacterial analysis and one on a Sabouraud dextrose agar with chloramphenicol plate for fungal analysis (NASA 2005d). Collected samples are incubated at ambient cabin temperature for 5 days (Pierson et al. 2012). In the Russian segment, the Ecosphera kit is used. Air samples (90 L) are collected by an aspiration/sedimentation method on TSA (bacteria) or Czapek agar (fungi) by a SAS air sampling device (PBI International, Italy). Collected samples are incubated at 37°C for 7 days. The ISS crew performs the onboard American and Russian procedures (comprising sample collection, incubation, data recording and interpretation). If microbial counts exceed the specifications, photographs will be downlinked and the microbial risk evaluated via visual inspection by NASA and RSA microbiologists.

25.2.2 Internal Surfaces

Again, to limit the introduction of contaminants in ISS, stringent pre-flight surface microbial limits, 500 CFU/100 cm^2 for bacteria and 10 CFU/100 cm^2 for fungi, have been established for cargo and vehicles traveling to the ISS (Pierson et al. 2012; ISS

MORD 2006). In-flight biocontamination acceptability limits for ISS internal surfaces are set to 10,000 CFU/100 cm^2 for bacteria and 100 CFU/100 cm^2 for fungi (ISS MORD 2006). On Earth, in general, standards or guidelines on what are acceptable levels of microbial contamination depend on the activity. For instance, cleaned and disinfected surfaces in establishments for the production and marketing of fresh meat can have a maximum total viable bacterial count of 10 CFU/cm^2 (equivalent to 1000 CFU/100 cm^2) (2001/471/EC). For hand contact surfaces in hospitals, the total viable bacterial count should be less than 2.5 CFU/cm^2 (equivalent to 250 CFU/100 cm^2) (Mulvey et al. 2011). Therefore, ISS surface standards for bacterial contamination are less stringent than standards for food contact surfaces or hand contact surfaces in hospitals. The ISS surface standard for fungal contamination is comparable to those described in the Australian Mold Guidelines, which rates the hygiene of indoor surfaces normal when the viable fungal concentration is between 50 and 105 CFU/100 cm^2 (Kemp and Neumeister-Kemp 2005; Van Houdt et al. 2012).

The hygiene level of internal surfaces is maintained by weekly housekeeping tasks, including vacuuming and cleaning with 0.4% benzalkonium chloride antiseptic towelettes (Satoh et al. 2011).

To assess the level and composition of the microbial surface contamination onboard ISS, different sites of the ISS are sampled regularly within the framework of the standard onboard procedure (ISS MORD 2006). In each module of the American segment, two sites are sampled with the American Surface Sampler Kit (SSK) (NASA 2005b) every 90 days, one for bacterial analysis (TSA) and one for fungal analysis (SDA) (Morris et al. 2012; NASA 2005d). Contact slides are used for flat surfaces, while saline-moistened swabs are used for nonflat surfaces (which are subsequently used to inoculate the contact slides). Collected samples are incubated at ambient cabin temperature for 5 days and analyses are performed by a crew member by comparing the amount of growth on each slide to a colony density chart included as part of the onboard procedure (Pierson et al. 2012). If the acceptability limit is exceeded, a digital image of the sample is downlinked and evaluated by NASA/JSC and RSA/IBMP microbiologists. In addition, samples are also returned to Earth and further analyzed post-flight to check for clinically significant organisms. In the Russian segment, the Test Tube Kit for Microbiological Sampling is used and 100 cm^2 surface areas are swabbed (1–2 days before completion of each mission). Swab samples are returned to Earth and processed post-flight by cultivation at 37°C (bacteria) or 28°C (fungi) and analyzed (total viable counts and identification of morphologically different isolates) (Novikova et al. 2006; Novikova 2004). Recommendations for surface cleaning are transmitted to the crew if acceptability limits are exceeded. Disinfection wipes containing either a mixture of hydrogen peroxide and a quaternary ammonium compound ("Fungistat" kit—supplied by Russian team) or a sole quaternary ammonium compound (supplied by US American team) are used if threshold limits are exceeded.

25.2.3 Drinking Water

The pre-flight and in-flight biocontamination acceptability limits for ISS drinking water are 50 CFU/mL for bacteria with no detectable coliforms (ISS MORD 2006;

Pierson et al. 2012). These are comparable with Earthly US standards (Dawson and Sartory 2000) and World Health Organization guidelines (WHO 2008).

Onboard ISS, ground supplied as well as recovered water is used. Ground supplied water is pre-flight sanitized, delivered on resupply vehicles to ISS and stored onboard ISS into storage tanks and containers (up to a year) (Straub et al. 2009; Bruce et al. 2005; James et al. 2008). Recovered water includes humidity condensate (produced by the Russian SRV-K system in the Service Module and the US Water Recovery System WRS) and urine distillate (produced by the US WRS) processed to potable water. A major difference between the US and Russian potable water is the used disinfectant, which is silver (i.e., colloidal and Ag^+; 0.5 mg/L) for the Russian type and iodine (2–4 mg/L) for the US type of water.

The quality of potable water from different dispensing ports in the American segment (WRS) and Russian segment (SRV-K, SVO-ZV, CWC) is checked monthly and every 3 months, respectively. The samples are processed and analyzed in-flight with the US-supplied Water Microbiology Kit (ISS MORD 2006; NASA 2005c). The kit collects 10 mL of water (1 mL of sample + 9 mL sterile water) on a filter membrane, which is subsequently wetted with modified R2A growth medium and incubated at ambient cabin temperature for 44 ± 4 h (Yamaguchi et al. 2014). After incubation, crew members evaluate the bacterial viable count by visual analysis (Pierson et al. 2012). Coliforms are detected by directly transferring about 100 mL of sample into a bag containing the Colisure reagent, incubation at ambient cabin temperature for 44 ± 4 h, and subsequent evaluation of a color change. A yellow color indicates the absence of coliforms, whereas a magenta color (metabolization of chlorophenyl red ß-D-galactopyranoside indicator) indicates the presence of coliforms (Pierson et al. 2012). In addition, samples are regularly returned to Earth for further analysis post-flight. The sampling schedule and frequency can be adjusted according to recommendation made by American and Russian experts to ensure water quality aboard the ISS. If unacceptable contamination is detected, these water supplies are not used for consumption to mitigate the potential health risks.

25.2.4 Food

Food processing (production and packaging) is rigorously tested and controlled pre-flight to guarantee that contamination levels comply with the implemented standards for spaceflight foods (ISS MORD 2006; NASA 2006; Perchonok and Douglas 2008). Actually, the currently globally applied Hazard Analysis Critical Control Point (HACCP) procedure, which is a systematic preventive approach assuring the safe production of foods, was developed by the Pillsbury Company while working on producing foods for NASA for spaceflights in the early 1960s (Lachance 1997). Commercially sterile food, which by definition is free of microorganisms capable of reproducing in the food under normal nonrefrigerated conditions of storage and distribution, is checked for package integrity. Foods for spaceflights that are not commercially sterile are analyzed at the stage of raw materials (before packaging) for certain specific microorganisms depending on the product (5 samples from each lot) and after flight packaging (finished goods) for total aerobic count (one sample from each daily production) (Table 25.2) (NASA 2006).

Table 25.2 Microbiological testing procedure for noncommercially sterile foods

Test	Pre-flight limit for rejection
Total aerobic count	If >20,000 CFU/g in a sample or If >10,000 CFU/g in more than 1 sample
Yeast and molds	If >1000 CFU/g in a sample or If >100 CFU/g in more than 1 sample
Coliform	If >100 CFU/g in a sample or If >10 CFU/g in more than 1 sample
Coagulase-positive staphylococci	If >100 CFU/g in a sample or If >10 CFU/g in more than 1 sample
Salmonella	If >0 CFU/g in a sample

CFU colony forming unit

25.3 The Environmental Microflora in the ISS

25.3.1 Air

Airborne microbial contamination can have diverse health effects such as irritation, respiratory infections, and allergic diseases (Husman 1996). Microorganisms can become airborne via diverse routes like talking, coughing, sneezing, and sewage removal. Their dispersion and deposition are directed by general physical principles with droplet size being one of the most important factors (Morawska 2006) and dust particles being an important carrier. In addition, humidity, oxygen concentration, and time affect the stability of airborne microorganisms and their survival in droplets or dust particles (Cox 1995). In microgravity, dust and microorganisms do not sediment, resulting in persistent airborne aerosols and high microbial densities in cabin air if the air filtering systems are not well maintained.

The airborne bacterial and fungal contamination levels were monitored during the occupation of the Mir station (1986–2001). While the bacterial population was fairly stable with 95% of the samples below 500 CFU/m^3 of air, the fungal contamination varied considerably from 2 up to 10,000 CFU/m^3 (Novikova 2004). A similar survey performed aboard ISS (from the year 1998 to 2005) indicated that both the airborne bacterial and fungal contamination were lower than 710 and 44 CFU/m^3, respectively (Novikova et al. 2006). The lower contamination levels in ISS compared to Mir were primarily because of the installation of the efficient Russian air disinfection and filtration system POTOK 150MK (POTOK Inter, Moscow, Russia) in April 2001. A similar decrease in contamination levels was also observed in the Mir station after installation of this system in January 1998 (Novikova 2004).

Although the airborne microbial contamination levels were generally below the limits (maximum 1000 CFU/m^3 for bacteria and 100 CFU/m^3 for fungi), these overall levels do not necessarily reflect the real associated risks since they do not provide information about the composition of the microbial population and the pathogenic potential of the contaminants. Therefore, it is valuable to include identification assays to detail the predominant microbial taxa. *Staphylococcus* and *Bacillus* spp. were found to be the dominant bacterial species in the air aboard ISS (Castro et al.

2004). In particular *Staphylococcus aureus* was observed frequently in ISS (in 3.2% of the cases) and exchange of *S. aureus* strains among crewmembers was already observed in previous missions (Berry 1973). *S. aureus* naturally colonizes the skin or nose of healthy people, but can also cause a range of illnesses. The dominance of *Staphylococcus* and *Bacillus* spp. in air is similar to other surveys in non-space-related confined environments, such as airplanes (Osman et al. 2008), polar stations (Van Houdt et al. 2009a), and during the Mars500 isolation campaign in which six crew members lived in a specifically designed spacecraft mock-up for 520 days (mimicking a crewed return flight to Mars) (Schwendner et al. 2017). The dominant fungal species were *Penicillum* and *Aspergillus*, with a high incidence of *Aspergillus flavus* (observed in 2.5% of the cases) (Novikova et al. 2006; Knox et al. 2016).

25.3.2 Internal Surfaces

Direct contact is another major route of transmission of microorganisms, both on Earth and in space. Most microorganisms are able to stick to surfaces and form biofilms. On Earth, it is well documented that this process of biofilm formation has major implications for many industrial and health-related processes (Van Houdt and Michiels 2010).

A large ISS monitoring campaign (from 1998 to 2005), including 243 surface swab samples, indicated that the bacterial and fungal contamination ranged from 25 to 43,000/100 cm^2 and from 25 to 300,000/100 cm^2, respectively (Novikova et al. 2006). Although the levels fluctuated within a broad range, surface contamination levels were in most cases low and below the international quality limits (i.e., 10,000 bacterial and 100 fungal CFU/100 cm^2). The dominant bacterial and fungal species isolated from surfaces were similar to those present in the airborne contamination, thus revealing predominant *Staphylococcus* and *Bacillus* bacterial species, and *Penicillum* and *Aspergillus* (e.g., *A. niger*; see Chap. 12) fungal species. On the equipment surfaces of the Japanese experimental module KIBO, *Staphylococcaceae*, *Enterobacteriaceae* and *Neisseriaceae*, and *Alternaria* sp. and *Malassezia* spp. were the predominant bacterial and fungal contaminants, respectively (Ichijo et al. 2016; Satoh et al. 2011).

The recent breakthrough in next-generation sequencing (NGS) in the last decade provided an unprecedented approach to investigate microbial communities and their dynamics in a culture independent manner. Not surprisingly, different approaches returned discrepancies in microbial abundances. For instance, during the Mars500 experiment *Staphylococcus* spp. and *Bacillus* spp. were found to be the most abundant surface contaminants in cultivation approaches, whereas they were not the most abundant via PhyloChip or NGS analysis (Schwendner et al. 2017). Likewise, the viable microbiomes from vacuumed dust particles (as well as filter element particles) collected in the ISS showed predominantly *Corynebacterium* spp. and *Propionibacterium* spp., which are also human-associated microorganisms, and not *Staphylococcus* (Checinska et al. 2015; Venkateswaran et al. 2014). Therefore, irrespective of variations in diversity and abundance, all studies indicated that the crew members are the main source contributing to the microbial communities onboard ISS.

Occasional increases in the bacterial and fungal surface contamination were registered for some ISS locations, such as on a panel of the ventilation screen and the table surface in the Service Module or behind panels of the Functional Cargo Blok (Novikova et al. 2006; James et al. 2008). Disinfectant wipes were used each time contamination levels were above the quality threshold and resulted systematically in a decrease of contamination below the acceptability limits. Nevertheless, the resilience of contamination at a certain location could indicate (a combination of) resistance to the used disinfectants, favorable survival/growth conditions (e.g., humidity) and a critical position in the design. The analysis of resilient microorganisms in the ISS indicated numerous resistance capabilities towards environmental stresses and diverse survival strategies including endospore formation, desiccation resistance and optimized DNA repair mechanisms (Mora et al. 2016).

25.3.3 Water

Microbiological contamination of drinking water is a well-known threat, not only from health perspective but also for microbial-mediated corrosion of the water piping system (Berry et al. 2006; Szewzyk et al. 2000). One of the important microbial characteristics in drinking water storage and distribution systems is biofilm development, which increases the persistence of pathogens and the resistance to disinfectants (Van Houdt and Michiels 2010; Wingender and Flemming 2004; Emtiazi et al. 2004; Stewart et al. 2001). This increased resistance can be attributed to different mechanisms such as a slow or incomplete penetration of the biocide into the biofilm, an altered physiology of the biofilm cells, expression of an adaptive stress response by some cells, or differentiation of a small subpopulation of cells into persister cells (Van Houdt and Michiels 2010). Adequate monitoring and disinfection methods are therefore needed to mitigate the risk to hardware disintegration and crew health.

A four-year monitoring campaign (2001–2004) was performed to identify the water contamination of ISS potable water. Shuttle potable water in Contingency Water Containers (CWCs) contained bacteria in 13 out of 52 samples (25%), Russian ground-supplied water contained bacteria in 2 out of 6 samples (33%) (Bruce et al. 2005). Seven out of ten US condensate samples (70%) from the US Lab system contained bacteria. No information was provided in how many of these samples the contamination level was above the former acceptability limit of 100 CFU/100 mL (Bruce et al. 2005). In addition, 4 out of 4 samples (100%) from SRV-K condensate contained bacteria in high amounts, ranging from 1.4×10^6 to 1.3×10^8 CFU/100 mL (Bruce et al. 2005). In-flight analyses (27 sampling times spread over the period 2000–2004) of the ISS SVO-ZV system, which dispenses at ambient temperature both the CWC and the Russian ground-supplied water for consumption, showed that bacterial contamination levels were above the former acceptability limit of 100 CFU/100 mL in 16 cases (60%) (Bruce et al. 2005). A series of remediation actions (disinfection with 10 mg/L silver and replacement of parts) were initiated to combat this recurrent water contamination, however, contamination levels increased again above the threshold limit shortly afterwards (Bruce et al. 2005).

Most of the isolates recovered from these samples were typical waterborne gram-negative bacteria from the genera *Methylobacterium*, *Ralstonia*, *Sphingomonas*, and *Pseudomonas*. The recurrence of the contamination problem indicated resilience and proliferation of the bacterial population in the water systems onboard ISS.

The resilience of this contamination could be caused by different events. The bacterial population could have adapted to survive in such oligotrophic environments. Indeed, *Cupriavidus metallidurans* and *Ralstonia pickettii* strains isolated from ISS water systems survived in mineral water for at least 2 years, even when supplemented with silver as disinfectant (Mijnendonckx et al. 2013). Correspondingly, *Ralstonia* spp. isolated from ultrapure water were previously shown to survive in such water for at least 6 months (McAlister et al. 2002) and *Ralstonia solanacearum* was even able to survive over 4 years in river water (Alvarez et al. 2008). In addition, the bacterial population could harbor resistance mechanisms, e.g., towards silver used as biocide in the ISS potable water (Straub et al. 2009). *Cupriavidus metallidurans* strains have been studied intensively for their resistance towards multiple heavy metals including silver (Janssen et al. 2010; Mijnendonckx et al. 2013). Most of these mechanisms are typically located on megaplasmids (Janssen et al. 2010; Monchy et al. 2007; Mijnendonckx et al. 2013), but some are also located on mobile genetic elements integrated in the chromosome (Van Houdt et al. 2009b) or on the chromosome itself (Janssen et al. 2010). The presence of similar megaplasmids as well as resistance determinants were also found in the *C. metallidurans* and *R. pickettii* strains isolated from the ISS water systems (Mijnendonckx et al. 2013), indicating that they are indeed equipped with resistance mechanisms to metals, including silver. Finally, certain areas in the water system could be prone to harbor and build up bacterial biofilms, which could augment the above described aspects (Bridier et al. 2011).

Although this contamination does not pose an immediate threat to the astronauts (only some species are recognized as opportunistic pathogens), high microbial concentrations aboard spacecraft prevents consumption of the water as currently no identification method is available during flight. Therefore, this type of contamination wastes large amounts of crew time and Earth-based resources. In addition, these water contamination events require more materials to be transferred both to and from the ISS and the development of adequate and efficient screening methods (Bechy-Loizeau et al. 2015).

25.4 Conclusions

The ISS is an artificial living and working habitat outside Earth, which has been in-flight and continuously occupied for 18 years, with an occupancy density higher than most homes or offices on Earth and in which crew members are confined indoors for 100% of their time, are limited in hygiene procedures, and are continuously exposed to a variety of health stressors. Microbial environmental quality and hazard control is therefore crucial to assure crew health. This might become even more important (1) when supplies from Earth are very restricted (cis-lunar stations or lunar habitats) or almost impossible while on a mission to Mars and (2) since the interactions between the microbial environment, the human microbiome and immune functions are receiving more attention and becoming more evidence-based (see also Chap. 34).

The collected data of microbiological contamination from different environmental sources indicates that the ISS is a microbiologically safe working and living habitat. Therefore, the measures in place shall be applied to a major degree when longer missions are envisaged. Nonetheless, occasional contamination hazard reports do indicate that the current prevention and monitoring strategies are a minimum to control the microbial burden in manned space stations. Fluctuations in microbial concentrations and contamination events suggest the need for continued diligence and evaluation as well as further improvements in engineering systems.

Prevention is an essential component in (future) orbital and planetary space stations and design should integrate, next to thermal, mechanical, and chemical resistance of equipment and utensils, also the hygienic properties (e.g., biofouling or antimicrobial surface properties). Rational habitat design, integrating both these constructional and health-related parameters, is essential and probably most cost-effective in the long term. Specific for the spread of biological aerosols, the development of a reliable bioaerosol dispersion and deposition model is important to pinpoint critical locations in a certain habitat design. Development of the latter has already been initiated in different projects including the FP7 project BIOSMHARS (http://www.biosmhars.eu; Salmela et al. 2018) and the ESA projects BIOSIS and BIOMODEXO.

In addition to prevention, monitoring tools could be optimized by developing online detection tools that can be used directly in flight and that are capable of simultaneous quantification and identification (see Chap. 24). These molecular, non-culture-dependent assays would be less time-consuming for the astronauts and circumvent the necessity for post-flight analyses, allowing quick and autonomous decisions by the crew for assessment and remediation of contamination problems in the air, on surfaces, or in water and food.

Acknowledgments This work was supported by the European Space Agency through the projects COMICS and EXANAM (together with the Belgian Science Policy), BIOSIS and BIOMODEXO, and by FP7 through the BIOSMHARS project.

References

Alvarez B, Lopez MM, Biosca EG (2008) Survival strategies and pathogenicity of *Ralstonia solanacearum* phylotype II subjected to prolonged starvation in environmental water microcosms. Microbiology 154(Pt 11):3590–3598

Aviles H, Belay T, Fountain K, Vance M, Sonnenfeld G (2003) Increased susceptibility to *Pseudomonas aeruginosa* infection under hindlimb-unloading conditions. J Appl Physiol 95(1):73–80

Bechy-Loizeau AL, Flandrois JP, Abaibou H (2015) Assessment of polycarbonate filter in a molecular analytical system for the microbiological quality monitoring of recycled waters onboard ISS. Life Sci Space Res (Amst) 6:29–35

Berry CA (1973) View of human problems to be addressed for long-duration space flights. Aerosp Med 44(10):1136–1146

Berry D, Xi C, Raskin L (2006) Microbial ecology of drinking water distribution systems. Curr Opin Biotechnol 17(3):297–302

Bridier A, Briandet R, Thomas V, Dubois-Brissonnet F (2011) Resistance of bacterial biofilms to disinfectants: a review. Biofouling 27(9):1017–1032

Brief RS, Bernath T (1988) Indoor pollution: guidelines for prevention and control of microbiological respiratory hazards associated with air conditioning and ventilation system. Appl Ind Hyg 3(1):5–10

Bruce RJ, Ott CM, Skuratov VM, Pierson DL (2005) Microbial surveillance of potable water sources of the international Space Station. SAE Trans 114(1):283–292

Castro VA, Thrasher AN, Healy M, Ott CM, Pierson DL (2004) Microbial characterization during the early habitation of the international Space Station. Microb Ecol 47(2):119–126

Checinska A, Probst AJ, Vaishampayan P, White JR, Kumar D, Stepanov VG, Fox GE, Nilsson HR, Pierson DL, Perry J, Venkateswaran K (2015) Microbiomes of the dust particles collected from the international Space Station and spacecraft assembly facilities. Microbiome 3(1):50

Cox CS (1995) Stability of airborne microbes and allergens. In: Cox CS, Wathes CM (eds) Bioaerosols handbook. CRC Press, Boca Rotan, Florida, pp 77–99

Dacarro C, Picco AM, Grisoli P, Rodolfi M (2003) Determination of aerial microbiological contamination in scholastic sports environments. J Appl Microbiol 95(5):904–912

Dawson DJ, Sartory DP (2000) Microbiological safety of water. Br Med Bull 56(1):74–83

Duncan JM, Bogomolov VV, Castrucci F, Koike Y, Comtois JM, Sargsyan AE (2008) Organization and management of the international Space Station (ISS) multilateral medical operations. Acta Astronaut 63(7–10):1137–1147

Emtiazi F, Schwartz T, Marten SM, Krolla-Sidenstein P, Obst U (2004) Investigation of natural biofilms formed during the production of drinking water from surface water embankment filtration. Water Res 38(5):1197–1206

Husman T (1996) Health effects of indoor-air microorganisms. Scand J Work Environ Health 22(1):5–13

Ichijo T, Yamaguchi N, Tanigaki F, Shirakawa M, Nasu M (2016) Four-year bacterial monitoring in the international Space Station-Japanese experiment module "Kibo" with culture-independent approach. NPJ Microgravity 2:16007

ISS MORD (2006) SSP 50260 Revision C: ISS medical operations requirement document. Houston

James JT, Parmet AJ, Pierson DL (2008) Aerospace toxicology and microbiology. In: Davis JR, Johnson R, Stepanek J, Fogarty JA (eds) Fundamentals of aerospace medicine, 4th edn. Lippincott, Williams & Wilkins, Philadelphia, pp 236–250

Janssen PJ, Van Houdt R, Moors H, Monsieurs P, Morin N, Michaux A, Benotmane MA, Leys N, Vallaeys T, Lapidus A, Monchy S, Medigue C, Taghavi S, McCorkle S, Dunn J, van der Lelie D, Mergeay M (2010) The complete genome sequence of *Cupriavidus metallidurans* strain CH34, a master survivalist in harsh and anthropogenic environments. PLoS One 5(5):e10433

Kemp PC, Neumeister-Kemp HG (2005) Australian mould guideline: AMG-2005-1. Myco Logia Australia Pty Ltd

Knox BP, Blachowicz A, Palmer JM, Romsdahl J, Huttenlocher A, Wang CC, Keller NP, Venkateswaran K (2016) Characterization of *Aspergillus fumigatus* isolates from air and surfaces of the international space station. mSphere 1(5):e00227–e00216

Lachance PA (1997) How HACCP started. Food Technol 51:35

McAlister MB, Kulakov LA, O'Hanlon JF, Larkin MJ, Ogden KL (2002) Survival and nutritional requirements of three bacteria isolated from ultrapure water. J Ind Microbiol Biotechnol 29(2):75–82

Mehta SK, Pierson DL, Cooley H, Dubow R, Lugg D (2000) Epstein-Barr virus reactivation associated with diminished cell-mediated immunity in antarctic expeditioners. J Med Virol 61(2):235–240

Mijnendonckx K, Provoost A, Ott CM, Venkateswaran K, Mahillon J, Leys N, Van Houdt R (2013) Characterization of the survival ability of *Cupriavidus metallidurans* and *Ralstonia pickettii* from space-related environments. Microb Ecol 65(2):347–360

Monchy S, Benotmane MA, Janssen P, Vallaeys T, Taghavi S, van der Lelie D, Mergeay M (2007) Plasmids pMOL28 and pMOL30 of *Cupriavidus metallidurans* are specialized in the maximal viable response to heavy metals. J Bacteriol 189(20):7417–7425

Mora M, Perras A, Alekhova TA, Wink L, Krause R, Aleksandrova A, Novozhilova T, Moissl-Eichinger C (2016) Resilient microorganisms in dust samples of the international Space Station-survival of the adaptation specialists. Microbiome 4(1):65

Morawska L (2006) Droplet fate in indoor environments, or can we prevent the spread of infection? Indoor Air 16(5):335–347

Morris HC, Damon M, Maule J, Monaco LA, Wainwright N (2012) Rapid culture-independent microbial analysis aboard the international space station (ISS) stage two: quantifying three microbial biomarkers. Astrobiology 12(9):830–840

Mulvey D, Redding P, Robertson C, Woodall C, Kingsmore P, Bedwell D, Dancer SJ (2011) Finding a benchmark for monitoring hospital cleanliness. J Hosp Infect 77(1):25–30

NASA (2005a) International Space Station bacteria filter element service life evaluation. Houston

NASA (2005b) MR050L, Microbial analysis of ISS surfaces Using the surface sampler kit (SSK). Houston

NASA (2005c) MR051L, Microbial analysis of ISS water using the water microbiology kit (WMK) and the microbiology water analysis kit. Houston

NASA (2005d) MR052L, Microbial analysis of ISS Air using the microbial air sampler (MAS). Houston

NASA (2006) SD-T-0251, Microbiological specification and testing procedure for foods which are not commercially sterile, Houston

Novikova ND (2004) Review of the knowledge of microbial contamination of the Russian manned spacecraft. Microb Ecol 47(2):127–132

Novikova N, De Boever P, Poddubko S, Deshevaya E, Polikarpov N, Rakova N, Coninx I, Mergeay M (2006) Survey of environmental biocontamination on board the international Space Station. Res Microbiol 157(1):5–12

Osman S, La Duc MT, Dekas A, Newcombe D, Venkateswaran K (2008) Microbial burden and diversity of commercial airline cabin air during short and long durations of travel. ISME J 2(5):482–497

Perchonok M, Douglas G (2008) Risk factor of an inadequate food system. In: Human research evidence book. National Aeronautics and Space Administration, Houston

Pierson DL, Botkin DJ, Bruce RJ, Castro VA, Smith MJ, Oubre CM, Ott CM (2012) Microbial monitoring of the international Space Station. In: Moldenhauer J (ed) Environmental monitoring: a comprehensive handbook, vol 6. PDA, Bethesda

Salmela A, Kokkonen E, Kulmala I, Veijalainen A-M, Van Houdt R, Leys N, Berthier A, Ilyin V, Kharin S, Morozova J, Tikhomirov A, Pasanen P (2018) Production and characterization of bioaerosols for model validation in spacecraft environment. J Environ Sci 69:227–238

Satoh K, Nishiyama Y, Yamazaki T, Sugita T, Tsukii Y, Takatori K, Benno Y, Makimura K (2011) Microbe-I: fungal biota analyses of the Japanese experimental module KIBO of the international Space Station before launch and after being in orbit for about 460 days. Microbiol Immunol 55(12):823–829

Schwendner P, Mahnert A, Koskinen K, Moissl-Eichinger C, Barczyk S, Wirth R, Berg G, Rettberg P (2017) Preparing for the crewed Mars journey: microbiota dynamics in the confined Mars500 habitat during simulated Mars flight and landing. Microbiome 5(1):129

Sessa R, Di PM, Schiavoni G, Santino I, Altieri A, Pinelli S, Del PM (2002) Microbiological indoor air quality in healthy buildings. New Microbiol 25(1):51–56

Stewart PS, Rayner J, Roe F, Rees WM (2001) Biofilm penetration and disinfection efficacy of alkaline hypochlorite and chlorosulfamates. J Appl Microbiol 91(3):525–532

Straub JE, Plumlee DK, Schultz JR (2009) Chemical analysis results for potable water returned from ISS expeditions 14 and 15. SAE Int J Aerosp 1(1):556–577

Szewzyk U, Szewzyk R, Manz W, Schleifer KH (2000) Microbiological safety of drinking water. Annu Rev Microbiol 54:81–127

Van Houdt R, Michiels CW (2010) Biofilm formation and the food industry, a focus on the bacterial outer surface. J Appl Microbiol 109:1117–1131

Van Houdt R, De Boever P, Coninx I, Le Calvez C, Dicasillati R, Mahillon J, Mergeay M, Leys N (2009a) Evaluation of the airborne bacterial population in the periodically confined Antarctic base Concordia. Microb Ecol 57(4):640–648

Van Houdt R, Monchy S, Leys N, Mergeay M (2009b) New mobile genetic elements in *Cupriavidus metallidurans* CH34, their possible roles and occurrence in other bacteria. Antonie Van Leeuwenhoek 96(2):205–226

Van Houdt R, Mijnendonckx K, Leys N (2012) Microbial contamination monitoring and control during human space missions. Planet Space Sci 60(1):115–120

Venkateswaran K, Vaishampayan P, Cisneros J, Pierson DL, Rogers SO, Perry J (2014) International Space Station environmental microbiome - microbial inventories of ISS filter debris. Appl Microbiol Biotechnol 98(14):6453–6466

Verhoeff A (1993) Biological particles in indoor environments. In: European collaborative action, indoor air quality and its impact on man, COST Project 613, Report N. 12, EUR 14988 EN1993. Luxembourg

Volodina E, Nagolkin A, Fedotov A (2003) Air cleaning device for destruction of microbes based on electroporation effect. In: Wirtanen G, Salo S (eds) 34th R3-Nordic contamination control symposium, Turku, Finland. pp. 199–204

WHO (2008) World Health Organization: Guidelines for drinking-water quality: incorporating 1st and 2nd addenda, vol 1, Recommendations. 3rd edn. Geneva

Wingender J, Flemming HC (2004) Contamination potential of drinking water distribution network biofilms. Water Sci Technol 49(11–12):277–286

Yamaguchi N, Roberts M, Castro S, Oubre C, Makimura K, Leys N, Grohmann E, Sugita T, Ichijo T, Nasu M (2014) Microbial monitoring of crewed habitats in space—current status and future perspectives. Microbes Environ 29(3):250–260

Monitoring of Core Body Temperature in Humans

26

Andreas Werner and Hanns-Christian Gunga

26.1 Introduction to the Assessment of Thermoregulation in Man

Humans have an endothermic metabolism, and the independent regulatory mechanisms ensure that the core body temperature of the vital organs in the body (brain, heart, liver, and kidney) is kept autonomously around 36.7°C with circadian variations of ±0.5°C in males. In females, the menstrual cycle additionally alters the core body temperature. Scientific investigations to monitor thermoregulation by the changes of core temperature are usually performed by placing a thermosensor in the esophagus, nasopharynx, rectum, or tympanum/auditory meatus. However, none of these methods are entirely applicable nor are they convenient during daily routines, especially during long-term recordings for chronobiological research. The requirements to measure core body temperature demands thermosensor properties to be noninvasive, easy to handle, and to fulfill basic hygiene standards and not to be biased towards various environmental conditions. Moreover, the quantitative changes should appropriately reflect small deltas of the arterial blood temperature changes and the response time of the thermosensor to temperature changes should be as short as possible (Cooper et al. 1964; Gundel et al. 1997). These requirements are essential because several studies in humans have shown if high environmental temperature and humidity prevail, the heat load will lead to a rapid rise in the core body temperature (Kirsch et al. 1996, 1999; Pandolf et al. 1988). Especially in combination with heavy physical workload and fluid loss (sweating) with inadequate rehydration (Gunga et al. 1991; Baartz 1994) it results in heat stress-related injuries such as heat stroke (Gunga 2008; Gunga et al. 1993, 2008b; Kirsch and Vogt-Kirsch 1985; Montain et al. 2001; Sawka and Wenger 1988; Taylor et al. 1998; Kirsch and

A. Werner (✉) · H.-C. Gunga
Institute of Physiology, Center for Space Medicine and Extreme Environments Berlin, University of Medicine, Charité Cross Over (CCO), Berlin, Germany
e-mail: andreas.werner@charite.de; hanns-christian.gunga@charite.de

© Springer Nature Switzerland AG 2020
A. Choukér (ed.), *Stress Challenges and Immunity in Space*,
https://doi.org/10.1007/978-3-030-16996-1_26

Gunga 1999). If thermoregulatory impairments due to fever or drugs prevail, the deleterious developments may occur even faster (Gunga et al. 2005; Hoyt and Friedl 2004; Hoyt et al. 2002; Clark and Lipton 1984; Imrie and Hall 1990).

The measurement of the circadian rhythm of the core body temperature in man requires a continuous data recording (Mendt et al. 2017). Up to now, the standard sensors to measure the "correct" core body temperature are either semi-invasive (e.g., rectal, oesophageal, or urine bladder probes) or invasive (e.g., pulmonary arterial or aortal sensors) and are restricted in their application and not suitable for everyday use. Therefore they are all limited to either innovative application or specific conditions during intensive care therapy.

Besides the handling and the applicability of thermal probes, it is of critical importance to estimate if the sites for measuring *core* body temperature in the past were the appropriate ones to assess core body temperature correctly. While the brain (hypothalamus) is the "command central" of thermoregulation, the standard in physiological and clinical research on thermoregulation is still the recording with a semi-invasive rectal probe. Other ways to measure core body temperature at the body's center are usually more invasive (see above) but more sensitive at the same time, while noninvasive means mostly remain closer to surfaces (e.g., skin) with a subsequent loss of sensitivity. Therefore, measuring the rectal temperature seems likely to be a compromise even though it is not convenient and lacks applicability as well as sensitivity under extreme conditions as in space or in coldness, respectively.

Additionally, temperature changes in the rectum are more variable due to gender differences. For example, in females, the assessment of the circadian temperature rhythm in the rectum is less reliable due to the menstrual cycle (28-day-rhythm) affecting core body temperature profiles with a variance of 0.5–1.0°C according to the day of the rhythm. This is a consequence of the anatomical proximity of the female genital organs to the rectum. The alternating uterine/vaginal blood flow, lasting until menopause, can locally affect the temperature in the rectum in the range of at least 1.0°C.

Another significant impact of the reliable measurements of the core body temperature results from the physio-physical effect of temperature regulation at different surrounding temperatures. This is explained by Aschoff saying that in hot (e.g., 35°C) environment the body shell is closed to reflect the core temperature. In contrast, in more ambient or colder climates (e.g., 20°C), the body shell is even colder, and the external measurement of the core temperature is biased (Aschoff et al. 1971). However, the head and hence the brain would not be included in these temperature changes because in this area the temperature will be stable, both under conditions of healthy life at ambient temperatures as well as in extreme and cold environments. In the last consequence blood from the lower extremities will also reduce the temperature in the pelvis and the rectal temperature accordingly. In conclusion, the head temperature (and particularly the brain) seems to reflect best the "real" core body temperature as the center of temperature regulation (hypothalamus). The rear hypothalamus as a temperature regulating center receives its information about peripheral and central temperature receptors. Cold receptors in the skin and the spinal cord and warmth receptors in the abdominal organs send signals about the lateral tractus spinothalamicus to the formatio reticularis and further to the rear hypothalamus (see Chap. 9). Warmth sensors in the anterior hypothalamus additionally measure the blood temperature (central receptors). In the hypothalamus, information concerning the normal core body temperature is “stored” (36.7°C). Divergences between that “set” temperature and external changes will be regulated from that location (e.g., increasing the metabolic rate in a cold environment) to hold the brain and consequently the core body temperature stable. Consequently, the ideal anatomical area would be in the head, which would obviously be impossible for ethical reasons. Therefore, the question is: “How to measure the brain’s, and at this moment the core body, temperature from outside the brain?” In this context, one important anatomical connection between the brain and skin by the venae emissariae may provide “access” to the brain and core temperature (Fig. 26.1). This network of veins, which is only found in the cranium of humans, is involved in the temperature regulation of the brain. These vessels have originated and further developed as a function of increasing brain volume and higher energy turnover together with the need to “remove” heat to maintain a stable brain temperature.

Briefly, the sensor for measuring core body temperature needs to fulfill the following requirements: 1. should not disturb the subject, 2. should be easily accessible, 3. must comply with basic hygiene standards, 4. should not be influenced by environmental conditions, 5. should be sensitive enough to measure small changes in the subject, 6. its response time has to represent quantitative results, and 7. should assess temperature (changes) as quickly as possible (Shiraki et al. 1986; Hoyt and Friedl 2004; Smith et al. 1980). These requirements will be discussed in the following sections, beginning with the gold standard and coming to a new device which is the so called Double Sensor.

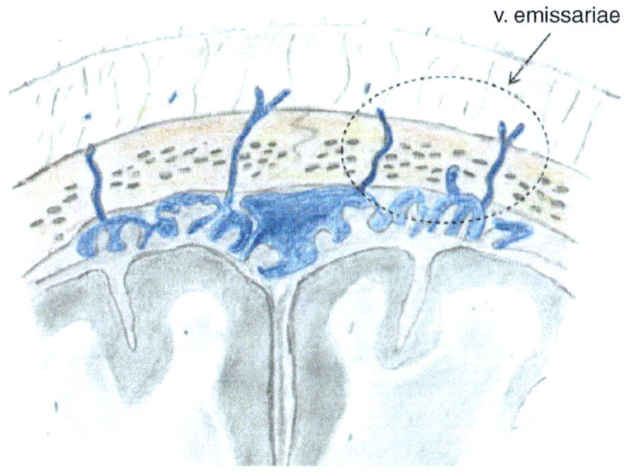

Fig. 26.1 Venae emissariae an evolutionary aspect in humans to regulate the core (brain) body temperature. Cross-section of parts of the left and right brain hemispheres. Cerebral vein and V. emissariae (blue), skull (bright brown) and skin, artwork by M. Hörl, inspired by the Sobotta Atlas of Descriptive Human Anatomy, Urban & Fischer

26.2 Current Status/Limitations

Currently, there is no accurate and easy method to measure core temperature in a field setting and to a certain extent in the laboratory as well, particularly during long-term (>24 h) core temperature recordings in chronobiology (see Chap. 9). The definitions of various temperature measurements used in wearable body activity monitors are summarized in Table 26.1.

The relative advantages and disadvantages of core temperature measurement sites including the time response of the different kind of sensors, have been intensively discussed since the first benchmark investigations on this topic by Claude Bernard in 1876 and will be briefly described in the following section.

26.2.1 Sensors to Measure Core Body Temperature

The most common places for measuring core body temperature are the rectum, esophagus, and tympanum/auditory meatus, or the gastrointestinal tract.

26.2.1.1 Esophageal and Rectal Temperature

Most thermal physiologists agree that the esophageal temperature is the best semi-invasive index of core body temperature for humans. It responds rapidly to changes in blood temperature elicited by extracorporeal circulation (Shiraki et al. 1986; Molnar and Read 1974) and body cooling by anesthesia (Cooper and Kenyon 1957). The esophageal temperature is obtained by inserting a catheter, containing a

Table 26.1 Exemplary types of sensors to measure core body temperature and their characteristics; Invasive temperature assessments in the aorta and pulmonary artery are not explained in the following text

Aorta ascendens	"Gold standard"—reference temperature, highly invasive
Pulmonary artery	As aorta ascendens, highly invasive
Esophageal	Good correlation to gold standard, semi-invasive
Urine bladder	Slow response time, very low compliance, semi-invasive
Rectal temperature	Good correlation to gold standard, slow reaction time, less compliance, semi-invasive
Tympanic	Infrared, dependent on user and environmental influences, low semi-invasive
"Temperature pill"	Dependent on position within the body (gastrointestinal tract), low semi-invasive
Axillary	"Old fashion," long calibration time (>30 min), noninvasive
Skin (Ramanathan)	Different places on the body to measure skin temperature and formular calculation (not possible in coldness), noninvasive
Skin – heat flux method	Good correlation to "gold-standard," highly dynamic, dependent on environmental conditions, noninvasive on the head

thermocouple or thermistor, through the nasal passage into the throat and then swallowing it. It is best used in clinical settings, but it is highly problematic in research or field assessments. This is also true for other devices to record core body temperatures as a rectal tube or tympanic/auditory meatus solutions; they are all impractical to use in the field (Steinman et al. 1987; Buller et al. 2005). The rectal temperature is obtained by inserting a temperature sensor a minimum of 5 cm past the anal sphincter, because temperature measurements are uniform within the rectum from 5 to 27 cm past the anal sphincter (Nielsen and Nielsen 1962; Greenleaf and Castle 1972). During exercise it takes approximately 25–40 min to achieve a steady-state rectal temperature value (Aikas et al. 1962; Nadel and Horvath 1970; Greenleaf and Castle 1972). These steady-state rectal temperatures are usually ~0.4°C (Mairiaux et al. 1983) higher than the mean skin temperatures (Wenger 2001) and 0.2–3°C higher than simultaneously measured nasopharyngeal and esophageal temperatures (Cranston et al. 1954; Saltin and Hermansen 1966; Saltin et al. 1970; Nadel and Horvath 1970).

Despite their limited reliability in specific conditions as described above, the esophageal and rectal temperatures have so far been considered to be largely independent of the environmental temperature (Strydom et al. 1965; Stolwijk et al. 1968; Saltin et al. 1970; Gunga et al. 2005). As a result, the steady-state rectal temperature provides a good index to assess body heat storage (Strydom et al. 1965; Saltin and Hermansen 1966). The main problem referred to rectal temperature measurements results from its slow response slope in comparison to the other measurement sites, an observation that has been proven again recently in 60 patients who underwent a postoperative rewarming (Melette 1950; Braeuer et al. 1977, 2000). The reason for the slow response is probably (1) a low rate of blood flow to the rectum compared to other measurement sites (Molnar and Read 1974; Aulick et al. 1981) and (2) the mass of organs located in the body cavity. This greater mass of tissue in the lower abdominal cavity requires a far greater amount of energy to cause a rapid temperature change.

26.2.1.2 Tympanic Temperature

Tympanic temperature is obtained by inserting a small temperature sensor into the ear canal and advancing it until it rests against the tympanic membrane. Proper placement is determined by the subject, hearing a sound when the temperature sensor touches the tympanic membrane. Some subjects find this contact to be uncomfortable (Brengelmann 1987). Also, there are reports of the temperature sensors perforating the tympanic membrane (Dickey et al. 1970; Wallace et al. 1974; Tabor et al. 1981). Because of the potential discomfort and trauma, as well as the placement problems associated with tympanic measurements, some investigators have chosen to measure the temperature of the external auditory meatus instead. For this measurement, a temperature sensor is placed in an earplug and inserted into the external auditory meatus. Placement of the temperature sensor is important since there is a substantial (~0.5°C) temperature gradient along the wall of the meatus. In addition, several studies have shown that tympanic or acoustic meatus temperature measurements do not provide a reliable index of the level of core temperature during either rest or exercise (Nadel and Horvath 1970; Greenleaf and Castle 1972; Marcus 1973a, b; McCaffrey et al. 1975; U.S. Army 2003). Depending upon the environmental conditions, tympanic or acoustic meatus temperature values can be lower or higher than simultaneously measured steady-state rectal (Greenleaf and Castle 1972) and esophageal temperature values (Marcus 1973a, b; McCaffrey et al. 1975). Moreover, local head heating and air flow to the face will bias the temperature of the external meatus (Nadel and Horvath 1970; Greenleaf and Castle 1972; Marcus 1973a, b; McCaffrey et al. 1975; Sawka and Wenger 1986). This method is still used for some reasons, but it is likely less good as the DoubelSensor (see below) and in recent publications some improvments ocured (Dakappa et al. 2017; Ota et al. 2017; Yeoh et al. 2017). Furthermore, it seems that this measured temperature probably doesn't display temperature under conditions of fluid shift as happen during weightlessness (Lorr et al. 2017).

26.2.1.3 Temperature Pill

The best and most reliable method of assessing thermal state in operational environments is a direct measurement of core body temperature by using the network-enabled ingestible core temperature sensor (TempPill™; Monnard et al. 2017). However, the use of a core temperature ingestible sensor is impractical for routine use, so these devices are reserved for use during high thermal stress missions, in case of heat injuries such as in hyper- and hypothermia (Weller and Withey 2005). The disadvantage of such a device is the movement of the temperature pill in the gastrointestinal tract. If the pill is located nearby the liver—a highly metabolic active organ—the temperature will be higher than in the jejunum, ileum or rectum and therefore the continuity of temperature recordings will be not accurate enough. On the other hand, in the distal intestinal tract, e.g., in the colon, the core body temperature might be underestimated. Therefore, it is assumed that different temperatures from the "core body" are recorded and the interpretation could be likely difficult.

26.2.1.4 Skin Temperature

Although the skin surface is more easily accessed than the other core body temperature sites mentioned above, this measurement site includes some important confounding factors, because the skin temperature is largely affected by cutaneous blood flow and sweat evaporation. Furthermore, environmental changes such as air temperatures, humidity, wind speed, and radiation will alter the skin temperature. That is why thermal physiologists prefer to determine mean skin temperature, a sum of weighted individual skin temperatures taken at up to 16 different skin surface sites, or for certain questions, even more (Ramanathan 1964; Mitchell and Wyndham 1969; Murgatroyd and Hardy 1970; Fox and Solman 1971). For the field purpose, such a complex, heavily wired temperature measurement setup is highly impractical. Other concepts have shown that it is complicated to use a single skin temperature sensor, even when it is combined with the body mass index (BMI), the knowledge of the clothes worn, to result in a reliable and accurate core body temperature (Yokota et al. 2005).

26.2.2 Heat Flux Sensor (Double Sensor)

26.2.2.1 Principles and Proof of Applications

The rectal temperature and the radio pill are the most widely used and accepted devices by physiologists for core body temperature recordings. In most cases, the core temperatures are obtained by inserting a temperature sensor a minimum of 15 cm past the anal sphincter (Abrams and Royston 1981; Benedict and Slack 1911; Karlberg 1949; Nielsen and Nielsen 1962; Aikas et al. 1962; Greenleaf and Castle 1972; Saltin et al. 1970; Saltin et al. 1972) or by swallowing a radio pill with a transmitter system, respectively. Nevertheless, the rectal and the radio pill temperature have some shortcomings as mentioned above. An advantage is that they are usually quite independent of the environmental temperature changes (Stolwijk et al. 1968; Nadel and Horvath 1970; Greenleaf and Castle 1972). As a result, the steady-state rectal temperature provides a good index to assess body heat storage (Strydom et al. 1965; Saltin and Hermansen 1966). The main disadvantage of the rectal temperature is the prolonged response time to changes in blood (Melette 1950) and other core temperatures. The reason for such a slow response is probably a lower rate of blood flow to the rectum compared to other recording sites (Molnar and Read 1974; Aulick et al. 1981; Sawka et al. 1996). The slow response time makes rectal temperature a poor core temperature index for estimating the input to the thermoregulatory controller (Eichna 1949; Gerbrandy et al. 1954; Saltin et al. 1970).

Gunga et al. (2005, 2008a, 2009a, 2009b) have introduced a combined skin temperature and heat flux sensor (Double Sensor) (Patent No. DE 100 38 247, DE 101 39 705, 2003). In contrast to similar methodological attempts in the past, this zero heat flux sensor principle (Teunissen et al. 2011) has been miniaturized and used without extra heating and has been specially sealed (Kimberger et al. 2009; Opatz et al. 2010). The sensor contains two temperature pills (Th1 and Th2) which are arranged one above the other separated by an insulating layer. With the direction of the skin to the environment the heat flux, which is an energy flow, passes through the pills (Fig. 26.2). The core body temperature can be calculated with the algorithm: $Tc = Th1 + Ks/Kg \times (Th1 - Th2)$ (Gunga 2008). The Double Sensor is placed at the vertex of the head or on the lateral forehead and has been tested under various physical and environmental conditions (i.e., changing workloads and ambient temperatures of 10°C, 25°C, and 40°C). A comparison of the new sensor with the rectal temperature revealed that the measurements of the Double Sensor differed between −0.16°C and 0.1°C from the average of the rectal temperature. With rising ambient temperatures an increasing concordance correlation coefficients (CCC) in the work periods was found (10°C = 0.49; 25°C – 0.69, 40°C = 0.75). Thirdly, the Double Sensor showed a more rapid response to the core body temperature changes for all resting periods at all ambient conditions, as compared to rectal temperature ($P < 0.01$) (Gunga 2008) (Fig. 26.3). The Double Sensor can be integrated into a helmet (Werner et al. 2009, 2010) or also into protective clothes and can measure the core body temperature accordingly. In the last years, the applicability and reliability of the Double Sensor could be presented under different physical and environmental conditions. When combining the data from the heat flux sensor with cardiovascular data (e.g., heart rate), a prediction of thermal physiological strain

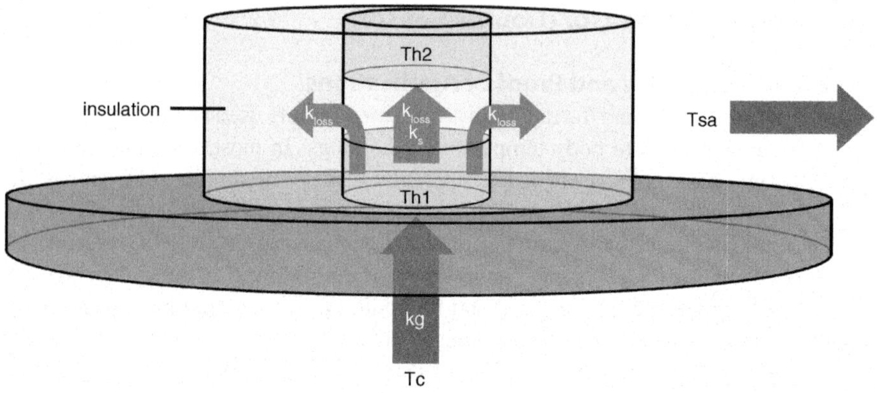

Fig. 26.2 Schematic side view of the Double Sensor. It contains two temperature sensors; an insulation disk placed in between which thermal conduction coefficient (*Ks*) is known (Gunga et al. 2008b). Core temperature (*Tc*), thermal conduction coefficient of human tissue (*Kg*), skin temperature (*Th1*), second temperature sensor for measuring the heat flux (*Th2*), lateral heat loss (K_{loss}), temperature on the outside area of the insulation (*Tsa*); middle gray—human skin (Dr. Koch, Dr. Sattler, Fa. Draeger Werke, Lübeck, Germany)

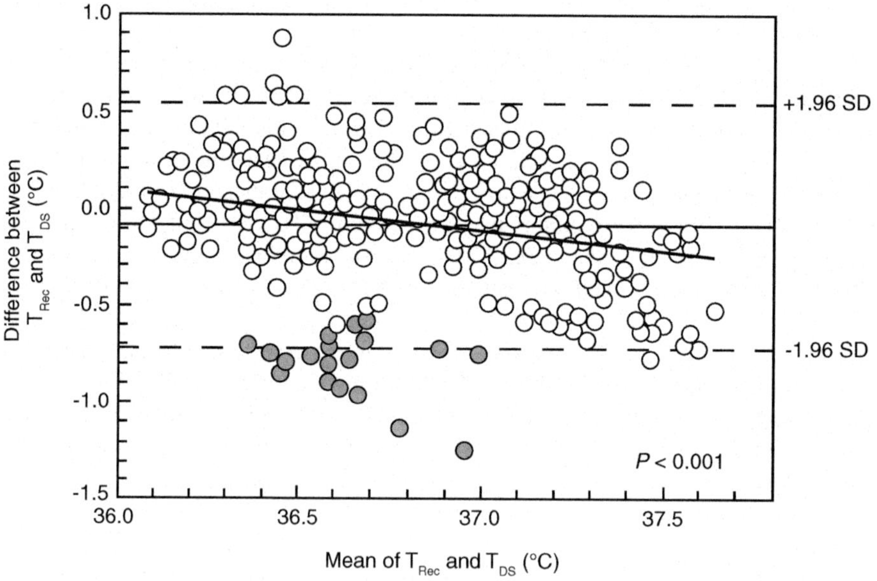

Fig. 26.3 Diagram of the difference between the methods against the averages of the methods (T_{REC} rectal temperature; T_{DS} Double Sensor). Statistical method according to Bland-Altman shows the evaluation of the Double Sensor underestimated compared with rectal temperature measurement at lower temperatures and overestimated at higher temperatures

(PSI) in humans, according to Moran et al. (1998) might be possible. However, they observed limitations of the heat flux sensor in cold environments that need to be addressed and investigated further. The main components of the Double Sensor unit hardware and the position of the Double Sensor are shown in Fig. 26.4.

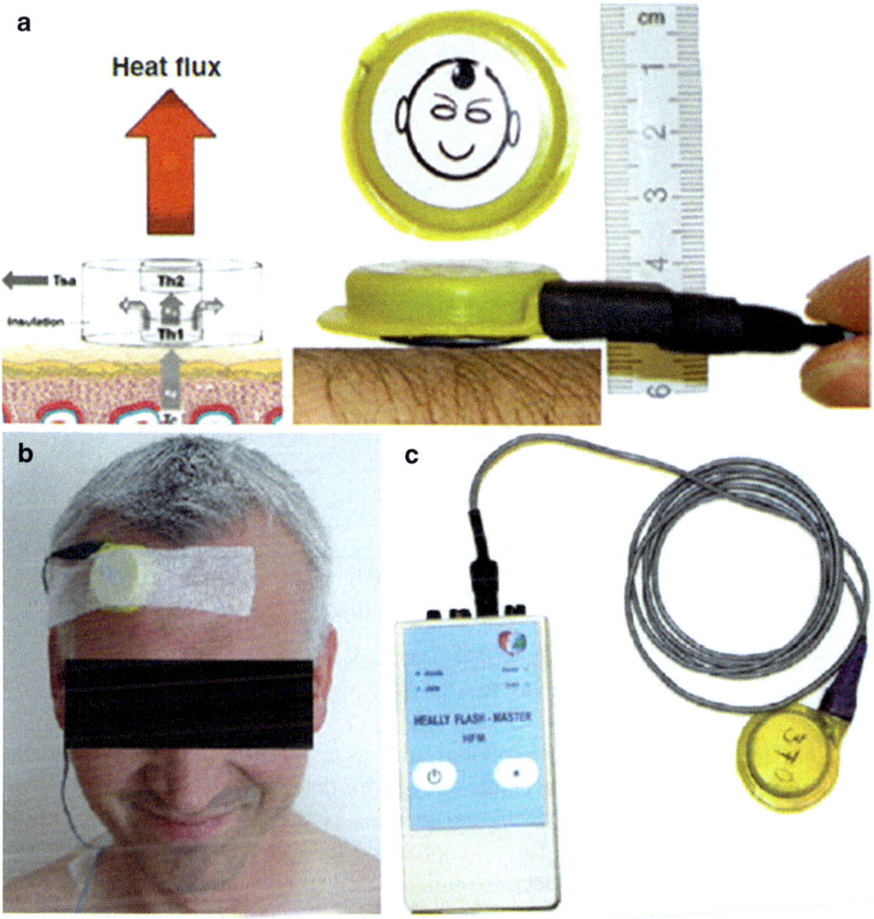

Fig. 26.4 Characteristics of the Double Sensor for the measurement of core body temperature (Draeger, Lübeck, Germany): (**a**) static display, (**b**) real dimensions, (**c**) clinical single use, self-adhesive sensor, (**d**) implementation in a health monitoring system, (**e**) healthLab hardware to operate the sensor (Koralewski, Hambühren and SpaceBit, Eberswalde, Germany), and (**f**) location of measurement stuck by tape on the forehead

In a bed rest study (Berliner bed rest study—BBR2) the Double Sensor was compared against the rectal temperature admission to examine the circadian rhythm profile (Gunga et al. 2009a). The temperature data with the Double Sensor were highly equivalent with the rectal temperature measurement. The dispersion of the measured values is shown in Fig. 26.5. In summary, about the great demand for developing an easy-to-operate and noninvasive technology to measure core body temperature in humans, we have evaluated the new skin temperature and heat flux measurement device to monitor core body temperature changes during 24 h at—6—head-down tilt bed rest. In this regard, it seems noteworthy that rectal temperature recordings themselves are not without error. Furthermore, it was found that circadian core body temperature profiles could be well approximated by the Double

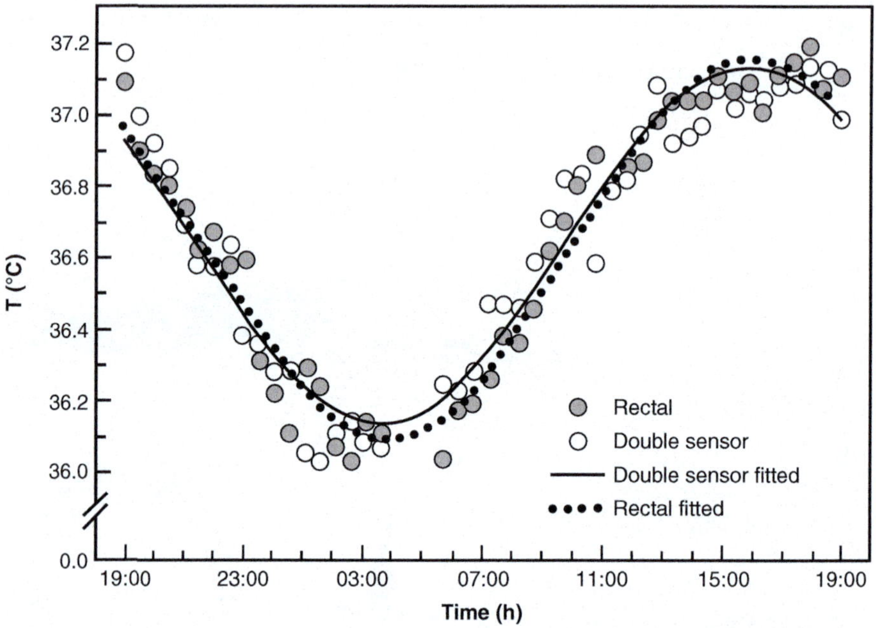

Fig. 26.5 Scatter plot of rectal temperature and core body temperature measured by Double Sensor under rest conditions (Berlin Bed Rest Study—BBR 2)

Sensor. Thus, even if the available pool of subjects is small, we are confident that the Double Sensor technology is accurate enough for detecting relatively small meaningful differences in circadian rhythm profiles compared to rectal temperature recordings. In addition to this, further approaches have been undertaken to evaluate the Double Sensor device in deep hypothermia (Werner et al. 2009, 2010), where promising results could be obtained. In conclusion, the knowledge and technological development gained from the study could promote applications of continuous core body temperature measurements in clinical, occupational, sports, and environmental medicine on Earth and in space (Werner and Gunga 2008). The Double Sensor device might be an effective, noninvasive and easy-to-operate technology to gain fundamental insights into cardiocirculatory regulation, thermoregulation, and circadian rhythms in humans (Werner et al. 2011a).

26.3 Thermoregulation in an Extreme Environment—Space

26.3.1 ThermoLab—Thermoregulation Under Exercise Conditions

Humans have an endothermic metabolism and autonomous regulation mechanisms to ensure that the core body temperature for the vital organs in the core body (brain, heart, liver, and kidneys) is approximately 36.7°C. In particular, different surrounding temperatures can challenge the temperature homeostasis and can alter the relationship of body core to body skin.

The thermoregulatory center functions act in the same way as a technical, regulatory system. It compares the information coming from the thermoreceptors with a preset target value, and, if there are variations, it activates the effector system, for example, the supply of blood to the skin (see below), to regain the intrinsic target values preset by the hypothalamus. The temperature homeostasis in humans is only ensured when the two decisive factors heat production, and heat loss, are balanced which is known as the so called thermal neutral zone (Hardy and DuBois 1938; Kirsch 1979; Jessen 2000; Gunga 2004). Only under this appropriate balance the human individual can feel comfortable and can perform adequately. This condition is achieved for a lightly clothed male adult at a temperature of 27°C (indifference temperature), under atmospheric conditions at sea level with 60% relative humidity and slight wind movement. In contradiction, water, with its high heat transmission capability and high heat capacity, has an indifference temperature of 33.5–34.5°C when compared to air. This means that the autonomous and morphological adaptation permits humans only a very narrow range of temperatures in which it is possible to remain unprotected.

All the intake energy (food) in the organism, unless it is needed for building new body substance or for physical work, is converted into heat. The following four mechanisms are available for heat transfer between the body and the environment: 1. radiation (emission of electromagnetic radiation), 2. convection (moving medium such as air), 3. conduction (objects that are in physical contact), and 4. evaporation (conversion of water to gas). If it is necessary to enhance heat loss, the peripheral blood circulation in the skin will be increased (vasodilatation); if it is needed to reduce heat loss, the blood circulation in the skin will be decreased (vasoconstriction). Vasodilatation and vasoconstriction are also described as "dry" heat loss, as opposed to the "wet," the evaporative heat loss (cooling by sweating). If heat loss is too high, the body temperature can be raised by increased metabolic heat production (e.g., shivering).

Thus, convective heat transfer makes up to 1/3 of the total heat loss mechanisms at environmental temperatures of 30°C. Convective heat transfer from the skin to the environment also affects, in particular, the buildup of the boundary layer between skin and the surrounding air. In this 4–8 mm thick boundary layer, there is usually a laminar air flow, which runs in an upright standing human from the feet along the body axis to the head. Approximately 600 l/min of air circulate in this way along the body; this upward streaming sheath of warmed air is called natural convection. At any skin area, the local heat loss depends on the temperature gradient and the thickness and velocity of the sheath. The structure and thickness of this boundary layer make a decisive contribution to the thermal well-being of the human (micro-climate).

In a sustained manner, a lack of gravity impairs (Yu and Yang 2000) the natural share of convective heat transfer from the body surface. Gravity, as the driving force for this convective heat transfer, ceases to apply at the body surface along the body

Fig. 26.6 A candle flame in Earth's gravity (*left*) and microgravity (*right*) showing that difference in the processes of combustion in microgravity. Image courtesy of NASA, Johnson Space Center (see http://www.nasa.gov/mission_pages/station/research/experiments/SAME.html)

axis under microgravity conditions (Blanc et al. 2000; Kuhlmann 2000; Yu et al. 2000; Zhang et al. 2000). The changes in convective heat transfer for a candle and the human body to the environment are shown in Fig. 26.6 and 26.7. These physical findings will affect the thermal comfort of the astronaut/cosmonaut under these specific conditions (Novak 1991; Qui et al. 1997, 2002), especially during extravehicular activities (EVA) (Clement 2003). Air temperatures in the Space Shuttle and the International Space Station (ISS) have been regulated between 18 and 29°C with a maximum humidity of 14 Torr and normobaric-normoxic conditions (Messerschmid and Bertrand 1999; Seibert 2001). It is known from the experience of the cosmonauts on the Space Station Mir that they complain of thermal discomfort at the extremities (Polyakov et al. 2001, Polyakov, personal communication; Clement 2003). This can result from increased catabolism in the muscular mass in this area (muscular metabolism), a vasoconstriction/vasodilatation, change in fluid shifts, energy balance and/or a change in heat transfer (Stein 2000).

Under these conditions the core body temperature can change rapidly, eventually reaching harmful levels. Core temperature recordings are quite difficult to achieve by inserting a thermosensor into the body until now. Furthermore, none of the methods apply to daily terrestrial routines in space. However, a thermosensor with its distinct advantageous, as described above, can be suitable for the following applications to

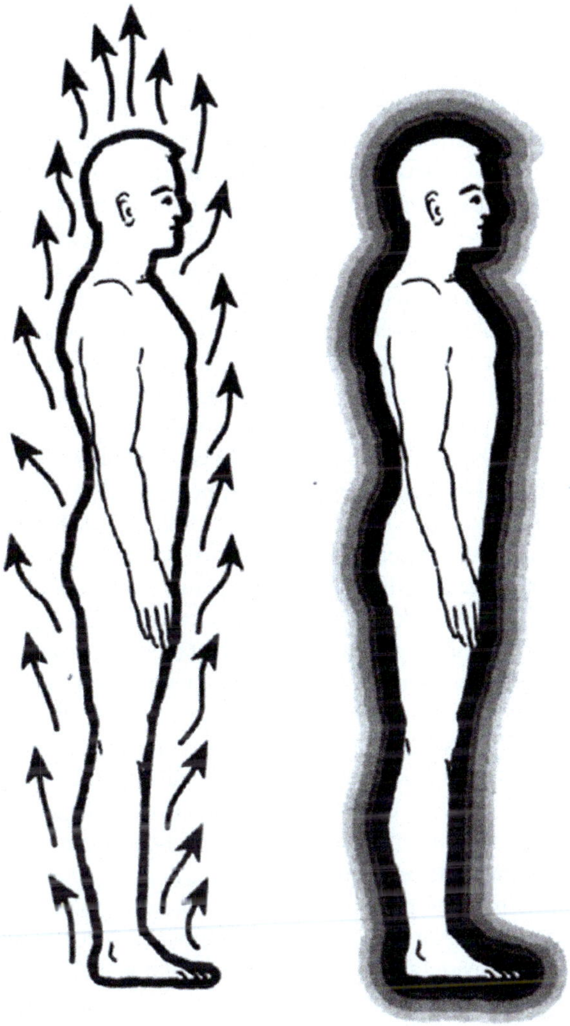

Fig. 26.7 The differences between being on Earth and in space. The *left figure* shows the airflow around the body under normal conditions (600 l/min of air flow); the *right figure* shows the condition under micro-g in space, the conductive heat transfer under these conditions is altered because of nonexistent air flow in space

assess stress effects in space: firstly, to assess and to monitor the heat homeostasis under stressful environmental conditions in space and secondly, to achieve reliable assessments of changes of the circadian rhythm during spaceflight.

26.3.2 ThermoLab—Affected Thermoregulation During Weightlessness

Based on the previous experiences (Werner et al. 2011b) it was approved to use the Double Sensor during long-term spaceflights on ISS to establish temperature recordings in humans. The purpose of the study was a better basic understanding of heat transfer and thermal regulation in humans under microgravity conditions. The findings might also be important for further long-term space explorations and extravehicular activities (EVA), or for future missions to the moon and Mars. Under the title: "Core temperature changes in humans before, during and after exercises performed on the International Space Station—Thermolab" has started on the ISS on February 2009 in combination with another experiment of NASA. Until the end of 2012 eleven astronauts participated in the study. Examination points were 90 days before launch, after 15, 45, 75, 105, 135, and 165 days of spaceflight, and 1, 10, and 30 days after the returning to Earth. The core body temperature was recorded with the Double Sensor on the head and was obtained at rest and during exercise on a bicycle ergometer which consisted of a submaximal protocol eliciting steady state cardiovascular and metabolic responses. Exercise intensity was based on an individual imputed protocol up to maximal oxygen uptake (VO2 max). The protocol consisted of 5 min resting period, followed by three continuous 5 min power levels prescribed to elicit 25%, 50%, and 75% of the individual's preflight VO2 max. These were immediately followed by 1-min stepwise increments of 25 W until subjects reached their symptom-limited maximum (Moore Jr et al. 2014). It was found that basal core body temperature increases during the first 45 days by about 1.01°C ($p < 0.001$) and was then stable the whole spaceflight at this temperature. For the temperature during the at VO2max peak, it was found an increase of body core temperature of nearly 40°C which means higher heat stress compared to the measurements preflight (max. core body temperature 38°C). Postflight, temperature regulation during the exercise was found altered until the last measurement point on day 30 (Fig. 26.8). These results reveal an incident of human temperature regulation during space habituation and could have impairment on physical and cognitive performance. Moreover, this elementary finding could have an impact on astronauts´ health by affecting the metabolic rate and cell stress (see Chap. 5), bone and muscle catabolism as well as on the immune systems' response by increasing the anti-inflammatory cytokines interleukin 1 receptor antagonist (IL1-RA) (Drescher et al. 2018; Stahn et al. 2017; Koch et al. 2016; Altrichter et al. 2014; Werner et al. 2010). Moreover, the permanent "resting" increase of the body's core temperature affects metabolism and functions continuously and may lead to a yet not fully understood "heat acclimation-mediated cross-tolerance" and further unknown adaptations of the organ systems to it.

The primary outcome of heat acclimation is increased thermotolerance, which stems from enhancement of innate cytoprotective pathways. These pathways produce "ON CALL" molecules that can combat stressors to which the body has never been exposed, via cross-tolerance mechanisms (heat acclimation-mediated cross-tolerance - HACT). The foundation of HACT lies in the sharing of generic stress signaling, combined with tissue/organ- specific protective responses. It becomes apparent when acclimatory homeostasis is achieved, lasts for several weeks, and has a memory. HACT differs from other forms of temporal protective mechanisms activated by exposure to lower "doses" of the stressor, which induce adaptation to higher "doses" of the same/different stressor; e.g., preconditioning, hormesis. These terms have been adopted by biochemists, toxicologists, and physiologists to describe the rapid cellular strategies ensuring homeostasis. HACT employs two major protective avenues: constitutive injury attenuation and abrupt post-insult release of help signals enhanced by acclimation. To date, the injury-attenuating features seen in all organs studied include fast-responding, enlarged cytoprotective reserves with HSPs, anti-oxidative, anti-apoptotic molecules, and HIF-1α nuclear and mitochondrial target gene products. Using cardiac ischemia and brain hypoxia models as a guide to the broader framework of phenotypic plasticity, HACT is enabled by a metabolic shift induced by HIF-1α and there are less injuries caused by Ca2+ overload, via channel or complex-protein remodeling, or decreased channel abundance. Epigenetic markers such as post-translational histone modification and altered levels of chromatin modifiers during acclimation and its decline suggest that dynamic epigenetic mechanisms controlling gene expression induce HACT and acclimation memory, to enable the rapid return of the protected phenotype. In this review the link between in vivo physiological evidence and the associated cellular and molecular mechanisms leading to HACT and its difference from short-acting cross-tolerance strategies will be discussed (Horowitz 2017).

In addition, the peak increases of temperature are adding to this or may act independently as either harmful stress peaks or could serve as "alert signals" and protective stress factors. Those are known as preconditioning which is an evolutionary very well preserved and clinically applicable tool to induce tissue protection (Thijssen et al. 2018). Those temporary heat peaks could be impacting the organ cell's metabolism, and e.g. inducing acute mitochondrial responses and translation of cyto-protective proteins (so called heat shock proteins HSP), altogether to increase tolerance of the heart muscle, brain and other cells (Ilievska et al. 2018; Tsai et al. 2016; Horowitz et al. 2015). Moreover, they seem to directly or indirectly affect immune functions as well (O'Neill and Hughes 2014) (Fig. 26.8).

26.3.3 Circadian Temperature Regulation During Weightlessness

After Thermolab, the Double Sensor was and is still in use on ISS for the study "core temperature and circadian rhythms in humans during long-term spaceflights—Circadian Rhythm." For a period of 36 h in 6 or 7 in-flight sessions during a six-month stay on ISS the temperature regulation is examined in astronauts. The light–dark rhythm on board is about 90 min, which means every 90 min there is a period of light and darkness. Therefore, it is expected that the circadian timing system (CTS), which is usually 24 h, will be affected by these rapid light–dark circuit. This may influence the coordinated daily variation of physiological and psychological behaviors. A synchronized circadian rhythm is fundamental to health, emotion, and cognitive function (Monk et al. 1998; Manzey 2001). In this context, human performance and symptoms of fatigue and concentration could be evaluated in further studies. Maybe one of the findings will be to bring adequate workplace illumination on ISS to simulate as far as possible normal circadian temperature rhythmicity under microgravity conditions. That will also have an impact on the planning of moon and Mars missions in the future. Temperature regulation during weightlessness: knowledge for application on Earth

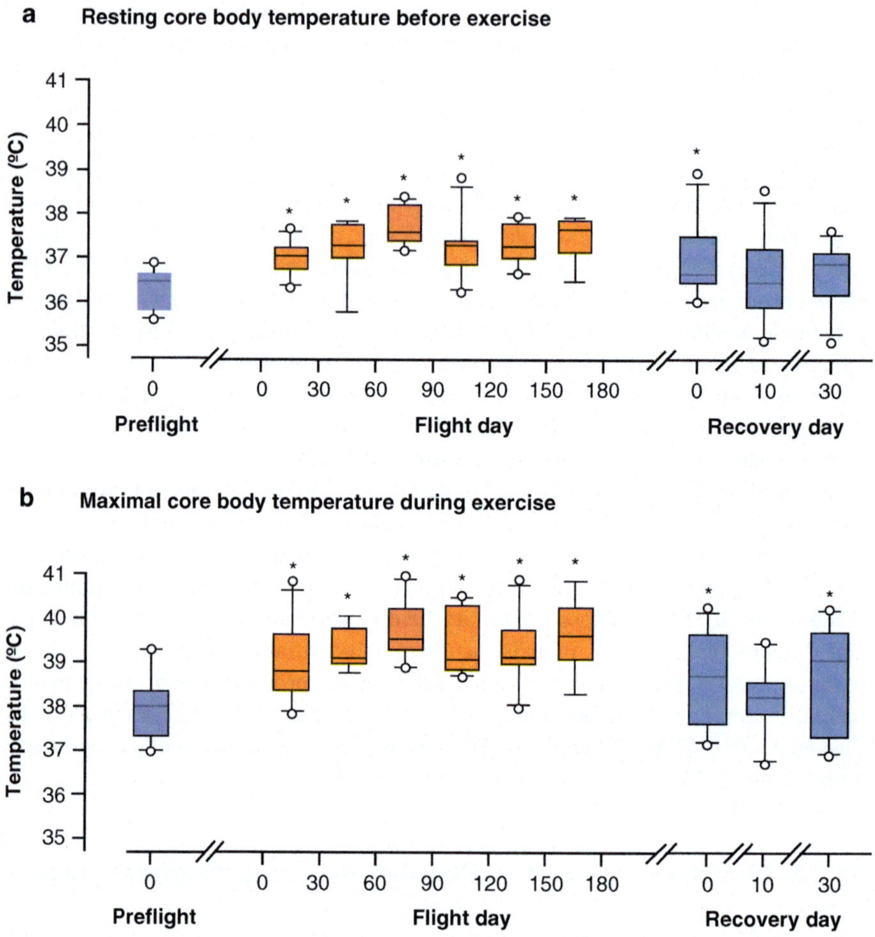

Fig. 26.8 Core body temperature before, during, and after a long-duration stay on the International Space Station (ISS). Panel (**a**) resting core body temperature before exercise; Panel (**b**) maximal core body temperature recorded during exercise

- Early prevention of hypothermia, perioperative monitoring (Desgranges et al. 2017; Opatz et al. 2013).
- Firefighters.
- Workers in temperature-challenged environments.

26.4 Summary

Humans have an endothermic metabolism which is regulated autonomously with a core body temperature of around 36.7°C ± 0.5°C. However, there is entirely a need for measuring core body temperature not only in patients but also in workers under extreme conditions during work. The measurement of core body temperature is not very easy. The more noninvasive you get, the more insufficient the sensors will be. Furthermore, the discussion where the core body is located is still an open issue. According to the center of thermoregulation, which is in the hypothalamus, the temperature should be measured around there. However, the invasiveness and therefore ethical reasons forbid implementing a sensor into the brain. Therefore, the apparent requirements for a sensor are to measure core body (brain) temperature noninvasively but very reliable, with hardware that is easy to use and which can be worn comfortably. To fulfill these requirements, a new heat-flux sensor was developed, the so called Double Sensor. In a series of studies, this sensor provided accurate and reliable core body temperature profiles by positioning the sensor on the forehead or vertex. Up to now, the sensor has an excellent correlation in hot environments. In the cold, however, there were some open questions. The sensor can be worn under the clothes and during a whole working day and is certified for clinical use as well. Because of this fact, the sensor was launched to the ISS to measure core body temperature before, during, and after long-term spaceflights (Stahn et al. 2017). The results show the reliability and fascinating insights into the thermoregulation under conditions of weightlessness, because body core temperature is increased and the regulation of this elementary physiological parameter is altered, and which can strongly affect—harm or protect—other organ systems including the immune functions sensitive to temperature changes.

Acknowledgments The acknowledgments go to the Deutsches Zentrum für Luft- und Raumfahrt (DLR) who has funded the Thermolab study (50WB0724).

References

Abrams RM, Royston JP (1981) Some properties of rectum and vagina as sites for basal body temperature measurement. Fertil Steril 35:313–316

Aikas E, Karvonen MJ, Piironen P, Ruosteenoja R (1962) Intramuscular, rectal and oesophageal temperature during exercise. Acta Physiol Scand 54:366–370

Altrichter S, Salow J, Ardelean E, Church MK, Werner A, Maurer M (2014) Development of a standardized pulse-controlled ergometry test for diagnosing and investigating cholinergic urticarial. J Dermatol Sci 75:88–93

Aschoff J, Günther B, Kramer K (1971) Energiehaushalt und Thermoregulation. In: Gauer O, Kramer K, Jung R (eds) Physiologie des Menschen Bd 2. Urban und Schwarzenberg, Berlin-München Wien

Aulick LH, Robinson S, Tzankoff S (1981) Arm and leg intravascular temperature of men during submaximal exercise. J Appl Physiol 51:1092–1097

Baartz F (1994) Die Hautwasserabgabe des Menschen unter extremen Umweltbedingungen. Dissertation, FU, Berlin

Benedict FG, Slack EP (1911) A comparative study of temperature fluctuations in different parts of the human body. Carnegie Institution of Washington, Washington, DC

Blanc S, Normand S, Pachiaudi C, Gauquelin-Koch G, Gharib C, Somody L (2000) Energy expenditure and blood flows in thermoregulatory organs during microgravity simulation in rat. Emphasis on the importance of the control group. Comp Biochem Physiol A Mol Integr Physiol 13:683–695

Braeuer A, Weyland W, Fritz U, Schuhmann MU, Schmidt JH, Braun U (1977) Bestimmung der Körpertemperatur. Anaesthesist 46:683–688

Braeuer A, Martin JD, Schuhmann MU, Braun U, Weyland W (2000) Genauigkeit der Blasentemperaturmessung bei intraabdominellen Eingriffen. Anasthesiol Intensivmed Notfallmed Schmerzther 35:435–439

Brengelmann GL (1987) Dilemma of body temperature measurement. In: Shiraki K, Yousef MK (eds) Man in a stressful environment; thermal and work physiology, vol 1. Thomas, C.C, Springfield, pp 5–22

Buller MJ, Hoyt RW, Ames JS, Latzka WA, Freund BJ (2005) Enhancing warfighter readiness through situational awareness–the warfighter physiologic monitoring–initial capability. 11th International Conference on Human-Computer Interaction Proceedings. Lawrence Erlbaum Associates Inc., Philadelphia, 2005

Clark WG, Lipton JM (1984) Drug-related heatstroke. Pharmacol Ther 26:345–388

Clement G (2003) Fundamentals of space medicine, Space technology library. Kluwer Academic Publishers, Dordrecht

Cooper KE, Kenyon JR (1957) A comparison of temperatures measured in rectum, oesophagus and on the surface of the aorta during hypothermia in man. Br J Surg 44:616–619

Cooper KE, Cranston WI, Snell S (1964) Temperature in the external auditory meatus as an index of central temperature changes. J Appl Physiol 19:1032–1035

Cranston WI, Gerbrandy J, Snell ES (1954) Oral, rectal and oesophageal temperatures and some factors affecting them in man. J Physiol 126:347–358

Dakappa PH, Prasad K, Rao SB, Bolumbu G, Bhat GK, Mahabala C (2017) A predictive model to classify undifferentiated fever cases based on twenty-four-hour continuous tympanic temperature recording. J Healthc Eng 2017:5707162

Desgranges FP, Bapteste L, Riffard C, Pop M, Cogniat B, Gagey AC, Boucher P, Bonnard C, Paturel B, Mullet C, Chassard D, Bouvet L (2017) Predictive factors of maternal hypothermia during Cesarean delivery: a prospective cohort study. Can J Anaesth 64:919–927

Dickey WT, Alhgren EW, Stephen CR (1970) Body temperature monitoring via the tympanic membrane. Surgery 67:981–984

Drescher U, Koschate J, Hoffmann U, Schneider S, Werner A (2018) Effect of acute ambient temperature exposure on cardio-pulmonary and respiratory kinetics in men. Int J Hyperth 34:442–454

Eichna LW (1949) Thermal gradients in man; comparison of temperatures in the femoral artery and femoral vein with rectal temperatures. Arch Phys Med Rehabil 30:584–593

Fox RH, Solman AJ (1971) A new technique for monitoring the deep body temperature in man from the intact skin surface. J Physiol 212:8P–10P

Gerbrandy J, Snell ES, Cranston WI (1954) Oral, rectal, and oesophageal temperatures in relation to central temperature control in man. Clin Sci 13:615–624

Greenleaf JE, Castle BL (1972) External auditory canal temperature as an estimate of core temperature. J Appl Physiol 32:194–198

Gundel A, Polyakov VV, Zulley J (1997) The alteration of human sleep and circadian rhythms during spaceflight. J Sleep Res 6:1–8

Gunga HC (2004) Physiologie. Urban & Fischer, München, pp 669–698

Gunga HC (2008) Wärmehaushalt und Temperaturregulation. In: Speckmann EJ, Hescheler J, Köhling R (eds) Physiologie. Urban & Fischer, München, Kap. 15, pp 615–641

Gunga H-C, Forson K, Amegby N, Kirsch K (1991) Lebensbedingungen und Gesundheitszustand von Berg - und Fabrikarbeitern im tropischen Regenwald von Ghana. Arbeitsmedizin Sozialmedizin Präventivmedizin 26:17–25

Gunga H-C, Maillet A, Kirsch K, Röcker L, Gharib C, Vaernes R (1993) European isolation and confinement study – water and salt turnover. In: Bonting SL (ed) Advances in space biology and medicine, vol 3. JAI Press Inc, Stamford, Conn, pp 185–200

Gunga H-C, Sandsund M, Reinertsen RE, Sattler F, Koch J (2005) The "Double sensor" – a new non-invasive device to measure continuously core temperature in humans. In: Environmental

Ergonomics XI, Holmér I, Kuklane K, Gao C (eds) Proceedings from the 11th international conference on environmental ergonomics. Ystad, pp 286–289

Gunga HC, Werner A, Sattler F, Koch J (2008a) Thermal monitoring systems. RTO Technical Report HFM-132 Real-Time Physiological and Psycho-Physiological Status Monitoring, 2008, Chapter 3

Gunga H-C, Sandsund M, Reinertsen RE, Sattler F, Koch J (2008b) A non-invasive device to continuously determine heat strain in humans. J Therm Biol 33:297–307

Gunga H-C, Steinach M, Stahn A, Werner A, Kirsch K (2009a) Handbook of space technology. In: Ley W, Wittmann K, Hallmann W (eds) Handbuch der Raumfahrttechnik. Hanser, München, pp 575–588. Chapter 7.6

Gunga HC, Werner A, Stahn A, Steinach M, Schlabs T, Koralewski E, Kunz D, Belavý DL, Felsenberg D, Sattler F, Koch J (2009b) The double sensor – a non-invasive device to continuously monitor core temperature in humans on earth and in space. Respir Physiol Neurobiol 169(Suppl 1):S63–S68

Hardy JD, DuBois EF (1938) Basal metabolism, radiation, convection and vaporization at temperatures of 22 to 35 °C. J Nutr 15:477–497

Horowitz M (2017) Heat acclimation-mediated cross-tolerance: origins in within-life epigenetics? Front Physiol 8:548

Horowitz M, Umschweif G, Yacobi A, Shohami E (2015) Molecular programs induced by heat acclimation confer neuroprotection against TBI and hypoxic insults via cross-tolerance mechanisms. Front Neurosci 9:256

Hoyt RW, Friedl KE (2004) Current status of field applications of physiological monitoring for the dismount soldier. In: Poos M (ed) Metabolic monitoring technologies for military field applications. National academy of sciences. National Academy Press, Washington, D.C., pp 247–257

Hoyt RW, Reifman J, Coster TS, Buller MJ (2002) Combat medical informatics: present and future. Proc AMIA Symp:335–339

Ilievska G, Dinevska-Kjovkarovska S, Miova B (2018) Effect of single and repeated heat stress on chemical signals of heat shock response cascade in the rat's heart. Cell Stress Chaperones 23:561–557

Imrie MM, Hall GM (1990) Body temperature and anaesthesia. Br J Anaesth 64:346–354

Jessen C (2000) Temperature regulation in humans and other mammals. Springer, Berlin

Karlberg P (1949) The significance of depth of insertion of the thermometer for recording rectal temperatures. Acta Paediatr 38;359–366

Kimberger O, Thell R, Schuh M, Koch J, Sessler DI, Kurz A (2009) Accuracy and precision of a novel non-invasive core thermometer. Br J Anaesth 103:226–231

Kirsch K (1979) Thermoregulation und cardiovasculäre adaptation. Der Kassenarzt 19:1–4

Kirsch KA, Gunga H-C (1999) Extreme Umwelten: Leben und Mobilität in Kälte. Flug- und Reisemedizin 6:36–38

Kirsch KA, Vogt-Kirsch C (1985) Die Leistungsgrenzen des Menschen beim Tragen von Atemschutz und Schutzanzug. Arbeitsmedizin Sozialmedizin Präventivmedizin 20:173–176

Kirsch K, Kaul A, Gunga H C, Roedler HD (1996) Physikalische Umweltfaktoren. In: Hierholzer K, Schmidt RF (eds) Pathophysiologie des Menschen, vol 40. VCH Verlag, Cambridge/New York, pp 1–40

Kirsch KA, Mendez-Gil A, Koralewski B, Johannes B, Bünsch B, Gunga H-C (1999) Probleme der thermoregulation während simulierter Schwerelosigkeit. Thermo Med 15:11–25

Koch K, Weller K, Werner A, Maurer M, Altrichter S (2016) Antihistamine updosing reduces disease activity in patients with difficult-to-treat cholinergic urticaria. J Allergy Clin Immunol 138:1483–1485

Kuhlmann H-C (2000) Transportprozesse unter Schwerelosigkeit. In: Keller MH, Sahm PR (eds) Bilanzsymposium Forschung unter Weltraumbedingungen. WPF, RWTH, Aachen, pp 31–41

Lorr D, Lund A, Fredrikson M, Secher NH (2017) Tympanic membrane temperature decreases during head up tilt: relation to frontal lobe oxygenation and middle cerebral artery mean blood flow velocity. Scand J Clin Lab Invest 77:587–591

Mairiaux P, Sagot J, Candas V (1983) Oral temperature as an index of core temperature during heat transients. Eur J Appl Physiol 50:331–341

Manzey D (2001) Limiting factors for human health and performance: psychological issues. In: HUMEX-Technical Note (TN)-002. Study on the survivability and adaptation of humans to long-duration interplanetary and planetary environments. Chapter 5, pp 1–45

Marcus P (1973a) Some effects of cooling and heating areas of the head and neck on body temperature measurement at the ear. Aerosp Med 44:397–402

Marcus P (1973b) Some effects of radiant heating of the head on body temperature measurements at the ear. Aerosp Med 44:403–406

McCaffrey TV, McCook RD, Wurster RD (1975) Effect of head skin temperature on tympanic and oral temperature in man. J Appl Physiol 39:114–118

Melette HC (1950) Skin, rectal and intravascular temperature adjustments in exercise. Am J Physiol 163:734. Abstract

Mendt S, Maggioni MA, Nordine M, Steinach M, Opatz O, Belavý D, Felsenberg D, Koch J, Shang P, Gunga HC, Stahn A (2017) Circadian rhythms in bed rest: Monitoring core body temperature via heat-flux approach is superior to skin surface temperature. Chronobiol Int 34(5):666–676

Messerschmid E, Bertrand R (1999) Space stations. Systems and utilization. Springer, New York, pp 240–244

Mitchell D, Wyndham CH (1969) Comparison of weighting formulas for calculating mean skin temperature. J Appl Physiol 26:616–622

Molnar GW, Read RC (1974) Studies during open-heart surgery on the special characteristics of rectal temperature. J Appl Physiol 36:333–336

Monk TH, Buysse DJ, Billy BD, Kennedy KS, Willrich LM (1998) Sleep and circadian rhythms in four orbiting astronauts. J Biol Rhythms 13:188–201

Monnard CR, Fares EJ, Calonne J, Miles-Chan JL, Montani JP, Durrer D, Schutz Y, Dulloo AG (2017) Issues in continuous 24-h core body temperature monitoring in humans using an ingestible capsule telemetric sensor. Front Endocrinol (Lausanne) 13(8):130

Montain SJ, Sawka MN, Wenger CB (2001) Hyponatremia associated with exercise: risk factors and pathogenesis. Exerc Sport Sci Rev 29:113–117

Moore AD Jr, Downs ME, Lee SM, Feiveson AH, Knudsen P, Ploutz-Snyder L (2014) Peak exercise oxygen uptake during and following long-duration spaceflight. J Appl Physiol (1985) 117:231–238

Moran DS, Shitzer A, Pandolf KB (1998) A physiological strain index to evaluate heat stress. Am J Physiol 275:R129–R134

Murgatroyd D, Hardy JD (1970) Central and peripheral temperatures in behavioral thermoregulation of the rat. In: Hardy JD, Gagge AP, Stolwijk JAJ (eds) Phys Behav Temp Reg, vol 58. Thomas, Springfield, Old Granville, pp 874–891

Nadel ER, Horvath SM (1970) Comparison of tympanic membrane and deep body temperatures in man. Life Sci 9:869–875

Nielsen B, Nielsen M (1962) Body temperature during work at different environmental temperatures. Acta Physiol Scand 56:120–129

Novak L (1991) Our experience in the evaluation of the thermal comfort during the space flight and in the simulated space environment. Astronaut 23:179–186

O'Neill S, Hughes J (2014) Heat-shock protein-70 and regulatory T cell-mediated protection from ischemic injury. Kidney Int 85:5–7

Opatz O, Stahn A, Werner A, Gunga HC (2010) Determining core body temperature via heat flux – a new promising approach. Resuscitation 81:1588–1589

Opatz O, Trippel T, Lochner A, Werner A, Stahn A, Steinach M, Lenk J, Kuppe H, Gunga HC (2013) Temporal and spatial dispersion of human body temperature during deep hypothermia. Br J Anaesth 111:768–775

Ota H, Chao M, Gao Y, Wu E, Tai LC, Chen K, Matsuoka Y, Iwai K, Fahad HM, Gao W, Nyein HYY, Lin L, Javey A (2017) 3D printed "earable" smart devices for real-time detection of core body temperature. ACS Sens 2:990–997

Pandolf KB, Sawka MN, Gonzales RR (1988) Thermoregulatory responses of middle-aged and young men during dry-heat acclimation. J Appl Physiol 65:65–71
Polyakov VV, Lacota NG, Gundel A (2001) Human thermohomeostasis on board "Mir" and stimulated microgravity studies. Acta Astronaut 49:137–143
Qui M, Liu W, Liu G, Wen J, Liu G, Chang S (1997) Thermoregulation under simulated weightlessness. Space Med Med Eng 10:210–213
Qui M, Wu JM, Gu DL, Yu XJ, Yuan XG, Chen JS (2002) Effects of head-down bedrest on surface temperature distribution and non-evaporative heat dissipation. Space Med Med Eng 12:93–97
Ramanathan NL (1964) A new weighting system for mean surface temperature of the human body. J Appl Physiol 19:531–533
Saltin B, Hermansen L (1966) Esophageal, rectal and muscle temperature during exercise. J Appl Physiol 21:1757–1762
Saltin B, Gagge AP, Stolwijk JAJ (1970) Body temperatures and sweating during thermal transients caused by exercise. J Appl Physiol 28:318–327
Saltin B, Gagge AP, Bergh U, Stolwijk JA (1972) Body temperatures and sweating during exhaustive exercise. J Appl Physiol 32:635–643
Sawka MN, Wenger C (1986) Physiological responses to acute exercise heat-stress. In: Pandolf KB, Sawka MN, Gonzalez RR (eds) Human performance physiology and environmental medicine at terrestrial extremes. Cooper Publishing Group, Traverse City, pp 97–152
Sawka MN, Wenger C (1988) Physiological responses to acute heat-exercise stress. In: Pandolf KB, Sawka MN' Gonzalez RR (eds) Human performance physiology and environmental medicine at terrestrial extremes. Cooper Publishing Group, Traverse City
Sawka MN, Wenger CB, Pandolf KB (1996) Thermoregulatory responses to acute exercise-heat stress and heat acclimation. In: Fregly MJ, Blatteis CM (eds) Handbook of physiology, vol I, Environmental physiology. Oxford University Press, New York
Seibert G (2001) A world without gravity. ESA SP-1251
Shiraki K, Konda N, Sagawa S (1986) Esophageal and tympanic temperature responses to core blood temperature changes during hyperthermia. J Appl Physiol 61:98–102
Smith P, Davies G, Christie MJ (1980) Continuous field monitoring of deep body temperature from the skin surface using subject-borne portable equipment: some preliminary observations. Ergonomics 23(1):85–86
Stahn AC, Werner A, Opatz O, Maggioni MA, Steinach M, von Ahlefeld VW, Moore A, Crucian BE, Smith SM, Zwart SR, Schlabs T, Mendt S, Trippel T, Koralewski E, Koch J, Choukèr A, Reitz G, Shang P, Röcker L, Kirsch KA, Gunga HC (2017) Increased core body temperature in astronauts during long-duration space missions. Sci Rep 7(1):16180
Stein TP (2000) The relationship between dietary intake, exercise, energy balance and the space craft environment. Pflugers Arch 441:R21–R31
Steinman AM, Hayward JS, Nemiroff MJ, Kubilis PS (1987) Immersion hypothermia: comparative protection of anti-exposure garments in calm versus rough seas. Aviat Space Environ Med 58(6):550–558
Stolwijk JA, Saltin B, Gagge AP (1968) Physiological factors associated with sweating during exercise. Aerosp Med 39:1101–1105
Strydom NB, Wyndham CH, Williams CG, Morrison JF, Bredell GAG, Joffe A (1965) Oral/rectal temperature difference during work and heat stress. J Appl Physiol 20:283–287
Tabor MW, Blaho DM, Schriver WR (1981) Tympanic membrane perforation: complication of tympanic thermometry during general anaesthesia. Oral Surg Oral Med Oral Pathol 51:581–583
Taylor N, Wilsmore B, Amos D, Takken T, Komen T, Cotter JD, Jenkins A (1998) Indirect measurement of core temperature during work: clothing and environmental influences. Abstracts of the 8th international conference on environmental ergonomics 97, San Diego
Teunissen LPJ, KlewerJ J, de Haan A, de Koning JJ, Daanen HAM (2011) Non-invasive continuous core temperature measurement by zero heat flux. Physiol Meas 32:559–570
Thijssen DHJ, Redington A, George KP, Hopman MTE, Jones H (2018) Association of exercise preconditioning with immediate cardioprotection: A Review. JAMA Cardiol 3:169–176

Tsai YC, Lam KK, Peng YJ, Lee YM, Yang CY, Tsai YJ, Yen MH, Cheng PY (2016) Heat shock protein 70 and AMP-activated protein kinase contribute to 17-DMAG-dependent protection against heat stroke. J Cell Mol Med 20:1889–1897

U.S. Army (2003) Heat-related injuries. MSMR Medical Surveillance Monthly Report 12(5) http://amsa.army.mil

Wallace CT, Marks WE, Adkins WY, Mahaffey JE (1974) Perforation of the tympanic membrane, a complication of tympanic thermometry during anesthesia. Anesthesiology 41:290–291

Weller AS, Withey WR (2005) Comparison of telemetry pill and rectal measurement of deep-body temperature during treadmill walking and running in the heat. ICEE, Ystad, p. 611ff

Wenger CB (2001) Human adaptation to hot environments. In: Pandolf KB, Burr RE (eds) Textbooks of military medicine, vol 1. USARIEM, Natick, pp 51–86

Werner A, Gunga HC (2008) Physiologie und Pathophysiologie des Wärmehaushalts und der Temperaturregulation des Menschen in extremen Umwelten und operationelle Konsequenzen für den militärischen Einsatz. Wehrmed Mschr 52(Heft 8):234–243

Werner A, Schlykowa L, Schlich G, Sattler F, Koch J, Koralewski HE, Gunga HC (2009) Preliminary data on a new non-invasive method to measure core temperature with the Double Sensor under cold conditions. RTO NATO Panel–HFM, Helsinki 2009

Werner A, Tiedemann J, Gunga HC, Falk M, Brugger H, Paal P (2010) Measurement of body core temperature by heat flux Double Sensor in hypothermic pigs during artificial avalanche burial. Poster in resuscitation congress, Porto 2010

Werner A, Lang V, Brix B, Krause W, Binnewies J, Gunga HC (2011a) Heat exposure of jet pilots during air traffic. In: Physiology of environmental stressors. AsMA, Anchorage 2011

Werner A, Fischer F, Stahn A, Gunga HC (2011b) Changes in body core temperature circadian rhythms during long-term isolation in MARS500. Project in Institute for Biological and Medical Problems–IBMP, Moscow 2011

Yeoh WK, Lee JKW, Lim HY, Gan CW, Liang W, Tan KK (2017) Re-visiting the tympanic membrane vicinity as core body temperature measurement site. PLoS One 12:e0174120

Yokota M, Moran DS, Berglund LG, Stephenson LA, Kolka MA (2005) Noninvasive warning indicator of the "Red Zone" of potential thermal injury and performance impairment: a pilot study. Proceedings for the 11th international conference of environmental ergonomics. Lund University, Sweden, pp 514–517

Yu XJ, Yang TD (2000) Ground-based studies on thermoregulation at stimulated microgravity by head-down tilt bed rest. Space Med Med Eng 13:382–385

Yu XJ, Yang TD, Pang C (2000) Weightlessness and heat stress on astronauts. Space Med Med Eng 13:70–73

Zhang WX, Chen JS, Li TQ (2000) A heat transfer model for liquid cooling garment (LCG) and its analysis. Space Med Med Eng 13:350–354

Flow Cytometry Methods to Monitor Immune Dysregulation Associated with Spaceflight

27

Brian Crucian, George Makedonas, and Clarence Sams

27.1 Technical Background

Flow cytometry is a laser-based technique for scanning and evaluation of suspended cells or other particles. A Nobel Prize was never awarded for this breakthrough; however, it has evolved into an extremely important technology in both research and clinical medicine. Flow cytometers have an extremely rapid throughput, and can analyze thousands of individual cells per minute. Typically, they have a laser excitation source (most commonly 488 nm (blue)) which can be augmented with additional lasers of 642 nm (red), 532 nm (green), or 405 nm (violet), and multiple photomultiplier tube detectors (PMT) for emitted light. Cell samples may be stained with various fluorescent antibodies to other dyes to identify cell surface proteins, intracellular markers, cell function, or other cellular targets. Upon cell excitation, the light emitted by the various fluorochromes is directed to the appropriate PMT via a network of long- and short-pass filters, dichroic mirrors, and band-pass filters. The number of colors visible to the cytometer (and thus the number of parameters that can be measured) is defined by the number of PMTs. The up multiple laser excitation sources have become common, and coupled with up to 10–15 PMTs to detect fluorescent dye emission. In addition, excitation laser light that has been "scattered" by the target cell, regardless of any fluorochromes present, may also be measured. Laser light scattered in near opposition to the excitation source is defined as "forward scatter," and generally corresponds to cell size. Light scattered at approximately 90° to the excitation source is defined as "side scatter", and is a measure of cellular granularity. Together, the scattered light parameters may be visualized in bivariate dot plots. This resolves the basic peripheral blood leukocyte

B. Crucian (✉) · C. Sams
NASA Johnson Space Center, Houston, TX, USA
e-mail: brian.crucian-1@nasa.gov

G. Makedonas
JES Tech, NASA-Johnson Space Center, Houston, TX, USA

© Springer Nature Switzerland AG 2020
A. Choukér (ed.), *Stress Challenges and Immunity in Space*,
https://doi.org/10.1007/978-3-030-16996-1_27

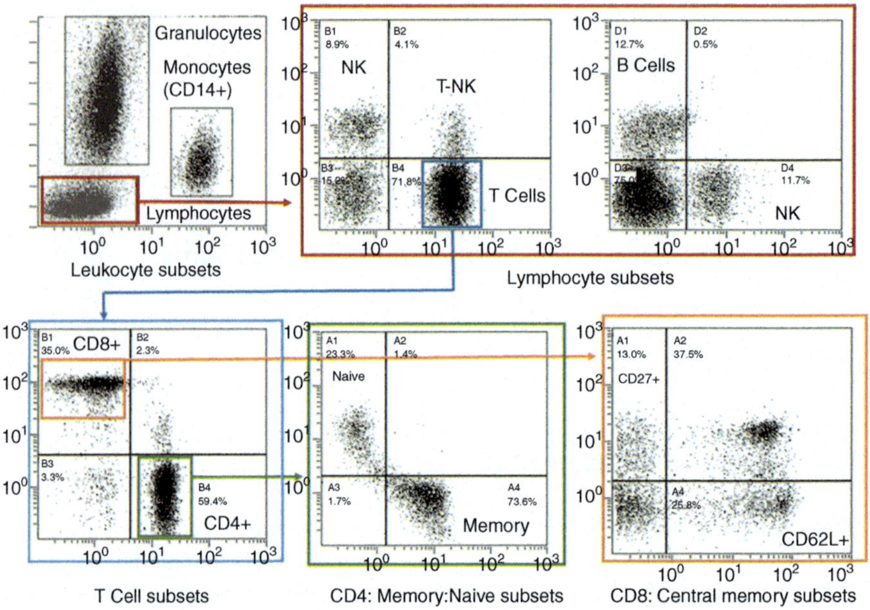

Fig. 27.1 Representative multicolor flow cytometry gating strategy. The bulk leukocyte subsets are resolved by both scatter properties and expression or absence of CD14. The lymphocyte population is then "gated," and further analyzed and quantified by resolving surface marker expression the T, B, and NK populations. Similarly, T cells and T cell subsets may be "gated," and subsequently further analyzed to resolve various memory:naive and central memory subsets

populations, granulocytes lymphocytes, and monocytes by scatter properties alone into distinctly visible populations prior to fluorescence measurements. This resolution allows identification, or "gating", of distinct leukocyte populations by scatter properties, to analyze their fluorescence emission properties separately (Fig. 27.1). Thus, a four color cytometer with forward and side scatter, essentially becomes a 6-parameter instrument, and allows measurements of all resolvable cell populations to occur simultaneously. Cell populations may also be gated off fluorescence expression, for populations with overlapping scatter properties. For example, T cells may be resolved from within the lymphocyte population (also containing B and NK) by expression of CD3. Gating by immunoscatter does reduce the remaining fluorescence channels available for data acquisition.

Samples for flow cytometry analysis must be prepared prior to running the samples on the instrument. For blood analysis, this usually includes the basic steps of WBC (white blood cell) staining, RBC (red blood cell) lysis, and sample fixation/stabilization. The quality of the sample preparation relates directly to the quality of data generated by the cytometer. The basic blood staining described above may be employed to resolve and determine relative percentages of all major peripheral blood leukocyte subsets ("phenotyping"). However, since classical flow cytometers suspend particles in sheath fluid, they do not determine absolute counts. This is one distinction from hematology instruments, which will derive absolute counts as part of a complete blood count (CBC), but will not stain cells or measure emitted light. Hematology instruments can also resolve bulk populations based on size/scatter or potentiometric measurements alone. Some newer miniaturized cytometers, including instruments from Guava and Partec, do not suspend cells in sheath fluid and thus can be used to determine absolute counts (Crucian and Sams 2005). By employing variation in sample preparation, including cell culture followed by analysis of the cultured cells, various measures of immune cell function may be measured by flow cytometry. Examples include the expression of cellular activation markers, determination of cytokine-secreting cells (permeabilization, intracellular cytometry), and measurement of secreted cytokine production. Measures of secreted analytes do not analyze cells, but use fluorescent bead technology as a capture platform, followed by staining with a fluorescent detection antibody. Various bead populations are "gated," and concentration of the secreted analyte may be determined.

Immune system dysregulation is associated with spaceflight, and a review of the phenomenon may be found elsewhere in this volume (see also Chaps. 11–17). A variety of flow cytometry techniques described above have been used to assess immune parameters during, or immediately following, spaceflight. Variations between the designs of these studies include simple postflight sampling of human or animal subjects, subject sampling during orbit for postflight cytometry analysis, and cell culture during orbit followed by postflight cytometry analysis. During one pilot study conducted during the Mir-18/STS-71 mission, staining, lysing, and preservation of astronaut whole blood samples was performed in-flight, and the stained cell samples were returned for postflight analysis (Sams et al. 1999). The relative ease and reliability of flow cytometry, and the large diversity of supported assays, make flow cytometers an ideal instrumentation platform for definition of spaceflight-associated immune dysregulation. This chapter discusses flow cytometry assays that have been applied to, and identify immune changes associated with, spaceflight, and the progress towards development and deployment of a microgravity-compatible flow cytometer for the International Space Station (ISS) and beyond.

27.2 Flow Cytometry Assays Used to Define Immune Changes Associated with Spaceflight in Humans or Animals

27.2.1 Peripheral Leukocyte Distribution

Measurements of the distribution of the peripheral blood leukocyte subsets (or splenocyte/marrow subsets, from mice or rats) is one of the most common flow cytometry techniques and has been used in many spaceflight studies. Alterations in distribution are frequently reported, and reflect in vivo immune changes that may have clinical significance, similar to that of the CBC measurement. Increases in immune cell populations may reflect immune activation or clonal expansion and proliferation. Decreases may reflect cell migration out of the blood to the periphery or a localized site of inflammation, or blockades in cell maturation processes. As the number of available fluorochromes and channel capabilities of flow cytometers has increased over the last two decades, the resolution of immune cell subsets has also increased. Earlier in the space program, simple "bulk" subsets were assessed on 2–3 color instruments. These may have included granulocytes, lymphocytes, monocytes, lymphocyte subsets, and the CD4:CD8 ratio. Today, 10–15 color instruments with distinct fluorochromes make possible the analysis of various activated populations, and "fine" T cell subsets, such as cytotoxic effector CD8+ and central/effector memory T cells. For example, T cells may be measured by simultaneously assessing CD45 and CD3, and memory T cells: naïve T cell subsets may be assessed by CD45RA/RO expression, whereas central memory T cells require CD3, CD8, CD45RA, and CD62L or CD11a. Studies by Roederer and others have indicated that diseases are often accompanied by changes in the number or function of "fine" lymphocyte subsets, even if changes in the bulk lymphocyte populations are not evident (De Rosa et al. 2001). Other "fine" leukocyte subsets are being identified, some with potential clinical significance, as the field of clinical immunology ever progresses. Specifically, there are also distinct subsets of B cells, NK, and monocytes, which all may have distinct physiological relevance. Some representative spaceflight studies employing phenotype analysis are discussed below.

In an example of an early assessment of bulk leukocyte subsets, Stowe et al. performed a postflight WBC and differential on short-duration Space Shuttle astronauts (Stowe et al. 1999). A postflight elevation in WBC and granulocyte percentage was observed. This is an almost universally reported phenomenon, widely believed to represent demargination associated with reentry and landing stress. A subsequent study from our laboratory assessed lymphocyte subsets, the T cell CD4:CD8 ratio, and basic memory naïve T cells as defined by CD45RA+/− following short-duration spaceflight (Crucian et al. 2000a). The data indicated a decreased percentage of T cells, an increased CD4:CD8 ratio, and variable changes in memory T cell levels.

Two subsequent studies used flow cytometry to assess leukocyte distribution and stress-related parameters in Space Shuttle astronauts on missions with varying durations to determine the effects of mission duration. Mills et al. reported that the WBC, granulocytes, monocytes, B cells, and CD4+ T cells were elevated, and NK were decreased postflight. The alterations were more pronounced following 1 week missions, as opposed to 2 week missions, and the authors concluded that shorter missions were associated with predominantly sympathetic responses, potentially attenuated following longer missions (Mills et al. 2001). Stowe et al. followed this study by assessing neuroimmune responses in astronauts participating in 9 or 16 day missions. The results confirmed that sympathetic nervous system responses dominate following short space missions, with longer missions characterized by glucocorticoid-mediated changes (Stowe et al. 2003).

During the STS-108 Space Shuttle mission in 2001, Peacut and Gridley extended the use of flow cytometry to analyze minute cell suspensions and blood from space-flown BALB/c mice, analyzing peripheral blood, marrow, and splenic leukocyte subset changes. Pecaut et al. indicated that no differences were observed in the general circulating lymphocyte proportions. However, changes in spleen leukocyte subsets were reported, including increased lymphocytes, decreased granulocytes, increased B cells, and decreased CD4+ T cells. In the same animals, there were proportional increases in marrow T cells and decreases in B cells (Pecaut et al. 2003). In the same animals, Gridley et al. reported percentages of CD25+ lymphocytes, CD25+ T cells, and NK1.1+/CD25+ NK were elevated in the STS-108 mouse subjects, whereas CD71 expression was decreased (Gridley et al. 2003). In these studies, the traditional bulk leukocyte distribution was expanded to include activation markers and adhesion markers, as well as organ-specific assessments of leukocyte distribution.

Subsequent to the 2001 studies, the authors performed a follow-up study during the STS-118 Space Shuttle mission in 2007. Gridley et al. showed that levels of T cells and B cells were reduced, whereas NK levels were elevated in the spleen (Gridley et al. 2009). Ortega et al. reported on a comprehensive marrow assessment, where flow cytometry was used to assess expression of molecules that define the maturation/activation state of cells in the granulocytic lineage. The molecules used were Ly6C, CD11b, CD31, Ly6G, F4/80, CD44, and c-Fos. There were no composite phenotypic differences between the total bone marrow cells isolated from spaceflight and ground control mice, however there were subpopulation differences in several markers. In particular, an elevation of CD11b suggested neutrophil activation in response to landing, and decreases in Ly6C, c-Fos, CD44, and Ly6G with an increase in F4/80 suggested marrow cells from the spaceflight mice were more differentiated (Ortega et al. 2009).

In a 2008 post-flight assessment of both short-duration and long-duration (6 month, ISS) astronaut crew members, our laboratory expanded the traditional "bulk" subset analysis to include "fine" T cell subsets. These included central memory/effector memory CD8+ T cells subsets, cytotoxic/senescent CD8+ T cell subsets, and various constitutively activated T cell subsets. These studies were conducted using 5-color cytometry. In addition to the commonly observed changes bulk subset alterations, crew members displayed elevated levels of both CD4 and CD8 memory T cells (defined as CD45RO+), reduced levels of early senescent (CD57+) and late senescent (CD28−) CD8+ T cells, and reduced levels of terminally differentiated CD8+ T cells (Crucian et al. 2008, Chap. 14).

Because postflight measurements may be altered by stressful effects of landing and readaptation, phenotype, T cell activation and virus-specific immunity were, after the construction and outfitting of the ISS, analyzed via flow cytometry from subjects who participated in long-duration orbital spaceflight. In particular, the NASA "Integrated Immune" investigation was the first to return ambient blood samples from spaceflight for subsequent terrestrial analysis. To enable functional cellular analyses, in addition to simple phenotyping, whole blood samples were collected near to undock and return of a visiting US or Russian spacecraft, followed by terrestrial cell culture and flow cytometry analysis (Crucian et al. 2010). The results from Integrated Immune demonstrated that crewmembers posessed altered leukocyte distribution, and specific reductions in both T cell function and mitogen stimulated cytokine production, that persisted during a 6-month orbital spaceflight. All of these findings were derived employing flow cytometry techniques.

In general, the studies described above demonstrate the utility of assessing leukocyte and lymphocyte subsets, as well as adhesion markers, activation markers, etc. for monitoring spaceflight associated immune dysregulation in humans and rodents. The resolution of subsets being monitored has advanced as cytometer technology and fluorochrome/dye development has advanced. It should be noted that changes in constitutive subset levels reflect in vivo immune changes, and thus may represent more individualized responses associated with adverse clinical reactions, than with generalized immune functional changes.

27.2.2 General Immune Cell Function

As several recent studies have demonstrated, measurements of immune function are proving important in defining what immune dysfunction exists during spaceflight. Leukocyte subset measurements assess in vivo distribution shifts, and by default in vivo immune responses. Typically, leukocyte populations become altered as part of an immune response related to illness. It is well established that stress may compromise the immune system (REFS). The stressors which influence human immunity in space: microgravity, physiological stress, isolation, altered nutrition, etc., are not necessarily related to pathological processes. Hence, they are much more likely to influence cellular function as opposed to cellular distribution. It is possible that crew member leukocyte distribution may be completely unaltered during flight, yet if immune cell function is (persistently) compromised, a clinical risk may still exist. Several investigators have reported altered granulocyte, monocyte, NK, and T cell function associated with spaceflight, and some of the current techniques for measuring cell function (and used in the cited spaceflight studies) employ flow cytometry. Some representative citations follow.

Kaur et al. performed a postflight assessment of monocyte oxidative burst capacity, a measure of monocyte function. Oxidative burst was determined by flow cytometry using the Fc OxyBURST Green assay reagent, which consists of bovine serum albumin covalently linked to dichlorodihydrofluorescein (H2DCF) and complexed with purified rabbit polyclonal anti-BSA IgG antibodies to yield a multivalent particulate immune complex. When this immune complex binds to the Fc receptors of monocytes, the nonfluorescent H2DCF molecules are internalized at 37° but not at 4° within the phagovacuole of the activated monocytes, and subsequently oxidized to green fluorescent dichlorofluorescein. The data indicated that following 5–11 days of spaceflight, astronauts' monocytes exhibited reduced capacity to elicit an oxidative burst (Kaur et al. 2005, see also Chaps. 11 and 12).

T cell activation initiates when the T cell receptor (TCR) is triggered in an antigen-specific fashion in the context of the major histocompatibility complex (MHC), with additional triggering of appropriate co-stimulatory molecules. During the process of full activation to clonal expansion, a >72 h series of events takes place. This process begins with nearly immediate transduction of intracellular signals to the T cell nucleus, involves changes in membrane potential and intracellular pH, and changes parameters such as receptor distribution and mobility, cytoskeletal

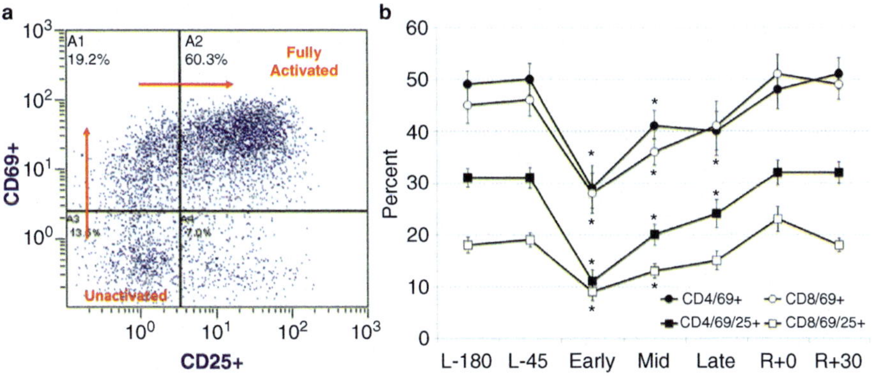

Fig. 27.2 T cell function by flow cytometry for ISS astronauts. Early blastogenesis of T cells, characterized by the expression of cell surface CD69 and CD25, may be measured via flow cytometry. This is a measure of the early events associated with T cell activation and may be considered a determination of cell function. (**a**) Representative dot plot following stimulation of T cells via culture in the presense of staphylococcus superantigens A and B; (**b**) Representative pre-, in-, and post-flight data for ISS astronauts (Crucian et al. 2015)

rearrangement, alterations in the molecular conformation of adhesion molecules, and results in activation of adenylate cyclase, ATPase, and other membrane-associated enzymes. Within the first few hours there is coalescence of patched receptors into a cap and expression of preformed CD69 molecules on the T cell surface. Within 24 h secretion of IL-2 begins, there is upregulation of the IL-2 receptor (CD25) on the cell surface and IL-2 mediated autocrine activation initiates (Fig. 27.2). By 36–72 h blast transformation begins, accompanied by expression of HLA-DR on the cell surface. Note that several of these changes may be detected by flow cytometry, especially tracking the progression of T cell activation via the expression of CD69, CD25, and HLA DR. In fact, such monitoring may essentially replace the previous gold standard of determination of T cell proliferation by [3H]thymidine incorporation. In a prior study, our laboratory monitored early T cell activation potential in US astronaut crew members, both short and long duration, immediately following spaceflight. The data showed that long-duration crew members had significantly blunted T cell function postflight, whereas T cell function was actually increased postflight for short-duration crew members (Crucian et al. 2008). The difference was attributed to the immunostimulatory effects of acute stress in less deconditioned crew members, versus the effect of chronic stress in severely deconditioned crew members. Since cultured cells may be analyzed by flow cytometry as easily as whole blood, there is no real limit to the functional measures that may be performed by flow cytometry.

In addition to postflight assessments of crew members, or postflight processing of in-flight samples, spaceflight effects may also be studied by culturing cells on-orbit. Following culture, astronauts may preserve the cultured cells by a variety of means, to be followed by ground-based cytometry analysis upon return to Earth. Hashemi et al. investigated T cell progression during the first 24 h of activation by culturing T cells during spaceflight onboard the Biorack facility, Space Shuttle missions STS-81 and 84. T cells were stimulated with soluble antibodies to CD3 to trigger the TCR. Similar to the postflight astronaut study described above, expression of both CD69 and CD25 were measured by flow cytometry. Culture during spaceflight showed a dramatic reduction in expression of both CD69 and CD25 following activation (Hashemi et al. 1999). Complementation of TCR-mediated signaling by phorbol ester restored the ability of the T cells to become fully activated, indicating a protein kinase C (PKC)-associated pathway could be compromised during microgravity conditions.

In an example of the high versatility of flow cytometers, our laboratory has developed an improved flow cytometry assay for NK function that is being used during a current ISS in-flight study. The previous standard assay for NK function involved coculturing "target cells" (K562 cell line) and human subject peripheral mononuclear cells (containing NK), to induce NK killing of the target cells. The target cells were radio-labeled, and killing was determined by measuring radioactivity released by the lysed target cells into the culture medium. Caveats to this assay include a varying number of NK within the peripheral blood mononuclear cell (PBMC) population, and the requirement for large numbers of subject cells to coculture varying ratios of PBMC:target cells in order to generate a kinetics curve for NK activity. For our current spaceflight study, the numbers of available astronaut cells are very low. For this reason, and to eliminate the requirement to work with radioactive materials, a nonradioactive flow cytometry method has utility for spaceflight studies. Nehlsen-Cannarella et al. developed an early nonradioactive flow cytometry assay, which used membrane dye labeling to identify target cells (Chang et al. 1993). Target cells were still incubated with subject cells; however, killing was measured by propidium iodide (PI) intercalation into target cell DNA via red fluorescence. To meet the requirements of our spaceflight study, we enhanced the assay by labeling NK with phycoerytherin (PE) labeled anti-CD16/56 antibodies, labeling the target cells with fluorescein isothiocyanate (FITC) labeled anti-CD71 antibodies. Killing is still determined by membrane permeation of PI via red fluorescence. Thus, a three color cytometry assay is created which can positively identify exact numbers of NK (not bulk PBMCs), live target cells and dead target cells (Crucian et al. 2006b ISAC, Fig. 27.3). This improved assay is appropriate for spaceflight studies, as minimal crewmember cells may be used. In addition, although measuring killing across multiple NK:target ratios is beneficial, it is not absolutely required since the precise number of astronaut NK and target cells is known. Using this method, an accurate killing percentage normalized to any NK value may be derived without assuming the PBMC population to accurately represent the NK population.

27.2.3 Cytokine Assays

A separate outcome from immune cell activation, distinct from the blastogenesis-related changes described above, is cytokine production. Cytokines are small

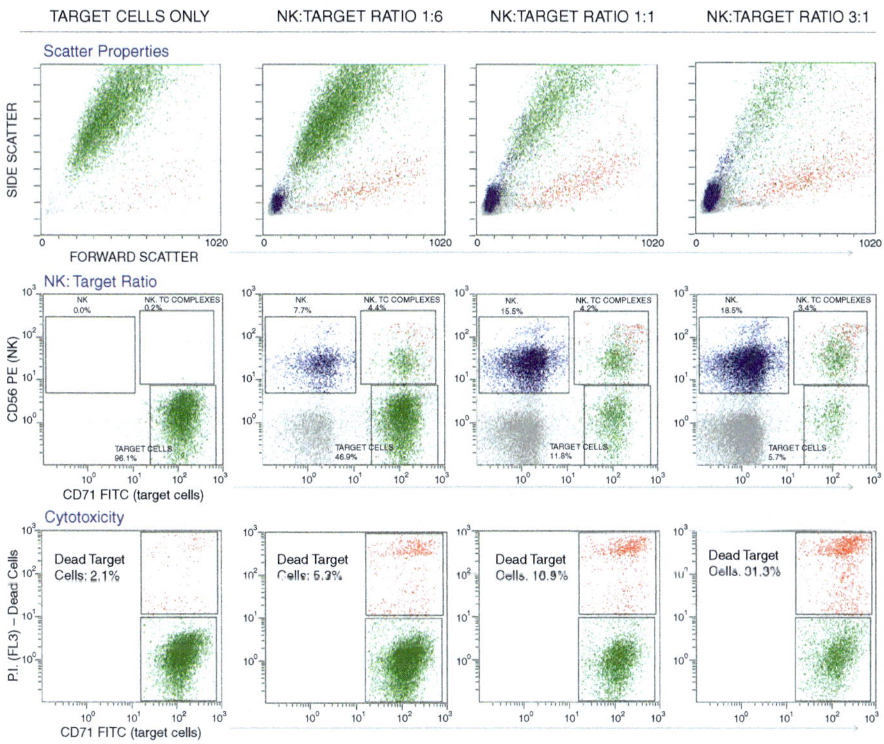

Fig. 27.3 A nonradioactive method for determining NK function by flow cytometry. Separate data are presented for four cell cultures containing free target cells and 3× increasing NK:target ratios (columns left to right, respectively). Fluorescent antibody staining is used to positively identify target cells (CD71-FITC/*green*) and NK (CD56-PE/*blue*). This allows a precise NK:target ratio to be determined for each coculture analyses (*middle row*). Note that free NK (*upper left gate*), free target cells (*bottom right gate*), and NK:target complexes (*upper right gate*) are all realized in the ratio scatter plots (*middle row*). Following culture and NK-mediated killing of target cells, dead target cells are identified by membrane disruption using the DNA intercalating dye PI (*red*). The percentage of dead target cells may be quantified (*bottom row*). Back-immunoscatter gating may be used to track live versus dead target cells via scatter properties, with the target cells manifesting altered scatter properties as they are killed (*top row*). This assay will be applied to a current study returning crewmember samples from the ISS

molecules secreted by immune cells which cause effects in other immune cells. Essentially they are the hormones, or communication messengers, of the immune system. Cytokine biology is incredibly complex, with dozens of cytokines secreted by a very diverse network of specific immune cell subsets. Examples of cytokine categories include the pro-inflammatory cytokines of the innate immune system, and the Th1, Th2, Th17 cytokines secreted by various T cell subsets. Following cell activation during mitogenic stimulation culture, techniques such as ELISA or cytometric bead array measure cytokine levels secreted into the culture supernatant. Secreted cytokine assays measure the net level following secretion, reabsorption, and molecule breakdown. Cytokine secretion may be halted however, by adding secretion blockers during activation such as monensin or brefeldin A. This allows intracellular cytokine accumulation in the cell cytoplasm. Following such culture, it is possible to detect intracellular cytokine production by flow cytometry. For this assay, surface markers are stained first, followed by cellular fixation, and then intracellular staining consisting of a permeabilization reagent and anti-cytokine antibodies. Preceding fixation is required to prevent the intracellular contents from leaking out of the cell following permeabilization. Intracellular cytokine assays, by flow cytometry, measure the percentage of cells capable of being stimulated to produce a particular cytokine, not the bulk supernatant level. Cytometric bead array technology allows detection of several secreted cytokine within a single flow cytometry assay. Essentially, fluorescent bead populations with varying fluorescence intensity along a single channel are created, with each population coated with "capture" antibodies for a specific cytokine. By incubating with a common secondary detection antibody, cytokine levels may be measured independently for each cytokine along a separate channel. As measures of cytokine production by specific, positively identified cell types (intracellular), or as a measurement of bulk secreted levels of multiple cytokines, these assays have clear utility for determining the effects of spaceflight on immune system cytokine regulation. Also, this technique is problematic for some cytokines. For example, IFNγ and IL-2 are readily detectable following activation; however, IL-4 and IL-10 are extremely difficult. This is likely due to the normal pre-programmed secretion patterns the cells will exhibit. Examining secretion by ELISA also yields low levels for the same cytokines that are difficult to detect by cytometry. Examples of utilization follow.

Our laboratory has investigated intracellular production of IFNγ and IL-2 in both CD4+ and CD8+ T cell subsets postflight for Shuttle and ISS astronauts. In general, following spaceflight crewmembers had reduced levels of both CD4+ and CD8+ T cells capable of producing both IFNγ and IL-2 (Crucian et al. 2008, see also Chap. 14). Flow cytometry scatter plots showing the gating strategy and analysis technique for a representative ISS crewmember are shown in Fig. 27.4. Kaur et al. examined monocyte intracellular cytokine production of IL-1β, IL-1ra, IL-6, and IL-8 postflight for Space Shuttle astronauts (Kaur et al. 2008). The stimulus for cytokine production was gram-negative bacterial endotoxin, known to be a powerful activator of monocytes via LPS binding to the CD14 cell surface antigen. Crewmember monocyte production of IL-6 and IL-1β were reduced compared to controls at all three mission associated timepoints. In contrast, crewmember production of IL-1ra and IL-8 were elevated as compared to controls.

In addition to measuring intracellular cytokines for the NASA study described above, secreted production of six cytokines was measured by cytometric bead array (CBA). CBA technology analyzes multiple secreted cytokines by using capture antibodies fixed to bead populations which vary in fluorescence along a single fluorescence channel. Detection antibodies then fluoresce along a different channel, allowing a 2 × 2 cytometry dot plot with cytokine specificity along the X axis, and cytokine production along the Y axis. The array used by NASA consisted of Th1/Th2 cytokines: IFNγ, TNF, IL-10, IL-5, IL-4, and IL-2. Crewmember cells were treated with either antibodies to the TCR (CD3/CD28) to stimulate T cells only, or

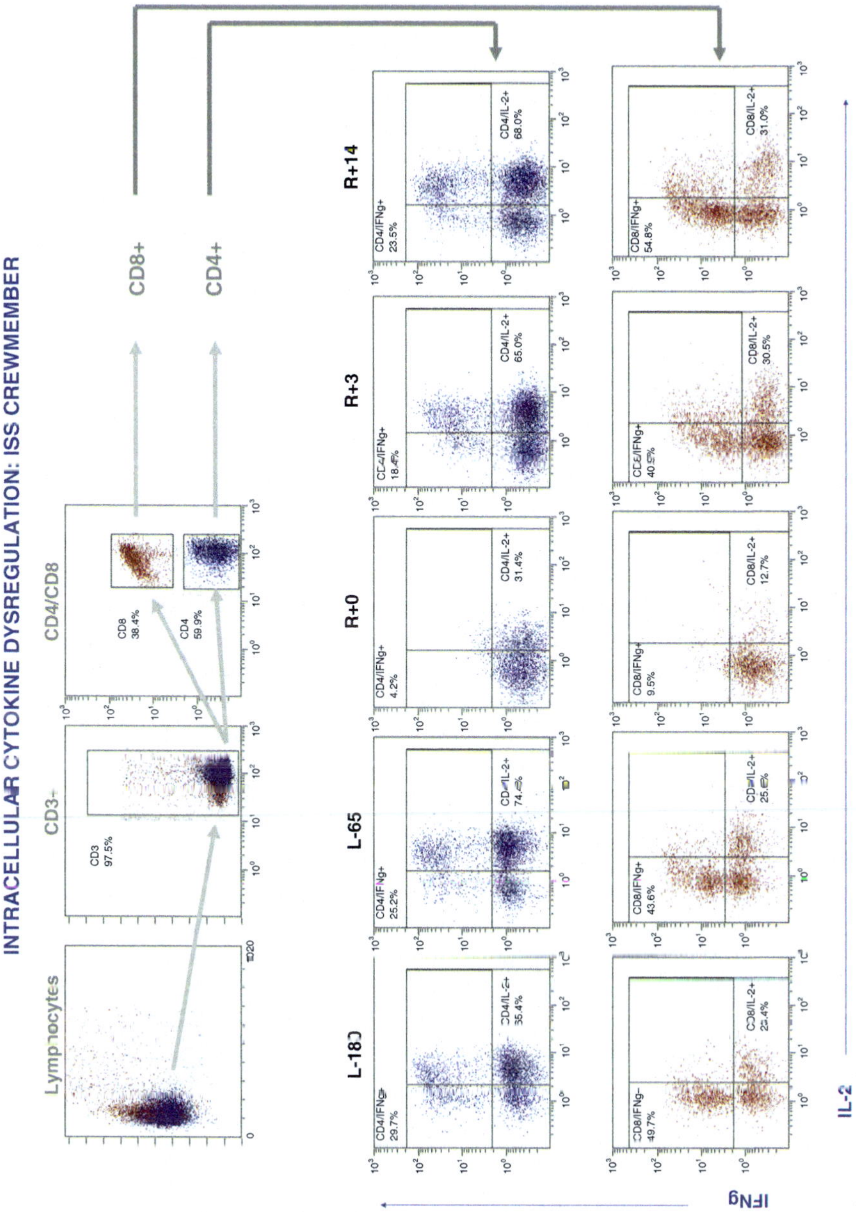

Fig. 27.4 Intracellular detection of IFNγ IL-2 in both CD4+ and CD8+ T cell subsets by flow cytometry. Representative postflight data from a single ISS crewmember. Following mitogenic stimulation in the presence of monensin to halt extracellular secretion, surface markers are (in order) stained, fixed, permeabilized, and followed by staining of intracellular cytokines

PMA + ionomycin to stimulate all leukocyte populations. Surprisingly, there were minimal postflight alterations for most cytokines, except a reduction in IFNγ for both Shuttle and ISS crewmembers (Crucian et al. 2008). Reduced production of in IFNγ associated with spaceflight has been reported by other (nonflow cytometry) techniques, and may be directly related to immune loss of control of latent viral reactivation. As important indicators of immune function in general, and Th1:Th2 or pro- /anti-inflammatory biases specifically, measurements of cytokine production have significant utility for defining spaceflight-associated immune dysregulation. This bead-based assay is performed on a traditional flow cytometer, however specialized "multiplex" cytometer-like instruments are now commonly used allowing much greater spectrum of simultaneous analyses. For example, the Luminex "Magpix" may analyze up to 44 cytokines using a bead platform. For these instruments, there is no manual gating or analysis, the process is automated.

27.2.4 Viral-Specific Immunity

Measurements of virus-specific T cells, either their direct enumeration from blood samples, or their functional capabilities, may have utility for monitoring spaceflight-associated immune dysregulation. There is ample evidence that reactivation of latent herpesviruses occurs during spaceflight (see Chap. 19). The reactivation of latent viruses is very likely a direct result of diminished immune function, particularly CD8+ cytotoxic T cells. Several of the viruses are known to reactivate during spaceflight, like EBV, CMV, and VZV, and have potential direct clinical consequences should reactivation persist for very long periods of time. Should spaceflight-associated immune compromise persist for the duration of a Mars mission, crews may be at risk for these adverse clinical events. For these reasons, flow cytometry assays to detect viral-specific T cells have utility for monitoring virus-specific immunity, as current studies define the in-flight phenomenon and potentially devise strategies to monitor countermeasures validation. Stowe et al. have monitored EBV and CMV specific immunity (both number and function) during a recent postflight NASA study. This study was conducted on both short- and long-duration US astronauts, and monitored numbers of virus-specific T cells via the tetramer method, and virus-specific T cell function via a peptide stimulation assay (Crucian et al. 2009). MHC tetramers consist of multimers of MHC molecules coupled to a particular viral protein and tagged with a fluorescent label. These reagents function similar to labeled antibodies, except that they bind to antigen-specific CD8+ T cells only. For that reason, the detection population is extremely small. For CMV peptide or EBV peptide-specific T cells, normal ranges in seropositive individuals may range from 0.1% to 0.6% of the total CD8+ T cell population, making this rare-event cytometry (Fig. 27.5). As such, a gating strategy employing multiple sequential gates with populations identified by immunoscatter to thoroughly resolve target cells from debris is essential. Peptide stimulation to measure viral-specific T cell function, consists of cell culture stimulation of astronaut peripheral blood cells with a particular viral peptide and antibodies to costimulatory molecules (Crucian et al.

2009 bed rest for method, Crucian et al. 2001 peptide article). Antigen presentation of the viral peptides occurs naturally in the culture, and only viral peptide-specific T cells are activated. Cultures are processed in the presence of either monensin or brefeldin. A similar to the intracellular cytokine assay. Following culture, a cytometry staining protocol consisting of cell surface CD3 and CD8, and intracellular IFNγ is performed to resolve activated peptide-specific T cells via production of cytokine (Fig. 27.5). When the viral peptide is a superantigen, as are several EBV peptides, concurrent staining of CD69 allows a monitoring of nonspecific superantigen activation versus legitimate viral-peptide activation (Crucian et al. 2001). The data to date indicate that, in general, numerous virus-specific T cells rise or are unchanged, whereas their functional capacity decreases (Stowe et al. 2007, HRP, Stowe et al. 2008, HRP). This at least partially explains the in-flight reactivation of latent viruses. These viral-specific assays are also being conducted as part of a current in-flight NASA study onboard the ISS (Crucian et al. 2010).

27.2.5 Bacterial Analysis

Flow cytometry is a versatile technique that may be applied to analyze nearly any particles in solution, provided the particles are in suspension and within the size limits for detection. The use of antibodies, fluorochromes, and membrane/functional dyes adds utility to the basic particle imaging via scattered laser light. Even bacteria, with most species within the 2–10 μm size range, may be detectable by some flow cytometers. For example, fully automated online cytometric approaches can evaluate, reliably, sensitive microbial dynamics in urine (Yang et al. 2017; Müller et al. 2018), as well as natural reservoirs of water (Besmer et al. 2014, 2016).

Several bacterial flow cytometry assays have been applied to spaceflight experiments. Nickerson et al. flew *Salmonella typhimurium* onboard Space Shuttle mission STS-115, and used flow cytometry to measure bacterial suspension counts postflight (Wilson et al. 2007). SYTO-BC dye, a membrane permeant dye which stains both gram+ and gram– bacteria was used for the assay. The most recent and comprehensive of the spaceflight bacterial studies which included cytometric measurements was performed by Leys et al. In this study, *Cupriavidus metallidurans* bacterium were grown for 10–12 days onboard the ISS. Upon return, flown bacterial cells were compared to ground control bacterial cells using a variety of flow cytometric techniques (Leys et al. 2009). Viability, membrane potential, intracellular pH, electron transport chain function, superoxide/peroxide detection, intracellular glutathione, intracellular calcium, DNA versus RNA concentration, and intracellular esterase activity were all measured by a standard flow cytometer equipped with a 15 mW 488 nm laser. Commonly, these assays employ dyes which manifest a color change in the presence or absence of the measured condition, or presence of the measured analyte. By employing multicolor cytometry, separate channels may be used to resolve bacteria from artifactual debris, and to measure the analyte of interest. A properly tuned optical path with very low cV values is necessary to resolve bacteria on a standard flow cytometer.

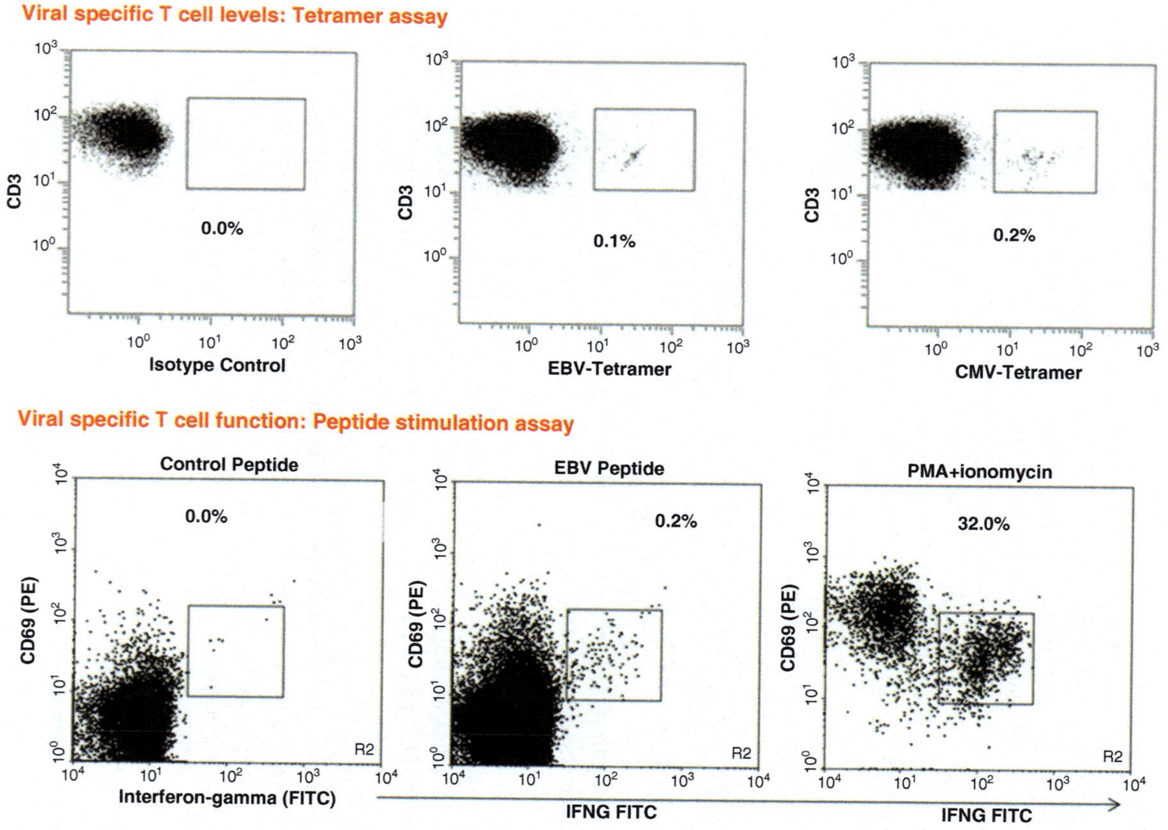

Fig. 27.5 Measurement of viral-specific immunity by flow cytometry. *Top row*: detection of CD8+ T cells specific for viral-peptides via the MHC-tetramer method. *Bottom row*: determination of viral-specific T cell function by viral-peptide stimulation followed by intracellular measurement of IFNγ production

27.3 Development of an In-Flight Flow Cytometer

Assuming the proven utility of some, or all, of the above described assays for monitoring immune dysregulation during spaceflight with flow cytometry, the entire development effort may be diminished without onboard monitoring hardware during exploration class space missions. Should the effect of spaceflight on immunity persist for very long periods of time, a spaceflight capable flow cytometer may be required to monitor crewmember immunity during missions beyond low Earth orbit. Unfortunately, standard commercially available flow cytometers are large, complex instruments that use high-energy lasers and require liters of liquid "sheath" fluid to operate. They also generate a significant amount of liquid biohazardous waste and require significant training to operate. Cytometers use the fluid mechanical property of "hydrodynamic focusing" to place stained blood cells in single-file (laminar flow) as they pass through a laser beam for scanning and evaluation. Many spaceflight experiments have demonstrated that fluid physics is dramatically altered in microgravity and previous studies have shown that sheath-fluid-based hydrodynamic focusing may also be altered during microgravity (Crucian et al. 2000b). For these reasons, it is likely that any spaceflight compatible design for a flow cytometer would abandon the sheath fluid requirement. The elimination of sheath fluid would remove both the problems of weight associated with large volumes of liquids as well as the large volume of liquid waste generated. It would also create the need for a method to create laminar particle flow distinct from the standard sheath-fluid-based method.

NASA has at several times investigated the possibility of an on-orbit functioning flow cytometer. In 1985, a panel convened at the Johnson Space Center (JSC) generated an extensive feasibility report, but no construction was initiated. However in the mid-2000s, an early prototype of a working on-orbit cytometer was created at the JSC. Fortuitously, a commercial cytometer was developed by Guava Technologies, which was extremely miniaturized and has eliminated the sheath fluid requirement. Although an innovative advance in technology, this instrument had primitive capabilities (2 color, 3 parameter) and was not spaceflight compatible due to fluidics handling, power requirements, sample delivery, size, and data handling. The JSC effort was designed to (1) engineer a prototype flight instrument based on the framework of the commerial off-the-shelf (COTS) cytometer; (2) perform ground-based and microgravity validation of the instrument; (3) design and validate a set of medical assays compatible with the prototype instrument; (4) design and validate a microgravity compatible cell staining device for sample processing that could interface with the instrument. Upon conclusion, in 2007, a miniaturized prototype was indeed developed (Fig. 27.7 left image) and validated during microgravity conditions (Fig. 27.6), using parabolic flight aircraft. The prototype flight cytometer was found to perform cytometry adequately in microgravity, moon, Mars, and even hypergravity conditions (Crucian and Sams 2005; Crucian et al. 2006a). In addition, the Whole Blood Staining Device (WBSD), designed to perform antibody staining, RBC lysis, and WBC fixation during microgravity conditions, was created. An early version of the WBSD was validated on-orbit during Space Shuttle mission STS-71

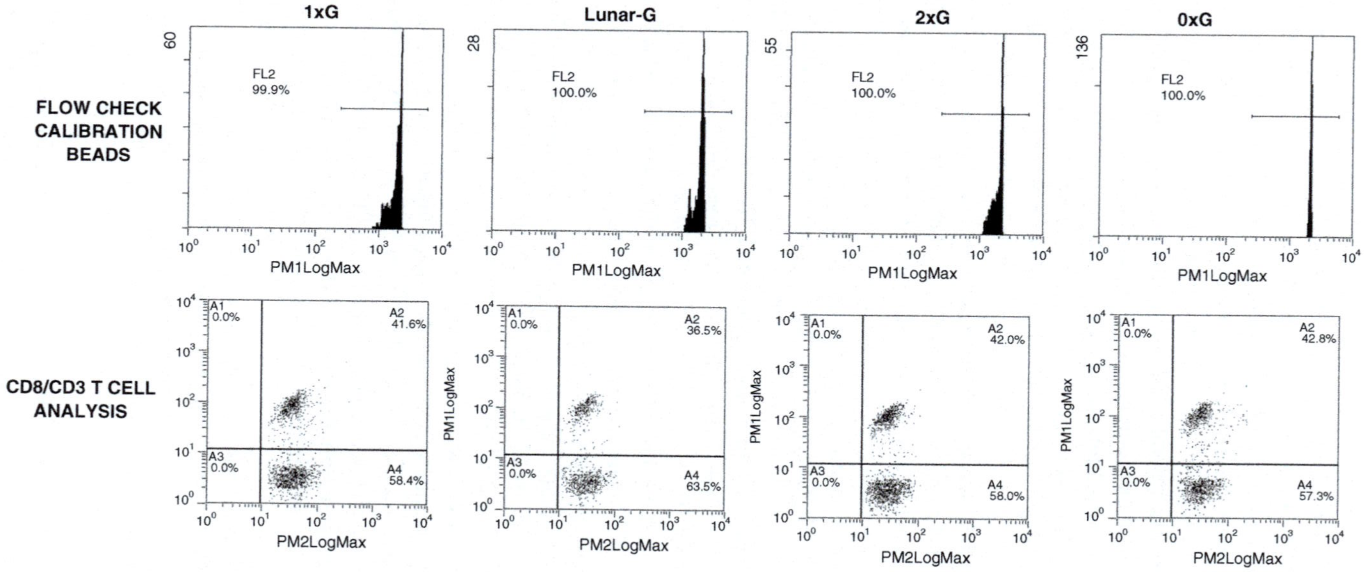

Fig. 27.6 Representative flow cytometry histograms and dot plots demonstrate performance of the NASA prototype flight cytometer in 1×g, lunar-g, 2×g, and microgravity conditions. *Top row*: assessment of flow cytometry calibration beads demonstrates detection stability and optical resolution; *bottom row*: detection of CD4+ T cells

(Sams et al. 1999). Subsequently, the WBSD was further modified to successfully interface with the JSC prototype flight cytometer and allow validation of blood-to-data sample processing and analysis during microgravity conditions.

Subsequent to the JSC effort, the Canadian Space Agency (CSA) developed a working demonstration unit flow cytometer "Microflow" which flew onboard ISS (Fig. 27.7 right image). Microflow was not a working unit such that samples could be introduced, it was a module designed to validate the fluidics in actual spaceflight conditions. For the demonstration, fluorescent beads and stained cells were already onboard. The unit was activated during flight to perform the pre-programmed analyses. In short, the Microflow demonstration was successful, and validated that the cytometer fluidics functioned appropriately during spaceflight. Unfortunately the planned fully usable flow cytometer envisioned to follow Microflow was not constructed.

NASA has funded several small business innovation grants for cytometer-like devices, many using microfluidics technology. Shi et al. (2009) are working to create a microgravity compatible cell analysis device. Their device uses a portable microflow cytometer system to create a two-part leukocyte differential and leukocyte count. Leukocytes are stained with Acridine Orange and blood is pumped through a disposable microfluidics device for fluorescence sensing. Under 460 nm LED excitation, DNA (green) and RNA (red) fluorescence are collected via two PMTs. Leukocytes are counted and differentiated into lymphocytes and non-lymphocytes based on fluorescence signatures. This technology has been microgravity validated and could serve as a platform for development of a capable in-flight cytometer instrument.

To date, there is no usable flow cytometer capability onboard ISS, but several efforts exist to deploy usable miniaturized laboratory instruments onboard ISS, either for cellular, protein or nucleic acid analysis. For example, CSA developed in 2017 an apparatus named The Bio-Analyzer, which quantifies both leukocyte

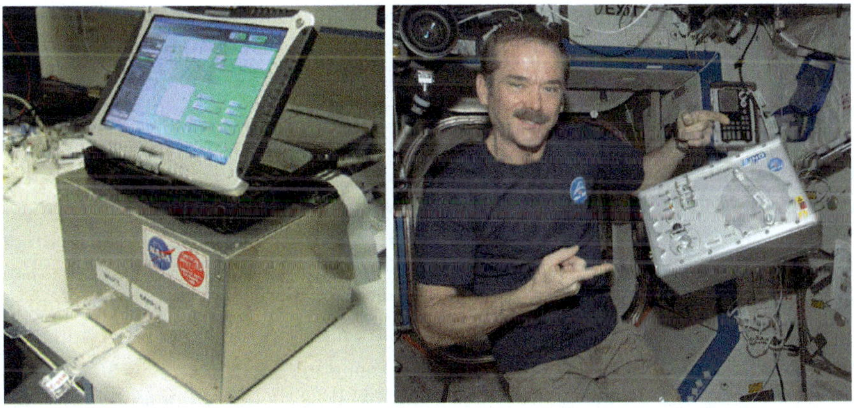

Fig. 27.7 NASA-JSC prototype flight cytometer with sample delivery WBSD units affixed to the sample delivery ports (left); and the Canadian Space Agencies "Microflow" demonstration unit onboard ISS (right)

subset percentages and soluble protein biomarkers using separate technology within a single instrument. The machine is not a flow cytometer per se; it uses microfluidic microchips to fluorescently label proteins of interest within a small-volume sample, and then quantifies them via imaging analysis. Still, the instrument has been commissioned for a CSA-NASA flight study to enumerate bulk leukocyte subsets, beginning with Expedition 57S in early 2019.

There are several recently developed commercial miniaturized traditional flow cytometers. In 2018 NASA performed a ground evaluation of a one-laser, 3-parameter, instrument from Orflo Technologies and found it likely well suited for use onboard ISS. A pilot parabolic flight microgravity evaluation of this technology indicated the fluidics were insensitive to gravity. This unit has been selected for a flight demonstration on ISS to occur in 2019 or 2020.

Thus, the long-standing goal of performing routine real-time flow cytometry in outer space may be at hand. This capability would significantly improve reasearch capabilities onboard ISS.

The advent of flow cytometry capacity in space would be revolutionary for biological surveillance of astronauts in real-time. Furthermore, this capability in space would open the door for more complex research endeavors—in theory—including those that require cell sorting. Modern advances in microfluidics have enabled portable cytometer devices with sorting capacity that exploit a variety of physical principles (reviewed aptly by Shields et al. (2015) and Yu et al. (2014)). For example, there is an instrument that uses 3D electrodes on a disposable chip; cells passes through whereas those of interest are retained and then recovered (Faraghat et al. 2017). While not an immediate priority, cell sorting in space may be very useful 1 day in the distant future.

27.4 Conclusion

As demonstrated by countless ground-based reports and the spaceflight-specific studies described above, flow cytometry is a versatile technique that has significant utility for both defining, and monitoring, spaceflight-associated immune dysregulation. This phenomenon, if found to persist during exploration class space missions, had the potential to be a significant clinical risk for crewmembers. Prolonged immune dysregulation could lead to infections, hypersensitivities, autoimmunity, malignancy, or other adverse events. The danger is likely heightened during deep space travel considering the high-energy radiation environment and lack of clinical care or return option. Cytometry-based assays, and a working in-flight flow cytometer, also in possible conjunction with other possible immune analyzers (e.g., multiplex-based, microfluidic chambers) using by individualized in vitro incubations platform (e.g. KUBIK), could altogether have significant utility for defining the on-orbit phenomenon, as well as the development and validation of countermeasures.

References

Besmer MD, Weissbrodt DG, Kratochvil BE, Sigrist JA, Weyland MS, Hammes F (2014) The feasibility of automated online flow cytometry for in-situ monitoring of microbial dynamics in aquatic ecosystems. Front Microbiol 5:265. https://doi.org/10.3389/fmicb.2014.00265. eCollection 2014. PubMed PMID: 24917858; PubMed Central PMCID: PMC4040452

Besmer MD, Epting J, Page RM, Sigrist JA, Huggenberger P, Hammes F (2016) Online flow cytometry reveals microbial dynamics influenced by concurrent natural and operational events in groundwater used for drinking water treatment. Sci Rep 6:38462. https://doi.org/10.1038/srep38462. PubMed PMID: 27924920; PubMed Central PMCID: PMC5141442

Chang L, Gusewitch GA, Chritton DB, Folz JC, Lebeck LK, Nehlsen-Cannarella SL (1993) Rapid flow cytometric assay for the assessment of natural killer cell activity. J Immunol Methods 166(1):45–54

Crucian B, Sams C (2005) Reduced gravity evaluation of potential spaceflight-compatible flow cytometer technology. Cytometry B Clin Cytom 66(1):1–9

Crucian BE, Cubbage ML, Sams CF (2000a) Altered cytokine production by specific human peripheral blood cell subsets immediately following space flight. J Interf Cytokine Res 20(6):547–556

Crucian B, Norman J, Brentz J, Pietrzyk R, Sams C (2000b) Laboratory outreach: student assessment of flow cytometer fluidics in zero gravity. Lab Med 31(10):569–573

Crucian BE, Stowe RP, Pierson DL, Sams CF (2001) Routine detection of Epstein-Barr virus specific T-cells in the peripheral blood by flow cytometry. J Immunol Methods 247(1–2):35–47

Crucian B, Guess T, Nelman-Gonzalez M, Sams C (2006a) C-9 and other microgravity simulations: prototype flight cytometer. NASA publication TM-2006-213727: 142–145. PFC online report. http://www.nasa.gov/centers/johnson/pdf/505835main_FY06_TM-2006-213727c.pdf

Crucian B, Nehlsen-Cannarella S, Sams C (2006b) An improved flow cytometry method for precise quantitation of natural-killer cell activity. In: ISAC International Congress, Quebec, 20–26 May, 2006

Crucian BE, Stowe RP, Pierson DL, Sams CF (2008) Immune system dysregulation following short – vs long-duration spaceflight. Aviat Space Environ Med 79(9):835–843

Crucian BE, Stowe RP, Mehta SK, Yetman DL, Leal MJ, Quiriarte HD et al (2009) Immune status, latent viral reactivation, and stress during long-duration head-down bed rest. Aviat Space Environ Med 80(5 Suppl):A37–A44

Crucian B, Mehta S, Stowe R, Uchakin P, Quiriarte H, Pierson D, et al (2010) Validation of procedures for monitoring crewmember immune function. In: NASA Human Research Program Investigators Workshop, 3–5 Feb 2010, Houston

Crucian B, Stowe RP, Mehta S, Quiriarte H, Pierson D, Sams C (2015) Alterations in adaptive immunity persist during long-duration spaceflight. NPJ Microgravity 1:15013. https://doi.org/10.1038/npjmgrav.2015.13. eCollection 2015. PubMed PMID: 28725716; PubMedCentral PMCID: PMC5515498

De Rosa SC, Herzenberg LA, Roederer M (2001) 11-color, 13-parameter flow cytometry: identification of human naive T cells by phenotype, function, and T-cell receptor diversity. Nat Med 7(2):245–248

Faraghat SA, Hoettges KF, Steinbach MK, van der Veen DR, Brackenbury WJ, Henslee EA, Labeed FH, Hughes MP (2017) High-throughput, low-loss, low-cost, and label free cell separation using electrophysiology-activated cell enrichment. Proc Natl Acad Sci U S A 114(18):4591–4596. https://doi.org/10.1073/pnas.1700773114. Epub 2017 Apr 13. PubMed PMID: 28408395; PubMed Central PMCID: PMC5422786

Gridley DS, Nelson GA, Peters LL, Kostenuik PJ, Bateman TA, Morony S et al (2003) Genetic models in applied physiology: selected contribution: effects of spaceflight on immunity in the C57BL/6 mouse. II. Activation, cytokines, erythrocytes, and platelets. J Appl Physiol 94(5):2095–2103

Gridley DS, Slater JM, Luo-Owen X, Rizvi A, Chapes SK, Stodieck LS et al (2009) Spaceflight effects on T lymphocyte distribution, function and gene expression. J Appl Physiol 106(1):194–202

Hashemi BB, Penkala JE, Vens C, Huls H, Cubbage M, Sams CF (1999) T cell activation responses are differentially regulated during clinorotation and in spaceflight. FASEB J 13(14):2071–2082

Kaur I, Simons ER, Castro VA, Ott CM, Pierson DL (2005) Changes in monocyte functions of astronauts. Brain Behav Immun 19(6):547–554

Kaur I, Simons ER, Kapadia AS, Ott CM, Pierson DL (2008) Effect of spaceflight on ability of monocytes to respond to endotoxins of gram-negative bacteria. Clin Vaccine Immunol 15(10):1523–1528

Leys N, Baatout S, Rosier C, Dams A, s'Heeren C, Wattiez R et al (2009) The response of *Cupriavidus metallidurans* CH34 to spaceflight in the international space station. Antonie Van Leeuwenhoek 96(2):227–245

Mills PJ, Meck JV, Waters WW, D'Aunno D, Ziegler MG (2001) Peripheral leukocyte subpopulations and catecholamine levels in astronauts as a function of mission duration. Psychosom Med 63(6):886–890

Müller M, Seidenberg R, Schuh SK, Exadaktylos AK, Schechter CB, Leichtle AB, Hautz WE (2018) The development and validation of different decision-making tools to predict urine culture growth out of urine flow cytometry parameter. PLoS One 13(2):e0193255. https://doi.org/10.1371/journal.pone.0193255. eCollection 2018. PubMed PMID: 29474463; PubMed Central PMCID: PMC5825091

Ortega MT, Pecaut MJ, Gridley DS, Stodieck LS, Ferguson V, Chapes SK (2009) Shifts in bone marrow cell phenotypes caused by spaceflight. J Appl Physiol 106(2):548–555

Pecaut MJ, Nelson GA, Peters LL, Kostenuik PJ, Bateman TA, Morony S et al (2003) Genetic models in applied physiology: selected contribution: effects of spaceflight on immunity in the C57BL/6 mouse. I. Immune population distributions. J Appl Physiol 94(5):2085–2094

Sams CF, Crucian BE, Clift VL, Meinelt EM (1999) Development of a whole blood staining device for use during space shuttle flights. Cytometry 37(1):74–80

Shi W, Zhenk S, Kasdan HL, Fridge A, Tai YC (2009) Leukocyte count and two-part differential in whole blood based on a portable microflow cytometer. In: IEEE, Transducers, Denver, 21–25 June 2009, pp 616–619

Shields CW 4th, Reyes CD, López GP (2015) Microfluidic cell sorting: a review of the advances in the separation of cells from debulking to rare cell isolation. Lab Chip 15(5):1230–1249. https://doi.org/10.1039/c4lc01246a. Review. PubMed PMID:25598308; PubMed Central PMCID: PMC4331226

Stowe RP, Sams CF, Mehta SK, Kaur I, Jones ML, Feeback DL et al (1999) Leukocyte subsets and neutrophil function after short-term spaceflight. J Leukoc Biol 65(2):179–186

Stowe RP, Sams CF, Pierson DL (2003) Effects of mission duration on neuroimmune responses in astronauts. Aviat Space Environ Med 74(12):1281–1284

Stowe R, Kozlova EV, Walling DM, Sams C, Pierson D (2007) Epstein-Barr virus gene expression in astronauts. In: NASA Human Research Program Workshop, Texas

Stowe R, Sams C, Pierson D (2008) Cytomegalovirus reactivation in astronauts. In: NASA Human Research Program Workshop, Texas

Wilson JW, Ott CM, Honer Zu Bentrup K, Ramamurthy R, Quick L, Porwollik S et al (2007) Space flight alters bacterial gene expression and virulence and reveals a role for global regulator Hfq. Proc Natl Acad Sci U S A 104(41):16299–16304

Yang CC, Yang SS, Hung HC, Chiang IN, Peng CH, Chang SJ (2017) Rapid differentiation of cocci/mixed bacteria from rods in voided urine culture of women with uncomplicated urinary tract infections. J Clin Lab Anal 31(5). https://doi.org/10.1002/jcla.22071. Epub 2016 Nov 15

Yu ZT, Aw Yong KM, Fu J (2014) Microfluidic blood cell sorting: now and beyond. Small 10(9):1687–1703. https://doi.org/10.1002/smll.201302907. Epub 2014 Feb 10. Review. PubMed PMID: 24515899; PubMed Central PMCID: PMC4013196

28 Assessment of Radiosensitivity and Biomonitoring of Exposure to Space Radiation

Roel Quintens, Sarah Baatout, and Marjan Moreels

28.1 Individual Radiosensitivity

Currently, the use of the terms radiation sensitivity and susceptibility is being debated (Foray et al. 2016; Wojcik et al. 2018). Some authors argue that radiation sensitivity should only be used for tissue reactions following cell death, while radiation susceptibility is the proneness to develop radiation-induced cancer (Foray et al. 2016). In the remainder of this chapter, we will use the term radiosensitivity for both cases.

It is well known from clinical practice as well as from radiobiological and toxicological experiments with high doses that each person reacts differently to radiation and drugs. The International Commission on Radiological Protection has estimated that between 5% and 15% of the population may be carriers of genetic mutations conferring them more radiosensitive (ICRP 2007). For manned spaceflight, and especially that beyond Low Earth Orbit, the individual biological responses to cosmic ray exposure may therefore be of critical concern for health-risk assessment of astronauts.

Interindividual differences in radiosensitivity have been reported with an increased incidence of both deterministic (e.g., tissue reactions like erythema or fibrosis) and stochastic effects (e.g., cancer, see also Chap. 20). Several studies demonstrated that the occurrence and severity of normal tissue reactions in response to ionizing radiation are influenced mainly by genetic susceptibility. In this context, variations in radiosensitivity between individuals may be a consequence of deficiencies in DNA repair capacity. This can be caused by specific mutations or polymorphisms in DNA repair genes such as ataxia telangiectasia mutated (*ATM*) (Taylor et al. 1975) or nibrin (*NBN* or *NBS1*), which is mutated in Nijmegen Breakage Syndrome (Chistiakov et al. 2008; Varon et al. 1998). They can also result from a

R. Quintens (✉) · S. Baatout · M. Moreels
Radiobiology Unit, Belgian Nuclear Research Centre, SCK•CEN, Mol, Belgium
e-mail: roel.quintens@sckcen.be

© The Author(s) 2020

A. Choukér (ed.), *Stress Challenges and Immunity in Space*,
https://doi.org/10.1007/978-3-030-16996-1_28

combination of small variations in multiple genes (Andreassen 2005), such as those functioning in specific DNA double-strand break (DSB) repair pathways including poly (ADP-ribose) polymerase (*PARP*), DNA-dependent protein kinase, catalytic subunit (*PRKDC*), *TP53*, or ATM and Rad3-related (*ATR*).

For a radiation-related health-risk assessment for astronauts, methods that can measure and/or predict individual radiosensitivity would be of great value. In the future such assays could be used as predictive tests to identify astronauts with a greater than average risk of developing radiation-induced health effects. Moreover, the identification of (predictive) biomarkers of radiation sensitivity could also be relevant for cancer patients treated with radiotherapy. With the growing interest in personalized medicine treament plans could be better tailored to the individual patients based on their personal radiation sensitivity.

28.2 Tools to Evaluate Radiosensitivity and Exposure Dose

Currently, a number of tests have been described and used to assess responses to radiation. In this light, peripheral blood lymphocytes are the preferred target for assessing radiosensitivity, as they are among the most radiosensitive cell types in humans (Venkatesulu et al. 2018). Biological endpoints such as chromosomal aberrations (e.g., dicentrics, translocations, micronuclei, DNA fragmentation) and the repair capacity of radiation-induced damage have been used to determine individual radiosensitivity in vitro. These methods mostly rely on previous observations that the occurrence of chromosomal aberrations is associated with increased radiosensitivity. Furthermore, novel methods such as gene expression profiles both at the mRNA and protein level, as well as epigenetic changes may be suitable as radiation sensitivity biomarkers. Besides this, these same methods can also be used as biodosimeters, i.e., as biological measurements to predict the dose to which an individual has been exposed.

Thus, different methods can be used for the detection of radiation-induced damage or biological effects and their choice depends on several parameters (exposure duration, estimated absorbed dose, type of exposure, and time since exposure). Importantly, most of these methods can be used for rather low doses from about 0.1 or 0.2 Sv and upward, which is within the range of what would be encountered by astronauts on interplanetary missions. Other important advantages of using biological markers for assessing radiation exposure over physical dosimeters are that they take into account shielding by the body itself, and that they are sensitive to all types of radiation, including neutrons.

28.2.1 Chromosomal Aberrations: Dicentrics and Stable Translocations

The use of cytogenetic tests to monitor the occurrence of chromosomal aberrations induced by ionizing radiation dates back to the 1960s when Tough et al. described chromosomal aberrations in the blood of patients who had undergone radiotherapy to treat ankylosing spondylitis (Tough et al. 1960). To date, chromosomal aberrations have been widely accepted as biomarkers of exposure to ionizing radiation. Moreover, the frequency of radiation-induced chromosomal aberrations is associated with overall cancer risk (Distel et al. 2006; Tucker 2008) suggesting that they may also be used as indicators for individual radiosensitivity.

Two of the most used cytogenetics assays are the Dicentric Chromosome Assay (DCA) and fluorescent in situ hybridisation (FISH) for the analysis of stable chromosomal translocations. The DCA usually uses metaphase spreads of lymphocytes to score for chromosomes with two centromeres. Advantages of the DCA are the low background, high specificity to radiation, low interindividual variability, and high correlation between in vitro and in vivo dose estimations (Hall et al. 2017). Their half-life is between 6 months to a year, which makes them suitable for retrospective analysis after most ISS missions, but not for long-duration missions spanning several years.

Chromosomal translocations refer to the occurrence of chromosome breakage, followed by transfer of the broken-off portion, often to a different chromosome. They are usually induced by the formation of DNA DSBs and subsequent interaction of different DSBs with each other. The scoring of chromosomal translocations is mostly performed using FISH painting which uses fluorescently labeled DNA probes to mark specific regions in the DNA after which they can be visualized using fluorescence microscopy. Translocations are in general very stable, with a half-life of 6–12 years (Durante 2005) and can be transmitted by mitosis making them very suitable for retrospective biodosimetry (Awa 1997). However, the analysis of chromosomal translocations is also useful in susceptibility studies to identify genetic polymorphisms of genes involved in the individual response to ionizing radiation (Rodrigues et al. 2005) and cancer. Currently, chromosomal translocations are used in the clinic as biomarkers for cancer risk and they have been validated as intermediate endpoints of cancer. This shows the potential of using chromosomal translocations as biomarkers for radiation sensitivity as well. One of the disadvantages of using FISH compared to the DCA is the rather large interindividual variability of chromosomal translocations. However, since it is easy to obtain preexposure (i.e. pre-flight) samples from astronauts to calculate their individual background levels, this is not really a concern (Beaton-Green et al. 2015).

The use of cytogenetics assays to assess radiation effects in astronauts during spaceflight, have been applied since the late 1990's (Durante et al. 2003; Fedorenko et al. 2001; George et al. 2004, 2001; Obe et al. 1997; Testard et al. 1996; Yang et al. 1997). In general, these studies and more recent ones (Beaton-Green et al. 2015; Cucinotta et al. 2008; George et al. 2010, 2013) have shown that the number of chromosomal aberrations significantly increased during the mission, although there were large interindividual differences, which could not be attributed to the measured dose. This variability could be a consequence of differences in radiation sensitivity, but the experimental uncertainties were so high that the result could also be explained by statistical fluctuations (Durante 2005) because the doses to which the astronauts were exposed were close to the sensitivity thresholds of the assays (Cologne et al. 1998). It may therefore be anticipated that these methods would have better sensitivity in the case of missions of longer duration, when astronauts are exposed to higher dose equivalents. In this respect, it was shown that the occurrence of translocations increased in astronauts after repeated missions (George et al. 2013).

28.2.2 Micronuclei

The cytokinesis-blocked micronucleus (CBMN) assay is one of the most commonly used methods for the measurement of chromosome loss and breakage in nucleated cells (Fenech 2010). Micronuclei are small nuclear membrane-bound structures. They originate from acentric chromosome fragments or complete chromosomes which are unable to attach to the mitotic spindle during cytokinesis, thereby resulting in exclusion from the daughter nuclei into the cytoplasm. Since micronuclei are present in cells that have completed nuclear division, they are ideally scored in the binucleated stage of the cell cycle. To distinguish between nondividing cells and cells undergoing mitosis, cytochalasin-B, an inhibitor of the mitotic spindle that prevents cytokinesis, is added to the cell cultures. As a consequence, cells that have completed one nuclear division can be identified by their binucleated appearance. This leads to cells which contain two nuclei and one or more micronuclei if chromosome breaks have occurred or the centromere is damaged (Fenech and Morley 1985). New methods of detection, using high resolution in combination with flow cytometry may significantly enhance the usefulness of this assay by facilitating automation and quantification of the assay (Rodrigues et al. 2018).

Currently, the CBMN assay is being used successfully for biodosimetry after occupational, medical and accidental radiation exposure, although it is only useful after exposures above 200–300 mGy (Feng et al. 2015; Liu et al. 2009; Thierens et al. 2014; Vral et al. 2011). Furthermore, since radiosensitive cells are more susceptible to radiation-induced micronucleus formation and because micronuclei originate from mis-repair or DSBs in DNA, the CBMN assay can be used to to assess individual in vitro radiosensitivity or cancer susceptibility (Vral et al. 2011). However, the CBMN assay seems less suitable for the assessment of exposure to energetic heavy ions (Wu et al. 2006), one of the major components of cosmic radiation, especially in terms of expected health effects.

28.2.3 Premature Chromosome Condensation (PCC) Assay

The problems that arise with the study of metaphases (e.g., radiation-induced mitotic delay) can be circumvented by directly studying interphase cell aberrations. The PCC assay is very powerful for detecting chromatin damage in G1 or G2-phase cells and it has the advantage of being used to compare responses to different radiation types such as heavy ions or X-rays (Suzuki et al. 2006). This is in contrast to chromosome aberrations in metaphase spreads, which are not always comparable when cells are exposed to different types of radiation. The PCC method is especially suitable for studying radiation-induced chromatid breaks in the G2-phase, which is the cell cycle phase in which cells are the most radiosensitive (Wang et al. 2006). Therefore, the frequencies of aberrations as determined via PCC are higher than those of metaphase spreads, indicating increased sensitivity. Another advantage of the PCC method is that it can be used for dosimetry purposes, within a very high range of doses (from 200 mGy to ~20 Gy) depending on the specific application (Pernot et al. 2012).

PCC occurs when mitotic cells, containing condensed chromosomes, fuse with interphase cells causing the interphase cells to produce condensed chromosomes prematurely. Although the initial protocol was very laborious and did not hold much promise to become a widely used method, a study by Gotoh and Asakawa showed that it was possible to induce PCC in G2-phase lymphocytes using the protein phosphatase inhibitor okadaic acid (Gotoh and Asakawa 1996). Later, other protein phosphatases were also applied (Durante et al. 1998), thereby significantly enhancing the suitability of the technique.

The PCC assay has also been applied for retrospective biodosimetry of astronauts, especially to compare results with those of metaphase analysis for complex chromosomal rearrangements. Some of these studies did not observe significant differences in the number of complex damages between pre- and post-flight samples (George et al. 2002; Greco et al. 2003) probably because exposures were too low. Another study evaluating translocations via the PCC assay did find increases in chromosome damage during flight, which was additive for repeated long-duration missions (George et al. 2013). In general, it is assumed that the PCC may be more accurate than metaphases for assessing complex chromosome damage because problems of chromosome damage underestimation due to cell cycle delays are avoided (George et al. 2002).

28.2.4 Comet Assay (Single-Cell Gel Electrophoresis)

The comet assay, also known as the single-cell gel electrophoresis assay, is a sensitive method for the detection of DNA damage and repair in individual cells (Singh et al. 1988). The size and shape of the comet and the distribution of DNA within the comet correlate with the extent of DNA damage (Fairbairn et al. 1995). For many years now, the comet assay has been utilized to study DNA damage induced by ionizing radiation. The dose range that can be investigated using the alkaline comet assay is 100 mGy to 8 Gy, although it has little potential as a radiation biodosimeter because of its lack of specificity for radiation-induced DNA damage (Pernot et al. 2012). However, this assay may be of interest because it can be used to evaluate DNA repair capacity which is a measure of individual radiosensitivity. A low DNA repair capacity is in general associated with increased (radiation-induced) cancer risk (Curtin 2012).

28.2.5 γ-H2AX Focus Assay

Upon DNA DSB induction in mammals, the histone H2A variant H2AX becomes rapidly phosphorylated at serine 139 in a region spanning several megabases around the initial lesion. The phosphorylated form of H2AX, termed γ-H2AX, facilitates the recruitment of other DNA repair factors to the damaged sites (Paull et al. 2000) and has an anchoring function to retain the broken chromosomal DNA ends in close proximity (Bassing and Alt 2004). As a result of the recruitment of DNA repair factors, a so-called focus is formed which can be easily visualized using light microscopy (Rogakou et al. 1999; Rogakou et al. 1998). As there is one focus formed per DSB, the number of DSBs can be directly determined from the number of foci that are present shortly after the induction of DNA damage.

The phosphorylation of H2AX occurs at a conserved carboxyl-terminal Ser-Gln-Glu (SQE) amino acid sequence and is catalyzed by members of the phosphoinositide 3-kinase (PI3K) family, such as DNA-PK catalytic subunit ATM, and ATR (Paull et al. 2000). The phosphorylation status of γ-H2AX is under tight control of the kinase activity of the abovementioned protein kinases, as well as the phosphatase activity of protein phosphatases such as WIP1/PPM1D (Cha et al. 2010). Another way of removing γ-H2AX, after DNA damage has been repaired, is by histone exchange. It has also been shown that the mechanism of γ-H2AX phosphorylation is dependent on the type of genotoxic stress. For example, at the site of ionizing radiation-induced DSBs, the phosphorylation of H2AX occurs mainly by ATM, whereas after UV irradiation the mechanism depends on the cell cycle and can occur in the absence of DSBs (Cha et al. 2010).

The γ-H2AX assay is probably the most sensitive method to detect radiation-induced DNA damage. DSBs can be visualized in cells after exposures to doses as low as 2 mGy (Rothkamm and Lobrich 2003). However, it does suffer from a number of drawbacks, including a lack of specificity for radiation-induced damage and the transient nature of the signal (depending on the dose and radiation quality, DSBs are repaired within hours after exposure). This precludes its usefulness as a biomarker of exposure, but like the comet assay, it could be an excellent method to assess individual radiosensitivity based on DNA repair capacity measurements (Hall et al. 2017).

28.3 Emerging Technologies for High-Throughput Screening

As previously mentioned, classical cytogenetic measurements such as dicentrics and chromosomal translocations have been mostly used to detect chromosomal aberrations in peripheral blood lymphocytes of astronauts returning from long-term space missions (Beaton-Green et al. 2015; Cucinotta et al. 2008; Durante et al. 2003; Fedorenko et al. 2001; George et al. 2001, 2004, 2010, 2013; Obe et al. 1997; Testard et al. 1996; Yang et al. 1997). In the next paragraphs, we will review some emerging technologies which might be useful for high-throughput screening of interindividual differences in radiosensitivity. So far, these methods have not yet been used to evaluate radiation exposure or sensitivity in astronauts.

28.3.1 Transcriptomic Profiling

Intrinsic radiosensitivity is correlated with the ability of the cell to detect and repair DNA damage (Hennequin et al. 2008), which is the ultimate factor for cell survival. To deal with radiation-induced DNA damage, cells have developed complex responses that rely on activation of genes that are involved in DNA damage repair and cell cycle arrest. However, when the cell damage is too severe, cellular apoptosis or cellular senescence can be induced, mainly via activation of p53 (Ou and Schumacher 2018), as a way to escape malignancy. Mechanistically, it is therefore likely that interindividual differences in DNA damage response gene activation may reflect individual radiation sensitivity. Thus, individuals having a genetic dysfunction of certain genes that are important for the DNA damage response display hypersensitivity toward ionizing radiation. In this context, not only the quality but also the quantity of changes in gene expression may differ greatly between individuals and contribute to the individual differences in response to radiation (Smirnov et al. 2009). For instance, in a screen of the radiation response between unrelated individuals and monozygotic twins it was shown that there was a very strong heritable component for the transcriptional response of the p53 target genes *FDXR* and *CDKN1A* to ionizing radiation. The variability in their postirradiation expression levels was much smaller amongst the monozygotic twins than the unrelated individuals (Correa and Cheung 2004).

Since the early 2000's, many studies have demonstrated the usefulness of radiation-induced changes in gene expression as signatures that could be used for biodosimetry purposes [reviewed in Hall et al. (2017), Lacombe et al. (2018), and Pernot et al. (2012)]. With time and the development of new methods for gene expression profiling, these studies have shown an increasing sensitivity of gene signatures towards prediction of ever lower doses, down to 100 mGy and below (Broustas et al. 2017; Knops et al. 2012; Macaeva et al. 2018; Macaeva et al. 2016; Nosel et al. 2013; Paul and Amundson 2011; Riecke et al. 2012), the dose range to which astronauts may be exposed. An important new application in this respect will be the identification of exon signatures, based on the observations of extensive radiation-induced alternative splicing (Macaeva et al. 2016; Sprung et al. 2011) which will benefit from advances in RNA sequencing methods. Another important finding from these studies is that these identified gene signatures are in general very robust, and independent of experimental conditions or differences related to the radiation itself (i.e., radiation quality, dose rate) (Hall et al. 2017). Finally, although most of these biomarkers have been investigated using ex vivo-irradiated blood samples, they have been proven to work also for in vivo-irradiated cancer patients (Abend et al. 2016; O'Brien et al. 2018; Paul and Amundson 2011). This is an important asset for their potential applicability for astronaut screening.

From these observations it can be concluded that the genes most appropriate for biodosimetry are those involved in the p53-regulated DNA damage response pathways. Interestingly, the same panel of genes has been proposed as potential biomarkers for studying radiation quality and dose rate effects, relevant for modeling cancer risk from space radiation (Sridharan et al. 2016). However, in contrast to gene expression signatures for individual radiosensitivity, genes that are useful for biodosimetry should ideally exhibit dose-dependent changes in expression with very little variation depending on age, gender, time, cell type, and/or other interindividual confounding factors (Pogosova-Agadjanyan et al. 2011).

28.3.2 Epigenetic Profiling

Besides individual differences in gene expression, differences in epigenetic marks may also influence the individual response to ionizing radiation. The term epigenetics has nowadays many different meanings. One classical definition is that epigenetics refers to changes in gene expression that do not involve changes in the DNA sequence but are nevertheless inherited (through mitosis and possibly meiosis) (Holliday 1987). Two classical epigenetic mechanisms are DNA methylation and posttranslational histone modification, both of which mainly affect gene expression by altering the chromatin structure, thereby making genes more or less accessible for transcription. Other epigenetic information carriers that have been proposed include transcription factors, prions, small RNAs (e.g., microRNAs), long intergenic noncoding RNAs and chromatin structure (Kaufman and Rando 2010; Rando 2016).

DNA methylation is a widely studied and best characterized epigenetic modification (Barros-Silva et al. 2018) and will therefore be mainly discussed here. It involves a covalent deposition of a methyl group, mostly catalyzed by DNA methyltransferases (DNMTs) at the 5′ position of a cytosine ring, which occurs mostly in the vicinity of a CpG dinucleotide (Taby and Issa 2010). The CpG dinucleotides tend to cluster in so-called CpG islands which are often located in promoter regions of genes (about 60% of the human gene promoters are associated with CpG islands). In a normal differentiated cell, most promoter CpG islands are unmethylated whereas the CpG islands that are distributed across the genome—mostly associated with repetitive elements—are methylated. Methylation of CpG islands is generally associated with gene silencing via various mechanisms such as recruitment of methyl-CpG-binding domain proteins, recruitment of histone-modifying and chromatin-remodeling complexes, or by precluding the recruitment of DNA-binding proteins (e.g., transcription factors) to their target sites (Portela and Esteller 2010).

The first observation that radiation affects DNA methylation was reported in *E. coli* in 1972 (Whitfield and Billen 1972) while the first report in mammals suggested both hypo- and hypermethylation in Wistar and outbred rats exposed to high doses of γ-radiation (Rakova 1979). Following in vivo studies indicated that exposure to radiation induces in general a loss of global DNA methylation in hematopoietic tissues as well as liver, but not in muscle and lung (Miousse et al. 2017). Nevertheless, alterations in DNA methylation of specific genes in response to radiation exposure may go in both directions, and one gene often found to be hypermethylated, for instance in the sputum of uranium miners (Su et al. 2006) or in lung-adenocarcinomas of Mayak nuclear facility workers (Belinsky et al. 2004) is the tumor-suppressor gene *CDKN2A*.

Of relevance for space exploration are results from experimental animal studies, showing global DNA hypermethylation in lungs of C57BL6 mice after exposure to low absorbed doses of ^{56}Fe ions (Nzabarushimana et al. 2014), which was corroborated in a number of other studies demonstrating hypermethylation in repetitive elements in heart and lung of ^{56}Fe-irradiated mice (Koturbash et al. 2016; Lima et al. 2014; Prior et al. 2016). Another study investigated the correlation between ^{56}Fe-induced cognitive decline and changes in DNA methylation and hydroxymethylation (hmC) in the hippocampus, showing that especially hmC could be correlated to cognitive impairments (Impey et al. 2016). Interestingly, treatment of mice with the DNA methylation inhibitor 5-iodotubercidin partially restored DNA methylation levels and could improve cognitive function of mice exposed to a space-relevant dose of ^{28}Si-ions (Acharya et al. 2017).

To link DNA methylation with other epigenetic markers, a recent study showed that the long noncoding RNA PARTICLE, which is specifically induced in response to low-dose radiation (O'Leary et al. 2015) links DNA and histone methylation and thereby affects gene expression at a genome-wide scale (O'Leary et al. 2017). Whether and how epigenetic mechanisms affect individual radiation sensititvity is currently under further investigation. In this respect, it is interesting to note that the NASA Twin Study, in which an astronaut spent 1 year at the ISS while his twin brother and fellow astronaut stayed on Earth, supposedly identified large changes in epigenetic modifications during spaceflight (Garrett-Bakelman et al. 2019).

28.3.3 Multiple Protein Expression Profiling

Although changes in gene transcription can serve as good candidate biomarkers, protein changes may be relevant as well (Hall et al. 2017). Differences in protein levels can occur between radioresistant and radiosensitive individuals, and therefore protein signatures might also have a predictive value (Chaze et al. 2013). Recent advances in proteomics might allow the identification of proteins associated with radiosensitivity (Leszczynski 2014; Smith et al. 2009; Turtoi et al. 2011).

28.4 Conclusion

For manned spaceflight, the biological effects induced by cosmic ray exposure on the immune and other organ systems are of critical concern to risk assessment for astronauts, especially since the orbiting (so-called (Deep Space) Gateway) and habitats on the moon are now clearly planned and envisaged for the near future. In this light, and when expanding further to the manned exploration of Mars, the increased risk of cancer associated with radiation exposure is widely considered to be the main obstacle to interplanetary travel (Chancellor et al. 2014; Cucinotta and Durante 2006; Durante and Cucinotta 2008). Shielding of the astronauts from space radiation is therefore a very important protective measure. Material shielding may only be partially effective against cosmic radiation in certain energy ranges, but may actually make the problem worse for some of the higher energy rays, as more shielding induces an increased amount of secondary radiation. Therefore, other protection measures such as onboard biodosimetry, or therapeutic measures, should also be considered, although their effectiveness in deep space is not yet established. Currently, several radioprotectors are available that can prevent and/or reduce radiation-induced health effects by enhancing the body's natural capacity to repair cell damage caused by radiation or by preventing DNA damage to occur (McLaughlin et al. 2017; Smith et al. 2017). Another possible preventive measure could be to include the individual's radiation resistance to the induction of early and late radiation effects in the medical assessment of the mission crew applicants. This has to be complemented by the provision of adequate onboard biodosimetry tools to guarantee the best possible diagnostics and personalized care in-flight.

References

Abend M, Badie C, Quintens R, Kriehuber R, Manning G, Macaeva E, Njima M, Oskamp D, Strunz S, Moertl S, Doucha-Senf S, Dahlke S, Menzel J, Port M (2016) Examining radiation-induced in vivo and in vitro gene expression changes of the peripheral blood in different laboratories for biodosimetry purposes: first RENEB gene expression study. Radiat Res 185:109–123

Acharya MM, Baddour AA, Kawashita T, Allen BD, Syage AR, Nguyen TH, Yoon N, Giedzinski E, Yu L, Parihar VK, Baulch JE (2017) Epigenetic determinants of space radiation-induced cognitive dysfunction. Sci Rep 7:42885

Andreassen CN (2005) Can risk of radiotherapy-induced normal tissue complications be predicted from genetic profiles? Acta Oncol 44:801–815

Awa A (1997) Analysis of chromosome aberrations in atomic bomb survivors for dose assessment: studies at the radiation effects research foundation from 1968 to 1993. Stem Cells 15(Suppl 2):163–173

Barros-Silva D, Marques CJ, Henrique R, Jeronimo C (2018) Profiling DNA methylation based on next-generation sequencing approaches: new insights and clinical applications. Genes (Basel) 9:E429

Bassing CH, Alt FW (2004) H2AX may function as an anchor to hold broken chromosomal DNA ends in close proximity. Cell Cycle 3:149–153

Beaton-Green LA, Lachapelle S, Straube U, Wilkins RC (2015) Evolution of the Health Canada astronaut biodosimetry program with a view toward international harmonization. Mutat Res Genet Toxicol Environ Mutagen 793:101–106

Belinsky SA, Klinge DM, Liechty KC, March TH, Kang T, Gilliland FD, Sotnic N, Adamova G, Rusinova G, Telnov V (2004) Plutonium targets the p16 gene for inactivation by promoter hypermethylation in human lung adenocarcinoma. Carcinogenesis 25(6):1063–1067

Broustas CG, Xu Y, Harken AD, Chowdhury M, Garty G, Amundson SA (2017) Impact of neutron exposure on global gene expression in a human peripheral blood model. Radiat Res 187:433–440

Cha H, Lowe JM, Li H, Lee JS, Belova GI, Bulavin DV, Fornace AJ Jr (2010) Wip1 directly dephosphorylates gamma-H2AX and attenuates the DNA damage response. Cancer Res 70:4112–4122

Chancellor JC, Scott GB, Sutton JP (2014) Space radiation: the number one risk to astronaut health beyond low earth orbit. Life (Basel) 4:491–510

Chaze T, Hornez L, Chambon C, Haddad I, Vinh J, Peyrat JP, Benderitter M, Guipaud O (2013) Serum proteome analysis for profiling predictive protein markers associated with the severity of skin lesions induced by ionizing radiation. Proteomes 1:40–69

Chistiakov DA, Voronova NV, Chistiakov PA (2008) Genetic variations in DNA repair genes, radiosensitivity to cancer and susceptibility to acute tissue reactions in radiotherapy-treated cancer patients. Acta Oncol 47:809–824

Correa CR, Cheung VG (2004) Genetic variation in radiation-induced expression phenotypes. Am J Hum Genet 75:885–890

Cologne JB, Pawel DJ, Preston DL (1998) Statistical issues in biological radiation dosimetry for risk assessment using stable chromosome aberrations. Health Phys 75:518–529

Cucinotta FA, Durante M (2006) Cancer risk from exposure to galactic cosmic rays: implications for space exploration by human beings. Lancet Oncol 7:431–435

Cucinotta FA, Kim MH, Willingham V, George KA (2008) Physical and biological organ dosimetry analysis for international space station astronauts. Radiat Res 170:127–138

Curtin NJ (2012) DNA repair dysregulation from cancer driver to therapeutic target. Nat Rev Cancer 12:801–817

Distel LV, Neubauer S, Keller U, Sprung CN, Sauer R, Grabenbauer GG (2006) Individual differences in chromosomal aberrations after in vitro irradiation of cells from healthy individuals, cancer and cancer susceptibility syndrome patients. Radiother Oncol 81:257–263

Durante M (2005) Biomarkers of space radiation risk. Radiat Res 164:467–473

Durante M, Cucinotta FA (2008) Heavy ion carcinogenesis and human space exploration. Nat Rev Cancer 8:465–472

Durante M, Furusawa Y, Gotoh E (1998) A simple method for simultaneous interphase-metaphase chromosome analysis in biodosimetry. Int J Radiat Biol 74:457–462

Durante M, Snigiryova G, Akaeva E, Bogomazova A, Druzhinin S, Fedorenko B, Greco O, Novitskaya N, Rubanovich A, Shevchenko V, Von Recklinghausen U, Obe G (2003) Chromosome aberration dosimetry in cosmonauts after single or multiple space flights. Cytogenet Genome Res 103:40–46

Fairbairn DW, Olive PL, O'Neill KL (1995) The comet assay: a comprehensive review. Mutat Res 339:37–59

Fedorenko B, Druzhinin S, Yudaeva L, Petrov V, Akatov Y, Snigiryova G, Novitskaya N, Shevchenko V, Rubanovich A (2001) Cytogenetic studies of blood lymphocytes from cosmonauts after long-term space flights on Mir station. Adv Space Res 27:355–359

Fenech M (2010) The lymphocyte cytokinesis-block micronucleus cytome assay and its application in radiation biodosimetry. Health Phys 98:234–243

Fenech M, Morley A (1985) Solutions to the kinetic problem in the micronucleus assay. Cytobios 43:233–246

Feng L, He L, Wang Y, Du L, Xu C, Liu Q, Fan F (2015) Eight-year follow-up study of three individuals accidentally exposed to (60)co radiation: chromosome aberration and micronucleus analysis. Mutat Res Genet Toxicol Environ Mutagen 784-785:10–14

Foray N, Bourguignon M, Hamada N (2016) Individual response to ionizing radiation. Mutat Res 770:369–386

Garrett-Bakelman FE (2019) The NASA Twins Study: A multidimensional analysis of a year-long human spaceflight. Science 12;364(6436)

George K, Durante M, Wu H, Willingham V, Badhwar G, Cucinotta FA (2001) Chromosome aberrations in the blood lymphocytes of astronauts after space flight. Radiat Res 156: 731–738

George K, Wu H, Willingham V, Cucinotta FA (2002) Analysis of complex-type chromosome exchanges in astronauts' lymphocytes after space flight as a biomarker of high-LET exposure. J Radiat Res 43(Suppl):S129–S132

George K, Durante M, Willingham V, Cucinotta FA (2004) Chromosome aberrations of clonal origin are present in astronauts' blood lymphocytes. Cytogenet Genome Res 104:245–251

George K, Chappell LJ, Cucinotta FA (2010) Persistence of space radiation induced cytogenetic damage in the blood lymphocytes of astronauts. Mutat Res 701:75–79

George K, Rhone J, Beitman A, Cucinotta FA (2013) Cytogenetic damage in the blood lymphocytes of astronauts: effects of repeat long-duration space missions. Mutat Res 756: 165–169

Gotoh E, Asakawa Y (1996) Detection and evaluation of chromosomal aberrations induced by high doses of gamma-irradiation using immunogold-silver painting of prematurely condensed chromosomes. Int J Radiat Biol 70:517–520

Greco O, Durante M, Gialanella G, Grossi G, Pugliese M, Scampoli P, Snigiryova G, Obe G (2003) Biological dosimetry in Russian and Italian astronauts. Adv Space Res 31:1495–1503

Hall J, Jeggo PA, West C, Gomolka M, Quintens R, Badie C, Laurent O, Aerts A, Anastasov N, Azimzadeh O, Azizova T, Baatout S, Baselet B, Benotmane MA, Blanchardon E, Gueguen Y, Haghdoost S, Harms-Ringhdahl M, Hess J, Kreuzer M, Laurier D, Macaeva E, Manning G, Pernot E, Ravanat JL, Sabatier L, Tack K, Tapio S, Zitzelsberger H, Cardis E (2017) Ionizing radiation biomarkers in epidemiological studies - an update. Mutat Res 771:59–84

Hennequin C, Quero L, Favaudon V (2008) Determinants and predictive factors of tumour radiosensitivity. Cancer Radiother 12:3–13

Holliday R (1987) The inheritance of epigenetic defects. Science 238:163–170

ICRP (2007) The 2007 recommendations of the International commission on radiological protection. ICRP publication 103. In Ann ICRP, 37

Impey S, Jopson T, Pelz C, Tafessu A, Fareh F, Zuloaga D, Marzulla T, Riparip LK, Stewart B, Rosi S, Turker MS, Raber J (2016) Short- and long-term effects of (56)Fe irradiation on cognition and hippocampal DNA methylation and gene expression. BMC Genomics 17:825

Kaufman PD, Rando OJ (2010) Chromatin as a potential carrier of heritable information. Curr Opin Cell Biol 22:284–290

Knops K, Boldt S, Wolkenhauer O, Kriehuber R (2012) Gene expression in low- and high-dose-irradiated human peripheral blood lymphocytes: possible applications for biodosimetry. Radiat Res 178:304–312

Koturbash I, Miousse IR, Sridharan V, Nzabarushimana E, Skinner CM, Melnyk SB, Pavliv O, Hauer-Jensen M, Nelson GA, Boerma M (2016) Radiation-induced changes in DNA methylation of repetitive elements in the mouse heart. Mutat Res 787:43–53

Lacombe J, Sima C, Amundson SA, Zenhausern F (2018) Candidate gene biodosimetry markers of exposure to external ionizing radiation in human blood: a systematic review. PLoS One 13:e0198851

Leszczynski D (2014) Radiation proteomics: a brief overview. Proteomics 14:481–488

Lima F, Ding D, Goetz W, Yang AJ, Baulch JE (2014) High LET (56)Fe ion irradiation induces tissue-specific changes in DNA methylation in the mouse. Environ Mol Mutagen 55:266–277

Liu Q, Cao J, Wang ZQ, Bai YS, Lu YM, Huang QL, Zhao WZ, Li J, Jiang LP, Tang WS, Fu BH, Fan FY (2009) Dose estimation by chromosome aberration analysis and micronucleus assays in victims accidentally exposed to (60)co radiation. Br J Radiol 82:1027–1032

Macaeva E, Saeys Y, Tabury K, Janssen A, Michaux A, Benotmane MA, De Vos WH, Baatout S, Quintens R (2016) Radiation-induced alternative transcription and splicing events and their applicability to practical biodosimetry. Sci Rep 6:19251

Macaeva E, Mysara M, De Vos WH, Baatout S, Quintens R (2018) Gene expression-based biodosimetry for radiological incidents: assessment of dose and time after radiation exposure. Int J Radiat Biol 95(1):64–75

McLaughlin MF, Donoviel DB, Jones JA (2017) Novel indications for commonly used medications as radiation protectants in spaceflight. Aerosp Med Hum Perform 88:665–676

Miousse IR, Kutanzi KR, Koturbash I (2017) Effects of ionizing radiation on DNA methylation: from experimental biology to clinical applications. Int J Radiat Biol 93:457–469

Nosel I, Vaurijoux A, Barquinero JF, Gruel G (2013) Characterization of gene expression profiles at low and very low doses of ionizing radiation. DNA Repair (Amst) 12:508–517

Nzabarushimana E, Miousse IR, Shao L, Chang J, Allen AR, Turner J, Stewart B, Raber J, Koturbash I (2014) Long-term epigenetic effects of exposure to low doses of 56Fe in the mouse lung. J Radiat Res 55:823–828

Obe G, Johannes I, Johannes C, Hallman K, Reitz G, Facius R (1997) Chromosomal aberrations in blood lymphocytes of astronauts after long-term space flights. Int J Radiat Biol 72:727–734

O'Brien G, Cruz-Garcia L, Majewski M, Grepl J, Abend M, Port M, Tichy A, Sirak I, Malkova A, Donovan E, Gothard L, Boyle S, Somaiah N, Ainsbury E, Ponge L, Slosarek K, Miszczyk L, Widlak P, Green E, Patel N, Kudari M, Gleeson F, Vinnikov V, Starenkiy V, Artiukh S, Vasyliev L, Zaman A, Badie C (2018) FDXR is a biomarker of radiation exposure in vivo. Sci Rep 8:684

O'Leary VB, Ovsepian SV, Carrascosa LG, Buske FA, Radulovic V, Niyazi M, Moertl S, Trau M, Atkinson MJ, Anastasov N (2015) PARTICLE, a triplex-forming long ncRNA, regulates locus-specific methylation in response to low-dose irradiation. Cell Rep 11:474–485

O'Leary VB, Hain S, Maugg D, Smida J, Azimzadeh O, Tapio S, Ovsepian SV, Atkinson MJ (2017) Long non-coding RNA PARTICLE bridges histone and DNA methylation. Sci Rep 7:1790

Ou HL, Schumacher B (2018) DNA damage responses and p53 in the aging process. Blood 131:488–495

Paul S, Amundson SA (2011) Gene expression signatures of radiation exposure in peripheral white blood cells of smokers and non-smokers. Int J Radiat Biol 87:791–801

Paull TT, Rogakou EP, Yamazaki V, Kirchgessner CU, Gellert M, Bonner WM (2000) A critical role for histone H2AX in recruitment of repair factors to nuclear foci after DNA damage. Curr Biol 10:886–895

Pernot E, Hall J, Baatout S, Benotmane MA, Blanchardon E, Bouffler S, El Saghire H, Gomolka M, Guertler A, Harms-Ringdahl M, Jeggo P, Kreuzer M, Laurier D, Lindholm C, Mkacher R, Quintens R, Rothkamm K, Sabatier L, Tapio S, de Vathaire F, Cardis E (2012) Ionizing radiation biomarkers for potential use in epidemiological studies. Mutat Res 751:258–286

Pogosova-Agadjanyan EL, Fan W, Georges GE, Schwartz JL, Kepler CM, Lee H, Suchanek AL, Cronk MR, Brumbaugh A, Engel JH, Yukawa M, Zhao LP, Heimfeld S, Stirewalt DL (2011) Identification of radiation-induced expression changes in nonimmortalized human T cells. Radiat Res 175:172–184

Portela A, Esteller M (2010) Epigenetic modifications and human disease. Nat Biotechnol 28:1057–1068

Prior S, Miousse IR, Nzabarushimana E, Pathak R, Skinner C, Kutanzi KR, Allen AR, Raber J, Tackett AJ, Hauer-Jensen M, Nelson GA, Koturbash I (2016) Densely ionizing radiation affects DNA methylation of selective LINE-1 elements. Environ Res 150:470–481

Rakova IA (1979) Methylation of newly synthesized DNA in rat bone marrow and thymus after irradiation. Radiobiologiia 19:413–416

Rando OJ (2016) Intergenerational transfer of epigenetic information in sperm. Cold Spring Harb Perspect Med 6:a022988

Riecke A, Rufa CG, Cordes M, Hartmann J, Meineke V, Abend M (2012) Gene expression comparisons performed for biodosimetry purposes on in vitro peripheral blood cellular subsets and irradiated individuals. Radiat Res 178:234–243

Rodrigues AS, Oliveira NG, Gil OM, Leonard A, Rueff J (2005) Use of cytogenetic indicators in radiobiology. Radiat Prot Dosim 115:455–460

Rodrigues MA, Beaton-Green LA, Wilkins RC, Fenech MF (2018) The potential for complete automated scoring of the cytokinesis block micronucleus cytome assay using imaging flow cytometry. Mutat Res 836:53–64

Rogakou EP, Pilch DR, Orr AH, Ivanova VS, Bonner WM (1998) DNA double-stranded breaks induce histone H2AX phosphorylation on serine 139. J Biol Chem 273:5858–5868

Rogakou EP, Boon C, Redon C, Bonner WM (1999) Megabase chromatin domains involved in DNA double-strand breaks in vivo. J Cell Biol 146:905–916

Rothkamm K, Lobrich M (2003) Evidence for a lack of DNA double-strand break repair in human cells exposed to very low x-ray doses. Proc Natl Acad Sci U S A 100:5057–5062

Singh NP, McCoy MT, Tice RR, Schneider EL (1988) A simple technique for quantitation of low levels of DNA damage in individual cells. Exp Cell Res 175:184–191

Smirnov DA, Morley M, Shin E, Spielman RS, Cheung VG (2009) Genetic analysis of radiation-induced changes in human gene expression. Nature 459:587–591

Smith L, Qutob O, Watson MB, Beavis AW, Potts D, Welham KJ, Garimella V, Lind MJ, Drew PJ, Cawkwell L (2009) Proteomic identification of putative biomarkers of radiotherapy resistance: a possible role for the 26S proteasome? Neoplasia 11:1194–1207

Smith TA, Kirkpatrick DR, Smith S, Smith TK, Pearson T, Kailasam A, Herrmann KZ, Schubert J, Agrawal DK (2017) Radioprotective agents to prevent cellular damage due to ionizing radiation. J Transl Med 15:232

Sprung CN, Li J, Hovan D, McKay MJ, Forrester HB (2011) Alternative transcript initiation and splicing as a response to DNA damage. PLoS One 6:e25758

Sridharan DM, Asaithamby A, Blattnig SR, Costes SV, Doetsch PW, Dynan WS, Hahnfeldt P, Hlatky L, Kidane Y, Kronenberg A, Naidu MD, Peterson LE, Plante I, Ponomarev AL, Saha J, Snijders AM, Srinivasan K, Tang J, Werner E, Pluth JM (2016) Evaluating biomarkers to model cancer risk post cosmic ray exposure. Life Sci Space Res (Amst) 9:19–47

Su S, Jin Y, Zhang W, Yang L, Shen Y, Cao Y, Tong J (2006) Aberrant promoter methylation of p16(INK4a) and O(6)-methylguanine-DNA methyltransferase genes in workers at a Chinese uranium mine. J Occup Health 48:261–266

Suzuki M, Tsuruoka C, Nakano T, Ohno T, Furusawa Y, Okayasu R (2006) The PCC assay can be used to predict radiosensitivity in biopsy cultures irradiated with different types of radiation. Oncol Rep 16:1293–1299

Taby R, Issa JP (2010) Cancer epigenetics. CA Cancer J Clin 60:376–392

Taylor AM, Harnden DG, Arlett CF, Harcourt SA, Lehmann AR, Stevens S, Bridges BA (1975) Ataxia telangiectasia: a human mutation with abnormal radiation sensitivity. Nature 258:427–429

Testard I, Ricoul M, Hoffschir F, Flury-Herard A, Dutrillaux B, Fedorenko B, Gerasimenko V, Sabatier L (1996) Radiation-induced chromosome damage in astronauts' lymphocytes. Int J Radiat Biol 70:403–411

Thierens H, Vral A, Vandevoorde C, Vandersickel V, de Gelder V, Romm H, Oestreicher U, Rothkamm K, Barnard S, Ainsbury E, Sommer S, Beinke C, Wojcik A (2014) Is a semi-automated approach indicated in the application of the automated micronucleus assay for triage purposes? Radiat Prot Dosim 159:87–94

Tough IM, Buckton KE, Baikie AG, Court-Brown WM (1960) X-ray-induced chromosome damage in man. Lancet 2:849–851

Tucker JD (2008) Low-dose ionizing radiation and chromosome translocations: a review of the major considerations for human biological dosimetry. Mutat Res 659:211–220

Turtoi A, De Pauw E, Castronovo V (2011) Innovative proteomics for the discovery of systemically accessible cancer biomarkers suitable for imaging and targeted therapies. Am J Pathol 178:12–18

Varon R, Vissinga C, Platzer M, Cerosaletti KM, Chrzanowska KH, Saar K, Beckmann G, Seemanova E, Cooper PR, Nowak NJ, Stumm M, Weemaes CM, Gatti RA, Wilson RK, Digweed M, Rosenthal A, Sperling K, Concannon P, Reis A (1998) Nibrin, a novel DNA double-strand break repair protein, is mutated in Nijmegen breakage syndrome. Cell 93:467–476

Venkatesulu BP, Mallick S, Lin SH, Krishnan S (2018) A systematic review of the influence of radiation-induced lymphopenia on survival outcomes in solid tumors. Crit Rev Oncol Hematol 123:42–51

Vral A, Fenech M, Thierens H (2011) The micronucleus assay as a biological dosimeter of in vivo ionising radiation exposure. Mutagenesis 26:11–17

Wang ZZ, Li WJ, Zhang H, Yang JS, Qiu R, Wang X (2006) Comparison of clonogenic assay with premature chromosome condensation assay in prediction of human cell radiosensitivity. World J Gastroenterol 12:2601–2605

Whitfield BL, Billen D (1972) In vivo methylation of Escherichia coli DNA following ultraviolet and x-irradiation. J Mol Biol 63:363–372

Wojcik A, Bouffler S, Hauptmann M, Rajaraman P (2018) Considerations on the use of the terms radiosensitivity and radiosusceptibility. J Radiol Prot 38:N25–N29

Wu H, Hada M, Meador J, Hu X, Rusek A, Cucinotta FA (2006) Induction of micronuclei in human fibroblasts across the Bragg curve of energetic heavy ions. Radiat Res 166:583–589

Yang TC, George K, Johnson AS, Tavakoli A, Durante M, Fedorenko BS (1997) Cytogenetic effects of space radiation in lymphocytes of MIR-18 crews. Aviakosm Ekolog Med 31:8–14

Open Access This chapter is licensed under the terms of the Creative Commons Attribution 4.0 International License (http://creativecommons.org/licenses/by/4.0/), which permits use, sharing, adaptation, distribution and reproduction in any medium or format, as long as you give appropriate credit to the original author(s) and the source, provide a link to the Creative Commons license and indicate if changes were made.

The images or other third party material in this chapter are included in the chapter's Creative Commons license, unless indicated otherwise in a credit line to the material. If material is not included in the chapter's Creative Commons license and your intended use is not permitted by statutory regulation or exceeds the permitted use, you will need to obtain permission directly from the copyright holder.

Analytical Approaches to the Quantitative Evaluation of Endocannabinoids and Glucocorticoids as Stress Markers: Growing Evidence for Hair Testing

29

Detlef Thieme, Patricia Anielski, Ann-Kathrin Helfers, and Aniko Krumbholz

29.1 Stress Markers

Cortisol represents a major tool to maintain homeostasis during times of stress and is hence one of the most efficient biochemical parameter to diagnose stress (see also Chaps. 4, 6 and 7). Cortisol (also "compound F"), like all rather lipophilic glucucorticoids (GCs), is secreted into and distributed via the blood stream and therefore easily accessible in most conventional matrices, i.e. plasma, oral fluid, and hair. The bioavailability of glucocorticoids is mainly limited by a significant protein-binding to albumin and cortisol-binding globulin (CBG) which amounts to 97% (F) or 80–85% (cortisone, also "compound E"), increasing the stability and detection window of GCs. Its biological activity –and its detectability- ranges from seconds to few hours and is hence much longer than those of short-acting catecholamines. Additionally to glucocorticoids as primary stress marker, the substance group of endocannabinoids (ECs) (see also Chap. 10) gained significant diagnostic attraction due to its ability to modulate GC response (Hill et al. 2010; Finn 2009). Respective agonists of the cannabinoid receptors are chemically defined as ethanolamides, e.g., N-arachidonoyl ethanolamine (anandamide, AEA), and its oleoyl, palmitoyl, or stearoyl analogs (OEA, PEA, SEA) or glycerol ester (arachidonoylglycerol, 2-AG). Endocannabinoids are locally synthesized—in particular in the brain—and systemic EC activity deems not likely. Nevertheless, the correlation between circulating plasma concentration—whether overflow from brain synthesis (Hillard et al. 2011) or peripherally produced—and stress is undisputed and well documented (Chouker et al. 2010; Dlugos et al. 2012; Hillard 2014). The biochemical mechanisms of

D. Thieme (✉) · P. Anielski · A.-K. Helfers · A. Krumbholz
Institut für Dopinganalytik und Sportbiochemie (IDAS), Kreischa, Germany
e-mail: detlef.thieme@idas-kreischa.de

© Springer Nature Switzerland AG 2020

A. Choukér (ed.), *Stress Challenges and Immunity in Space*,
https://doi.org/10.1007/978-3-030-16996-1_29

habituation to repeated stress seem to differ between the chemical classes (Sharkey and Wiley 2016). AEA and 2-AG were both found to be elevated during acute stress but respective concentration did not correlate. AEA was postulated to represent a "tonic-like" mechanism, whereas 2-AG acts as a "burst-like" modulator of the hypothalamic–pituitary–adrenal (HPA) axis (see also Chap. 10).

Besides the activation of the EC system, e.g., by stimulated synthesis or administration of agonists, the prevention of its deactivation plays an important diagnostic and therapeutic role. Ethanolamides are rapidly cleaved by fatty acid amide hydrolase (FAAH, mainly AEA) or NAE-Hydrolyzing Acid Amidase (NAAA, e.g., OEA, PEA); 2-AG is predominantly hydrolyzed by monoacylglycerol lipase (MAGL). The inhibition of respective enzymes; e.g., 4-nitrophenyl-4-[bis(1,3-benzodioxol-5-yl)(hydroxy) methyl]piperidine-1-carboxylate (JZL 184), (4-nitrophenyl) 4-[(3-phenoxyphenyl) methyl]piperazine-1-carboxylate (JZL 195) or methoxyarachidonyl fluorophosphonate (MAFP) is suitable for in vivo or in vitro stabilization ECs in biological specimens (Long et al. 2009). There are numerous therapeutic concepts, e.g., in treatment of pain, depression, or maladaptive behaviors to acute and chronic stress (Griebel et al. 2018; Jiang et al. 2018; Ratano et al. 2018) based on this indirect enhancement of EC levels.

The different kinetics of EC-hydrolysis, e.g., its dependence on unsaturation and length of the fatty acids (Vandevoorde et al. 2005), will positively affect its concentrations (and corresponding ratios) in various specimens and need to be taken into account in its interpretation.

The choice of the most efficient stress testing strategy depends therefore on multiple parameters, i.e. the duration and intensity of triggers. Availability of minimal invasive biological matrices (e.g., hair) and feasibility of appropriate collection and stabilization requirements need to be granted. Rapid collection, centrifugation, and freezing of plasma samples appear to be most efficient as a diagnostic tool but extremely challenging to accomplish under stressful conditions.

29.2 Suitable Testing Matrices

The distribution of glucocorticoids and cannabinoids follows a common concept of many drugs and hormones. The majority of bioactive substances is rather lipophilic and is therefore rapidly absorbed, transported and bound to the respective receptors. The process of a quick deactivation of the agonists is similarly important for a quick and reversible stress response.

All endocannabinoids are easily cleaved by enzymatic hydrolysis in blood, the remaining fatty acids, alcylalcohols, and amides are identical to endogenous compounds or their metabolites and hence of little diagnostic relevance.

Glucocorticoids are synthesized via a cascade of enzymatic reactions, stimulated by the adrenocorticotropic hormone (ACTH). Cortisol, the most relevant agonist in stress response is further oxidized to cortisone. Precursors (e.g., ACTH) as well as biotransformation products (cortisone) are highly specific and may well contribute to the GC-based stress investigation.

The choice of suitability specimens is closely related to the chemical and pharmacokinetic behavior of respective markers. Moreover the access to noninvasive matrices is of paramount importance when dealing with stress. Any potentially stressful manipulation (phlebotomy, liquor punctuation) will positively influence stress levels of tested individuals.

Due to the systemic mechanism of GCs there is no need for local sampling at the target receptor. Blood concentrations well represent bioavailable amounts at any target organ. Protein-binding represents the main interfering factor, because only free, i.e., protein-unbound portions of GCs are available for biological effects, biotransformation, migration into other compartments, incorporation into hair or excretion via sweat, saliva or urine. In case of the locally synthesized and bioactive ECs the collection of brain, brain areas (hypothalamus) or at least liquor appears clearly preferential over blood and other secondary matrices.

29.2.1 Blood (Plasma)

Finally, target concentrations and diagnostic value of drugs in different matrices, e.g., brain, blood (plasma), urine, oral fluid or hair depend on a couple of physicochemical parameters. Transfer from brain to blood (relevant for ECs) requires moderate molecular masses and only limited numbers of polar substituents (e.g., hydroxyl groups, both prerequisites certainly given). The detection window in blood or plasma is mainly restricted by the half-life and stability of the markers. In particular the potential hydrolysis, isomerization (Vogeser and Schelling 2007; Zoerner et al. 2018) or in vitro formation (Vogeser et al. 2006) of ECs post sampling requires substantial effort and consideration. The bioavailability and detection period of glucocorticoids is prolonged by its significant protein-binding.

Sample preprocessing includes rapid centrifugation of blood samples and immediate freezing of the plasma, which requires significant logistic and instrumental requirements, which is difficult to achieve under stressful conditions. Following thawing, the stored samples are purified by ether extraction (Hauer et al. 2014; Thieme et al. 2014).

29.2.2 Hair

An old and probably nonproven report that Marie Antoinette's hair turned white the night before she was guillotined (16th October 1793) and presumably the stress of impending decapitation caused her locks to lose color within hours. This time course seems difficult to prove and one of the reasons discussed are that "the stress of the situation could have generated a swarm of free radicals in her hair follicles, which traveled along the hair shafts, destroying pigment and creating a bleaching effect" (Ballantyne 2007). The biology of incorporation into hair is mainly defined by the physicochemical conditions at the hair bulb. Basic lipophilic compounds are preferentially incorporated into the hair fiber due to the slightly acidic condition around the hair bulb. Bioactive parent drugs, which are typically lipophilic, are generally better incorporated into hair than polar metabolites. This is a major diagnostic advantage compared to blood and urine analysis. Direct detection of heroin is almost impossible in blood due to its short half-life (20 min) and pointless in urine because of its high lipophilicity. In spite of its quick hydrolysis, heroin is significantly incorporated into hair and available as long-term marker for its abuse. Similarly all slightly basic lipophilic ethanolamides will be well transferred into hair while its acidic hydrolysis products (fatty acids) cannot be detected.

Moreover, basic compounds are often tightly bound to melanin in pigmented hairs. Therefore hair concentrations of basic compounds (e.g., cocaine) are known to depend significantly on hair color. Equal blood concentrations will result in highest hair concentration in black hair followed by brown, red, blond and gray hair. Once incorporated into hair, basic lipophilic drugs are well immobilized and only slowly washed out. Segmental analysis of hair strands, i.e., quantitation of drugs in adjacent hair segments can often provide retrospective information on administration (or production) profiles of basic drugs (Fig. 29.1). Owing to an average hair growth of 1 cm/month (human scalp hair (Pötsch and Skopp 2004; Cooper 2015)), a retrospection period of several months is mostly given. The active growing cycle of hair lasts ca. 3–5 years (anagen phase) followed by a transition (catagen) phase and a resting period of the nongrowing (telogen) hair. Owing to this fact, the amount of anagen hairs, which are kinetically relevant, is about 90–92% (Musshoff et al. 2013).

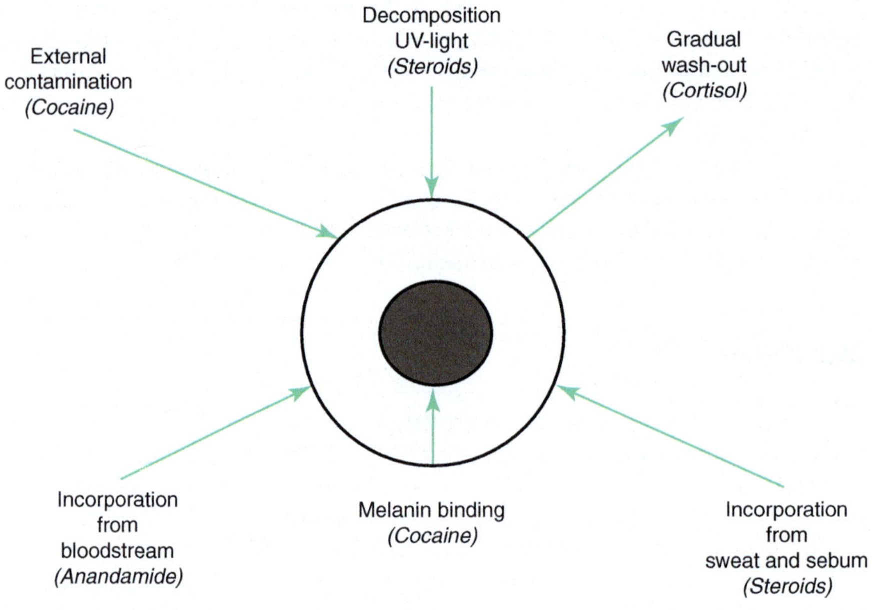

Fig. 29.1 Incorporation and elimination of compounds into hair is mainly controlled by its chemical structure. Lipophilic basic compounds (e.g., anandamide) are preferentially incorporated from blood into the hair fiber and may be tightly bound to hair pigments, e.g., melanin. Neutral compounds (steroids) migrate typically into hair via sweat and sebum and are potential subjects of elimination (wash-out following cosmetic treatment)

Neutral drugs (steroids, including GCs, probably 2-AG) are incorporated into hair via sweat and sebum into the keratinizing zone of hair formation. Those drugs are no subject of melanin binding, and wash-out is much more relevant when quantifying hair concentrations for retrospective considerations. A reduction of hair concentration following regular cosmetic treatment of 16% per month was estimated for cortisol (Krumbholz et al. 2013). Therefore, hair segmentation (Fig. 29.2) and long-term interpretation of respective concentration is much more critical for neutral glucocorticoids than for basic compounds (Duvivier et al. 2016; Thieme and Sachs 2007), e.g., endocannabinoids. Metabolite ratios of cortisol to cortisone differ significantly compared to blood. The original F/E ratio in blood of ~8:1 changes to ~1:4 as a consequence of different protein-binding rates (only the protein-unbound portion of GCs is available for hair incorporation) (Krumbholz et al. 2018), slightly different polarity (E > F) and potential local conversion of F to E (Zhang et al. 2016).

Long-term storage of hair samples does not require any precaution. Hair strands are typically stored at room temperature in aluminum foil following fixation of hair strands, which appear to be the easiest option for diagnostic purposes under crucial conditions, e.g., space missions (Fig. 29.3). Hair sample preparation includes a methanol extraction followed by hair digestion (Krumbholz et al. 2013).

29.2.3 Oral Fluid

The quantification of GCs in oral fluid (saliva) became a standard diagnostic tool in clinical chemistry and endocrinology(Perogamvros et al. 2009; Inder et al. 2012;

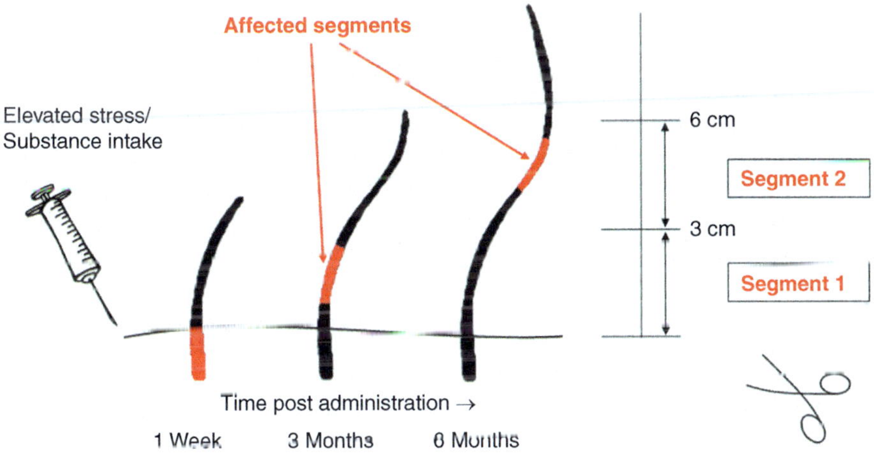

Fig. 29.2 Hair segmentation provides access to retrospective testing. Following a latency period of several days, hair segments affected by substance incorporation (administration or endogenous reaction, e.g. stress) reach the body surface and become available for testing until falling out. The live cycle, i.e. percentage of nongrowing hair, is critically dependent on its location and species, human scalp hair is known to growth ~1 cm/month

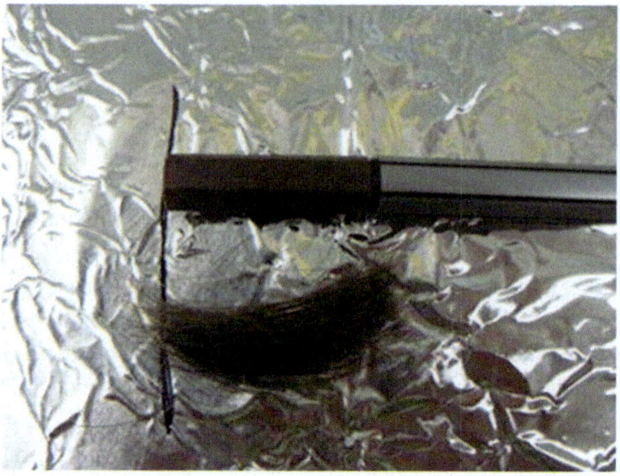

Fig. 29.3 Hair sample placed on line marked aluminum foil next to a pencil to illustrate the required thickness of the sample and indication of the cutting edges (vs. scull)

Bastin et al. 2018). Its noninvasive collection, commercial availability of robust collection devices—which use stabilization buffers to easily avoid decomposition—and the extensive availability of testing laboratories are strong arguments in favor of saliva testing. There is little information of testing endocannabinoids in oral fluids. Owing to the similarity with tetrahydrocannabinol (THC), which is not sufficiently incorporated into oral fluid (Moore et al. 2007), it seems to be unlikely that EC testing in saliva may become a suitable option.

29.2.4 Urine

Urine represents generally a preferential matrix for clinical and forensic tests due to the noninvasive collection mode and comparatively high substance concentration of xenobiotics. However, lipophilic compounds as GCs as well as ECs are not sufficiently excreted into urine without prior biotransformation. GCs are mainly excreted into urine as glucuroconjugates, whereas ECs are hydrolyzed into unspecific metabolites. Inter-individual variations of these biotransformation reactions and subsequent renal excretion as well as the temporal delay of urinary collection discriminate against the use of urine for stress detection.

29.3 Analytical Methods

29.3.1 Immunoassays

There are numerous immunoassays available to test cortisol in different matrices, e.g., serum and oral fluid (El-Farhan et al. 2013). Well described systematic deviations between those assays could be easily avoided by using identical assays within the same study. Therefore, there still is a justification for using these tests due to their sensitivity, simplified sample preparation and stabilization, low investments. The use of immunoassays in hair testing for cortisol is widely used and requires special consideration due to cross reactivity with other steroids, e.g., cortisone which is predominantly incorporated into hair. The cross reactivity to other steroids, e.g., cortisone or progesterone, represents the most crucial aspect whenever matrices with other composition than default (the matrix in which typical reference concentrations were tested, i.e. usually blood) were used. Moreover changes in the endogenous steroid pattern, i.e. its dramatic variation during pregnancy, may considerably interfere cortisol profiling. At present, there is no immunoassay available for quantification of endocannabinoids and its feasibility is unlikely due to the structural similarity between all ECs and with other, highly abundant endogenous substances.

29.3.2 Gas Chromatography–Mass Spectrometry (GC-MS)

The general analytical approach consists in a combination of a separation technique (gas chromatography) with the selective detection technique of mass spectrometry, which is monitoring masses of ionized molecules and corresponding fragments. GS-MS requires an initial stabilization of most analytes (derivatization) because gas chromatography requires the evaporation of molecules prior to its separation in chromatographic columns. This includes an extra procedure of sample preprocessing, e.g., formation of pentafluoropropionyl- (Zoerner et al. 2009) dimethyl isopropylsilyl- (Obata et al. 2003) or trimethylsilyl derivatives of target compounds (Amendola et al. 2003) representing a significant extra effort which devaluates this approach.

29.3.3 Liquid Chromatographic–Mass Spectrometry (LC-MS)

The combination of liquid chromatography and mass spectrometry represents the default technique for both substance groups, endocannabinoids (Lam et al. 2008, 2010; Ozalp and Barroso 2009; Zhang et al. 2009; Zoerner et al. 2011) and glucocorticoids. The general advantage over gas-chromatography coupling consists in reduced sample preparation and enhanced comprehensiveness of screening procedures. Hyphenated techniques need to be applied to obtain a sufficient analytical selectivity and sensitivity. This improvement is nowadays technically achieved by either a combination of two mass spectrometers (tandem mass spectrometry) and/or the elevation of mass resolution. Tandem mass spectrometer uses the concept of monitoring fragmentation reactions instead of detecting ions. Anandamide for instance (Fig. 29.4) may be tested in single or tandem mode at unit (one mass unit) or high (Res = $M/\Delta M$ ~16,000) resolution. A combination of tandem and high-resolution mass spectrometry (Ottria et al. 2014) proved to be the most robust and sensitive option. Respective analytical parameters are shown in Table 29.1, results from a representative plasma sample are shown in Fig. 29.5.

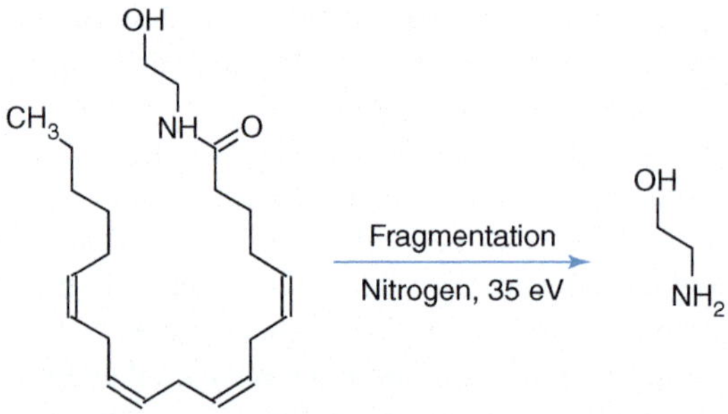

Anandamide: MS Mode	Single	Tandem
Low resolution	348.3 ± 0.5 Da	348.3 → (62 ± 0.5) Da
High resolution	348.289 ± 0.006 Da	348.3 → (62.064 ± 0.002) Da

Fig. 29.4 Potential mass spectrometric detection modes in LC-MS monitoring of anandamide. The intact molecule or its fragmentation reaction—i.e., the selective formation of ethanolamide by fragmentation of anandamide—can be monitored either in unit resolution (1 Da mass window) or focusing on a narrow mass detection window in high-resolution MS

Table 29.1 Modified (so far unpublished) experimental parameters of the LC-MS/HRMS assay

Compound	Precursor *m/z* [Da]	CE [eV]	Product *m/z* [Da]
2-AG	379.2	21	269.2264 287.2369
AEA	348.2	35	62.064
OEA	328.2	22	62.064
SEA	326.2	25	62.064
PEA	300.2	35	62.064
F	363.2	24	309.1853 345.2066 327.1955
E	361.2	38	163.111

Details on sample preprocessing and LC parameters and published elsewhere (Krumbholz et al. 2013, 2018)

29.3.3.1 Potential Implementation of MS Assays in Space

Any determination of biochemical parameters in isolated environments, e.g., space missions, needs to balance corresponding efforts between either sample return—facing logistic and stability issues—or an implementation of suitable techniques on site. Technically, a clear shift of biochemical detection techniques from sensitive target testing by immunoassays to flexible multitarget testing procedures using mass spectrometry is generally observed in clinical and forensic analysis. This is due to the fact, that mass spectrometers became recently more sensitive, selective, and robust. Mass spectrometry consists in accurate measurement of molecular masses, which necessarily requires their ionization and subsequent acceleration. The main technical prerequisite of mass spectrometry—besides high acceleration voltages—is the generation of high vacuum to allow undisturbed separation of ions.

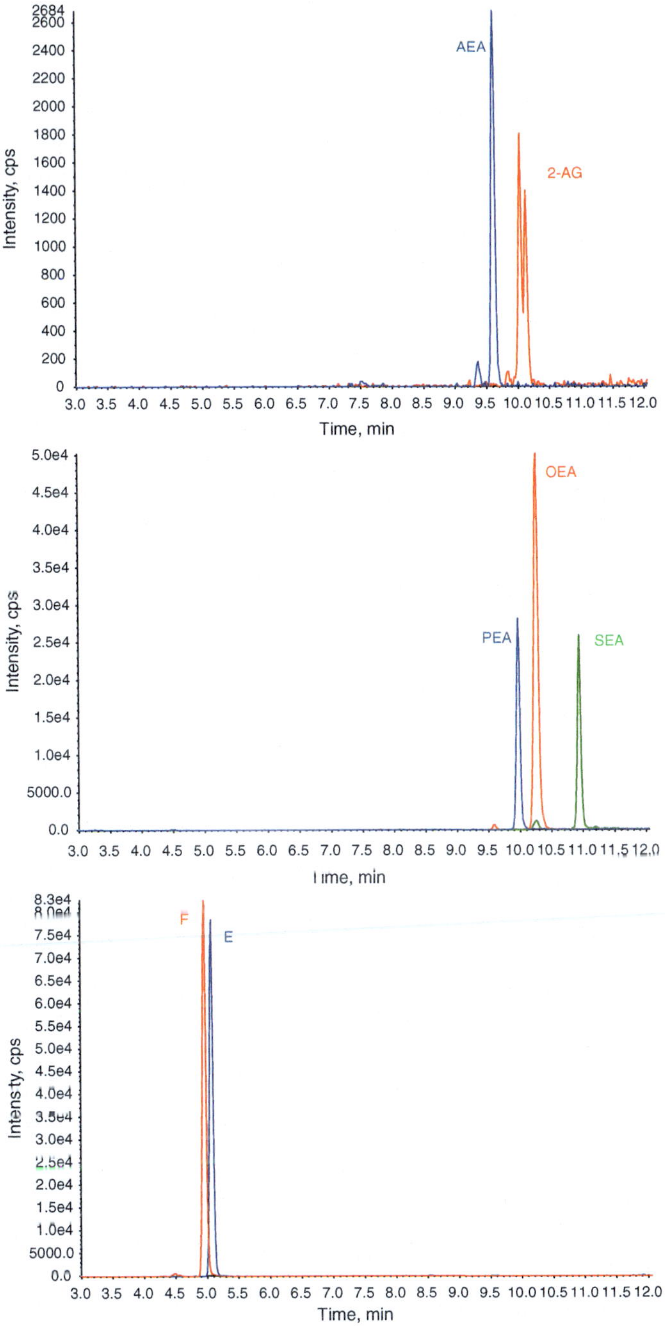

Fig. 29.5 Simultaneous monitoring of ECs (AEA, 2-AG, OEA, PEA, SEA) and glucocorticoids (cortisol (F) and cortisone (E), deuterated standards of AEA, 2-AG, PEA, F and E not shown) in an ether extract obtained from a plasma sample permits a direct access to corresponding stress markers from both substance groups

Advanced mass spectrometers typically generate vacuum better than 10^{-5} Pa and corresponding pumps cover the majority of mass, volume and energy consumption of current instruments. Compared to the situation at sea level, the effort of evacuation may significantly be reduced in space. Therefore, numerous attempts of utilizing MS in space missions were e.g., dedicated to measure gas composition, search for organic matter or quantify stable isotope compositions (Biemann et al. 1976; Evans-Nguyen et al. 2008). Figure 29.6 illustrates the pressure reduction at elevated altitude, describing a rather complex variation due to temperature fluctuations within the thermosphere. However, pressure values at an altitude >350 km appear to be compatible with requirements of mass spectrometry. This would enable to apply powerful and versatile MS detectors with low consumption of energy and limited space requirement. Moreover, investigation of biochemical matrices always includes an initial chromatographic separation, which means typically liquid chromatography. Neither respective LC-instruments nor the application of solvents constitute relevant safety hazards or unconquerable technical challenges. It appears likely that utilization of atmospheric vacuum and further miniaturization of MS detectors will soon become efficient and flexible alternatives to an elaborate storage and return of samples.

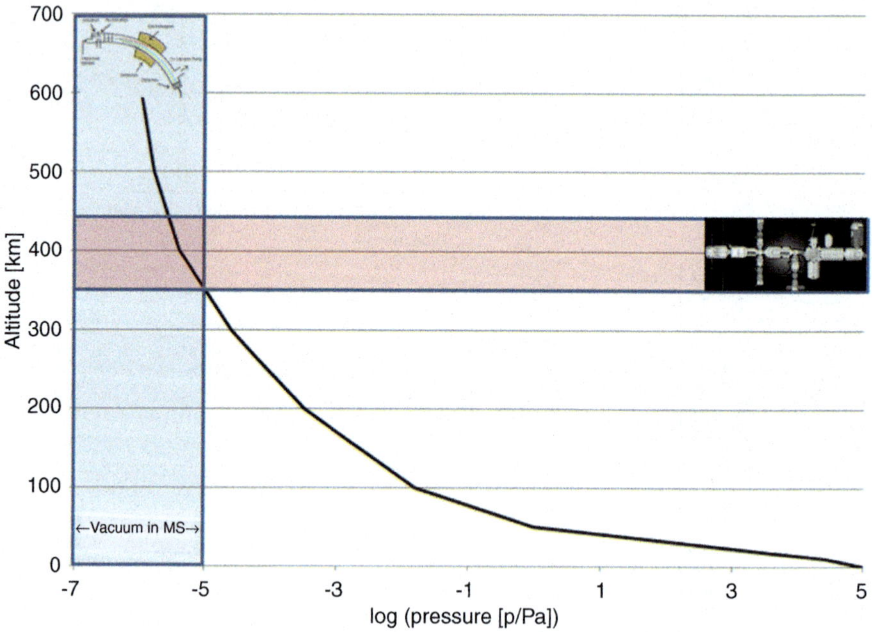

Fig. 29.6 Reduction of atmospheric pressure at elevated altitude. Vacuum levels at an altitude >350 km provide mean free path lengths of molecules larger than 100 m and are hence compatible with requirements of mass spectrometry

29.4 Analytical Results

Plasma concentrations of all relevant compounds do not pose an analytical challenge, reproducibility and detection limits of the established procedure proved to be more than adequate. Significant deviations from expected behavior and potential outliers are due to sample preprocessing issues, in particular the critical time prior to centrifugation and freezing of plasma samples. In particular the excessively high 2-AG concentrations up-to 400 ng/ml, i.e. 40 times the median value, could be analytically verified, but could not be attributed to exceptional situations and were attributed to storage limitations. The association of EC pairs was in good accordance with theoretical considerations, in particular the high correlation of ethanolamide pairs (Fig. 29.7, top) demonstrates a gratifying analytical robustness. The moderate correlation between AEA and cortisol would be in good accordance

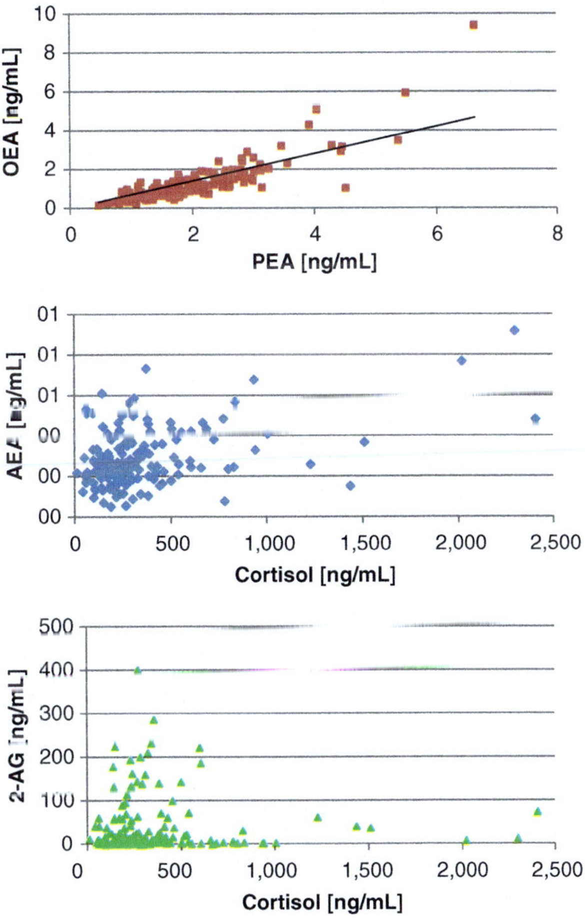

Fig. 29.7 The highest correlation ($p = 0.851$) was observed between different ethanolamides (PEA and OEA) concentrations in plasma, compared to low correlation between AEA and F ($p = 0.440$) and uncorrelated levels of 2-AG and cortisol (0.006). Patients and volunteers from different clinical and scientific studies were summarized to this statistical evaluation (Atypical high values of F are thought to be due to infections of patients or its treatment in a respective inflammation study.)

with an anandamide-modulation of chronic stress causing elevated cortisol levels (Fig. 29.7, middle), whereas 2-AG do apparently not correlate with corresponding cortisol concentrations (Fig. 29.7, bottom).

Hair testing includes higher analytical requirements, because of the limited amount of sample material. The available amount of sample remained often below the desirable amount of 50 mg hair. The resulting amount of biological material is therefore 10–30 times lower than the mass of corresponding plasma samples. The limit of quantitation (LOQ) of the most crucial compound AEA was found to be 0.3 pg/mg (Krumbholz et al. 2013) and is hence in the order of magnitude of the median value of all samples whereas LOQs of other ethanolamides (OEA and PEA) were way below basal hair concentrations. This is due to the fact that relative concentrations differ significantly between hair and plasma. Concentration ratios of PEA to AEA were found to be 7 in plasma but ~1000 in hair (Table 29.2), demonstrating different stability and/or hair incorporation behavior. The latter deems unlikely, because neither the basicity (governed by the ethanolamide group) nor the lipophilicity (controlled by structure of respective fatty acids) suggests significant physicochemical variations. It is therefore likely, that pharmacokinetic variations, in particular a limited temporal plasma stability of OEA and PEA may cause their increased hair concentrations.

However, it needs to be acknowledged that plasma and hair concentrations cannot be easily converted from hair to plasma, not even concentration ratios are suitable for direct comparison. Any quantitative evaluation requires highest standardization of all relevant constraints. The benefit of plasma samples consists in the access to short term variations, e.g. in cases of acute stress. This can turn out to be a disadvantage, if these short term variations are overlaid by endogenous (circadian) or pathological fluctuations. The logic benefit of hair testing is the averaging of high frequency fluctuations to quarterly (if 3 cm segments were collected) averages and the access to long-term trends. Short term variations are in return not accessible, the shortest feasible segmentation of hair samples amounts to 1 mm segments, corresponding to an average time window of 3 days.

Exemplary data from few participants of an overwintering mission to CONCORDIA Station in the high Antarctic at Dome C (see also Chap. 16)

Table 29.2 Comparison of endocannabinoid concentrations in blood and hair

	2-AG	AEA	OEA	PEA
Plasma concentration [ng/mL]				
Min	0.80	0.05	0.20	0.47
Max	402.3	0.92	9.42	6.64
Median	10.82	0.26	1.01	1.79
Hair concentration [pg/mg]				
Min	181.5	0.20	359.6	897.8
Max	4476	33.16	2041	3929
Median	1142	2.0	884.6	2032

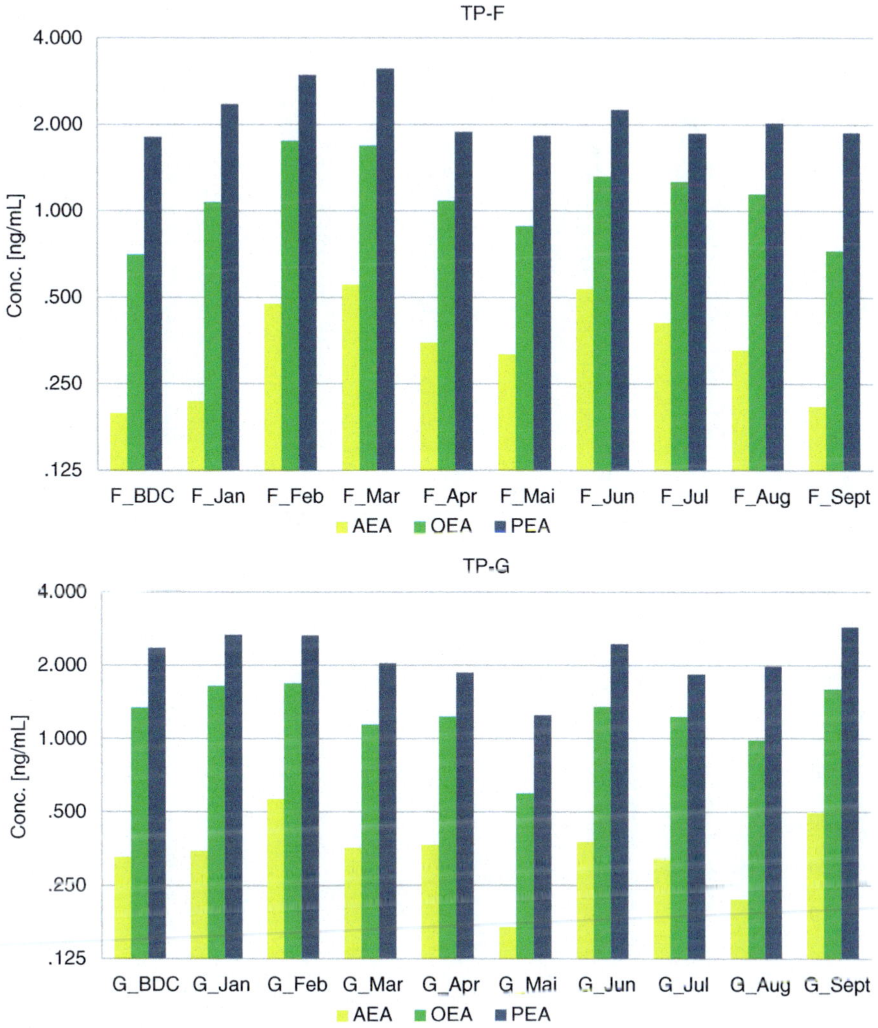

Fig. 29.8 Synchronized variations of the ethanolamide-type endocannabinoids AEA, OEA, and PEA in two of a 9-month isolation study in plasma, compared to baseline data collection (BDC)

(Figs. 29.8 and 29.9) demonstrate the tendency of moderate increase of endocannabinoids during isolation on the Antarctica station. This effect is observed for different individuals and markers (AEA, OPE, PEA) in plasma and was in principle confirmed in hair samples, keeping in mind that there is a temporal offset of 3 month, according to segmentation length. However, individual interpretations require a comprehensive analysis of all corresponding data and information.

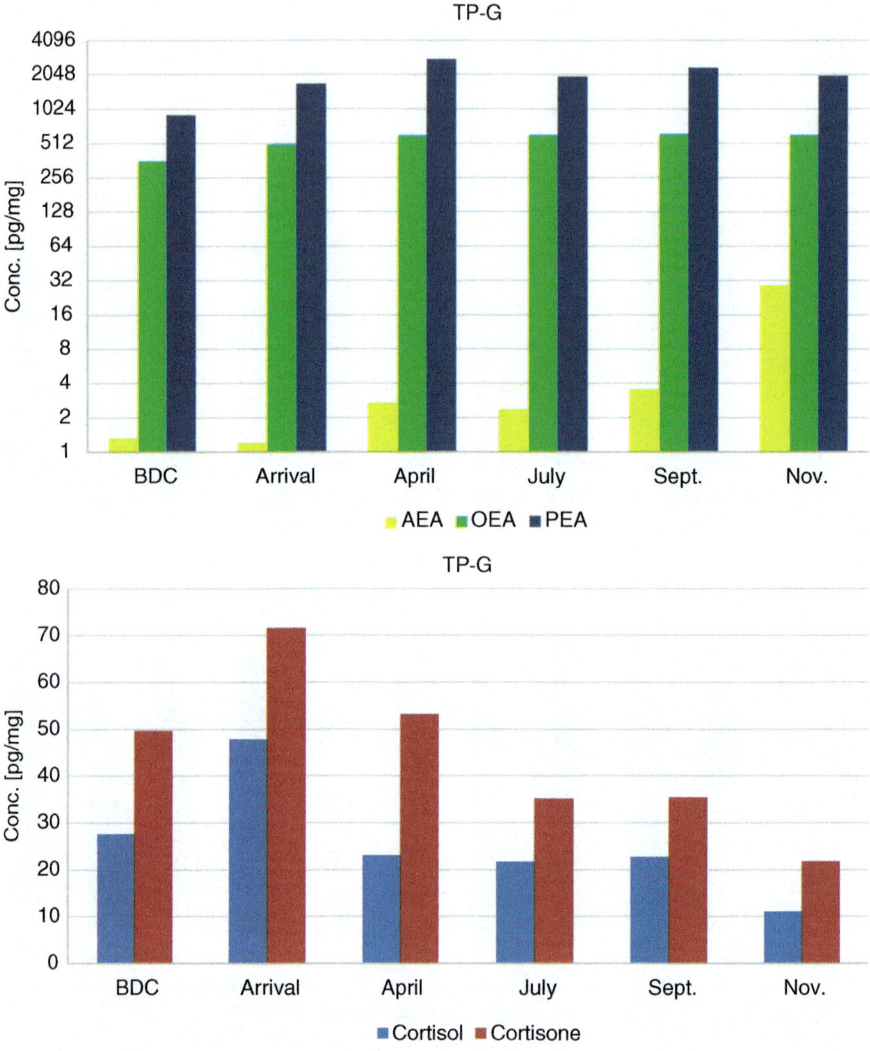

Fig. 29.9 Variations of the ethanolamide-type endocannabinoids AEA, OEA, PEA, and of glucocorticoids in two participants of a 9-month isolation study in hairs, compared to baseline data collection (BDC). The length of each proximal hair sample was 3 cm representing the recent quarterly period before collection

29.5 Discussion

Space is a stressful environment and any analytical evaluation of biochemical stress markers needs to account for numerous confounding factors which may significantly affect its interpretation. Appropriate biochemical markers need to be defined to reflect the relevant stress kinetics, e.g., acute vs. chronic mechanisms to cope external influences. These markers—or corresponding diagnostic biotransformation products—need to be sufficiently stable in the relevant timeframe of stress adaptation and differ significantly from basal values. Another critical parameter is the selection of suitable biological matrices which reflects stress-related variations of its markers in a diagnostic way and permits retrospective diagnostics. Blood represents the most suitable and logical compromise between diagnostic value and practical feasibility (i.e., low-invasiveness) of sample collection. However, also hair samples start to serve for certain parameters as a very reliable biospecimen. Because of the growth rate they may not allow high time resolution of days or hours, but permit an integration of stress responses over several weeks and months, as a retrospective calendar of stress. Any other common specimen is affected by additional biotransformation reactions (in particular urine) and by substance-specific incorporation or excretion mechanisms and lacks specificity and sensitivity.

Interpretation of presumptive stress as based on monitoring of its biochemical markers is always a complex challenge, because respective markers may be affected by the variables and their combinations as listed (see box below):

- Stress,
- Chemical instability (isomerization of 2-AG, hydrolysis of ECs) in plasma, wash-out of glucocorticoids from hair,
- Influence of nutrition and exposure to sunlight,
- Temporal, e.g., circadian variations (Valenti et al. 2004),
- Exceptional metabolic statuses, e.g., pregnancy (Abbassi-Ghanavati et al. 2009), potentially leading to duplication of cortisol levels in third trimester,
- Acute diseases (e.g., infections) causing metabolic syndromes or influencing the endocrine system,
- Use (or abuse) of any steroid involved in the endogenous synthesis of glucocorticoids including its endogenous precursors (Hengevoss et al. 2015).

Therefore, it deems helpful to combine different markers and matrices to verify the consistency of analytical results. Plasma and hair samples cover different time windows and can provide complementary data on acute vs. chronic stress. Introduction of apparently redundant (i.e., highly correlated) markers may help to identify analytical issues.

From an analytical point of view, respective data exhibits a high robustness and reproducibility. Finally, analytical techniques for the quantitation of respective markers need to be considered. Immunoassays and other specific low-cost screening methods are often well established under well-controlled default conditions, e.g., clinical chemistry, but cannot be easily extended to different matrices and situations. Therefore hyphenated techniques, in particular the use of mass spectrometry in combination to chromatic separations, became a golden standard in stress diagnostic. This can typically not be accomplished in a local manner including the problem of proper stabilization and transporting of sample material.

The high correlation of ethanolamides, i.e., anandamide and its analogs OEA, PEA and SEA demonstrate their synchronized variation under various conditions whereas 2-AG acts apparently more independently. A moderate correlation between AEA and cortisol would be in line with the concept of a modulation of chronic stress causing elevated cortisol levels by ethanolamides.

Topical data from small cohorts involved in a 9-month isolation study (Antarctica Concordia station on the high Antarctic plateau) suggested mild variations of endocannabinoids plasma concentrations relative to the basal levels, whereas this was more pronounced and elevated during overwintering in coastal stations (Neumayer III base). Moreover there were seen in preliminary analyzed samples variations of endocannabinoid and particularly glucocorticoid concentrations in hair samples. However, a final assessment of the data requires the statistical evaluation of the total cohort and the consideration of individual confounding factors.

29.6 Summary

Monitoring stress is challenged by the variability of the compliance when psychosocial stress is assessed by questionnaires and requires new and innovative tool (see Chaps. 22 and 23). Monitoring the extent and time slope of the adaptive responses of the human body to extreme environments seems to be very compelling. To assess the degree of allostatic load (Chap. 4) in a low-invasive or noninvasive fashion can be of importance to maintain crew health and to include and apply appropriate countermeasures. Moreover, ground analogs to mirror exploration class mission scenarios, such as winterover deployments to Antarctica, could also help to identify the effects of living conditions and to modulate the latter to reduce overspill of stress hormone release with all its effects as described in this volume on mood, cognitive performance, memory, metabolism, and immunity, to name a few.

Analytically there are two technical options, i.e., using low invasive specimens, e.g., saliva, to specifically determine selected biochemical markers (e.g., cortisol) on site with minimum technical effort for sample preparation and quantitation. Respective assays are highly selective and sensitive but focused on very few biochemical targets. Any comprehensive "stress screening" including acute and chronic markers and modulators is technically more challenging and requires thoroughly sample preprocessing to preserve the mostly unstable biochemical markers and subsequent return of the samples under controlled conditions. Both options include significant logistic efforts. An alternative application of hair testing was considered as a compromise, because of the high stability of compounds incorporated in the hair matrix. Target concentrations of stress marker—whether glucocorticoids or endocannabinoids—were found to be sufficient for combination to multitarget quantitation procedures based on high-resolution mass spectrometry. Moreover, segmentation of hairs permits a retrospective evaluation of concentration profiles which may give rise to estimation of stress variations. Concentration profiles of cortisol, cortisone and 5 endocannabinoids could be determined. A clear limitation of the procedure consists in its limited temporal resolution. Segment lengths of ~2 mm corresponding to an average growth cycle of a week are supposed to represent the minimum time span which could be tackled.

References

Abbassi-Ghanavati M, Greer LG, Cunningham FG (2009) Obstet Gynecol 114:1326

Amendola L, Garribb F, Botrè F (2003) Anal Chim Acta 489:233

Ballantyne C (2007) Sci Am Vol. 22.09.2018, www.scientificamerican.com/article/fact-or-fiction-stress-causes-gray-hair/

Bastin P, Maiter D, Gruson D (2018) Ann Biol Clin (Paris)

Biemann K, Oro J, Toulmin P 3rd, Orgel LE, Nier AO, Anderson DM, Simmonds PG, Flory D, Diaz AV, Rushneck DR, Biller JA (1976) Science 194:72

Chouker A, Kaufmann I, Kreth S, Hauer D, Feuerecker M, Thieme D, Vogeser M, Thiel M, Schelling G (2010) PLoS One 5:e10752

Cooper G (2015) In: Kintz P, Salomone A, Vincenti M (eds) Hair Analysis. Elsevier, Amsterdam, p 1

Dlugos A, Childs E, Stuhr KL, Hillard CJ, de Wit H (2012) Neuropsychopharmacology 37:2416

Duvivier WF, van Putten MR, van Beek TA, Nielen MW (2016) Anal Chem 88.2489

El-Farhan N, Pickett A, Ducroq D, Bailey C, Mitchem K, Morgan N, Armston A, Jones L, Evans C, Rees DA (2013) Clin Endocrinol 78:673

Evans-Nguyen T, Becker L, Doroshenko V, Cotter RJ (2008) Int J Mass Spectrom 278:170

Finn DP (2009) Immunobiology 215:629

Griebel G, Stemmelin J, Lopez-Grancha M, Fauchey V, Slowinski F, Pichat P, Dargazanli G, Abouabdellah A, Cohen C, Bergis OE (2018) Sci Rep 8:2416

Hauer D, Schelling G, Gola H, Campolongo P, Morath J, Roozendaal B, Hamuni G, Karabatsiakis A, Atsak P, Vogeser M, Kolassa IT (2014) PLoS One 8:e62741

Hengevoss J, Piechotta M, Muller D, Hanft F, Parr MK, Schanzer W, Diel P (2015) J Steroid Biochem Mol Biol 150:86

Hill MN, McLaughlin RJ, Bingham B, Shrestha L, Lee TT, Gray JM, Hillard CJ, Gorzalka BB, Viau V (2010) Proc Natl Acad Sci U S A 107:9406

Hillard CJ (2014) Semin Immunol 26:380

Hillard CJ, Weinlander KM, Stuhr KL (2011) Neuroscience 204:207

Inder WJ, Dimeski G, Russell A (2012) Clin Endocrinol 77:645

Jiang HX, Ke BW, Liu J, Ma G, Hai KR, Gong DY, Yang Z, Zhou C (2018) Anesth Analg

Krumbholz A, Anielski P, Reisch N, Schelling G, Thieme D (2013) Ther Drug Monit 35:600

Krumbholz A, Schonfelder M, Hofmann H, Thieme D (2018) Forensic Sci Int 286:23

Lam PM, Marczylo TH, El-Talatini M, Finney M, Nallendran V, Taylor AH, Konje JC (2008) Anal Biochem 380:195

Lam PM, Marczylo TH, Konje JC (2010) Anal Bioanal Chem 398:2089

Long JZ, Nomura DK, Cravatt BF (2009) Chem Biol 16:744

Moore C, Rana S, Coulter C (2007) J Chromatogr B Analyt Technol Biomed Life Sci 852:459

Musshoff F, Arrey T, Strupat K (2013) Drug Test Anal (5):361

Obata T, Sakurai Y, Kase Y, Tanifuji Y, Horiguchi T (2003) J Chromatogr B Analyt Technol Biomed Life Sci 792:131

Ottria R, Ravelli A, Gigli F, Ciuffreda P (2014) J Chromatogr B Analyt Technol Biomed Life Sci 958:83

Ozalp A, Barroso B (2009) Anal Biochem 395:68

Perogamvros I, Owen LJ, Keevil BG, Brabant G, Trainer PJ (2009) Clin Endocrinol 72:17

Pötsch L, Skopp G (2004) In: Madea B, Mußhoff F (eds) Haaranalytik. Deutscher Ärzte-Verlag GmbH, Köln, p 29

Ratano P, Petrella C, Forti F, Passeri PP, Morena M, Palmery M, Trezza V, Severini C, Campolongo P (2018) Neuropharmacology 138:210
Sharkey KA, Wiley JW (2016) Gastroenterology 151:252
Thieme D, Sachs H (2007) Forensic Sci Int 166:110
Thieme U, Schelling G, Hauer D, Greif R, Dame T, Laubender RP, Bernhard W, Thieme D, Campolongo P, Theiler L (2014) Drug Test Anal 6:17
Valenti M, Vigano D, Casico MG, Rubino T, Steardo L, Parolaro D, Di Marzo V (2004) Cell Mol Life Sci 61:945
Vandevoorde S, Saha B, Mahadevan A, Razdan RK, Pertwee RG, Martin BR, Fowler CJ (2005) Biochem Biophys Res Commun 337:104
Vogeser M, Schelling G (2007) Clin Chem Lab Med 45:1023
Vogeser M, Hauer D, Azad SC, Huber E, Storr M, Schelling G (2006) Clin Chem Lab Med 44:488
Zhang MY, Gao Y, Btesh J, Kagan N, Kerns E, Samad TA, Chanda PK (2009) J Mass Spectrom 45:167
Zhang Q, Chen Z, Chen S, Xu Y, Deng H (2016) Steroids 118:61
Zoerner AA, Gutzki FM, Suchy MT, Beckmann B, Engeli S, Jordan J, Tsikas D (2009) J Chromatogr B Analyt Technol Biomed Life Sci 877:2909
Zoerner AA, Batkai S, Suchy MT, Gutzki FM, Engeli S, Jordan J, Tsikas D (2011) J Chromatogr B Analyt Technol Biomed Life Sci 883-884:161
Zoerner AA, Gutzki FM, Batkai S, May M, Rakers C, Engeli S, Jordan J, Tsikas D (2018) Biochim Biophys Acta 1811:706

Part V

Therapeutic Strategies

Preventive and Therapeutic Strategies to Counter Immune System Dysfunctioning During Spaceflight

30

Jean-Pol Frippiat, Sergey A. Ponomarev, Martina Heer, Brian Crucian, and Alexander Choukér

30.1 Considerations for Preventive and Therapeutic Strategies

Because about half of the astronauts have experienced altered immune functions during spaceflight (Kimzey 1977; Crucian et al. 2016a), one might consider robotic missions to circumvent any further risk to humans on mission to space. However, even though robotic mission will undoubtedly precede and further accompany manned exploration mission ("human robotic exploration teams"), the capabilities of man to support space exploration missions with his capacity to estimate situation by previous experiences, to change plans considering complex conditions, develop and communicate ideas and not just data, and react intuitively makes him irreplaceable for now. It is therefore crucial to develop preventive and therapeutic strategies to preserve astronaut health.

J.-P. Frippiat (✉)
Stress Immunity Pathogens Laboratory, EA 7300, Faculty of Medicine, Lorraine University, Vandœuvre-lès-Nancy, France
e-mail: jean-pol.frippiat@univ-lorraine.fr

S. A. Ponomarev
Institute of the Russian Academy of Sciences, Moskva, Russia

M. Heer
Institute of Nutritional and Food Sciences, University of Bonn, Bonn, Germany
e-mail: martina.heer@uni-bonn.de

B. Crucian
NASA Johnson Space Center, Houston, TX, USA
e-mail: Brian.Crucian-1@nasa.gov

A. Choukér
Laboratory of Translational Research Stress and Immunity, Department of Anaesthesiology, Hospital of the University of Munich (LMU), Munich, Germany
e-mail: Alexander.Chouker@med.uni-muenchen.de

© Springer Nature Switzerland AG 2020
A. Choukér (ed.), *Stress Challenges and Immunity in Space*,
https://doi.org/10.1007/978-3-030-16996-1_30

Maintaining crew health requires the development of efficient measures to prevent from and combat the deleterious effects of spaceflight on the immune system before undertaking prolonged space voyages (Brumfiel 2007; Heppener 2008; Frippiat et al. 2016; Crucian et al. 2018). Indeed, as described elsewhere in this volume, the effect of spaceflight stressors on the immune system are manifold, especially since the "mobile" immune system is embedded and any targeted organ will likely affect cells of the immune system, and vice versa, and a countermeasure targeting the immune system will affect the other organs (see Fig. 30.1).

These preventive or even therapeutic measures include psychological (Chap. 31), physical (Chap. 32), nutritional (Chap. 33), microbiome (Chap. 34) as well as pharmacological (Chap. 35) approaches, alone or in combination, as summarized in Fig. 30.2. Several investigations have been conducted to investigate the role of psychological measures to mitigate the likelihood of psychological problems among the crew during extended space missions, the effectiveness of physical exercise to maintain and ameliorate mood, to maintain adequate immune allostasis, and thereby reduce health risks under those extreme conditions of life. Moreover, laboratories have undertaken studies to understand the impact of malnutrition (often reported in astronauts and cosmonauts (Smith et al. 2009) that may exacerbate immune function alterations in long-term spaceflight) and nutritional and pharmacological approaches to combat the deleterious effects of spaceflight on the immune system. Despite the increasing knowledge in the field, an appropriate understanding of the complex nutrition/immune interactions is required to find an approach to "the right" balance for nutrient supply or even supplementation. Gut microbiota is also important to consider because its equilibrium and composition are important factors for host defense against infection. Sympathetic nervous system (SNS) and hypothalamic-pituitary-adrenal (HPA) axis are strongly involved in the brain-gut axis that drives communication between the central nervous system and the gastrointestinal tract, including microbiota (Dinan and Cryan 2017) (see also Chap. 34). As an imbalance in gut microbiota could be correlated with a shift from a healthy to a diseased state, it is important to evaluate the status of astronauts' gut microbiomes for long-duration space missions. Indeed, spaceflight and terrestrial models of microgravity showed that the integrity of rodent intestinal microbiota and epithelium could be affected (Li et al. 2015; Ritchie et al. 2015; Shi et al. 2017; Alauzet et al. 2019).

Finally, it is essential to understand the need, timing, and efficiency in space for all kind of interventions triggered by adequate immune read-out parameters. This will require achieving a better understanding of stress challenges during space missions and the description of the environmental effects which can add also to operational reasons (e.g. lowering oxygen tension, see Chap. 16) and purely space environment (e.g. radiation, see Chap. 20).

Testing of an efficient and balanced set of nutrients and pharmacologics together with physical and psychosocial measures will open up new preventive and therapeutic tools helpful to counterbalance immune dysfunctions encountered in space and on Earth, such as those induced by aging or acute and chronic stress exposures (Godbout and Glaser 2006) resulting in immunosenescence (Benjamin et al. 2016; Crucian et al. 2016a, b; Pereira and Akbar 2016; Bektas et al. 2017; Mehta et al. 2017; Feuerecker et al. 2019; Hirano and Matsunaga 2018).

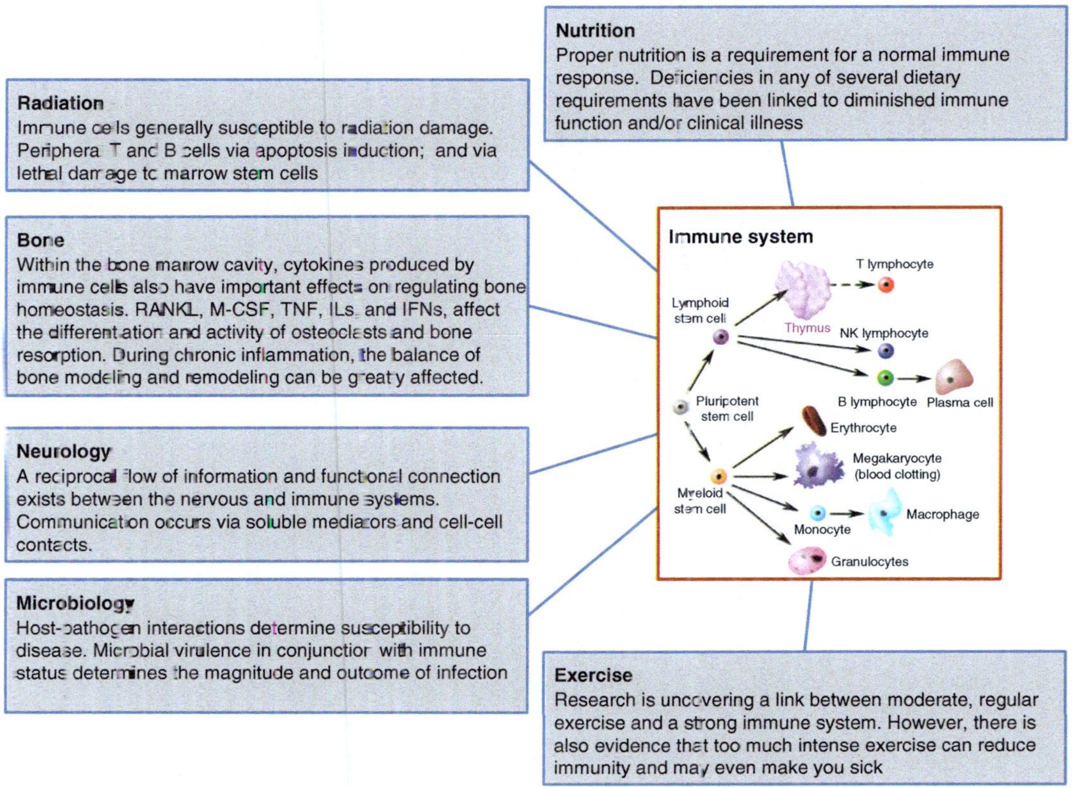

Fig. 30.1 Cells of the immune system, located virtually everywhere throughout the body, exchange reciprocal information and influence/are influenced by most of the other physiological systems in the human body. Therefore, immune countermeasures may influence other body systems, and conversely any biomedical countermeasures implemented (e.g., for bone) during flight may similarly have bystander effects on immunity. Cellular graphic sourced from "The Biology Project," University of Arizona

Detail slides

Nutrition

Vitamin D important promoter of T cell regulation:

↓ Vitamin D = Th1 shift,↓ Treg (Hamzaoui 2010)
T cell signaling/induction of PLC isoenzymes Vitamin D dependent (von Essen 2009)
Vit D enables early innate response to microbes (Schauber 2007)
Vit D deficiency = increased latent viral reactivation during Antarctic WO (Zwart 2010)

Other:

Protein energy malnutrition = impairment of CMI, Phagocyte function, cytokine production, etc.
Deficiency of zinc, selenium, glutathione or vitamins C, E, B6 (and others) correlates with reduction in T cell function
Antioxidants prevent free radical damage, maintain immune function
Omega 3: generally anti-inflammatory

Neurology (1)

•Neural targets that control thermogenesis, behavior, sleep, and mood can be affected by pro-inflammatory cytokines, which are released by activated macrophages and monocytes during infection. Within the central nervous system production of cytokines has been detected as a result of brain injury, during viral and bacterial infections, and in neurodegenerative processes
•Pathfinding molecules creating the complex pattern of neuronal connectivity in the brain and nervous system, the semaphorins and their receptors, Netrins, are expressed also in cells of the immune system and as such are components of a complex circuitry of positive and negative signals controlling neuroimmune functions.
•innervation of immune organs such as spleen, thymus, and lymph nodes by sympathetic and parasympathetic neurons
•participation of the sympathetic nervous system in immune functions is evidenced also by the observation that sympathectomy and lesioning of specific regions of the brain can both enhance and/or suppress immune responses.
•Firing rates of hypothalamic neurons have been shown to be altered during immune responses.
•Monocytes, macrophages and T-cells are able to cross the blood brain barrier. Macrophages can persist for very long intervals as resident microglial cells of the brain and constitute approximately 10 % of the total glial cell population. Activated T-cells are retained for days if they react specifically with central nervous system antigens. A variety of stimuli have been shown to induce expression of MHC molecules on astrocytes, microglial cells, and oligodendrocytes, which then can function to present antigens and to become targets for cytotoxic T-cells. Functionally significant concentrations of some neuropeptides are found also at sites of immune and inflammatory reactions.
•Many different classes of molecules, including cytokines, neurohormones, neurotransmitters, and many non-peptide mediators are involved in the amplification, coordination, and regulation of communication pathways within the neuroimmune system. Moreover, many classical cytokines of the immune system have been shown to be produced by a variety of brain cells, including neurons and glial cells.
•Neurohormonal involvement in immune reactions has been known for some time, in particular through the immunosuppressive effects of glucocorticoid hormones. Pituitary and/or hypothalamic hormones in turn are usually controlled negatively by end products of the particular neuroendocrine cascade; glucocorticoid hormones, for example, suppress ACTH production (see also: GIF, glucocorticoid increasing factor). These interactions form the basis of the physiologically important regulatory entity known as the **Hypothalamic-pituitary-adrenocortical axis**

Neurology (2)

•Many of these neuromediators and neurohormones have been shown to be released also by cells of the immune system in response to cell activation. Immune cells have been shown to produce authentic neuromodulatory molecules identical with those produced also by brain and nerve cells. Precisely how these factors can modulate immunity and/or neuronal processes (cell growth, survival, and differentiation) remains to be determined.
•classical cytokine mediators including, for example, IL1, IL2, IL6, TNF, LIF, IFN(interferons), thymic hormones, and bFGF, also have potent neuroendocrine activities
•IL1 is produced by pituitary cells. It has been shown to be as potent as corticotropin releasing hormone in some systems to induce the production of ACTH
•It has been shown that IL6, which is produced also by pituitary cells, is also a more potent secretagogue for ACTH than corticotropin releasing hormone in some systems. Intravenous injection of IL6 causes a dose-dependent increase in plasma ACTH levels.
•TNF-alpha is produced by astrocytes and probably also by pituitary cells. It stimulates adrenal and inhibits thyroid functions. All interferons (see: IFN) have been shown to act within the hypothalamic-pituitary-adrenal axis. IFN-gamma has been shown to inhibit corticostatin releasing hormone induced ACTH secretion by cultured pituitary cells.
•One of the interferons (see: IFN), IFN-beta, has been found to be an unconditioned stimulus signal responsible for the bidirectional communication which links the central nervous system with the immune system (augmentation of natural killer cell activity).
•It has been suggested that the immune system possesses sensory functions (Blalock, 1984). Leukocytes are capable of identifying stimuli/stressor signals that are not recognizable by the central and peripheral nervous system (see also: acute phase reaction). Cytokine receptors on cells of the immune and nervous system seem to play a sensory and regulatory role enabling the brain to monitor the progress of immune responses. The brain may be able also to modulate immune responses, for example, by using its neuroimmunomodulatory factors to alter the functional capacities of immune cells.
•The complexity of the system suggests that there will be no single unifying role of one individual factor in nervous and immune system interactions.

Fig. 30.1 (continued)

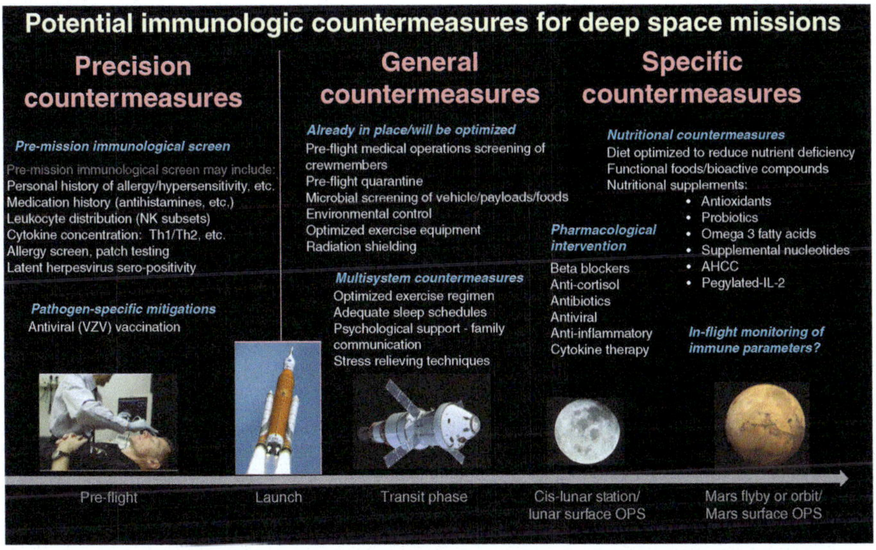

Fig. 30.2 Listing of likely biomedical and/or operational countermeasures for deep space exploration missions with the potential to benefit spaceflight-associated dysregulation of the human immune system. Broadly, the countermeasures may be considered general/operational multisystem influencing the entire crew, or specific biomedical treatments (generally ingestible). A third category, precision or personalized, describes the possibility to "screen" crewmembers pre-flight to determine their unique immunological biases or susceptibilities, against which specific in-flight countermeasures may be tailored for individual crewmembers. Reprinted from Crucian, Choukèr et al. (2018)

The most promising psychological, physical, nutritional, and pharmacologic approaches deduced from human investigations as well as from experimental setting are presented in the following chapters. However, this overview can neither cover all detailed strategies of combined action of tentative countermeasures, nor tentative concepts of immune modulation by e.g., boosting immunity through vaccinations in space. This last strategy may be less efficient than originally expected since somatic hypermutation, which diversifies antibody-binding sites to improve their quality, occurs in space following immunization but at a frequency that is two times lower than the frequency observed when animals are immunized on Earth (Bascove et al. 2011).

Acknowledgments The authors wish to thank all participants of the 2017 meeting of the European Space Agency's "Topical Team Stress Challenges and Immunity in Space." This international meeting was funded by the European Space Agency and focused on countermeasures for immune dysregulation during space missions. The authors highly appreciate also all the discussions and inputs that cannot be reflected or attributed to each meeting participant in this introduction but are reflected due to given limitations. The authors also wish to acknowledge support provided by the NASA Human Research Program, Human Health and Countermeasures Element to BC. SP is supported by the program of fundamental research (theme 65.1) of the Institute for Biomedical Problems (IBMP), JPF and his team are supported by CNES (DAR 4800000894 and DAR 4800000950), AC and MH are funded under the German National Space Program (DLR) on behalf of the German Ministry of Economics and Energy (for AC BMWi 50WB1622).

References

Alauzet C, Cunat L, Wack M, Lozniewski A, Busby H, Agrinier N, Cailliez-Grimal C, Frippiat JP (2019) Hypergravity disrupts murine intestinal microbiota. Sci Rep 9(1):9410. https://doi.org/10.1038/s41598-019-45153-8

Bascove M, Guéguinou N, Schaerlinger B, Gauquelin-Koch G, Frippiat JP (2011) Decrease in antibody somatic hypermutation frequency under extreme, extended spaceflight conditions. FASEB J 25:2947–2955. https://doi.org/10.1096/fj.11-185215

Bektas A, Schurman SH, Sen R, Ferrucci L (2017) Human T cell immunosenescence and inflammation in aging. J Leukoc Biol 102:977–988. https://doi.org/10.1189/jlb.3RI0716-335R

Benjamin CL, Stowe RP, St John L, Sams CF, Mehta SK, Crucian BE, Pierson DL, Komanduri KV (2016) Decreases in thymopoiesis of astronauts returning from space flight. JCI Insight 1:e88787. https://doi.org/10.1172/jci.insight.88787

Brumfiel G (2007) Space exploration: where 24 men have gone before. Nature 445:474–478. https://doi.org/10.1038/445474a

Crucian B, Babiak-Vazquez A, Johnston S, Pierson DL, Ott CM, Sams C (2016a) Incidence of clinical symptoms during long-duration orbital spaceflight. Int J Gen Med 9:383–391. https://doi.org/10.2147/IJGM.S114188

Crucian B, Johnston S, Mehta S, Stowe R, Uchakin P, Quiriarte H, Pierson D, Laudenslager ML, Sams C (2016b) A case of persistent skin rash and rhinitis with immune system dysregulation onboard the International Space Station. J Allergy Clin Immunol Pract 4:759–762.e8. https://doi.org/10.1016/j.jaip.2015.12.021

Crucian BE, Choukèr A, Simpson RJ, Mehta S, Marshall G, Smith SM, Zwart SR, Heer M, Ponomarev S, Whitmire A, Frippiat JP, Douglas GL, Lorenzi H, Buchheim JI, Makedonas G, Ginsburg GS, Ott CM, Pierson DL, Krieger SS, Baecker N, Sams C (2018) Immune system dysregulation during spaceflight: potential countermeasures for deep space exploration missions. Front Immunol 9:1437. https://doi.org/10.3389/fimmu.2018.01437

Dinan TG, Cryan JF (2017) The microbiome-gut-brain axis in health and disease. Gastroenterol Clin North Am 46:77–89. https://doi.org/10.1016/j.gtc.2016.09.007

Feuerecker M, Crucian BE, Quintens R, Buchheim JI, Salam AP, Rybka A, Moreels M, Strewe C, Stowe R, Mehta S, Schelling G, Thiel M, Baatout S, Sams C, Choukèr A (2019) Immune sensitization during one year in the Antarctic high altitude Concordia Environment. Allergy 74:64. https://doi.org/10.1111/all.13545

Frippiat JP, Crucian BE, de Quervain DJF, Grimm D, Montano N, Praun S et al (2016) Towards human exploration of space: the THESEUS review series on immunology research priorities. NPJ Microgravity. 2:16040. https://doi.org/10.1038/npjmgrav.2016.40

Godbout JP, Glaser R (2006) Stress-induced immune dysregulation: implications for wound healing, infectious disease and cancer. J Neuroimmune Pharmacol 1:421–427. https://doi.org/10.1007/s11481-006-9036-0

Heppener M (2008) Spaceward ho! The future of humans in space. EMBO Rep 9:S4–S12. https://doi.org/10.1038/embor.2008.98

Hirano T, Matsunaga K (2018) Late-onset asthma: current perspectives. J Asthma Allergy 11:19–27. https://doi.org/10.2147/JAA.S125948. eCollection 2018

Kimzey SL (1977) Hematology and immunology studies. In: Johnson RS, Dietlein LF (eds) Biomedical results from Skylab. National Aeronautics and Space Administration, U.S. Government Printing Office, Washington, DC, pp 249–282

Li P, Shi J, Zhang P, Wang K, Li J, Liu H, Zhou Y, Xu X, Hao J, Sun X, Pang X, Li Y, Wu H, Chen X, Ge Q (2015) Simulated microgravity disrupts intestinal homeostasis and increases colitis susceptibility. FASEB J 29:3263–3273. https://doi.org/10.1096/fj.15-271700

Mehta SK, Laudenslager ML, Stowe RP, Crucian BE, Feiveson AH, Sams CF, Pierson DL (2017) Latent virus reactivation in astronauts on the international space station. NPJ Microgravity 3:11. https://doi.org/10.1038/s41526-017-0015-y

Pereira BI, Akbar AN (2016) Convergence of innate and adaptive immunity during human aging. Front Immunol 7:445. https://doi.org/10.3389/fimmu.2016.00445
Ritchie LE, Taddeo SS, Weeks BR, Lima F, Bloomfield SA, Azcarate-Peril MA, Zwart SR, Smith SM, Turner ND (2015) Space environmental factor impacts upon murine colon microbiota and mucosal homeostasis. PLoS One 10:e0125792. https://doi.org/10.1371/journal.pone.0125792
Shi J, Wang Y, He J, Li P, Jin R, Wang K, Xu X, Hao J, Zhang Y, Liu H, Chen X, Wu H, Ge Q (2017) Intestinal microbiota contributes to colonic epithelial changes in simulated microgravity mouse model. FASEB J 31:3695–3709. https://doi.org/10.1096/fj.201700034R
Smith SM, Zwart SR, Kloeris V, Heer M (2009) Nutritional biochemistry of space flight. In: Nicogossian AE, Williams RS, Huntoon CL, Doarn CR, Polk JD, Schneider VS (eds) Space physiology and medicine: from evidence to practice. Nova Science Publishers Inc, New York, NY

31 Psychological Countermeasures

Gro Mjeldheim Sandal and Gloria R. Leon

31.1 Introduction

Psychological research has been recognized as a major part of understanding challenges associated with human space exploration. Long-duration missions may involve chronic exposure to many stressors that can negatively impact health, performance, and even safety. A growing body of research on crews working for extended periods in different kinds of isolated, confined environments reveals the existence of psychological and performance problems in varying degrees of magnitude. Psychological reactions have included depression, emotional lability, fatigue, sleep difficulties, decrements in cognitive functioning, psychosomatic symptoms, irritability toward crewmates, and/or mission control staff, and a considerable decline in vigor and motivation (Kanas 2015).

In the book "Dragonfly: NASA and the Crisis Aboard Mir," Bryan Burrough (1998) described many of these reactions among astronauts and cosmonauts who served aboard the Russian space station. For example, at one stage, communication between one of the NASA astronauts and the ship's cosmonaut commander was so limited as to be almost nonexistent. A veteran astronaut returned from Mir suffering from exhaustion and depression, blaming both on NASA's lack of ground support. Clashes of cultures were evident within the crew as well as among the personnel in the mission control particularly in terms of handling of safety-sensitive information.

G. M. Sandal (✉)
Department of Psychosocial Science, University of Bergen, Bergen, Norway
e-mail: gro.sandal@psysp.uib.no

G. R. Leon
Department of Psychology, University of Minnesota, Minnesota, MN, USA

© Springer Nature Switzerland AG 2020
A. Choukér (ed.), *Stress Challenges and Immunity in Space*,
https://doi.org/10.1007/978-3-030-16996-1_31

Together with other spaceflight factors, such as microgravity, exposure to prolonged stress can alter physiological responses, including the immune system, and thereby pose increased health risks for crewmembers during long-duration spaceflights (Pagel and Chouker 2016; Choukèr et al. 2001). The purpose of this chapter is to discuss countermeasures that may act to reduce or even prevent psychological and interpersonal problems during long-duration missions, and more generally to support the psychological resilience of crewmembers, which will have a positive impact on organ function, health, and mission success. In psychological science the concept of resilience refers to the capacity to cope successfully and even experience personal growth under significant adverse conditions (Fletcher and Sarkar 2013). Resources associated with psychological resilience in space crews go beyond individual traits and abilities, and also involve how the crew functions as a team, family characteristics, and organizational features (see Fig. 31.1). Actions to support or increase resilience may focus on different components of these aspects.

Kanas and Manzey (2008, p. 161) defined psychological countermeasures as including "all actions and measures that alleviate the effects of the extreme living and working conditions of space flight on crew performance and behavior." They distinguished between two complementary approaches.

The *first approach* focuses on the accommodation of the working and living conditions during space missions to the psychological capabilities and needs of humans, for example, ergonomics, work design, and work–rest scheduling.

The *second approach* focuses on adapting crewmembers to the psychological vicissitudes of space missions. The latter can be achieved by psychological selection of astronauts, crew composition based on interpersonal compatibility, preflight

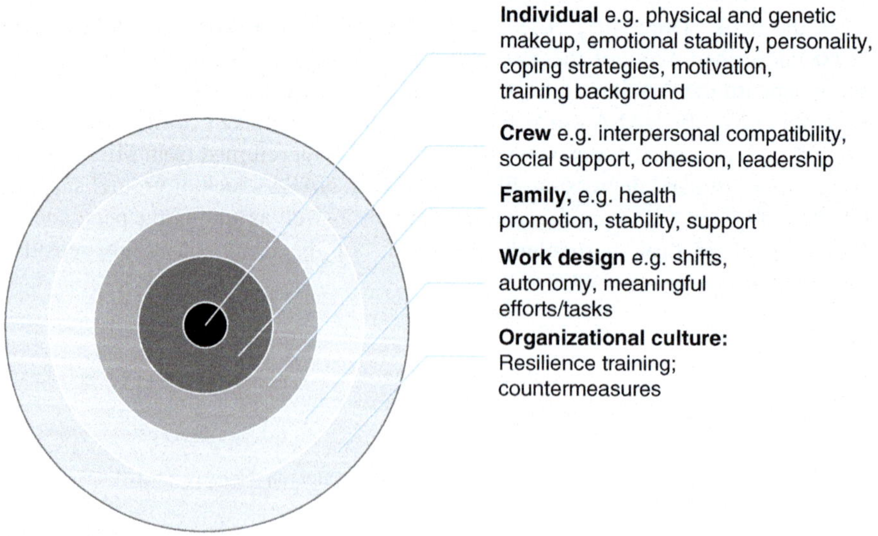

Fig. 31.1 Resource factors associated with psychological resilience can be represented by successive layers, from individual characteristics to organizational features

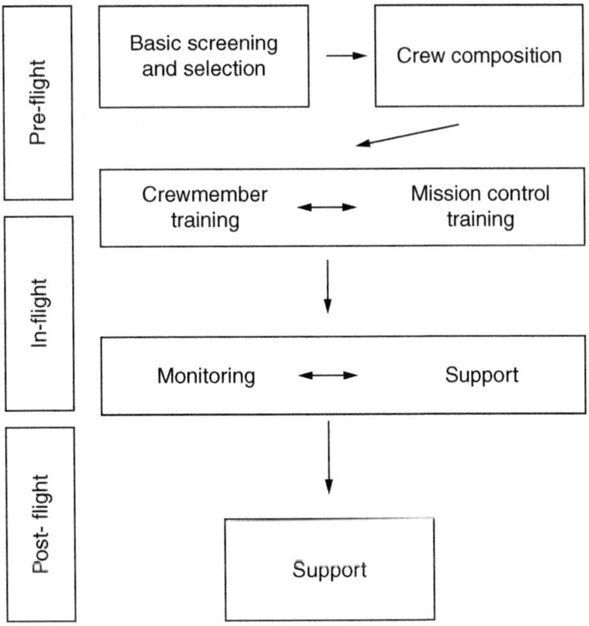

Fig. 31.2 Elements of psychological countermeasures (from Kanas and Manzey 2008. Reproduced with permission)

psychological training, and psychological support to individual astronauts and crews while they are on-orbit as well as to their families. In addition, postflight support activities assisting with the readjustment of astronauts to life on Earth after a space mission is also a psychological countermeasure, since they help to prevent adverse psychological aftereffects of spaceflight. As the former approach typically represents multidisciplinary efforts and approaches, the focus of this chapter will be on countermeasures falling into the latter category (see Fig. 31.2).

A review of the history of human spaceflight reveals large differences between national agencies in the extent to which psychological challenges have been officially recognized and addressed by appropriate countermeasures. From the beginning of the Soviet/Russian space program psychological countermeasures have been an important element. In contrast, attention to psychological factors in research and operations were not a high priority for NASA over a considerable period of time (Harrison and Fiedler 2011). Many observed that the NASA culture discouraged questions about the behavioral health of astronauts since they were assumed to have the "Right Stuff." The importance of psychological factors in mission success has become more salient as a result of experiences during the Russian/American Mir/Shuttle and the International Space Station (ISS) programs. Psychological countermeasures are now being used by all agencies in current operations on ISS and they will be an indispensable factor during future long-duration exploration space missions that go beyond Low Earth Orbit and eventually, a mission to Mars.

31.2 Selection

Selection is the process of identifying from a pool of applicants those who are best fit to work as astronauts (see also Chap. 3). At the individual level, the objective of selection strategies is twofold: to eliminate unfit or potentially unfit applicants, and to select from otherwise qualified candidates those who will adapt and perform optimally. A distinction is therefore made between "select-out" and "select-in" criteria.

Select-out: The select-out process most often involves psychiatric and psychological screening procedures designed to reduce the likelihood of psychological dysfunction and impaired performance. Psychiatric screening, interviews and standardized tests (in particular, the Minnesota Multiphasic Personality Inventory) have been used by the U.S., Russian, and European space agencies. Some mental disorders, for example, certain types of depression and anxiety disorders, as well as liability for schizophrenia have a considerable genetic component that might be detected by inquiries about family psychopathology as part of the interviews. As the field of epigenetics, genetic testing, and other psychometrically based predictors of dysfunction develop over time, this information will be used in screening, selection, and follow-up evaluations.

A challenging issue in selecting crews for future long-duration and planetary missions is the prediction of the later development of psychopathology in initially healthy individuals (Leon 2009; Sandal and Leon 2011). The onset of a serious disorder during a space mission would affect not only the disabled person but would also have a deleterious impact on the performance of the full crew, their safety, and potentially could jeopardize the mission. Passing a psychological/psychiatric screening examination at initial selection does not necessarily predict the absence of psychological/behavioral problems that might occur at a future time. Therefore, it is imperative that assessment of the psychological fitness of the astronauts is carried out on a regular basis and in particular before and after participation in space missions.

Select-in: Establishing assessment tools to select the best individuals for the astronaut corps from a pool of healthy applicants also represents a challenge. The psychological *select-in* process focuses on identifying candidates who, according to their operational aptitudes, attitudes, and personality characteristics, appear to be best suited for spaceflights. Best practices from organizational research emphasize that essential precursors for efficient selection involve three steps: (a) a job analysis identifying essential competencies; (b) ratings of criticality and importance of competencies; and (c) validation of assessment methods (Landon et al. 2017). Kanas (2015) pointed out that attempts to define select-in factors for astronaut selection have been based primarily on expert judgments, which may seem plausible but usually lack scientific validation. To date, the absence of formal criteria for assessing astronaut performance and the limited research opportunities for longitudinal studies have made it difficult to evaluate the predictive power of individual traits and capabilities in the context of space missions. However, relevant research has been conducted during shorter missions and in analog environments on Earth that entails some similarities to the conditions that astronauts experience in space. For example, a study of military units undertaking survival training identified strong associations between cortisol levels and personality traits that have been suggested as criteria for astronaut selection (Sandal et al. 1998). The stress hormone cortisol may have a negative impact on performance through the generation of anxiety and fatigue states, and may also affect health parameters though suppression of the immune system (see Chaps. 6 and 7).

There is growing interest in the study of resilience and hardiness (the latter defined more broadly as dispositional resilience, maintaining health while undergoing psychological stress) to identify individual characteristics associated with sustaining optimal performance in difficult conditions. For example, studies have demonstrated large individual differences in tolerance for sleep deprivation and shift work that in part are linked with personality traits that might be viewed as hardiness characteristics (for review, see Saksvik et al. 2011). Other researchers have demonstrated associations between hardiness and healthy immune and neuroendocrine responses to stress (Sandvik et al. 2013). Such research represents steps in providing empirical evidence on personality traits or styles that can be applied in identifying the "Right Stuff" in astronauts, and assist in defining scientifically sound select-in criteria.

NASA historically has focused its selection process on select-out procedures based on psychological tests and psychiatric interviews. The Soviet/Russian program, in addition, has included careful assessment of psychophysiological responses in induced high stress situations. Comparing the different select-in approaches applied in Russia, Japan, Canada, Europe, and the USA, there seems to be overall agreement that at least the following aspects need to be considered in evaluating psychological fitness for spaceflight: motivation, relevant biographical experiences, cognitive and psychomotor capabilities, stress-coping, and interpersonal and teamwork skills. Despite this consensus, differences in the national and organizational cultures of the space agencies are probably reflected in the choice of methods used as part of the selection process.

For the European Space Agency (ESA), representing 22 member states, an important consideration is that psychological selection methods should not favor or disfavor applicants from certain countries or language groups. In the selection round performed in 2008, a total of 8413 valid applications were received. After psychological, medical and professional screening (Maschke et al. 2011), six new ESA astronauts were selected. From an assessment point of view, assuring a fair and accurate evaluation in a multicultural and multinational pool of applicants raises a number of challenges. One issue is whether candidates should complete the same or different language versions of a psychological measure. Another issue is whether the results from multinational candidates should be compared using a common norm or the local norms of their countries. At present, there are no definite answers to these questions.

It is important to distinguish between select-in criteria to enter the astronaut- or cosmonaut corps and crew assignment for specific missions, for example essential individual characteristics for short-term missions may be completely different from those needed for long-duration missions.

31.3 Crew Composition

Psychological guidance in crew composition for specific missions involves consideration of the *interpersonal compatibility* of crewmembers assigned to particular missions. Interpersonal compatibility is a concept that describes the long-term interaction among individuals in relation to ease and comfort of communication and behavior. It assumes greater importance with increasing mission duration as long-term social monotony and confinement are likely to increase the risk for interpersonal tension that may negatively impact on psychological health and the ability of crewmembers to collaborate efficiently. While no general theory of interpersonal compatibility exists within psychology, research on groups operating in confined and isolated environments suggests that compatibility is determined by compatibility in personality characteristics, needs, values, and motivation (Sandal et al. 1995, 2011); expectations about specific work performance roles; and congruence of abilities with task requirements (Leon et al. 2011).

An important question is the selection of the most efficient tools for assessing interpersonal compatibility. Because congruent personality traits and values seem to play an important role, comparing individual scores on standardized instruments assessing such characteristics represents one approach. Russian space psychologists have based their compatibility assessments in part on psychophysiological testing undertaken during group exercises (e.g., the "Homoestat," Gazenko 1980). Yet, very little empirical research has been published about these methods in the accessible literature, and the theoretical basis of at least some of these approaches has been called into question (Manzey 2003; Santy 1994). At present, careful observation of group interactions in high-fidelity analogs or simulations, that may include behavioral exercises combined with mutual peer ratings, seems to be the most accepted basis for assessment of interpersonal compatibility across agencies (Inoue and Tachibana 2013; Maschke et al. 2011; Slack et al. 2014). Since the beginning of long-term flights in the Salyut 6 orbital station, the Russian space program has strongly emphasized interpersonal compatibility of crewmembers. In these earlier days, crew composition was decided well in advance of the mission, which allowed for observation of the interactions of crewmembers as they trained together on Earth. In the Mir program, the addition of astronauts from foreign agencies, whose flight assignment was not under the full responsibility of the Russians, added a new dimension of complexity to maintaining crew compatibility. Suddenly, the procedure of rotating one crewmember from a flight no longer allowed for the well-established practice of replacing the entire prime crew with the backup crew if the prime crew was found to be interpersonally incompatible. Within the ISS program, the shift from three to six crewmember operations, the rotation of individual crewmembers at different time intervals, and the required handover of leadership have injected a change. As a result of these constraints, it is hardly feasible to compose a psychologically guided crew based on observational data. Also, in practice, each partner agency nominates its own candidates for certain missions, and social, political, and cultural forces will always be contributing factors. On the other hand, it will be even more important for the success of long-duration exploration missions to use a psychologically informed method of composing the crew. While this already has been ascribed some significance for orbital space missions (Gazenko 1980; Manzey 2003), it will become a pivotal element for missions involving high levels of crew autonomy, such as a mission to Mars.

31.4 Pre-mission Training

Psychological training programs aim to prepare the crew for individual, psychosocial, and cultural challenges they may experience during the mission. Promoting the ability of crewmembers to cope with their individual reactions is one important aspect of pre-mission training. Many stressors involved in space missions are chronic and beyond the control of crewmembers. Under such conditions, there is a risk that they may experience sustained activation and an inability to relax. Ongoing activation in the skeletomuscular, vegetative, and endocrine systems may result in immunosuppression (see Chaps. 8–11) and represents a health risk. The negative effects of prolonged microgravity on the astronaut's immune system is well documented (Sonnenfeld 2002). It is therefore crucial that astronauts and cosmonauts learn to use efficient strategies to reduce activation during spaceflights. For example, learning principles for optimizing the quality of sleep ("sleep hygiene" and relaxation techniques) may be of great value based on the frequency of sleeping problems reported by crews in space. Equally essential with regard to long-duration missions, is to prepare crewmembers to cope with hypostimulation resulting from monotony and boredom. Given the psychosocial demands of long-duration, international missions and the interdependence between crewmembers for task accomplishment, conflict management and team building activities are also important aspects of pre-mission training. In addition, addressing language issues and cultural differences during training are particularly important for multinational missions. These activities aim to build an optimal crew cohesion, which is known to have a strong influence not only on the impact on performance but also on the psychological resilience of group members in stressful situations.

Of special concern for long-duration missions to Mars and beyond is the retention of optimal cognitive and perceptual-motor skills over long periods without "on-the-job" practice. To compensate for this problem, crewmembers and mission control personnel may receive computer-based psychosocial education refresher courses to remind them of key issues discussed prior to launch. Such programs were tested during a Mars mission simulation study at the Institute of Biomedical Problems (IBMP) in Moscow.

The Russian, American, and European space programs all offer team building and group activities as part of pre-mission training. However, published information about the contents and methods of training has been sparse. According to Santy (1994), Russian psychological training focused mainly on stress management and the familiarization with stressful events during field exercises like survival training, parachute jumping, or periods in isolation chambers. American approaches during the Mir/Shuttle program were limited to theoretical briefings to crewmembers and their families about psychiatric and psychological issues of long-duration spaceflight (Ritsher 2005). Now that the Russian, European, American, Canadian and Japanese space agencies are partners on joint projects such as the ISS, they are faced with challenges related to collaboration of multinational teams representing different national and organizational cultures. This situation has increased the need for empirically tested standardized training programs. In 2008, a working group that included experts and specialists from different disciplines and all major ISS international partners was tasked to develop the "International Space Station Human Behavior and Performance Competency Model" (Bessone et al. 2008) specifying training requirements to help ensure competencies that are critical for the success of long-duration missions. For example, the guidelines call for the early and frequent exposure of astronauts and cosmonauts to relevant ISS languages and cultures through professional trips and full-immersion language courses. This training aspect has also been emphasized by crewmembers who have experience with multinational missions.

Crewmembers and mission control personnel have emphasized that prelaunch psychosocial training should involve key members of both groups to prepare them for coping with psychological issues occurring during long-duration missions and to establish mutual trust. A frequent observation during actual as well as simulated space missions is that crewmembers express frustration about lack of empathy from mission control personnel (Gushin et al. 1997; Sandal et al. 1995; Kanas et al. 2000). Although this tendency to some extent has been interpreted to be a displacement of tension, it is extremely important that the space crew and the mission control are able to understand each other's perspectives. When training for a specific ISS mission, joint training of crewmembers and flight control teams is often not possible due to scheduling constraints. However, it is still possible to perform joint training at an earlier stage, namely, during basic and preassigned crew training. Exposing the crew and mission control teams to the same behavioral training, using similar case studies, situations, and role-play scenarios creates a basic, common culture and language for all operational personnel throughout the program.

31.5 In-Flight Support

Countermeasures necessary for long-duration space missions can be described as targeted at two interrelated actions. The first involves monitoring the mental, emotional, and psychophysiological state of crewmembers, prevention of and intervention when problems are evident; the second action involves several in-flight support measures to prevent feelings of monotony, boredom, and social isolation. In addition, countermeasures are needed to enhance the positive experiences of spaceflight, as well as mitigating interpersonal tensions among crewmembers and between space crews and mission control.

The purpose of behavioral *monitoring* is to identify the possible emergence of psychological dysfunction among crewmembers. Early, accurate detection of even mild symptoms and the rapid implementation of countermeasures to deal with these problems are extremely important in long-duration space missions because of the extraordinarily high level of sustained performance required for safety and mission success. Methods for psychological monitoring are dealt with in Chap. 22.

Basic elements of *in-flight support* involve helping astronauts to stay connected to life or Earth. In-flight support has been carried out in Russia since the earliest days of human spaceflight. Learning from this experience, a similar system has been established by other agencies. Activities have included frequent communication with families via two-way video, radio, and television programs, and delivery of gifts, letters, and favorite food via the cargo vehicles. In addition, the arrival of visiting astronauts and cosmonauts on Mir and ISS has helped break the monotony and provided stimulation and assistance in performing mission activities (Kanas 2015).

The spaceflight experience affects individual astronauts differently. The enhancement of individually tailored leisure time activities that takes into account changing interests and needs over the course of the mission is an important element of psychological support systems. As an illustration, clear differences in leisure pursuits were described among crewmembers on Mir: Shannon Lucid read books; John Blaha watched videos, and Andy Thomas sketched. The U.S. astronaut Jerry Linenger (2002) published a book "Letters from MIR: An astronaut's letters to his son" that he wrote during his period on the ISS. Chris Hadfield recorded a music video in space with his guitar; Scott Kelly took pictures that he posted on Twitter; and in 2016 Tim Peake completed a run that coincided with the London marathon (see also Chap. 32).

Family separation and concerns about beloved ones on Earth may represent a significant source of mood and morale change in people who work in extreme environments (Johnson 2012). Nonetheless, some astronauts indicate that they are able to set aside these concerns during training and flight through so-called "compartmentalization." Astronaut Daniel Tani was living on the space station, with no way to get back to Earth quickly, when his mother died in a car crash in 2007. He declined his bosses' offer to cut his workload and continued to work productively for the rest of his stay in orbit. Astronaut Mark Kelly decided to continue as Commander of a space shuttle mission even as his wife recovered from being shot. Providing psychological and social support to families during the mission can be helpful in maintaining the crewmembers' concentration on the objectives of the mission by relieving them of considerations about possible problems at home and feelings of responsibility.

Families need to be coached in interacting with their in-space family member and be prepared for possible psychological changes during and following the mission. Regularly scheduled private psychological conferences between crewmembers and ground consultants, using two-way audio or video transmissions, is routinely offered to crewmembers on the ISS. For this countermeasure to be effective, it remains essential that information is treated in a highly confidential manner. During international flights, crewmembers should be given the opportunity to speak with consultants who share the same cultural background and in their native language. On future missions to Mars, access to expertise on Earth will become more limited as most of the mission will occur at so great a distance that the communication delay will impair the effectiveness of the consultation. These communication challenges will mandate the use of methods for self-assessment to identify and deal with psychological/psychiatric or interpersonal disorders. Such systems are currently being developed with clinical populations and during simulated space missions. For example, computer-interactive instruction programs for the treatment of conflict, stress, and depression have been developed by investigators funded through the NASA-supported National Space Biomedical Research Institute (Anderson et al. 2016). The feasibility of an autonomous computer-based Virtual Space Station encompassing psychological training and treatment programs was recently evaluated in a high-fidelity analog environment (Anderson et al. 2016). The goal of these programs is to enable astronauts to overcome emotional and behavioral problems through teaching modules they can access in private.

Other types of virtual reality (VR) and robotic programs are being developed and tested with a focus on psychological and biological support for long-duration exploration missions to Mars. These include immersive scenes from the natural Earth environment to alleviate the feeling of distance from Earth and loved ones; however, questions have been raised whether Earth scenes might have a negative effect when turned off in increasing the sense of isolation from one's home environment. A ground-based study found promising results evaluating whether a software generated exercise partner (avatar) that accompanied/competed with a crewmember during exercise stints was effective as a motivational device (Feltz et al. 2016). A NEEMO analogue exercise study with positive findings assessed the motivational potential of running with a virtual coach avatar in a virtual environment (Hanson et al. 2017). In addition, a robotic crew companion/assistant with artificial intelligence is currently being tested on the ISS. The robot, named CIMON, was designed for the DLR to provide in a very first proof of concept verbal commands to ESA astronaut Alexander Gerst to facilitate his efforts conducting several onboard experiments (see also Chap. 21). The use of VR technology is still in its infancy; it opens new possibilities for psychological as well as technical support for astronauts during long-duration missions.

Psychoactive drugs always need to be part of the onboard medication kit as a countermeasure for emergency and chronic psychological/psychiatric problems developing during long-duration missions and have to complement other concepts of countermeasures (see Chap. 30). However, the choice of specific mood/behavioral altering medications needs to be based on research findings regarding their effectiveness in the space environment.

31.6 Postflight Support

Space voyagers have to deal with the return to the Earth environment and the adjustment to a more usual routine. Personal and marital problems that the crewmember tried to ignore during the mission may become more evident upon return, and behavioral health problems may be intensified by the external pressure and media attention postflight (Abraham and Aldrin 2009). Moreover, family problems may become exacerbated while the voyager is in space, particularly as the length of the mission increases. Access to and encouragement in making use of individual and family counseling for at least 1 year post mission could be needed, either directly through the particular space agency, or else through referral sources outside of the agency. The latter may be more comfortable for astronauts and their families who desire privacy and to alleviate concerns that the disclosure of problems could jeopardize assignment to future missions.

31.7 Concluding Remarks

The success of most human space missions and the numerous examples of ambitious tasks that have been accomplished are generally taken as evidence of the ability of most astronauts to perform and cope in space, both as individuals and as teams. Despite this, psychological and interpersonal challenges associated with long-duration missions must not be ignored. A positive development in the history of space exploration is that greater attention is being paid to the importance of psychological factors in mission success by all national agencies. Along with this recognition, psychological countermeasures have been more accepted as part of mission planning. Such countermeasures have the potential not only to help astronauts maintain performance and well-being, but also mitigate the potential for health problems due to immune dysfunctional states and other physiological alterations induced by the exposition to stress of the space environment (see Crucian et al. 2018). Evidence-based research is required to support the use of countermeasures, as well as the most effective methods of crew selection, training, and in-flight and postflight psychological support. For this purpose, high-fidelity ground-based models of spaceflight are of great value to supplement the limited flight studies in this area, also to test and validate the benefits and risks of new technologies involved, such as virtual reality or crew supporting artificial intelligence robotic entities. In addition, while human spaceflight used to be exclusively the domain of the massive governmental programs, the challenging era of space tourism is about to begin. The involvement of commercial entities flying people in space, even for short durations, raises questions regarding guidelines for pilot training, passenger selection and psychological and medical support needing attention by mission planners.

References

Abraham K, Aldrin B (2009) Voices from the Moon. Nature 460:328

Anderson AP, Fellows AM, Binsted KA et al (2016) Autonomous, computer-based behavioral health countermeasure evaluation at HI_SEAS Mars analog. Aerosp Med Hum Perform 87(11):912–920

Bessone L, Coffey E, Filippova N, Greenberg E, Inoue N, Gittens M et al (2008) International Space Station human behavior & performance competency model. NASA/TM-2008-214775, vol 2. NASA, Washington, DC

Burrough B (1998) Dragonfly: NASA and the crisis aboard MIR. Fourth Estate Limited, London
Choukèr A, Thiel M, Baranov V, Meshkov D, Peter K, Messmer K, Christ F (2001) Simulated microgravity, psychic stress, and immune cells in men: observations during 120-day 6° HDT. J Appl Physiol 90:1736–1743
Crucian BE, Choukèr A, Simpson RJ, Mehta S, Marshall G, Smith SM, Zwart SR, Heer M, Ponomarev S, Whitmire A, Frippiat JP, Douglas GL, Lorenzi H, Buchheim JI, Makedonas G, Ginsburg GS, Ott CM, Pierson DL, Krieger SS, Baecker N, Sams C (2018) Immune system dysregulation during spaceflight: potential countermeasures for deep space exploration missions. Front Immunol 28(9):1437. https://doi.org/10.3389/fimmu.2018.01437
Feltz DL, Ploutz-Snyder L, Winn B, Kerr NL, Pivarnik JM, Ede A, Hill C, Samendinger S, Jeffery W (2016) Simulated partners and collaborative exercise (SPACE) to boost motivation for astronauts: study protocol. BMC Psychol 4(1):54
Fletcher D, Sarkar M (2013) Psychological resilience. A review and critique of definitions, concepts, and theory. Eur J Psychol 18:12–23
Gazenko OG (1980) Psychological compatibility on Earth and in outer space. Aviat Space Environ Med 51:622–623
Gushin VI, Zaprisa NS, Kolinitchenko TB et al (1997) Content analysis of the crew communication with external communicants under prolonged isolation. Aviat Space Environ Med 68:1093–1098
Hanson AM, Kalogera K, Sandor A, Hardy M, Frank A, Amonette W, English K, Williams T, Perera J (2017) Evaluation of the next-gen exercise software interface in the NEEMO analog. NASA Human Research Program Investigators' Workshop, Galveston, TX. NASA, Washington, DC
Harrison AAF, Fiedler ER (2011) Introduction. Psychology and the U.S. space program. In: Vakoch DA (ed) Psychology of space exploration: contemporary research in historical perspective. National Aeronautics and Space Administration, Washington, DC, pp 1–16
Inoue N, Tachibana S (2013) An isolation and confinement facility for the selection of astronaut candidates. Aviat Space Environ Med 84(8):867–871
Johnson PJ (2012) The roles of NASA, U.S. astronauts, and their families in long-duration missions. In: Vakoch DA (ed) On orbit and beyond. Springer, New York, NY, pp 69–89
Kanas N (2015) Humans in space. The psychological hurdles. Springer, Cham
Kanas N, Manzey D (2008) Space psychology and psychiatry, 2nd edn. Springer, Dordrecht
Kanas N, Salnitskiy V, Grund E et al (2000) Interpersonal and cultural issues involving crews and ground personnel during Shuttle/Mir space missions. Aviat Space Environ Med 7:A11–A16
Landon LB, Rokholt C, Slack KJ, Pecena Y (2017) Selection best prectices, critical factors, and skills. In: Sgobba BKT, Clervoy JF, Sandal GM (eds) Space safety and human performance. Elsevier. Butterworth-Heinemann, Oxford, pp 731–764
Leon GR (2009) Strategies to optimize individual and team performance. In: Building a safer space together. The 3rd International Association for the Advancement of Space Safety Conference (IAASS), Rome 2009. European Space Agency Communication Production Office, Paris
Leon G, Sandal GM, Fink B et al (2011) Positive experiences and personal growth in a two-man North Pole expedition team. Environ Behav 43(5):710–731
Linenger JM (2002) Letters from MIR: an astronaut's letters to his son. McGraw-Hill Companies, New York, NY
Manzey D (2003) Study of the survivability and adaptation of humans to long-duration interplanetary and planetary environments. ESA/ESTEC, Noordwijk
Maschke P, Oubaid V, Pecena Y (2011) How do astronaut candidate profiles differ from airline pilot profiles? Results from the 2008/2009 ESA astronaut selection. Aviat Psychol Appl Hum Fact 1(1):38–44
Pagel JI, Chouker A (2016) Effects of isolation and confinement on humans-implications for manned space explorations. J Appl Physiol 120(12):1449–1457
Ritsher JB (2005) Cultural factors and the International Space Station. Aviat Space Environ Med 76(6, Suppl):B135–B144
Saksvik IB, Bjorvatn B, Hetland H, Sandal GM, Pallesen S (2011) Individual differences in tolerance to shift work--a systematic review. Sleep Med Rev 15(4):221–235

Sandal GM, Leon GR (2011) From the past to the future. In: Vakoch DA (ed) Psychology of space exploration: contemporary research in historical perspective (The NASA History Series). National Aeronautics and Space Administration, Washington, DC, pp 195–204
Sandal GM, Værnes R, Ursin H (1995) Interpersonal relations during simulated space missions. Aviat Space Environ Med 66:617–624
Sandal GM, Groenningsaeter H, Eriksen HR, Gravraakmo A, Birkeland K, Ursin H (1998) Personality and endocrine activation in military stress situations. Mil Psychol 10(1):45–61
Sandal GM, Bye HH, van de Vijver FJR (2011) Personal values and crew compatibility during a simulated space mission, vol 69, pp 141–149
Sandvik AM, Bartone PT, Hystad SW, Phillips TM, Thayer JF, Johnsen BH (2013) Psychological hardiness predicts neuroimmunological responses to stress. Psychol Health Med 18(6):705–713
Santy P (1994) Choosing the right stuff: the psychological selection of astronauts and cosmonauts. Praeger, Westport/London
Slack KJ, Holland A, Sipes W (2014). Selecting astronauts: the role of psychologists. Paper presented at the annual convention of the American Psychological Association, Washington, DC.
Sonnenfeld G (2002) The immune system in space and microgravity. Med Sci Sports Exerc 34(12):2021–2027

32 Physical Countermeasures to Stress

Vera Abeln, Alexander Choukér, and Stefan Schneider

32.1 Introduction

Since the first Apollo missions, exercise in space has been an important field of research (Halberg et al. 1970; Rummel et al. 1973; Rummell et al. 1975). For scientists it has always been a major interest how fundamental physiological systems as the cardiovascular and the musculoskeletal system are affected by missing gravity; how they adapt, how they degrade. A number of exercise ideas and concepts have evolved since that time, to minimize both, the effects of microgravity on degenerative processes as well as the amount of time that needs to be spent on exercise.

Besides its impact on the cardiovascular and musculoskeletal system, physical activity is also known to affect brain cortical function and to have a positive impact on psychophysiological parameters (Schneider et al. 2009b). Especially endurance training is regarded as a pertinent compensatory activity for different kinds of stress (Hollmann and Strüder 2000). So, in addition to the positive effects of exercise on the peripheral physiological system (Convertino 1996, 2002), it is assumed that adequate, individually specified exercise programs will help to maintain possible neurocognitive decrements of long-term confinement (Lipnicki and Gunga 2009) as well as counteract the effects of confinement on the mental health status in space (Palinkas and Suedfeld 2008) and result in an improvement of general well-being and mood (see Fig. 32.1).

Although in general positive effects of exercise on musculoskeletal and neurocognitive processes predominate, for further space missions it needs to be taken into consideration that specific problems might evolve out of different exercise

V. Abeln · S. Schneider (✉)
Institute for Movement and Neurosciences, German Sport University Cologne, Cologne, Germany
e-mail: schneider@dshs-koeln.de

A. Choukér
Laboratory of Translational Research Stress and Immunity, Department of Anaesthesiology, Hospital of the University of Munich (LMU), Munich, Germany

© Springer Nature Switzerland AG 2020
A. Choukér (ed.), *Stress Challenges and Immunity in Space*,
https://doi.org/10.1007/978-3-030-16996-1_32

Fig. 32.1 Astronaut Tim Peake during his marathon run in the microgravity condition at the International Space Station (© **ESA**)

programs. Besides the fact that individual exercise preferences need to be considered (Schneider et al. 2009a), exercise, if not adequately applied, might provoke additional stress and, therefore exercise recommendations for further space missions need to be carefully defined.

A Holistic Approach to Exercise: The Balance of Activity/Inactivity (ora et labora)

An important number of papers in the last decades have addressed the benefit of exercise on physiological and psychological well-being. Nevertheless, apart from some very detailed and complex physiological mechanisms, there are some anthropological mechanisms that might explain the power of exercise.

Looking back, physical habits of the human being underwent a dramatic change within the last 30 years. Throughout the evolution of mankind, until the late 1970s when the industrial age ended, working life was widely categorized by physical work (from hunters to pitmen). Unsurprisingly recreational exercise and sports were not very common before the 1970s as the physical work load during a working day did not allow for much further energy expenditure. In contrast, today working life is widely characterized by nonphysical work. Accordingly exercise has developed as a key variable to stay physically and mentally fit. This has been acknowledged by a recent initiative of the American Society of Sports Medicine (ACSM) "Exercise is medicine."

With regard to the underlying metabolic and hormonal changes caused by exercise, it can be assumed that a balance of activity and inactivity is important for a general well-being. This has been identified already by the regula benedicti, which sums up in the traditional "ora et labora," pray and work. Whereas "labora" was physical work, the prayer was characterized by physical recovery. So one might postulate that exercise itself is not important for a mental well-being but it is the adequate alteration between inactivity and activity. Therefore it might be regarded as counterproductive and stress-evoking when people, living in space are forced to exercise beside a physical stressful working day (see concrete recommendations).

32.2 Exercise in Space: Ideas and Concepts to Prevent and to Counteract Physical Stress

With the end of the industrial age and the beginning of the information age in the late 1970s, physicians realized that regular exercise and an active lifestyle is a helpful tool to substitute missing physical workload and to prevent civilization diseases as diabetes, cardiovascular disease, and an increasing obesity among all age groups.

Since then a tremendous amount of research activities and publications has validated the positive effects of exercise on the cardiovascular system, the musculoskeletal system, on the immune system, as well as the metabolic system. With the beginning of manned space flight in the late 1960s it was realized that the loss of physical work load caused by weightlessness results in a decrease of muscle mass, bone density/stability, and cardiovascular changes. A new area of exercise science was born in this time, and research has very well documented the positive effects of exercise on the cardiovascular and the musculoskeletal system. This research has helped scientists to understand fundamental principles of musculoskeletal and cardiovascular deconditioning and to develop countermeasures (Richter et al. 2017), which are nowadays used, for example, in the rehabilitation of patients suffering from the negative effects of immobilization after surgery. However, their interaction with other organ systems (e.g., immune system) especially in space are warranted in future investigations.

In the first decades mainly standardized gym equipment such as treadmill and bike ergometry have been used to exercise in space. However, this exercise takes up to 2 h off of the precious work time every day. Therefore in the recent years more efficient, new devices have been developed and evaluated (e.g., Fly Wheel, Short Arm Human Centrifuge (SAHC), vibration training) in order to increase efficiency by maximizing the output and minimizing the time spend for exercising.

Today an "exercise in space" program is guided by three fundamental principles, which all are aiming to counteract the physiological stress of weightlessness. Exercise in space aims

1. To counteract a loss of muscle mass/muscle force in order (a) to protect work capabilities and (b) to accelerate postflight recovery.
2. To counteract cardiovascular deconditioning as to prevent postflight orthostatic intolerance.
3. To counteract osteoporotic deconditioning to facilitate postflight recovery (Fig. 32.2).

32.2.1 Exercise, Weightlessness, and the Muscle

Despite a well-designed and intensive (up to 2 h/day) exercise program, it still seems impossible to counteract the loss of muscle mass and muscle strength during long-term space flights. In a recent study, Gopalakrishnan et al. (2010), report a decrease of 15% muscle mass and approximately 20% of isokinetic strength in the plantar flexors of four subjects although they were regular exercisers. This deconditioning is more emphasized in the early phases of spaceflights, which exponentially decreases in the following weeks. So far it is not clear whether this deconditioning levels out and shows no further decline. Recreation follows a similar pattern with rapid improvements in the early phases postflight. It needs to be mentioned that muscular deconditioning mainly affects the locomotor system. Muscle mass and strength for the upper extremities show no major changes and also the strength losses in the ankle dorsiflexor group was negligible, probably due to the regular use of foot loops, which are used to align and move the body against the resistance of inertial forces.

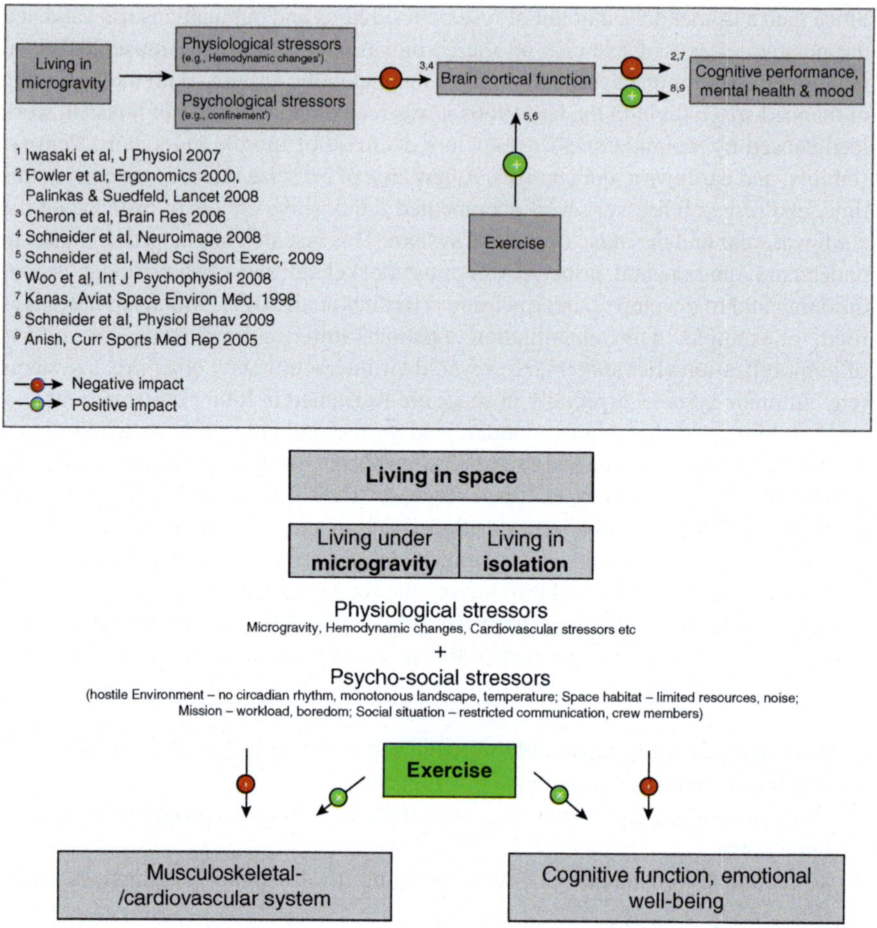

Fig. 32.2 Living in microgravity is known to result in a multitude of stressors resulting from physiological changes (e.g., hemodynamic changes in the brain) in combination with psychological effects caused by the confined situation (e.g., workload, limited nutritional supply, isolation) exist while living in space. This has a negative impact on brain cortical activity, cognitive performance, mental health, and mood. In contrast, exercise is known to have a positive impact on brain cortical function, performance, mental health, and mood. It is therefore proposed that, individually specified exercise programs might help to counteract the negative impacts of microgravity on brain function

32.2.2 Exercise, Weightlessness, and the Cardiovascular System

As recent study by Verheyden et al. (2010) has shown that under weightless conditions cardiovascular parameters (heart rate and mean arterial pressure) are similar to those of being in a supine position on Earth. It is concluded that the cardiovascular system is chronically relaxed in space due to an increase of venous return and cardiac output (Norsk et al. 2006; Verheyden et al. 2010). Instead of deconditioning they name it neural adaptations to changed gravity conditions and in doing so it is indirectly marked that, despite postflight orthostatic problems, these adaptations have no negative impact on in-flight health.

Unfortunately, exercise itself is not able to provide a blood pressure gradient from head to feet in space, which would be helpful to counteract orthostatic intolerances postflight. As a result, a combination of low body negative pressure (LBNP) and exercise was proposed, and showed promising results. The data suggest that treadmill exercise using a graded lower body compression suit, and 100 mmHg lower body negative pressure provides equivalent or greater physiologic stress than similar upright exercise on Earth (Hargens 1994).

32.2.3 Exercise, Weightlessness, and the Skeletal System

Exposure caused by routine exercise stimulates bone formation and prevents bone loss, whereas unload causes a loss in bone density. Although, as compared to the muscle system, the decline in bone markers during spaceflight and recovery postflight occurs much slower (recovery may require 1–3 years (LeBlanc et al. 2000)) several similarities between muscle and bone exist. Those similarities include a very large inter- and intra-subject variability, a distinct localization with significant losses in the lower extremities and no significant changes measured in the upper body (e.g., arms) (LeBlanc et al. 2000). Most importantly it is worth to note that a loss of bone density AND muscle mass/strength cannot be completely prevented by exercise.

In summary, it can be concluded that exercise in space is simply used to facilitate postflight recovery processes on the organ systems as described above. Although some authors account for the necessity to exercise in order to maintain work productivity in space or to promote a physical healthy lifestyle, the main focus today is to counteract musculoskeletal and cardiovascular deconditioning in order to prevent postflight physical stress. As unfortunately no tool, no exercise routine, is able to completely prevent physical deconditioning so far, a number of new ideas are currently implemented in space exercise research, such as artificial gravity produced by a short arm human centrifuge (SAHC) or vibration augmented resistive exercise. The main attempt is to reduce precious work time spent on exercise by increasing exercise effectiveness. Although in general, one might support those approaches, we need to take into consideration that during long-term trips to Mars and beyond, time limitations seem to be negligible and a packed schedule will probably be replaced by boredom and the need to entertain oneself. Moreover, these attempts so far are guided by physiological definitions and rarely take into account that there is a need to exercise beyond counteracting physical stress and facilitating postflight recovery processes.

32.3 The Effects of Exercise on Brain Cortical Function, Cognitive Performance Mood and (Mental) Health

While regular physical activity is primarily recommended for its beneficial effects on cardiorespiratory health and fitness, habitual exercisers often cite the positive effect of exercise and physical activity on mood and general well-being. There is a growing body of evidence to support the positive psychological effects of exercise, and this is acknowledged in a recent report of the US Surgeon General on Physical

Activity and Health. Yeung (1996) reviewed 81 studies that investigated the influence of a single bout of exercise on mood and mental state. The vast majority (85%) of these studies found an improvement in mood and mental state with exercise, and this benefit seems dependent on the duration and intensity of the exercise (Lind et al. 2005; Ekkekakis and Lind 2006). The mechanisms underlying the link between exercise and acute changes in mood are not clear, but it is likely that changes in the concentration of different neurotransmitters (Buckworth and Dishman 2002; Hollmann and Strüder 2003) and alterations in central neural activity (Hall et al. 2007) play a role.

Also the effects of physical exercise on cognitive performance have been studied throughout the last 30 years. While research concentrating on simple motor performance tasks after exercise provided inconsistent results (for review see Tomporowski 2003) recent studies provided clear support for an improvement of cognitive performance and emotional well-being during and after exercise (Schneider et al. 2013; Abeln et al. 2015; Wollseiffen et al. 2016). One of the main contributing factors for enhancement of cognitive performance seems to be the increase in the level of arousal related to physical exercise, which seems to affect central nerval processes in those brain areas responsible for emotional and cognitive processing (see also Chap. 7).

In spite of a general agreement that exercise improves mood and cognitive performance, there is an ongoing debate about dose–response effects of exercise (Ekkekakis and Petruzzello 1999; Ekkekakis et al. 2000). Recent evidence suggests that the transition from aerobic to anaerobic exercise metabolism seems to have an impact on mood (Hall et al. 2002). Furthermore, there is a consensus that changes in mood should be monitored on an individual level as experiments on a group level might blur important variations (Ekkekakis and Petruzzello 1999; Schneider et al. 2009b; Brümmer et al. 2011).

In the past two decades, there has been an increased interest in detecting the underlying neurophysiological processes of these behavioral findings (Boecker et al. 2012). However, research is still in its early stages as standardized brain imaging methods as positron emission tomography (PET) or functional magnetic resonance imaging (fMRI) are limited in an exercise setting. An overview of the applicability of different imaging techniques during and after exercise was presented in an edition of the Methods Journal from 2008, Volume 45, Issue 4. So far we can conclude from several studies that exercise does not only affect cortical activity in regions that have been identified to control motor activity like the sensory-motor cortex or supplementary motor areas (SMA) but also areas that have recently been connected with emotional and cognitive processes, especially (pre-)frontal cortex (PFC) areas. From affective neuroscience and exercise psychobiology, it is known that the PFC plays a major role in regulating and coping with emotions such as fatigue, distress, and tension (Faw 2003). Today, two main theories exist which describe and explain the effects of exercise on PFC areas: (1) the transient hypofrontality hypothesis and (2) the dual mode theory.

32.3.1 The Transient Hypofrontality Hypothesis

Dietrich's (2006) transient hypofrontality theory assumes that exercise reduces activity within the prefrontal cortex. This idea is based on the assumption that cerebral blood-flow remains stable during exercise and that an increase in oxygen consumption in regions that are directly involved in motor control (e.g., motor cortex, sensory cortex, SMA, etc.) coercively results in a decrease in cortical activity in regions that are not directly involved with exercise, like the PFC. The hypofrontality theory is supported by PET findings (Tashiro et al. 2008) showing a decreased prefrontal cortex activity during exercise. A similar theory (the dual mode theory) has been raised by Ekkekakis (Ekkekakis and Lind 2006; Ekkekakis 2009). Several studies in the recent years brought empirical evidence for this theory, which today is commonly accepted (Boecker et al. 2012).

Following both theories of a shift of cortical resources away from regions responsible for cognitive and emotional processing *during* exercise, studies have shown that also *after* exercise, previously "overloaded" brain regions are now downregulated due to the increased computational demand of brain regions associated with exercise (Schneider et al. 2009, 2013; Vogt et al. 2012). If we regard the brain as a multiprocessor unit, then exercise is able to reduce clock frequency in specific processors (because calculating capacity is needed elsewhere) and this seems to reset cognitive as well as emotional processors and might explain positive effects of exercise on emotional as well as neurocognitive performance. A less technical, more resource-orientated model would assume that, during exercise, the brain is, with increasing intensity and duration, more and more concerned about keeping the physical system running. If there are limited resources available this will prevent from dealing with emotions (you will definitely not worry any more about the question whether your wife still loves you or not while spending the 10th minute at 90% VO_{2max} or with cognitive processes you won't be able to worry about your company's next year's business plan at this intensity). After exercise, the tremendous workload will result in physical relaxation, which is accompanied by neurophysiological relaxation. Body and mind need to recover from an exceptional physical experience and from organizing this experience, and there is some experimental evidence to confirm this understanding: Tashiro et al. were able to demonstrate that the adjusted regional metabolic rate ratio is increased in sensorimotor and premotor but decreased in temporal and prefrontal cortex areas while running (Tashiro et al. 2008). Within two recent studies, Schneider et al. reported a decrease of cortical high-frequency activity (beta-activity, desynchronized state of the brain, indicator for excitatory brain activity) in areas that are responsible for language processing (Brodmann areas 21/22) (Schneider et al. 2009c, 2010), as well as an increase in alpha activity (synchronized state of the brain, indicator for decrease of brain cortical activity) in prefrontal cortex areas (Schneider et al. 2010). Apart from that, an increase in cortical alpha frequency was reported in somatosensory areas after exercise (Schneider et al. 2009c). This indicates on a cortical level that exercise results in physical relaxation, which is definitely going along with cortical relaxation.

32.3.2 A Neurotransmitter Theory

Besides these theories that are more focused on changes in regional brain cortical activity, changes within the neurotransmitter system, especially dopamine and serotonin might be considered to be involved in the improvement of mood and cognitive performance. There is good evidence that the neurotransmitter serotonin plays an important role in the etiology of mood changes. The precursor of serotonin Tryptophan (TRP), but not serotonin (5-HT), can pass the blood–brain barrier (BBB) via a carrier system, in which TRP and large neutral amino acids (LNAA), especially the branched-chain amino acids (BCAA), compete for transport into the brain (Fernstrom and Wurtman 1971; Pardridge 1977). If TRP entry from blood across the BBB into the brain is attenuated, central TRP concentration as well as consecutive 5-HT biosynthesis is increased (Chaouloff 1989; Strüder and Weicker 2001a). Experimental depletion of TRP induces a transient lowering of mood in control subjects (Klaassen et al. 1999), and can induce an acute depressive syndrome in patients actually remitted from major depression (Delgado et al. 1990). In contrast, a sufficient amount of aerobic physical exercise >30 min is known to affect the TRP uptake across the BBB positively, which consecutively increases serotonin biosynthesis (Strüder and Weicker 2001a, b). In parallel, previously reported effects of exercise on cognitive performance, mood, and mental health might be correlated with a release of neurotrophic factors like brain-derived neurotrophic factor (BDNF), insulin-like growth factor (IGF-1), vascular endothelial growth factor (VEGF), and prolactin (PRL) (Rojas Vega et al. 2006, 2008). In those studies, an increase of BDNF during physical exercise was shown and from animal studies it is well known that the release of brain BDNF triggers neurogenetic processes in different brain areas such as the hippocampus and prefrontal cortex areas (van Praag et al. 1999). A recent meta-analytic review targeted the effects of exercise on BDNF in more than 1000 humans (Szuhany et al. 2015) indicating that exercise indeed improves cognition and mood and it is giving more evidence suggesting that BDNF could mediate these effects of exercise. Interestingly, the authors concluded that "the magnitude of these effects may be lower in females relative to males."

The effects of exercise on brain functions and mood can of course also be discussed in the light of other phenomena like the *runners high* as caused by opioidergic mechanisms (Boecker et al. 2008) or endocannabinoids (Butterly et al. 2010) (see also Chaps. 10 and 29). Unfortunately it seems—at least from individual reports—that the *runners high* is hardly predictable nor reproducible but seems to be connected to a number of framework conditions which are hard to entitle ("sometimes it works—sometimes not"). Moreover a *runners high* seems to be connected to a personal bias to running. At least first findings indicate that a *runners high* is limited to exercise intensities and durations far beyond what is used for health aspects.

To sum up, for now we can only speculate about the underlying neurophysiological processes caused by exercise. Existing approaches to access to the brains' electrical and neurochemical (sub)systems during exercise are restricted. This includes limited access by MRI imaging techniques and the fact that the analysis of *peripheral* blood is not of much support due to the impenetrability of the BBB by several

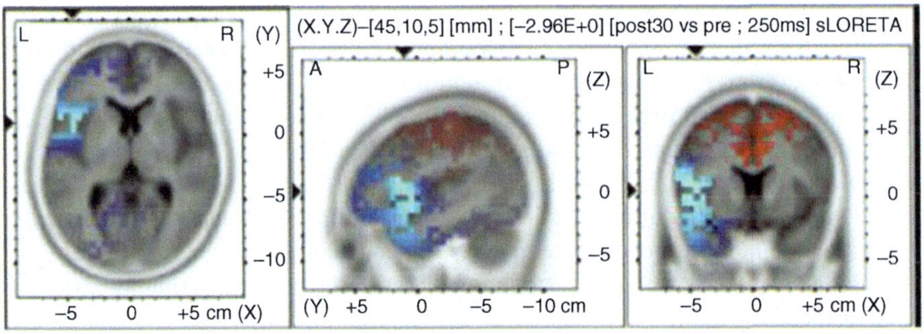

Fig. 32.3 Statistical parametric maps (SPM) of standardized low resolution brain electromagnetic tomography (*sLORETA*) differences in the EEGs high-frequency activity (13–35 Hz) comparing 30 min postexercise at 80%VO_{2max} with pre-exercise (n = 12). *Red* and *yellow* indicate increased activity and *blue* indicates decreased activity. Several studies have shown that a significant decrease of activity occurs in frontal and left temporal lobes, including Wernickes' Area, which is well known to be connected to language processing (Schneider et al. 2009c, 2010, 2013). It is concluded that a postexercise decrease in these areas might facilitate cognitive processing

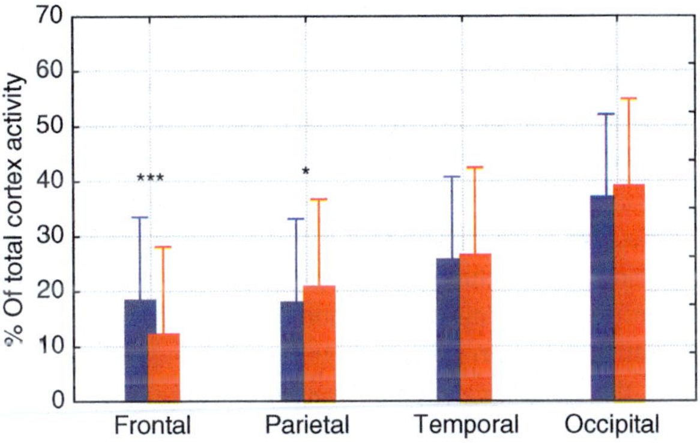

Fig. 32.4 Changes in cortical current density pre (blue) and post (red) exercise in the frontal, parietal, temporal, and occipital lobe. Data is averaged over several training sessions during the Mars500 study and clearly shows a decrease in frontal cortex activity

of the mentioned neurotransmitters and neurotrophic factors. However, new technologies have provided modern tools to get more insight into neurophysiological changes of the brain in the course of exercise. Just recently, software-based electrotomography, in combination with recently developed active EEG systems, allows to localize exercise-induced changes in brain cortical activity, supporting a transient hypofrontality theory. In contrast to standard brain imaging techniques, electrotomography therefore allows an easy to use tool for brain imaging even under extreme conditions like exercise or in weightlessness (Figs. 32.3 and 32.4).

32.3.3 Exercise Immunology

As physical health plays a major role on mental well-being, this last paragraph is briefly dedicated to the effect of exercise on the immune system. Shephard has reported the need for a more integrative view of exercise and its positive and negative effects on the immune system functions and disease (Shephard 2010). This is likely because exercise can affect different immune responses as a function of the duration, degree, and type of the exercise. This includes several organs as well as a redistribution of immune cells (Adams et al. 2011) and direct functional control of innate and adaptive immune cells (see Chaps. 11–16) through complexly orchestrated levels of interaction including (1) direct brain region and nerval structures (e.g., the autonomous nerve systems (ANS), see Chap. 8) (2) neurohumoral (e.g., catecholamines) or hormone-like stress mediators (e.g., endocannabinoids, NPY) as well as metabolites released from either degradation of energy rich phosphates during strenuous exercise or from direct release from stressed and damaged cells (see Chap. 16) (Choukèr et al. 2005). These and other yet not fully understood interactions of mechanisms affect immune cell proliferation and function (Giraldo et al. 2009) hereby finely tuning immune functions and either resulting in higher resistance to infection or tumor growth, or inversely to higher risk of disease (Kakanis et al. 2010). The U-shaped or inverted J hypotheses have been discussed to describe the incremental effects of exercise on immune (dys-)function. Here it is stated that moderate exercise improves immune functional states, hereby increasing health by decreasing the susceptibility for disease (Woods et al. 1999). Moreover, the immune system functions, the percentages of immune cells in circulating blood are considered to be adapted as function of the intensity of activity, as well as the duration over which such activities are elevated. E.g. acute physical activity and longer-term, also incremental changes can have dissimilar effects on immunity. From an ecological and evolutionary point of view, the energy to support immune answers is often traded off to the use of energy to migrate (e.g. birds) or to mate (van Dijk and Matson 2016) which may guide to the conclusion that long duration and exhaustive exercise is limiting energy sources for appropriate immune answers, and hence limit immune cells' proliferation. In a meta-analyses some more evidence was provided, that the lymphocyte proliferation is suppressed even more when exercise is prolonged for more than 1 h (Siedlik et al. 2016). In contrast very short, 20 min of strenuous physical exercise (80% VO_2 Max) was shown to mobilize plasmacytoid dendritic cells (DCs) which are playing a critical role in the presentation of antigens to naïve T cell (at the immunologic synapse). Brown et al. (2018) reported, that because of the functional repertoire of these cells, their role in orchestrating also antiviral and antibacterial pathogens, it is considered that short bouts of exercise increase "immune-surveillance by preferentially mobilising effector cells." Similar effects were seen also for virus specific T-cell that are mobilized and increased in their activitiy after excerice (Kunz et al. 2018). But not only circulating immune cells but also resting immune functions and tissue microenvironments are affected. Recently, Koelwyn et al. (2017) reviewed the beneficial effects of physical exercise on e.g. the metabolic properties and immune responses that are all in all resulting in an anti-tumor reprogramming of the tumor microenvironment.

This ambiguity of immune activation and suppression became evident in a study reported after strenuous endurance exercise showing that both pro-inflammatory and anti-inflammatory pathways are activated (Ostrowski et al. 1999). How this immune balance turns into a not favorable direction and which molecular mechanisms are involved in the immune regulation in response to exercise and the subsequent neural, hormonal, or metabolic pathway remains unclear but certainly involves expression of genes which are differentially regulated by different transcription factors, alternative splicing, gene silencing by small RNAs (miRNAs or miRs), or others (Wessner et al. 2010). More recent mechanistic investigations are further needed to understand these ambiguous effects if and how exercise may increase or impair the immune dysfunctional states. "The immune system shows remarkable plasticity in response to exercise" which "may be used as a tool to optimize the immune system for patients receiving cancer immunotherapy and other patients exhibiting abnormal immunity caused by obesity, infections, or other diseases" (Gustafson et al. 2017) but could serve along this understanding to overcome or prevent from immune dysfunctional states during space flight.

In brief, this chapter tried to identify the positive effects of exercise on emotional and cognitive processes as well as the underlying neurophysiological processes. How exactly and to which degree this is affecting immune function and health remains to be understood. However, it is a major contribution of exercise science together with other fields, to provide a more detailed and more precise descriptions of the functionality of the (neuro-) physiological system and its implication on the behavioral system and "downstream" pathways affecting the immune homeostasis.

32.4 Recommendations

Exercise might be used as a suitable "global tool," not only for counteracting physiological stress but also to mitigate psychological stress during long-term space missions. Exercise might therefore be defined as one key factor for crew performance and mission success. The following paragraph will try to identify exercise recommendations in order to maximize a psychophysiological health outcome. These recommendations are based on simple principles how exercise affects psychophysiological health and a general well-being.

1. The current exercise recommendations on board of the International Space Station (ISS) foresee a daily exercise routine of up to 2 h/day. With respect to the immense workload and the tight schedule that astronauts and cosmonauts experience while in space, it needs to be questioned whether a strict exercise schedule provokes additional stress physiologically as well as psychologically.
2. As described before, the balance of physical activity and inactivity is a major determinant for the well-being and health. Physically demanding activities (e.g., extravehicular activity, EVA) should not be followed by additional exercise routines the same or following day as they are likely to result in immune imbalance (as seen in athletes).
3. To further minimize the impact of exercise routines on the time management while in-flight, new time-saving approaches like vibration training, or artificial gravity provided, for example, by a short arm human centrifuge should be

broadly evaluated. This evaluation must include effects on cognitive performance, mood, and immune function. For example, vibration training might be a time-saving method to counteract muscle loss and bone stability, but perhaps duration and not intensity of training has a major impact on neurocognitive and mental processes. Current research (unpublished) shows that artificial gravity levels, which have been shown to be tolerated over a longer time frame (30 min) on a centrifuge ($\leq 2 \times g$) can not provoke cardiovascular reactions (e.g. increase in heart rate) similar to moderate exercise. Higher g-loads ($>2.5 \times g$) are necessary to gain cardiovascular reactions similar to exercise, but obviously this goes along with an increased risk of syncope. Cognitive impacts have to be explored.
4. In general, healthy individuals will feel a need to exercise if they are physically underchallenged. For mission success, it is of utter importance not only to physically train astronauts/cosmonauts but also to sensitize them in their predeparture training years to the effects of exercise on psychophysiological health and mental performance.
5. Besides all this, we need to keep in mind that an individual dose–response relationship as well as an exercise preference hypothesis exists. If we aim to optimize exercise, we need to aim for individual exercise prescriptions, taking into account individual habits and an individual bias rather than generalizing exercise on a physiological level.
6. The individual preferences and needs of a "personalized exercise suite" might become even expanded by adjusted regimen when it comes to very long space flights, when continuous excerice programs are in place throughout the entire mission but which might further be deepened by special training programs in the months before return to the gravitational pull. This is yet very speculative and will be much dependent on the individual again, and on the mission duration and other (gravitational) countermeasures.

32.5 Concluding Remarks

In the last 20 years, an increasing number of studies from exercise science were devoted to the positive effects of exercise on cognitive performance, mental health, and a general well-being. Exercise is a "global tool" that might help to facilitate the individual's psychophysiological adaptation to extreme conditions of life, to microgravity, and therefore to act positively on the crews' performance level, mission safety, and mission success.

However, when the role of exercise in space is discussed, there is a need to distinguish physiological aspects from health aspects. One example for this differential view, is the applicaton of artificial gravity. Although one might state that if research within the next years will show that artificial gravity is the non plus ultra to prevent musculoskeletal deconditioning, too high g-levels might have an overall negative effect on the individuals health status, especially if individuals would prefer "real" exercise. While physiological reactions to a given exercise stimulus are supposed to be widely similar, the emotional reactions hereto are not. This reflects well, that the general idea of health does not equal physical health alone.

Throughout the last decades exercise has been regarded as a countermeasure to physiological deconditioning in space. Within this chapter, it is proposed to identify exercise not only as a countermeasure solving problems, but as a preventive tool to avoid problems inter- as well as intra-individually. While developing tools and programs to fragmentarily prevent musculoskeletal deconditioning as an organ-directed approach, we should also aim to enhance crew performance level as well as individual mental and physical health in a more holistic view. This integrated approach is needed because mission success is endangered by a multitude of stressors (social, workload, environment, confinement, and others) that can be likely avoided by regular and individually adapted exercise. The degree and the consequences of exercise, respectively, have to be assessed also in the light of their effects on other general health relevant conditions, such as "healthy ageing" as well as on distinct organ entities such as on to the immune system. A good reason for exercising however was already acknowledged in the first century by Juvenal "*mens sana in corpore sano*."[1] And finally, what we currently see in the new generation of European Astronauts: The best thing to avoid physical degeneration in space is to start the mission in a fit state!

References

Abeln V, MacDonald-Nethercott E, Piacentini MF, Meeusen R, Kleinert J, Strueder HK, Schneider S (2015) Exercise in isolation--a countermeasure for electrocortical, mental and cognitive impairments. PLoS One 10(5):e0126356

Adams GR, Zaldivar FP, Nance DM, Kodesh E, Radom-Aizik S, Cooper DM (2011) Exercise and leukocyte interchange among central circulation, lung, spleen, and muscle. Brain Behav Immun 25(4):658–666

Boecker H, Henriksen G, Sprenger T, Miederer I, Willoch F, Valet M, Berthele A, Tolle TR (2008) Positron emission tomography ligand activation studies in the sports sciences: measuring neurochemistry in vivo. Methods 45:307–318

Boecker H, Hillman CH, Scheef L, Strüder HK (2012) Functional neuroimaging in exercise and sport sciences. Springer, New York, NY

Brown FF, Campbell JP, Wadley AJ, Fisher JP, Aldred S, Turner JE (2018) Acute aerobic exercise induces a preferential mobilisation of plasmacytoid dendritic cells into the peripheral blood in man. Physiol Behav 194:191–198. https://doi.org/10.1016/j.physbeh.2018.05.012

Brümmer V, Schneider S, Abel T, Vogt T, Strüder HK (2011) Brain cortical activity is influenced by exercise type and intensity. Med Sci Sports Exerc 43:1863

Buckworth J, Dishman RK (2002) Exercise psychology. Human Kinetics, Champaign

Butterly LF, Goodrich M, Onega T, Greene MA, Srivastava A, Burt R, Dietrich A (2010) Improving the quality of colorectal cancer screening: assessment of familial risk. Dig Dis Sci 55:754–760

Chaouloff F (1989) Physical exercise and brain monoamines: a review. Acta Physiol Scand 137:1–13

[1] Although a traditional translation to "a healthy mind in a healthy body" does not reflect the meaning of the context which is more: "we should pray to the gods that there should be a healthy mind in (=and) a healthy body."

Choukèr A, Demetz F, Martignoni A, Smith L, Setzer F, Bauer A, Hölzl J, Peter K, Christ F, Thiel M (2005) Strenuous physical exercise inhibits granulocyte activation induced by high altitude. J Appl Physiol 98:640–647
Convertino VA (1996) Exercise as a countermeasure for physiological adaptation to prolonged spaceflight. Med Sci Sports Exerc 28:999–1014
Convertino VA (2002) Planning strategies for development of effective exercise and nutrition countermeasures for long-duration space flight. Nutrition 18:880–888
Delgado PL, Charney DS, Price LH, Aghajanian GK, Landis H, Heninger GR (1990) Serotonin function and the mechanism of antidepressant action. Reversal of antidepressant-induced remission by rapid depletion of plasma tryptophan. Arch Gen Psychiatry 47:411–418
Dietrich A (2006) Transient hypofrontality as a mechanism for the psychological effects of exercise. Psychiatry Res 145:79–83
van Dijk JG, Matson KD (2016) Ecological immunology through the lens of exercise immunology: new perspective on the links between physical activity and immune function and disease susceptibility in wild animals. Integr Comp Biol 56(2):290–303. https://doi.org/10.1093/icb/icw045. Review
Ekkekakis P (2009) Illuminating the black box: investigating prefrontal cortical hemodynamics during exercise with near-infrared spectroscopy. J Sport Exerc Psychol 31:505–553
Ekkekakis P, Lind E (2006) Exercise does not feel the same when you are overweight: the impact of self-selected and imposed intensity on affect and exertion. Int J Obes (Lond) 30: 652–660
Ekkekakis P, Petruzzello SJ (1999) Acute aerobic exercise and affect: current status, problems and prospects regarding dose-response. Sports Med 28:337–374
Ekkekakis P, Hall EE, VanLanduyt LM, Petruzzello SJ (2000) Walking in (affective) circles: can short walks enhance affect? J Behav Med 23:245–275
Faw B (2003) Pre-frontal executive committee for perception, working memory, attention, long-term memory, motor control, and thinking: a tutorial review. Conscious Cogn 12(1):83–139
Fernstrom JD, Wurtman RJ (1971) Brain serotonin content: increase following ingestion of carbohydrate diet. Science 174:1023–1025
Giraldo E, Garcia JJ, Hinchado MD, Ortega E (2009) Exercise intensity-dependent changes in the inflammatory response in sedentary women: role of neuroendocrine parameters in the neutrophil phagocytic process and the pro-/anti-inflammatory cytokine balance. Neuroimmunomodulation 16:237–244
Gopalakrishnan R, Genc KO, Rice AJ, Lee SM, Evans HJ, Maender CC, Ilaslan H, Cavanagh PR (2010) Muscle volume, strength, endurance, and exercise loads during 6-month missions in space. Aviat Space Environ Med 81:91–102
Gustafson MP, DiCostanzo AC, Wheatley CM, Kim CH, Bornschlegl S, Gastineau DA, Johnson BD, Dietz AB (2017) A systems biology approach to investigating the influenceof exercise and fitness on the composition of leukocytes in peripheral blood. J Immunother Cancer 5:30. https://doi.org/10.1186/s40425-017-0231-8. PubMed PMID: 28428879; PubMed Central PMCID: PMC5394617
Halberg F, Vallbona C, Dietlein LF, Rummel JA, Berry CA, Pitts GC, Nunneley SA (1970) Human circadian circulatory rhythms during weightlessness in extraterrestrial flight or bedrest with and without exercise. Space Life Sci 2:18–32
Hall EE, Ekkekakis P, Petruzzello SJ (2002) The affective beneficence of vigorous exercise revisited. Br J Health Psychol 7:47–66
Hall EE, Ekkekakis P, Petruzzello SJ (2007) Regional brain activity and strenuous exercise: predicting affective responses using EEG asymmetry. Biol Psychol 75:194–200
Hargens AR (1994) Recent bed rest results and countermeasure development at NASA. Acta Physiol Scand Suppl 616:103–114
Hollmann W, Strüder HK (2000) Brain, psyche and physical activity. Orthopade 29:948–956
Hollmann W, Strüder HK (2003) Körperliche Aktivität fördert Gehirngesundheit und -leistungsfähigkeit: Übersicht und eigene Befunde. Nervenheilkunde 9:467–474

Kakanis MW, Peake J, Brenu EW, Simmonds M, Gray B, Hooper SL, Marshall-Gradisnik SM (2010) The open window of susceptibility to infection after acute exercise in healthy young male elite athletes. Exerc Immunol Rev 16:119–137
Klaassen T, Riedel WJ, Deutz NE, van Someren A, van Praag HM (1999) Specificity of the tryptophan depletion method. Psychopharmacology (Berl) 141:279–286
Koelwyn GJ, Quail DF, Zhang X, White RM, Jones LW (2017) Exercise-dependent regulation of the tumour microenvironment. Nat Rev Cancer 17(10):620–632. https://doi.org/10.1038/nrc.2017.78
Kunz HE, Spielmann G, Agha NH, O'Connor DP, Bollard CM, Simpson RJ (2018) A single exercise bout augments adenovirus-specific T-cell mobilization and function. Physiol Behav 194:56–65. https://doi.org/10.1016/j.physbeh.2018.04.035
LeBlanc A, Schneider V, Shackelford L, West S, Oganov V, Bakulin A, Voronin L (2000) Bone mineral and lean tissue loss after long duration space flight. J Musculoskelet Neuronal Interact 1:157–160
Lind E, Joens-Matre RR, Ekkekakis P (2005) What intensity of physical activity do previously sedentary middle-aged women select? Evidence of a coherent pattern from physiological, perceptual, and affective markers. Prev Med 40:407–419
Lipnicki DM, Gunga HC (2009) Physical inactivity and cognitive functioning: results from bed rest studies. Eur J Appl Physiol 105:27–35
Norsk P, Damgaard M, Petersen L, Gybel M, Pump B, Gabrielsen A, Christensen NJ (2006) Vasorelaxation in space. Hypertension 47:69–73
Ostrowski K, Rohde T, Asp S, Schjerling P, Pedersen BK (1999) Pro- and anti-inflammatory cytokine balance in strenuous exercise in humans. J Physiol 15:287–291
Palinkas LA, Suedfeld P (2008) Psychological effects of polar expeditions. Lancet 371:153–163
Pardridge WM (1977) Kinetics of competitive inhibition of neutral amino acid transport across the blood-brain barrier. J Neurochem 28:103–108
van Praag H, Christie BR, Sejnowski TJ, Gage FH (1999) Running enhances neurogenesis, learning, and long-term potentiation in mice. Proc Natl Acad Sci U S A 96:13427–13431
Richter C, Braunstein B, Winnard A, Nasser M, Weber T (2017) Human biomechanical and cardiopulmonary responses to partial gravity - a systematic review. Front Physiol 8:583
Rojas Vega S, Strüder HK, Vera Wahrmann B, Schmidt A, Bloch W, Hollmann W (2006) Acute BDNF and cortisol response to low intensity exercise and following ramp incremental exercise to exhaustion in humans. Brain Res 1121:59–65
Rojas Vega S, Abel T, Lindschulten R, Hollmann W, Bloch W, Strüder HK (2008) Impact of exercise on neuroplasticity-related proteins in spinal cord injured humans. Neuroscience 153:1064–1070
Rummel JA, Michel EL, Berry CA (1973) Physiological response to exercise after space flight–Apollo 7 to Apollo 11. Aerosp Med 44:235–238
Rummell JA, Sawin CF, Buderer MC, Mauldin DG, Michel EL (1975) Physiological response to exercise after space flight–Apollo 14 through Apollo 17. Aviat Space Environ Med 46:679–683
Schneider S, Vogt T, Frysch J, Guardiera P, Struder HK (2009) School sport - a neurophysiological approach. Neurosci Lett 467(2):131–134
Schneider S, Brümmer V, Abel T, Askew CD, Strüder HK (2009a) Changes in brain cortical activity measured by EEG are related to individual exercise preferences. Physiol Behav 98:447–452
Schneider S, Mierau A, Diehl J, Askew CD, Strüder HK (2009b) EEG activity and mood in health orientated runners after different exercise intensities. Physiol Behav 96:706–716
Schneider S, Vogt T, Frysch J, Guardiera P, Strüder HK (2009c) School sport–a neurophysiological approach. Neurosci Lett 467:131–134
Schneider S, Askew CD, Abel T, Mierau A, Strüder HK (2010) Brain and exercise: a first approach using electro tomography. Med Sci Sports Exerc 42:600–607
Schneider S, Abeln V, Popova J, Fomina E, Jacubowski A, Meeusen R, Struder HK (2013) The influence of exercise on prefrontal cortex activity and cognitive performance during a simulated space flight to Mars (MARS500). Behav Brain Res 236(1):1–7

Shephard RJ (2010) Development of the discipline of exercise immunology. Exerc Immunol Rev 16:194–222

Siedlik JA, Benedict SH, Landes EJ, Weir JP, Vardiman JP, Gallagher PM (2016) Acute bouts of exercise induce a suppressive effect on lymphocyte proliferation in human subjects: a meta-analysis. Brain Behav Immun 56:343–351. https://doi.org/10.1016/j.bbi.2016.04.008. Review

Strüder HK, Weicker H (2001a) Physiology and pathophysiology of the serotonergic system and its implications on mental and physical performance Part I. Int J Sports Med 22:467–481

Strüder HK, Weicker H (2001b) Physiology and pathophysiology of the serotonergic system and its implications on mental and physical performance. Part II. Int J Sports Med 22:482–497

Szuhany KL, Bugatti M, Otto MW (2015) A meta-analytic review of the effects of exercise on brain-derived neurotrophic factor. J Psychiatr Res 60:56–64

Tashiro M, Itoh M, Fujimoto T, Masud MM, Watanuki S, Yanai K (2008) Application of positron emission tomography to neuroimaging in sports sciences. Methods 45:300–306

Tomporowski PD (2003) Effects of acute bouts of exercise on cognition. Acta Psychol (Amst) 112:297–324

Verheyden B, Liu J, Beckers F, Aubert AE (2010) Operational point of neural cardiovascular regulation in humans up to 6 months in space. J Appl Physiol 108:646–654

Vogt T, Schneider S, Abeln V, Anneken V, Struder HK (2012) Exercise, mood and cognitive performance in intellectual disability - a neurophysiological approach. Behav Brain Res 226(2):473–480

Wessner B, Gryadunov-Masutti L, Tschan H, Bachl N, Roth E (2010) Is there a role for microRNAs in exercise immunology? A synopsis of current literature and future developments. Exerc Immunol Rev 16:22–39

Wollseiffen P, Ghadiri A, Scholz A, Struder HK, Herpers R, Peters T, Schneider S (2016) Short bouts of intensive exercise during the workday have a positive effect on neuro-cognitive performance. Stress Health 32:514

Woods JA et al (1999) Exercise and cellular innate immune function. Med Sci Sports Exerc 31(1):57–66. https://doi.org/10.1097/00005768-199901000-00011

Yeung RR (1996) The acute effects of exercise on mood state. J Psychosom Res 40:123–141

Nutritional Countermeasures for Spaceflight-Related Stress

33

Martina Heer, Natalie Baecker, Scott M. Smith, and Sara R. Zwart

33.1 Introduction

Every cell, from a single-cell organism to the most complex biological system as we learn in humans, requires energy to maintain the cell's function and homeostasis (see also Chap. 5). Optimal function of the cells of the immune system—with a mass of all in all about 4 kg for all immune organs and cells depends like all other organs upon adequate nutrient supply. This led to the concept of "Immunonutrition," which describes diets that are specifically designed to enhance immune functions, and without adequate nutrition the immune system is clearly deprived of the components needed to generate an effective immune response (Beisel 1992; Mizock 2010) and support host defense mechanisms (McCarthy and Martindale 2018). Nutrients act as both antioxidants and as cofactors at the level of cytokine regulation (Cunningham-Rundles et al. 2005). Crews during spaceflight generally have lower dietary intake than their requirement (requirement is based on the World Health Organization equation for energy or on individual measurements of resting metabolic rate before flight) (Smith et al. 2014, 2009b). It is well known from ground research that a lack of macronutrients or selected micronutrients (e.g., zinc, selenium, and the antioxidant vitamins) can have profound effects on immune function (Chandra 1992; Chandra and Kumari 1994; Keith and Jeejeebhoy 1997). Micronutrient deficiency suppresses immune functions by affecting the innate and

M. Heer (✉) · N. Baecker
Institute of Nutritional and Food Sciences, University of Bonn, Bonn, Germany

IUBH International University, Bad Reichenhall, Germany
e-mail: martina.heer@uni-bonn.de

S. M. Smith
Human Health and Performance Directorate, NASA Lyndon B. Johnson Space Center, Houston, TX, USA

S. R. Zwart
University of Texas Medical Branch, Galveston, TX, USA

© Springer Nature Switzerland AG 2020

A. Choukér (ed.), *Stress Challenges and Immunity in Space*,
https://doi.org/10.1007/978-3-030-16996-1_33

T-cell-mediated immune response and adaptive antibody response. It leads also to a dysregulation of the balanced host response (Wintergerst et al. 2007) and seems to be rather important to maintain immune homestasis than to reestablish it in the course of severe inflammation, where, e.g., intravenously applied selenium could not improve the outcome in humans (Bloos et al. 2016). Disruption of dietary intake and nutritional balance of astronauts and cosmonauts during spaceflight, which is often accompanied by a stress response, might influence their immune response as the nutrition status can influence host immune responses and resistance to infection (Sonnenfeld 2002; Sonnenfeld and Shearer 2002). Some of the immune function alterations documented in astronauts (see Chaps. 11–15) have been reported also in states of nutrient deficiencies, and could potentially be counteracted by improving crew nutrition (Crucian et al. 2018). Detailed information on the absorption, metabolism, and excretion of many micronutrients during spaceflight are required before general nutritional recommendations can be made, and especially with regard to their relationship with the immune system function as well as other newly discovered thermoregulatory dysfunction (Chap. 26). It was recently quantified on board the International Space Station (ISS) that crews' body core temperature increased in space by 1 °C at rest and even higher during exercise which can further aggravate catabolism and may result in adjustments of the nutritional regimen (Stahn et al. 2017).

33.2 Energy Intake

During brief space shuttle flights of 8–14 days duration, energy expenditure was measured using doubly labeled water (Lane et al. 1997). This study demonstrates that energy expenditure for 13 men during short spaceflights was nearly identical to that measured in a 5-day ground-based baseline period. Another study, on a similar mission with intensive exercise, actually documented increased energy expenditure during flight (Stein et al. 1999). Therefore, human energy requirements during spaceflight are—at least in short-term missions—similar to those on Earth, and increased depending on exercise. Whether this also holds true for long-term space missions is the aim of an ESA-sponsored study on ISS, being completed as of this writing.

Despite the data suggesting no change in energy requirements during (short-term) flights, energy intake during flight is often lower than the estimated requirements for individual crew members (Fig. 33.1). Many astronauts could be considered malnourished during their mission, meaning their appetite is decreased and thereby their nutrient supply is inadequate (Bourland et al. 2000) leading to body mass loss during flight (Fig. 33.2, and Zwart et al. 2014). Although permissive underfeeding in critically ill patients rather than aggressive nutritional support might be evolutionarily conserved and beneficial, a continuous reduced energy intake in space travelers may not (van et al. 2016). There is anecdotal data from recently flown astronauts who were able to maintain body mass suggesting adequate energy intake, but that is still not true for the all the space travelers.

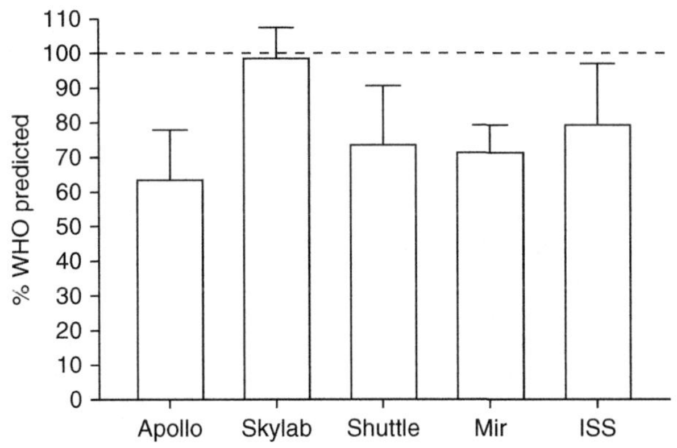

Fig. 33.1 In-flight dietary intake of crew members in different space programs. Data are expressed as percentage of energy requirements predicted by the World Health Organization (WHO) (WHO (World Health Organization) 1985). Apollo $n = 33$, Skylab $n = 9$, Shuttle $n = 32$, Mir $n = 7$, ISS $n = 70$. Apollo and Skylab data are from Bourland et al. (2000). Figure is adapted from Smith and Lane (2008), with additional data from Smith et al. (2005b) and Smith and Zwart (2008)

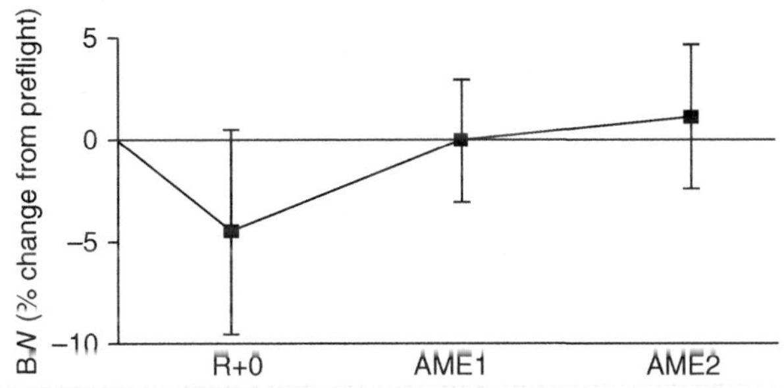

Fig. 33.2 Postflight body weight (*BW*) of Mir and ISS crew members ($n = 20$). Data are expressed as mean ± SD of the percent change from preflight body weight. *R* + 0 landing day, *AME1* first annual medical exam after return from the mission, and *AME2* second exam (Smith et al. 2009b)

A study on the Russian Mir space station assessed the urinary excretion of 8-iso-prostaglandin-F2 alpha and 8-oxo-7,8-dihydro-2 deoxyguanosine (8-OH-dG), markers for oxidative damage in six subjects during and after long-duration spaceflight (90–180 days) (Stein and Leskiw 2000). Although urinary isoprostane decreased significantly during flight, 8-OH-dG excretion tended to increase. A similar observation was made on the 'Life and Microgravity Science Spacelab Shuttle' mission. The regression of the excretion rate of 8-OH-dG against the energy deficit of either Mir or Spacelab showed that the greater the deficit in energy intake, the more extensive free radical propagation occurs because of the diminished protein-based antioxidant defense mechanisms (Stein 2002).

33.3 Protein and Amino Acids

Protein and amino acid deficiencies can have profound effects on a variety of immune system functions (Chandra et al. 1984; Guadagni and Biolo 2009). When the body is deprived of energy, protein is one of the most important limiting factors, because indispensable amino acids are not stored in the body. There is evidence that inactivity amplifies the catabolic response of skeletal muscle to inflammatory mediators (Bosutti et al. 2008). Guadagni and Biolo (2009) concluded that optimal protein (amino acid) intake could be much greater than the minimum amount of protein required to mitigate whole body protein wasting. This concept of optimal protein intake to promote non-anabolic actions of specific amino acids could also be applied to microgravity. Physical inactivity is associated with a low-grade inflammatory condition, as Steffen and Musacchia (1986) demonstrated in a study of experimental bed rest in healthy volunteers. Results from this bed rest study showed that physical inactivity might also be associated with increased protein requirement. Microgravity is generally a state of physical inactivity, more so on some missions than others. Shuttle astronauts did not routinely have access to exercise equipment, and future missions will depend on small spacecraft. ISS missions have multiple exercise devices, and crews typically exercise nearly every day. Whether or not those transient periods of exercise are enough is not yet known.

Moreover, Stein and Gaprindashvili (1994) used the ^{15}N-glycine method to measure whole body protein syntheses in-flight. The studies showed that whole body protein synthesis rates were increased by approximately 30% on the second and eighth day of spaceflight, with decreased nitrogen balance. At the same time, urinary cortisol, fibrinogen, and IL-2 levels were elevated, suggesting that the increased protein synthesis was due to stress. Stein et al. suggest that spaceflight triggers a stress response similar to injury-induced stress.

The defined spaceflight requirement for protein intake is 0.8 g/kg per day, not to exceed 35% of the total daily energy intake (Smith and Zwart 2008; Smith et al. 2009b). About 2/3 of the total amount of protein is to be provided in the form of animal protein, and 1/3 of the total should be in the form of vegetable protein. A study by Heer et al. showed that, on long missions, reaching (or exceeding) nominal protein intakes is common, but that on short flights (shuttle missions) protein intake is less than the recommended amount because of insufficient food intake (Heer et al. 2000). An energy deficit, as seen during short-duration spaceflight (Lane et al. 1997), will also lead to loss of protein.

Altering dietary amino acid composition or concentration of certain amino acids in food products may differentially affect immune system function. Supplementing whey protein, which has a high leucine content, has been shown to enhance natural killer (NK) function and IL-12 concentration (Kang et al. 2017) and increase plasma glutathione concentrations in HIV-infected patients (Micke et al. 2001). While further study is required, provision of protein- and/or amino acid-rich foods and/or supplements might be a means to maintain immune system function on exploration missions.

Glutathione is the most abundant endogenous intracellular antioxidant and plays a central role in antioxidant defenses. Glutathione concentrations decline with aging (al-Turk et al. 1987; Lang et al. 1992; Matsubara and Machado 1991; Samiec et al. 1998). Glutathione is synthesized from the amino acids cysteine and glycine; and in older adults, increasing dietary intake of these amino acids brought reduced glutathione levels back to the concentration in young controls (Sekhar et al. 2011). In a limited number of mice flown to ISS for 3 months, red blood cell glutathione concentrations were higher than in ground controls, an effect suggesting an adaptation to increased oxidative stress (Rizzo et al. 2012). One could speculate that providing food products rich in cysteine and glycine may support glutathione synthesis to help mitigate oxidative stresses during spaceflight, but this has not been tested up to date.

In general, caution must always be taken in interpreting protein or amino acid supplementation studies, given the inherent energy content of protein. Experimental designs often fail to capture the effects of energy supplementation alone, and rather compare an energy restricted group to a protein supplemented group, with inherently confounded outcomes. This point was highlighted in the review by Stein and Blanc (2011).

33.3.1 Arginine

Arginine is a necessary amino acid for normal T-cell function and may become essential in catabolic states. Many studies have shown improved immune response following arginine supplementation as part of immunonutrition (Bronte and Zanovello 2005; Joyner 2005; Popovic et al. 2007; Ralph et al. 2008; Van Buren 2004). Supplementary dietary arginine has been shown to have useful effects on cellular immunity in animal studies, showing an increased thymic size, enhanced lymphocyte proliferation to mitogen and alloantigen, augmented macrophage and killer cell lysis, and increased lymphocyte interleukin-2 production and receptor activity (Kirk and Barbul 1990). In mice, arginine supplementation induced changes of the intestinal microbiota which contributed to the activation of the innate immune system through the reduced expression of NF-κB, MAPK, and PI3K-Akt signaling pathways (Ren et al. 2014). In another study, in a rodent model for mammalian tumor, L-Arginine supplementation inhibited the growth of breast cancer by enhancing innate and adaptive immune responses mediated by suppression of MDSCs in vivo (Cao et al. 2016). In humans, supplementation of arginine led to improved wound healing and immune responses in elderly subjects (Kirk et al. 1993) and to lower C-reactive protein levels and earlier discharge from the hospital following total hip arthroplasty when an immune supplement also containing arginine was supplied (Alito and de Aguilar-Nascimento 2016). Based on these observations, it seems promising to supplement arginine during long-term missions; however, no studies have been carried out up to now testing arginine as a measure to improve immune response during space missions or in ground analog studies with humans.

33.3.2 Glutamine

Glutamine is the most abundant free amino acid in the body. It can inhibit NF-κB activation and cytokine expression following sepsis (Singleton and Wischmeyer 2008). Beneficial effects of glutamine include: antioxidant effects, serving as a precursor to glutathione, for arginine through inter-organ transport of citrulline, as an energy substrate for lymphocytes and neutrophils, and stimulation of nucleotide synthesis (Wischmeyer 2007, 2008). Glutamine supplements seem to have a significant benefit on mortality, length of hospitalization, and infectious morbidity in critical illness (Wischmeyer 2008). Positive results of glutamine supplementation have been shown patients undergoing thoracic surgery (Brinkmann et al. 2016) and in critically ill patients in whom supplemental glutamine reduced complications, mortality rates (Mondello et al. 2010; Novak et al. 2002), and increased gut barrier and lymphocyte function, and preserves lean body mass (Singleton et al. 2005). On the other hand, supplementation of glutamine during strenuous exercise in hypoxic conditions did not mitigate the stress effects of strenuous exercise on the immune system as demonstrated by oral mucosal immunity (Caris et al. 2017). Up to now, however, the use of glutamine as a pharmaconutrient has not been tested in spaceflight or spaceflight analogs (i.e., bed rest).

33.4 Vitamins and Minerals

33.4.1 Vitamin D

Although Vitamin D can be derived from the diet, the main supply of vitamin D_3 (cholecalciferol) for humans on Earth is obtained through endogenous synthesis in the skin following exposure to ultraviolet (UV) light (270–300 nm). Once in the circulation, to attain the biologically active metabolite $1,25(OH)_2D_3$, vitamin D_3 must first be hydroxylated in the liver into 25-hydroxyvitamin D_3 ($25(OH)D_3$) and then in the kidney by the $25(OH)D_3$-1-hydroxylase into $1,25(OH)_2D_3$, vitamin D_3 (calcitriol). The latter step is under control of parathyroid hormone; however, in case of vitamin D deficiency/insufficiency, renal hydroxylation becomes substrate dependent, that is, dependent on the circulating 25(OH)D (Schleithoff et al. 2008). Vitamin D status can best be assessed by measuring circulating 25(OH)D concentrations (Zittermann 2010). The 2011 Dietary Reference Intake (DRI) report from the Institute of Medicine (Institute of Medicine 2011) states that 50 nmol/L is optimal for serum 25(OH)D concentration to maximize benefits for bone. There continues to be debate regarding non-skeletal benefits of vitamin D, and many have argued for potential benefits of vitamin D for immune function, which may require higher serum concentrations. To this end, many maintain that 75–80 nmol/L is optimal, as this is the point of maximum suppression of parathyroid hormone. Cut-offs for defining vitamin D status are: <30 nmol/L for deficiency, <50 nmol/L as insufficiency, >125 nmol/L as of concern, and ≥200 nmol/L for intoxication (Institute of Medicine 2011; Zittermann and Gummert, 2010).

Most of the astronauts in space stations like Mir or ISS do have low 25(OH)D levels due to almost absent natural UVB from sunlight and limited food sources (Rettberg et al. 1998; Smith et al. 2009b). Indeed, it was reported in 2005 that decreased vitamin D status was one of the most striking nutritional changes that occurs during spaceflight (Smith et al. 2005a, b). The mean serum 25(OH)D concentration for US ISS crew members was 62 ± 14 nmol/L (Smith et al. 2009b), given the initial ISS supplement provision of 400 IU vitamin D per day. Serum 25(OH)D concentrations have typically been about 30% lower after 4–6 month spaceflights, and in several crewmembers, even decreased to levels considered clinically significant (i.e., <25 nmol/L) (Smith et al. 2005b). The current documented spaceflight requirement for dietary intake of vitamin D is 25 μg per day (National Aeronautics and Space Administration Johnson Space Center 2005), an increase from the original ISS recommendation of 10 μg per day (National Aeronautics and Space Administration Johnson Space Center 1996). The ISS food system provides less than half of this amount (4 μg per day on average) (Smith et al. 2009b). Given this shortfall, and because astronauts in space are shielded from sunlight, considering them to be in a high-risk group seems appropriate. Based on early ISS crew data, in 2006 the recommended was increased for ISS crew members take 800 IU of vitamin D per day during long-duration spaceflight (Smith et al. 2009b), and this resulted in crews maintaining serum vitamin D in the "optimal" range of 75–80 nmol/L (Smith et al. 2012). Investigations in the Antarctic overwintering crew are very useful as they mimic the effects of sunlight deprivation on the vitamin D status during long-duration spaceflight. Here it was observed that overwintering is associated with a decrease in 25-OH D levels, where only the baseline 25-OH D concentrations impacted the concentrations after 13 months of overwintering (Steinach et al. 2015). Another study carried out in the Antarctic winter aiming at the optimal level of vitamin D supplementation revealed that 1000 IU per day is a sufficient to maintain adequate blood concentrations (Smith et al. 2009a).

The classical function of vitamin D is to regulate calcium homeostasis and thus bone formation and resorption. However, in the last three decades a considerable body of evidence confirms the link and causal interaction between vitamin D and immune functions as identified in different innate antigen presenting cells (macrophages and dendritic cell), and adaptive immune cells (the T and B cell subsets), respectively (Vanherwegen et al. 2017). Vitamin D plays an important role in modulating the immune response to infections and exerts also other biological activities including immunomodulation (Lang and Aspinall 2017; Pincikova et al. 2017). The latter seems to be mediated via the (nuclear) vitamin D receptor (VDR) expressed in antigen-presenting cells and activated T-cells (van Etten and Mathieu 2005). Vitamin D and the VDR are required for normal numbers of regulatory T-cells. The discovery that VDR is inducibly expressed by lymphocytes following activation suggests a role for $1,25(OH)_2D_3$ in the immune system (Mathieu and Adorini 2002). Moreover, even the enzyme 25(OH)D3-1-α-hydroxylase is expressed by active macrophages, enabling to synthesis and secretion of calcitriol (Hewison et al. 2003). However, here the enzyme is mainly activated by immune signals such as interferon (IFN)-γ rather than the parathyroid hormone as the mediator in the kidney (van Etten and Mathieu 2005). Moreover, the active vitamin D metabolite 1,25$(OH)_2D_3$ can also modulate by alternative mechanisms to increase the ability of PBMC from sensitized human donors to resist to microbes (here mycobacteria). Martineau et al. found that 1,25$(OH)_2D_3$ suppressed both bacillus Calmette Guérin (BCG-) infected cell cultures and *Mycobacterium tuberculosis* likely through "non-classical" mechanisms including the induction of antimicrobial peptides (Martineau et al. 2007a, b). The effect of vitamin D on the "switch" of transition of information from adaptive to innate immunity as occurring by the antigen presenting cells (see Chap. 12) is of critical interest to enable adequate immune response and to prevent from a failure of immune tolerance resulting in allergic immune response (Pfeffer and Hawrylowicz 2018). Moreover, in tolerogenic dendritic cells metabolic reprogramming could be induced by a shift of heir intracellular metabolism under influence of 1,25$(OH)_2D_3$ (Vanherwegen et al. 2017), it also affects cells of the adaptive immune system stimulating the function of T-cells and activated B-cells (reviewed in Lang and Aspinall 2017).

Studies during or after spaceflight have shown numerous changes in astronauts' immune status, including altered distribution of circulating leukocytes, altered production of cytokines, decreased activity of natural killer cells, decreased function of granulocytes, decreased activation of T-cells, altered levels of immunoglobulins, latent viral reactivation, altered virus specific immunity, expression of Epstein–Barr virus immediate early/late genes, and altered neuroendocrine responses (see Chap. 19). Furthermore, there is an interactive effect between vitamin D status, stress (serum cortisol), and viral reactivation identified in crews wintering-over in the Antarctic (Zwart et al. 2011). Hence, vitamin D insufficiency of astronauts during space missions might impact the immune status of astronauts. Further studies are mandatory to distinguish between the effects of vitamin D and microgravity effects.

33.4.2 Vitamin B_6

Vitamin B_6 is essential in the biosynthesis of nucleic acid and protein by providing one-carbon units used in the production of deoxythymidylate and purines affecting DNA and mRNA synthesis. The active form of vitamin B_6, pyridoxal 5′-phosphate, is known to be a cofactor for many enzymatic reactions. Pyridoxal 5′-phosphate in plasma is reported to be inversely associated with inflammatory markers and a supplementation of vitamin B_6 improves some immune functions in humans who are deficient of vitamin B_6 (Ueland et al. 2017). Antibodies and cytokines are built up from amino acids and therefore require vitamin B_6 as a coenzyme in their metabolism. A deficiency in vitamin B_6 impairs lymphocyte maturation and growth in human, as well as antibody production and T-cell activity (Wintergerst et al. 2007) while supplementation could provide also benefits to autoimmune compromised patients (Huang et al. 2010).

Weightlessness has been shown to reduce the cross-sectional area of muscle fibers and is associated with a change from type I to type II muscle fibers

(Kraemer et al. 2000). Since vitamin B_6 is stored mainly in muscle tissue (Coburn et al. 1988), a decrease in muscle cross-sectional area could reduce the amount of the vitamin that is stored. Increased excretion of 4-pyridoxic acid (4-PA) during bed rest, a finding observed in short- (Zwart et al. 2009a) and long-duration bed rest studies (Coburn et al. 1995), likely reflects this loss of muscle stores of vitamin B_6.

33.4.3 Vitamin B_{12}

Vitamin B_{12} functions in many enzymatic reactions, and deficiencies result in anemia and neurological disorders. Vitamin B_{12} functions as a coenzyme in two metabolic forms: adenosylcobalamin and methylcobalamin. Vitamin B_{12} works as a cofactor for three different enzymatic reactions: (1) the conversion of homocysteine to methionine, (2) the conversion of L-methylmalonyl-coenzyme A (CoA) to succinyl-CoA, and (3) the isomerization of L-leucine and β-leucine. Vitamin B_{12} deficiency may cause the accumulation of folate in the serum because of a reduction in B_{12}-dependent methyltransferase, also known as the methyl-folate trap (Shin et al. 1975). Vitamin B_{12} also functions in the synthesis of choline, which can be converted to the neurotransmitter acetylcholine.

Unlike other water-soluble vitamins, vitamin B_{12} can be stored in the body for years. It is stored predominantly in the liver, but smaller amounts can also be found in the muscles, kidneys, bones, heart, brain, and spleen. About 2–5 mg of vitamin B_{12} is stored in the body (Institute of Medicine 1998). The size of B_{12} stores remains relatively stable, partly because urinary and fecal excretion decrease in direct relationship to decreases in the body pools. The half-life of vitamin B_{12} in humans is 350–400 days.

A human study in vitamin B_{12} deficient patients evaluated the alterations of immunological indicators following administration of vitamin B_{12}. In these patients, the number of lymphocytes was significantly decreased and NK activity was suppressed. Supplementation with vitamin B_{12} reversed these effects indicating that it may act as a modulatory agent for cellular immunity (Tamura et al. 1999). In elderly subjects (aged 70 years) who received over 4 months in addition to the regular diet a special nutritional formula providing, among other nutrients, 120 IU vitamin E, 3.8 mg vitamin B_{12}, and 400 mg folic acid, NK cytotoxic activity increased in supplemented subjects, indicating increased innate immunity in elderly people (Bunout et al. 2004). These few studies demonstrate the importance of a sufficient vitamin B_{12} status to maintain an adequate immune response (Maggini et al. 2007).

The current documented spaceflight requirement for dietary intake of vitamin B_6 is 1.7 mg/day (National Aeronautics and Space Administration Johnson Space Center 2005) and vitamin B_{12} is 2.4 µg/day. The interaction between the immune system and vitamin B_6 and B_{12} status has not been investigated up to now in spaceflight.

33.4.4 Sodium

Sodium is the major cation of the extracellular space and as such responsible for water and electrolyte metabolism. Sodium intake, mainly as sodium chloride, is very high in the Western world, being 2379 mg/day for women and 3216 mg/day in men in Germany, which is above the recommended intake per day of 550 mg/day (Max-Rubner-Institut 2008). Sodium intake on Skylab and shuttle missions averaged 4000–5000 mg/day, and were similar to preflight values (Bourland et al. 2000). In the ISS missions, typical intakes have been in excess of 4500 mg/day and in some cases intake as high as 10,000–12,000 mg/day have been observed (Smith et al. 2009b).

High sodium intake is correlated with development of hypertension in sodium sensitive people. We have shown in spaceflight as well as in ambulatory conditions that already at a level of about 4000 mg/day sodium is retained without accompanying fluid retention (Drummer et al. 2000; Heer et al. 2000, 2009). The hypothesis how sodium can be bound in an osmotically inactive way has been brought forward by Titze et al. (2003, 2004) and proposes that sodium can be stored on proteoglycans in interstitial sites. This uniquely bound sodium can induce a state of local hypertonicity in the skin interstitium. In a further study, they suggest that the local hypertonicity is sensed by macrophages which then activate a transcription factor (tonicity-enhanced binding protein (TonEBP)) which, in turn, induces VEGF-C (Machnik et al. 2010, 2009; Marvar et al. 2009) signaling. It seems that in mice the hyperosmotic stress induced by high salt intake leads to enhanced induction of Th17 response upon immunization suggesting an unknown function of the inflammasome (Ip and Medzhitov 2015). In vitro high salt concentrations, comparable to concentrations found in animals fed a high salt diet, dramatically boost the induction of murine and human Th17 cells (Kleinewietfeld et al. 2013). In a study with different level of sodium intake in humans, a high intake level (12 g NaCl/day) displayed a higher number of immune cell monocytes while low intake (6 g NaCl/day) was accompanied by reduced production of proinflammatory cytokines IL-6 and IL-23 together with enhanced production of anti-inflammatory cytokine IL-10 (Yi et al. 2016). Macrophages play a key role in innate immunity and therefore further studies in microgravity should distinguish between the effects of microgravity and high sodium intake on the immune system and to the pathways of how sodium induces bone cell turnover and demineralization (Frings-Meuthen et al. 2011).

33.5 Antioxidants

Endogenous antioxidants play an important role in minimizing cellular damage (Stein 2002) (see also Chap. 5). The antioxidant vitamins A, C, and E are cofactors in the immune response. Vitamin A deficiency impairs mucosal barriers (Zeng et al. 2016) and diminishes the function of neutrophils, macrophages, and NK (Stephensen 2001a). Vitamin E is a strong antioxidant that can support monocyte/macrophage-mediated responses (Park et al. 2003) and vitamin C is a regulator of redox and metabolic checkpoints controlling activation and survival of immune cells (Wintergerst et al. 2006).

Selenoproteins are an important component of the antioxidant host defense system affecting leukocyte and NK function (Ferencik and Ebringer 2003).

33.5.1 Vitamin A

Vitamin A plays a well-known role in immune function and protection against infections (Ross 1992; Stephensen 2001b). Vitamin A deficiency can affect host defenses directly through its essential functions in metabolism in the various immune cells (Semba 1998) or indirectly through its role in epithelial cell differentiation and host barrier function (Ross 1992; Stephensen 2001b). The considerable immune benefits, which would contribute in reducing the risk of various pathogen-mediated diseases, warrant a recommendation to supplement individuals with minimal or poor vitamin A status. Supplementing vitamin A after training in rats led to increased total serum antioxidant capacity, but concurrently expression of superoxide dismutase-1 was downregulated and upregulation of superoxide dismutase-2 induced by exercise was blunted by vitamin A (Petiz et al. 2017). Additionally, IL-10 and heat shock protein 70 expression, which are both positive for tissue damage protection after exercise, were decreased (Petiz et al. 2017). In a study in healthy aged subjects, vitamin A supplementation did not affect lymphocyte proliferation (Bouamama et al. 2017).

The current documented spaceflight requirement for vitamin A intake is 700–900 μg/day. Vitamin A content and stability in the space food supply should be determined. The role of vitamin A as an antioxidant in spaceflight has not been investigated up to now. Whether vitamin A supplementation would be a desired measure to improve immune response in spaceflight needs to be thoroughly thought through.

33.5.2 Vitamin C

Ascorbic acid (vitamin C) is an essential component of every living cell. The concentration of vitamin C is very high in leukocytes and is used rapidly during infection to prevent oxidative damage. A deficiency in vitamin C status is associated with reduced immune function (Schwager and Schulze 1998). The immune-enhancing role of vitamin C has been reviewed (Carr and Maggini 2017; Wintergerst et al. 2006). Vitamin C has been shown to stimulate the immune system by enhancing T-lymphocyte proliferation in response to infection increasing cytokine production and synthesis of immunoglobulins (Jeng et al. 1996). Vitamin C stimulates neutrophil migration to the site of infection, enhances phagocytosis and oxidant generation, and microbial killing (Carr and Maggini 2017).

Spaceflight requirement for dietary intake of vitamin C is 90 mg/day. Vitamin C status of crew members has not been investigated to date. However, in analog studies like short-duration bed rest, no significant change in vitamin C, but a trend for an increase, could be shown (Zwart et al. 2009b). This might be related to dietary vitamin C intake during the study compared to the intake before the study (Zwart et al. 2009b).

33.5.3 Vitamin E

Vitamin E and selenium have synergistic functions in tissues to reduce damage to lipid membranes by the formation of reactive oxygen species (ROS) during infections. The ability of vitamin E to scavenge lipid soluble-free radicals is dependent to some extent on the status of two other antioxidant compounds, vitamin C and glutathione, which are involved in reducing oxidized vitamin E back to a reusable (i.e., able to be oxidized) form (Carmeliet et al. 2001). Additionally, vitamin E may improve T-cell function by decreasing macrophage prostaglandin E2 production by modulating the amino acid cascade initiated by lipoxygenase and/or cyclooxygenase (Chapkin et al. 2007). Furthermore, vitamin E influences lymphocyte maturation, possibly by stabilizing membranes and allowing enhanced binding of antigen-presenting cells to immature T cells via increased expression of intercellular adhesion molecule-1. Although a vitamin E deficiency is not common, intake above currently recommended levels may help restore T cell function as seen in an aging population were T-cell functions are significantly impaired. "This effect of vitamin E can be accomplished by directly impacting T cells as well as indirectly, by inhibiting production of prostaglandin E2, a T cell-suppressing lipid mediator known to increase with aging" as reviewed by Wu and Meydani (2014).

After early ISS crew members had spent 4–6 months in space, their plasma γ-tocopherol was 50% less than preflight levels (Smith et al. 2005b). No change in α-tocopherol occurred in these subjects.

Although no striking changes occurred in plasma vitamin E concentrations, the spaceflight menus provide only about 60% of the documented requirement for vitamin E (Smith et al. 2009b). Antioxidant properties of vitamin E may help to counteract the free-radical damage caused by high-linear energy transfer radiation in space. Pretreatment with antioxidants may help decrease radiation damage during mission (Pence and Yang 2000), and it may be necessary to provide enough vitamin E so that astronauts' blood levels of the vitamin is higher during spaceflight than on Earth. However, knowledge gaps weaken the evidence for use of vitamin E as a countermeasure.

33.5.4 Copper

Copper is an important mineral, and has wide-ranging functions, including many that are considered vital for spaceflight (Borchers et al. 2002; Crucian et al. 2008; Levine and Greenleaf 1998; Taylor et al. 1997; Tipton et al. 1996). This might have direct or indirect (when alterations are induced by psychological stress or radiation stress) implications for nutrition and nutritional status as possible causes or effects on immune system function (Smith et al. 2009b). Nonetheless, to date, little/no information is available about copper metabolism during spaceflight, especially with regard to immune system function.

33.5.5 Zinc

In addition to its many essential functions in growth and development, zinc is essential for the function of cells of the immune system and has been described as a gatekeeper of the immune system. It has an important role in promotion of wound healing and in maintenance of intestinal integrity. A deficiency of zinc is pro-inflammatory and it is also associated with reduced concentrations of IGF 1 and reduced rates of protein synthesis (Wessels et al. 2017). Therefore, zinc deficiency could be especially detrimental during immobility. However, zinc status of astronauts, as assessed by mean serum zinc and urinary zinc excretion (admittedly, not the best markers of zinc status), did not change after long-duration spaceflight (Smith et al. 2009b). There is no knowledge available on the use of zinc supplementation as a countermeasure in spaceflight.

33.5.6 Polyphenols (Resveratrol, Quercetin, Curcumin, Catechins)

Naturally occurring polyphenols like resveratrol, quercetin, curcumin, and catechins have shown antioxidant and anti-inflammatory effects (Chung et al. 2010). These effects seem to be modulated via different pathways such as NF-κB- and mitogen-activated protein kinase-dependent pathways, as well as preventing the generation of reactive oxygen species by binding iron (Perron and Brumaghim 2009). Additionally, polyphenols seem to activate sirtuin 1 (SIRT1) directly or indirectly and thereby are beneficial—besides others—for regulation of oxidative stress, inflammation, and autoimmunity. Accumulating evidence has shown that polyphenols such as resveratrol, curcumin, catechins, and quercetins, have a regulatory role in immune function in vitro and in vivo (Devine et al. 2007; Gao et al. 2003; Park et al. 2009; Sharma et al. 2007; Shim et al. 2008, Singh et al. 2007; Song et al. 2003; Vasamsetti et al. 2016). Therefore, they might also have beneficial effects in prevention of immune dysfunction during long-term missions, particularly because body iron stores are higher during spaceflight. However, the role of polyphenols in SIRT1-mediated or iron-related regulation in immune function remains still to be studied.

There is an increasing amount of studies investigating the effect of effective candidates of natural compounds such as polyphenols on the immune system. As summarized in a recent review, different immune cells express polyphenols receptors and the polyphenols may activate signaling pathways in the cells and subsequently initiate immune response (Ding et al. 2018). Specific polyphenols, such as curcumin and epigallocatechingallate may induce epigenetic changes within cells

(Ding et al. 2018). An impact on epigenetic mechanisms such as DNA methylation or histone modification may also be induced by quercetin (Cuevas et al. 2013). Polyphenols are also the most studied natural compounds with regard to their capability to prevent allergies. They seem, for instance, to dampen the onset of allergic inflammation by inhibiting microsomal prostaglandin E2 synthase-1 or inhibit the expression of pro-inflammatory TH-2-cytokines IL-4 and IL-13 cytokine-producing signaling factors, and prostaglandin-endoperoxidase 2, indicating that specific polyphenol candidates broadly inhibit allergic inflammation (Chirumbolo 2014). The main mechanism of polyphenol inhibitory effects on allergy development seems to be the interference with T-helper 2 cell activation (Magrone and Jirillo 2012). Although there is some evidence for polyphenols as a promising tool to modulate immune function, further studies are mandatory to demonstrate positive effects in situations where inflammation and oxidative stress is induced, for instance by exercise. In cyclists a blend of freeze-dried juice powder or a watermelon puree did not alter inflammation and oxidative stress or immune function during or after intensified exercise sessions at the end of the supplemental period (Knab et al. 2014; Shanely et al. 2016). In endurance athletes, myeloid DCs increase and plasmazytoid DCs decrease during exercise, while levels of viral antigen presenting toll-like receptor (TLR) 7 messenger RNA are declined. Providing a polyphenol-rich beverage for 3 weeks before a marathon did not affect the changes in the DCs; however, it supported the regeneration of the viral antigen presenting TLR7 (Lackermair et al. 2017). To prove any effects of an antioxidant cocktail during inactivity such as bed rest, the European Space Agency sponsored a 60-day bed rest study which has been finished recently. The antioxidant cocktail consisted of an antioxidant and anti-inflammatory food supplement (a mixture of natural polyphenolic extracts from edible plants) which was supplied in the form of capsules. The respective experiments will address changes in effects of the cocktail on metabolism, the cardiovascular system, muscles, bones, immunology, the neurosensory system, and sleep.

No results are obtained during spaceflight up to date. Therefore, it is very obvious that further studies are also warranted to examine polyphenols as potential dietary measures to counteract immune dysfunction in microgravity.

33.5.7 Iron

Iron is a critical micronutrient involved in many cellular processes. Functions include oxygen binding, electron transport, and serving as a catalyst for literally hundreds of enzymes. Iron deficiency can have irreversible consequences, but excess iron can be toxic through the formation of oxygen free radicals. Iron overload can increase cardiovascular disease risk and cancer risk, impair immune function to activated or inhibited states (Kernan and Carcillo 2017), and contribute to eye diseases such as cataracts and age-related macular degeneration (Crichton 2009; Loh et al. 2009). Thus, maintenance of iron homeostasis is extremely important for human health.

During spaceflight, it is well established that iron homeostasis is altered (Smith et al. 2009b). There is a decreased red blood cell mass, increased serum ferritin, decreased transferrin receptors, and increased serum iron—all of which document increased iron storage during spaceflight. Furthermore, the space food system provides almost three times the recommended intake (Smith et al. 2009b), at a time when dietary iron recommendations for the general population being lowered. This can in turn bring iron into an even more ambiguous "role," as it is not only necessary for host survival, but microorganisms require iron acquisition from the environment for survival as well. Generally, during an infection, iron in the body is made less available for invading microorganisms by increasing iron uptake into cells or by increasing protein-bound iron. Cells of the innate immune system have genes which regulate proteins that can modulate iron homeostasis at the cellular and systemic level in order to restrict iron availability to invading microorganisms. One such protein is hepcidin, a key regulator of iron homeostasis and critical factor in the anemia of inflammation (Ganz 2006). Hepcidin has been shown to be endogenously expressed by innate immune cells—macrophages and neutrophils, and it plays a role in making iron less available by increasing intracellular iron sequestration and decreasing circulating iron concentrations and it is influenced by cytokines IL-6 and IL-1 (Lee et al. 2005; Nemeth et al. 2004) and represents one component of the innate immune cell's response to acute infections (Armitage et al. 2011). Iron is also associated with certain optic neuropathies and retinal degeneration (Theriot et al. 2016). Studies suggest that some types of radiation exposure, which is increased in spaceflight, and oxidative stress can release ferrous iron (Fe^{2+}) from ferritin (Aubailly et al. 1991), further adding to the load of free iron in the body. Independently, both radiation exposure and high dietary iron load promote a state of oxidative stress with increased risk of pathophysiological outcomes (Sannita et al. 2004; Stevens et al. 2000). In a recent study in rats the effect of high dietary iron intake and whole body radiation exposure was analyzed (Theriot et al. 2016). They found increased levels of 8-OH-dG in the high iron intake-, radiation and the combined group. An attenuation of radiation-induced DNA oxidation in the retina of animals under the high-iron diet was observed (Theriot et al. 2016).

Further research is warranted to determine what role increased iron storage during spaceflight plays on changes in immune function, viral reactivation, microbial growth, and virulence.

33.6 Polyunsaturated Fatty Acids (PUFAS)

Omega-3 fatty acids are long-chain, polyunsaturated fatty acids (PUFA). Extensive documentation exists showing that omega-3 fatty acids provide protection to the general population for the cardiovascular and immune systems (Fernandes et al. 2008; Kang and Weylandt 2008; Yaqoob and Calder 2003), for psychiatric diseases (Pusceddu et al. 2016), stress-related cognitive dysfunction in experimental model (Pusceddu et al. 2015) and even protection from oxidative damage and radiation-related risks (Turner et al. 2002; Vanamala et al. 2008), all of which are concerns for space travelers. Moreover, a summary on human and animal studies strongly indicates the potential action of omega-3 PUFAs to affect "the gut-brain axis, acting through gut microbiota composition" (Costantini et al. 2017) (see also Chap. 34), but also to directly affect inflammatory states (Michalak et al. 2016).

The mechanism of action of omega-3 fatty acids on these systems is likely related to multiple pathways, but there is evidence that the nuclear transcription factor, NF-κB, is affected differently by omega-3 or omega-6 fatty acids (Camandola et al. 1996). This transcription factor affects transcription of genes involved in cell cycle regulation and inflammatory processes. Not only is NF-κB activated by arachidonic acid and specifically by prostaglandin E2, but Camandola and colleagues have also found that eicosapentaenoic acid (EPA) inhibits NF-κB activation (Camandola et al. 1996). Interestingly experimental data indicate the potential to block autoimmune response and to hereby regulate autoimmunity (Bi et al. 2017).

Cytokines such as tumor necrosis factor (TNF) can activate NF-κB, and subsequently stimulate other processes such as muscle protein loss through upregulation of mRNA and expression of various ubiquitin-proteasome protein subunits (Li and Reid 2000). Not surprisingly, conditions associated with a chronic elevation of TNF are also commonly associated with muscle atrophy. We have reported elevated NF-κB after short-duration spaceflight (Zwart et al. 2010). The effects of omega-3 fatty acids on inflammatory cytokines, and specifically TNF, are well documented on the ground (Kang and Weylandt 2008; Kim et al. 2008; Magee et al. 2008; Zwart et al. 2010), but warrant further studies during spaceflight.

33.7 Prebiotics

Intestinal microbiota is a major player in immune tolerance and inflammation. Its crosstalk with the innate and adaptive immune cells is involved in the regulation of immunomodulatory effects and disrupted communication seems to negatively influence the intestinal immune homoeostasis (Frei et al. 2015). For an appropriate immune function, early bacterial colonization seems to be very important to develop respective immune regulatory networks, which may influence disease risks in later life (Frei et al. 2015). Prebiotics in mothers' milk, such as certain oligosaccharides, will support the development of a healthy gut microbiota composition. Even supplementing prebiotics, such as Xylo-Oligosaccharides to healthy adults for a 3-week period increased bifidobacterial counts, as well as changes in the cell-surface markers on NKT cells and lowered IL-10 secretion, suggesting an immunomodulatory effect of Xylo-Oligosaccharides (Childs et al. 2014). Supplementing prebiotics might be a promising tool to improve immune function in space exploration missions and warrants further investigation space exploration.

33.8 Conclusion

The understanding of the impact of nutrition on cell/organ functions is continuously increasing. Also the interactions of direct effects of nutritional compounds/nutraceuticals or indirect effects have come more to our attention. The field of immunonutrition has emerged as a scientific topic critical for improving outcomes in a wide range of patients, particularly when subject to distinct stressors, such as trauma or inflammation, and as in under/malnourished individuals. Since astronauts in space are generally not optimally nourished, dietary formulas tailored to the astronauts needs may be beneficial for immune system function. Furthermore, the environmental stress of spaceflight (see Chap. 3) can altogether lead to changes in immune response as well as the individual needs of the respective astronaut. These and other factors are required to support optimal astronaut health during long-duration missions.

However, it is important to be aware that "one size does *not* fit all." This implies that genetics and the immune nutrient profile that is appropriate for one astronaut or one condition may be of minimal benefit for another one and could be potentially harmful in other settings. Moreover, further understanding is needed to which degree changes in the thermoregulatory functions of crew subject, the microbiome or microbiota composition in the gut and new opportunities of grown fresh food in the space vessel or habitat during a long-duration space missions might strongly affect the strategy of a personalized and adequate nutrition. Making evidence-based decisions in choosing the optimal diet or formula will minimize adverse effects. In order to reach that level of individualized immunonutrition for astronauts, further research is needed to assess whether immune function is benefited by individually tailored immunonutrient formulas.

The use of basic clinical pharmacology, genetics, molecular biology, and clinical research principles in the study of nutritional therapy during spaceflight and analog studies will lead to answers on how to administer the right nutrients, in the right amounts, at the right time during astronauts' exploration missions. In summary, we have to provide the "Right Stuff" for exploration missions after all the research and the new challenges ahead (see also Chap. 3). This means that out of the mix of things we need to do everything right, may be starting with an optimized and individualized nutrition at the forefront.

References

Alito MA, de Aguilar-Nascimento JE (2016) Multimodal perioperative care plus immunonutrition versus traditional care in total hip arthroplasty: a randomized pilot study. Nutr J 15:34

Armitage AE, Eddowes LA, Gileadi U, Cole S, Spottiswoode N, Selvakumar TA, Ho LP, Townsend AR, Drakesmith H (2011) Hepcidin regulation by innate immune and infectious stimuli. Blood 118(15):4129–4139

Aubailly M, Santus R, Salmon S (1991) Ferrous ion release from ferritin by ultraviolet-A radiations. Photochem Photobiol 54(5):769–773

Beisel WR (1992) History of nutritional immunology: introduction and overview. J Nutr 122(3 Suppl):591–596

Bi X, Li F, Liu S, Jin Y, Zhang X, Yang T, Dai Y, Li X, Zhao AZ (2017) omega-3 polyunsaturated fatty acids ameliorate type 1 diabetes and autoimmunity. J Clin Invest 127(5):1757–1771

Bloos F et al (2016) Effect of sodium selenite administration and procalcitonin-guided therapy on mortality in patients with severe sepsis or septic shock: a randomized clinical trial. JAMA Intern Med 176(9):1266–1276

Borchers AT, Keen CL, Gershwin ME (2002) Microgravity and immune responsiveness: implications for space travel. Nutrition 18(10):889–898

Bosutti A, Malaponte G, Zanetti M, Castellino P, Heer M, Guarnieri G, Biolo G (2008) Calorie restriction modulates inactivity-induced changes in the inflammatory markers C-reactive protein and pentraxin-3. J Clin Endocrinol Metab 93(8):3226–3229

Bouamama S, Merzouk H, Medjdoub A, Merzouk-Saidi A, Merzouk SA (2017) Effects of exogenous vitamins A, C, and E and NADH supplementation on proliferation, cytokines release, and cell redox status of lymphocytes from healthy aged subjects. Appl Physiol Nutr Metab 42(6):579–587

Bourland CT, Kloeris V, Rice BL, Vodovotz Y (2000) Food systems for space and planetary flights. In: Lane HW, Schoeller DA (eds) Nutrition in spaceflight and weightlessness models. CRC press, Boca Raton, FL, pp 19–40

Brinkmann SJ, Buijs N, Vermeulen MA, Oosterink E, Schierbeek H, Beishuizen A, de Vries JP, Wisselink W, van Leeuwen PA (2016) Perioperative glutamine supplementation restores disturbed renal arginine synthesis after open aortic surgery: a randomized controlled clinical trial. Am J Physiol Renal Physiol 311(3):F567–F575

Bronte V, Zanovello P (2005) Regulation of immune responses by L-arginine metabolism. Nat Rev Immunol 5(8):641–654

Bunout D et al (2004) Effects of a nutritional supplement on the immune response and cytokine production in free-living Chilean elderly. JPEN J Parenter Enteral Nutr 28(5):348–354

Camandola S, Leonarduzzi G, Musso T, Varesio L, Carini R, Scavazza A, Chiarpotto E, Baeuerle PA, Poli G (1996) Nuclear factor kB is activated by arachidonic acid but not by eicosapentaenoic acid. Biochem Biophys Res Commun 229(2):643–647

Cao Y, Feng Y, Zhang Y, Zhu X, Jin F (2016) L-Arginine supplementation inhibits the growth of breast cancer by enhancing innate and adaptive immune responses mediated by suppression of MDSCs in vivo. BMC Cancer 16:343

Caris AV, Da Silva ET, Dos Santos SA, Tufik S, Dos Santos RVT (2017) Effects of carbohydrate and glutamine supplementation on oral mucosa immunity after strenuous exercise at high altitude: a double-blind randomized trial. Nutrients 9(7):E692

Carmeliet G, Vico L, Bouillon R (2001) Space flight: a challenge for normal bone homeostasis. Crit Rev Eukaryot Gene Expr 11(1–3):131–144

Carr AC, Maggini S (2017) Vitamin C and immune function. Nutrients 9(11):E1211

Chandra RK (1992) Nutrition and immunoregulation. Significance for host resistance to tumors and infectious diseases in humans and rodents. J Nutr 122(3 Suppl):754–757

Chandra RK, Kumari S (1994) Nutrition and Immunity: an overview. J Nutr 124(8):1433S–1435S

Chandra RK, Chandra S, Gupta S (1984) Antibody affinity and immune complexes after immunization with tetanus toxoid in protein-energy malnutrition. Am J Clin Nutr 40(1):131–134

Chapkin RS, Davidson LA, Ly L, Weeks BR, Lupton JR, McMurray DN (2007) Immunomodulatory effects of (n-3) fatty acids: putative link to inflammation and colon cancer. J Nutr 137(1 Suppl):200S–204S

Childs CE et al (2014) Xylo-oligosaccharides alone or in synbiotic combination with Bifidobacterium animalis subsp. lactis induce bifidogenesis and modulate markers of immune function in healthy adults: a double-blind, placebo-controlled, randomised, factorial cross-over study. Br J Nutr 111(11):1945–1956

Chirumbolo S (2014) Dietary assumption of plant polyphenols and prevention of allergy. Curr Pharm Des 20(6):811–839

Chung S, Yao H, Caito S, Hwang JW, Arunachalam G, Rahman I (2010) Regulation of SIRT1 in cellular functions: role of polyphenols. Arch Biochem Biophys 501(1):79–90

Coburn SP, Lewis DL, Fink WJ, Mahuren JD, Schaltenbrand WE, Costill DL (1988) Human vitamin B-6 pools estimated through muscle biopsies. Am J Clin Nutr 48(2):291–294

Coburn SP et al (1995) Pyridoxic acid excretion during low vitamin B-6 intake, total fasting, and bed rest. Am J Clin Nutr 62(5):979–983

Costantini L, Molinari R, Farinon B, Merendino N (2017) Impact of omega-3 fatty acids on the gut microbiota. Int J Mol Sci 18(12):E2645

Crichton RR (2009) Iron metabolism: from molecular mechanisms to clinical consequences. John Wiley & Sons Ltd, West Sussex

Crucian BE, Stowe RP, Pierson DL, Sams CF (2008) Immune system dysregulation following short- vs long-duration spaceflight. Aviat Space Environ Med 79(9):835–843

Crucian BE et al (2018) Immune system dysregulation during spaceflight: potential countermeasures for deep space exploration missions. Front Immunol 9:1437

Cuevas A, Saavedra N, Salazar LA, Abdalla DS (2013) Modulation of immune function by polyphenols: possible contribution of epigenetic factors. Nutrients 5(7):2314–2332

Cunningham-Rundles S, McNeeley DF, Moon A (2005) Mechanisms of nutrient modulation of the immune response. J Allergy Clin Immunol 115(6):1119–1128

Devine A, Hodgson JM, Dick IM, Prince RL (2007) Tea drinking is associated with benefits on bone density in older women. Am J Clin Nutr 86(4):1243–1247

Ding S, Jiang H, Fang J (2018) Regulation of immune function by polyphenols. J Immunol Res 2018:1264074

Drummer C, Hesse C, Baisch F, Norsk P, Elmann-Larsen B, Gerzer R, Heer M (2000) Water and sodium balances and their relation to body mass changes in microgravity. Eur J Clin Invest 30(12):1066–1075

van Etten E, Mathieu C (2005) Immunoregulation by 1,25-dihydroxyvitamin D3: basic concepts. J Steroid Biochem Mol Biol 97(1-2):93–101

Ferencik M, Ebringer L (2003) Modulatory effects of selenium and zinc on the immune system. Folia Microbiol (Praha) 48(3):417–426

Fernandes G, Bhattacharya A, Rahman M, Zaman K, Banu J (2008) Effects of n-3 fatty acids on autoimmunity and osteoporosis. Front Biosci v. 13:4015–4020

Frei R, Akdis M, O'Mahony L (2015) Prebiotics, probiotics, synbiotics, and the immune system: experimental data and clinical evidence. Curr Opin Gastroenterol 31(2):153–158

Frings-Meuthen P, Buehlmeier J, Baecker N, Stehle P, Fimmers R, May F, Kluge G, Heer M (2011) High sodium chloride intake exacerbates immobilization-induced bone resorption and protein losses. J Appl Physiol 111(2):537–542

Ganz T (2006) Hepcidin--a peptide hormone at the interface of innate immunity and iron metabolism. Curr Top Microbiol Immunol 306:183–198

Gao X, Deeb D, Media J, Divine G, Jiang H, Chapman RA, Gautam SC (2003) Immunomodulatory activity of resveratrol: discrepant in vitro and in vivo immunological effects. Biochem Pharmacol 66(12):2427–2435

Guadagni M, Biolo G (2009) Effects of inflammation and/or inactivity on the need for dietary protein. Curr Opin Clin Nutr Metab Care 12(6):617–622

Heer M, Boerger A, Kamps N, Biener C, Korr C, Drummer C (2000) Nutrient supply during recent European missions. Pflugers Arch 441(2–3):R8–R14

Heer M, Frings-Meuthen P, Titze J, Boschmann M, Frisch S, Baecker N, Beck L (2009) Increasing sodium intake from a previous low or high intake affects water, electrolyte and acid-base balance differently. Br J Nutr 101:1286–1294

Hewison M, Freeman L, Hughes SV, Evans KN, Bland R, Eliopoulos AG, Kilby MD, Moss PA, Chakraverty R (2003) Differential regulation of vitamin D receptor and its ligand in human monocyte-derived dendritic cells. J Immunol 170(11):5382–5390

Huang SC, Wei JC, Wu DJ, Huang YC (2010) Vitamin B(6) supplementation improves pro-inflammatory responses in patients with rheumatoid arthritis. Eur J Clin Nutr 64(9):1007–1013

Institute of Medicine (1998) Dietary reference intakes for thiamin, riboflavin, niacin, vitamin B6, folate, vitamin B12, pantothenic acid, biotin, and choline. National Academic Press, Washington DC

Institute of Medicine (2011) Dietary reference intakes for calcium and vitamin D. National Academies Press, Washington DC

Ip WK, Medzhitov R (2015) Macrophages monitor tissue osmolarity and induce inflammatory response through NLRP3 and NLRC4 inflammasome activation. Nat Commun 6:6931

Jeng KC, Yang CS, Siu WY, Tsai YS, Liao WJ, Kuo JS (1996) Supplementation with vitamins C and E enhances cytokine production by peripheral blood mononuclear cells in healthy adults. Am J Clin Nutr 64(6):960–965

Joyner MJ (2005) Glutamine and arginine: immunonutrients and metabolic modulators? Exerc Sport Sci Rev 33(3):105–106

Kang JX, Weylandt KH (2008) Modulation of inflammatory cytokines by omega-3 fatty acids. Subcell Biochem 49:133–143

Kang M et al (2017) Supplementation of fermented Maillard-reactive whey protein enhances immunity by increasing NK cell activity. Food Funct 8(4):1718–1725

Keith ME, Jeejeebhoy KN (1997) Immunonutrition. Baillieres Clin Endocrinol Metab 11(4):709–738

Kernan KF, Carcillo JA (2017) Hyperferritinemia and inflammation. Int Immunol 29(9):401–409

Kim HH, Lee Y, Eun HC, Chung JH (2008) Eicosapentaenoic acid inhibits TNF-alpha-induced matrix metalloproteinase-9 expression in human keratinocytes, HaCaT cells. Biochem Biophys Res Commun 368(2):343–349

Kirk SJ, Barbul A (1990) Role of arginine in trauma, sepsis, and immunity. JPEN J Parenter Enteral Nutr 14(5 Suppl):226S–229S

Kirk SJ, Hurson M, Regan MC, Holt DR, Wasserkrug HL, Barbul A (1993) Arginine stimulates wound healing and immune function in elderly human beings. Surgery 114(2):155–159

Kleinewietfeld M, Manzel A, Titze J, Kvakan H, Yosef N, Linker RA, Muller DN, Hafler DA (2013) Sodium chloride drives autoimmune disease by the induction of pathogenic TH17 cells. Nature 496(7446):518–522

Knab AM, Nieman DC, Gillitt ND, Shanely RA, Cialdella-Kam L, Henson D, Sha W, Meaney MP (2014) Effects of a freeze-dried juice blend powder on exercise-induced inflammation, oxidative stress, and immune function in cyclists. Appl Physiol Nutr Metab 39(3):381–385

Kraemer WJ et al (2000) The effects of 10 days of spaceflight on the shuttle Endeavor on predominantly fast-twitch muscles in the rat. Histochem Cell Biol 114(5):349–355

Lackermair K et al (2017) Influence of polyphenol-rich diet on exercise-induced immunomodulation in male endurance athletes. Appl Physiol Nutr Metab 42(10):1023–1030

Lane HW, Gretebeck RJ, Schoeller DA, Davis-Street J, Socki RA, Gibson EK (1997) Comparison of ground-based and space flight energy expenditure and water turnover in middle-aged healthy male US astronauts. Am J Clin Nutr 65(1):4–12

Lang PO, Aspinall R (2017) Vitamin D status and the host resistance to infections: what it is currently (not) understood. Clin Ther 39(5):930–945

Lang CA, Naryshkin S, Schneider DL, Mills BJ, Lindeman RD (1992) Low blood glutathione levels in healthy aging adults. J Lab Clin Med 120(5):720–725

Lee P, Peng H, Gelbart T, Wang L, Beutler E (2005) Regulation of hepcidin transcription by interleukin-1 and interleukin-6. Proc Natl Acad Sci U S A 102(6):1906–1910

Levine DS, Greenleaf JE (1998) Immunosuppression during spaceflight deconditioning. Aviat Space Environ Med 69(2):172–177

Li YP, Reid MB (2000) NF-kappaB mediates the protein loss induced by TNF-alpha in differentiated skeletal muscle myotubes. Am J Physiol Regul Integr Comp Physiol 279(4):R1165–R1170

Loh A, Hadziahmetovic M, Dunaief JL (2009) Iron homeostasis and eye disease. Biochim Biophys Acta 1790(7):637–649

Machnik A et al (2009) Macrophages regulate salt-dependent volume and blood pressure by a vascular endothelial growth factor-C-dependent buffering mechanism. Nat Med 15(5):545–552

Machnik A et al (2010) Mononuclear phagocyte system depletion blocks interstitial tonicity-responsive enhancer binding protein/vascular endothelial growth factor C expression and induces salt-sensitive hypertension in rats. Hypertension 55(3):755–761

Magee P, Pearson S, Allen J (2008) The omega-3 fatty acid, eicosapentaenoic acid (EPA), prevents the damaging effects of tumour necrosis factor (TNF)-alpha during murine skeletal muscle cell differentiation. Lipids Health Dis 7:24

Maggini S, Wintergerst ES, Beveridge S, Hornig DH (2007) Selected vitamins and trace elements support immune function by strengthening epithelial barriers and cellular and humoral immune responses. Br J Nutr 98(Suppl 1):S29–S35

Magrone T, Jirillo E (2012) Influence of polyphenols on allergic immune reactions: mechanisms of action. Proc Nutr Soc 71(2):316–321

Martineau AR et al (2007a) IFN-gamma- and TNF-independent vitamin D-inducible human suppression of mycobacteria: the role of cathelicidin LL-37. J Immunol 178(11):7190–7198

Martineau AR et al (2007b) A single dose of vitamin D enhances immunity to mycobacteria. Am J Respir Crit Care Med 176(2):208–213

Marvar PJ, Gordon FJ, Harrison DG (2009) Blood pressure control: salt gets under your skin. Nat Med 15(5):487–488

Mathieu C, Adorini L (2002) The coming of age of 1,25-dihydroxyvitamin D(3) analogs as immunomodulatory agents. Trends Mol Med 8(4):174–179

Matsubara LS, Machado PE (1991) Age-related changes of glutathione content, glutathione reductase and glutathione peroxidase activity of human erythrocytes. Braz J Med Biol Res 24(5):449–454

Max-Rubner-Institut (2008) Nationale Verzehrsstudie II Karlsruhe. Bundesforschungsinstitut für Ernährung und Lebensmittel, Karlsruhe

McCarthy MS, Martindale RG (2018) Immunonutrition in critical illness: what is the role? Nutr Clin Pract 33(3):348–358

Michalak A, Mosinska P, Fichna J (2016) Polyunsaturated fatty acids and their derivatives: therapeutic value for inflammatory, functional gastrointestinal disorders, and colorectal cancer. Front Pharmacol 7:459

Micke P, Beeh KM, Schlaak JF, Buhl R (2001) Oral supplementation with whey proteins increases plasma glutathione levels of HIV-infected patients. Eur J Clin Invest 31(2):171–178

Mizock BA (2010) Immunonutrition and critical illness: an update. Nutrition 26(7-8):701–707

Mondello S, Italiano D, Giacobbe MS, Mondello P, Trimarchi G, Aloisi C, Bramanti P, Spina E (2010) Glutamine-supplemented total parenteral nutrition improves immunological status in anorectic patients. Nutrition 26(6):677–681

National Aeronautics and Space Administration Johnson Space Center (1996) Nutritional requirements for International Space Station (ISS) missions up to 360 days. NASA, Houston, TX

National Aeronautics and Space Administration Johnson Space Center (2005) Nutrition requirements, standards, and operating bands for exploration missions. NASA, Houston, TX

Nemeth E, Rivera S, Gabayan V, Keller C, Taudorf S, Pedersen BK, Ganz T (2004) IL-6 mediates hypoferremia of inflammation by inducing the synthesis of the iron regulatory hormone hepcidin. J Clin Invest 113(9):1271–1276

Novak F, Heyland DK, Avenell A, Drover JW, Su X (2002) Glutamine supplementation in serious illness: a systematic review of the evidence. Crit Care Med 30(9):2022–2029

Park OJ, Kim HY, Kim WK, Kim YJ, Kim SH (2003) Effect of vitamin E supplementation on antioxidant defense systems and humoral immune responses in young, middle-aged and elderly Korean women. J Nutr Sci Vitaminol (Tokyo) 49(2):94–99

Park HJ et al (2009) Quercetin regulates Th1/Th2 balance in a murine model of asthma. Int Immunopharmacol 9(3):261–267

Pence BC, Yang TC (2000) Antioxidants: radiation and stress. In: Lane HW, Schoeller DA (eds) Nutrition in spaceflight and weightlessness models. CRC press, Boca Raton, FL, pp 233–251

Perron NR, Brumaghim JL (2009) A review of the antioxidant mechanisms of polyphenol compounds related to iron binding. Cell Biochem Biophys 53(2):75–100

Petiz LL, Girardi CS, Bortolin RC, Kunzler A, Gasparotto J, Rabelo TK, Matte C, Moreira JC, Gelain DP (2017) Vitamin A oral supplementation induces oxidative stress and suppresses IL-10 and HSP70 in skeletal muscle of trained rats. Nutrients 9(4):E353

Pfeffer PE, Hawrylowicz CM (2018) Vitamin D in asthma: mechanisms of action and considerations for clinical trials. Chest 153(5):1229–1239

Pincikova T, Paquin-Proulx D, Sandberg JK, Flodstrom-Tullberg M, Hjelte L (2017) Vitamin D treatment modulates immune activation in cystic fibrosis. Clin Exp Immunol 189(3):359–371

Popovic PJ, Zeh HJ III, Ochoa JB (2007) Arginine and immunity. J Nutr 137(6 Suppl 2):1681S–1686S

Pusceddu MM, Kelly P, Ariffin N, Cryan JF, Clarke G, Dinan TG (2015) n-3 PUFAs have beneficial effects on anxiety and cognition in female rats: effects of early life stress. Psychoneuroendocrinology 58:79–90

Pusceddu MM, Kelly P, Stanton C, Cryan JF, Dinan TG (2016) N-3 Polyunsaturated fatty acids through the lifespan: implication for psychopathology. Int J Neuropsychopharmacol 19(12):pyw078

Ralph AP, Kelly PM, Anstey NM (2008) L-arginine and vitamin D: novel adjunctive immunotherapies in tuberculosis. Trends Microbiol 16(7):336–344

Ren W et al (2014) Dietary arginine supplementation of mice alters the microbial population and activates intestinal innate immunity. J Nutr 144(6):988–995

Rettberg P, Horneck G, Zittermann A, Heer M (1998) Biological dosimetry to determine the UV radiation climate inside the MIR station and its role in vitamin D biosynthesis. Adv Space Res 22(12):1643–1652

Rizzo AM, Corsetto PA, Montorfano G, Milani S, Zava S, Tavella S, Cancedda R, Berra B (2012) Effects of long-term space flight on erythrocytes and oxidative stress of rodents. PLoS One 7(3):e32361

Ross AC (1992) Vitamin A status: relationship to immunity and the antibody response. Proc Soc Exp Biol Med 200(3):303–320

Samiec PS, Drews-Botsch C, Flagg EW, Kurtz JC, Sternberg P Jr, Reed RL, Jones DP (1998) Glutathione in human plasma: decline in association with aging, age-related macular degeneration, and diabetes. Free Radic Biol Med 24(5):699–704

Sannita WG et al (2004) Effects of heavy ions on visual function and electrophysiology of rodents: the ALTEA-MICE project. Adv Space Res 33(8):1347–1351

Schleithoff SS, Zittermann A, Tenderich G, Berthold HK, Stehle P, Koerfer R (2008) Combined calcium and vitamin D supplementation is not superior to calcium supplementation alone in improving disturbed bone metabolism in patients with congestive heart failure. Eur J Clin Nutr 62(12):1388–1394

Schwager J, Schulze J (1998) Modulation of interleukin production by ascorbic acid. Vet Immunol Immunopathol 64(1):45–57

Sekhar RV, Patel SG, Guthikonda AP, Reid M, Balasubramanyam A, Taffet GE, Jahoor F (2011) Deficient synthesis of glutathione underlies oxidative stress in aging and can be corrected by dietary cysteine and glycine supplementation. Am J Clin Nutr 94(3):847–853

Semba RD (1998) The role of vitamin A and related retinoids in immune function. Nutr Rev 56(1 Pt 2):S38–S48

Shanely RA, Nieman DC, Perkins-Veazie P, Henson DA, Meaney MP, Knab AM, Cialdell-Kam L (2016) Comparison of watermelon and carbohydrate beverage on exercise-induced alterations in systemic inflammation, immune dysfunction, and plasma antioxidant capacity. Nutrients 8(8):E518

Sharma S, Chopra K, Kulkarni SK, Agrewala JN (2007) Resveratrol and curcumin suppress immune response through CD28/CTLA-4 and CD80 co-stimulatory pathway. Clin Exp Immunol 147(1):155–163

Shim JH, Choi HS, Pugliese A, Lee SY, Chae JI, Choi BY, Bode AM, Dong Z (2008) (-)-Epigallocatechin gallate regulates CD3-mediated T cell receptor signaling in leukemia through the inhibition of ZAP-70 kinase. J Biol Chem 283(42):28370–28379

Shin YS, Beuhring KU, Stokstad EL (1975) The relationships between vitamin B12 and folic acid and the effect of methionine on folate metabolism. Mol Cell Biochem 9(2):97–108

Singh NP, Hegde VL, Hofseth LJ, Nagarkatti M, Nagarkatti P (2007) Resveratrol (trans-3,5,4′-trihydroxystilbene) ameliorates experimental allergic encephalomyelitis, primarily via induction of apoptosis in T cells involving activation of aryl hydrocarbon receptor and estrogen receptor. Mol Pharmacol 72(6):1508–1521

Singleton KD, Wischmeyer PE (2008) Glutamine attenuates inflammation and NF-kappaB activation via Cullin-1 deneddylation. Biochem Biophys Res Commun 373(3):445–449

Singleton KD, Serkova N, Beckey VE, Wischmeyer PE (2005) Glutamine attenuates lung injury and improves survival after sepsis: role of enhanced heat shock protein expression. Crit Care Med 33(6):1206–1213

Smith SM, Lane HW (2008) Spaceflight metabolism and nutritional support. In: Barratt MR, Pool SL (eds) Principles of clinical medicine for spaceflight. Springer, New York, NY, pp 559–576

Smith SM, Zwart SR (2008) Nutritional biochemistry of spaceflight. Adv Clin Chem 46:87–130

Smith SM, Wastney ME, O'Brien KO, Morukov BV, Larina IM, Abrams SA, Davis-Street JE, Oganov V, Shackelford LC (2005a) Bone markers, calcium metabolism, and calcium kinetics during extended-duration space flight on the mir space station. J Bone Miner Res 20(2):208–218

Smith SM, Zwart SR, Block G, Rice BL, Davis-Street JE (2005b) The nutritional status of astronauts is altered after long-term space flight aboard the International Space Station. J Nutr 135(3):437–443

Smith SM, Gardner KK, Locke J, Zwart SR (2009a) Vitamin D supplementation during Antarctic winter. Am J Clin Nutr 89(4):1092–1098

Smith SM, Zwart SR, Kloeris V, Heer M (2009b) Nutritional biochemistry of space flight. Nova Science Publishers, Inc., New York, NY

Smith SM, Heer MA, Shackelford LC, Sibonga JD, Ploutz-Snyder L, Zwart SR (2012) Benefits for bone from resistance exercise and nutrition in longduration spaceflight: evidence from biochemistry and densitometry. J Bone Miner Res 27(9):1896–1906

Smith SM, Abrams SA, Davis-Street JE, Heer M, O'Brien KO, Wastney ME, Zwart SR (2014) Fifty years of human space travel: implications for bone and calcium research. Annu Rev Nutr 34:377–400

Song EK, Hur H, Han MK (2003) Epigallocatechin gallate prevents autoimmune diabetes induced by multiple low doses of streptozotocin in mice. Arch Pharm Res 26(7):559–563

Sonnenfeld G (2002) The immune system in space and microgravity. Med Sci Sports Exerc 34(12):2021–2027

Sonnenfeld G, Shearer WT (2002) Immune function during space flight. Nutrition 18(10):899–903

Stahn AC et al (2017) Increased core body temperature in astronauts during long-duration space missions. Sci Rep 7(1):16180

Steffen JM, Musacchia XJ (1986) Spaceflight effects on adult rat muscle protein, nucleic acids, and amino acids. Am J Physiol 251(6 Pt 2):R1059–R1063

Stein TP (2002) Space flight and oxidative stress. Nutrition 18(10):867–871

Stein TP, Gaprindashvili T (1994) Spaceflight and protein metabolism, with special reference to humans. Am J Clin Nutr 60(5):806S–819S

Stein TP, Leskiw MJ (2000) Oxidant damage during and after spaceflight. Am J Physiol Endocrinol Metab 278(3):E375–E382

Stein TP, Schluter MD, Leskiw MJ (1999) Cortisol, insulin and leptin during space flight and bed rest. J Gravit Physiol 6(1):85–86

Stein TP, Blanc S (2011) Does protein supplementation prevent muscle disuse atrophy and loss of strength? Crit Rev Food Sci Nutr 51(9):828–834

Steinach M, Kohlberg E, Maggioni MA, Mendt S, Opatz O, Stahn A, Tiedemann J, Gunga HC (2015) Changes of 25-OH-vitamin D during overwintering at the German antarctic stations neumayer II and III. PLoS One 10(12):e0144130

Stephensen CB (2001a) Examining the effect of a nutrition intervention on immune function in healthy humans: what do we mean by immune function and who is really healthy anyway? Am J Clin Nutr 74(5):565–566

Stephensen CB (2001b) Vitamin A, infection, and immune function. Annu Rev Nutr 21:167–192

Stevens RG, Morris JE, Anderson LE (2000) Hemochromatosis heterozygotes may constitute a radiation-sensitive subpopulation. Radiat Res 153(6):844–847

Tamura J, Kubota K, Murakami H, Sawamura M, Matsushima T, Tamura T, Saitoh T, Kurabayshi H, Naruse T (1999) Immunomodulation by vitamin B12: augmentation of CD8+ T lymphocytes and natural killer (NK) cell activity in vitamin B12-deficient patients by methyl-B12 treatment. Clin Exp Immunol 116(1):28–32

Taylor GR, Konstantinova I, Sonnenfeld G, Jennings R (1997) Changes in the immune system during and after space flight. Adv Space Biol Med 6:1–32

Theriot CA, Westby CM, Morgan JLL, Zwart SR, Zanello SB (2016) High dietary iron increases oxidative stress and radiosensitivity in the rat retina and vasculature after exposure to fractionated gamma radiation. NPJ Microgravity 2:16014

Tipton CM, Greenleaf JE, Jackson CG (1996) Neuroendocrine and immune system responses with spaceflights. Med Sci Sports Exerc 28(8):988–998

Titze J, Lang R, Ilies C, Schwind KH, Kirsch KA, Dietsch P, Luft FC, Hilgers KF (2003) Osmotically inactive skin Na+ storage in rats. Am J Physiol Renal Physiol 285:F1108–F1117

Titze J, Shakibaei M, Schafflhuber M, Schulze-Tanzil G, Porst M, Schwind KH, Dietsch P, Hilgers KF (2004) Glycosaminoglycan polymerization may enable osmotically inactive Na+ storage in the skin. Am J Physiol Heart Circ Physiol 287(1):H203–H208

al-Turk WA, Stohs SJ, el-Rashidy FH, Othman S, Shaheen O (1987) Changes in glutathione, glutathione reductase and glutathione-S-transferase as a function of cell concentration and age. Pharmacology 34(1):1–8

Turner ND, Braby LA, Ford J, Lupton JR (2002) Opportunities for nutritional amelioration of radiation-induced cellular damage. Nutrition 18(10):904–912

Ueland PM, McCann A, Midttun O, Ulvik A (2017) Inflammation, vitamin B6 and related pathways. Mol Aspects Med 53:10–27

Van Buren CT (2004) Arginine immunonutrition in critically ill patients. Am J Crit Care 13(4):290

Vanamala J et al (2008) Dietary fish oil and pectin enhance colonocyte apoptosis in part through suppression of PPARdelta/PGE2 and elevation of PGE3. Carcinogenesis 29(4):790–796

Vanherwegen AS, Gysemans C, Mathieu C (2017) Vitamin D endocrinology on the cross-road between immunity and metabolism. Mol Cell Endocrinol 453:52–67

Vasamsetti SB, Karnewar S, Gopoju R, Gollavilli PN, Narra SR, Kumar JM, Kotamraju S (2016) Resveratrol attenuates monocyte-to-macrophage differentiation and associated inflammation via modulation of intracellular GSH homeostasis: relevance in atherosclerosis. Free Radic Biol Med 96:392–405

Wessels I, Maywald M, Rink L (2017) Zinc as a Gatekeeper of Immune Function. Nutrients 9(12):E1286

WHO (World Health Organization) (1985) Energy and protein requirements. Report of a joint FAO/WHO/UNU expert consultation. 724. Technical report series. World Health Organization, Geneva

Wintergerst ES, Maggini S, Hornig DH (2006) Immune-enhancing role of vitamin C and zinc and effect on clinical conditions. Ann Nutr Metab 50(2):85–94

Wintergerst ES, Maggini S, Hornig DH (2007) Contribution of selected vitamins and trace elements to immune function. Ann Nutr Metab 51(4):301–323

Wischmeyer PE (2007) Glutamine: mode of action in critical illness. Crit Care Med 35(9 Suppl):S541–S544

Wischmeyer PE (2008) Glutamine: role in critical illness and ongoing clinical trials. Curr Opin Gastroenterol 24(2):190–197

Wu D, Meydani SN (2014) Age-associated changes in immune function: impact of vitamin E intervention and the underlying mechanisms. Endocr Metab Immune Disord Drug Targets 14(4):283–289

Yaqoob P, Calder PC (2003) N-3 polyunsaturated fatty acids and inflammation in the arterial wall. Eur J Med Res 8(8):337–354

Yi B, Nichiporuk I, Nicolas M, Schneider S, Feuerecker M, Vassilieva G, Thieme D, Schelling G, Chouker A (2016) Reductions in circulating endocannabinoid 2-arachidonoylglycerol levels in healthy human subjects exposed to chronic stressors. Prog Neuropsychopharmacol Biol Psychiatry 67:92–97

Zeng R, Bscheider M, Lahl K, Lee M, Butcher EC (2016) Generation and transcriptional programming of intestinal dendritic cells: essential role of retinoic acid. Mucosal Immunol 9(1):183–193

Zittermann A (2010) The estimated benefits of vitamin D for Germany. Mol Nutr Food Res 54(8):1164–1171

Zittermann A, Gummert JF (2010) Sun, vitamin D, and cardiovascular disease. J Photochem. Photobiol.B 101(2):124–129

Zwart SR et al (2009a) Effects of 21 days of bed rest, with or without artificial gravity, on nutritional status of humans. J Appl Physiol 107(1):54–62

Zwart SR, Oliver SA, Fesperman JV, Kala G, Krauhs J, Ericson K, Smith SM (2009b) Nutritional status assessment before, during, and after long-duration head-down bed rest. Aviat Space Environ Med 80(5 Suppl):A15–A22

Zwart SR, Pierson D, Mehta S, Gonda S, Smith SM (2010) Capacity of omega-3 fatty acids or eicosapentaenoic acid to counteract weightlessness-induced bone loss by inhibiting NF-kappaB activation: from cells to bed rest to astronauts. J Bone Miner Res 25:1049–1057

Zwart SR, Mehta SK, Ploutz-Snyder R, Bourbeau Y, Locke JP, Pierson DL, Smith SM (2011) Response to vitamin D supplementation during Antarctic winter is related to BMI, and supplementation can mitigate Epstein-Barr Virus Reactivation. J Nutr 141(4):692–697

Zwart SR, Launius RD, Coen GK, Morgan JL, Charles JB, Smith SM (2014) Body mass changes during long-duration spaceflight. Aviat Space Environ Med 85(9):897–904

van NG, Isaacs AW, Nell T, Engelbrecht AM (2016) Sickness-associated anorexia: mother nature's idea of immunonutrition? Mediators Inflamm 2016:8071539

Microbiome and Immunity: A Critical Link for Long-Duration Space Exploration Missions

34

Hernan Lorenzi

34.1 Introduction

Everyday humans intimately interact with trillions of microbes that live and thrive across every inner and outer surface of the human body. This huge collection of microorganisms, known as the human microbiome, is mostly composed of bacteria, viruses, archaea, and unicellular eukaryotes that have coevolved with humans over millions of years, establishing a mutualistic, beneficial relationship of such depth that it would not be possible for humans to survive without them. Indeed, it is known that the human microbiome performs many functions that are essentials to humans including the processing and absorption of otherwise indigestible complex nutrients, the synthesis of essential compounds such as vitamins and antioxidants, the maturation and modulation of the immune system, the biotransformation of xenobiotics and the prevention of infections by pathogenic organisms (Ottman et al. 2012).

It is estimated that the microbiome contains as many bacterial cells as the number of cells in the human body (Sender et al. 2016). This enormous microbial population performs a remarkable metabolic activity driven by a microbial genetic pool (known as metagenome) whose size is a hundred times larger than the number of genes encoded by the human genome. During the first weeks of life, the composition of the gastrointestinal microbiome is heavily influenced by the maternal vaginal microbiota, mode of delivery and feeding (Backhed et al. 2015). As we become older, the microbiome gains in complexity by the incorporation of new microbial species acquired from the environment and the interaction with other animals and its composition is strongly influenced by eating habits. Even though the microbiome composition varies from person to person, its diversity is greater between body sites of the same person than across individuals. This observation indicates that, across the human body, the microbiome is

H. Lorenzi (✉)
Department of Infectious Diseases, The J. Craig Venter Institute, Rockville, MD, USA
e-mail: hlorenzi@jcvi.org

© Springer Nature Switzerland AG 2020

A. Choukér (ed.), *Stress Challenges and Immunity in Space*,
https://doi.org/10.1007/978-3-030-16996-1_34

a highly variable ecosystem as a result of the different environmental properties specific to each body site. In spite of this variation, it is still possible to define a core human microbiome associated with a "healthy state" across different body sites and people, mainly composed of different combinations of four major bacterial phyla, *Firmicutes* (Gram positive), *Bacteroidetes* (Gram negative), *Actinobacteria* (Gram positive), and *Proteobacteria* (Gram negative) (Costello et al. 2009). For example, while more than 95% of the gastrointestinal tract (GIT) microbiota belongs to the phyla *Firmicutes* and *Bacteroidetes*, the skin microbiome contains approximately 45% *Actinobacteria*, 30% *Firmicutes* and 10% *Proteobacteria* (Costello et al. 2009).

Host–microbiome interactions are complex and fluid, capable of adjusting to physiological perturbations that are encountered on a daily basis and are further subject to myriads of environmental stressors and interactions (see Chap. 3). However, large or selective shifts in the gut microbiota (known as dysbiosis) as a consequence of host pathobiology, alterations of diet, medications, and other environmental triggers can upset critical inter-microbe as well as host–microbe relationships to initiate pathophysiological processes leading to disease. During a space mission, astronauts are constantly exposed to a series of stressors (microgravity, sleep deprivation, radiation, dietary changes, etc.) that are likely to affect the composition and dynamic of the astronauts' microbiome. Several foundational studies on specific culturable commensal and opportunistic pathogenic bacteria performed under real or simulated microgravity conditions suggest that space travel can disturb the composition and function of the microbiome, including bacterial virulence, antibiotic resistance, and growth (Ott et al. 2016).

In order to maintain health, the immune system plays an essential role in maintaining a delicate balance between eliminating invading pathogens and keeping the homeostatic relationship with beneficial resident microorganisms of the GIT. At the same time, resident bacteria have a significant immunomodulatory activity that profoundly shape mammalian immunity. This chapter will focus on what is currently known about the interaction between the human intestinal microbiota and the immune response and how conditions found in space might alter this interaction and pose a risk to astronauts' health.

34.2 The Role of the Gut Microbiome in Immunity

By far the most heavily colonized organ is the gastrointestinal tract; the colon alone is estimated to contain over 70% of all the microbes in the human body (Ley et al. 2006; Whitman et al. 1998). The human GIT has an estimated surface area of 200 m^2 (Gebbers and Laissue 1989) and, as such a large organ, represents a major surface for microbial colonization. Therefore, the intestinal immune system faces unique challenges relative to other organs, as it must continuously confront an enormous microbial load and avoid a systemic infection by keeping commensal and pathogenic microorganisms from breaking through the intestinal epithelium barrier. At the same time, it is necessary to avoid pathologies arising from exacerbated innate immune signaling or from microbiota alterations that disturb essential metabolic functions.

A first mechanism to minimize bacterial access to internal host tissues consists of a viscous coating made out of mucin glycoproteins secreted by goblet cells that covers the luminal surface of the intestinal epithelium and that works as a first barrier of the immune defense. In the colon, this mucus layer is very thick and poses a physical barrier that significantly reduces the number of bacteria that can access and penetrate the colonic endothelium.

A second immune mechanism that limits bacteria–epithelial cell contact is the secretion of antimicrobial proteins (AMPs) by intestinal epithelial cells. AMPs span a number of diverse protein families, including defensins, cathelicidins, and C-type lectins. The expression of most AMPs is controlled by bacterial signals through the activation of pattern recognition receptors (PPRs), which recognize microorganism-associated molecular patterns (MAMPs) that are species-specific and essential for their viability. For example, in the small intestine the secretion to the luminal space of the C-type lectin REGIIIγ is under the control of a subfamily of PRRs named Toll-like receptors (TLRs) (Brandl et al. 2007; Vaishnava et al. 2008). REGIIIγ binds to the peptidoglycan layer located on the surface of gram-positive bacteria. On the other hand, the expression of a subset of α-defensins and defensin-related cryptdins is controlled by another PRR named nucleotide-binding oligomerization domain-containing protein 2 (NOD2) (Wehkamp et al. 2005; Kobayashi et al. 2005), located in the cytoplasm of intestinal Paneth cells. NOD2 recognizes muramyl dipeptide, a constituent of gram-positive and gram-negative bacterial peptidoglycan (Ogura et al. 2003; Inohara et al. 2003). Also, it has been shown that molecules of meso-diamino pimelic acid, a degradation product of peptidoglycans from gram-negative bacteria, can translocate the epithelial barrier and reach the blood stream where they activate neutrophils via intracellular NOD1 receptors (Clarke et al. 2010). These activated neutrophils can kill bacteria more efficiently, suggesting that the microbiome can directly influence neutrophil activity and prepares them for possible invading pathogenic bacteria (Clarke et al. 2010).

Keeping intestinal bacteria on the luminal side of the epithelial barrier also depends on secreted immunoglobulin A (IgA). IgA specific for intestinal bacteria is produced with the help of intestinal dendritic cells (DCs) that sample bacteria at various sites along the GIT. DCs located beneath the epithelial dome of Peyer's patches take up the small number of bacteria that penetrate the overlying mucosal epithelium. In addition, lamina propria DCs extend dendrites between epithelial cells to monitor bacteria that associate with the mucosal surface. These bacteria-laden DCs migrate to the mesenteric lymph nodes where they induce B cells to differentiate into IgA+ plasma cells that translocate to the intestinal lamina propria. From there, secreted IgA is transcytosed across the mucosal epithelium and deposited on the apical surface. Transcytosed IgA binds to luminal bacteria, limiting bacterial associations with the epithelium and preventing microbial translocations across the epithelial barrier (Macpherson et al. 2000). In addition, IgA secretion helps to contrast the inflammatory activity of bacteria through opsonization and to reduce the expression of inflammatory epitopes on the microbiota (Peterson et al. 2007).

In spite of the aforementioned immune mechanisms to protect the intestinal epithelium from direct bacterial contact, occasionally some bacteria translocate through the intestinal epithelial barrier. Microorganisms that penetrate the mucosal epithelium are phagocytosed and eliminated by lamina propria macrophages (Kelsall 2008) through mechanisms that include AMPs and reactive oxygen species (ROS) (see also Chap. 5). Intestinal macrophages also contribute to the restoration of the physical integrity of the epithelial cell barrier by migrating to damaged areas and secreting growth factors that interact with the epithelium, promoting enterocyte proliferation to replace damaged epithelial cells (Pull et al. 2005). This restorative process is activated by bacterial MAMPs recognized by TLRs expressed by intestinal macrophages (Pull et al. 2005; Rakoff-Nahoum et al. 2004). In addition, innate lymphoid cells that reside in the lamina propria and produce IL-22 are also essential for containment of lymphoid resident bacteria to the intestine, thus preventing their spread to systemic sites (Sonnenberg et al. 2012).

One important property of the intestinal immune system is its capacity to keep a balanced response able to eradicate pathogenic organisms and simultaneously tolerate the presence of a huge number of commensal microorganisms and their derivatives without triggering an exacerbated inflammatory response that could eventually lead to immunopathology. CD4+ regulatory T (Treg) cells play an essential role keeping immune tolerance and homeostasis in the intestine. In the GIT, commensal microbiota and dietary metabolites promote specialized host immune responses through differentiation of appropriate effector and regulatory T-cell populations. Segmented filamentous bacteria, that directly interact with the apical side of the mucosal epithelium, promote effector T-cell accumulation, particularly of pro-inflammatory Th17 cells, within the intestine. In the colon, some specific groups of bacteria, such as Clostridiales IV and XIVa, or *Bacteroides fragilis* promote accumulation of Foxp3+ Treg populations and induction of IL-10. In addition, production of cytokines and chemokines by antigen-presenting cells, such as DCs, heavily influences the balance between effector and regulatory T-cell responses. Secretion of STAT3-activating cytokines, IL-6, IL-23, and IL-12, favors effector T-cell differentiation. For example, IL-6 and IL-23 promote Th17 cell-mediated pathogenic effector responses as well as restrain Treg differentiation and IL-10 production in the intestine (Ahern et al. 2010), while IL-12 is required for Th1 differentiation (Coghill et al. 2011). DC-secreted retinoic acid and TGF-β production, on the other hand, modulate the balance between effector and regulatory responses, depending on the inflammatory microenvironment and associated cytokine production. Foxp3+ Treg secretes anti-inflammatory IL-10 and TGF-β, which reciprocally inhibits Th17 and Th1 cells. In the small intestine, another Foxp3− Treg cell subset, referred to as T_R1 cells, is also stimulated by IL-6 to express IL-10. Th17 cells secretes IL-17A and IL-17F that have pro-inflammatory effects and mediate neutrophil chemotaxis. Th17 cells also express IL-22, which contributes to epithelial homeostasis and stimulates the secretion of AMPs. The mechanisms through which the microbiota modulates Treg differentiation have recently started to be unraveled. *B. fragilis* releases outer membrane vesicles that contain the capsular polysaccharide (PSA) that is involved in Treg differentiation (Shen et al. 2012). PSA targets Toll-like receptor

TLR2 on intestinal DCs and induces their differentiation to Treg-promoting DCs. In addition, bacteria-derived short chain fatty acids, produced during the fermentation of dietary fiber by commensal anaerobic bacteria, are also involved in Treg differentiation (Arpaia et al. 2013; Furusawa et al. 2013; Smith et al. 2013). It has been proposed that metabolite-sensing G-protein-coupled receptors GPR43 and GPR109a, located at the apical surface of the GIT epithelium, are involved in this process (Macia et al. 2015).

The immune system plays an important role shaping the composition of the GIT microbiome by the production and secretion of AMPs into the lumen of the gastrointestinal tract. For example, in graft-versus-host disease (GVHD), disruption of Paneth cell function results in reduced luminal secretion of α-defensins altering intestinal microbiota causing dysbiosis, which further accelerates the underlying diseases (Eriguchi et al. 2015). Therapeutic treatment with recombinant α-defensins restore GVHD-mediated dysbiosis in mice (Hayase et al. 2017). In addition, the pro-inflammatory response is able to control the microbial population of the GIT. Mice lacking the bacterial flagelin Toll-like receptor TLR5 develop altered GIT microbiota, spontaneous colitis and metabolic syndrome. Similarly, mice with intestinal epithelial cells unable to express the inflammasome component NLRP6 develop GIT dysbiosis, increased recruitment of inflammatory cells to the intestine and susceptibility to chemically induced colitis (Elinav et al. 2011). All this evidence suggests that the immune system provides the mammalian host some control over the composition of the microbial communities of the GIT and that immune dysregulation can alter host-microbial homeostasis in the gut leading to dysbiosis-associated diseases.

34.3 Dysbiosis in Outer Space

Dysbiosis in the GIT microbiome has been implicated in auto-immune and inflammatory diseases, some cancers and mental disorders. To date, several space-related studies have identified shifts in the microbial composition of the oral, intestinal and nasal microbiome using culture-based methods on samples collected from astronauts before and after spaceflights of up to 2 months of duration (Decelle and Taylor 1976; Lencner et al. 1984; Brown et al. 1976; Lizko et al. 1984). Although it is likely that the observed compositional changes were caused, at least in part, by the stress associated with the flight back to Earth, some moderate inflight increases of dental plaque and gum inflammation were observed (Brown et al. 1976). Among the observed compositional changes in cosmonauts' GIT were decreased post-flight levels of lactobacilli, some of which are beneficial to human health (Shao et al. 2017; Ganji-Arjenaki and Rafician-Kopaei 2018), and increased enterobacteria and clostridia. Another study looking at changes in the intestinal bifidoflora of eight astronauts before and after flight detected a significant drop in the abundance of bifidobacteria before flight but no significant post-flight changes. It was suggested that the observed reduction of bifidoflora before the mission could be a result of the effect of nervous and emotional tension during the period of training and preparation for the incoming flight (Goncharova et al. 1981). Several species of bifidobacteria produce anti-microbial molecules that prevents intestinal colonization by know gram-negative pathogens, including *Salmonella enterica*, *Shigella*, and *Vibrio cholera* (Lievin et al. 2000; Gibson and Wang 1994). Also, studies looking at the effect of short-term space travel on the upper respiratory tract of astronauts identified a significant post-flight increase and potential inflight cross-contamination with *Staphylococcus aureus* (Decelle and Taylor 1976) but no changes in other microorganisms were detected.

It has been proposed that the number of bacterial species that populate the different ecological niches of the human body are likely to decrease during long space missions (Taylor and Sommer 2005; Hales et al. 2002) due to the environmental conditions astronauts are exposed to in space. However, a comprehensive 16S analysis of the intestinal flora from mice flown on the Space Shuttle Atlantis mission STS-135 for a period of 13 days did not revealed any post-flight drop in bacterial diversity. Moreover, the same study noticed a reduction of lactobacillales with a concomitant increase in clostridiales (Ritchie et al. 2015), as observed in post-flight gut microbiome samples from crew members (Lizko et al. 1984). The absence of a drop in bacterial diversity associated with space travel is also supported by preliminary post-flight results from our astronauts' Microbiome project, that investigated the effect of long-term space travel (6 months or longer) on the astronauts' gut microbiome (Voorhies et al. 2017).

In spite of all the existing evidence suggesting the composition of the human microbiome changes in space, it is important to emphasize that most of these studies rely on the contrast between the microbial content of samples collected pre- and post-flight and therefore, extrapolations to what happens in space should be done with caution. Both, astronauts and animal models are subject to highly stressful situations during takeoff and landing. It is known that the stress response is coordinated by the central nervous system through the hypothalamic-pituitary-adrenal (HPA) axis, which in turn maintains a complex cross talk with the GIT microbiota and the immune system (Rea et al. 2016). It has been shown that through the HPA axis different types of psychological stressors can alter the composition of the GIT microbiome and lead to bacterial translocations through the intestinal epithelial barrier inducing the production of immunomodulatory cytokines such as IL-6 and CCL2/MCP-1 (Rea et al. 2016; Bailey 2014; De Palma et al. 2015; Bailey et al. 2011, 2006). Therefore, it is likely that the microbial composition of samples collected immediately after a space mission are highly influenced by the acute stress experienced by the crew or animal models during the flight back to Earth. Moreover, it is possible that during short space missions, the proximity of two very stressful situations, takeoff and landing, might synergistically impact the microbial composition of samples collected immediately after flight.

Two more recent studies made use of 16S analysis and metagenomic sequencing to characterize the influence of long space missions (of 6 months or longer) on the astronauts' microbiome by sampling the stool, skin, nose, and tongue of crew members before, during and after flight (Voorhies et al. 2017; Turek et al. 2017). Although these studies are still ongoing at the time of writing this book edition, preliminary results from inflight samples revealed that the overall composition of bacterial communities from the GIT, skin, and nose significantly change in space. Moreover, for some crew members, intestinal microbial diversity increased under microgravity conditions (Voorhies et al. 2017) most likely promoted by the new spaceflight diet and/or the crowded living conditions of the International Space Station (ISS) that might favor a more fluid interchange of microbial flora among crew members. In addition, increases in plasma levels of CCL2/MCP-1 suggest that inflight microbial dysbiosis may lead to intestinal bacteria translocations, as reported by Bailey et al. (Voorhies et al. 2017; Bailey et al. 2011, 2006).

34.4 Microbial Physiological Changes Associated with the Space Environment

Over the past 50 years a significant amount of microbial research has been done to investigate the effects of space travel on microbial organisms. Although much of the accumulated data correspond to studies focused on a limited number of opportunistic pathogenic and probiotic organisms (Ott et al. 2016), their conclusions may well provide hints about the way commensals from the human microbiota respond to stressors to which they are exposed during a space mission. It is important to consider, however, that the main conclusion derived from these studies is that there exists no common, predictable bacterial response to Low Shear Modeled Microgravity (LSMMG) and spaceflight and therefore, any extrapolation of these studies to the microbial communities that conform the human microbiome must be taken with caution.

It has been shown that exposure of several bacteria species to LSMMG or the spaceflight environments results in altered bacterial behaviors including increased final population density, thicker cellular envelope, reduced sensitivity to antibiotics, improved resistance to acidic environments, reduced cellular volume, bacterial aggregation in culture and enhanced biofilm formation, increased conjugation efficiency, higher specific productivity of secondary metabolites and increased virulence (Shao et al. 2017; Zea et al. 2017; Klaus and Howard 2006; Nickerson et al. 2000; Ciferri et al. 1986; Benoit et al. 2006) (see also Chaps. 17 and 18). Microgravity may also influence the cell interaction between mutualistic bacteria and their host, accelerating bacteria-induced apoptosis in host tissues (Foster et al. 2013; Ilyin 2005). It has been proposed that these bacterial responses are the result of the low fluid share of liquid cultures associated with the microgravity environment. In addition, new transcriptomics data suggest that the microgravity-associated altered microbial behavior observed in nonmotile bacteria is the result, at least in part, of the lack of gravity-dependent forces and flows that limits the extracellular mass transport to diffusive processes only (Zea et al. 2017, 2016). However, some experimental evidence shows that not all bacteria species respond in a similar way to microgravity and that other environmental factors, such as oxygen or nutrient availability, may influence bacterial responses to the space environment. For example, induction of acid tolerance by simulated microgravity in *Salmonella typhimurium* is dependent on the presence of phosphate ions in the growth medium (Wilson et al. 2008). Also, liquid cultures of *Lactobacillus acidophilus* grown in aerobic conditions under LSMMG showed a shorter lag phase compared with 1 × *g* controls and increased resistance to acid and bile-juice stressors (Shao et al. 2017). Nevertheless, under anaerobic conditions, similar to those found in the gut, *L. acidophilus* show no difference in growth rate nor in its resistance to simulated gastric and intestinal juices (Castro-Wallace et al. 2017). Thus, the absence of a *L. acidophilus* response to LSMMG when cultured under anaerobic conditions indicates the potential role of oxygen in the response to low fluid shear, which has been previously suggested (Shao et al. 2017; Crabbe et al. 2011). Contrary to *L. acidophilus*, this oxygen-dependent differential response to microgravity is not observed in *Escherichia coli* that grow to higher cell densities in space compared to ground controls in anaerobic cultures (Zea et al. 2017).

Virulence is another bacterial response that varies across bacteria species. Simulated and actual microgravity induces changes in the virulence potential of several gram-negative enteric bacteria. For example, the adherent-invasive *E. coli* strain O83:H1 increased it's adherence and infectivity to Caco-2 gastrointestinal epithelial cells (Allen et al. 2008) and the pathogen *S. typhimurium* presents enhanced invasion of both epithelial and macrophage cells and has a hypervirulence phenotype in BALB/c mice when infected orally (Wilson et al. 2007). Simulated microgravity also enhanced production of TNF in murine macrophages infected with an enteropathogenic strain of *E. coli*. Also, both simulated and real microgravity increased resistance to thermal and oxidative stressors in *Pseudomonas aeruginosa* and enhanced the production of biofilm and alginate, which play an important role in protecting mucoid *P. aeruginosa* biofilm bacteria from the human immune system (Crabbe et al. 2011, 2010; Kim et al. 2013a, b; McLean et al. 2001; Leid et al. 2005).

Also, there are instances where simulated microgravity or spaceflight induces an hypovirulence phenotype, such as *Yersinia pestis* in cell culture infections (Lawal et al. 2010, 2013) and methicillin-resistant *Staphylococcus aureus*, *Enterococcus faecalis*, and *Listeria monocytogenes* in a *Caenorhabditis elegans* infection model (Hammond et al. 2013). *S. aureus* exposed to LSMMG also presented increased sensitivity to oxidative stress with a concomitant downregulation of *Hfq* expression, an RNA-binding factor that seems to play a major role in gram-negative response to low fluid share (Castro et al. 2011).

34.5 Bacterial Response and Infectious Diseases During Spaceflight

How these space-associated microbial responses may affect host–microbiota homeostasis and the interaction with pathogenic bacteria? From the many studies looking at microbial responses to spaceflight environment it is clear that bacterial behaviors in space do not follow a one-size-fits-all paradigm. These differential responses may alter the way subgroups of commensal and opportunistic pathogenic bacteria interacts with the host immune system making them more or less susceptible to host defenses and imposing a differential selective pressure that may lead to microbial dysbiosis.

Enhanced biofilm formation is one of the most consistently observed microbial changes induced by simulated or real microgravity and may explain, at least in part, many other physiological changes associated to spaceflight and/or LSMMG (e.g. increased resistance to oxidative stress, production of secondary metabolites, antibiotic resistance and increased conjugation efficiency). At the epithelial mucosa of the gastrointestinal and respiratory tract or gums, enhanced biofilm formation can promote bacterial adhesion to the epithelium (Zea et al. 2017; Allen et al. 2008) attracting and activating cellular components of the innate immune response. Two of the main cellular components, neutrophils and macrophages, deliver defensive actions against pathogenic bacteria through the release of reactive oxygen species (ROS) and oxygen-independent enzymes such as lysozymes and lactoferrin and the engulfment and killing of bacteria via phagocytosis. Although these actions are very efficient eliminating planktonic, single-celled microbes, they are not as effective against biofilm bacteria. In consequence, biofilms form persistent infections that are recalcitrant to the action of activated and opsonized phagocytes of the innate response, antibody responses, other components of the host response (e.g. complement) and to the action of antibiotics. As the highly active immune response cannot effectively eliminate the biofilm infection, the surrounding epithelium mucosa is exposed to deleterious oxidative radicals and enzymes released from inflammatory cells that may lead to a leaky epithelial barrier and translocation of microorganisms and/or their products into the host contributing to systemic inflammation. This is in agreement with changes observed in plasma cytokine profiles from astronauts' peripheral blood in space. Comparison of cytokine levels before and during long-duration space missions revealed increased levels of TNF, IL-8, IL-1ra, CCL2, CCL4, and CXCL5, suggesting systemic mild inflammation and leukocyte recruitment (Crucian et al. 2014a). In addition, a more recent study of the leukocyte population in space showed an increase of neutrophils in peripheral blood of astronauts during long-stays in the ISS (Crucian et al. 2015). The same study also identified a significant inflight reduction in the production of anti-inflammatory IL-4, IL-5, and IL-10 and of INFγ, which may impact the secretion of IgA in the intestinal lumen and reduced local and systemic tolerance as well as protection against opportunistic pathogens.

34.6 Immune Dysregulation in Spaceflight

Immunological studies of animal models and astronauts during spaceflight have identified important alterations of several immune parameters. Rats flown on 9- and 14-day missions had reduced total white blood cell (WBC) counts, lymphocytes, monocytes, and eosinophils and an increase in neutrophils (Allebban et al. 1994; Ichiki et al. 1996). Both studies also reported a drop in the absolute number of CD4+ and CD8+ T cells as well as B lymphocytes. Depressed cell-mediated immunity (delayed-type hypersensitivity), was observed in astronauts during long-duration flight determined in vivo (Taylor and Janney 1992; Cogoli 1993; Gmunder et al. 1994) and supports the hypothesis that the immune system is functionally altered during spaceflight. Another study looking at changes in the human immune response during 6-month spaceflight missions reported a significant inflight increase in total WBC counts and granulocyte levels whereas levels of lymphocytes, monocytes, CD4+ and CD8+ T cells subsets, memory T cells, and B cells did not change in space (Crucian et al. 2015). The same study detected an upward inflight trend, that became significant at some point during the mission, in the levels of active CD8+ T cytotoxic (Tc) cells (CD28+/CD244+) and Natural Killer (NK) cells (CD45+/CD16+/CD56+). Even though the detection of higher NK cells in space contradicts findings from other investigations, it has been proposed that this disparity could be due to the use of different experimental setups (Crucian et al. 2015). It is likely that the elevated inflight levels of NK and Tc cells are in response to the virus reactivation phenomena associated to spaceflight (Crucian et al. 2015; Mehta et al. 2014, 2013).

Even though neutrophil counts are consistently increased in space (Crucian et al. 2015; Allebban et al. 1994; Ichiki et al. 1996), lower phagocytosis, oxidative burst capacity and reduced chemotaxis have been reported after 9-day or longer missions (Stowe et al. 1999; Kaur et al. 2004). Similarly, a reduction in oxidative burst, degranulation and phagocytosis capacity were noted in monocytes collected from astronauts after short-duration space missions with a concomitant drop in the expression of CD32 and CD64, two surface markers involved in phagocytosis (Kaur et al. 2005). However, inflight cell culture experiments revealed that human bone-marrow macrophages stimulated with lipopolysaccharide (LPS) during spaceflight secretes higher levels of IL-1 and TNF compared to ground controls and agrees with results from monocytes exposed to simulated microgravity (Chapes et al. 1992; Batkai et al. 1999; Saei and Barzegari 2012). Even though macrophages seem capable of producing an enhanced response to LPS by secreting higher levels of IL-1 and INF-α, the reduced phagocytic and oxidative burst capacity observed in macrophages and neutrophils under microgravity conditions strongly suggests that in space cosmonauts have a higher risk of contracting bacterial infections due to a weakened, less efficient, innate immune response against planktonic and biofilm bacteria. Whether this impaired response against biofilms also increases the risk of chronic bacterial infections merits further investigations.

In addition, another study revealed that the density of ICAM-1, a surface adhesion molecule that participates in the stabilization of leukocyte cell–cell interactions and endothelial transmigration, increases at the surface of macrophage-like differentiated human U937 cells under microgravity conditions, as opposed to murine BV-2 microglial cells where ICAM-1 density drops during spaceflight (Paulsen et al. 2015). A similar investigation determined that primary human M1 macrophages cultured in the ISS for 11 days had lower density of ICAM-1 molecules compared to ground controls, proliferated faster under microgravity and did not exhibit any quantitative or structural change of the F-actin and vimentin cytoskeleton (Tauber et al. 2017). A reduced surface expression of the adhesion molecule

ICAM-1 may affect cell migration and result in disturbed activation of CD4+ T lymphocytes and the specific immune response. Interestingly, the same study also identified more de-fucosylated proteins on the cell surface of M1 macrophages associated with the spaceflight environment suggesting that under microgravity M1 macrophages could become less responsive to LPS stimulation or other activation mechanisms upon microbial exposure (Tauber et al. 2017; Cameron 1985). It is intriguing why the effect of microgravity on ICAM-1 densities at the surface of monocyte/macrophage cells is not consistent. It has been proposed that it could be the consequence of different cell systems (tumor cell lines in various states of differentiation versus primary cells) and experimental conditions (Paulsen et al. 2015; Tauber et al. 2017). The lack of cytoskeletal changes associated with real microgravity in primary M1 macrophages (Tauber et al. 2017) also contradicts the results of other studies, mostly based on tumor cells as model systems. There exists a significant parallelism between cytoskeletal alterations reported in microgravity and in tumor cells regarding the loss of actin filaments (Yamaguchi and Condeelis 2007) and the disorganization of microtubules (Pachenari et al. 2014), and the potential crucial function of Rho-GTPases both in tumor cell biology (Gomez del Pulgar et al. 2005; Jaffe and Hall 2005; Vega and Ridley 2008) and in microgravity-induced cellular reorganization (Louis et al. 2015). Therefore, the results from studies using tumorigenic cell lines to study the microgravity effects on cytoskeleton organization should be taken with caution.

Dendritic cells (DCs) play a fundamental role in the processing and presentation of antigens to T lymphocytes during the initiation of both innate and acquired immune responses. It has been shown that the generation and function of DCs are perturbed in cell cultures under simulated microgravity (Savary et al. 2001). DCs grown under these conditions presented reduced phagocytosis capacity and density of HLA-DR at the cell surface, suggesting a diminished capacity of antigen presentation compared to control DCs grown in static cultures. In addition, a smaller proportion of microgravity-grown DCs expressed CD80 at the cell surface and produced lower levels of IL-12 (Savary et al. 2001). These results seem to indicate that the generation of DCs as well as some of the functions required to mount an effective immune response to pathogens may be disturbed in the space environment.

A study of cosmonauts aboard the ISS for 6 months registered a sustained and significant increase in blood of the endocannabinoid (EC) anandamide (see also Chap. 10) while no significant changes were registered for 2-arachidonoylglycerol. After the mission, the blood concentrations for both ECs returned to baseline. In the same study, the authors hypothesized that enhanced EC signaling could be required for adaptation and tolerance to the microgravity environment (Strewe et al. 2012). Interestingly, it has been shown that the EC system regulates immune homeostasis in the GIT, intestinal barrier permeability and has been associated with obesity, insulin resistance and type II diabetes (Muccioli et al. 2010; Acharya et al. 2017). Moreover, GIT microbial composition modulates EC system activity (Muccioli et al. 2010; Mehrpouya-Bahrami et al. 2017). Anandamide and 2-arachidonoylglycerol have antagonistic effects on the gut epithelium, respectively enhancing or diminishing gut barrier permeability by altering the distribution and localization of tight-junction proteins. It is possible that the observed inflight increase of anandamide could lead to enhanced gut epithelium barrier permeability to LPS and other microbiome products and contribute, at least partially, to the mild inflammatory response observed in cosmonauts during long-term space missions. If that is the case, then treatments known to block the effect of anandamide and inflammation in the GIT such as the administration of prebiotics or treatment with the probiotic bacteria *Akkermansia muciniphila* could be implemented in space to reduce the risks of diseases associated with chronic inflammatory responses (Cani et al. 2016).

34.7 The Microbiome, Immune Dysregulation in Spaceflight Analogs

Further studies have been conducted or initiated in confined and high-fidelity stressful space analog environments such as in the Antarctic or in controlled confined conditions (e.g. HERA facility in the US, the Mars500/SIRIUS facility in Russia or the envihab at DLR in Germany, as examples, see also Chap. 36). Here, some specific conditions associated with long-duration and exploration space missions can be reproduced and investigated, allowing to explore more holistically the interactions of the space analog stressors in the future with the microbiome, and how the immune system shapes the composition of the GIT microbiome and vice versa. This is possible since the immune system is affected in such analogs at different slopes and intensities, indicating the development of an immune hypersensitivity as a function of these stressors (Feuerecker et al. 2019; Yi et al. 2015; Crucian et al. 2014b). Since allergies have also been reported by crew members during a space mission (Crucian et al. 2016), and because of the emerging knowledge on the link between allergies and the (GIT) microbiome, these space analog environments will contribute to a better understanding of the impact of space-associated stressors on the crew microbiota and its interaction with the immune response.

34.8 Conclusion

The human microbiome is essential to human health. Over millions of years the human immune system has learnt how to tolerate commensal beneficial microorganisms that conform the microbiome while keeping out opportunistic pathogens. Central to this "peaceful" coexistence is an intense cross-talk between the human microbiome and the immune system. Current experimental evidence from studies performed during space missions or under simulated microgravity strongly suggest that space travel disrupts host–microbiome homeostasis causing dysbiosis and dysregulation of the immune system that, in the long run, might lead to immunopathologies and infectious diseases. Higher inflight concentrations of pro-inflammatory cytokines and the endocannabinoid anandamide together with increased numbers of neutrophils are all compatible with a potential increase in the permeability of the intestinal epithelial barrier associated with space travel. Increased anandamide and a chronic leaky gut have been found associated to a number of inflammatory disorders such as insulin resistance, type II diabetes and inflammatory bowel disease. Lower phagocytosis and oxidative burst capacity of macrophages and neutrophils might also contribute to a less efficient inflight clearance of microbial infections and the establishment of systemic diseases. In addition, the potential space-associated alteration of host–microbiome homeostasis together with the enhanced virulence phenotype observed in some opportunistic pathogens suggest cosmonauts may have an increased risk of contracting infectious diseases.

Although in most cases immune dysregulation remains subclinical in space (Crucian and Sams 2009), there is a possibility that alterations in host–microbiome homeostasis, immune response, microbial physiology, virulence and antimicrobial resistance as well as dysbiosis become more exacerbated during longer space journeys, such as future missions to Mars or long stays on the moon surface. Thus, in deep space, all these factors could converge synergistically into a "perfect storm" situation facilitating the establishment and speeding up the development of microbial infection or immune-related diseases as well as reducing the efficacy of planned therapeutic treatments. Therefore, in the near future it will be essential to use model organisms (either wild-type or susceptible for specific immune diseases) to perform inflight experiments that allow to assess all these factors together, simulating infections and investigating the effectivity of countermeasures such as the usage of pre- and probiotics, enriched diets and antimicrobial/immunomodulatory therapies (Crucian et al. 2018).

References

Acharya N, Penukonda S, Shcheglova T, Hagymasi AT, Basu S, Srivastava PK (2017) Endocannabinoid system acts as a regulator of immune homeostasis in the gut. Proc Natl Acad Sci U S A 114:5005–5010

Ahern PP, Schiering C, Buonocore S, McGeachy MJ, Cua DJ, Maloy KJ, Powrie F (2010) Interleukin-23 drives intestinal inflammation through direct activity on T cells. Immunity 33:279–288

Allebban Z, Ichiki AT, Gibson LA, Jones JB, Congdon CC, Lange RD (1994) Effects of spaceflight on the number of rat peripheral blood leukocytes and lymphocyte subsets. J Leukoc Biol 55:209–213

Allen CA, Niesel DW, Torres AG (2008) The effects of low-shear stress on Adherent-invasive Escherichia coli. Environ Microbiol 10:1512–1525

Arpaia N, Campbell C, Fan X, Dikiy S, van der Veeken J, deRoos P, Liu H, Cross JR, Pfeffer K, Coffer PJ, Rudensky AY (2013) Metabolites produced by commensal bacteria promote peripheral regulatory T-cell generation. Nature 504:451–455

Backhed F, Roswall J, Peng Y, Feng Q, Jia H, Kovatcheva-Datchary P, Li Y, Xia Y, Xie H, Zhong H, Khan MT, Zhang J, Li J, Xiao L, Al-Aama J, Zhang D, Lee YS, Kotowska D, Colding C, Tremaroli V, Yin Y, Bergman S, Xu X, Madsen L, Kristiansen K, Dahlgren J, Wang J (2015) Dynamics and stabilization of the human gut microbiome during the first year of life. Cell Host Microbe 17:852

Bailey MT (2014) Influence of stressor-induced nervous system activation on the intestinal microbiota and the importance for immunomodulation. Adv Exp Med Biol 817:255–276

Bailey MT, Engler H, Sheridan JF (2006) Stress induces the translocation of cutaneous and gastrointestinal microflora to secondary lymphoid organs of C57BL/6 mice. J Neuroimmunol 171:29–37

Bailey MT, Dowd SE, Galley JD, Hufnagle AR, Allen RG, Lyte M (2011) Exposure to a social stressor alters the structure of the intestinal microbiota: implications for stressor-induced immunomodulation. Brain Behav Immun 25:397–407

Batkai L, Varkonyi A, Minarovits J (1999) The effect of simulated microgravity conditions on the TNF-alpha production by human PBMCS. J Gravit Physiol 6:P109–P110

Benoit MR, Li W, Stodieck LS, Lam KS, Winther CL, Roane TM, Klaus DM (2006) Microbial antibiotic production aboard the International Space Station. Appl Microbiol Biotechnol 70:403–411

Brandl K, Plitas G, Schnabl B, DeMatteo RP, Pamer EG (2007) MyD88-mediated signals induce the bactericidal lectin RegIII gamma and protect mice against intestinal Listeria monocytogenes infection. J Exp Med 204:1891–1900

Brown LR, Fromme WJ, Handler SF, Wheatcroft MG, Johnston DA (1976) Effect of Skylab missions on clinical and microbiologic aspects of oral health. J Am Dent Assoc 93:357–363

Cameron DJ (1985) Specificity of macrophage-mediated cytotoxicity: role of target and effector cell fucose. Immunol Lett 11:39–44

Cani PD, Plovier H, Van Hul M, Geurts L, Delzenne NM, Druart C, Everard A (2016) Endocannabinoids--at the crossroads between the gut microbiota and host metabolism. Nat Rev Endocrinol 12:133–143

Castro SL, Nelman-Gonzalez M, Nickerson CA, Ott CM (2011) Induction of attachment-independent biofilm formation and repression of Hfq expression by low-fluid-shear culture of Staphylococcus aureus. Appl Environ Microbiol 77:6368–6378

Castro-Wallace S, Stahl S, Voorhies A, Lorenzi H, Douglas GL (2017) Response of Lactobacillus acidophilus ATCC 4356 to low-shear modeled microgravity. Acta Astronaut 139:463–468

Chapes SK, Morrison DR, Guikema JA, Lewis ML, Spooner BS (1992) Cytokine secretion by immune cells in space. J Leukoc Biol 52:104–110

Ciferri O, Tiboni O, Di Pasquale G, Orlandoni AM, Marchesi ML (1986) Effects of microgravity on genetic recombination in Escherichia coli. Naturwissenschaften 73:418–421

Clarke TB, Davis KM, Lysenko ES, Zhou AY, Yu Y, Weiser JN (2010) Recognition of peptidoglycan from the microbiota by Nod1 enhances systemic innate immunity. Nat Med 16:228–231

Coghill JM, Sarantopoulos S, Moran TP, Murphy WJ, Blazar BR, Serody JS (2011) Effector CD4+ T cells, the cytokines they generate, and GVHD: something old and something new. Blood 117:3268–3276

Cogoli A (1993) The effect of space flight on human cellular immunity. Environ Med 37:107–116

Costello EK, Lauber CL, Hamady M, Fierer N, Gordon JI, Knight R (2009) Bacterial community variation in human body habitats across space and time. Science 326:1694–1697

Crabbe A, Pycke B, Van Houdt R, Monsieurs P, Nickerson C, Leys N, Cornelis P (2010) Response of Pseudomonas aeruginosa PAO1 to low shear modelled microgravity involves AlgU regulation. Environ Microbiol 12:1545–1564

Crabbe A, Schurr MJ, Monsieurs P, Morici L, Schurr J, Wilson JW, Ott CM, Tsaprailis G, Pierson DL, Stefanyshyn-Piper H, Nickerson CA (2011) Transcriptional and proteomic responses of Pseudomonas aeruginosa PAO1 to spaceflight conditions involve Hfq regulation and reveal a role for oxygen. Appl Environ Microbiol 77:1221–1230

Crucian B, Sams C (2009) HRP evidence report: risk of crew adverse health event due to altered immune response. HRP evidence report. NASA, Washington, DC

Crucian BE, Zwart SR, Mehta S, Uchakin P, Quiriarte HD, Pierson D, Sams CF, Smith SM (2014a) Plasma cytokine concentrations indicate that in vivo hormonal regulation of immunity is altered during long-duration spaceflight. J Interferon Cytokine Res 34:778–786

Crucian B, Simpson RJ, Mehta S, Stowe R, Chouker A, Hwang SA, Actor JK, Salam AP, Pierson D, Sams C (2014b) Terrestrial stress analogs for spaceflight associated immune system dysregulation. Brain Behav Immun 39:23–32

Crucian B, Stowe RP, Mehta S, Quiriarte H, Pierson D, Sams C (2015) Alterations in adaptive immunity persist during long-duration spaceflight. NPJ Microgravity 1:15013

Crucian B, Johnston S, Mehta S, Stowe R, Uchakin P, Quiriarte H, Pierson D, Laudenslager ML, Sams C (2016) A case of persistent skin rash and rhinitis with immune system dysregulation onboard the International Space Station. J Allergy Clin Immunol Pract 4:759–762e8

Crucian BE, Chouker A, Simpson RJ, Mehta S, Marshall G, Smith SM, Zwart SR, Heer M, Ponomarev S, Whitmire A, Frippiat JP, Douglas GL, Lorenzi H, Buchheim JI, Makedonas G, Ginsburg GS, Ott CM, Pierson DL, Krieger SS, Baecker N, Sams C (2018) Immune system dysregulation during spaceflight: potential countermeasures for deep space exploration missions. Front Immunol 9:1437

De Palma G, Blennerhassett P, Lu J, Deng Y, Park AJ, Green W, Denou E, Silva MA, Santacruz A, Sanz Y, Surette MG, Verdu EF, Collins SM, Bercik P (2015) Microbiota and host determinants of behavioural phenotype in maternally separated mice. Nat Commun 6:7735

Decelle JG, Taylor GR (1976) Autoflora in the upper respiratory tract of Apollo astronauts. Appl Environ Microbiol 32:659–665

Elinav E, Strowig T, Kau AL, Henao-Mejia J, Thaiss CA, Booth CJ, Peaper DR, Bertin J, Eisenbarth SC, Gordon JI, Flavell RA (2011) NLRP6 inflammasome regulates colonic microbial ecology and risk for colitis. Cell 145:745–757

Eriguchi Y, Nakamura K, Hashimoto D, Shimoda S, Shimono N, Akashi K, Ayabe T, Teshima T (2015) Decreased secretion of Paneth cell alpha-defensins in graft-versus-host disease. Transpl Infect Dis 17:702–706

Feuerecker M, Crucian BE, Quintens R, Buchheim JI, Salam AP, Rybka A, Moreels M, Strewe C, Stowe R, Mehta S, Schelling G, Thiel M, Baatout S, Sams C, Chouker A (2019) Immune sensitization during one year in the Antarctic high altitude Concordia Environment. Allergy 74:64. https://doi.org/10.1111/all.13545

Foster JS, Khodadad CL, Ahrendt SR, Parrish ML (2013) Impact of simulated microgravity on the normal developmental time line of an animal-bacteria symbiosis. Sci Rep 3:1340

Furusawa Y, Obata Y, Fukuda S, Endo TA, Nakato G, Takahashi D, Nakanishi Y, Uetake C, Kato K, Kato T, Takahashi M, Fukuda NN, Murakami S, Miyauchi E, Hino S, Atarashi K, Onawa S, Fujimura Y, Lockett T, Clarke JM, Topping DL, Tomita M, Hori S, Ohara O, Morita T, Koseki H, Kikuchi J, Honda K, Hase K, Ohno H (2013) Commensal microbe-derived butyrate induces the differentiation of colonic regulatory T cells. Nature 504:446–450

Ganji Arjenaki M, Rafieian Kopaei M (2018) Probiotics are a good choice in remission of inflammatory bowel diseases: a meta analysis and systematic review. J Cell Physiol 233:2091–2103

Gebbers JO, Laissue JA (1989) Immunologic structures and functions of the gut. Schweiz Arch Tierheilkd 131:221–238

Gibson GR, Wang X (1994) Regulatory effects of bifidobacteria on the growth of other colonic bacteria. J Appl Bacteriol 77:412–420

Gmunder FK, Konstantinova I, Cogoli A, Lesnyak A, Bogomolov W, Grachov AW (1994) Cellular immunity in cosmonauts during long duration spaceflight on board the orbital MIR station. Aviat Space Environ Med 65:419–423

Gomez del Pulgar T, Benitah SA, Valeron PF, Espina C, Lacal JC (2005) Rho GTPase expression in tumourigenesis: evidence for a significant link. Bioessays 27:602–613

Goncharova GI, Liz'ko NN, Lianmaia AM, Shilov VM, Spitsa TI (1981) Bifidobacterium flora status of cosmonauts before and after completing space flights. Kosm Biol Aviakosm Med 15:14–18

Hales NW, Yamauchi K, Alicea A, Sundaresan A, Pellis NR, Kulkarni AD (2002) A countermeasure to ameliorate immune dysfunction in in vitro simulated microgravity environment: role of cellularnucleotide nutrition. In Vitro Cell Dev Biol Anim 38:213–217

Hammond TG, Stodieck L, Birdsall HH, Becker JL, Koenig P, Hammond JS, Gunter MA, Allen PL (2013) Effects of microgravity on the virulence of Listeria monocytogenes, Enterococcus faecalis, Candida albicans, and methicillin-resistant Staphylococcus aureus. Astrobiology 13:1081–1090

Hayase E, Hashimoto D, Nakamura K, Noizat C, Ogasawara R, Takahashi S, Ohigashi H, Yokoi Y, Sugimoto R, Matsuoka S, Ara T, Yokoyama E, Yamakawa T, Ebata K, Kondo T, Hiramine R, Aizawa T, Ogura Y, Hayashi T, Mori H, Kurokawa K, Tomizuka K, Ayabe T, Teshima T (2017) R-Spondin1 expands Paneth cells and prevents dysbiosis induced by graft-versus-host disease. J Exp Med 214:3507–3518

Ichiki AT, Gibson LA, Jago TL, Strickland KM, Johnson DL, Lange RD, Allebban Z (1996) Effects of spaceflight on rat peripheral blood leukocytes and bone marrow progenitor cells. J Leukoc Biol 60:37–43

Ilyin VK (2005) Microbiological status of cosmonauts during orbital spaceflights on Salyut and Mir orbital stations. Acta Astronaut 56:839–850

Inohara N, Ogura Y, Fontalba A, Gutierrez O, Pons F, Crespo J, Fukase K, Inamura S, Kusumoto S, Hashimoto M, Foster SJ, Moran AP, Fernandez-Luna JL, Nunez G (2003) Host recognition of bacterial muramyl dipeptide mediated through NOD2. Implications for Crohn's disease. J Biol Chem 278:5509–5512

Jaffe AB, Hall A (2005) Rho GTPases: biochemistry and biology. Annu Rev Cell Dev Biol 21:247–269

Kaur I, Simons ER, Castro VA, Mark Ott C, Pierson DL (2004) Changes in neutrophil functions in astronauts. Brain Behav Immun 18:443–450

Kaur I, Simons ER, Castro VA, Ott CM, Pierson DL (2005) Changes in monocyte functions of astronauts. Brain Behav Immun 19:547–554

Kelsall B (2008) Recent progress in understanding the phenotype and function of intestinal dendritic cells and macrophages. Mucosal Immunol 1:460–469

Kim W, Tengra FK, Shong J, Marchand N, Chan HK, Young Z, Pangule RC, Parra M, Dordick JS, Plawsky JL, Collins CH (2013a) Effect of spaceflight on Pseudomonas aeruginosa final cell density is modulated by nutrient and oxygen availability. BMC Microbiol 13:241

Kim W, Tengra FK, Young Z, Shong J, Marchand N, Chan HK, Pangule RC, Parra M, Dordick JS, Plawsky JL, Collins CH (2013b) Spaceflight promotes biofilm formation by Pseudomonas aeruginosa. PLoS One 8:e62437

Klaus DM, Howard HN (2006) Antibiotic efficacy and microbial virulence during space flight. Trends Biotechnol 24:131–136

Kobayashi KS, Chamaillard M, Ogura Y, Henegariu O, Inohara N, Nunez G, Flavell RA (2005) Nod2-dependent regulation of innate and adaptive immunity in the intestinal tract. Science 307:731–734

Lawal A, Jejelowo OA, Rosenzweig JA (2010) The effects of low-shear mechanical stress on Yersinia pestis virulence. Astrobiology 10:881–888

Lawal A, Kirtley ML, van Lier CJ, Erova TE, Kozlova EV, Sha J, Chopra AK, Rosenzweig JA (2013) The effects of modeled microgravity on growth kinetics, antibiotic susceptibility, cold growth, and the virulence potential of a Yersinia pestis ymoA-deficient mutant and its isogenic parental strain. Astrobiology 13:821–832

Leid JG, Willson CJ, Shirtliff ME, Hassett DJ, Parsek MR, Jeffers AK (2005) The exopolysaccharide alginate protects Pseudomonas aeruginosa biofilm bacteria from IFN-gamma-mediated macrophage killing. J Immunol 175:7512–7518

Lencner AA, Lencner CP, Mikelsaar ME, Tjuri ME, Toom MA, Valjaots ME, Silov VM, Liz'ko NN, Legenkov VI, Reznikov IM (1984) The quantitative composition of the intestinal lactoflora before and after space flights of different lengths. Nahrung 28:607–613

Ley RE, Peterson DA, Gordon JI (2006) Ecological and evolutionary forces shaping microbial diversity in the human intestine. Cell 124:837–848

Lievin V, Peiffer I, Hudault S, Rochat F, Brassart D, Neeser JR, Servin AL (2000) Bifidobacterium strains from resident infant human gastrointestinal microflora exert antimicrobial activity. Gut 47:646–652

Lizko NN, Silov VM, Syrych GD (1984) Events in he development of dysbacteriosis of the intestines in man under extreme conditions. Nahrung 28:599–605

Louis F, Deroanne C, Nusgens B, Vico L, Guignandon A (2015) RhoGTPases as key players in mammalian cell adaptation to microgravity. Biomed Res Int 2015:747693

Macia L, Tan J, Vieira AT, Leach K, Stanley D, Luong S, Maruya M, Ian McKenzie C, Hijikata A, Wong C, Binge L, Thorburn AN, Chevalier N, Ang C, Marino E, Robert R, Offermanns S, Teixeira MM, Moore RJ, Flavell RA, Fagarasan S, Mackay CR (2015) Metabolite-sensing receptors GPR43 and GPR109A facilitate dietary fibre-induced gut homeostasis through regulation of the inflammasome. Nat Commun 6:6734

Macpherson AJ, Gatto D, Sainsbury E, Harriman GR, Hengartner H, Zinkernagel RM (2000) A primitive T cell-independent mechanism of intestinal mucosal IgA responses to commensal bacteria. Science 288:2222–2226

McLean RJ, Cassanto JM, Barnes MB, Koo JH (2001) Bacterial biofilm formation under microgravity conditions. FEMS Microbiol Lett 195:115–119

Mehrpouya-Bahrami P, Chitrala KN, Ganewatta MS, Tang C, Murphy EA, Enos RT, Velazquez KT, McCellan J, Nagarkatti M, Nagarkatti P (2017) Blockade of CB1 cannabinoid receptor alters gut microbiota and attenuates inflammation and diet-induced obesity. Sci Rep 7:15645

Mehta SK, Crucian BE, Stowe RP, Simpson RJ, Ott CM, Sams CF, Pierson DL (2013) Reactivation of latent viruses is associated with increased plasma cytokines in astronauts. Cytokine 61:205–209

Mehta SK, Laudenslager ML, Stowe RP, Crucian BE, Sams CF, Pierson DL (2014) Multiple latent viruses reactivate in astronauts during Space Shuttle missions. Brain Behav Immun 41:210–217

Muccioli GG, Naslain D, Backhed F, Reigstad CS, Lambert DM, Delzenne NM, Cani PD (2010) The endocannabinoid system links gut microbiota to adipogenesis. Mol Syst Biol 6:392

Nickerson CA, Ott CM, Mister SJ, Morrow BJ, Burns-Keliher L, Pierson DL (2000) Microgravity as a novel environmental signal affecting Salmonella enterica serovar Typhimurium virulence. Infect Immun 68:3147–3152

Ogura Y, Lala S, Xin W, Smith E, Dowds TA, Chen FF, Zimmermann E, Tretiakova M, Cho JH, Hart J, Greenson JK, Keshav S, Nunez G (2003) Expression of NOD2 in Paneth cells: a possible link to Crohn's ileitis. Gut 52:1591–1597

Ott CM, Oubre C, Wallace S, Mehta SK, Pierson DL (2016) Risk of adverse health effects due to host-microorganisms interactions. NASA, Human Health Countermeasures (HHC) Element, Washington, DC

Ottman N, Smidt H, de Vos WM, Belzer C (2012) The function of our microbiota: who is out there and what do they do? Front Cell Infect Microbiol 2:104

Pachenari M, Seyedpour SM, Janmaleki M, Babazadeh Shayan S, Taranejoo S, Hosseinkhani H (2014) Mechanical properties of cancer cytoskeleton depend on actin filaments to microtubules content: investigating different grades of colon cancer cell lines. J Biomech 47:373–379

Paulsen K, Tauber S, Dumrese C, Bradacs G, Simmet DM, Golz N, Hauschild S, Raig C, Engeli S, Gutewort A, Hurlimann E, Biskup J, Unverdorben F, Rieder G, Hofmanner D, Mutschler L, Krammer S, Buttron I, Philpot C, Huge A, Lier H, Barz I, Engelmann F, Layer LE, Thiel CS, Ullrich O (2015) Regulation of ICAM-1 in cells of the monocyte/macrophage system in microgravity. Biomed Res Int 2015:538786

Peterson DA, McNulty NP, Guruge JL, Gordon JI (2007) IgA response to symbiotic bacteria as a mediator of gut homeostasis. Cell Host Microbe 2:328–339

Pull SL, Doherty JM, Mills JC, Gordon JI, Stappenbeck TS (2005) Activated macrophages are an adaptive element of the colonic epithelial progenitor niche necessary for regenerative responses to injury. Proc Natl Acad Sci U S A 102:99–104

Rakoff-Nahoum S, Paglino J, Eslami-Varzaneh F, Edberg S, Medzhitov R (2004) Recognition of commensal microflora by toll-like receptors is required for intestinal homeostasis. Cell 118:229–241

Rea K, Dinan TG, Cryan JF (2016) The microbiome: a key regulator of stress and neuroinflammation. Neurobiol Stress 4:23–33

Ritchie LE, Taddeo SS, Weeks BR, Lima F, Bloomfield SA, Azcarate-Peril MA, Zwart SR, Smith SM, Turner ND (2015) Space environmental factor impacts upon murine colon microbiota and mucosal homeostasis. PLoS One 10:e0125792

Saei AA, Barzegari A (2012) The microbiome: the forgotten organ of the astronaut's body--probiotics beyond terrestrial limits. Future Microbiol 7:1037–1046

Savary CA, Grazziuti ML, Przepiorka D, Tomasovic SP, McIntyre BW, Woodside DG, Pellis NR, Pierson DL, Rex JH (2001) Characteristics of human dendritic cells generated in a microgravity analog culture system. In Vitro Cell Dev Biol Anim 37:216–222
Sender R, Fuchs S, Milo R (2016) Revised estimates for the number of human and bacteria cells in the body. PLoS Biol 14:e1002533
Shao D, Yao L, Riaz MS, Zhu J, Shi J, Jin M, Huang Q, Yang H (2017) Simulated microgravity affects some biological characteristics of Lactobacillus acidophilus. Appl Microbiol Biotechnol 101:3439–3449
Shen Y, Giardino Torchia ML, Lawson GW, Karp CL, Ashwell JD, Mazmanian SK (2012) Outer membrane vesicles of a human commensal mediate immune regulation and disease protection. Cell Host Microbe 12:509–520
Smith PM, Howitt MR, Panikov N, Michaud M, Gallini CA, Bohlooly YM, Glickman JN, Garrett WS (2013) The microbial metabolites, short-chain fatty acids, regulate colonic Treg cell homeostasis. Science 341:569–573
Sonnenberg GF, Monticelli LA, Alenghat T, Fung TC, Hutnick NA, Kunisawa J, Shibata N, Grunberg S, Sinha R, Zahm AM, Tardif MR, Sathaliyawala T, Kubota M, Farber DL, Collman RG, Shaked A, Fouser LA, Weiner DB, Tessier PA, Friedman JR, Kiyono H, Bushman FD, Chang KM, Artis D (2012) Innate lymphoid cells promote anatomical containment of lymphoid-resident commensal bacteria. Science 336:1321–1325
Stowe RP, Sams CF, Mehta SK, Kaur I, Jones ML, Feeback DL, Pierson DL (1999) Leukocyte subsets and neutrophil function after short-term spaceflight. J Leukoc Biol 65:179–186
Strewe C, Feuerecker M, Nichiporuk I, Kaufmann I, Hauer D, Morukov B, Schelling G, Chouker A (2012) Effects of parabolic flight and spaceflight on the endocannabinoid system in humans. Rev Neurosci 23:673–680
Tauber S, Lauber BA, Paulsen K, Layer LE, Lehmann M, Hauschild S, Shepherd NR, Polzer J, Segerer J, Thiel CS, Ullrich O (2017) Cytoskeletal stability and metabolic alterations in primary human macrophages in long-term microgravity. PLoS One 12:e0175599
Taylor GR, Janney RP (1992) In vivo testing confirms a blunting of the human cell-mediated immune mechanism during space flight. J Leukoc Biol 51:129–132
Taylor PW, Sommer AP (2005) Towards rational treatment of bacterial infections during extended space travel. Int J Antimicrob Agents 26:183–187
Turek FW, Vitaterna MH, Jiang P, Keshavarzian A, Green SJ (2017) Metagenomic sequencing of the bacteriome in GI tract of twin astronauts on ground and on one-year ISS mission. In: Space life & physical sciences research & applications division task book. https://taskbook.nasaprs.com/Publication/index.cfm?action=public_query_taskbook_content&TASKID=11124
Vaishnava S, Behrendt CL, Ismail AS, Eckmann L, Hooper LV (2008) Paneth cells directly sense gut commensals and maintain homeostasis at the intestinal host-microbial interface. Proc Natl Acad Sci U S A 105:20858–20863
Vega FM, Ridley AJ (2008) Rho GTPases in cancer cell biology. FEBS Lett 582:2093–2101
Voorhies AA, Mehta SK, Crucian BE, Torralba M, Moncera K, Feiveson A, Zurek-Varela EE, Pierson DL, Ott CM, Lorenzi HA (2017) Study of the impact of long-term space travel on the astronaut's microbiome. In: Space life & physical sciences research & applications division task book. https://taskbook.nasaprs.com/Publication/index.cfm?action=public_query_taskbook_content&TASKID=11370
Wehkamp J, Salzman NH, Porter E, Nuding S, Weichenthal M, Petras RE, Shen B, Schaeffeler E, Schwab M, Linzmeier R, Feathers RW, Chu H, Lima H Jr, Fellermann K, Ganz T, Stange EF, Bevins CL (2005) Reduced Paneth cell alpha-defensins in ileal Crohn's disease. Proc Natl Acad Sci U S A 102:18129–18134
Whitman WB, Coleman DC, Wiebe WJ (1998) Prokaryotes: the unseen majority. Proc Natl Acad Sci U S A 95:6578–6583
Wilson JW, Ott CM, Honer zu Bentrup K, Ramamurthy R, Quick L, Porwollik S, Cheng P, McClelland M, Tsaprailis G, Radabaugh T, Hunt A, Fernandez D, Richter E, Shah M, Kilcoyne M, Joshi L, Nelman-Gonzalez M, Hing S, Parra M, Dumars P, Norwood K, Bober R, Devich J, Ruggles A, Goulart C, Rupert M, Stodieck L, Stafford P, Catella L, Schurr MJ, Buchanan K,

Morici L, McCracken J, Allen P, Baker-Coleman C, Hammond T, Vogel J, Nelson R, Pierson DL, Stefanyshyn-Piper HM, Nickerson CA (2007) Space flight alters bacterial gene expression and virulence and reveals a role for global regulator Hfq. Proc Natl Acad Sci U S A 104:16299–16304
Wilson JW, Ott CM, Quick L, Davis R, Honer zu Bentrup K, Crabbe A, Richter E, Sarker S, Barrila J, Porwollik S, Cheng P, McClelland M, Tsaprailis G, Radabaugh T, Hunt A, Shah M, Nelman-Gonzalez M, Hing S, Parra M, Dumars P, Norwood K, Bober R, Devich J, Ruggles A, CdeBaca A, Narayan S, Benjamin J, Goulart C, Rupert M, Catella L, Schurr MJ, Buchanan K, Morici L, McCracken J, Porter MD, Pierson DL, Smith SM, Mergeay M, Leys N, Stefanyshyn-Piper HM, Gorie D, Nickerson CA (2008) Media ion composition controls regulatory and virulence response of Salmonella in spaceflight. PLoS One 3:e3923
Yamaguchi H, Condeelis J (2007) Regulation of the actin cytoskeleton in cancer cell migration and invasion. Biochim Biophys Acta 1773:642–652
Yi B, Rykova M, Jager G, Feuerecker M, Horl M, Matzel S, Ponomarev S, Vassilieva G, Nichiporuk I, Chouker A (2015) Influences of large sets of environmental exposures on immune responses in healthy adult men. Sci Rep 5:13367
Zea L, Prasad N, Levy SE, Stodieck L, Jones A, Shrestha S, Klaus D (2016) A molecular genetic basis explaining altered bacterial behavior in space. PLoS One 11:e0164359
Zea L, Larsen M, Estante F, Qvortrup K, Moeller R, Dias de Oliveira S, Stodieck L, Klaus D (2017) Phenotypic changes exhibited by E. coli cultured in space. Front Microbiol 8:1598

Pharmacological Countermeasures to Spaceflight-Induced Alterations of the Immune System

35

Thi Tuong Le, Armelle Ropars, Alamelu Sundaresan, Brian Crucian, Alexander Choukér, and Jean-Pol Frippiat

35.1 Introduction

Opportunities for microbes to establish infections are enhanced under spaceflight conditions because space travel can stimulate their growth (Fig. 35.1) and has a negative impact on immune cells (Chap. 18). Although a well-characterized post-flight phenomenon, immune dysregulation occurs and persists during spaceflight, confirming in-flight dysregulation distinct from the influences of landing and read-aptation following deconditioning (Crucian et al. 2015, 2016a, b).

Spaceflight affects lymphoid organs (Gridley et al. 2003; Baqai et al. 2009) and induces variations in peripheral blood leukocyte subsets (Chaps. 12–14). Several studies were undertaken to understand how spaceflight environment impairs natural immunity and T cell responses (for review see Frippiat et al. 2016; Guéguinou et al. 2009). It has been shown that the phagocytic and oxidative functions of neutrophils are affected by spaceflight conditions (Kaur et al. 2004; Rykova et al. 2008) and that

T. T. Le · A. Ropars · J.-P. Frippiat (✉)
Stress Immunity Pathogens Laboratory, EA 7300, Faculty of Medicine, Lorraine University, Vandœuvre-lès-Nancy, France
e-mail: thi-tuong.le@univ-lorraine.fr; armelle.ropars@univ-lorraine.fr; jean-pol.frippiat@univ-lorraine.fr

A. Sundaresan
Department of Biology, U.T. Medical School, Texas Southern University, Houston, TX, USA
e-mail: sundaresana@tsu.edu

B. Crucian
NASA Johnson Space Center, Houston, TX, USA
e-mail: Brian.Crucian-1@nasa.gov

A. Choukér
Laboratory of Translational Research Stress and Immunity, Department of Anaesthesiology, Hospital of the University of Munich (LMU), Munich, Germany
e-mail: alexander.chouker@med.uni-muenchen.de

© Springer Nature Switzerland AG 2020

A. Choukér (ed.), *Stress Challenges and Immunity in Space*,
https://doi.org/10.1007/978-3-030-16996-1_35

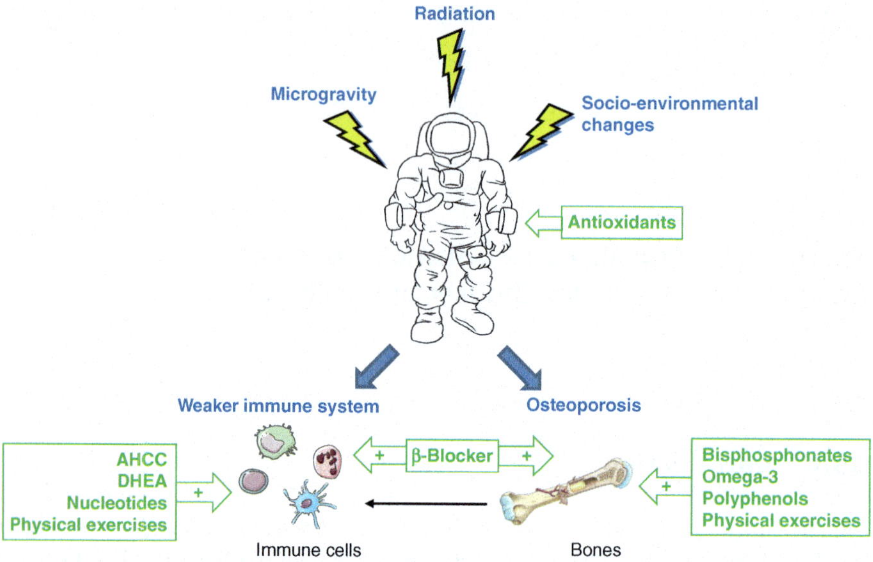

Fig. 35.1 Microgravity, radiation, and modifications of socio-environmental factors encountered during spaceflight weaken the immune system and induce osteoporosis. Cells involved in natural immunity and immune-competent B and T lymphocytes derive from hematopoietic stem cells that reside in the bone marrow in specialized niches. Changes in bone microstructure could therefore contribute to variations in peripheral blood leukocyte subsets observed at landing. Promising countermeasures are indicated in green. Antioxidants have a broad action, other countermeasures are effective on the immune system, bones or both

astronauts' monocytes exhibit phenotypic and cytokine-production deregulations, a reduced ability to engulf *E. coli*, elicit an oxidative burst, and degranulate (Crucian et al. 2011; Kaur et al. 2005, 2008; Rykova et al. 2008). Low natural killer cell cytotoxicity and a delay in responses to hypersensitivity skin tests were observed (Meshkov and Rykova 1995; Taylor and Janney 1992). Reactivation of latent herpes viruses has frequently been reported and can be considered as a good biomarker of spaceflight-induced weakening of cell-mediated immunity (Cohrs et al. 2008; Mehta et al. 2000; Pierson et al. 2005). Numerous studies did also investigate reduced T cell activation under low gravity conditions (Cogoli et al. 1984; Gridley et al. 2009) and highlighted that almost all cellular parameters can be affected such as: (1) gene expression, as shown by lower expressions of interleukin-2 (IL-2) and IL-2 receptor alpha chain (Walther et al. 1998); (2) cell–cell interactions and cytoskeleton structure, as T lymphocytes were found to be highly motile under microgravity while the motility of monocytes was severely reduced and the structure of their cytoskeleton was modified (Sciola et al. 1999; Meloni et al. 2006, 2011); (3) signal transduction, as PKA and NF-κB signaling pathways were shown to contribute to T cell dysfunction under altered gravity (Boonyaratanakornkit et al. 2005; Chang et al. 2012; Martinez et al. 2015) and (4) disturbed expression of cell cycle regulatory proteins (Thiel et al. 2012) (see also Chap. 17). Studies on plasma antibody levels did not reveal significant changes after short spaceflights (Rykova et al. 2008), but contradictory results were reported after long missions. Indeed, several studies (Konstantinova et al. 1993; Bascove et al. 2011, 2009; Guéguinou et al. 2009) reported changed in immunoglobulin production while Rykova et al. (2008) reported normal amounts of antibodies after prolonged space missions.

Spaceflight conditions also have a negative impact on immune cell production. Indeed, the in vitro culture of human CD34$^+$ bone marrow progenitors during spaceflight revealed the inhibitory effect of microgravity on erythropoiesis and myelopoiesis (Davis et al. 1996). Changes in the maturation of granulocytic cells in murine bone marrow were also reported after a 13-day spaceflight (Ortega et al. 2009). Finally, both T and B lymphopoiesis were shown to be reduced under altered gravity conditions (Woods et al. 2003, 2005; Huin-Schohn et al. 2013; Lescale et al. 2015; Ghislin et al. 2015).

Taken together, these data demonstrate that spaceflight-induced modifications of immune cell function and production could have an immediate impact on mission objectives.

The development of efficient countermeasures to combat the deleterious effects of spaceflight on the immune system is therefore required before we undertake prolonged space voyages. Furthermore, some of the observations presented above are also found in the elderly and people subjected to chronic or acute stress (for review see Weiskopf et al. 2009; Glaser and Kiecolt-Glaser 2005) (see Chap. 6). Finding countermeasures to spaceflight-induced immune alterations are therefore of interest to counter immunosenescence and the effects of stress-inducing situations on Earth.

35.2 Effects of Combined Antioxidant Treatment

Increased oxidative stress, which is harmful to cells and can induce many disorders, has been observed after radiation exposure and is associated with spaceflight (Stein and Leskiw 2000; Wan et al. 2005; Barrila et al. 2016). Indeed, lipopolysaccharide (LPS)-activated splenocytes from mice that flew onboard the space shuttle during mission STS-118 produced more interleukin-6 (IL-6) and interleukin-10 (IL-10) and less tumor necrosis factor (TNF) than control mice (Baqai et al. 2009). The same study showed that many of the genes responsible for scavenging reactive oxygen species (ROS) were upregulated after the flight, suggesting that cells attempted to scavenge ROS produced during spaceflight. An increase in the superoxide response by murine polymorphonuclear neutrophils was also reported even after short periods of microgravity (Fleming et al. 1991). Furthermore, it was shown that the urinary concentration of 8-hydroxy-2′-deoxyguanosine, a marker of oxidative damage to DNA, was higher and that red blood cell superoxide dismutase, an antioxidant enzyme that functions as a superoxide radical scavenger, was lower in astronauts after long-duration spaceflight (Smith and Zwart 2008). A downregulation of *GPX1*, which encodes glutathione peroxidase (GPX1), an enzyme that protects cells from oxidative damage and modulates the immune response, was also noted in astronaut's blood (Barrila et al. 2016). In the same way, it has been shown that spaceflight can downregulate antioxidant defense capacity and elicits an oxidative stress in rat liver (Hollander et al. 1998). Spaceflight significantly decreased catalase, glutathione (GSH) reductase and GSH sulfur-transferase activities in the liver of flown rats. It did also induce a decrease of liver GSH, GSH disulfide and total GSH contents which were accompanied by a lower gamma-glutamyl transpeptidase activity. Finally, it has been shown that microgravity generates a pro-oxidative environment that activates inflammatory responses in cultured human umbilical vein endothelial cells (Versari et al. 2013). These two last examples indicate that oxidative stress is a general feature.

Research was undertaken to determine if antioxidants could protect organisms from radiation-induced oxidative stress. Three studies showed that a mixture of L-selenomethionine (SeM), vitamin C, vitamin E succinate, alpha-lipoic acid, and *N*-acetyl cysteine improved the survival of mice after exposure to protons or to a potentially lethal dose of X-rays (Wambi et al. 2008, 2009) (Table 35.1). Pretreatment of mice with this mixture of antioxidants resulted in significantly higher total white blood cell and neutrophil counts in the peripheral blood and increased bone marrow cell counts after irradiation. Antioxidants increased Bcl-2 (B cell lymphoma-2, a protein regulating anti-apoptotic mechanisms) and decreased Bax (Bcl-associated X protein, promoting apoptosis), caspase 9, and TGF (transforming growth factor)-β1 mRNA expression in the bone marrow after X-ray irradiation (Wambi et al. 2008). This diet altered the expression pattern of several pro- and anti-apoptotic genes (Finnberg et al. 2013). In mice or rats exposed to high-energy particle radiation, D- or L-SeM or a combination of selected antioxidant agents, which included SeM, could also prevent the decrease in total antioxidants by regulating the expression of genes involved in the repair of radiation-induced DNA damage (Kennedy et al. 2004, 2007). Besides, it has been shown that a diet containing a mixture of antioxidant agents reduced the risk of developing malignant lymphoma in mice exposed to space radiation. It reduced the yields of a variety of different rare tumor types (Kennedy et al. 2008).

Taken together, these results suggest that antioxidant dietary supplements could be useful in the prevention of malignancies and other neoplastic lesions developing from exposure to space radiation.

35.3 Nucleotides

Nucleotides are beneficial for health because they positively influence lipid metabolism, immunity, and tissue growth, development and repair (Gil 2002). Rapidly proliferating tissues, such as those of the immune system, are not able to fulfill the needs of cell nucleotides exclusively by de novo synthesis and consequently use the salvage pathway that recovers nucleotides from the blood and diet. Nucleotides modulate the immune system (Nagafuchi et al. 1997; Holen et al. 2006). They influence lymphocyte maturation, activation, and proliferation. Likewise, they affect lymphocyte subset populations in the blood and are involved in enhancing macrophage phagocytosis and delayed hypersensitivity as well as allograft and tumor

Table 35.1 Examples of promising countermeasures to protect astronaut immune system. Arrows indicate up and down modulations

Countermeasure	Experiment performed	Results	References
Antioxidants	Irradiated mice (protons or X-rays) + antioxidants	– ↑ survival – ↑ white blood cells – ↑ bone marrow cells – Changes in apoptotic gene expression: ↑ Bcl-2 mRNA expression and ↓ Bax, caspase 9 and TGF-β1 mRNA expression in bone marrow	Wambi et al. (2008, 2009), Finnberg et al. (2013)
	Irradiated mice/rats (high-energy particles) + antioxidants	– Antioxidants prevented the decrease of the antioxidant status of animals	Kennedy et al. (2004, 2007)
Nucleotides	In vitro Mouse splenocytes cultured under simulated microgravity conditions and stimulated with PHA (phytohemagglutinin) + nucleotides	– Nucleoside-nucleotide mixture and uridine restored splenocyte proliferation – Nucleoside-nucleotide mixture ↑ IL-1β, IL-2 and IFN-γ	Hales et al. (2002)
	In vitro Mouse splenocytes cultured under simulated microgravity conditions and stimulated with PHA + nucleotides	– PHA-induced proliferation of splenocytes restored by uridine and nucleoside-nucleotide mixture	Kulkarni et al. (2002, 2005)
	In vivo Hindlimb-unloaded mice + nucleotides	– RNA and uracil restored popliteal lymph node proliferation, PHA-induced splenocyte proliferation, IL-2 and IFN-γ production	Kulkarni et al. (2002, 2005)
	In vivo Hindlimb-unloaded mice + nucleotides	– ↑ proliferation in lymph nodes – ↑ IL-2 and IFN-γ production in lymph nodes – ↓ corticosterone plasma level	Yamauchi et al. (2002)

(continued)

Table 35.1 (continued)

Countermeasure	Experiment performed	Results	References
AHCC	T cells + AHCC	– ↑ expression of LAT involved in TCR signaling.	Olamigoke et al. (2015)
	Hindlimb-unloaded mice infected with *K. pneumoniae* + AHCC	– ↓ mortality – ↑ time to death – ↑ bacteria clearance – ↑ anti-*K. pneumoniae* IgG levels	Aviles et al. (2003)
	Normally housed mice + AHCC	– ↑ spleen cell proliferation induced by Con-A (concanavalin A) or LPS – ↑ IL-2 and IFN-γ after Con-A stimulation – ↑ anti-*K. pneumoniae* IgG levels – ↑ IL-4, IL-6 and IL-10 after LPS stimulation – ↑ nitric oxide production in peritoneal cells	Aviles et al. (2003, 2004)
	Hindlimb-unloaded mice + AHCC.	– No effect on splenocyte proliferation induced by Con-A or LPS – ↑ IL-2 and IFN-γ after Con-A stimulation of splenocytes – ↑ nitric oxide production in peritoneal cells – Restored peritoneal cell function	Aviles et al. (2004)

DHEA	In vitro KLH-primed mouse splenocytes stimulated with KLH + DHEA	*TH2 favored* – ↑ IL-4 – ↓ IFN-γ	Du et al. (2001)
	In vitro Mouse splenocytes stimulated with ConA and LPS + DHEA	*TH2 favored* – ↓ IL-1, IL-2 and IFN-γ – ↑ IL-10 – IL-4, IL-6 and TNF not affected	Powell and Sonnenfeld (2006)
	In vivo Retrovirus infected mice + DHEA	*TH1 favored* – ↑ IL-2 and IFN-γ – ↓ IL-6 and TNF	Araghi-Niknam et al. (1997)
	In vivo Old female mice + DHEA	*TH1 favored* – ↑ IL-2 and IFN-γ – ↓ IL-6 and IL-10	Inserra et al. (1998)

responses. In addition, they contribute to the immunoglobulin response (Navarro et al. 1996; Nagafuchi et al. 1997; Maldonado et al. 2001), which has a positive effect on clearing infection. The molecular mechanisms by which nucleotides modulate the immune system are still largely unknown. Nucleotides may influence protein biosynthesis as well as signal membrane transduction mediated by the interaction of exogenous nucleosides and their receptors. They may also contribute to modulate the expression of a number of genes, including those involved in the immune system.

Because nutrient absorption and metabolism appear to be altered under spaceflight conditions (see Chap. 33), several studies have analyzed the effects of an exogenous source of nucleotides on immune function using ground-based models of microgravity. Hales et al. (2002) and Kulkarni et al. (2002, 2005) have shown that the decreased splenocyte proliferation in response to phytohemagglutinin (PHA) under simulated microgravity can be restored by a nucleoside-nucleotide mixture and uridine but not by inosine. This observation indicates that pyrimidines are more effective for immunoprotection of the hosts (Table 35.1). In vitro studies also revealed that cultured splenocytes secreted more IL-1β, IL-2, and interferon (IFN)-γ in the presence of a nucleoside-nucleotide mixture. In addition, Kulkarni et al. (2002, 2005) performed in vivo studies that demonstrated that popliteal lymph node proliferation, PHA-induced splenocyte proliferation, and IL-2 and IFN-γ production, which are significantly suppressed in hindlimb-unloaded mice (a ground-based model of choice for simulating spaceflight conditions on Earth (Globus and Morey-Holton 2016)), are restored by RNA and uracil. Similarly, Yamauchi et al. (2002) showed that in hindlimb-unloaded mice, nucleotides significantly increased in vivo lymph node proliferation and ex vivo lymphoproliferation response to alloantigen and mitogens, respectively, and IL-2 and IFN-γ production. Moreover, a lower plasma corticosterone level was observed in hindlimb-unloaded mice with RNA and uracil-supplemented diet. Furthermore, it has been shown that adenosine limits the oxidative stress response of polymorphonuclear leukocytes after parabolic flight through an upregulation of the adenosine A2(A) receptor function. This stop signal on inflammation is stronger than that under normal physiologic states and may limit further cytotoxic damage (Kaufmann et al. 2011). Thus, nucleotides possess immune-protective effects. These molecules are therefore potential countermeasures for the observed immune dysfunction associated with space travel.

35.4 Active Hexose-Correlated Compound

Another interesting compound is the active hexose-correlated compound (AHCC). AHCC is an extract prepared from cocultured mycelia of several species of *Basidiomycete* mushrooms that contains 40% of polysaccharides (β-glucan and acetylated α-glucan which are known to have immune-stimulating effects), amino acids, and minerals. AHCC is a popular complementary and alternative medicine used by cancer patients in Japan. It is available to the public without a prescription.

AHCC may help in the treatment of cancer. Indeed, a cohort study showed a significantly longer no recurrence period and an increased overall survival rate in 113 postoperative liver cancer patients taking AHCC (Matsui et al. 2002). Another study showed that AHCC significantly enhanced cisplatin-induced antitumor effect (Hirose et al. 2007). In a murine engraftment model of acute myeloid leukemia, AHCC led to significantly increased survival time and decreased blast counts (Fatehchand et al. 2017). As a last example, it has been shown that ovarian cancer cell viability was significantly reduced through treatment with AHCC (Choi et al. 2017).

Several studies have shown that AHCC has also a positive effect on human and rodent immune systems, including the enhancement of host resistance to influenza and West Nile viruses, improved protective antibody titers to influenza B vaccine, the prevention of thymic apoptosis induced by dexamethasone, the increase of natural killer cell activity, the enhancement of $CD4^+$ and $CD8^+$ T cell immune responses in healthy elderly persons and the induction of IL-12 production (Burikhanov et al. 2000; Matsui et al. 2002; Yagita et al. 2002; Nogusa et al. 2009; Wang et al. 2009; Yin et al. 2010; Roman et al. 2013).

AHCC was shown to induce the overexpression of the Linker for activated T cells (LAT) gene by nearly fourfold (Olamigoke et al. 2015) (Table 35.1). LAT, which is the primary activator after TCR engagement, increased by 2.7-fold in AHCC-treated lymphocytes compared to control even at 360 h post-treatment. As an adaptor protein, the function of LAT in TCR signaling centers upon its tyrosine phosphorylation and subsequent recruitment of other signaling proteins. Upon TCR engagement, phosphorylation of LAT allows it to interact with several SH2 domain-containing proteins, such as Grb2, Gads, and PLC-γ1. Interestingly, inhibitions of LAT and PLC-γ1, related to reduced T cell activation, were noted when lymphocytes are placed in microgravity (unpublished data). Hence, AHCC might be able to directly activate LAT and thus might be a possible future countermeasure candidate to restore T cell activation in immunosuppressive scenarios.

AHCC was also tested on hindlimb-unloaded mice that present decreased resistance to bacterial infections (*Klebsiella pneumoniae* and *Pseudomonas aeruginosa*) (Belay et al. 2002; Aviles et al. 2003). Hindlimb-unloaded mice showed significantly increased mortality and reduced mean time to death, increased levels of corticosterone, reduced ability to clear bacteria from their organs, and delayed production of anti-*P. aeruginosa* IgG antibodies, by comparison with controls. Aviles et al. (2003) showed that the administration of AHCC for 1 week before suspension and throughout the 10-day suspension period yielded significant beneficial effects for hindlimb-unloaded mice infected with *K. pneumoniae*, including decreased mortality, increased time to death, and increased ability to clear bacteria. Furthermore, mice receiving AHCC independent of the type of treatment (hindlimb-unloaded or normally caged) had higher anti-*K. pneumoniae* IgG antibody levels. The same team later demonstrated that AHCC significantly enhanced the function of the immune system in normally housed mice but only enhanced the TH1 response in mice under hindlimb-unloading conditions (Aviles et al. 2004) (Table 35.1). Interestingly, TH1 cytokine production has been shown to be depressed after short- and long-duration missions on the International Space Station (Crucian et al. 2008). Indeed, both groups of astronauts had a low IFN-γ to IL-10 secretion ratio on the day of landing after activation of peripheral blood T cells with anti-CD3 and anti-CD28 antibodies. This observation was confirmed by another study performed on PHA-stimulated splenocytes from mice flown on STS-108, which revealed that both IL-2 and IFN-γ were significantly lower after the flight (Gridley et al. 2003) indicating that a shift toward the TH2 subset is associated with spaceflight. This shift could represent a significant clinical risk for TH2-related autoimmune diseases, allergies, hypersensitivities, and disease susceptibility related to diminished cell-mediated immunity. AHCC also restored peritoneal cell functions that are suppressed by hindlimb-unloading and increased nitric oxide production in peritoneal cells isolated from hindlimb-unloaded mice. Other studies confirmed that AHCC enhances resistance to infection. In a mouse model of surgical wound infection, mice receiving AHCC were better able to clear bacteria than control animals (Aviles et al. 2006). AHCC increased immune function that resulted in a lower bacterial load in a murine model of intramuscular infection (Aviles et al. 2008) and in an increased resistance of mice to chlamydia genital infection (Belay et al. 2015). AHCC was also able to decrease LPS-induced inflammatory markers (cytokines, nitric oxide) and edema formation in the gut of rats, and diminished lymphocyte infiltration restoring gut architecture (Doursout et al. 2016). Finally, and interestingly, it was suggested that enhanced immune function observed with AHCC could be caused by attenuated concentrations of stress hormones and catecholamine (Love et al. 2013).

In conclusion, AHCC is an immunoenhancer that restores innate immunity, and a good candidate to lower stress hormones and restore T cell activation in immunosuppressive scenarios on Earth and in space.

35.5 DHEA

Dehydroepiandrosterone (DHEA) is one of the major circulating adrenal cortical hormones in humans and many other warm-blooded animals. This hormone is secreted by the adrenal cortex in response to stress (Kroboth et al. 1999). In the plasma, DHEA is predominantly present as DHEA-S that generates DHEA after cleavage of the sulfate group. For many years, the physiological significance of DHEA remained elusive. However, many studies have now shown that DHEA has significant immune modulatory functions, exhibiting both immune stimulatory and anti-glucocorticoid effects (for review see Hazeldine et al. 2010). The DHEA response to acute stress appears to be an important factor in stress-mediated immunological responses, with differential effects on immunity depending upon the presence of other hormones (Prall et al. 2017).

DHEA-S increases superoxide generation in primed human neutrophils in a dose-dependent fashion, thereby impacting a key bactericidal mechanism (Radford et al. 2010). In the same way, DHEA treatment has been shown to promote autophagy in *Mycobacterium tuberculosis*-infected human macrophage-like THP-1 cells

(Bongiovanni et al. 2015). In murine models, exogenous DHEA counteracts stress-induced glucocorticoid immunosuppression and increases the resistance of mice to viral and bacterial infections (Ben et al. 1999; Zhang et al. 1999). DHEA-S supplemented pigs, immunized against KLH (keyhole limpet hemocyanin) and ovalbumin, had greater concentrations of IgG and relative concentrations of antigen-specific IgG compared to control pigs indicating that DHEA-S supplementation increases responsiveness to an antigenic challenge (Burdick et al. 2009). Administration of DHEA has been demonstrated to improve survival and cellular immune functions in a murine model of sepsis (Schmitz et al. 2010). It also suppressed airway hyperresponsiveness and decreased eosinophil infiltration in the lungs to improve the symptoms of asthma in ovalbumin-sensitized mice (Liou and Huang 2011). In murine model systems of aging, DHEA appears to reverse the immunological defects seen as a consequence of aging. In particular, DHEA increased the ability of old mice to resist experimental viral and bacterial disease (Daynes et al. 1993; Straub et al. 1998). DHEA administration also restored immune function after thermal and trauma-hemorrhage injury and reduced mortality rates from septic challenge (Knoferl et al. 2003). In addition, DHEA provides protection against several diseases, including diabetes, oncological disorders, autoimmune disease, and chronic inflammatory illness. DHEA appears to be a potent regulator of cytokine production supporting the idea that this molecule acts on T cells, which is the lynch pin of the adaptive immune response. However, conflicting results on cytokine production in the presence of DHEA have been reported (Table 35.1). In vitro studies (Du et al. 2001; Powell and Sonnenfeld 2006) showed that DHEA might be an important factor for increasing TH2 cytokine synthesis, which encourage vigorous antibody production and are commonly associated with antibody responses important for resisting infection, and decreasing TH1 and pro-inflammatory cytokine production. However, DHEA has shown an opposite effect in vivo in which a TH1 upregulation associated with DHEA administration has been found in old or retrovirus-infected mice (Inserra et al. 1998; Zhang et al. 1999; Araghi-Niknam et al. 1997). These discrepancies may reflect differences in assays used to determine DHEA effects on cytokine production or differences in animal models used. Additionally, whereas in vitro DHEA is protected from biomodifications, in vivo DHEA administration could lead to rapid clearance from the blood and conversion to other steroids in peripheral tissue, which can affect T cells differently from DHEA. Despite these contradictory data, DHEA and DHEA-S seem beneficial to immune function and disease resistance, and are therefore promising countermeasures to fight the effects of spaceflight-associated stress on the immune system.

35.6 Pharmacological Approaches to Reduce Stress Hormone Effects

Targeting sympatho-adrenergic activation could be a pharmacological countermeasure to be envisaged as reviewed in Crucian et al. (2018). Sympatho-adrenergic activation in space may be strongly affecting the immune functions (see also Chaps.

6 and 8). As a surrogate marker of sympatho-adrenergic activation, it has been shown that the systolic arterial pressures as well as the heart rates were higher with dysfunction of vagal baroreflex control in space. Also, sympathetic nerve activity was observed to be elevated during spaceflight as compared to conditions on ground and microgravity exposure seems hence to induce a prevalence of sympathetic and decreased vagal cardiovascular control.

The key hormones of the sympathetic nerve system (SNS), the catecholamines epinephrine (E) and norepinephrine (NE), are potent immune modulators as shown in vitro (1) for E to negatively downregulate LPS-related TNF and IL-1β release in healthy subjects and ill patients, (2) to alter the TH1/TH2 immune balance and hereby impact the likelihood of hypersensitive immune reactions and (3) for norepinephrine to suppress oxidative burst reactions of granulocytes in a dose-dependent fashion. Interestingly, elevations of plasma E and NE levels in astronauts positively correlated with high adaptive immune impairments and higher herpes viral shedding (see Chap. 19). Not surprisingly, this virus directed effect can be expanded on other viruses, because catecholamine-mediated stress responses have the potential to differentially impact herpes simplex viruses (HSV) HSV-1 and HSV-2 expression and theoretically affect clinical outcomes of an infection. Thus, exploration crew might be taking advantage from selected antiadrenergic therapies since excessive sympathetic activity can suppress antiviral $CD4^+$ T cell responses. Also, it was reported that antiviral $CD8^+$ T cell responses can be enhanced by the administration of a β2-adrenergic antagonist in mice (Grebe et al. 2009). Here, β-blockers, a medication commonly used to treat cardiac arrhythmias and high blood pressure and to prevent from migraine, could represent a safe tool to counterbalance spaceflight related sympathetic nervous system activation that could lead to adverse immune consequences.

β-blockers bind to and block the ARs with no intrinsic secondary activation. One of the most widely used, peripherally and centrally acting β-blocker is propranolol. The expression and function of the target ARs have been studied extensively in dendritic cells, lymphocytes, and monocytes, and have recently been characterized in PMN as well and have shown beneficial effect by immune modulating properties. Therefore, also the evidence has been growing that propranolol could become a drug with anti-cancer properties. For example, propranolol improved the recurrence-free survival rates in mice undergoing primary tumor excision and has been reported to limit the recurrence of melanoma in humans. A combination of cyclooxygenase-2 and β-adrenergic blockade was seen to reduce liver metastasis of colon cancer and improved metastatic biomarkers in breast cancer patients (De Giorgi et al. 2017; Glasner et al. 2010; Killock 2017; Shaashua et al. 2017; Sorski et al. 2016). However, these effects cannot yet be extrapolated to other solid tumors.

Targeting HPA activation in space could be another complimentary tool to reduce the action of the increased concentrations of free cortisol, which is known to be of strong immune modulator function. Mifepristone, a competitive glucocorticoid receptor antagonist, and ketoconazole, a steroidogenesis inhibitor, are used in Cushing patients suffering from the consequences of hypercortisolism but the side effects, especially for ketoconazole, can be very considerable and preclude from a further application. New orally active and high-affinity selective antagonist of the glucocorticoid receptor are now tested and first results on dose-related safety, tolerability, pharmacokinetics, and pharmacological effects of e.g. CORT125134 have become available and look promising. Hunt et al. (2017) investigated CORT125134 in a cohort of 81 subjects and demonstrated that this antagonist was able to prevent several effects of prednisone, a strong agonist of the glucocorticoid receptor. Further studies are warranted and the clinical effects on immune functions and recovery have to be further quantified in clinical trials and high fidelity ground-based models. These results can probably suggest in the future also their evaluation for counteracting the immune suppression as related to chronically elevated corticoid levels in selected crewmembers in space.

35.7 Osteoporosis Treatment

Cells involved in natural immunity (e.g. granulocytes, monocytes) and immune-competent B and T lymphocytes derive from hematopoietic stem cells (HSC) that reside in the bone marrow within specialized niches made up of bone and vascular structures, including bone forming osteoblasts and bone resorbing osteoclasts (Mercier et al. 2011). Interactions between HSC and bone marrow niches control the balance between quiescence, self-renewal and differentiation of HSC (Calvi et al. 2003; Wang and Wagers 2011; Xie et al. 2009). Given that prolonged exposure to microgravity induces osteopenia, with decreased bone formation and mineralization and increased bone resorption (LeBlanc et al. 2000, 2007; Lang et al. 2004), changes in bone microstructure during spaceflights could contribute to variations in peripheral blood leukocyte subsets. Preserving bone structure could consequently help maintaining host immunity.

Bisphosphonates are largely used to preserve bone. Interestingly, transient changes in the numbers of murine hematopoietic stem cells, myeloid-biased progenitor cells and lymphoid-biased cells concurrent with changes to hematopoietic stem cell niches were reported following zoledronic acid (a bisphosphonate) administration (Ubellacker et al. 2017). Amino bisphosphonates are potent inhibitors of bone resorption and were shown to increase the number of granulocytes, indicating that they have an effect on murine hematopoiesis (Nakamura et al. 1999; Otsuka et al. 2011). Supplementation with omega-3 or polyphenols (see Chap. 33) also has a positive impact on bone mineral density and affects hematopoiesis. Indeed, it was shown that consumption of fish (a rich source of omega-3 fatty acids) was associated with reduced loss of bone mineral density in astronauts returning from a flight (Zwart et al. 2010), that a fish oil-rich diet promotes hematopoiesis in murine bone marrow and spleen (Xia et al. 2015), and that omega-3 fatty acids impact murine hematopoietic differentiation (Varney et al. 2009). In the same way, it was reported that curcumin improves anemia and extramedullary murine hematopoiesis (Fu et al. 2015). Another example is *Amaranthus cruentus* extract, rich in polyphenols. This extract significantly aided in restoring the levels of red blood cells, white blood cells, and hemoglobin in rats treated with phenylhydrazine to induce anemia (Pandey et al. 2016). Physical exercises can also countermeasure disuse-induced bone loss (Shackelford et al. 2004) and were shown to positively affect antibody production following immunization (Shearer et al. 2009). Another promising product is nacre, or mother-of-pearl, an acellular calcium carbonate composite produced by mollusks. Oral administration of nacre powder has a positive impact on murine and human osteoporosis (Kim et al. 2012; Vujasinović-Stupar et al. 2009) but no studies have yet addressed its effects on hematopoiesis. Finally, some studies suggest that β-blocker might also prove useful in the prevention of osteoporosis and the risk of fractures (Bonnet et al. 2007; Yang et al. 2011) (see also Sect. 35.6).

These examples show that bone countermeasures can likely contribute to preserve astronauts from spaceflight-associated immune alterations.

35.8 Conclusion

This chapter shows that the combination of antioxidants and the pharmacologic, immune-directed action of nucleotides, AHCC and DHEA show various degrees of efficiency to restore immune system alterations (Fig. 35.1). Furthermore, it appears that some drugs used to reduce the effects of stress hormones, and multidisciplinary bone countermeasures, could very likely contribute to preserve astronauts from spaceflight-induced immune alterations. Research in that direction could also help countering the age-associated decline in immune function on Earth because alterations of bone structure and immune system that manifest under reduced gravity conditions has features of accelerated aging. Finally, this chapter highlights the importance of taking into account connections between organs and physiological systems.

Acknowledgments JPF team was supported by CNES and the "Impact Biomolecules" project of the "Lorraine Université d'Excellence" (Investissements d'avenir—ANR 15-004), AS team by Amino Up Chemical, NASA, NuVue therapeutics and National Science Foundation, BC team by NASA and AC team by the DLR on behalf of the Federal Ministry of Economics and Technology (BMWi 50WB0719 and 50WB0919, 50WB1319) and the European Space Agency (ESA's ELIPS 3 and 4 programs, and E3P).

References

Araghi-Niknam M, Zhang Z, Jiang S, Call O, Eskelson CD, Watson RR (1997) Cytokine dysregulation and increased oxidation is prevented by dehydroepiandrosterone in mice infected with murine leukemia retrovirus. Proc Soc Exp Biol Med 216:386–391

Aviles H, Belay T, Fountain K, Vance M, Sun B, Sonnenfeld G (2003) Active hexose correlated compound enhances resistance to *Klebsiella pneumoniae* infection in mice in the hindlimb-unloading model of spaceflight conditions. J Appl Physiol 95:491–496. https://doi.org/10.1152/japplphysiol.00259.2003

Aviles H, Belay T, Vance M, Sun B, Sonnenfeld G (2004) Active hexose correlated compound enhances the immune function of mice in the hindlimb-unloading model of spaceflight conditions. J Appl Physiol 97:1437–1444. https://doi.org/10.1152/japplphysiol.00259.2004

Aviles H, O'Donnell P, Sun B, Sonnenfled G (2006) Active hexose correlated compound (AHCC) enhances resistance to infection in a mouse model of surgical wound infection. Surg Infect (Larchmt) 7:527–535. https://doi.org/10.1089/sur.2006.7.527

Aviles H, O'Donnell P, Orshal J, Fujii H, Sun B, Sonnenfeld G (2008) Active hexose correlated compound activates immune function to decrease bacterial load in a murine model of intramuscular infection. Am J Surg 195:537–545. https://doi.org/10.1016/j.amjsurg.2007.05.045

Baqai FP, Gridley DS, Slater JM, Luo-Owen X, Stodieck LS, Ferguson V et al (2009) Effects of spaceflight on innate immune function and antioxidant gene expression. J Appl Physiol 106:1935–1942. https://doi.org/10.1152/japplphysiol.91361.2008

Barrila J, Ott CM, LeBlanc C, Mehta SK, Crabbé A, Stafford P et al (2016) Spaceflight modulates gene expression in the whole blood of astronauts. NPJ Microgravity 2:16039. https://doi.org/10.1038/npjmgrav.2016.39

Bascove M, Guéguinou N, Schaerlinger B, Gauquelin-Koch G, Frippiat JP (2011) Decrease in antibody somatic hypermutation frequency under extreme, extended spaceflight conditions. FASEB J 25(9):2947–2955. https://doi.org/10.1096/fj.11-185215

Bascove M, Huin-Schohn C, Guéguinou N, Tschirhart E, Frippiat JP (2009) Spaceflight-associated changes in immunoglobulin VH gene expression in the amphibian *Pleurodeles waltl*. FASEB J 23:1607–1615. https://doi.org/10.1096/fj.08-121327

Belay T, Aviles H, Vance M, Fountain K, Sonnenfeld G (2002) Effects of the hindlimb-unloading model of spaceflight conditions on resistance of mice to infection with Klebsiella pneumoniae. J Allergy Clin Immunol 110:262–268

Belay T, Fu CL, Woart A (2015) Active hexose correlated compound activates immune function to decrease chlamydia trachomatis shedding in a murine stress model. J Nutr Med Diet Care 1(1)

Ben ND, Padgett DA, Loria RM (1999) Androstenediol and dehydroepiandrosterone protect mice against lethal bacterial infections and lipopolysaccharide toxicity. J Med Microbiol 48:425–431. https://doi.org/10.1099/00222615-48-5-425

Bongiovanni B, Mata-Espinosa D, D'Attilio L, Leon-Contreras JC, Marquez-Velasco R, Bottasso O, Hernandez-Pando R et al (2015) Effect of cortisol and/or DHEA on THP1-derived macrophages infected with Mycobacterium tuberculosis. Tuberculosis 95:562–569. https://doi.org/10.1016/j.tube.2015.05.011

Bonnet N, Gadois C, McCloskey E, Lemineur G, Lespessailles E, Courteix D et al (2007) Protective effect of β blockers in postmenopausal women: influence on fractures, bone density, micro and macroarchitecture. Bone 40:1209–1216. https://doi.org/10.1016/j.bone.2007.01.006

Boonyaratanakornkit JB, Cogoli A, Li CF, Schopper T, Pippia P, Galleri G et al (2005) Key gravity-sensitive signaling pathways drive T-cell activation. FASEB J 19:2020–2022. https://doi.org/10.1096/fj.05-3778fje

Burdick NC, Dominguez JA, Welsh TH Jr, Laurenz JC (2009) Oral administration of dehydroepiandrosterone-sulfate (DHEAS) increases in vitro lymphocyte function and improves in vivo response of pigs to immunization against keyhole limpet hemocyanin (KLH) and ovalbumin. Int Immunopharmacol 9:1342–1346. https://doi.org/10.1016/j.intimp.2009.07.007

Burikhanov RB, Wakame K, Igarashi Y, Wang S, Matsuzaki S (2000) Suppressive effect of active hexose correlated compound (AHCC) on thymic apoptosis induced by dexamethasone in the rat. Endocr Regul 34:181–188

Calvi LM, Adams GB, Weibrecht KW, Weber JM, Olson DP, Knight MC et al (2003) Osteoblastic cells regulate the haematopoietic stem cell niche. Nature 425:841–846. https://doi.org/10.1038/nature02040

Chang TT, Walther I, Li CF, Boonyaratanakornkit J, Galleri G, Meloni MA et al (2012) The Rel/NF-κB pathway and transcription of immediate early genes in T cell activation are inhibited by microgravity. J Leukoc Biol 92:1133–1145. https://doi.org/10.1189/jlb.0312157

Choi JY, Lee S, Yun SM, Suh DH, Kim K, No JH et al (2017) Active hexose correlated compound (AHCC) inhibits the proliferation of ovarian cancer cells by suppressing signal transducer and activator of transcription 3 (STAT3) activation. Nutr Cancer 7:1–7. https://doi.org/10.1080/01635581.2018.1380203

Cogoli A, Tschopp A, Fuchs-Bislin P (1984) Cell sensitivity to gravity. Science 225:228–230
Cohrs RJ, Mehta SK, Schmid DS, Gilden DH, Pierson DL (2008) Asymptomatic reactivation and shed of infectious varicella zoster virus in astronauts. J Med Virol 80:1116–1122. https://doi.org/10.1002/jmv.21173
Crucian BE, Stowe RP, Pierson DL, Sams CF (2008) Immune system dysregulation following short- vs long-duration spaceflight. Aviat Space Environ Med 79:835–843
Crucian B, Stowe R, Quiriarte H, Pierson D, Sams C (2011) Monocyte phenotype and cytokine production profiles are dysregulated by short-duration spaceflight. Aviat Space Environ Med 82:857–862
Crucian B, Stowe RP, Mehta S, Quiriarte H, Pierson D, Sams C (2015) Alterations in adaptive immunity persist during long-duration spaceflight. NPJ Microgravity 1:15013. https://doi.org/10.1038/npjmgrav.2015.13
Crucian B, Babiak-Vazquez A, Johnston S, Pierson DL, Ott CM, Sams C (2016a) Incidence of clinical symptoms during long-duration orbital spaceflight. Int J Gen Med 9:383–391. https://doi.org/10.2147/IJGM.S114188
Crucian B, Johnston S, Mehta S, Stowe R, Uchakin P, Quiriarte H et al (2016b) A case of persistent skin rash and rhinitis with immune system dysregulation onboard the International Space Station. J Allergy Clin Immunol Pract 4:759–762. https://doi.org/10.1016/j.jaip.2015.12.021
Crucian BE, Choukèr A, Simpson RJ, Mehta S, Marshall G, Smith SM et al (2018) Immune system dysregulation during spaceflight: potential countermeasures for deep space exploration missions. Front Immunol 9:1437. https://doi.org/10.3389/fimmu.2018.01437
Davis TA, Wiesmann W, Kidwell W, Cannon T, Kerns L, Serke C et al (1996) Effect of spaceflight on human stem cell hematopoiesis: suppression of erythropoiesis and myelopoiesis. J Leukoc Biol 60:69–76
Daynes RA, Araneo BA, Ershler WB, Maloney C, Li GZ, Ryu SY (1993) Altered regulation of IL-6 production with normal aging: possible linkage to the age-associated decline in dehydroepiandrosterone and its sulfated derivate. J Immunol 150:5219–5230
De Giorgi V, Grazzini M, Benemei S, Marchionni N, Botteri E, Pennacchioli E et al (2017) Propranolol for off-label treatment of patients with melanoma. JAMA Oncol 4:e172908. https://doi.org/10.1001/jamaoncol.2017.2908
Doursout MF, Liang Y, Sundaresan A, Wakame K, Fujii H, Takanari J et al (2016) Active hexose correlated compound modulates LPS-induced hypotension and gut injury in rats. Int Immunopharmacol 39:280–286. https://doi.org/10.1016/j.intimp.2016.07.023
Du C, Guan Q, Khalil MW, Sriram S (2001) Stimulation of TH2 response by high doses of dehydroepiandrosterone in KLH-primed splenocytes. Exp Biol Med 226:1051–1060
Fatehchand K, Santhanam R, Shen B, Erickson EL, Gautam S, Elavazhagan S et al (2017) Active hexose-correlated compound enhances extrinsic-pathway-mediated apoptosis of Acute Myeloid Leukemic cells. PLoS One 12:e0181729. https://doi.org/10.1371/journal.pone.0181729
Finnberg N, Wambi C, Kennedy AR, El-Deiry WS (2013) The effects of antioxidants on gene expression following gamma-radiation (GR) and proton radiation (PR) in mice in vivo. Cell Cycle 12:2241–2247. https://doi.org/10.4161/cc.25324
Fleming SD, Edelman LS, Chapes SK (1991) Effects of corticosterone and microgravity on inflammatory cell production of superoxide. J Leukoc Biol 50:69–76
Frippiat JP, Crucian BE, de Quervain DJ, Grimm D, Montano N, Praun S et al (2016) Towards human exploration of space: the THESEUS review series on immunology research priorities. NPJ Microgravity 2:16040. https://doi.org/10.1038/npjmgrav.2016.40
Fu Z, Chen X, Guan S, Yan Y, Lin H, Hua ZC (2015) Curcumin inhibits angiogenesis and improves defective hematopoiesis induced by tumor-derived VEGF in tumor model through modulating VEGF-VEGFR2 signaling pathway. Oncotarget 6:19469–19482. https://doi.org/10.18632/oncotarget.3625
Ghislin S, Ouzren-Zarhloul N, Kaminski S, Frippiat JP (2015) Hypergravity exposure during gestation modifies the TCRβ repertoire of newborn mice. Sci Rep 5:9318. https://doi.org/10.1038/srep09318

Gil A (2002) Modulation of the immune response mediated by dietary nucleotides. Eur J Clin Nutr 56:S1–S4. https://doi.org/10.1038/sj.ejcn.1601475

Glaser R, Kiecolt-Glaser JK (2005) Stress-induced immune dysfunction: implications for health. Nat Rev Immunol 5:243–251. https://doi.org/10.1038/nri1571

Glasner A, Avraham R, Rosenne E, Benish M, Zmora O, Shemer S et al (2010) Improving survival rates in two models of spontaneous postoperative metastasis in mice by combined administration of a β-adrenergic antagonist and a cyclooxygenase-2 inhibitor. J Immunol 184:2449–2457. https://doi.org/10.4049/jimmunol.0903301

Globus RK, Morey-Holton E (2016) Hindlimb unloading: rodent analog for microgravity. J Appl Physiol 120:1196–1206. https://doi.org/10.1152/japplphysiol.00997.2015

Grebe KM, Hickman HD, Irvine KR, Takeda K, Bennink JR, Yewdell JW (2009) Sympathetic nervous system control of anti-influenza CD8+ T cell responses. Proc Natl Acad Sci U S A 106:5300–5305. https://doi.org/10.1073/pnas.0808851106

Gridley DS, Nelson GA, Peters LL, Kostenuik PJ, Bateman TA, Morony S et al (2003) Genetic models in applied physiology: selected contribution: effects of spaceflight on immunity in the C57BL/6 mouse. II. Activation, cytokines, erythrocytes, and platelets. J Appl Physiol 94:2095–2103. https://doi.org/10.1152/japplphysiol.01053.2002

Gridley DS, Slater JM, Luo-Owen X, Rizvi A, Chapes SK, Stodieck LS et al (2009) Spaceflight effects on T lymphocyte distribution, function and gene expression. J Appl Physiol 106:194–202. https://doi.org/10.1152/japplphysiol.91126.2008

Guéguinou N, Huin-Schohn C, Bascove M, Bueb JL, Tschirhart E, Legrand-Frossi C et al (2009) Could spaceflight-associated immune system weakening preclude the expansion of human presence beyond Earth's orbit? J Leukoc Biol 86:1027–1038. https://doi.org/10.1189/jlb.0309167

Hales NW, Yamauchi K, Alicea A, Sundaresan A, Pellis NR, Kulkarni AD (2002) A countermeasure to ameliorate immune dysfunction in in vitro simulated microgravity environment: role of cellularnucleotide nutrition. In Vitro Cell Dev Biol Anim 38:213–217. https://doi.org/10.1290/1071-2690(2002)038<0213:ACTAID>2.0.CO;2

Hazeldine J, Arlt W, Lord JM (2010) Dehydroepiandrosterone as a regulator of immune cell function. J Steroid Biochem Mol Biol 120:127–136. https://doi.org/10.1016/j.jsbmb.2009.12.016

Hirose A, Sato E, Fujii H, Sun B, Nishioka H, Aruoma OI (2007) The influence of active hexose correlated compound (AHCC) on cisplatin-evoked chemotherapeutic and side effects in tumor-bearing mice. Toxicol Appl Pharmacol 222:152–158. https://doi.org/10.1016/j.taap.2007.03.031

Holen E, Bjørge ÖA, Jonsson R (2006) Dietary nucleotides and human immune cells. II. Modulation of PBMC growth and cytokine secretion. Nutrition 22:90–96

Hollander J, Gore M, Fiebig R, Mazzeo R, Ohishi S, Ohno H et al (1998) Spaceflight downregulates antioxidant defense systems in rat liver. Free Radic Biol Med 24:385–390

Huin-Schohn C, Guéguinou N, Schenten V, Bascove M, Gauquelin-Koch G, Baatout S et al (2013) Gravity changes during animal development affect IgM heavy-chain transcription and probably lymphopoiesis. FASEB J 27:333–341. https://doi.org/10.1096/fj.12-217547

Hunt H, Donaldson K, Strem M, Zann V, Leung P, Sweet S et al (2017) Assessment of safety, tolerability, pharmacokinetics, and pharmacological effect of orally administered CORT125134: an adaptive, double-blind, randomized, placebo-controlled phase 1 clinical study. Clin Pharmacol Drug Dev 7:408–421. https://doi.org/10.1002/cpdd.389

Inserra P, Zhang Z, Ardestani SK, Araghi-Niknam M, Liang B, Jiang S et al (1998) Modulation of cytokine production by dehydroepiandrosterone (DHEA) plus melatonin (MLT) supplementation of old mice. Proc Soc Exp Biol Med 218:76–82

Kaufmann I, Feuerecker M, Salam A, Schelling G, Thiel M, Choukèr A (2011) Adenosine A2(A) receptor modulates the oxidative stress response of primed polymorphonuclear leukocytes after parabolic flight. Hum Immunol 72:547–552. https://doi.org/10.1016/j.humimm.2011.03.021

Kaur I, Simons ER, Castro VA, Ott CM, Pierson DL (2004) Changes in neutrophil functions in astronauts. Brain Behav Immun 18:443–450. https://doi.org/10.1016/j.bbi.2003.10.005

Kaur I, Simons ER, Castro VA, Ott CM, Pierson DL (2005) Changes in monocyte functions of astronauts. Brain Behav Immun 19:547–554. https://doi.org/10.1016/j.bbi.2004.12.006
Kaur I, Simons ER, Kapadia AS, Ott CM, Pierson DL (2008) Effect of spaceflight on ability of monocytes to respond to endotoxins of gram-negative bacteria. Clin Vaccine Immunol 15(10):1523–1528. https://doi.org/10.1128/CVI.00065-08
Kennedy AR, Ware JH, Guan J, Donahue JJ, Biaglow JE, Zhou Z et al (2004) Selenomethionine protects against adverse biological effects induced by space radiation. Free Radic Biol Med 36:259–266
Kennedy AR, Guan J, Ware JH (2007) Countermeasures against space radiation induced oxidative stress in mice. Radiat Environ Biophys 46:201–203. https://doi.org/10.1007/s00411-007-0105-4
Kennedy AR, Davis JG, Carlton W, Ware JH (2008) Effects of dietary antioxidant supplementation on the development of malignant lymphoma and other neoplastic lesions in mice exposed to proton or iron-ion radiation. Radiat Res 169:615–625. https://doi.org/10.1667/RR1296.1
Killock D (2017) Skin cancer: propranolol limits melanoma recurrence. Nat Rev Clin Oncol 14:714–714. https://doi.org/10.1038/nrclinonc.2017.170
Kim H, Lee K, Ko CY, Kim HS, Shin HI, Kim T et al (2012) The role of nacreous factors in preventing osteoporotic bone loss through both osteoblast activation and osteoclast inactivation. Biomaterials 33:7489–7496. https://doi.org/10.1016/j.biomaterials.2012.06.098
Knoferl MW, Angele MK, Catania RA, Diodato MD, Bland KI, Chaudry IH (2003) Immunostimulatory effects of dehydroepiandrosterone in proestrus female mice after trauma-hemorrhage. J Appl Physiol 95:529–535. https://doi.org/10.1152/japplphysiol.01201.2002
Konstantinova IV, Rykova MP, Lesnyak AT, Antropova EA (1993) Immune changes during long-duration missions. J Leukoc Biol 54:189–201
Kroboth PD, Salek FS, Pittenger AL, Fabian TJ, Frye RF (1999) DHEA and DHEA-S: a review. J Clin Pharmacol 39:327–348
Kulkarni AD, Yamauchi K, Hales NW, Ramesh V, Ramesh GT, Sundaresan A et al (2002) Nutrition beyond nutrition: plausibility of immunotrophic nutrition for space travel. Clin Nutr 21:231–238
Kulkarni AD, Yamauchi K, Sundaresan A, Ramesh GT, Pellis NR (2005) Countermeasure for space flight effects on immune system: nutritional nucleotides. Gravit Space Biol Bull 18:101–102
Lang T, LeBlanc A, Evans H, Lu Y, Genant H, Yu A (2004) Cortical and trabecular bone mineral loss from the spine and hip in long-duration spaceflight. J Bone Miner Res 19:1006–1012. https://doi.org/10.1359/JBMR.040307
LeBlanc A, Schneider V, Shackelford L, West S, Oganov V, Bakulin A et al (2000) Bone mineral and lean tissue loss after long duration space flight. J Musculoskelet Neuronal Interact 1:157–160
LeBlanc AD, Spector ER, Evans HJ, Sibonga JD (2007) Skeletal responses to space flight and the bed rest analog: a review. J Musculoskelet Neuronal Interact 7:33–47
Lescale C, Schenten V, Djeghloul D, Bennabi M, Gaignier F, Vandamme K et al (2015) Hind limb unloading, a model of spaceflight conditions, leads to decreased B lymphopoiesis similar to aging. FASEB J 29:455–463. https://doi.org/10.1096/fj.14-259770
Liou CJ, Huang WC (2011) Dehydroepiandrosterone suppresses eosinophil infiltration and airway hyperresponsiveness via modulation of chemokines and Th2 cytokines in ovalbumin-sensitized mice. J Clin Immunol 31:656–665. https://doi.org/10.1007/s10875-011-9529-3
Love KM, Barnett RE, Holbrook I, Sonnenfeld G, Fujii H, Sun B et al (2013) A natural immune modulator attenuates stress hormone and catecholamine concentrations in polymicrobial peritonitis. J Trauma Acute Care Surg 74:1411–1418. https://doi.org/10.1097/TA.0b013e31829215b1
Maldonado J, Navarro J, Narbona E, Gil A (2001) The influence of dietary nucleotides on humoral and cell immunity in the neonate and lactating infant. Early Hum Dev 65:S69–S74
Martinez EM, Yoshida MC, Candelario TL, Hughes-Fulford M (2015) Spaceflight and simulated microgravity cause a significant reduction of key gene expression in early T-cell activation. Am J Physiol Regul Integr Comp Physiol 308:R480–R488. https://doi.org/10.1152/ajpregu.00449.2014

Matsui Y, Uhara J, Satoi S, Kaibori M, Yamada H, Kitade H et al (2002) Improved prognosis of postoperative hepatocellular carcinoma patients when treated with functional foods: a prospective cohort study. J Hepatol 37:78–86

Mehta SK, Stowe RP, Feiveson AH, Tyring SK, Pierson DL (2000) Reactivation and shedding of cytomegalovirus in astronauts during spaceflight. J Infect Dis 182:1761–1764. https://doi.org/10.1086/317624

Meloni MA, Galleri G, Pippia P, Cogoli-Greuter M (2006) Cytoskeleton changes and impaired motility of monocytes at modelled low gravity. Protoplasma 229:243–249. https://doi.org/10.1007/s00709-006-0210-2

Meloni MA, Galleri G, Pani G, Saba A, Pippia P, Cogoli-Greuter M (2011) Space flight affects motility and cytoskeletal structures in human monocyte cell line J-111. Cytoskeleton (Hoboken) 68:125–137. https://doi.org/10.1002/cm.20499

Mercier FE, Ragu C, Scadden DT (2011) The bone marrow at the crossroads of blood and immunity. Nat Rev Immunol 12:49–60. https://doi.org/10.1038/nri3132

Meshkov D, Rykova M (1995) The natural cytotoxicity in cosmonauts on board space stations. Acta Astronaut 36:719–726

Nagafuchi S, Katayanagi T, Nakagawa E, Takahashi T, Yajima T, Yonekubo A et al (1997) Effects of dietary nucleotides on serum antibody and splenic cytokine production in mice. Nutr Res 17:1163–1174

Nakamura M, Yagi H, Endo Y, Kosugi H, Ishi T, Itoh T (1999) A time kinetic study of the effect of aminobisphosphonate on murine haemopoiesis. Br J Haematol 107:779–790

Navarro J, Ruiz-Bravo A, Jiménez-Valera M, Gil A (1996) Modulation of antibody-forming cell and mitogen-driven lymphoproliferative responses by dietary nucleotides in mice. Immunol Lett 53:141–145

Nogusa S, Gerbino J, Ritz BW (2009) Low-dose supplementation with active hexose correlated compound improves the immune response to acute influenza infection in C57BL/6 mice. Nutr Res 29:139–143. https://doi.org/10.1016/j.nutres.2009.01.005

Olamigoke L, Mansoor E, Mann V, Ellis I, Okoro E, Wakame K et al (2015) AHCC activation and selection of human lymphocytes via genotypic and phenotypic changes to an adherent cell type: a possible novel mechanism of T cell activation. Evid Based Complement Alternat Med 2015:508746. https://doi.org/10.1155/2015/508746

Ortega MT, Pecaut MJ, Gridley DS, Stodieck LS, Ferguson V, Chapes SK (2009) Shifts in bone marrow cell phenotypes caused by spaceflight. J Appl Physiol 106:548–555. https://doi.org/10.1152/japplphysiol.91138.2008

Otsuka H, Yagi H, Endo Y, Nonaka N, Nakamura M (2011) Kupffer cells support extramedullary erythropoiesis induced by nitrogen-containing bisphosphonate in splenectomized mice. Cell Immunol 271:197–204. https://doi.org/10.1016/j.cellimm.2011.06.025

Pandey S, Ganeshpurkar A, Bansal D, Dubey N (2016) Hematopoietic effect of amaranthus cruentus extract on phenylhydrazine-induced toxicity in rats. J Diet Suppl 13:607–615. https://doi.org/10.3109/19390211.2016.1155685

Pierson DL, Stowe RP, Phillips TM, Lugg DJ, Mehta SK (2005) Epstein-Barr virus shedding by astronauts during space flight. Brain Behav Immun 19:235–242. https://doi.org/10.1016/j.bbi.2004.08.001

Powell JM, Sonnenfeld G (2006) The effects of dehydroepiandrosterone (DHEA) on in vitro spleen cell proliferation and cytokine production. J Interferon Cytokine Res 26:34–49. https://doi.org/10.1089/jir.2006.26.34

Prall SP, Larson EE, Muehlenbein MP (2017) The role of dehydroepiandrosterone on functional innate immune responses to acute stress. Stress Health 33:656–664. https://doi.org/10.1002/smi.2752

Radford DJ, Wang K, McNelis JC, Taylor AE, Hechenberger G, Hofmann J et al (2010) Dehdyroepiandrosterone sulfate directly activates protein kinase C-beta to increase human neutrophil superoxide generation. Mol Endocrinol 24:813–821. https://doi.org/10.1210/me.2009-0390

Roman BE, Beli E, Duriancik DM, Gardner EM (2013) Short-term supplementation with active hexose correlated compound improves the antibody response to influenza B vaccine. Nutr Res 33:12–17. https://doi.org/10.1016/j.nutres.2012.11.001

Rykova MP, Antropova EN, Larina IM, Morukov BV (2008) Humoral and cellular immunity in cosmonauts after the ISS missions. Acta Astraunaut 63:697–705. https://doi.org/10.1016/j.actaastro.2008.03.016

Schmitz D, Kobbe P, Wegner A, Hammes F, Oberbeck R (2010) Dehydroepiandrosterone during sepsis: does the timing of administration influence the effectiveness. J Surg Res 163:e73–e77. https://doi.org/10.1016/j.jss.2010.05.017

Sciola L, Cogoli-Greuter M, Cogoli A, Spano A, Pippia P (1999) Influence of microgravity on mitogen binding and cytoskeleton in Jurkat cells. Adv Space Res 24:801–805

Shaashua L, Shabat-Simon M, Haldar R, Matzner P, Zmora O, Shabtai M et al (2017) Perioperative COX-2 and β-adrenergic blockade improves metastatic biomarkers in breast cancer patients in a phase-II randomized trial. Clin Cancer Res 23:4651–4661. https://doi.org/10.1158/1078-0432.CCR-17-0152

Shackelford LC, LeBlanc AD, Driscoll TB, Evans HJ, Rianon NJ, Smith SM et al (2004) Resistance exercise as a countermeasure to disuse-induced bone loss. J Appl Physiol 97:119–129. https://doi.org/10.1152/japplphysiol.00741.2003

Shearer WT, Ochs HD, Lee BN, Cohen EN, Reuben JM, Cheng I et al (2009) Immune responses in adult female volunteers during the bed-rest model of spaceflight: antibodies and cytokines. J Allergy Clin Immunol 123:900–905. https://doi.org/10.1016/j.jaci.2008.12.016

Smith SM, Zwart SR (2008) Nutrition issues for space exploration. Acta Astronaut 63:609–613. https://doi.org/10.1016/j.actaastro.2008.04.010

Sorski L, Melamed R, Matzner P, Lavon H, Shaashua L, Rosenne E et al (2016) Reducing liver metastases of colon cancer in the context of extensive and minor surgeries through β-adrenoceptors blockade and COX2 inhibition. Brain Behav Immun 58:91–98. https://doi.org/10.1016/j.bbi.2016.05.017

Stein TP, Leskiw MJ (2000) Oxidant damage during and after spaceflight. Am J Physiol Endocrinol Metab 278:E375–E382. https://doi.org/10.1152/ajpendo.2000.278.3.E375

Straub RH, Konecna L, Hrach S, Rothe G, Kreutz M, Schölmerich J et al (1998) Serum dehydroepiandrosterone (DHEA) and DHEA sulfate are negatively correlated with serum interleukin-6 (IL-6), and DHEA inhibits IL-6 secretion from mononuclear cells in man in vitro: possible link between endocrinosenescence and immunosenescence. J Clin Endocrinol Metab 83:2012–2017. https://doi.org/10.1210/jcem.83.6.4876

Taylor GR, Janney RP (1992) In vivo testing confirms a blunting of the human cell-mediated immune mechanism during space-flight. J Leukoc Biol 51:129–132

Thiel CS, Paulsen K, Bradacs G, Lust K, Tauber S, Dumrese C et al (2012) Rapid alterations of cell cycle control proteins in human T lymphocytes in microgravity. Cell Commun Signal 10:1. https://doi.org/10.1186/1478-811X-10-1

Ubellacker JM, Haider MT, DeCristo MJ, Allocca G, Brown NJ, Silver DP et al (2017) Zoledronic acid alters hematopoiesis and generates breast tumor-suppressive bone marrow cells. Breast Cancer Res 19:23. https://doi.org/10.1186/s13058-017-0815-8

Varney ME, Hardman WE, Sollars VE (2009) Omega 3 fatty acids reduce myeloid progenitor cell frequency in the bone marrow of mice and promote progenitor cell differentiation. Lipids Health Dis 8:9. https://doi.org/10.1186/1476-511X-8-9

Versari S, Longinotti G, Barenghi L, Maier JA, Bradamante S (2013) The challenging environment on board the International Space Station affects endothelial cell function by triggering oxidative stress through thioredoxin interacting protein overexpression: the ESA-SPHINX experiment. FASEB J 27:4466–4475. https://doi.org/10.1096/fj.13-229195

Vujasinović-Stupar N, Novković S, Jezdić I (2009) Supplementation with bio-calcium from shells Pinctada maxima in postmenopausal women with decreased mineral bone density--pilot study. Srp Arh Celok Lek 137:518–523

Walther I, Pippia P, Meloni MA, Turrini F, Mannu F, Cogoli A (1998) Simulated microgravity inhibits the genetic expression of interleukin-2 and its receptor in mitogen-activated T lymphocytes. FEBS Lett 436:115–118

Wambi C, Sanzari J, Wan XS, Nuth M, Davis J, Ko YH et al (2008) Dietary antioxidants protect hematopoietic cells and improve animal survival after total-body irradiation. Radiat Res 169:384–396. https://doi.org/10.1667/RR1204.1
Wambi CO, Sanzari JK, Sayers CM, Nuth M, Zhou Z, Davis J et al (2009) Protective effects of dietary antioxidants on proton total-body irradiation-mediated hematopoietic cell and animal survival. Radiat Res 172:175–186. https://doi.org/10.1667/RR1708.1
Wan XS, Bloch P, Ware JH, Zhou Z, Donahue JJ, Guan J et al (2005) Detection of oxidative stress induced by low- and high-linear energy transfer radiation in cultured human epithelial cells. Radiat Res 163:364–368
Wang LD, Wagers AJ (2011) Dynamic niches in the origination and differentiation of haematopoietic stem cells. Nat Rev Mol Cell Biol 12:643–655. https://doi.org/10.1038/nrm3184
Wang S, Welte T, Fang H, Chang GJ, Born WK, O'Brien RL et al (2009) Oral administration of active hexose correlated compound enhances host resistance to West Nile encephalitis in mice. J Nutr 139:598–602. https://doi.org/10.3945/jn.108.100297
Weiskopf D, Weinberger B, Grubeck-Loebenstein B (2009) The aging of the immune system. Transpl Int 22:1041–1050. https://doi.org/10.1111/j.1432-2277.2009.00927.x
Woods CC, Banks KE, Gruener R, DeLuca D (2003) Loss of T cell precursors after spaceflight and exposure to vector-averaged gravity. FASEB J 17:1526–1528. https://doi.org/10.1096/fj.02-0749fje
Woods CC, Banks KE, Lebsack TW, White TC, Anderson GA, Maccallum T et al (2005) Use of a microgravity organ culture dish system to demonstrate the signal dampening effects of modeled microgravity during T cell development. Dev Comp Immunol 29:565–582. https://doi.org/10.1016/j.dci.2004.09.006
Xia S, Li XP, Cheng L, Han MT, Zhang MM, Shao QX et al (2015) Fish oil-rich diet promotes hematopoiesis and alters hematopoietic niche. Endocrinology 156:2821–2830. https://doi.org/10.1210/en.2015-1258
Xie Y, Yin T, Wiegraebe W, He XC, Miller D, Stark D et al (2009) Detection of functional haematopoietic stem cell niche using real-time imaging. Nature 457:97–101. https://doi.org/10.1038/nature07639
Yagita A, Maruyama S, Wakasugi S, Sukegawa Y (2002) H-2 haplotype-dependent serum IL-12 production in tumor-bearing mice treated with various mycelial extracts. In Vivo 16:49–54
Yamauchi K, Hales NW, Robinson SM, Niehoff ML, Ramesh V, Pellis NR et al (2002) Dietary nucleotides prevent decrease in cellular immunity in ground based microgravity analog. J Appl Physiol 93:161–166. https://doi.org/10.1152/japplphysiol.01084.2001
Yang S, Nguyen ND, Center JR, Eisman JA, Nguyen TV (2011) Association between beta-blocker use and fracture risk: the Dubbo Osteoporosis Epidemiology study. Bone 48:451–455. https://doi.org/10.1016/j.bone.2010.10.170
Yin Z, Fujii H, Walshe T (2010) Effects of active hexose correlated compound on frequency of CD4+ and CD8+ T cells producing interferon-γ and/or tumor necrosis factor-α in healthy adults. Hum Immunol 71:1187–1190. https://doi.org/10.1016/j.humimm.2010.08.006
Zhang Z, Araghi-Niknam M, Liang B, Inserra P, Ardestani SK, Jiang S et al (1999) Prevention of immune dysfunction and vitamin E loss by dehydroepiandrosterone and melatonin supplementation during murine retrovirus infection. Immunology 96:291–297
Zwart SR, Pierson D, Mehta S, Gonda S, Smith SM (2010) Capacity of omega-3 fatty acids or eicosapentaenoic acid to counteract weightlessness-induced bone loss by inhibiting NF-kappaB activation: from cells to bed rest to astronauts. J Bone Miner Res 25:1049–1057. https://doi.org/10.1359/jbmr.091041

Part VI

Perspectives for Manned Space Exploration—from Visions to Realities

36 Platforms for Stress and Immune Research in Preparation for Long-Duration Space Exploration Missions

Thu Jennifer Ngo-Anh and Andrea Rossiter

36.1 Introduction

Future human space exploration missions will be challenging endeavors on many different levels. When considering possible immunological changes and their implications for such missions, it is helpful to understand the characteristics that exploration endeavors entail, and that are likely to affect the functions of the immune system.

These characteristics include, for example,

- Long-term isolation and confinement (a small crew living within one limited habitat for many months or even years in the case of a mission to Mars)
- Unloading of the body in freefall (or partial gravity on another planet)
- Hypobaric hypoxic conditions (under discussion for future planetary habitats to facilitate frequent extra-habitat activities)

It appears obvious that the effects of the many challenges in spaceflight are best investigated in the real space environment. With the International Space Station, orbiting in Low Earth Orbit for more than a decade, an excellent permanently manned laboratory is available. The Chinese Space Station will also offer unique opportunities to expand this research. However, due to issues like limited resources for up-/download and crew time, operational constraints, and limited sample sizes, it is beneficial and complementary to also utilize terrestrial analogs for preparatory research. Analogs also allow differentiating the effects of different spaceflight factors, which can be helpful when making projections for new mission scenarios (Pagel and Choukèr 2016). Some examples of useful analogs are described in the

T. J. Ngo-Anh (✉) · A. Rossiter
Human and Robotic Exploration Programmes (HRE-RS), European Space Agency, Noordwijk, The Netherlands
e-mail: Jennifer.Ngo-Anh@esa.int

© Springer Nature Switzerland AG 2020

A. Choukér (ed.), *Stress Challenges and Immunity in Space*,
https://doi.org/10.1007/978-3-030-16996-1_36

following sections. While helpful results may also be collected using cell, tissue, or animal models, the descriptions here will focus on analogs where humans can be the test subjects.

36.2 Bed Rest and Dry Immersion

Bed rest studies are a well-established and frequently used model for many of the physiological effects related to the unloading of the human body from the pull of gravity in spaceflight. Healthy volunteers are confined to bed in a −6° head down tilt position for periods ranging from a few days to months, depending on the research objectives. During the study period the test subjects do not leave the supine position. Any activity, from eating and reading to personal hygiene and exercise are performed lying down (Fig. 36.1). The resulting muscle atrophy, bone mass loss, cardiovascular deconditioning, and other changes mimic the adaptations observed in astronauts, and thus this model can be used for research into the mechanisms behind these changes, as well as the evaluation of countermeasures. All major space agencies are supporting bed rest studies through their own mechanisms, and some bed rest studies have even been organized independently by investigators themselves. One critical issue is, however, the comparability of results or, in other words, the standardization of the studies. In an effort to improve that aspect, an international study group under the umbrella of the International Academy of Astronautics (IAA) has worked in recent years to develop guidelines for standardized conditions in bed rest studies (Sundblad and Orlov 2015).

Bed rest studies are, due to their high fidelity and standardization, an important environment to test the effectiveness of measures to counteract the unfavorable consequences of unloading conditions, including with respect to the immune

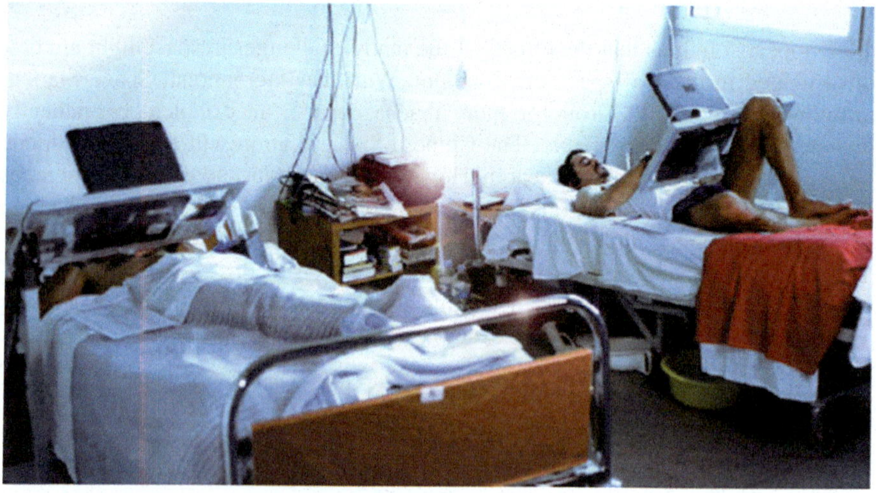

Fig. 36.1 Bed rest: Test subjects in an ESA long-duration bed rest study (©ESA)

system. Short-term (5 days; Caiani et al. 2014; Clément et al. 2015, 2016; Kos et al. 2014; Li et al. 2017; Provost et al. 2015), mid-term (21 days; Kelsen et al. 2012) and long-term (60 days; Koschate et al. 2018; Kramer et al. 2017; Schoenrock et al. 2018) bed rest periods have been established. Also, an intermediate duration of, for instance, 14 days (Buehlmeier et al. 2017) or longer durations (70, 90, and up to 120 days; Belavý et al. 2017; Cromwell et al. 2018; Dillon et al. 2018; Ploutz-Snyder et al. 2018) have been realized. Exercise, gravitational and nutritional countermeasures have been tested among other studies implemented under these standardized condition. Female bed rest studies have also been conducted to test for gender effects which are important in space research with mixed crews (Beller et al. 2011; Evans et al. 2018; Holt et al. 2016; Klassen et al. 2018; Lee et al. 2014).

Similar to bed rest, the dry immersion model aims at reproducing unloading effects. Here, the test subjects are immersed in big water tanks, comparable to oversized bath tubs. They are enveloped in non-water-permeable blankets, and as such, it is a dry immersion. This approach, which is especially popular for space research in Russia but is planned to be used in Europe as well, leads to an even-faster deconditioning (Demangel et al. 2017; Navasiolava et al. 2010; Treffel et al. 2017). However, practical aspects of hygiene and care limit the duration for which such studies can be performed.

36.3 Isolation and Confinement Studies

Isolating/confining small crews in simulation chambers has been done for different purposes in the past. Sometimes this isolation was a by-product of the testing of spacecraft systems to validate them for an upcoming space mission. At other times, the aim was the testing of life support systems or elements thereof.

A further common motivation for organizing isolation studies is to investigate operations concepts and study the psychological effects of long-duration isolation. During the latter type of study especially, research from other scientific fields is also implemented.

Typical crew sizes are 3–6 persons, though for special purposes or scenarios, higher or lower numbers have also been used.

Many differences exist between these studies in duration, target scenario, crew selection and training procedures, and other characteristics, creating complications in drawing overall conclusions. Nevertheless, valuable data has been collected on a variety of topics, including the reactions of the physiological systems (including the immune system), psychological issues, the evolution of the microbial communities inside the confined space, circadian changes, and some countermeasure concepts.

Hopefully, in the future this approach will be further standardized and more repetitive to enhance the results. A special challenge for the future will be the integration of interacting/interfering factors like psychology, medicine, astrobiology, and life support, for example: when a crew is isolated in autonomous mode in a habitat with an advanced, closed-loop life support system.

36.3.1 Mars500: SIRIUS

An example of current isolation and confinement studies is the Mars500 study performed in 2010/2011. This study, conducted in a facility of the Russian Institute for BioMedical Problems (IMBP) in Moscow, aimed at the replication of a full Mars mission scenario of 520 days, with a transit phase to Mars, a stay in Mars orbit including a simulated landing of half of the crew, and a transit phase back to Earth. Six crew members (three Russian, two European, one Chinese) conducted a very extensive international experiment program (see Chap. 37). Nowadays this program is run under the programme name SIRIUS (www.sirius.imbp.info).

36.3.2 :envihab

What happens to the human body on a flight to Mars? How does being confined to bed after a serious illness impact the body? How does the lack of daylight affect mood? Are there any measures to counteract these adverse effects? These basic questions need to be answered for us on Earth, to understand the effects of ageing, bedriddenness, immobilisation, and isolation, to name but a few. The DLR Institute of Aerospace Medicine is a world leader in aviation and space medicine. With its one-of-a-kind, highly sophisticated medical research facility, :envihab, this institute is taking a step forward in its ground-breaking research into the ways in which people adjust to extreme environments and other stressful situations. :envihab (from the words "Environment" and "Habitat"), a one-story, 3500-square-meter, state-of-the-art laboratory, will be used to explore the effects of extreme environmental conditions on humans and to determine possible countermeasures. It is ideally suited for exploring the future challenges of human spaceflight. Made up of eight separate modules, built according to a "house-in-house" design, which can be combined within one facility, it includes a short arm human centrifuge, several laboratories, MRI/PET analysis facilities, psychological stress simulations and rehabilitations, microbiological and molecular biological research tools, and places to house and monitor test subjects for analog research.

:envihab consists of:

- A Psychology Lab to test the psychological effects of confinement over a longer duration of time. Up to six test subjects (similar to a crew size) can be isolated, immobilized, and exposed to targeted stress situations.
- A Sleep and Physiology Lab that can facilitate 12 test subjects for research under different analog aspects (e.g. bed rest studies) and under highly controlled conditions of: ambient light, temperature, humidity, oxygen, and nitrogen levels. The lab also has a kitchen for metabolic nutrition, and the capacity to carry out: blood sampling during sleep without disturbance of the test subject, isolation for crew simulation, a reduction of oxygen down to 12% within 5 h (15,000 ft), enrichment with CO_2 up to 3% with a 1% change per hour, and variations of individual lighting between 0.5 and 1500 Lux.

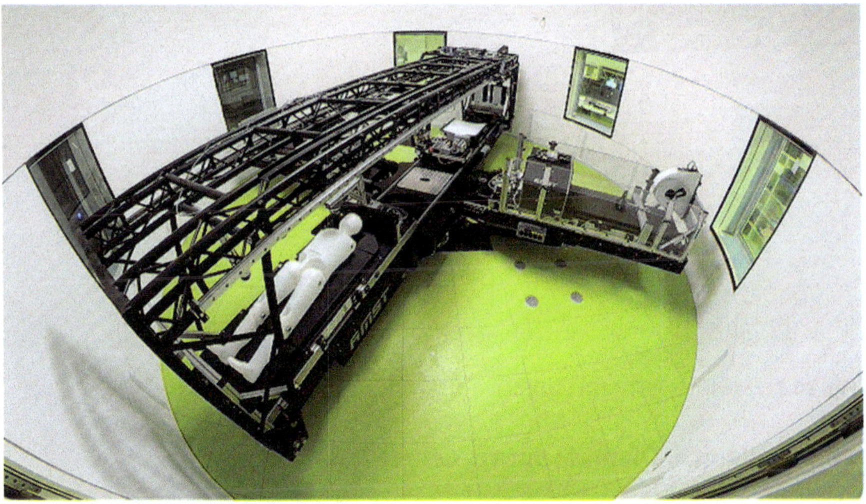

Fig. 36.2 :envihab's short-arm human centrifuge (©DLR)

- A Short-Arm Human Centrifuge (Fig. 36.2): advanced functional possibilities for the development of a countermeasure against physiological deconditioning under weightlessness. It has: a radius of 3.80 m, max. radial acceleration of 6 g at outer perimeter, acceleration from 0 g to 6 g within 30 s, nacelles are moveable when centrifuge is running, and simultaneous testing of up to four test subjects.
- A Prevention and Rehabilitation Lab: assessment of health parameters and physical performance capacity (cardiovascular and locomotor system) of astronauts during and following space missions to develop countermeasures. The lab also facilitates: tilt testing, Lower Body Negative Pressure (LBNP), a treadmill, an isokinetic dynamometer, and a barocomplex to simulate spaceflight scenarios. Ambient pressure, oxygen levels, temperature, and humidity can be adjusted independently, also.
- Positron Emission Tomography/Magnetic Resonance Imaging (PET-MRI): a whole body MRI and MR spectroscopy machine with integrated PET to monitor the physiological changes that could occur during long term missions (Fig. 36.3).
- A Biology Lab: for microbiological research and biological space experiments.

At :envihab, the institute will conduct supertargeted research in space and flight physiology, radiation biology, space psychology, operational medicine, biomedical research, and analogous terrestrial situations. A major emphasis of :envihab is to form a closely interrelated network of scientists with industry and the general public. In addition to its cutting-edge facilities, :envihab will serve as a communications centre, focused on outreach and inspiring the next generation of scientists. With its wide range of trend-setting research opportunities, :envihab is ideally suited for exploring the future challenges of human spaceflight, as well as for discovering a host of new applications for improving life on Earth.

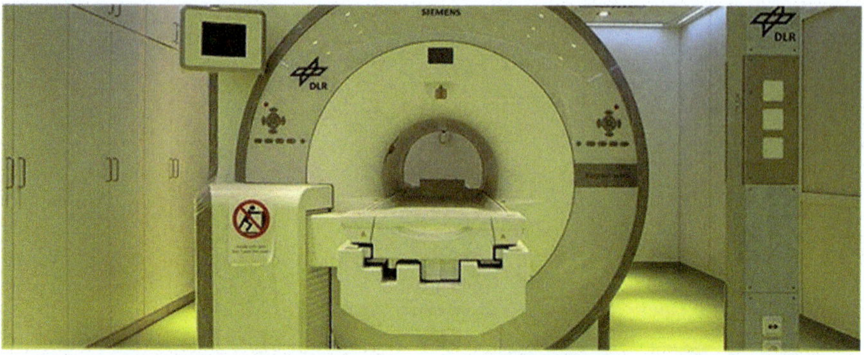

Fig. 36.3 :envihab's PET-MRI (©DLR)

36.3.3 Planica, Olympic Sports Centre

The Planica Nordic Centre is a skiing complex situated at 940 m above sea level. This facility was converted or upgraded to a clinical research facility with support from the EU Regional Fund, and is now owned and maintained by the Slovenian Ministry of Education, Science and Sport. The most prominent feature of the facility is that the ambient air composition on the entire bed rest study floor can be regulated by reducing oxygen levels via a special oxygen dilution system. With that system, any target normobaric hypoxic condition simulating up to 6000 m in altitude can be achieved.

Planica has already hosted multiple bed rest studies in the past, e.g. ESA PECS (Plan for European Cooperating States) supported LunHab and FemHab studies, as well as the FP7 European Commission PlanHab bed rest study (see also Chap. 16). In the framework of these projects, a total of nine bed rest studies were coordinated and successfully completed. A large number of international science teams participated in these studies and have published impressive results in highly ranked journals. The Planica facility maintains a full range of special research equipment, which is available to all scientists involved in studies that are conducted there. The Olympic Sport Centre Planica meets the technical and logistical requirements for conducting bed rest studies, and as such, ESA plans to conduct bed rest studies there in the future and—by introducing physical deconditioning in a hypoxic environment—add a new dimension to "classical" bed rest studies.

36.3.4 Eurac/terraXcube Bolzano/Italy

The terraXcube is a research facility in Bolzano/Bozen, Italy, that has been designed to simulate the Earth's most extreme climatic conditions. Here, the intrinsic relationship between environmental stresses and the physiological and ecological responses of humans and nature, as well as the effects of these stresses on material and product performance can be repeatedly investigated in an experimental setting.

Climate variations can range from storms atop the Himalayas, to extreme arctic cold, to the heat of the North African deserts. The scale of the climactic chambers is such that they can house equipment, machinery, plants and other organisms, as well as products both small and large for extended periods of time. This platform can therefore suit the needs of simulating exploration conditions, while testing gear and human–robotic interactions (text reported and adjusted from the facilities website at https://terraxcube.eurac.edu/).

36.3.5 HERA-Facility

The Human Exploration Research Analog (HERA, see Figs. 36.4 and 36.5) plans for campaigns of incremental duration began in 2014 with four 7-day missions. A campaign is defined as one integrated protocol with one primary mission scenario, consisting of multiple missions in order to meet study subject requirements. Studies designed to utilize the capabilities of HERA are integrated with other investigations on a noninterference basis and run together as one integrated campaign.

Planned mission durations may range from 7 days up to 45 days. The HERA planning schedule currently anticipates 4 missions per year (one per quarter) of 45-day duration in 2018. The HERA is a two-story, four-port habitat unit residing in Building 220 at NASA Johnson Space Center (JSC). It is cylindrical with a vertical axis, and connects to a simulated airlock and hygienemodule.

Fig. 36.4 NASA's Human Exploration Research Analog (© NASA)

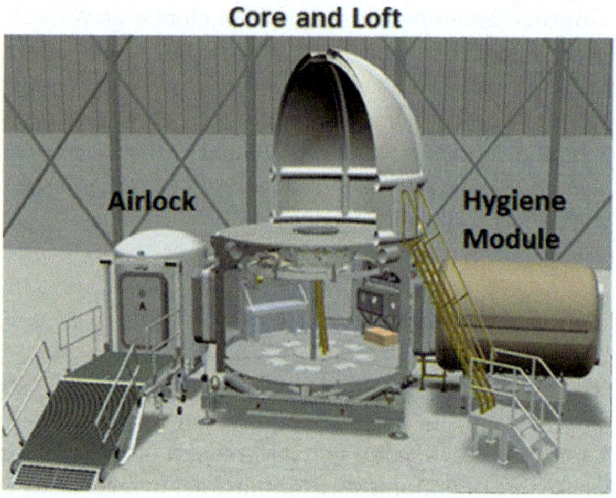

Fig. 36.5 Schematic Representation of HERA (© NASA)

HERA facility capabilities include a network that allows electronic research data and voice communications to be exchanged between the crew and ground controllers. The research data can be securely accessed by remote investigators real-time or near real-time through the JSC Telescience Center (TSC). HERA has a surveillance video and audio system, flight-like timeline and procedure viewer to provide a space mission experience.

Currently, the HERA represents an analog for simulation of isolation, confinement, and remote conditions of exploration mission scenarios. Studies suitable for this analog may include, but are not limited to: behavioral health and performance assessments, communication and autonomy studies, human factors evaluations, human health countermeasures, and exploration medical capabilities assessments and operations (see also https://www.nasa.gov/sites/default/files/files/HRP-HERA-Experiment-Information-Package.pdf).

36.4 Studies in Antarctica

Dozens of countries maintain stations on the Antarctic continent. Most of the stations are located in the coastal areas, only a few deeper on the continent itself. Many are only operated during the Antarctic summer; others are occupied all year-round, which, due to the harsh environmental conditions, typically implies a multiple-month period during which the crew is fully isolated from resupply or evacuation. While the coastal bases are at sea level, inland stations are on higher altitudes. Crew sizes range from less than 10 to more than 1000 (like at the US base McMurdo during summer).

Accordingly, the research questions that can be addressed are diverse, ranging from studies of small autonomous crews to physiological, epidemiological investigations.

In the spaceflight context, factors that make some Antarctic stations very interesting analogs for exploration missions are the changed day/night cycle, the partially extreme isolation and therefore the need for autonomy of the crew, the hostile environment with very limited sensory stimulation, and the realistic operational situation with real dangers, which can never be perfectly simulated. Further, the effects of long-term hypobaric conditions can be tested and controlled by implementing research protocols at inner continental bases (e.g., Concordia station at 3200 m, see Chaps. 16 and 38) and at sea level, as at the Halley VI-station (British Antarctic Survey, BAS, Fig. 36.6), Neumayer III (German Alfred Wegener Institute, AWI, Fig. 36.7) and others which are at intermediate altitude, such as the Belgian

Fig. 36.6 Halley VI Research Station (© BAS)

Fig. 36.7 Neumayer III Research Station (© AWI)

Princess Elisabeth station. This is "bought" through some operational constraints, and through the fact that often the "test subjects" are volunteering for experiments on top of their normal main tasks, so that the overall experimental load and the level of invasiveness is less than in dedicated simulations. Therefore, these two approaches are complementary, and in fact both are required to adequately address this research.

36.5 Underwater Habitats

Diving stations, submarines, and similar underwater habitats are another classical analog for space missions. While, like in any analog or model, limitations to their use exist, they also offer realistic challenges and natural constraints and dangers. In some cases, especially for shorter campaigns, it can be easier in these settings to adopt an operational setup that mimics those used in space missions.

One underwater habitat that has been especially used by NASA is the Aquarius laboratory off the Florida coast. Aquarius is located 5.6 km (3.5 miles) off Key Largo and is deployed at 19 m (62 ft) below the surface (see https://www.nasa.gov/mission_pages/NEEMO/about_neemo.html). In the frame of NEEMO (NASA Extreme Environment Mission Operations, see Fig. 36.8), regular campaigns of a couple of weeks are conducted. These missions are operationally conducted like a space mission, and normally include multiple extra habitat excursions. Aims range from aspects of astronaut training, tests of technologies and tools (e.g., for telemedicine) to tests of operational concepts and provide a very interesting opportunity for research on some physiological immune adaption processes under realistic operational conditions (Strewe et al. 2015).

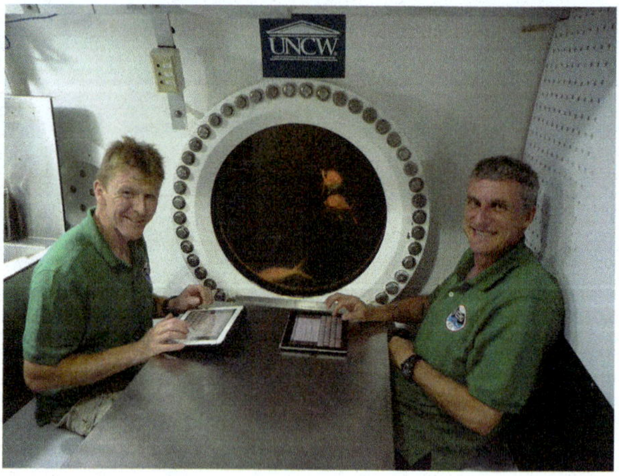

Fig. 36.8 Tim Peake (ESA Astronaut) and Steve Squyres (Cornell University) in the Aquarius Underwater Laboratory (©NASA)

Fig. 36.9 Alexander Gerst (ESA astronaut) during a parabolic flight (©ESA)

36.6 Parabolic Flights

If studying short-term effects of true weightlessness in human beings is of interest, then parabolic flights can be a useful platform. By flying a parabola-shaped trajectory with an airplane, circa 20 s periods of free fall can be produced (see Fig. 36.9). For some studies it can be of additional interest that this is combined with periods of hypergravity of 1.8–2 g. Specifically, with a view to exploration, more opportunities are now becoming available to also perform "lunar" or "Martian" parabolas, which reproduce 0.16 g or 0.38 g, respectively, for up to 30 s, as well as hyper-G parabolas. This is useful as a general gravity-related stress model, for investigating fundamental mechanisms of acute immune cell changes (see Chap. 17; Adrian et al. 2013; Kaufmann et al. 2009, 2011; Tauber et al. 2015), as well as for testing of stress and immune research relevant technologies, procedures before being considered for space application (see also Chap. 27; Sams et al. 1999).

36.7 International Space Station

The International Space Station (ISS) is a huge international laboratory, with major contributions from NASA, ROSKOSMOS, ESA, JAXA, and CSA, orbiting the Earth at an altitude of 350–400 km (Fig. 36.10). The station is permanently occupied since the year 2000, and with six crew members since 2009. American, Russian, European, and Japanese laboratory modules allow a variety of research, including biological and physiological. For that purpose, hardware like ECG, blood pressure instruments, sampling kits, a pulmonary function system, EEG, incubators, freezers, and much more are available as research payloads on board the ISS. In addition to the regular 5–6 months stays on the station—as called increments—double increments have been implemented, first in 2015 with Cosmonaut M. Kornienko and US-astronaut S. Kelly to investigate expanded mission durations on immune (see Chaps. 11–16) and other physiological systems.

Fig. 36.10 ISS: International Space Station as seen from space shuttle Endeavour in February 2010 (©NASA)

Generally life science research on the ISS is very well-coordinated through a working group (International Space Life Science Working Group, ISLSWG) that encompasses all ISS partners except Russia. Regular Announcements of Opportunity are published to gather experiment proposals, and the evaluation process is common for all ISLSWG partners. This coordination and cooperation makes it possible for researchers to rely on any life science payload in orbit, and not be limited to the hardware owned by their respective agency. Thus, like the ISS itself, the life science program represents a major accomplishment in international cooperation.

Next to operational issues, which may constrain specific experiments (e.g., when crews need to sleep shift in preparation for an upcoming vehicle docking), crew time as well as up- and download mass still represent the major constraints for research. Also, the environment in Low Earth Orbit and the current operational and logistical setup are different from what can be expected for exploration missions. Nevertheless, the ISS represents an excellent tool for preparatory research, as it is the only available platform in which humans are living and working in actual weightlessness for extended periods of time, within a real spacecraft.

36.8 Chinese Space Station (CSS)

China is going to start sending parts of its future space station into space as soon as 2020, with the aim of having it up and running by 2022. It will be made of three modules joined in a T shape and will weigh some 66 metric tons (72 tons), which is around one-sixth of the International Space Station. When docked with manned spaceships and cargo vehicles, the station may reach around 100 metric tons (110 tons). The station builds on the knowledge China gathered from its first space station prototype, the space lab Tiangong-1, which came crashing back to Earth in April 2018, after spending 6 years in space.

On the 21st September 1992, the China Manned Space Program (CMSP) was officially approved by the Government of China, a three-step program designed to build a permanently manned Earth-orbiting Space Station.

The first step of this program was to launch a manned spaceship with the aim of building up fundamental capability in human space exploration and space experiments.

The second step was to launch a space laboratory tasked with making technological breakthroughs for extravehicular activities, space rendezvous and spacecraft docking procedures, as well as providing a solution for man-tended space utilization on a certain scale and short-term basis.

The third step was to establish a Space Station with the aim of providing a solution for man-tended-space utilization on a larger scale and longer-term basis.

At present, the first and second steps have been achieved, while the third step of setting up a manned Space Station is well underway.

The Tiangong 3 station is expected to be complete by 2022 with three modules: one core module (launch planned for 2019) plus two science modules in a T-configuration. During the assembly period 2–3 months' missions to Tiangong will be conducted with three crewmembers. Once completed, the Tiangong will be permanently inhabited by three crewmembers, on a 6-month crew rotation (long duration missions).

China expects to have its station in operation for at least 10 years. Up to six astronauts will be able to stay at a time for at least 180 days, and will explore topics such as how space living affects humans, microgravity physics, and material science. Beijing is already portraying its station as a beacon of global cooperation—it has already said it will welcome other countries who want to conduct experiments aboard its space station.

ESA has initiated discussions to determine potential partnership scenarios with China in the area of astronaut training, scientific utilization, and infrastructure. Some initial activities have successfully been conducted.

Engaging with CMSA in a partnership in Low Earth Orbit (LEO) research has the potential to open access to affordable astronaut and payload flight opportunities. In addition, expanding the agency's partnership portfolio answers directly to strategic goals and objectives related to peaceful international cooperation. It is assumed that astronaut and/or payload flights and operational cost will be compensated on a no-exchange-of-funds basis by contributions to the China Manned Space Agency (CMSA) programme. The associated development activities create opportunities for European stakeholders to forge new partnerships in China. In particular a substantial level of collaboration between ESA and CMSA for the Chinese Space Station (CSS) could ensure Europe has continued access to LEO in a post-ISS timeframe—access which could be institutional or commercial.

36.9 Conclusions

This brief description of research platforms is intended to illustrate the many available opportunities for performing exploration preparation science. Crucial for the success of the investigations is to select the most suitable analog for a given research question. If, for example, primarily musculoskeletal unloading is of interest and related countermeasures are evaluated, bed rest or dry immersion studies may be very suitable. If effects related to long-term confinement and isolation are a topic, then isolation simulation studies or analog environments like Antarctic stations may be useful settings. Though simulation studies with less realistic dangers have greater overall control and space fidelity, they are not representative of real-life situations and dangers, and operational constraints, particularly in comparison with analog environments. As a result, it will often be sensible to employ both settings, in order to collect complementary data.

Ultimately, most lines of investigation will require use of the unique space laboratory ISS or CSS or new orbiters (e.g. lunar) for validation or basic data collection. However, as the cost and effort for any flight activity is high and resources like crew time are limited, it is essential that any research project is very well prepared and optimized through the use of adequate ground-based analogs by the time the "ultimate" step to a flight experiment is made. In this way, all the platforms mentioned in this chapter, and the international groups taking advantage of them are all crucial and play an important role toward enabling our next step further out into the solar system.

References

Adrian A, Schoppmann K, Sromicki J, Brungs S, von der Wiesche M, Hock B, Kolanus W, Hemmersbach R, Ullrich O (2013) The oxidative burst reaction in mammalian cells depends on gravity. Cell Commun Signal 11(1):98

Belavý D, Ohshima H, Rittweger J, Felsenberg D (2017) High-intensity flywheel exercise and recovery of atrophy after 90 days bed-rest. BMJ Open Sport Exerc Med 3(1):e000196

Beller G, Belavý D, Sun L, Armbrecht G, Alexandre C, Felsenberg D (2011) WISE-2005: bed-rest induced changes in bone mineral density in women during 60days simulated microgravity. Bone 49(4):858–866

Buehlmeier J, Frings-Meuthen P, Mohorko N, Lau P, Mazzucco S, Ferretti J, Biolo G, Pisot R, Simunic B, Rittweger J (2017) Markers of bone metabolism during 14 days of bed rest in young and older men. Musculoskelet Neuronal Interact 17(1):399–408

Caiani E, Massabuau P, Weinert L, Vaïda P, Lang R (2014) Effects of 5 days of head-down bed rest, with and without short-arm centrifugation as countermeasure, on cardiac function in males (BR-AG1 study). J Appl Physiol 117(6):624–632

Clément G, Bareille M, Goel R, Linnarsson D, Mulder E, Paloski W, Rittweger J, Wuyts F, Zange J (2015) Effects of five days of bed rest with intermittent centrifugation on neurovestibular function. J Musculoskelet Neuronal Interact 15(1):60–68

Clément G, Paloski W, Rittweger J, Linnarsson D, Bareille M, Mulder E, Wuyts F, Zange J (2016) Centrifugation as a countermeasure during bed rest and dry immersion: what has been learned? J Musculoskelet Neuronal Interact 16(2):84–91

Cromwell R, Scott J, Downs M, Yarbough P, Zanello S, Ploutz-Snyder L (2018) Overview of the NASA 70-day Bed Rest Study. Med Sci Sports Exerc 50(9):1909–1919

Demangel R, Treffel L, Py G, Brioche T, Pagano A, Bareille M, Beck A, Pessemesse L, Candau R, Gharib C, Chopard A, Millet C (2017) Early structural and functional signature of 3-day human skeletal muscle disuse using the dry immersion model. J Physiol 595(13):4301–4315

Dillon E, Sheffield-Moore M, Durham W, Ploutz-Snyder L, Ryder J, Danesi C, Randolph K, Gilkison C, Urban R (2018) Efficacy of testosterone plus NASA exercise countermeasures during head-down bed rest. Med Sci Sports Exerc 50(9):1929–1939

Evans J, Knapp C, Goswami N (2018) Artificial gravity as a countermeasure to the cardiovascular deconditioning of spaceflight: gender perspectives. Front Physiol 9:716

Holt J, Macias B, Schneider S, Watenpaugh D, Lee S, Chang D, Hargens A (2016) WISE 2005: aerobic and resistive countermeasures prevent paraspinal muscle deconditioning during 60-day bed rest in women. J Appl Physiol 120(10):1215–1222

Kaufmann I, Schachtner T, Feuerecker M, Schelling G, Thiel M, Choukèr A (2009) Parabolic flight primes cytotoxic capabilities of polymorphonuclear leucocytes in humans. Eur J Clin Invest 39(8):723–728

Kaufmann I, Feuerecker M, Salam A, Schelling G, Thiel M, Choukèr A (2011) Adenosine A2A receptor modulates the oxidative stress response of primed polymorphonuclear leukocytes after parabolic flight. Hum Immunol 72(7):547–552

Kelsen J, Bartels L, Dige A, Hvas C, Frings-Meuthen P, Boehme G, Thomsen M, Fenger-Grøn M, Dahlerup J (2012) 21 Days head-down bed rest induces weakening of cell-mediated immunity – some spaceflight findings confirmed in a ground-based analog. Cytokine 59(2):403–409

Klassen S, De Abreu S, Greaves D, Kimmerly D, Arbeille P, Denise P, Hughson R, Normand H, Shoemaker J (2018) Long-duration bed rest modifies sympathetic neural recruitment strategies in male and female participants. J Appl Physiol 124(3):769–779

Kos O, Hughson R, Hart D, Clément G, Frings-Meuthen P, Linnarsson D, Paloski W, Rittweger J, Wuyts F, Zange J, Gorczynski R (2014) Elevated serum soluble CD200 and CD200R as surrogate markers of bone loss under bed rest conditions. Bone 60:33–40

Koschate J, Thieschäfer L, Drescher U, Hoffmann U (2018) Impact of 60 days of 6° head down tilt bed rest on muscular oxygen uptake and heart rate kinetics: efficacy of a reactive sledge jump countermeasure. Eur J Appl Physiol 118(9):1885–1901

Kramer A, Kümmel J, Mulder E, Gollhofer A, Frings-Meuthen P, Gruber M (2017) High-intensity jump training is tolerated during 60 days of bed rest and is very effective in preserving leg power and lean body mass: an overview of the cologne RSL study. PLoS One 12(1):e0169793

Lee S, Schneider S, Feiveson A, Macias B, Smith S, Watenpaugh D, Hargens A (2014) WISE-2005: countermeasures to prevent muscle deconditioning during bed rest in women. J Appl Physiol 116(6):654–667

Li X, Yang C, Zhu Y, Sun J, Shi F, Wang Y, Gao Y, Zhao J, Sun X (2017) Moderate exercise based on artificial gravity preserves orthostatic tolerance and exercise capacity during short-term head-down bed rest. Physiol Res 66(4):576–580

Navasiolava N, Custaud M, Tomilovskaya E, Larina I, Mano T, Gauquelin-Koch G, Gharib C, Kozlovskaya I (2010) Long-term dry immersion: review and prospects. Eur J Appl Physiol 111(7):1235–1260

Pagel J, Choukèr A (2016) Effects of isolation and confinement on humans-implications for manned space explorations. J Appl Physiol 120(12):1449–1457

Ploutz-Snyder L, Downs M, Goetchius E, Crowell B, English K, Ploutz-Snyder R, Ryder J, Dillon E, Sheffield-Moore M, Scott J (2018) Exercise training mitigates multisystem deconditioning during bed rest. Med Sci Sports Exerc 50(9):1920–1928

Provost R, Zuj K, Arbeille P (2015) 5-Day bed rest: portal and lower limb veins with and without artificial gravity countermeasures. Aerosp Med Hum Perform 86(6):524–528

Sams C, Crucian B, Clift V, Meinelt E (1999) Development of a whole blood staining device for use during Space Shuttle flights. Cytometry 37(1):74–80

Schoenrock B, Zander V, Dern S, Limper U, Mulder E, Veraksitš A, Viir R, Kramer A, Stokes M, Salanova M, Peipsi A, Blottner D (2018) Bed rest, exercise countermeasure and reconditioning effects on the human resting muscle tone system. Front Physiol 9:810

Strewe C, Crucian B, Sams C, Feuerecker B, Stowe R, Choukèr A, Feuerecker M (2015) Hyperbaric hyperoxia alters innate immune functional properties during NASA Extreme Environment Mission Operation (NEEMO). Brain Behav Immun 50:52–57

Sundblad P, Orlov O (eds) (2015) Guidelines for standardization of bed rest studies in the spaceflight context. International Academy of Astronautics, Paris. ISBN: 9782917761342. June

Tauber S, Hauschild S, Paulsen K, Gutewort A, Raig C, Hürlimann E, Biskup J, Philpot C, Lier H, Engelmann F, Pantaleo A, Cogoli A, Pippia P, Layer L, Thiel C, Ullrich O (2015) Signal transduction in primary human T lymphocytes in altered gravity during parabolic flight and clinostat experiments. Cell Physiol Biochem 35(3):1034–1051

Treffel L, Massabuau N, Zuj K, Custaud M, Gauquelin-Koch G, Blanc S, Gharib C, Millet C (2017) Pain and vertebral dysfunction in dry immersion: a model of microgravity simulation different from bed rest studies. Pain Res Manag 2017:1–10

37 Mars500: The First Preparation of Long-Duration Space Exploration Missions—Results and Implications for a Holistic Stress and Immune Research Approach

Thu Jennifer Ngo-Anh, Andrea Rossiter, Alexander Suvorov, Galina Vassilieva, and Vadim Gushin

The opportunities given in this study as experts investigating different organ systems in humans exposed to the same environmental challenge in a coordinated fashion allowed a more holistic view in humans' adaptation. The results obtained thus far highlight the importance of the link of stress, neuro(-humoral) and immune reaction on the one hand, and identifying behavioral, psychological, and biological markers of characteristics that predispose prospective crewmembers to both effective and ineffective behavioral reactions during the confinement of prolonged spaceflight, to inform crew selection, training, and individualized countermeasures.

This chapter will give an overview of the results obtained thus far.

37.1 Introduction

Human exploration and colonization of the solar system is an important focus for the European Space Agency (ESA), which has started on the path to making this a reality in the future. ESA was one of the 14 space agencies that signed up to the Global Exploration Strategy, which emphasizes ESA's and its partners' vision and plan to return to the moon and beyond with the goal of sustained and ultimately self-sufficient human presence beyond Earth. This is an enormous challenge, and one that no nation can undertake on its own.

Within this cooperative framework ESA has developed its strategic goals, which fulfill an important role within global exploration activities whilst at the same time

T. J. Ngo-Anh (✉) · A. Rossiter
Human and Robotic Exploration Programmes (HRE-RS), European Space Agency, Noordwijk, The Netherlands
e-mail: Jennifer.Ngo-Anh@esa.int

A. Suvorov · G. Vassilieva · V. Gushin
State Scientific Center of Russian Federation, Institute of Biomedical Problems, Russian Academy of Sciences, Russian Federation, Moscow, Russia

© Springer Nature Switzerland AG 2020

A. Choukér (ed.), *Stress Challenges and Immunity in Space*,
https://doi.org/10.1007/978-3-030-16996-1_37

providing ESA with the opportunity to make significant advancements in its own science and technology research areas.

The success of any human long-duration exploration spaceflight mission, which is anticipated to be realized in this century, will depend on the ability of crewmembers to remain confined and isolated from Earth much longer than previous missions or simulations, while maintaining the intensity and timing of behavioral activity necessary to accomplish the mission and mitigate the effects of microgravity. To date, a total of six people have spent >1 year in space, with the record of 437 consecutive days on the Mir space station set by Valery Polyakov. The longest Earth-based spaceflight simulation involved four Russians confined in connected hyperbaric chambers for 240 consecutive days (SFINCSS 99 *Simulation of a Flight of International Crew on Space Station—'99*). Antarctic winter-over missions have extended up to 363 days. The most confined habitat available today is the International Space Station, which has been constantly inhabited since November 2000.

Predictions on how prolonged confinement will affect future crewmembers is needed to advance knowledge on spacecraft habitability requirements, crew selection, and behavioral countermeasures during interplanetary missions. Making sure that future international astronaut/cosmonaut/taikonauts crews are prepared mentally and physically for the demands of long-duration exploration missions is imperative to a mission's success.

37.2 The Mars500 Program

In light of this, ESA embarked on an international cooperative project called the Mars500 program. The Mars500 program was financed by ESA's European Programme for Life and Physical Sciences in Space (ELIPS) and involved scientists from across Europe. It was jointly implemented with the State Scientific Center of the Russian Federation—Institute of Biomedical Problems (IBMP) of the Russian Academy of Sciences. Preparations for this study date back to 2004, when IBMP started considering using its facilities for a simulation of a full Martian mission profile (i.e. 520–700 days of isolation) and invited ESA to participate in the program. In the end, this consisted of three isolation studies: a 14-day pilot study (completed in November 2007), a 105-day pilot study (completed in July 2009), and the main 520-day study, simulating a complete human mission to Mars. A multinational crew, composed of six adult male volunteers (three selected by the Russian Federation, two by the European Space Agency, and one by the China National Space Administration; mean age, 31.8 years; range, 27–38 years), entered the isolation facility in the IBMP, Moscow, on June 3, 2010, where they remained in continuous spatial confinement until November 4, 2011 (Fig. 37.1).

During the stay in the spacecraft-like habitat, consisting of four hermetically sealed interconnected modules and one external module to simulate the Martian surface, they performed realistic activities of a round-trip mission to Mars following a weekly work schedule, including, among others, operational and maintenance

Fig. 37.1 Exterior view of the Mars500 facility at the IBMP in Moscow (© ESA)

work and meetings, exercise, as well as scientific experiments covering the areas and subareas of physiology, psychology, biochemistry, immunology, biology, and microbiology, and even simulated emergency events.

During the isolation period the candidates only had personal contact within the group, in addition to voice contact with a simulated control center and family and friends—as would happen in a future human spaceflight mission to Mars. All of their communications with the outside world were subjected to a continuously increasing delay (adjusted by "distance" from Earth) up to 20 min, which is the time it takes for signals to get from Earth to Mars and vice versa. Food and water, as well as any other consumables, supply was carried out in the same way as for space crews, namely in limited amounts. As with a human spaceflight mission, the chosen candidates were free to take certain personal items, in addition to being supplied with books, movies and personal laptops and kept themselves busy with physical exercise or self-studies.

The crew was subjected to a 7-day week, with 2 days off and a rotational system in place to account for night shifts. Nonstandard and emergency situations were simulated to determine the effect of a decrease in work capability, sickness, and failures of the on-board systems and equipment.

During "Mars surface operations" the crew was divided into two groups of three people each. Once the first group exited to the Martian surface, the hatch between the Martian simulation module and the rest of the facility was closed by the second group and only opened again when the Mars surface stay simulation ended (Fig. 37.2).

Fig. 37.2 A typical day for the Mars500 crew during the transit period (© ESA)

37.3 First Results Stemming from the Mars500 Program

ESA's science program during the Mars500 program covered all systems of the human body and demonstrated convincingly that isolation and confinement lead to whole-body effects requiring an all-encompassing holistic view on potential countermeasures. The Mars500 crewmembers acted as subjects in scientific investigations to assess the effect that isolation has on various psychological and physiological aspects, such as stress, hormone regulation and immunity, sleep quality, mood and the effectiveness of dietary supplements. This chapter will give an overview of the results obtained thus far.

37.3.1 Psychology and Performance

In the last few decades, the interest in the effects of stress on mood and performance has increased. One measure of stress is, for instance, loneliness (see also Chaps. 22 and 31). Research on Earth has shown that loneliness is related to social, physical and personal factors. It is not known, however, whether astronauts or cosmonauts will experience loneliness during a long-term spaceflight, and what the effects may be on their mental processes. The Mars500 program confirmed earlier findings on the effect of isolation and confinement on mood, emotional responses and psychological adaptation by use of an objective system (the so-called "wireless group structure monitoring system" which was integrated into a mobile psycho-physiological system) to unobtrusively monitor crew cohesion and possible individual stress reactions (Johannes et al. 2015). Two days per week, each crewmember wore a small sensor that registered the presence and distance of the sensors either worn by the other subjects or strategically placed throughout the isolation facility. Results confirm the initial hypothesis that the registered amount of time spent together during free time was associated with the intensity of personal relationships, resulting in varying levels of team work and crew cohesion.

In addition, long-duration isolation and confinement led to stronger feelings of stress, including loneliness, and had effects on performing and controlling professional tasks. The crew exhibited a stage-changing pattern of psychological adaptation during 520 days of confinement, which is similar to the well-described third-quarter phenomenon (Wang et al. 2014). Of course, there were substantial inter-individual differences within the crew itself, as well as inter-group differences between the crew and mission control (Basner et al. 2014); conflicts with mission control were reported five times more often than conflicts amongst crewmembers (Fig. 37.3).

The results of the Mars500 experiment generally confirmed N. Kanas and D. Manzey's hypothesis about the possibility of the psychological "break-off" phenomenon during a long autonomous mission in the conditions of isolation with loss of direct visual contact with the native planet (Kanas and Manzey 2008). In the Mars500 project, this phenomenon manifested itself in the decreasing dependence of the crew on MCC ("Mission Control") decisions and recommendations, and the increasing number of independent decisions based on the knowledge, values, and goals of the isolated small group (Ushakov et al. 2014). A subsequent decrease of crew motivation after the middle of the Mission (after "Mars surface operations") was also defined, especially in the third quarter. Crewmembers felt increasingly distant from their usual social surroundings, which caused mood deterioration,

Fig. 37.3 Mars500 crewmember Diego Urbina with a computer simulation (© ESA)

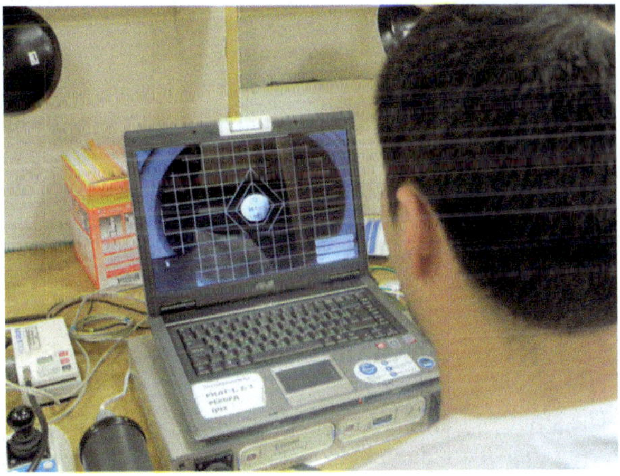

feelings of boredom and loneliness. This phenomenon was aggravated by the influence of communication delays caused by the simulated signal lag effect. Both parties (MCC and the crew) under communication delay not only developed a sense of dissatisfaction with the contact and felt it to be deficient but also were worse at making subsequent decisions. These were sometimes based on conjectures, rather than an understanding of the mutual positions. The lack of the usual encouragement from outside had an adverse effect on the crew's motivation to perform the scheduled operations and, gradually, to communicate with the MCC. It is necessary not only to further investigate this phenomenon but also to seek and develop effective means to mitigate the impact of communication delays on decision making during the execution of space and ground based operations.

The results of this study will help to improve our understanding of the concerns and attitudes of space crew and personnel towards prevention and control of stress and stress-related problems. It will also benefit ongoing research by providing data which could lead to the development of new procedures for crew selection, routine and emergencies operations, and countermeasures for use in future long-duration spaceflight and Antarctic programs.

37.3.2 Hormones, Brain, and Sleep

Previous studies have shown that the ability of a human being to defend oneself against infections exhibits clear changes in response to simulated weightlessness or confinement on Earth. Interestingly, these changes appeared to be associated with the mental stress experienced by the participating individuals. Schneider et al. (2013) demonstrated during the Mars500 program that stress caused by isolation and confinement manifested itself in decreased cortical activity and increased cortisol levels and that exercise was efficient as countermeasure, leading to an increase in cortical activity (Fig. 37.4).

These data are complemented by other investigation and findings (Schneider et al. 2010, 2013, examining prefrontal cortex activity after endurance exercise). It was found that both mood and cognitive performance were increased after a running exercise. Additionally, Gemignani et al. (2014) showed that the increased cortisol levels altered sleep structure and sleep EEG spectral content (indicators for pathological conditions such as primary insomnia or insomnia associated with depression)—one crewmember developed a persistent sleep onset insomnia with ratings of poor sleep quality, which resulted in chronic partial sleep deprivation, elevated ratings of daytime tiredness, and frequent deficits in behavioral alertness (Basner et al. 2014).

37.3.3 Salt, Hormones, and the Immune System

The strictly controlled conditions of the Mars500 program also allowed for the carrying out of fundamental metabolic studies (supported by IBMP and the German

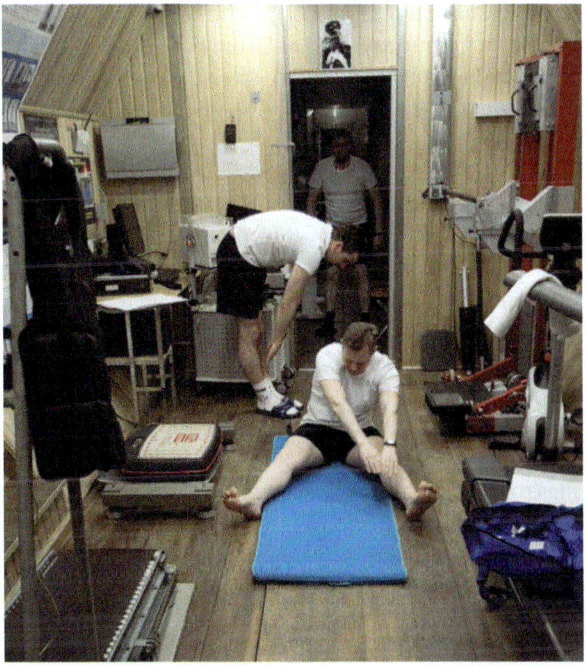

Fig. 37.4 Crewmembers performing their daily exercise (© ESA)

Space Life Sciences Programme) (Lerchl et al. 2015; Yi et al. 2015a, b, c; Birukov et al. 2016). Salt-driven changes in mineralocorticoid and glucocorticoid urinary excretion on day-to-day osmolyte and water balance were measured by exposing participating volunteers to three different regimes of salt intake levels (12, 9, or 6 g/day) via strictly controlled dietary menus, while maintaining all other nutrients constant. Across all three levels of salt intake, half-weekly and weekly rhythmical mineralocorticoid release promoted free water reabsorption via the renal concentration mechanism. Mineralocorticoid-coupled increases in free water reabsorption were counterbalanced by rhythmical glucocorticoid release, with excretion of endogenous osmolyte and water surplus by relative urine dilution. A 6-g/day increase in salt intake decreased the level of rhythmical mineralocorticoid release and elevated rhythmical glucocorticoid release. The projected effect of salt-driven hormone rhythm modulation corresponded well with the measured decrease in water intake and an increase in urine volume with surplus osmolyte excretion. The study therefore showed that humans regulate osmolyte and water balance by rhythmical mineralocorticoid and glucocorticoid release, endogenous accrual of surplus body water, and precise surplus excretion.

Interestingly, this controlled dietary study also showed that immune function was changed; subjects on the high-salt diet of 12 g/day displayed a significantly higher number of immune cell monocytes compared with the same subjects on a lower-salt diet, and correlation test revealed a strong positive association between salt-intake levels and monocyte numbers. The decrease in salt intake was accompanied by reduced production of proinflammatory markers (Yi et al. 2015a, b, c).

These results suggest that in healthy humans high-salt diet has a potential to elicit excessive immune responses, which can be damaging to immune homeostasis, and a reduction in habitual dietary salt intake may induce potentially beneficial immune alterations.

37.3.4 Confinement and Hypersensitivity

The immune system is always shaped by its environment. Getting exposed to such environments as the Mars500 one can lead to immediate reactions, but with increasing duration of exposure, the effects of a return to the normal environment seem to be a strong stressor to the immune system. Since epidemiologic studies have reported the association between exposure to a new environment, as in the course of migration, and an increased risk of atopy/asthma (Bråbäck et al. 2011; Cabieses et al. 2014; Gold and Acevedo-Garcia 2005), and reports from space crews returning from long-duration space missions have reflected some similar patterns (Crucian et al. 2016, 2018), the long-duration isolation and confinement condition of the Mars500 setting has provided a unique setting to gather information on the role of environmental (re-)exposures in the immune responses toward hypersensitivity reactions, and the potential development of immune- allergic or autoimmune type diseases.

It was observed that cellular immune responses toward known and so-called recall antigens were affected in this special migration study—"migration from space to Earth" (Yi et al. 2015a, b, c)—especially when the crew completed their mission and the hatches were reopened after 520 days. These responses might be either protective and an appropriate alert response when reexposed to "Earths' normal environment" or might be triggering a damaging role, thereby enhancing the risk to overshooting immune responses as seen in autoimmune and/or allergic diseases. These effects lasted up to 3 months. Interestingly, comparable results had been observed in the SFINCSS 99 study with elevated skin inflammatory responses to similar recall antigens placed intra-cutaneously (Choukèr et al. 2002) after long-duration isolation and confinement. These observations were complemented in studies ran in Antarctica, mirroring immune-sensitization in crew during deployment at the Concordia station at higher altitude, though a follow up when they were reexposed to "normal" environments after isolation was not investigated at that time (see also Chap. 16).

37.3.5 Confinement and Crew Microbiota

The Mars500 program, the longest ground-based space simulation ever, provided scientists with a unique opportunity to trace the crew microbiota over 520 days of isolated confinement.

Wherever humans live, microorganisms populate the habitat. Not only does a specific microbe community develop but also a particular 'Mobile Genetic Elements' pool through which the microbial population can exchange information. Several studies with cosmonauts have shown that during long-term stays on the International Space Station (ISS), the number of opportunistic pathogens increased while the population of certain protective micro-organisms decreased in the skin and intestinal microflora (see also Chaps. 25 and 34).

Prolonged human confinement in isolated habitats, often combined with particular features of waste disposal, personal hygiene, weightlessness, high oxygen content, and conditions such as localised high temperature, humidity, and concentrations of metabolites influence the microflora population of the crewmembers and the habitat. This situation can potentially result in the undesired accumulation and proliferation of microbes on structural materials of the interior of the habitat (metals, polymers) and systems of life support (water tanks, air filters, etc.), which may cause a risk for crew infection and material deterioration, possibly resulting in hardware and equipment malfunctioning.

In the study from the group of Turroni et al. (2017), the microbial population present and developing in the Mars500 habitat was monitored. Microbial samples were taken from all possible reservoirs of microbes such as the atmosphere, surface, water, food, waste and greenhouse reservoirs and the crewmembers, at several time points during the isolation period (see also Fig. 37.5). Data showed that even under the strictly controlled conditions of an enclosed environment, the human gut

Fig. 37.5 Mars500 crewmember Romain Charles taking an air sample for the microbial experiments (© ESA)

microbiota is inherently dynamic, capable of shifting between different steady states, typically with rearrangements of autochthonous members. Notwithstanding a strong individuality in the overall gut microbiota trajectory, some key microbial components showed conserved temporal dynamics, with potential implications for the maintenance of a health-promoting, mutualistic microbiota configuration. Sharing life in the confined habitat of Mars500 does therefore not affect the resilience of the individual gut microbial ecosystem, even in the long term. However, the temporal dynamics of certain microbiota components should be monitored when programing future mission simulations and real space flights, to prevent breakdowns in the metabolic and immunological homeostasis of the crewmembers.

The results from the psychological, neurohumoral, and immune studies and in conjunction with research in other isolation conditions (e.g. Antarctica, see also Chaps. 16, 36 and 38) will help to provide a better estimate of the overall effects of extreme long-term confinement on brain functions, sleep disturbances and immune responses and provide the basis for the development of pharmacological tools to counter unwanted immunological side effects during long-duration space missions. The stress due to isolation and confinement is expected to cause an increase in cortisol and a disturbance of circadian rhythms, eventually creating an impact upon appetite-regulating hormones, the immune system and the so-called hypothalamic-pituitary-gonadal axis, which plays a critical part in reproductive and immune system regulation.

37.3.6 Confinement, Activity and Muscle

The Mars500 program confirmed that prolonged confinement is a form of physical inactivity and is associated with adaptations in the neuromuscular system (Belavý et al. 2013). Noticeable deconditioning was expected for muscle function, mental and physical health, and well-being of the crew. To this end, the physical fitness of the candidates was monitored before, during, and after the isolation period within the chamber. They followed a defined set of test procedures at regular intervals which monitored the development of cardiovascular and pulmonary function, muscle function, force, and power as well as bone mass and strength. The Mars500 crewmembers experienced significant loss of quadriceps'/hamstring's maximal voluntary isokinetic force but not calf maximal voluntary contraction (Gaffney et al. 2017). Collectively, these data suggest that muscles with predominantly type I fibres were affected less by isolation compared to type II dominant muscles. Multifunctional Dynamometer for space and whole-body vibration afforded the best protection against isolation-induced loss of strength and thus may have virtue in exploration class missions, these findings are underlined by a study conducted by Meigal and colleagues (2016) for which six countermeasures to reduce muscle atrophy were shown to be successful (Fig. 37.6).

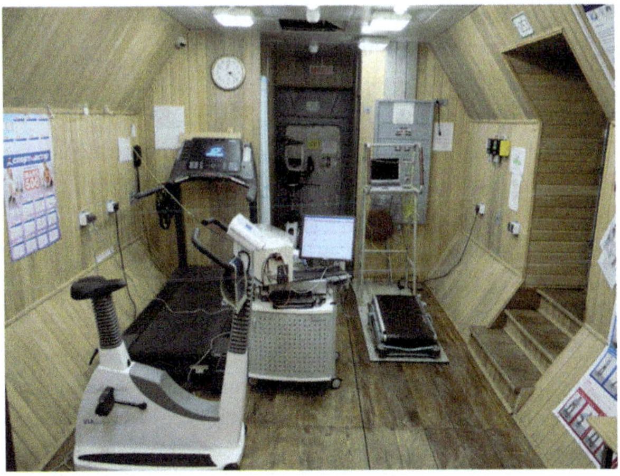

Fig. 37.6 Mars500's gym facilities (© ESA)

37.3.7 Cardiovascular System

Space missions have shown to induce a series of slowly reversible physiological adaptations resembling aging. It is not yet clear how much is caused by exposure to weightlessness and how much is due to environmental stress. It is crucial to know the extent to which long-term isolation/confinement can cause adverse adaptations in the immune system, hormone regulation and metabolism especially since chronic stress can cause physiological changes leading to such conditions as insulin resistance, inflammation, and clogged arteries (atherosclerosis). The Mars500 program offered the possibility of evaluating the principal effects of hormonal and cardiovascular control, for example, by controlling dietary salt, as realized by Professor Titze and colleagues (see also Sect. 37.3.3), and supporting the concept that salt is causally responsible for increases in blood pressure in healthy humans (Rakova et al. 2013, Yi et al. 2015a, b, c). Moreover, cardiovascular changes were also affected by isolation and confinement (Arbeille et al. 2014). Under normal circumstances heart rate is influenced by the sympathetic and parasympathetic nervous systems, which in basic terms influence accelerated and decelerated bodily activities respectively. They determine heart rate and strength of contraction and adapt these to different needs during our daily activities. The control of heart rate receives important feedback information via the baroreflex mechanism (the relationship between heart rate and blood pressure). This mechanism monitors blood pressure and adjusts heart rate to maintain a stable blood pressure within healthy limits. The results from the study led by A. Aubert evaluated the effect of confinement and isolation on changes in the psychological wellbeing of the Mars500 crew and correlated mood changes with changes in cardiac regulation and cardiopulmonary function. Vessel images from the carotid artery as well as the femoral artery obtained by Doppler sonography showed *an increase* in main peripheral arterial diameter and wall thickness and main vein size. In the absence of other influencing factors such as radiation, microgravity, temperature and air composition, those cardiovascular changes can therefore be attributed to stress due to isolation and confinement and need to be taken into account when planning for long-duration human exploration missions. Interestingly, circadian heart rate variability was shown to progressively decrease and dampen during confinement (Vigo et al. 2013).

Also the rhythms of the human autonomous nervous system and its seasonal dynamics have been investigated and stationary and nonstationary intervals of tonicity of the autonomous nervous system within the first year of the experiment. The maximal parasympathetic tonus of the autonomous nervous system was observed in the spring and/or fall seasons (Demin et al. 2013). Additional impact on the cardiorespiratory system was exerted by the physical activity of testers. In the Mars500 program, various exercise machines changed every 2 months and there were two periods of 1 month when exercise machines were not used. As a result changes of physical working capacity (PWC170) were observed, reflecting their state of health.

37.3.8 Metabolism and Gastrointestinal System

The impact of confinement and isolation on body composition, glucose metabolism/insulin (HOMA-IR) resistance and adipokine levels was assessed in the precursor study (Mars105) and during the Mars500 experiment. The results of these studies indicate what countermeasures will be necessary and can be used for the development of a well-focused exercise program (Strollo et al. 2014, 2018).

Additionally, the effect of isolation conditions on gastrointestinal (GI) motility and inflammation state of the GI tract was assessed by means of the developed integrated approach. This involves the use of breath tests based on the oral administration of labeled or hydrogenous substrates containing ^{3}CC, followed by the detection of their metabolites ($^{33}CO_2$ or H_2) during breathing and the measurement of fecal calprotectin by a cassette-type lateral flow immunoassay. The obtained data points to the fact that a combination of isolation, stress and dietary factors (i.e., prolonged nutrition with canned and preserved foods) can contribute to the onset of pathological processes (Roda et al. 2013).

37.4 Conclusion

The Mars500 program offered and yielded remarkable opportunities for science. It focused on the first, basic, human factor-oriented steps on a sustainable pathway to Mars. The Mars500 program and its scientific outcomes have been recognized as an important milestone of Europe's pathway to exploration as it gathered unique, high-fidelity data, knowledge and experience to help prepare one day for a real mission to Mars.

The participants acted as subjects in scientific investigations to assess the effect that isolation has on various psychological and physiological aspects. The unique opportunities given in this study as by experts investigating different organ systems in humans exposed to such a standardized environmental challenge (the "Exposome") in a coordinated fashion, have allowed a more holistic view in humans' adaptation to such a simulated very long-duration mission to Mars.

The knowledge gained during the Mars500 program on the effect of this stressful environment, the consecutive hormone regulation and changes of immunity, sleep quality, mood and the effectiveness of dietary supplements have altogether become invaluable in providing the basis for the potential development of countermeasures. The latter can deal with any unwanted side effects of such a mission—and also help in astronaut selection procedures, and at a modest expense. Obviously the Mars500 program had its limitations, naturally, microgravity, radiation, and threat-to-life, three important physiological and psychological stressors that will be encountered during exploration-type missions—could not be simulated in the Mars500 program, which restricts the generalizability of the findings to long-duration space-exploration missions, but makes it still very complimentary when the study protocols are replicated and seen together with other Earth analog environments.

The crew was male only, so one cannot make inferences about female-only or mixed crews. However, the study did help to determine key psychological and physiological effects of being in such an enclosed environment for such an extended period of time. For instance, the effect of isolation and confinement on daily crew life and operational capabilities, the need for full autonomy and resourcefulness, the isolation, the interaction with fellow crewmembers and other aspects.

37.5 Outlook

ESA and other space agencies have a long history of conducting and participating in isolation and confinement studies. Substantial experience and achievements in assessing the risks for humans in these space analog environments have been gained. Nevertheless, the currently available database is too small and anecdotal to derive definite risk assessments, risk preventions and mitigation strategies. Further research is needed to assess the psychological and physiological issues associated with human long-duration space exploration missions in order to best select and prepare future crews and to develop appropriate countermeasures.

Exploratory missions to the moon and Mars, including the establishment of a permanently crewed base on the lunar surface, will add a new dimension to human spaceflight. It will require taking into account the distance of travel and subsequent remoteness of the crew, the radiation environment, the gravity levels, the duration and mission scenario, and the level of confinement and isolation the crews will be exposed to. For such long-duration exploration missions, the physical, psychological, and physiological challenges presented by the long distances of travel, the duration of time spent dependent on automated life-support systems, the degree of isolation and confinement, and the lack of short-term rescue possibilities in case of emergencies, will be unlike any that humans have ever undertaken before. This will raise the importance of addressing several health issues, including those related to individual and crew performance and well-being, as well as psychological and physiological issues. These factors are expected to become possible limiting factors to human adaptability during these missions and therefore need to be efficiently prevented, mitigated and counteracted.

In order to protect astronauts from the negative effects of isolation and confinement and in order to ensure mission success, the factors that promote or threaten crew cohesion, as well as individual and crew performance and well-being need, to be identified. Since the questions that need to be addressed are manifold and multifactorial, a structured step-by-step approach will need to be put in place to retrieve valid and comparable scientific data within a reasonable and realistic time frame in preparation for future human long-duration exploration missions.

Isolation and confinement studies conducted by ESA and its partners will allow the development of integrative tools and concepts to prevent and mitigate health risks related to performance, emotional, cognitive, sensory, and psychosocial degradation of the crewmembers, as well as to maintain the crew's cohesion and their professional and social skills during the mission.

We want to learn more about how humans are physically able to explore the Solar System, and reduce risks that humans face in space. We also want to use scientific results to determine how explorers will use local resources, to maximize the use of what we find around us by "living off the land." This would be a shift from the current paradigm of taking everything with us from Earth. Additionally, these resources have high intrinsic scientific value. The key disciplines involved in this theme include human physiology, behavior and performance, life sciences, planetary geology, astrobiology, and engineering sciences related to life support and spaceflight.

The search for knowledge is an essential part of why humans explore. Scientific exploration addresses overarching questions that society seeks to answer: "Where did we come from?", "Are we alone in the Universe?", "What will happen to us in the future?", "How far can humans safely travel in space?" (see also Chap. 39). This search for knowledge is best achieved by a combination of research platforms and opportunities for performing exploration preparation science, including isolation and confinement studies. The value of data obtained from these studies to future mission planning cannot be overstated.

References

Arbeille P, Provost R, Vincent N, Aubert A (2014) Adaptation of the main peripheral artery and vein to long term confinement (Mars 500). PLoS One 9(1):e83063

Basner M, Dinges D, Mollicone D, Savelev I, Ecker A, Di Antonio A, Jones C, Hyder E, Kan K, Morukov B, Sutton J (2014) Psychological and behavioral changes during confinement in a 520-day simulated interplanetary mission to Mars. PLoS One 9(3):e93298

Belavý DL, Gast U, Daumer M, Fomina E, Rawer R, Schießl H, Schneider S, Schubert H, Soaz C, Felsenberg D (2013) Progressive adaptation in physical activity and neuromuscular performance during 520d confinement. PLoS One 8(3):e60090

Birukov A, Rakova N, Lerchl K, Engberink RH, Johannes B, Wabel P, Moissl U, Rauh M, Luft FC, Titze J (2016) Ultra-long-term human salt balance studies reveal interrelations between sodium, potassium, and chloride intake and excretion. Am J Clin Nutr 104(1):49–57

Bråbäck L, Vogt H, Hjern A (2011) Migration and asthma medication in international adoptees and immigrant families in Sweden. Clin Exp Allergy 41(8):1108–1115

Cabieses B, Uphoff E, Pinart M, Antó J, Wright J (2014) A systematic review on the development of asthma and allergic diseases in relation to international immigration: the leading role of the environment confirmed. PLoS One 9(8):e105347

Choukèr A, Smith L, Christ F, Larina I, Nichiporuk I, Baranov V, Bobrovnik E, Pastushkova L, Messmer K, Peter K, Thiel M (2002) Effects of confinement (110 and 240 days) on neuroendocrine stress response and changes of immune cells in men. J Appl Physiol 92(4):1619–1627

Crucian B, Babiak-Vazquez A, Johnston S, Pierson D, Ott C, Sams C (2016) Incidence of clinical symptoms during long-duration orbital spaceflight. Int J Gen Med 9:383–391

Crucian B, Choukèr A, Simpson R, Mehta S, Marshall G, Smith S, Zwart S, Heer M, Ponomarev S, Whitmire A, Frippiat J, Douglas G, Lorenzi H, Buchheim J, Makedonas G, Ginsburg G, Ott C, Pierson D, Krieger S, Baecker N, Sams C (2018) Immune system dysregulation during spaceflight: potential countermeasures for deep space exploration missions. Front Immunol 9:1437

Demin A, Dyachenko A, Ivanov A, Orlov O, Suvorov A (2013) Instrumental monitoring of the autonomous nervous system in the Mars-520 experiment. Biomed Eng 47(2):86–90

Gaffney C, Fomina E, Babich D, Kitov V, Uskov K, Green D (2017) The effect of long-term confinement and the efficacy of exercise countermeasures on muscle strength during a simulated mission to Mars: data from the Mars500 study. Sports Med Open 3(1):40

Gemignani A, Piarulli A, Menicucci D, Laurino M, Rota G, Mastorci F, Gushin V, Shevchenko O, Garbella E, Pingitore A, Sebastiani L, Bergamasco M, L'Abbate A, Allegrini P, Bedini R (2014) How stressful are 105days of isolation? Sleep EEG patterns and tonic cortisol in healthy volunteers simulating manned flight to Mars. Int J Psychophysiol 93(2):211–219

Gold D, Acevedo-Garcia D (2005) Immigration to the United States and acculturation as risk factors for asthma and allergy. J Allergy Clin Immunol 116(1):38–41

Johannes B, Sitev A, Vinokhodova A, Salnitski V, Savchenko E, Artyukhova A, Bubeev Y, Morukov B, Tafforin C, Basner M, Dinges D, Rittweger J (2015) Wireless monitoring of changes in crew relations during long-duration mission simulation. PLoS One 10(8):e0134814

Kanas N, Manzey D (2008) Space psychology and psychiatry, 2nd edn. Springer Science + Business Media; Microcosm Press, New York, NY; El Segundo, CA

Lerchl K, Rakova N, Dahlmann A, Rauh M, Goller U, Basner M, Dinges DF, Beck L, Agureev A, Larina I, Baranov V, Morukov B, Eckardt KU, Vassilieva G, Wabel P, Vienken J, Kirsch K, Johannes B, Krannich A, Luft FC, Titze J (2015) Agreement between 24-hour salt ingestion and sodium excretion in a controlled environment. Hypertension 66(4):850–857

Meigal A, Fomina E (2016) Electromyographic evaluation of countermeasures during the terrestrial simulation of interplanetary spaceflight in Mars500 project. Pathophysiology 23(1):11–18

Rakova N, Jüttner K, Dahlmann A, Schröder A, Linz P, Kopp C, Rauh M, Goller U, Beck L, Agureev A, Vassilieva G, Lenkova L, Johannes B, Wabel P, Moissl U, Vienken J, Gerzer R, Eckardt K, Müller D, Kirsch K, Morukov B, Luft F, Titze J (2013) Long-term space flight simulation reveals infradian rhythmicity in human Na+ balance. Cell Metab 17(1):125–131

Roda A, Mirasoli M, Guardigli M, Simoni P, Festi D, Afonin B, Vasilyeva G (2013) Non-invasive panel tests for gastrointestinal motility monitoring within the MARS-500 Project. World J Gastroenterol 19(14):2208

Schneider S, Brümmer V, Carnahan H, Kleinert J, Piacentini MF, Meeusen R, Strüder HK (2010) Exercise as a countermeasure to psycho-physiological deconditioning during long-term confinement. Behav Brain Res 211(2):208–214

Schneider S, Abeln V, Popova J, Fomina E, Jacubowski A, Meeusen R, Strüder H (2013) The influence of exercise on prefrontal cortex activity and cognitive performance during a simulated space flight to Mars (MARS500). Behav Brain Res 236:1–7

Strollo F, Vassilieva G, Ruscica M, Masini M, Santucci D, Borgia L, Magni P, Celotti F, Nikiporuc I (2014) Changes in stress hormones and metabolism during a 105-day simulated Mars mission. Aviat Space Environ Med 85(8):793–797

Strollo F, Macchi C, Eberini I, Masini M, Botta M, Vassilieva G, Nichiporuk I, Monici M, Santucci D, Celotti F, Magni P, Ruscica M (2018) Body composition and metabolic changes during a 520-day mission simulation to Mars. J Endocrinol Invest 41(11):1267–1273

Turroni S, Rampelli S, Biagi E, Consolandi C, Severgnini M, Peano C, Quercia S, Soverini M, Carbonero F, Bianconi G, Rettberg P, Canganella F, Brigidi P, Candela M (2017) Temporal dynamics of the gut microbiota in people sharing a confined environment, a 520-day ground-based space simulation, MARS500. Microbiome 5(1):39

Ushakov I, Vladimirovich M, Bubeev Y, Gushin V, Vasil'eva G, Vinokhodova A, Shved D (2014) Main findings of psychophysiological studies in the Mars 500 experiment. Herald Russ Acad Sci 84(2):106–114

Vigo D, Tuerlinckx F, Ogrinz B, Wan L, Simonelli G, Bersenev E, Van den Bergh O, Aubert A (2013) Circadian rhythm of autonomic cardiovascular control during Mars500 simulated mission to Mars. Aviat Space Environ Med 84(10):1023–1028

Wang Y, Jing X, Lv K, Wu B, Bai Y, Luo Y, Chen S, Li Y (2014) During the long way to Mars: effects of 520 days of confinement (Mars500) on the assessment of affective stimuli and stage alteration in mood and plasma hormone levels. PLoS One 9(4):e87087

Yi B, Matzel S, Feuerecker M, Hörl M, Ladinig C, Abeln V, Choukèr A, Schneider S (2015a) The impact of chronic stress burden of 520-d isolation and confinement on the physiological response to subsequent acute stress challenge. Behav Brain Res 281:111–115

Yi B, Rykova M, Jäger G, Feuerecker M, Hörl M, Matzel S, Ponomarev S, Vassilieva G, Nichiporuk I, Choukèr A (2015b) Influences of large sets of environmental exposures on immune responses in healthy adult men. Sci Rep 5(1):13367

Yi B, Titze J, Choukèr A (2015c) Dietary sodium intake and risk of cardiovascular disease. JAMA Intern Med 175(9):1578

Exploration Class Missions on Earth: Lessons Learnt from Life in Extreme Antarctic Isolation and Confinement

38

Alex P. Salam

38.1 Introduction

Sometime this century humans will attempt to set foot on the planet Mars. This will mark a defining moment in the history of human exploration. Never before will a journey of this magnitude or complexity have been attempted. Apart from the technical challenges, the physical and psychological threats to the crew will be without precedent. Severe derangements of numerous physiological processes will occur. The psychological state of the crew will, without a doubt, also heavily influence the mission. Prolonged separation from loved ones and our home, planet Earth, will place immense emotional strains on the crew. A life devoid of Earth's colors, sounds, and smells, will lead to sensory deprivation and monotony. Altered circadian rhythms and disturbed sleep patterns will result in cognitive and motor impairment. Endless months confined to a small habitable volume will test the patience, stability, and diplomacy of even the most highly trained astronauts. Overall, a state of constant physical and psychological stress will loom over the crew, and might indeed put their lives at risk.

I spent a year at the Concordia Antarctic research station in 2009 as a researcher in human biology and medicine for the European Space Agency, investigating the consequences of chronic stress (and chronic hypoxia) on immunity and sleep. Concordia is one of the most isolated research stations on the planet and shares many stressor characteristics with long duration–deep space missions (LDDS). I discuss, in brief, some of the stressors present at Concordia and their consequences and similarities with LDDS missions, and potential countermeasures to disturbed behavioral health and performance.

A. P. Salam (✉)
Nuffield Department of Medicine, University of Oxford, Oxford, UK

Institut Polaire Emile Victor (French Polar Institute), Plouzane, France (former)
e-mail: alexsalam@doctors.org.uk

© Springer Nature Switzerland AG 2020

A. Choukér (ed.), *Stress Challenges and Immunity in Space*,
https://doi.org/10.1007/978-3-030-16996-1_38

38.2 Concordia Station

Concordia station is located at Dome Charlie (Dome C) at an elevation of 3232 m, 1000 km inland from the Antarctic coast, on the high Antarctic plateau. Dome C is a vast barren ice desert that stretches, uninterrupted, for hundreds of kilometers in all directions. It is one of the coldest, driest, most inhospitable, and inaccessible regions on the planet. This alien world was my home for 379 days. The station serves as a research platform for glaciology, astronomy, atmospheric chemistry, seismology, geomagnetism, and human biology. It is jointly operated by the French Polar Institute (IPEV) and the Italian Polar Institute (PNRA). There are only two other permanent stations on the high Antarctic plateau, the Russian Vostok station and the American South Pole station. During the summer season (mid-November to mid-February) the station is lively and there are on average 50 international scientists and technical personnel. Planes come and go, transporting people to and from the station. Snow tractors and snow ploughs arrive from the coastal station, Dumont D'Urville (distance 1200 km), bringing supplies in preparation for the winter. The outdoor temperatures are relatively mild, averaging −30 °C, and it is possible to explore the surroundings of the station before the bitter winter begins; the seismology ice cave, the astronomy platforms, the glaciology shelter, and the climatology tower. The 24-h intense sunlight encourages late "night" gatherings and the ambience is social. People are busy, occupied with station and equipment repairs, logistics, and science (Fig. 38.1).

Fig. 38.1 Concordia station. The two principal towers can be seen; the "noisy" tower which houses life support, recreation, food storage and preparation, and the mess; the "quiet" tower which houses sleeping quarters, laboratories, and medical

Throughout the last week of January and the first week of February, the summer crew begin to leave and numbers dwindle until the last flight approaches. The mood and feel of the station changes significantly, becoming much more still and composed. On February 8th 2009, at 09h02 UTC, the 12 of us staying for the fifth Concordia winter-over, a mix of French, Italians, and myself, all stood side by side, waiving hesitantly as the last Douglas DC3 revved into action, kicking a mist of snow and fine ice particles towards us. A few minutes later the plane roared over our heads and rushed away into the vast expanse. The buzz of the engines dissipated, the defined form of the plane became a blur, until eventually it disappeared in an instant. This was the start of 9 months of isolation during which there was no possibility of evacuation or deliveries. The next nearest human presence was the crew of the Russian Vostok station, 560 km away. Because of its location and the environmental extremes, Concordia station is completely isolated during the winter season from mid-February to mid-November. It is simply too dangerous to attempt airplane landings or pass-overs, and in all likelihood technically impossible during the depths of winter. Telecommunications were limited and consisted of 2–3 satellite connections during the day for emails to be sent and received. Our personal email accounts were limited to 1 Mb per connection. There was no real-time data transfer, although we did have occasional use of satellite telephone at our own expense. As individuals and as a crew, we had to be self-reliant and semiautonomous during the winter period.

On February 14th, the sun set for the first time in 3 months, marking the end of the polar summer and the start of the polar winter. It was a fleeting moment; within a few minutes the sun was back above the horizon. Over the course of several weeks, however, night and day equalized and temperatures began to drop. By March, outside temperatures were averaging −45 °C. By April, night had overtaken day and outside temperatures were averaging −60 °C. On May 6th the sun finally disappeared beneath the horizon for three long, dark, cold months. Concordia experiences cycles of total light or total darkness because of its location deep within the polar circle. Mid-June was the coldest and darkest period of the winter, in many ways. Temperatures regularly approached −80 °C (not counting wind-chill) and people were generally confined to the station for long stretches of time. On the morning of August the 11th, the sun shot a crimson flame low into the sky as it made its first, but short-lived, appearance in 3 months. By November it had returned to its 24-h blinding fury. On November 17th, early afternoon, the exact same plane that had left us 9 months ago returned, slowly transforming from a mere glistening speckle in the desolate distance to a wonderful, thundering, colorful beast arriving just a few meters from our feet

38.3 Stressors, Consequences, and Parallels with Long Duration–Deep Space Missions

The main stressors present at Concordia station are shown in Fig. 38.2a. The list is not exhaustive. Apart from some of the miscellaneous stressors, all of the stressors listed will also occur during LDDS missions, although their intensity and the level

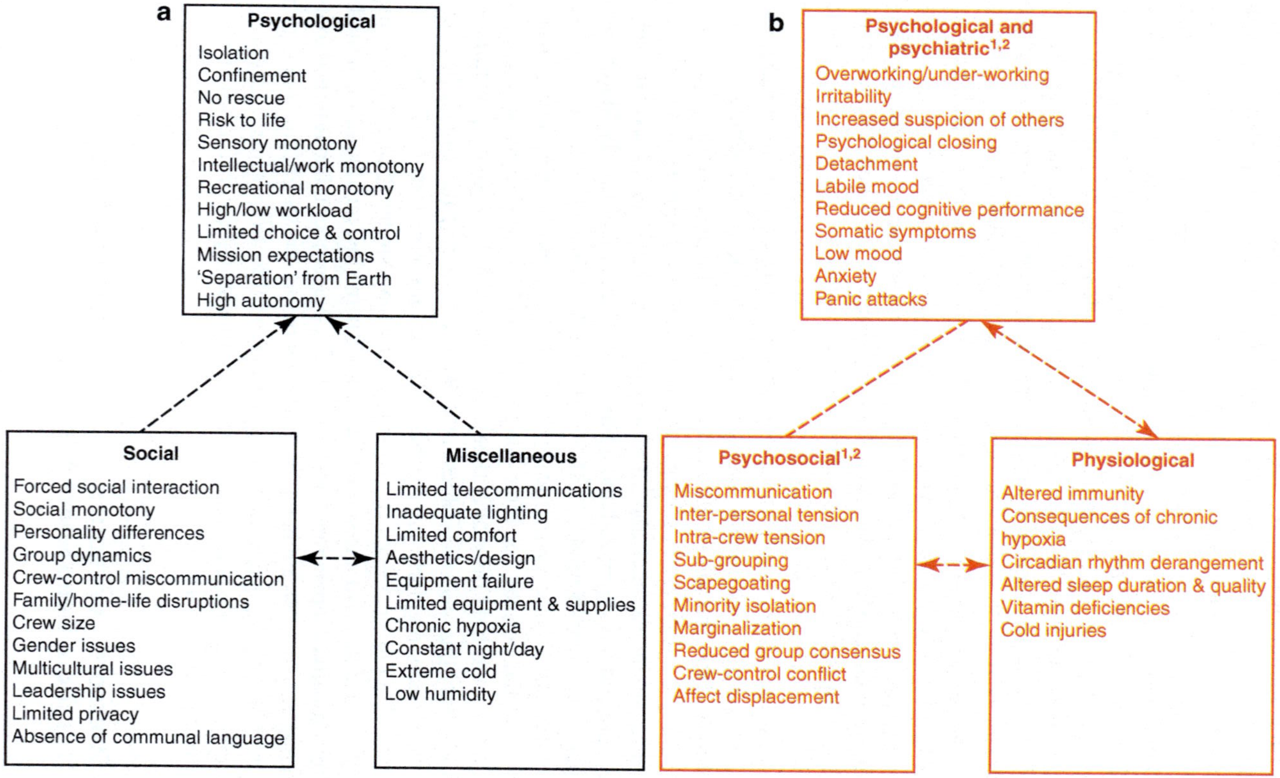

Fig. 38.2 (**a**) Stressors present during a Concordia winter-over; not listed in any particular ranking order. Many of the stressors are interrelated. Miscellaneous includes human factors, environmental factors, and habitat factors. (**b**) Consequences of the stressors; not listed in any particular ranking order. Many of the consequences are interrelated. Stressors and their sequelae feedback, in many cases, in a bidirectional manner (Palinkas and Suedfeld 2008; Lugg 2005)

of risk associated with them will vary. The stressors are listed under specific categories but in reality some could be listed under more than one category, and their classification is fluid. Many of the stressors are interrelated. For example, some of the social stressors can influence the psychological stressors, and vice-versa. The intensity and numbers of stressors that an individual can tolerate is time dependent. The main consequences of the stressors are shown in Fig. 38.2b. Again, many of the sequelae are interrelated. Importantly, some of the stressors and sequelae feedback in a bidirectional manner. As a result, a single sequela experienced by a single individual can eventually have a profound impact on the group. For example, an individual feels low in mood due to their extreme isolation from society. A vicious cycle occurs as the low mood is exacerbated without access to the support of friends and family. The individual begins to withdraw from the group, and eventually forgoes some of their professional or communal duties. This can lead to the individual being made a scapegoat and marginalization from the group (see Chap. 19), which in turn amplifies the individual's mood disorder and even arouses feelings of suspicion towards others. If this individual is part of a minority group, subdivision of the group may occur. In the worst-case scenario, all these events lead to reduced group consensus, and possibly risk to the mission (Fig. 38.3).

38.3.1 Psychological Stressors

Living at Concordia is in many ways like living on a distant desolate planet. When I stepped out of the Douglas DC3 twin-engine propeller aircraft onto the ice on December 5th 2008, what struck me the most was the vast, bleak, expanse of the high Antarctic plateau. The landscape has no topography, and there is not a single drop of life or natural color in the endless ocean of listless ice that surrounds the station for as far as the eye can see. There is nothing in the environment to remind you that you are still on Earth, a planet filled to the brim with flowing water, lush vegetation, geological wonders, and wildlife. A sun that never sets and then subsequently never rises adds to this surreal sense of detachment from Earth. In addition Dome C experiences, essentially, no weather system as there are almost no clouds and there is almost no precipitation and no wind. Therefore, one feels not only very physically isolated due to the remote location of Concordia and distance from civilization but also emotionally isolated and detached from Earth. This is compounded by the limited telecommunications and generally poor access to ample and personalized information from the outside world, including communication with friends and family. The extreme separation and isolation from close relatives and friends places a significant emotional strain on many crew members and can lead to low mood, emotional liability, or sometimes even emotional detachment. Such emotional detachment is both a positive and negative consequence of the physical and emotional isolation. It allows one to adapt to the high isolation, but when experienced to the extreme can result in emotional blunting and apathy, which in turn can result in disruptions to personal and family life, relationships that are already strained by the immense physical separation (Fig. 38.4).

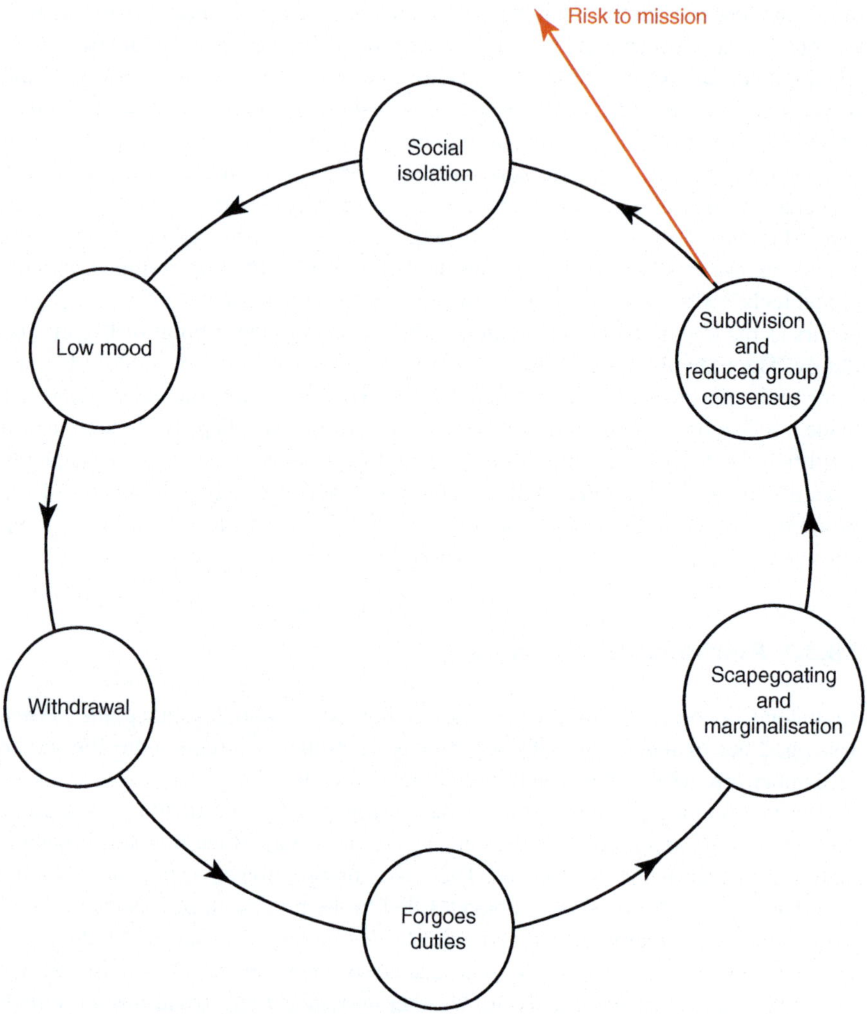

Fig. 38.3 An individual reacting negatively to a stressor can eventually have a profound influence on the group

Life at Concordia is, perhaps surprisingly, very monotonous. It is not a "polar adventure" and in this respect is sharply different to many of the Antarctic coastal stations that have wildlife, weather patterns, dynamic landscapes, as well as shorter periods of total darkness/daylight and shorter periods of confinement. The monotony is present in several forms including sensory, intellectual/work, recreational, and social. Sensory monotony is profound. There are very few colors, sounds, or smells present in the environment or the station, and those that are present are unchanging. Olfactory and gustatory senses are probably, in addition, blunted due to the chronic hypoxia. Although food is ample, it is frozen and dried and highly

Fig. 38.4 Fish-eye view from the roof of Concordia station, taken during the winter, looking out onto the Dome C plateau

dependent on the quality of preparation. The limited access to substantive and personalized information, in particular real-time data transfer, amplifies the general monotony. Although friends and family could assist in retrieving information, one was wary of hassling them on a daily basis for this purpose. I managed to find a crude way round this issue by designing, with the help of a colleague back in Europe, a program that retrieved Pubmed and Google searches through an automated email process. I would send an email to my colleague's account in Europe with an identifying title and my search request in specific syntax. At the next satellite connection, usually 8 h later, I would receive an automated email response with the search results. I would then send the http links for the abstracts/articles I wanted at the next connection, and finally, usually 24–48 h after my initial email request, I would receive an automated response with the abstracts/articles I desired. This was an incredibly laborious and slow process, but the only option available. I highlight this point to illustrate the extent to which we found the lack of access to scientific knowledge and information frustrating. Work was interesting for many of the scientists, but due to limited amounts of reagents and equipment, the ability to be scientifically creative or "play" when on site was restricted. Between the biomedical experiments and my own personal goals and projects, I didn't have time for much else. My routine consisted of work, lunch, work, dinner, gym, work, socialize, and bed. I, and others, followed this similar pattern almost unchanged for 279 days straight during the winter. Days all seemed to merge into one. Unexpected events and pleasures were welcome but very rare. The major external events that occurred were the first and last sunset and sunrise, but even these events were predictable. Work served as an important adaptation method to the isolation and sensory void, but was in itself a monotonous routine for many. Again, if used to the extreme, this adaptation process had negative consequences including overworking, burnout, and detachment.

Leisure activities in the station were sparse. Some of us brought recreational materials, but these were limited by transport weight and dimension restrictions. It is interesting, but perhaps obvious, to note that the motivation to pursue goals dropped soon after arrival at Concordia if these goals and pastimes were not related to Antarctica directly, to work or to preexisting hobbies. The gym consisted of a few weight machines and a bicycle. Several months into the winter though, using the same machines and repeating the same movements was dry, and a strain on motivation. Midway through the winter I welded some steel pipes together to make a balancing beam. This helped break the repetitiveness of the gym machines and even just this small amount of variety in my exercise routine and motor functions was a huge sensory rush. Although leisure activities such as table football and a pool table were available, these were less frequently used than one might expect. This may sound surprising, but was related to choice. People didn't necessarily want to play pool or table football simply because they were available. These were not recreational activities that people had interest in prior to arriving. This lack of choice and control was noticeable in many other areas including food, work, and most importantly personal and family life. It was very difficult for those who experienced negative relationship and family events whilst at Concordia to feel as though they had any control or interventional influence over such situations.

The no abort/rescue scenario and high autonomy are subtle stressors at Concordia and not necessarily ones that are obviously felt until system failures or major medical emergencies occur. One major technical event in particular during our winter-over reminded us of how fragile we were. Around mid-winter, as endless night surrounded the station, we heard a menacing, twisting groan escape from the depths of the station in the early afternoon. The lights went out and we were plunged into darkness. The station had lost power and all systems were down. The power failure lasted over an hour whilst the technical crew worked to repair the generator systems. Human presence at Dome C is completely dependent on life support systems; electricity, heat and water principally. The summer camp, 1 km away, serves as an evacuation destination but it consists of very basic huts, simple radio communication, and one very cold, unused, generator. This event served to remind us that Concordia station is a vulnerable speck of artificially supported life in a seemingly endless expanse of uninhabited and deadly ice, and several of the crew felt a deep sense of lasting anxiety and fear as a result. Although the International Space Station (ISS) is inherently more dangerous than Concordia, the crew can evacuate and be back on Earth within 24 h. This possibility does not exist at Concordia and we are completely at the mercy our life support systems and health during the 9-month winter. The dangers at Concordia are minimal compared to spaceflight, but it is of note that this event, in the context of no rescue possible, caused a deep sense of fear and unease in some of the crew members. It is not unreasonable to expect that such thoughts may weigh heavily on the crew of LDDS missions as, unlike during current spaceflight, they will clearly not have the ability to abort or be rescued when millions of kilometers away from Earth.

38.3.2 Psychosocial Stressors

It can be difficult to remain socially engaged and active within a group that has not been selected as a team unit. Socializing within such a small unchanging social sphere can be repetitive after many months. Forced interpersonal contact can easily reveal, sometimes at an early stage, personality incompatibilities. In a group size of 12, in a closed and confined environment, minor disagreements can alter the mood and feel of the group significantly. Individual stress can lead to affect displacement towards other individuals, the group, and even ground control. In the case of Concordia station, the two nationality groups and two command structures can be complex and can lead to subgrouping, which can be amplified by the absence of a communal language. The role of the leader becomes increasingly important and influential as the winter progresses, as a mediator between both individuals and subgroups and as a figurehead. After 9 months of isolation, some individuals had seemingly irrational and suspicious negative emotions towards others and the group. As mentioned before, changes in mood of even just a single individual can theoretically, depending on the role of the individual and their interaction with the group, eventually have a profound impact. The presence of a mixed gender crew was generally a positive and calming influence, but this is clearly dependent on the personality types and the inverse can arise. Relationships often develop in isolated and monotonous environments. There is no reason to expect that this will not occur during LDDS missions when the crew is even more physically and emotionally detached from Earth. This has potentially very significant social and health implications, and is something that needs to be addressed openly when planning long duration missions.

38.3.3 Environmental Stressors

My first night's sleep at Concordia station was terrible. Dome C is one of the driest places on Earth. I woke up several times with my lips cracked and stuck together, raw, almost bleeding. On my first night, at two in the morning I stirred from sleep and noticed the sun blazing fiercely through a tiny crack in the window blind as though it were midday. The light pounded my retina and suddenly I was awake and viciously alert. I couldn't get back to sleep despite the fact that after days of traveling to get to Concordia, my body and mind craved rest desperately. Within a couple of weeks I had got used to the 24 h sunlight but I couldn't shake off the lack of humidity, and this was something that affected my sleep for several months. Cheap and rusty humidifiers, and buckets of recycled water, had little effect. By the time most of us had got used to the dryness, darkness arrived. The 3 months of total darkness did not affect everyone equally, however. Some noticed little difference in their sleep patterns and mood, whilst others required sleeping aids. Overall, people felt more fatigued, with diminished abilities to concentrate for sustained periods. During the winter, at temperatures below −60 °C outside, my eyelashes would freeze together and I would have to pry them apart to be able to see. My fingers would ping with pain as well, no matter how many gloves or hand warmers I used. Even the best polar gloves and boots offered minimal protection from the cold. Below −50 °C to −60 °C, people spent very little time outside and were confined to the station for long stretches of time. Time spent outside at such temperatures was usually a maximum of 20–30 min and mostly for technical reasons rather than recreation.

All these environmental stressors, cold, hypoxia, lack of humidity, and altered day/night cycles, result in significant physical stress. These stressors are experienced most forcefully during summer, as the change in physiological homeostasis is very acute. By the time winter begins, people have had time to adapt somewhat to the environmental stressors. Most people arrive direct from sea level and the adaptation to an altitude equivalent to 3800 m at the equator is sudden and abrupt. Despite living at such an altitude for over a year, and many of us developing hemoglobin levels above 20 g/dL, exercise tolerance never quite matched that at sea level. It is interesting to note that during the summer, approximately 50% of the entire crew developed one or more infections, most commonly upper respiratory tract infections. This high incidence of infection was probably multifactorial and due to: a reduced ability of dry mucous membranes to act as barriers to microorganisms; the effects of hypoxia on immunity and inflammation; the consequences of sleep deprivation and fatigue on immunity; the fact that Concordia is a dynamic and high population density environment during the summer (up to 70 individuals); and the fact that people arrive on a regular basis from the outside world bringing new microorganisms with them. During the winter there were only two incidences of symptomatic infections despite the much higher levels of psychic stressors; a soft tissue hand infection that responded to antibiotics promptly, and a 24 h febrile illness of unclear etiology that spontaneously resolved. Clearly the fact that Concordia is a closed system during the winter is a major factor with respect to this lower incidence of symptomatic infection. Wound healing however seemed, subjectively, slow. Results of the immune study CHOICE (Consequences of Hypobaric hypOxia on Immunity in the antarctic Concordia Environment) have shed light on the interaction between hypoxia, stress and immunity at Concordia (see Chap. 16). Chronic moderate hypoxia at Concordia is likely to influence not just immunity and inflammation, but many other systems and processes including, for example, cardiovascular function, cognitive abilities and muscle metabolism. Vitamins and minerals are not routinely supplemented at Concordia, and it is therefore likely that the wintering crews experience nutritional deficiencies, as well as presumably vitamin D deficiency due to the lack of sunlight exposure. Overall therefore, Concordia is not only a psychologically and socially challenging environment to live in but also a physically stressing environment.

38.4 Adaptation and Countermeasures to Stress During Long Duration–Deep Space Missions

38.4.1 Personality Characteristics and Adaptation to Stress

Many of the sequelae listed in Fig. 38.2b were experienced by one or more individuals, and the group, during the winter. Some of us successfully dealt with the issues and stressors present and were fortunate enough to remain professionally and personally motivated throughout the winter. Others unfortunately suffered psychological disturbances as described. Of these, the most serious were anxiety and panic attacks for which several individuals required benzodiazepines at one stage or another. Such psychological imbalances, perhaps paradoxically, can sometimes continue, and indeed develop de novo, upon return to civilization. The abrupt cessation of life in isolation and confinement and the subsequent readaptation to normality can be stressful and lengthy, and individuals are unfortunately often unprepared for this. Readjusting to social rules, routine employment and relationships can be difficult and some people experience feelings of confusion, uneasiness, disinterest and social claustrophobia. With time however, many people who have spent a winter in Antarctica feel positive effects associated with the privilege of experiencing one of the planet's most spectacularly vast and daunting environments, such as: a profound sense of accomplishment; increased personal and professional confidence; a better tolerance and adaptation to stress; a clearer vision of one's personal needs, limits and ambitions; a deeper appreciation of personal freedoms and the natural environment. Although there are many stressors present at Concordia, there is an absence of some of the stressors present in "everyday" life, such as commuting, shopping, queues, bills, excessive choice (both restricted choice and excessive choice can, in my opinion, function as stressors), advertising and information overload, rules and regulations and so on. And although everyone feels some of the psychological and social stressors to a certain degree, some experience the absence of routine life stressors very positively. Indeed, this is probably why some people subsequently return to Antarctica for a further winter-over, even to Concordia!

Based on my observations of the crew, several personality characteristics and attitudes appeared regularly in those who remained consistently in good psychological health during the winter: detachment, lack of sentimentality, relatively unemotional yet still empathetic, stoicism, pragmatism, determination, flexibility, diplomacy, resourcefulness and creativity, and a low need for social stimulation. Admittedly, these observations were personal and possibly biased. These are not all inherent personality characteristics and some represent personality adaptations to specific stressors. In my opinion, the most critical frame of mind and approach in ensuring a successful and healthy winter-over were: having a sharply defined and almost unquestionable comprehension of one's reasons for doing a winter stay, including an obvious understanding and acceptance of the risks and sacrifices involved; the ability to keep oneself occupied and motivated, both professionally and personally, whilst ensuring time and methods for effective relaxation and recuperation; having clear and realistic personal and work goals relevant to Antarctica, careers, or previous past-times and hobbies; a willingness to help others through personal and professional issues including the ability to appropriately re-prioritize one's own time, needs and beliefs for the benefit of other individuals and the group; resourcefulness and creativity in making the most out of the environment and limited materials available; and a certain detachment and lack of sentimentality with regards to the outside world.

38.4.2 Effective Adaptation and Countermeasures

It is unrealistic to expect humans not to suffer psychological distress when exposed to the prolonged intense and novel conditions that will occur during LDDS missions. After all, we evolved in the lush, open, vast plains of Africa, and not in a small, dark, noisy, dangerous spacecraft. That kind of environment is not, and will never be, a natural environment for humans to live in for years at a time. Expecting to identify "super-human" astronauts and teams in advance who will successfully adapt to the stressors of LDDS missions is not realistic (see Chap. 2). Rather, helping highly trained and well-balanced individuals and teams to adapt through appropriate interventions and countermeasures seems more appropriate. The general aim of countermeasures should be to increase individual physical and psychological well-being and performance and minimize downside risks, increase group cohesion and teamwork, minimize the disruptions to families and relatives back on Earth, maximize the chances of mission success, and ensure successful readaptation upon return to Earth. Briefly, countermeasures can be classified along the lines of crew composition, crew training, habitability, social, recreational, psychological, sensory, work design and pharmacological (see also Chaps. 30–35). Complimentary to countermeasures is the monitoring of stress via physiological (e.g., autonomic responses, biological stress parameters), psychological (e.g., questionnaires, video analysis, sociometry) and performance parameters and markers. Early identification of the consequences of stressors allows for early implementation of countermeasures. Indeed, on site analysis of physiological markers of stress (e.g., salivary amylase, neuropeptide-Y) would allow astronauts to monitor stress levels objectively themselves on a regular basis and instigate self-directed countermeasures appropriately. Some of the social and sensory countermeasures available to the crew of the ISS presently will not be present during LDDS missions. The ISS experiences stunning and ever-changing views of Earth. Visiting crews bring fresh foods, recreational materials, and private packages from friends and family. The ISS crew also has access to real-time data transfer including on demand communication with relatives and psychological counseling, and live video links. None of these stress-reducing factors will be present during LDDS missions.

Effective adaptation and countermeasures should start early, at the time of selection and training. Although all candidates for LDDS missions will undergo extensive psychological profiling and testing, and perhaps genetic characterization, the exact personality traits and attitudes needed will be extremely difficult to correctly identify in advance, for the simple reason that no one will have experienced such high levels of novel stress before. Selecting the right candidates for spaceflight is a difficult process at the moment but will represent an even bigger challenge in LDDS missions. Testing and training of individuals and crews in advance in analogous extreme situations and environments will, without a doubt, be necessary. The Concordia winter crews are not selected as a team unit, undergo little training together prior to deployment, and do not share a specifically defined common goal. Clearly this is not ideal with regards to maximizing work and social cohesion. Crews of future LDDS missions will have to spend many years training together and, importantly, also socializing together.

There are a number of habitability factors and social and sensory countermeasures that, although limited by engineering and financial constraints, could however significantly ease the stressors during LDDS missions. For example, at Concordia station, the windows are relatively small. One large, possibly panoramic window, in the central living area, even at the expense of other windows, would have made a significant difference to the feeling of confinement. Simply being able to clearly and comfortably visualize the outside world, the Antarctic plateau, would have a calming and comforting effect. The illusion of not being physically limited by our internal environment is critical. There is psychological strength and meaning in the fact that there is a world outside the winter confines of the station. In the case of LDDS missions, Earth will disappear from view relatively soon but depending on the location of windows, and orientation with regards to the sun, stars and galaxies should still be visible. In any case, the view of Earth, slowly increasing in size as the mission end approaches, will also have powerful positive psychological connotations at a time point that may prove one of the most psychologically demanding and difficult. Indeed, the view of the approaching mission destination, whether it be an asteroid, moon, or Mars, is equally psychologically important.

A minimum amount of privacy is also needed for personal recreation, sleep, hygiene, and communication. At Concordia station the communication facilities were poorly sound insulated, which meant that passersby could easily hear conversations. The ability to communicate in complete confidence with friends and family, whether real time or not, allows for the dissolution of concerns, fears, and anger, which in turn helps to minimize interpersonal conflicts. Admittedly, the crew of LDDS missions will not be able to communicate in real time during most of the journey, but they will have the ability to send and receive files, and video and audio messages. The use of short messaging services and social media to communicate with friends and family will take on increasing importance the further the crew travel from Earth. Short messaging services in particular give somewhat the illusion of instant communication and would probably be very useful yet fairly simple to implement. Our private and sleeping quarters at Concordia were small and basic but thankfully solitary, which allowed us the chance to retreat when necessary. Even just a few months into the winter, I was acutely aware of the importance of being able to retreat to private space every now and then, either one's bedroom or laboratory, as a necessary adaptation aid. It is comforting to know that there is a space within your cocooned environment that belongs to your own little private world. Volume will clearly be a limiting factor during LDDS missions but it is absolutely critical that each crew member has a small but private space that can be used for personal use and sleep, and can be defined as their own for the entire journey. The lack of humidity was a significant contributor to poor sleep at Concordia and this, and other environmental characteristics (e.g., noise, temperature, vibration, and lighting), should be taken into account during the design of life support systems for LDDS habitats.

Concordia station is decorated with few colors. The water treatment unit, diesel engines, and ventilation are responsible for most of the ambient background noise. Although the most prevalent smell throughout the station was that of the water treatment unit, we were fortunate to have cooking facilities, which did thankfully provide us with pleasant aromas and a degree of sensory stimulation. Indeed the quality of food preparation was absolutely paramount. Food was the single most important pleasure we had at Concordia, and meals were one of the few occasions when the entire crew came together. The quality and variety of the food had a significant impact on the mood of individuals as well as the group dynamics. When the food was well prepared, moods and tensions improved, whereas they significantly worsened when the food was disliked. Having the opportunity to actually cook for ourselves, every now and again, and create a meal that we specifically desired was also very important. It was fun and challenging to try and make interesting and varied meals from frozen, dried, and limited ingredients. When confined and isolated for long periods of time there are very few choices allowed or available, yet even a small amount of choice can significantly alter the perception of personal freedom. In addition, in an environment where many of the sensory and social pleasures that we normally take for granted are absent, the few that are available become increasingly important and necessary.

Countermeasures to dulled senses could be relatively simple to implement during LDDS missions and might include colors, sounds, smells and possibly tactile sensations such as video projections of Earth, virtual windows, virtual reality, small plants, bottled aromas, sounds of nature, different textures, etc. Variety in motor function should also be provided, and perhaps this is best included as part of the training systems that will be necessary to keep the crew's technical skills up to date en route to their destination. It will be difficult, probably impossible, to accommodate any form of cooking during LDDS missions. It is therefore critical that adequate and personal choice be allowed for in meal selection, together with odor and flavor enhancers. There may not be enough volume to accommodate a separate relaxation area, but there should at least be a defined dining and communal area, which ideally should be separate from work areas.

Finally, the journey to and from any deep space destination will present ample opportunities for boredom and motivational insecurity to creep into the minds of the crew. Providing the crew with work and occupations during this period is vital. Above all, any work or occupation should be meaningful and relevant to the mission or the individual's needs, experience and goals. At Concordia there was a marked difference in the ability of individuals to withstand stressors between those individuals who had meaningful and plentiful work and those who did not. Due to volume and technical constraints, it will not be possible to have large volume, well-equipped and sophisticated laboratories on board LDDS missions. Any laboratory will be basic and therefore the scope for scientific creativity will be limited, much like at Concordia. Individuals who can creatively exploit the scientific wonders of microgravity and space through their own devise with the limited equipment and materials available, or keep themselves intellectually stimulated through the use of electronic information, will be better suited to the monotonous and confined environment.

38.5 Conclusion

The stressors, consequences, and countermeasures mentioned in this chapter have been discussed briefly, hoping to give an insight into some of the behavioral health issues that will affect the crews of LDDS missions. Interplanetary missions will be expensive, and physically, psychologically, and technically dangerous endeavors. The stressors and consequences that have been observed during spaceflight and analogous environments, such as Concordia, will be markedly amplified during LDDS missions. Investing significant amounts of time and finance into training and countermeasures that will improve the well-being of the crew is a small fraction to pay when one considers the vast expense, expectations, and risks that are at stake.

Acknowledgment Special thanks go to the DC5 winter crew, my friends, and colleagues during the winter. Loredana Bessone was hugely helpful in providing support to the entire crew during the DC5 winter-over, and the crew of the ISS expedition 20 mission were also incredibly kind in sharing their experiences with the DC5 winter crew. I would also like to acknowledge the NASA Behavioral Health and Performance Group, in which my involvement has helped clarify and crystallize some of the thoughts documented in this chapter. Finally I would like to thank Dr. Alexander Choukér, Dr. Brian Crucian, Dr. Clarence Sams, and Dr Oliver Angerer for having faith in my scientific and technical abilities in isolation and confinement.

References

Lugg DJ (2005) Behavioral health in Antarctica: implications for long-duration space missions. Aviat Space Environ Med 76(6 Suppl):B74–B77

Palinkas LA, Suedfeld P (2008) Psychological effects of polar expeditions. Lancet 371(9607):153–163

Moon, Mars and Beyond

39

Marc Heppener

39.1 Introduction

This book deals with stress and the immune system in astronauts with a link to the importance of this research to Earth application and patients benefits. However, without astronauts and space travel, the book would be devoid of practical purpose for space missions. The same is true for space travel in itself because, one might ask, what is the purpose of travel if it is not a destination? Yet, today, most spaceflights seem to consist of endless turns around the Earth, without ever leaving the so-called Low Earth Orbit (LEO). This is the orbit of most inhabited spacecraft, including the International Space Station (ISS), at an altitude of ca. 400 km (roughly the distance Paris-Lyon, Bonn-Berlin, London-Newcastle, Rome-Venice, or Houston-Dallas etc.). Only 24 human beings, almost half a century ago, have ever have travelled beyond.

In this chapter therefore we will look at the future perspectives of human spaceflight beyond the Earth and its very direct surroundings. Space is vast and the definition of 'direct surroundings of the Earth' can range from the moon (380,000 km), our Solar System (distance of termination shock: ca. 25 billion km or 23 light-hours), our Galaxy (100,000 light years), our 'local group' of some 30 nearby galaxies (10 million light years). Wherever we want to go, we can make a few general statements on the nature of such a trip:

- It takes a long to very long time to reach our destination
- The travel itself and life at the respective destination will both challenge humans by a series of known and unknown stressors related to the hostile environment of space.

M. Heppener (✉)
Directorate of Human Spaceflight and Robotic Exploration,
European Space Agency (Retired), Paris, France

© Springer Nature Switzerland AG 2020

A. Choukér (ed.), *Stress Challenges and Immunity in Space*,
https://doi.org/10.1007/978-3-030-16996-1_39

- The effects of these environmental factors on humans are not always well known or quantified. Yet, we need to ensure that the crew will survive the trip and is able to perform useful tasks.
- For almost all conceivable destinations, new technology will need to be developed for the travel itself, life-support, physical and mental health.

This chapter will endeavour to give a brief summary of the current understanding of and most important stumbling blocks towards extended space mission durations, as well as possible technological solutions to overcome these. The further we go—in terms of distance from Earth, and also in terms of the build-up of this chapter—the less facts are available, and the more speculative the text will become. The final paragraphs are bordering on science fiction, although still based on current research.

39.2 Consideration for Mission Planning

39.2.1 Going Places

First, the question needs to be answered what time will be required to reach our future destinations. To determine the boundary conditions of such travel, most importantly the basic orbital physics have to be taken into account.

To characterise the trip a bit better, let us examine two factors in a bit more detail. In order not to get too stressed already from reading this chapter, we restrict ourselves in this discussion first to targets within our own Solar System. The energy required to go from Earth to another planet in our Solar System is basically determined by

- The need to leave the potential energy well of the Earth (leading to a certain so-called escape velocity)
- The need to reach a stable transfer orbit to the destination planet
- The need to decelerate again to reach the surface of the destination planet again with a safe velocity.
- The need to retrace our steps in reverse order, to get safely back to Earth

Each of these steps are characterised by a parameter called delta-v, describing the velocity change required to carry out orbital manoeuvres in space, and is expressed in units of m/s. Obviously, such a velocity change can only be achieved through acceleration of the vehicle, and is therefore a direct function of the efficiency of the propulsion system and the total mass of the spacecraft. It is important to realise that any increase in delta-v will have a significant impact on the mass of the space vehicle. Simply put, increasing the total velocity of the vehicle will require additional fuel, and that fuel, including its containment, will have to be accelerated also, leading to even further fuel requirements, etc. The net result is that there is an exponential relation between the ratio of useful to total vehicle mass and the effective delta-v. This relation was first derived by the Russian pioneer of spaceflight Tsiolkovsky in

1903 and carries his name. As an illustration, the take-off mass of the Apollo-16 mission was 3 million kg, whereas the mass of the lunar module was only 5000 kg; a ratio of 600.

It therefore makes sense, certainly with the currently used chemical propulsion technology, to try to minimize energy requirements whilst keeping travel times reasonable. The orbit with the most interesting characteristics is the one in which the space vehicle makes optimum use of the relative orbital velocities of the two planets, getting a boost from the departure planet to arrive with just about the right velocity at the destination, as depicted in Fig. 39.1. It can be shown mathematically that the transfer time of such a so-called Hohmann transfer orbit is a function only of the orbital radii and the mass of the sun. Going faster is possible, but at the costs of a significant amount of additional delta-v, whereas trajectories consuming less energy will lead immediately to longer travel times.

Table 39.1 gives some parameters for destinations in our Solar System. From this it can be seen that, with our current technological knowledge, destinations within reach of humans are restricted to Mercury, Venus, moon, Mars, and possibly Jupiter (and in particular its moons). Anything further away is prohibitive in terms of travel times relative to a normal human lifespan.

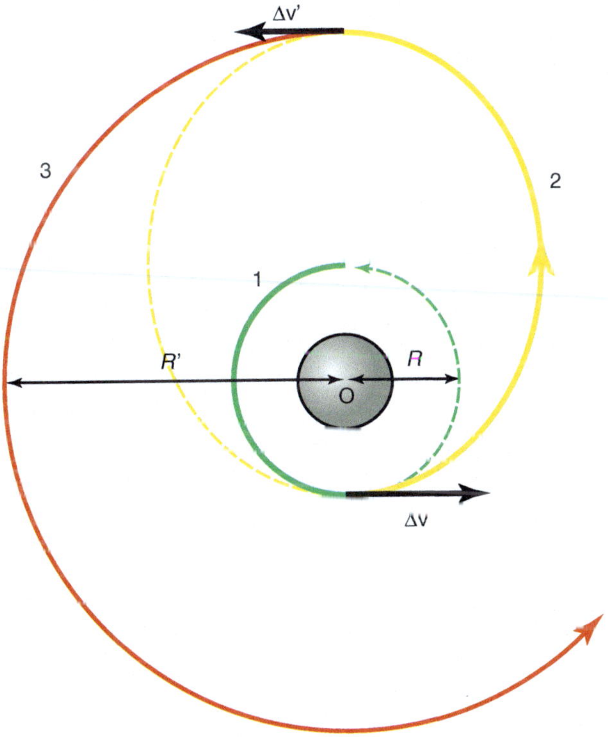

Fig. 39.1 A Hohmann transfer from a planet of origin with orbital radius R to a destination planet with orbital radius R' (source: Wikimedia)

Table 39.1 Some specific parameters of the Solar System

	Mercury	Venus	Earth	Moon	Mars	Jupiter	Saturn	Uranus	Neptune
Distance from Sun (million km)	57.9	108.2	149.6	[0.38][a]	227.9	778.6	1433.5	2872.5	4495.1
Diameter (km)	4879	12,104	12,756	3475	6792	142,984	120,536	51,118	49,528
Density (kg/m^3)	5427	5243	5514	3340	3933	1326	687	1271	11,638
Orbital period (days/years)	88.0	224.7	365.2	27.3	687.0	11.9 years	29.5 years	84 years	165 years
Rotation period (days)	58.7	−243.0	1.0	27.30	1.03	0.41	0.45	−0.72	0.67
Gravity at equator (*g*)	0.38	0.91	1.00	0.16	0.38	2.35	0.92	0.89	1.12
Mean surface temperature (°C)	167	464	15	−153 – +107	−65	−110	−140	−195	−200
Surface atmospheric pressure (bar)	–	92	1.01	–	0.01	0.7	1.4	1.2	1.3
Surface composition	Basaltic	Basaltic	Basaltic/H_2O	Basaltic	Basaltic				
Atmosphere composition	Na	CO_2	N_2/O_2	–	$CO_2/N_2/Ar$	H_2/He	H_2/He	$H_2/He/CH_4$	$H_2/He/CH_4$
Number of moons	0	0	1		2	67	62	27	14
Hohmann transfer time (days)	104.62	144.87		[3]	256.71	989.28	2196.05	5810.42	11,110.96

[a]Relative to the Earth

More distant destinations would be possible if we would use different propulsion technologies, which would allow us to deviate from the Hohmann orbit. Even then, however, we have to realise that the fundamental laws of nature put limits on the velocities we can reach. Assuming that 10% of the speed of light will be a very optimistic long-term achievable goal, it would still take 15 years to reach the boundaries of our Solar system, and 44 years to travel to the nearest sunlike star (Alpha Centauri), where one might actually find a habitable planet (Kopparapu et al. 2013).

The universe is immense, and it will take a really long time to reach even the closest interesting destinations. Will humans ever survive such a trip?

39.2.2 The Dragons on the Way

According to our current knowledge the most dangerous physical stress of long-distance and long-duration spaceflight is radiation. Until this very day, almost all experience in long-duration spaceflight has been restricted to the first 500 km above the Earth surface. This fact is significant because in LEO an important fraction of the harmful radiation present in space is shielded by the Earth's magnetic field. In deep space, defined for the moment as beyond the Van Allen Belts (Fig. 39.2), two sources of radiation can be identified:

- Solar radiation, which mainly consists of protons. Their intensity shows a long-term variation with the solar cycle (11 years), but can increase rapidly during periods of solar flares and reach levels that, without proper protection, can be lethal. To illustrate the importance of this, in August 1972, just a few months after the Apollo 16 mission, a particular vehement solar eruption took place.

Fig. 39.2 Graphic representation of the Van Allen radiation belts generated by the Earth's magnetic field and protecting the Earth from a large part of cosmic radiation (source: Wikimedia)

Would this solar flare have happened a few weeks earlier during a lunar extra-vehicular activities (EVA), the Apollo16 crew would likely have received a lethal radiation dose (Parsons and Townsend 2000) and their perspectives of returning safely to Earth would have been bleak. Fortunately, shielding against proton radiation is relatively easy, and a relatively thin layer of water, or lunar or Martian soil, would already provide sufficient protection. Combined with the transient nature of solar flares, it is assumed that efficient protection strategies can be devised, although much further work is still needed.

- Galactic cosmic radiation, which is present throughout our Galaxy, consists of nuclei of heavier atoms (with higher atomic number *Z*, such as iron) originating from stellar nuclear fusion processes and supernova explosions. Whilst the intensity of cosmic radiation is not particularly high, shielding is much more complicated and hence the exposure will be almost continuous. The effect of a high-*Z* particle impinging on a cell can also be much more dramatic than a proton event, since it creates a track of ionisation of up to millimeters in length (Cucinotta and Durante 2006).

Current knowledge of the radiation effects of long-duration spaceflight is generally restricted to LEO and in particular based on measurements on the ISS (Narici et al. 2015). It is important to note that, contrary to Earth, both the moon and Mars have no magnetic field or dense atmosphere that could attenuate the radiation effects while on the surface. In the vicinity of Jupiter, the radiation environment is even more hostile as a result of the high concentration of cosmic radiation by the planet's radiation belts.

The uncertainties of the effect of all forms of space radiation, both in LEO and beyond, on long-term risks of cancer are high. Best available estimates put the risk of cancer death at 0.1–2% for ISS missions and 1–20% for Mars missions (Durante and Cucinotta 2008). Certainly the higher-end ranges would put such missions at or beyond acceptable occupational hazard levels (generally set at 3%). More knowledge is therefore required on the effects of long-term continuous exposure to cosmic radiation and possible countermeasures (shielding or other). The use of ground-based experiments using particle beams such as the 'Investigations into Biological Effects of Radiation (IBER)' program is particularly important since it can offer more controlled conditions and improve statistics complementing space experiments (Durante et al. 2010).

39.2.3 Floating Free, Falling Deep

The most visible aspect of current day spaceflight is weightlessness. This phenomenon is inherent to the fact that a vehicle in space that is only subject to the force of gravity (i.e. without engines burning or resting on a planetary surface) is in fact in free fall. Such conditions exist in orbit around a planet, or during interplanetary travel using a Hohmann transfer.

What are the effects of microgravity on biological systems? This has been an important area of scientific research in space during the past decades. With the advent of the ISS access to extended periods of microgravity for research purposes is now available for almost two decades. The ISS has been used to continue the study of biological effects on cells, plants, animals and humans. To increase the statistical validity, bed rest studies form a very useful analogue for studying the effects of prolonged weightlessness on the human body (see Chap. 36).

Initial assumptions, based on the very small difference in gravitational energy over distances of several microns, predicted that at the cellular level biological processes would not be affected by weightlessness (Pollard 1965). However, when results from early Russian and US (Gemini and Biosatellite) experiments became available, it became clear that weightlessness did affect individual cells (Young and Tremor 1968; Jenkins 1968). The first case where the pathways of gravity sensing was unraveled was in plant-roots, where it was shown that the mechanism of sensing gravity is based on the presence of a heavy cell element called statolith which is connected to the cytoskeleton, whose deformation trigger local changes in growth rate of the root walls, leading in the end to the curvature of the root in the direction of the gravity vector (Friml et al. 2002; Perbal and Driss-Ecole 2003).

One can now consider it proven that gravity has a measurable influence on cellular molecular processes that determine the functionality of virtually all types of cells, ranging from simple algae to mammalian cells. Through the multitude of experiments being carried out to date it is now emerging that, while the nature of the gravity-sensing element in a cell may differ, the underlying molecular mechanisms are very similar in different cell types (Häder et al. 2017), indicating that responding to gravity is a very fundamental process for life as we know it on Earth.

For mammalian cells in particular, the stress effects of weightlessness on signal transduction, gene expression and subsequent effects on cell division or functionality of, for example immuno cells, have been well documented (see Chap. 11). More than being of mere theoretical interests, these results have direct relevance for humans living under conditions of weightlessness. Other relevant effects in this context are for example the increased proliferation rate of several bacteria (Chap. 18; Häder et al. 2005), the reduced proliferation of lymphocytes in the human body during spaceflight (Cogoli 2002), or the change in bone turnover as a result of exposure to weightlessness (LeBlanc et al. 2000; Marie et al. 2000; Scheld et al. 2001).

From the description of the effects of weightlessness at the cellular level, it is already clear that the state of the human body will change during spaceflight. However, beyond the effects on our cells, the human body is large enough to react macroscopically to changes in gravity (both size and direction), as anybody can witness in a rapidly accelerating/decelerating elevator or by standing on one's head.

The most visible and common of these effects are known already since the early days of spaceflight and are by now extremely well documented. They include:

- Effects on the neurovestibular and proprioceptive system:

  The equilibrium system and in particular the semicircular canals comprise a motion-detection system that is based on the detection of fluid motion resulting from the vector-product of the gravity force and angular acceleration of the head. Clearly, without gravity vector, the output of this system is completely different and at odds with other visual or proprioceptive clues (Clarke et al. 2000). The net result can range from mild disorientation to severe space sickness in the first days of spaceflight. Interestingly, also here the human body is capable of adapting to the new environment and the effects in general disappear after a few days. In the new configuration, visual inputs have an even higher importance than under Earth gravity, and many studies take use of this effect to look into the specific mechanisms of image processing, orientation and the interaction with other proprioceptive signals. Of practical importance are tests that look into the capability of the crew in performing complex tasks such as manipulating equipment or carrying out docking or landing procedures (Viguier et al. 2001).

- Fluid shifts and subsequent neurological, cardiovascular and hormonal adaptation processes:

  Very visible is the so-called puffy face in astronauts in the first days after launch, resulting from the redistribution of fluid from the lower extremities to the head. Interestingly, adaptation processes will set in rather soon, and after a few days a new equilibrium will be reached, characterised by different hormonal levels and functioning of the cardiovascular system (cardiac output, arterial friction, etc.). However, fluid shift and other not yet fully understood mechanisms can result in 'space headaches (Feuerecker et al. 2016)', and long lasting visual impairments of the crew (see below).
- Bone and muscle mass loss:

  Loss of calcium and overall bone-mass loss sets in almost immediately after exposure to weightlessness and at a very important rate of up to 1–2% of total bone mass per month (which is 10–15 times faster than in the most severe case of osteoporotic patients). At the same time the very long recovery time of more than a year after return to normal conditions was observed in bed rest models investigating this pathology on Earth (Grimm et al. 2016). Similar effects arise in muscle mass (DiPrampero and Narici 2003), although here the recovery rate is faster.
- Lung ventilation and related processes:

  Detailed studies have been performed on lung ventilation in weightlessness by monitoring the composition of exhaled air with or without premixed gaseous additions. The results have led to a change in the understanding of the process of gas exchange in the lungs, in particular by invalidating the assumed effect of gravity on the ventilation/perfusion ratio in the lungs (Verbanck et al. 1996). Further studies look into the potential increased risk of lung inflammation due to inhalation of floating particles, or exposure to reduced pressure environment such as in space suits during extra-vehicular activities (EVA) (Karlsson et al. 2009).

With the start of the ISS era long-duration flights have become more common and the number of test subjects for studies increased importantly. This has allowed more in-depth research, identifying a few less known effects that emerged in the last decade:

- It is now demonstrated that the immune system is influenced by spaceflight. The effect goes beyond the effect of weightlessness on the proliferation of lymphocytes mentioned before and is probably due to a combination of weightlessness and stress associated with spaceflight. There is evidence for a role of stress hormones being (partially) responsible for this effect (this book; Strewe et al. 2012). For long-term space missions this aspect may have important consequences, both in terms of the design of life support systems, countermeasures and medication.
- A surprise finding, first identified in 2005, is that during spaceflight a significant fraction of astronauts experience a noticeable impairment of their eyesight. Upon closer examination, in some 60–80% of the cases investigated, astronaut's eyes are altered in a more or less serious way during spaceflight (Mader et al. 2011). The effect, which seems to be permanent, is not yet fully understood but thought to be related to changes in intra-cranial pressure and plasticity of the brain. MRI scans of volunteers during bed-rest studies seem to confirm this hypothesis (Gerlach et al. 2017). Not much is known on the long-term evolution of the phenomenon nor on possible countermeasures, but it clearly deserves attention when discussing perspectives of longer duration missions.
- The study of the human body protein system (proteomics) is a current topic in research and it is therefore not surprising that it is also extended to space research. Indeed, it appears that changes in the protein system in the human body under spaceflight are different from those seen at individual cellular level and that there are important interaction effects. The impact of these findings for long-duration spaceflight is under study (Grimm et al. 2014).

At first sight the effects of weightlessness on the human body seem impressive and could constitute a potential major showstopper for human spaceflight in general. However, the most remarkable result of these studies is the apparent capability of the body's organ systems to adapt to the new stressful environmental conditions. After a few weeks of spaceflight the most notable adverse effects seem to be compensated and humans are able to survive well in space for prolonged periods of time. In fact, after a few weeks, the situation of weightlessness has almost become as natural to body and mind as the normal Earth environment and new methods of orientation and locomotion (resembling dolphin-like swimming) are adopted.

Humans have survived in space for extensive periods of time by now, and records include:

- Longest continuous duration in space, men: Valery Poliakov, 438 days, 1994/5
- Longest continuous duration in space, women: Peggy Whitson, 290 days, 2017.
- Longest total duration in space Gennady Padalka, 879 days, 1998–2015.
- Longest total duration of EVA activity, Anatolyi Solovyov, 82 h, 1988–1998.
- The ISS represents 19 years of continuous human occupation of space. It has been visited by more than 230 astronauts, of whom roughly half was long-term crew, i.e. staying on board for approximately 6 months. More recently also a 1-year stay for two astronauts was implemented in 2015.

Suitable countermeasures to mitigate the consequences of life under the condition of weightlessness are also under development. Two approaches are under consideration. One is based on physical countermeasures such as exercise machines and treadmills (Petersen et al. 2016) or centrifugation (Iwasaki et al. 2001; Clément and Pavy-Le Traon 2004; Linnarsson et al. 2015). Another approach is looking at metabolism, nutrition or food supplements (see also Chap. 33) (Vermeer et al. 1998; Stein et al. 2003; Heer et al. 2004) and pharmaceutical countermeasures (Chap. 35) (Grimm et al. 2016). One of the more fascinating results of such studies indicate that bone-mass loss may be correlated with an increased sodium intake by astronauts (Frings-Meuthen et al. 2008). Such research and its outcome transcends the space domain and has direct relevance for people on Earth (Bühlmeier et al. 2016). Also for the immune system's dysfunctions in space international and interdisciplinary groups of scientists have discussed and proposed some helpful measures to mitigate these immune related risks (Crucian et al. 2018, see also Chap. 35). A different aspect in this context is how the crew cope with sudden changes in gravity level, such as occur not only directly after launch, but also directly after landing on another planetary surface. As can be seen at any return of astronauts after a long stay in microgravity, the adaptation to Earth's gravity is stressful, cumbersome and takes roughly the same time as the adaptation to microgravity at the beginning of the flight. Some of the already mentioned countermeasure strategies may be also effective to counteract this effect. However, certainly for the immediate availability of astronauts to carry out critical tasks further research is required (Bles and Groen 2009).

Finally, once on a planetary surface, the body will be exposed to longer times of non-zero gravity different from Earth's gravity. Here very little research is available. As long as the gravity level is between zero and one no major effects are expected. This is in a way evidenced by the limited experience of the Apollo astronauts on the moon. In the near future, the European Space Agency (ESA) plans to carry out parabolic flight campaigns simulating lunar and Martian gravity levels to study this further. However, prolonged exposure to higher gravity levels has never been tested in a serious way, to the knowledge of the author. Certainly an upper limit must exist for the gravity level that a human body can sustain in a prolonged way. Astronaut training includes testing in a long-arm centrifuge operating at up to ~6 g, but the duration of this does not exceed more than several minutes. In any case, it is unlikely that an astronaut can do useful physical activities at permanent g-levels exceeding ~2 g.

In conclusion, it can be stated that the impact of weightlessness on the human body—in spite of its effects on almost every organ system—most likely is not THE most critical problem for the voyage to Mars.

39.2.4 Stressed, Bored and Lonely

Psychological constraints are the least known of all factors that determine whether humans will be able to sustain very long-duration space trips. For example, a mission to Mars and back following a Hohmann transfer orbit will take some 520 days, of which roughly 1 month will be spent on the Martian surface and the rest in transit.

At its largest distance, the crew will be some 360 million km from home, more than 1000 times farther away than the Apollo lunar astronauts. From this distance, the Earth will only be a faint dot in the sky that cannot be distinguished among the stars without a good telescope, and communication delays may run up to 20 min one way, that is almost three quarters of an hour before a reply can be expected. In some configurations, the sun will be in the line of sight between Mars and Earth and visual contact and communications are even virtually non-existent. Naturally, for farther destinations the numbers are correspondingly higher.

During all this time, the crew may experience various forms of physical and emotional stressors. These include not only increased noise levels, limited privacy and contact with family or friends, prolonged confinement, small crew-size, increased expectations as to performance and significant risks of equipment failure or fatal mishaps, but also long periods of boredom. Prolonged exposure to such situations may affect individual psychological health, interpersonal relations in a small group, including effects related to different cultural backgrounds and language problems, leadership and teamwork, problems between the crew ('us') and ground-control ('them'), monotony, loneliness, etc. (see Chaps. 22, 31, and 37). Of particular importance are the possibility for the crew to maintain not only their mental health, but also crew coherence and cognitive capabilities. Finally, as argued in other chapters and elsewhere, physical and mental health cannot be seen as separated items but as different and interacting aspects of overall crew health and performance.

To study these aspects further, long-term isolation studies are carried out. ESA is performing studies in ground-based analogue environments. Important research is, for example carried out in the French-Italian Antarctic station Concordia (see also Chaps. 36 and 38). Due to its very isolated location (more than 1000 km from the nearest coastline) and small crew size of 12–16, the conditions in this base are quite comparable to those on a Martian mission. Nine months of the year the base cannot be reached by any vehicle and the crew is hence totally isolated. Several medical and psychological studies are carried out here.

Another study was the so-called Mars500 isolation study, organised and carried out by the Russian IBMP institute in Moscow together with ESA between June 2010 and November 2011. The study simulated a complete Mars mission with six crew members, enclosed for 520 days in a spacecraft-like complex. With exception of real space travel, weightlessness and radiation, all important aspects of a Mars mission, including communication delays, were simulated and a multitude of medical and psychological protocols was carried out to study the effects on the crew. It is interesting to note that the Concordia and Mars500 studies are in a way complementary to each other. Concordia is a real research environment with real physical isolation. The situation is therefore on the one hand more realistic, but the environment is not well suited for testing different protocols and simulated emergencies, because these might interfere with the operational aspects of the base. Mars500 on the other hand has been a simulation, and the participants will to a certain extent stay aware of that. While this may influence the study, the environment is more controlled, and for that reason the data will be easier to interpret (Chap. 37).

The results of these isolation studies confirm that psychological effects during very long and distant space missions can be expected. The effects can exist at the level of the individual such as autonomy, motivation and decision making (Van Baarsen 2013), performance (Schneider et al. 2013), crew interaction processes (Sandal and Bye 2015) but also be of a more physiological nature such as changes in metabolism (Rakova et al. 2013) or changes in the immune system (Yi et al. 2014). Recommendations are being formulated to alleviate these effects, both during crew selection and training and during mission definition and operations (De la Torre et al. 2012).

39.3 Technological Leaps

The conclusion of the above sections can be summarised as: from a human health perspective, with technology that exists today human spaceflight to destinations like moon or Mars could be feasible, but not much farther away. This chapter investigates a series of technological developments that are under consideration to secure this endeavour and to also extend these frontiers.

39.3.1 Health-Care in Space

During spaceflight a multitude of health risks exist. These can be related to spaceflight, but can also include very common diseases which need treatment. Whereas for example on the ISS in case of serious medical emergency, a crew member can be transported back to Earth in a matter of hours, during a mission to Mars there is no return possibility other than the nominal flight schedule. The consequences of this will have to be examined in every stage of the mission, from the crew selection (from risk screening to preventive medical treatment), crew training (proficiency in routine anamnesis and treatment should be present in at least two of the crew members), selection and preservation of medical supplies (even the number of painkillers, so to speak, will have to be determined in advance), diagnostic tools of minimal or non-invasive nature (see Chaps. 21–29) with adequate and personalised countermeasure algorithms up to the definition of a small operation theatre including the possibility of tele-assisted diagnosis and surgery (Haidegger et al. 2011).

39.3.2 Closed-Loop Life Support Systems

Space travelers are very human, 'normal' people, and are not different in their need for some very basic consumables, such as water, oxygen and food. Although this sounds trivial, it is in fact something that has far-reaching implications. Table 39.2

Table 39.2 Daily consumption of consumables on the ISS per crewmember

Item	Mass (kg/crewmember/day)
Drinking water	1.5
Food water	0.5
Metabolic water	0.35
Hygiene water	1.0
Water for oxygen supply	1.0
Cooling water	0.85
Food	2
Crew items	1.0
Other	0.4
Total	8.6

shows the daily requirements of water, food and other supplies on the ISS. In this table it is assumed that all oxygen will be generated from electrolysis of water.

For a space mission of 520 days, which is the realistic travel time to Mars, this would add up to approximately 27,000 kg for a crew of six. This is totally incompatible with the capabilities of current or future launchers. Clearly, recycling of waste back into air, water and food is mandatory to keep the total mass of a Mars mission under control. There are several approaches to this problem. Currently on the ISS the Environmental Control and Life Support System (ECLSS) is taking care of the physico-chemical recovery of waste water (atmospheric water vapor as well as urine). The throughput of the urine system is maximum 9 kg/day with a recovery level of 80%. In the words of astronaut Frank de Winne: 'I have seen this cup of coffee before'. More advanced systems are being developed and tested to improve on these numbers. ESA developed a waste-water recycling system based on nano-filtration and reverse osmosis that is currently being employed in the realistic environment of the Concordia base on Antarctica. Its capacity is some 40,000 L of water per year, with a recycling efficiency of 90%.

The next and ultimate step would be to create a complete bioregenerative cycle, in which all water, air waste and food channels are combined (Mergeay et al. 1988). Such a system, under development by ESA under the name Melissa (Micro-ecological Life Support System Alternative) is depicted in Fig. 39.3. Of particular importance is the Photosynthesis Compartment 4, where the products of other compartments are used to produce higher plants for consumption. In 2009, the Melissa Pilot Plant (Fig. 39.4) was activated that will serve to provide actual data on performance (Lasseur et al. 2010).

Edible plants (lettuce, radish, peas) have been produced in various experiments dating back to the early 70s and up to the ISS (Zabel et al. 2014). Experiments on the ISS are carried out to study the fundamental aspects of growing plants for food in space (Kittang et al. 2014). Results of these experiments will be used in the future to improve the efficiency of food production on Earth.

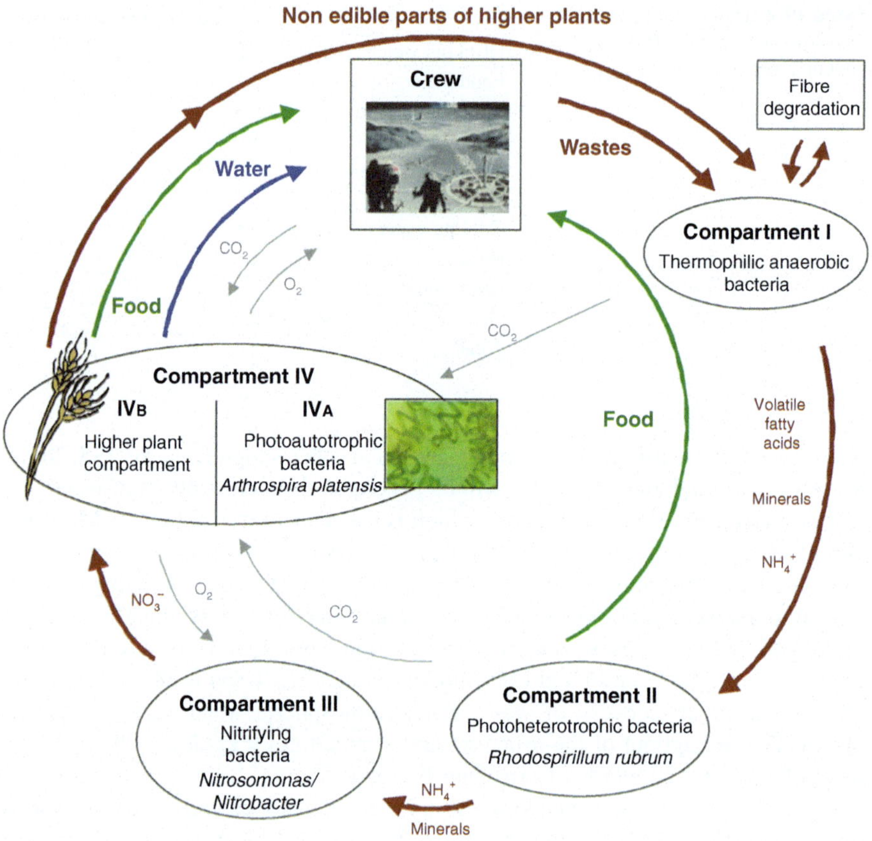

Fig. 39.3 The MELISSA advanced loop concept, showing the different compartments and their connections, forming an artificial micro-ecosystem (image ESA)

39.3.3 Chemical and Biological Dangers

Both in the spacecraft and on a planetary surface, astronauts may be exposed to an environment that may be hazardous for other reasons than just radiation (Horneck et al. 2006). Equipment has to be developed such that no off-gassing of toxic material occurs, and that dangerous chemicals will remain contained. A particular health risk results from the development of pathogens in a closed environment. Experience shows that bacteria and fungi can grow on surfaces or in hidden niches on board manned spacecraft, posing a potential risk both to crew health and equipment depending on the degree of contamination (Lauber and Ullrich 2016). More specifically, in systems where waste, water, air and food will be recycled to a large extent preferential growth of certain bacteria cannot be excluded. Also the possibility of genetic mutation of bacteria by radiation or other environmental factors has to be taken into account (Baatout et al. 2007).

Fig. 39.4 The MELISSA pilot plant at the University Autònoma of Barcelona (image ESA)

On the planetary surface, even if the soil would not contain toxic materials, the mere inhalation of dust particles, particularly in a reduced gravity environment, can pose health hazards. Free radicals, salts and oxidants can be aggressive both for human tissue and for equipment, in particular in humid conditions. Toxic materials and potential organics add to the dangers. In extremis, even unknown biohazards may be encountered such as mutant or fully extraterrestrial viruses, yeasts or bacteria (Cousins and Cockell 2016). New monitoring and protection techniques will have to be developed to keep such risks under control.

39.3.4 Future Permanent Habitats Beyond LEO

Since the first edition of this book the situation regarding the perspectives of future human space habitats has evolved, although not always in a clear direction. Within the context of an international coordination group (ISECG), 14 global Space Agencies are formulating an approach towards the 'next destination' in human space exploration. However, the final choice will stay a matter of national priorities and funding and is therefore, as so many political issues, subject to rather frequent changes in orientation. At the moment of writing no clear approved and funded projects are available that foresee creating a permanently inhabited space base. What seems to emerge, however, is a sense that a base on the lunar surface or in the vicinity of the moon most likely will be the first step, to be expected towards the end of the next decade. Many important milestones are still to be reached, in particular related to radiation protection, habitat design and supply possibilities. It is certain that these will need further study, including looking into possibilities of subsurface construction and using local resources for supply (see below). Therefore, initially such bases may not be permanently crewed, to allow for milder radiation regimes and supply requirements. A next step may then see the (robotic?) construction of a shielded lunar base that would allow for permanent habitation. The technological solutions for such an approach are within reach and it is more a question of finding the proper financing and international collaboration scheme that will determine the timescale.

For Mars, the main reasons for making the trip are the sense of exploration and the gathering of scientific knowledge, in particular related to the possibility of finding extinct or even extant non-terrestrial life. Also for Mars, however, there are some strong voices arguing for commercial endeavours. What is slightly disturbing in these discussions is that most of the fundamental issues such as radiation protection and resource scenarios are not really addressed. Whatever the objective, it should be realised that for a nominal flight to Mars the available time on the surface cannot be changed at will. It is either roughly 1 month, or almost 1.5 years, determined by the relative position of Earth and Mars allowing for a Hohmann transfer orbit. In the first case the amount of time spent on the Martian surface is less than 10% of the total mission duration, which can be considered wasteful. The second scenario will require something that could only be described as a permanent base. Almost certainly the first flight to Mars will be of the first type. Based on all these arguments, this author does not expect a permanent Martian base to happen in the first half of this century. However, the recently emerging private interest in lunar or Martian missions may change or even accelerate progress in this area.

39.3.5 Living of the Land

The greater the distance, delta-v and mission duration, the more supplies will have to be carried along. Some gains can be made by recycling technologies such as described above. However, major reductions in launch mass could be achieved if part of the consumables could be found at the place of destination. Possibilities include:

- Building material. Lunar or Martian regolith can be used, either in its pure state or after processing, as material for covering habitation modules for protection against radiation or meteorites. More advanced use could include regolith-base concrete for construction purposes. Local soil may also be used as substrate for food production (Wamelink et al. 2014).

Fig. 39.5 Mars Express image, showing water ice in an impact crater on Vastitas Borealis, an area near the Martian Polar regions (image ESA)

- Water. The presence of water ice in Martian polar regions has first been demonstrated in 2005 from images from the ESA Mars Express mission (Fig. 39.5) and confirmed in 2008 by the NASA Phoenix Lander (Smith et al. 2009). There is also speculation on the existence of underground reservoirs of (frozen) water based on analysis of Martian meteorites and direct observation (Stuurman et al. 2016).
- More recently, the presence of water ice on the moon was inferred from spectroscopic measurements of dust plumes originating from the lunar south pole region (Colaprete et al. 2010) and water traces have been identified in lunar soil samples returned during the Apollo missions (Hauri et al. 2011).
- Oxygen. Whereas pure oxygen cannot be found on either moon or Mars, it can be produced relatively easily from water using electrolysis. In absence of water, oxygen may also be produced from carbon dioxide, which is relatively abundant in the Martian atmosphere, or from the regolith itself. In the latter case, significant amounts of energy will be required, however (Zubrin and McKay 1997). In order to reduce weight and energy requirements further (also in the sense of allowing less sturdy and hence massive constructions) it is not unlikely that early habitats will use hypobaric conditions which in itself may impact physiological adaptation and health risk (see Chap. 16).
- Propellant. The mass of propellant required for the return trip to Earth is quite significant. If one bases oneself on a simple hydrogen oxygen engine, in particular the mass of the oxygen will be important. As seen in the previous section, production of oxygen is feasible on Mars and maybe on the moon as well. More advanced ideas are also studied, such as the production of methane and oxygen from Martian carbon dioxide and hydrogen (Meier et al. 2017).
- Metals, plastic, etc. Various studies have looked into chemical processes to produce iron, aluminum, copper and even ethylene from locally available resources. Given the presence on Mars of carbon and hydrogen, Earth-based technologies could be applicable there (Zubrin and McKay 1997). On the moon this may be more complicated, in view of the low abundance of these two elements.

The more complex the above procedures become in general, the more energy intensive they will be. Therefore, energy production, either solar or nuclear energy, will be an issue that needs to be resolved first. Another aspect that requires consideration is that if mission-critical consumables are supposed to be generated in this way, it is most likely wise that a mission be split in two parts, and that the human crew will only be sent on its way when positive confirmation has been obtained that the required oxygen, propellant, etc. is already available.

39.3.6 Terraforming

An more drastic approach would be the creation of a breathable atmosphere and Earth-like water-based ecosystem on Mars in which humans could live without spacesuits or other protection, and where standard food production techniques can be employed.

The details of such a procedure have been studied in some detail (McKay and Marinova 2001; Beech 2009). A first step in this process would be raising the Martian temperature. On Mars, solid carbon dioxide is present at the poles. It can be shown that raising the temperature by only a few degrees could lead to a runaway effect, starting with the release of gaseous carbon dioxide which in turn, through the greenhouse effect, will lead to higher temperatures and release of even more carbon dioxide. In this way, the surface temperature on Mars could approach the melting temperature of water. There would be several ways to achieve the initial raise of temperature required for this process. Under more or less serious consideration are the use of large solar mirrors, redirecting meteors towards Mars and using the energy released on their impact, and adding strong greenhouse gases like Chlorofluorocarbons (CFC's) to the Martian atmosphere. Needless to say all of these solutions require massive technological developments and will not be available in this century.

A next step in terraforming Mars would be the activation of an artificial ecosystem. Suitable bacteria, most likely selected extremophiles, could initiate this, followed by plant life once sufficient water and organics will be available. This will then raise the oxygen concentration in the atmosphere, further increase of the temperature and finally the creation of a more or less Earth-like environment. It is estimated that the total process will require hundreds to thousands years as a minimum.

Of course, such a drastic development is at odds with current ethical values and principles of planetary protection, which say that mankind should leave other planetary systems pristine and free from Earth contamination (Sparrow 2015).

39.3.7 Nuclear Propulsion

To a large extent, the discussions before on the possibility to reach Mars or other destinations were based on current propulsion technologies, which are basically

Table 39.3 Performance characteristics of some current and future propulsion systems

Propulsion type	Specific impulse (s)	Thrust/weight
Chemical	200–400	0.1–10
Ion propulsion	1200–5000	10^{-4}–10^{-3}
Nuclear fission	500–3000	10^{-4}–10
Nuclear fusion	10^4–10^5	10^{-5}–10^{-2}
Antimatter annihilation	10^3–10^6	10^{-3}–1

chemical in nature. Using different, more efficient propulsion techniques, trajectories could be chosen that are faster than the Hohmann transfer orbit, since the higher energy requirements could be met without serious mass penalties. The two key parameters to define a propulsion system are the thrust (in Newton) and the so-called specific impulse I_{sp} (in seconds), representing the achieved impulse per unit of weight of propellant. Specific impulse determines how efficient a propulsion system is (the higher I_{sp}, the less propellant is required to achieve a certain delta-v), the thrust determines how fast such delta-v will be achieved. In Table 39.3, some examples are given of I_{sp}, and thrust (or rather thrust per weight) can be achieved with current and future propulsion systems (Mallove and Matloff 1989; Sutton 1992).

From the table it can be seen that chemical propulsion is the most reasonable solution today, even if it is not very efficient. Ion thrusters, in which ionised heavy gases (most often xenon is used) are accelerated in an electric field, are much more efficient, but have very low thrust and are therefore not practical fur human spaceflight purposes. As an example, the ESA Smart-1 mission in 2004 used only ca. 60 kg of xenon to reach the moon from LEO, but it took 409 days to get there (Rathsman et al. 2005).

For the very long future, propulsion systems for human spaceflight will therefore be based likely on nuclear technologies. The first concrete designs for this were in the late 1940s, and assumed a series of nuclear bombs to be detonated some 50 ft behind the rocket, with the blast being caught by a pusher plate (Dyson 2002). Since then, 'more elegant' designs have become available with less obvious drawbacks. These designs are based on nuclear fission reactors in which the energy is either used to generate power for electric propulsion or transferred to inert gas that is used as propellant (Bruno 2008). Designs on nuclear fusion are also proposed.

Even more promising, on paper at least, are engines in which anti-protons (such as are generated for example in CERN), would be stored in magnetic flasks and brought to annihilation through contact with normal matter. The energy released in such a process is enormous and can be used to generate huge exhaust velocities at the engine nozzle (Forward 1985). Designs for such engines do exist (Fig. 39.6) and prototype antiproton storage systems are in existence. Nevertheless, such systems will require amounts of anti-protons that are beyond the current capabilities, and it cannot be expected that within the century this technology will become mature.

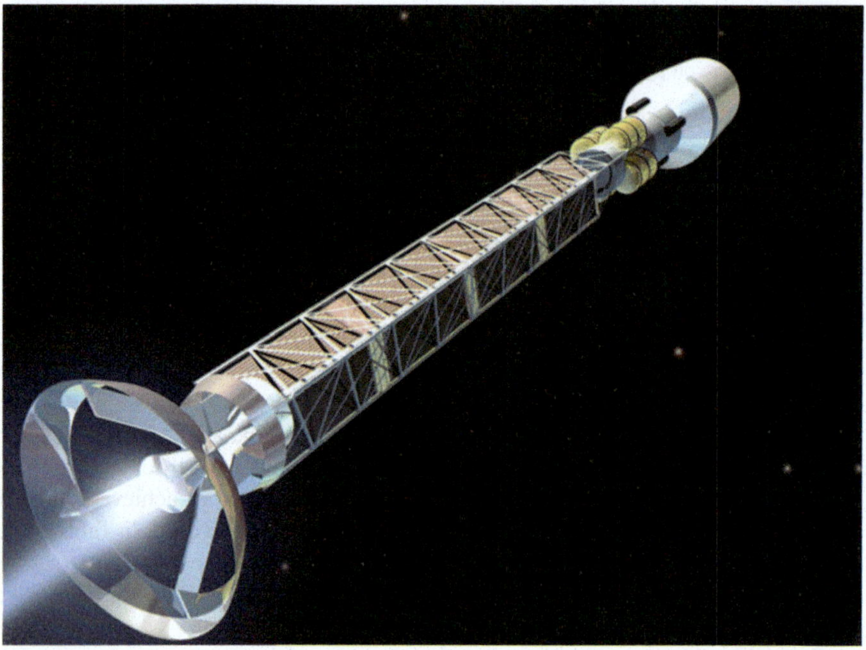

Fig. 39.6 Artist impression of a rocket using antimatter propulsion (source: Wikimedia)

39.3.8 Hibernation

Even when much more efficient engines would be produced routinely, the time of a space trip will remain enormous. As discussed in Sect. 39.2.1, the travel time (one way!) to even the closest neighbouring stars would take several tens of years in the most optimistic of all cases. Science Fiction literature and movies present us with many elegant solutions to this problem. Unfortunately, most of them involve transportation means ('hyperspace', 'warp') that have no connection with physical reality. Only one of these concepts has realistic merit. This is the possibility to extend our lifetime and overcome boredom through hibernation. Hibernation is a physiological process in which the body reduces its metabolic rate for prolonged periods of time, allowing some animals to overcome periods of extreme conditions (mostly cold). Specific chemicals triggering hibernation have been identified and synthetic derivatives have been shown to be effective in non-hibernating animals (Oeltgen et al. 1988; Vecchio et al. 2003; Blackstone et al. 2005; Hamid et al. 2009), although as yet a generic mechanism to induce hibernation is not identified. One, or much more likely many triggers and distal mechanisms to control the cells' metabolism need to be identified—and to be modulated in the right sequence and intensity (see Chap. 5). The question whether also humans could be brought in a hibernation is not yet answered, but from a biological perspective there are no reasons why this would not be possible. In fact, studies on the physiology of fetuses, breath-hold divers or persons using deep meditation techniques, suggest that they are able to enter a state of slowed vital functions (Singer and Mühlfeld 2007; Lindholm and Lundgren 2009; Tyagi and Cohen 2013). A specific advantage of hibernation for space applications may be found in the general observation that hibernating animals seem to be more robust with respect to external damage to organs and even cells (Fleck et al. 2005; Bouma et al. 2013). It could be hypothesised that this would also hold for effects of weightlessness and possibly even radiation damage (Cerri et al. 2016). This field of research has recently attracted further attention, not because of its interest for space exploration, but from the medical research community (Bouma et al. 2012), in view of the potential advantages for medical applications. Many open questions still exist, but research programs are in progress to address these. If it becomes practically feasible to bring astronauts in a hibernating state, this will be a true game-changer for future exploration missions.

39.4 Summary

The purpose of this chapter has not only been to list the challenges and caveats but also to demonstrate that the drive to enable human space travel beyond LEO is strong. The feasibility to travel to the moon has already been proven, but is still attracting renewed interest from space agencies across the world. Much research is now needed to tackle potential showstoppers to prolong human presence on or near the lunar surface, and to reach out farther, in particular to Mars. That amazing adventure will happen, although possibly only later in this century.

What is even more amazing is the capability and willingness of humans to think of the impossible and try to make it real. Several of the ideas presented above sound very fantastic. Yet, creative minds are busy to push the boundaries. In writing an update to this chapter for the second edition, I was amazed to see how much progress was made in the cosmologically negligible timespan of 5 years. History has demonstrated that humans have a huge drive to explore new horizons. The question is hardly 'how,' but 'when.'

References

Baatout S, Leys N, Hendrickx L, Dams A, Mergeay M (2007) Physiological changes induced in bacteria following pH stress as a model for stress research. Acta Astronaut 60:451

Beech M (2009) Terraforming. Springer, New York, NY

Blackstone E, Morisson M, Roth MB (2005) H2S induces a suspended animation-like state in mice. Science 59:308

Bles W, Groen E (2009) The DESDEMONA motion facility: applications for space research. Microgravity Sci Technol 21:281

Bouma HR, Verhaag EM, Otis JP, Heldmaier G, Swoap SJ, Strijkstra AM, Henning RH, Carey HV (2012) Induction of torpor: mimicking natural metabolic suppression for biomedical applications. J Cell Physiol 227(4):1285–1290

Bouma HR, Henning RH, Kroese FG, Carey HV (2013) Hibernation is associated with depression of T-cell independent humoral immune responses in the 13-lined ground squirel. Dev Comp Immunol 39(3):154–160

Bruno C (ed) (2008) Progress in astronautics and aeronautics. AIAA, Reston, VA, p 225

Bühlmeier J, Frings-Meuthen P, Maser-Gluth C, Heer M (2016) Glucocorticoid activity and metabolism with NaCl-induced low-grade metabolic acidosis and oral alkalization: results of two randomized controlled trials. Endocrine 52:139

Cerri M, Tinganelli W, Negrini M, Helm A, Scifoni E, Tommasino F, Sioli M, Zoccoli A, Durante M (2016) Hibernation for space travel: impact on radioprotection. Life Sci Space Res (Amst) 11:1–9. https://doi.org/10.1016/j.lssr.2016.09.001

Clarke AH, Grigull J, Müller R, Scherer H (2000) The three-dimensional vestibuloocular reflex during prolonged microgravity. Exp Brain Res 134:322

Clément G, Pavy-Le Traon A (2004) Centrifugation as a countermeasure during actual and simulated spaceflight: a review. Eur J Appl Physiol 92:235

Cogoli A (ed) (2002) Cell biology and biotechnology in space. Elsevier Publishing, Amsterdam. ISBN 0-444-50735-3

Colaprete A, Schultz P, Heldmann J, Wooden D, Shirley M, Ennico K, Hermalyn B, Marshall W, Ricco A, Elphic RC, Goldstein D, Summy D, Bart GD, Asphaug E, Korycansky D, Landis D, Sollitt L (2010) Detection of water in the LCROSS ejecta plume. Science 330:463

Cousins CR, Cockell CS (2016) An ESA roadmap for geobiology in space exploration. Acta Astronaut 118:286

Crucian BE, Choukèr A, Simpson RJ, Mehta S, Marshall G, Smith SM, Zwart SR, Heer M, Ponomarev S, Whitmire A, Frippiat JP, Douglas GL, Lorenzi H, Buchheim JI, Makedonas G, Ginsburg GS, Ott CM, Pierson DL, Krieger SS, Baecker N, Sams C (2018) Immune system dysregulation during spaceflight: potential countermeasures for deep space exploration missions. Front Immunol 9:1437. https://doi.org/10.3389/fimmu.2018.01437

Cucinotta FA, Durante M (2006) Cancer risk from exposure to galactic cosmic rays: implications for space exploration by human beings. Lancet Oncol 7:431

De la Torre GG, Van Baarsen B, Ferlazzo F, Kanas N, Weiss K, Schneider S, Whiteley I (2012) Future perspectives on space psychology: recommendations on psychosocial and neurbehavioural aspects of human spaceflight. Acta Astronaut 81:587

DiPrampero PE, Narici MV (2003) Muscles in microgravity: from fibres to human motion. J Biomech 36:403

Durante M, Cucinotta FA (2008) Heavy ion carcinogenesis and human space exploration. Nat Rev Cancer 8:456

Durante M, Reitz G, Angerer O (2010) Space radiation research in Europe: flight experiments and ground-based studies. Radiat Environ Biophys 49(3):295

Dyson G (2002) Project orion – the atomic spaceship 1957–1965. Penguin, London

Feuerecker M, van Oosterhout WP, Feuerecker B, Matzel S, Schelling G, Rehm M, Vein AA, Choukèr A (2016) Headache under simulated microgravity is related to endocrine, fluid distribution, and tight junction changes. Pain 157(5):1072–1078. https://doi.org/10.1097/j.pain.0000000000000481. PubMed PMID: 26761382

Fleck CC, Carey HV (2005) Modulation of apoptotic pathways in intestinal mucosa during hibernation. Am J Physiol Regul Integr Comp Physiol 289(2):R586–R595

Forward RL (1985) Antiproton annihilation propulsion. J Propuls 1:370

Friml J, Wisniewska J, Benkova E, Mendgen K, Palme K (2002) Lateral relocation of auxin efflux regulator PIN3 mediates tropism in Arabidopsis. Nature 415:806

Frings-Meuthen P, Baecker N, Heer M (2008) Low-grade metabolic acidosis may be the cause of sodium chloride-induced exaggerated bone resorption. J Bone Miner Res 23(4):517–524

Gerlach DA, Marshall-Goebe K, Hasan KM, Kramer LA, Alperin M, Rittweger J (2017) MRI-derived diffusion parameters in the human optic nerve and its surrounding sheath during head-down tilt. NPJ Microgravity 3:18

Grimm D, Pietsch J, Wehland M, Richter P, Strauch MS, Lebert M, Magnusson NE, Wise P, Bauer J (2014) The impact of microgravity-based proteomics research. Expert Rev Proteomics 11:465

Grimm D, Grosse J, Wehland M, Mann V, Reseland JE, Sundares A, Corydon TJ (2016) The impact of microgravity on bone in humans. Bone 87:44

Häder D, Hemmersbach R, Lebert M (2005) Gravity and the behaviour of unicellular organisms. Cambridge University Press, Cambridge. ISBN 0-521-82059-9

Häder D, Braun M, Grimm D, Hemmersbach R (2017) Gravireceptors in eukaryotes – a comparison of case studies on the cellular level. NPJ Microgravity 3:13
Haidegger T, Sándor J, Benyó Z (2011) Surg Endosc 25:681
Hamid A, Schultz MJ, Juffermans NP (2009) Potential applications of hydrogen sulfide-induced suspended animation. Curr Med Chem 16:1295
Hauri EH, Weinreich T, Saal AE, Rutherford MC, Van Orman JA (2011) High pre-eruptive water contents preserved in lunar melt inclusions. Science 333:213
Heer M, Boese A, Bäcker N, Zittermann A, Smith SM (2004) Moderate hypocaloric nutrition does not exacerbate bone resorption during bed-rest. FASEB J 18:478
Horneck G, Facius R, Reichert M, Rettberg P, Seboldt W, Manzey D, Comet B, Maillet A, Preiss H, Schauer L, Dussap CG, Poghdon L, Belyavin A, Reitz G (2006) HUMEX, a study on the survivability and adaptation of human long-duration exploratory missions, part II: Mission to mars. Adv Space Res 38:752
Iwasaki KI, Sasaki T, Hirayanagi K, Yajima K (2001) Usefulness of daily +2Gz load as a countermeasure against physiological problems during weightlessness. Acta Astronaut 49:227
Jenkins DW (1968) USSR and US Bioscience. Bioscience 18:543
Karlsson LL, Kerckx Y, Gustafsson LE, Hemmingsson TE, Linnarsson D (2009) Microgravity decreases and hypergravity increases exhaled nitric oxide. J Appl Physiol (1985) 107(5):1431–1437. https://doi.org/10.1152/japplphysiol.91081.2008
Kittang AI, Iversen TH, Fossum KR, Mazars C, Carnero-Diaz E, Boucheron-Dubuisson e LDI, Legue V, Herranz R, Pereda-Loth V, Medina FJ (2014) Exploration of plant growth and development using the European Modular Cultivation System facility on the International Space Station. Plant Biol 16:528
Kopparapu RK, Ramirez R, Kasting JF, Eymet V, Robinson TD, Mahadevan S, Deshpande R (2013) Habitale zones around main-sequence stars: new estimates. Astrophys J 765:2
Lasseur C, Brunet J, de Weever H, Dixon M, Dussap G, Godia N, Leys N, Mergeay M, Van der Straeten D (2010) MELISSA: the European project of closed life support system. Gravitat Space Res 23:3
Lauber BA, Ullrich O (2016) Spacecraft contamination monitoring and control. In: Chouker A, Ullrich O (eds) The immune system: are we prepared? Springer, New York, NY, p 89
LeBlanc A, Schneider V, Shakelford L, West S, Oganov V, Bakulin A, Varonin L (2000) Bone mineral and lean tissue loss after long duration space flight. J Musculoskelet Neuronal Interact 1:157
Lindholm P, Lundgren CEG (2009) The physiology and pathophysiology of human breath hold diving. J Appl Physiol 106:284–292
Linnarsson D, Hughson RL, Fraser KL, Clément G, Karlson LL, Mulder E, Paloski WH, Rittweger J, Wuyts FL, Zange J (2015) Effects of an artificial gravity countermeasure on orthostatic tolerance, blood volumes and aerobic power after short-term bed rest. J Appl Physiol 118:29
Mader TH, Gibson R, Pass AF, Kramer LA, Lee AG, Fogarty J, Tarver WJ, Phillips JL, Tran D, Lipsky W, Cho J, Stern C, Kuyumjian R, Polk JD (2011) Optic disc edema, globe flattening, choroidal folds, and hyperopic shifts observed in astronauts after long-duration space flight. Ophthalmology 118:2058
Mallove EF, Matloff GL (1989) The starflight handbook. John Wiley & Sons, New York, NY
Maric PJ, Jones D, Vico L, Zallone A, Hinsenkamp M, Cancedda R (2000) Osteobiology, strain and microgravity. Part I: studies at the cellular level. Calcif Tissue Int 67:2
McKay CP, Marinova MM (2001) The physics, biology, and environmental ethics of making mars habitable. Astrobiology 1:89
Meier AJ, Shah MG, Hintze PE, Muscatello AC, Petersen E (2017) Mars atmospheric conversion to methane and water: an engineering model of the Sabatier reactor with characterization of Ru/Al_2O_3 for long duration use on Mars. 47th Int Conf on Environmental Systems (ICES), p 161
Mergeay M, Verstraete W, Dubertret G, Lefort-Tran M, Chipaux C, Binot R (1988) Proceedings 3rd European symposium on space thermal control and life support systems, Noordwijk
Narici L, Berger T, Matthiä D, Reitz G (2015) Radiation measurements performed with active detectors relevant for human exploration. Front Oncol 5:273

Oeltgen PR, Nuchols PA, Nilekani SP, Spurrier WA, Su T-P (1988) Further studies on opioids and hibernation: delta opioid receptor ligand selectively induced hibernation in summer-active ground squirrels. Life Sci 43:1565

Parsons JL, Townsend LW (2000) Interplanetary crew dose rates for the August 1972 solar particle event. Radiat Res 153(6):729

Perbal G, Driss-Ecole D (2003) Mechanotransduction in gravisensing cells. Trends Plant Sci 8:398

Petersen N, Jaekel P, Rosenberger A, Weber T, Scott J, Castrucci F, Lambrecht G, Ploutz-Snyder L, Damann V, Kozlovskaya I, Mester J (2016) Exercise in space: the European Space Agency approach to in-flight countermeasures for long-duration missions on the ISS. Extr Phys Med 5:9

Pollard EC (1965) Theoretical studies on living systems in the absence of mechanical stress. J Theor Biol 8:113

Rakova N, Jüttner K, Dahlmann A, Schröder A, Linz P, Kopp C, Rauh M, Goller U, Beck L, Agureev A, Vassilieva G, Lenkova L, Johannes B, Wabel P, Moissel U, Vienken J, Gerzer R, Eckardt KU, Müller DN, Kirsch K, Morukov B, Luft FC, Titze J (2013) Long-term space flight simulation reveals infradian rhythmicity in human Na^+ balance. Cell Metab 17:125

Rathsman P, Kugelberg J, Bodin P, Racca GD, Foing B, Stangaro L (2005) SMART-1: development and lessons learnt. Acta Astronaut 57:455

Sandal GM, Bye HH (2015) Value diversity and crew relationships during a simulated space flight to Mars. Acta Astronaut 114:164

Scheld K, Zimmermann A, Heer M, Herzog B, Mika C, Drummer C, Stehle P (2001) Nitrogen metabolism and bone metabolism markers in healthy adults during 16 weeks of bed-rest. Clin Chem 47:1688

Schneider S, Abeln V, Popova J, Fomina E, Jacubowski A, Meeusen R, Strüder HK (2013) The influence of exercise on prefrontal cortex activity and cognitive performance during a simulated spaceflight to Mars (MARS500). Behav Brain Res 236:1

Singer D, Mühlfeld C (2007) Perinatal adaptation in mammals: the impact of metabolic rate. Comp Biochem Phys A 148:780–784

Smith PH, Tamppari LK, Arvidson RE, Bass D, Blaneym D, Boyntonm WV, Carswell A, Catling DC, Clark BC, Duck T, Dejong E, Fisher D, Goetz W, Gunnlaugsson HP, Hecht MH, Hipkin V, Hoffman J, Hviid SF, Keller HU, Kounaves SP, Lange CF, Lemmon MT, Madsen MB, Markiewicz WJ, Marshall J, McKay CP, Mellon MT, Ming DW, Morris RV, Pike WT, Renno N, Staufer U, Stoker C, Taylor P, Whiteway JA, Zent AP (2009) H_2O at the phoenix landing site. Science 325(5936):58

Sparrow R (2015) Terraforming, vandalism and virtue ethics. In: Galliott J (ed) Commerical space exploration. Ashgate, London, p 161

Stein TP, Donaldson MR, Leskiw MJ, Schluter MD, Baggett DW, Boden G (2003) Branched-chain amino acid supplementation during bed-rest: effect on recovery. J Appl Physiol 94:1345

Strewe C, Feuerecker M, Nichiporuk I, Kauffman I, Hauer D, Morukov B, Schelling G, Choukèr A (2012) Effects of parabolic flight and spaceflight on the endocannabinoid system in humans. Rev Neurosci 23:673

Stuurman CM, Osinski GR, Holt JW, Levy JS, Brothers TC, Kerrigan M, Campbell BA (2016) SHARAD detection and characterization of subsurface water ice deposits in Utopia Planitia, Mars. Geophys Res Lett 43:9484

Sutton GP (1992) Rocket propulsion elements. John Wiley & Sons, New York, NY

Tyagi A, Cohen M (2013) Oxygen consumption changes with Yoga practices: a systematic review. J Evid Based Compl Alter Med 18:290–308

Van Baarsen B (2013) Person autonomy and voluntariness as important factors in motivation, decision making and astronaut safety; first results from the Mars500 LODGEAD study. Acta Astronaut 87:139

Vecchio L, Baldelli B, Malatesta M, Biggiogera M (2003) Eur J Clin Invest 33(Suppl):49

Verbanck S, Linnarsson D, Prisk GK, Paiva M (1996) Specific ventilation distribution in microgravity. J Appl Physiol 80:458

Vermeer K, Wolf J, Craciun AM, Knapen MH (1998) Bone markers during a 6-month space flight: effects of vitamin K supplementation. J Gravit Physiol 5:65
Viguier A, Clément G, Trotter Y (2001) Distance perception within near visual space. Perception 30:115
Wamelink GWW, Frissel JY, Krijnen WHJ, Verwoert RM, Goedhart PW (2014) Can plants grow on Mars and the Moon: a growth experiment on Mars and Moon soil simulants. PLoS One 9:e103138
Yi B, Rykova M, Feuerecker M, Jager B, Lading C, Basner M, Horl M, Matzel S, Kaufmann I, Strewe C, Nichiporuk I, Vassilieva G, Rinas K, Baatout S, Schelling G, Thiel M, Dinges DF, Morukov B, Choukèr A (2014) 520-d isolation and confinement simulating a flight to Mars reveals heightened immune responses and alteration of leukocyte phenotype. Brain Behav Immun 40:20
Young RS, Tremor JW (1968) Weightlessness and the developing frog egg. Proceeding of the 10th COSPAR meeting, London, p 87
Zabel P, Bamsey M, Schubert D, Tajmar M (2014) Review and analysis of plant growth chambers and greenhouse modules for space. 44th Int Conf on Environmental Systems (ICES)
Zubrin R, McKay CP (1997) Technological requirements for terraforming Mars. JBIS 50:83

Part VII

Synopsis

Summary and Outlook 40

Dominique Moser, Alexander Choukér,
and for all authors...

40.1 Stress and the Effect of Stress Mediators

40.1.1 Effects of Stress on Human Physiology

The term *Stress* denotes the event of unknown or known but challenging situations, individuals have to cope with and to adapt to. This event may be of physiological or psychological nature or a combination of both and endangers the inner equilibrium (homeostasis). Regain of homeostasis and maintenance of stability is reached by allostasis which is described as physiological or psychological changes in response to disturbance. Dependent of the kind of stress, such adaptations may occur already in cellular compartments like mitochondria, at single-cell level or physiologically like in neurobiological systems or the autonomic nervous system (ANS) (see Chaps. 4–6).

At the neurobiological level, stressful situations lead to the activation of the hypothalamic-pituitary adrenal (HPA) and the sympathetic-adrenal medullary axis, resulting in a *stress response* or *fight-or-flight* response, which is characterized by the release of stress hormones like cortisol, catecholamines or ligands to the peripheral or central endocannabinoid (EC) receptor system. Adaptation to stress is common throughout all organisms and crucial to survive everyday life as well as extreme situations. An appropriate concentration of stress mediators or an acute short concentration peak is beneficial for performance or outcome; however, chronic stress has opposite effects. An overuse or dysregulation of stress mediators is likely to result in allostatic load or overload and the ability to cope with these situations

D. Moser (✉) · A. Choukér (✉) · for all authors...
Laboratory of Translational Research Stress and Immunity, Department of Anaesthesiology, Hospital of the University of Munich (LMU), Munich, Germany
e-mail: Dominique.Moser@med.uni-muenchen.de; Alexander.Chouker@med.uni-muenchen.de

© Springer Nature Switzerland AG 2020

A. Choukér (ed.), *Stress Challenges and Immunity in Space*,
https://doi.org/10.1007/978-3-030-16996-1_40

strongly depends on the individual variations in perceiving and handling stress as well as support from outside (see Chaps. 4, 6, and 10).

Stress hormones affect diverse physiological systems and in extremely stressful and/or chronic situations the effect is negative for cognitive and physical performance, emotional regulation, mood and health.

Findings in the research field of psychoneuroimmunology state, that chronic stress exposure affects negatively inflammatory processes and wound healing and increases the susceptibility to illness. Furthermore chronic stress is associated with an increased production of reactive oxygen species (ROS), a reduced telomerase activity and shortened telomeres, factors which induce and indicate accelerated aging. Premature aging (senescence) can take place in several cell types in the body; however, a senescent immune system (immunosenescence) is associated with an increased susceptibility to infection and a greater risk for cancer development (see Chaps. 4 and 6).

40.1.2 Stress Mediators

40.1.2.1 Cortisol

The glucocorticoid cortisol is a powerful stress hormone involved in essential mechanisms like metabolism, memory, and functionality of the immune system (see Chaps. 4, 7, and 12). Cortisol is a widely used pharmacologic agent in psychiatrics since its regulatory function on memory includes consolidation of emotionally arousing experiences and the impairment of memory retrieval. Though chronically enhanced cortisol levels are believed to be a psychological adaptive response to help the organism coping with stressful events, it is also considered to play a role in pathogenesis and symptomatology of anxiety disorders (see Chap. 7). Moreover, cortisol has a known and very potent anti-inflammatory function and is used in a variety of inflammatory diseases as an antiphlogistic drug. However, the strong suppressive properties of cortisol severely impair immune cell functions (see Chap. 12) at the transcriptional level and by direct inhibition of signalling pathways. The prominent role of cortisol in stress-induced immunosuppression encouraged scientists to focus on the identification of naturally occurring and synthetically produced glucocorticoid receptor (GR) antagonists. The adrenal cortical hormone DHEA (dehydroepiandrosterone), which is secreted in response to stress, showed promising anti-glucocorticoid and immune-stimulatory effects. Similarly, pharmacological GR antagonists like Mifepristone demonstrated anti-glucocorticoid effects and further substances are currently under evaluation (see Chap. 35).

40.1.2.2 Catecholamines

The ANS acts mostly involuntarily and regulates essential body functions such as heart rate or respiratory rate and is the primary mechanism in control of the *fight-or-flight* response. Under stressful situations the sympathetic activity rises, which is mirrored by an increased concentration of the catecholamines epinephrine and norepinephrine. Dysregulation of the ANS or sympathetic over-activity can lead over time to cardiovascular diseases like hypertension or myocardial infarction (see Chap. 8). Furthermore norepinephrine has direct inhibitory effects on the immune system via action on α- and β-adrenergic receptors, which are expressed on the surface of immune cells. Immunosuppression, which is induced by sympathetic over-activity, has been shown to be counteracted by β-blockers through binding and thereby blocking adrenergic receptors on immune cells (see Chap. 35).

40.1.2.3 The Endocannabinoid System

The endocannabinoid (EC) system, which has a highly specific negative feedback control of the HPA-axis, is the third neurobiological mechanism introduced in this book and acts also as a regulator under stressful conditions. The EC system is highly conserved and controls key elements of physiological and psychological homeostasis like metabolism, neurobehavioral changes during stress and anxiety, and the regulation of cognition and memory. Ligands of the EC system bind to their respective receptors (CB1 and CB2) which are primarily located in the brain and also on immune cells where ECs display immunosuppressive properties (see Chap. 10).

40.2 The "Space Exposome" as a Collection of Stressors

The space exposome comprises several conceivable stressors that occur in space and in and around space stations. At a first glance this includes entirely new conditions like permanent microgravity (μg) and not unknown factors like isolation, confinement, and sleep disturbances; however, extent and intensity exceed levels which we are accustomed to on Earth. Since almost two decades, the International Space Station (ISS) represents a platform for on-site research in space and a habitat for astronauts. With the onset of mission, the body is forced to adapt to this new environment, which seriously impacts homeostasis and allostasis. The duration of mission, conditions of isolation, confinement and the high pressure to succeed have a negative impact on the cognitive performance and mood of astronauts (see Chaps. 12 and 22).

40.2.1 Lack of Natural *Zeitgeber*

The ISS orbits the Earth every 90 minutes, which means that astronauts undergo 16 day/night cycles in 24 hours. As a consequence of lack of a natural *zeitgeber* and high ambient noise, circadian rhythm and sleep behaviour is disrupted. This does not only lead to reduced concentration and bad mood, insufficient and unsatisfactory sleep also leads to an increase of cortisol levels and this in turn further worsens sleep quality, increases risks of metabolic syndromes, and has negative impact on the immune system (see Chaps. 4 and 9).

40.2.2 Microgravity

In space, the body is exposed to permanent μg, which leads to a dramatically reduced strain of muscles and bones, especially in the legs. The adaptive response to this situation is deconditioning, a consequence which is not fully realized until return to Earth (see Chap. 32). Furthermore body fluids are redistributed in μg to the upper part of the body including the head. This often leads within the first days in space to headaches and later to visual impairments. In the case of blood redistribution, a tendency to higher systolic arterial blood pressure and heart rate was observed during space missions (see Chap. 8). In addition, lack of gravity leads to a reduction of convective heat transfer, resulting in an increase in core body temperature (CBT) by about 1.01 °C during the first 45 days of deployment. Until the end of mission, CBT does not rise further but remains constant at an elevated level (see Chaps. 9 and 26). Single cells sense changed gravity conditions by altered extracellular matrix mechanics, cell shape, and cytoskeletal organization. Since all cells and tissues represent highly dynamic systems, ultra-fast adaptation processes take place in the sense of gene expression and cytoskeletal stability (see Chap. 17). Aside from μg, hypobaric hypoxia prevails in the ISS (see Chap. 16).

40.2.3 Radiation

An additional potential stressor in space is radiation, which is also considered to be one of the most critical and limiting factors for long-term space missions beyond low Earth orbit. Radiation in space is different from radiation on Earth, and on board the ISS 200 times higher than on Earth with strong mutagenic properties. Radiation-induced cell damage happens either directly by ionization of DNA, proteins, or lipids or indirectly by the production of harmful free oxygen radicals. Both factors increase ultimately the risk of developing malignancies (see Chap. 20).

40.2.4 Microbes

Besides the human inhabitants, the ISS is also colonized by diverse strains of microbes. The main source of the different microbial communities is undoubtedly the members of the crew themselves. However, under conditions of chronic stress even symbiotic and commensal microorganisms can turn out to be harmful to the host. Moreover, external bacteria e.g. on surfaces also hold a potential health risk. Bacteria which were exposed to spaceflight conditions display an increased virulence and altered susceptibility to antibiotics (see Chaps. 18 and 25).

In brief, the space exposome comprises conditions of isolation, a misaligned external *zeitgeber*, μg, high radiation, an increased microbioal load, and hypobaric hypoxia with distinct effects of every single stressor. However, all of these stressors have in common, that they have a negative impact on the immune system (see Fig. 40.1) which rises concerns about the feasibility of long-duration interplanetary

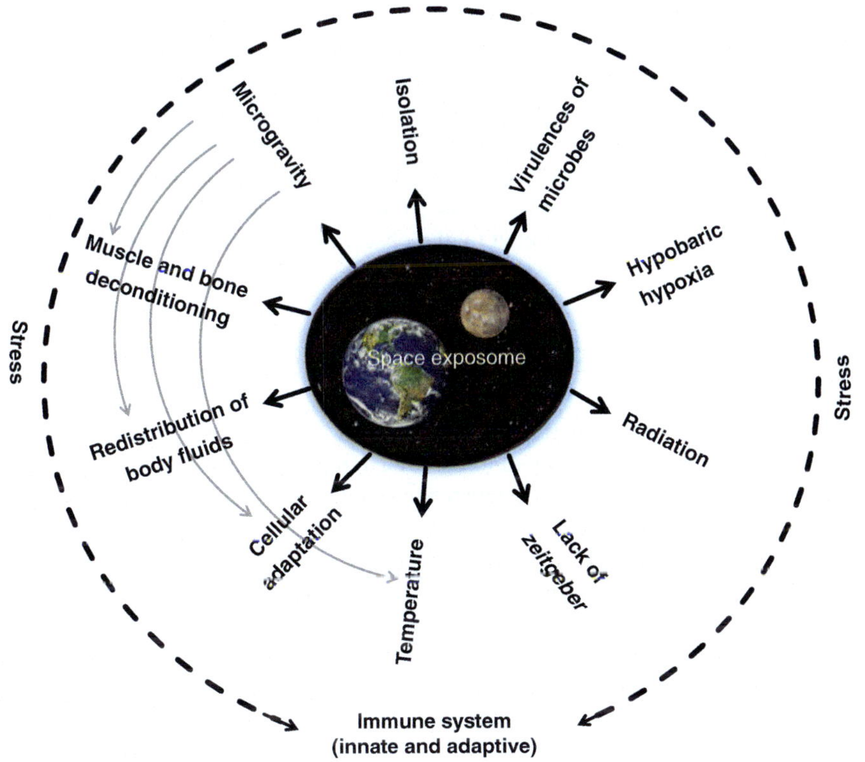

Fig. 40.1 The special living conditions on board a space station are summarized as "Space Exposome". These stressors are of physical (radiation, microgravity), biological (bone and muscle loss, misaligned circadian rhythm), microbial (increased virulence of pathogens), and psychological (isolation, confinement) nature and have all a negative impact on the immune system

space missions due to health issues. An impaired immune system may lead to an increased susceptibility to bacterial and viral infections, favors reactivation of latent viruses and enhances the risk of cancer development.

40.3 Effects of Spaceflight on the Immune System

Due to technical limitations, the majority of space-related immune studies gain results from the comparison of the immune state pre-flight and after return to Earth (see Chap. 11).

40.3.1 Alterations of the Innate Immune System

Spaceflight has proven effects on different subsets of the innate immune system. Neutrophils for instance of which the main functions are induction of tissue inflammation and eliminating pathogens by the production of ROS and phagocytosis are highly activated directly after mission, which is most probably due to hypergravity-induced stress during landing. Though, the ability to phagocytose is reduced and the cytokine profile is pushed in an inflammatory state (see Chap. 12). NK display changes in percentage in peripheral blood as well as in crucial functions like cytotoxicity (see Chap. 13). Flow-cytometry based inflight measurements revealed an altered distribution of peripheral blood leukocytes, which indicates a hampered immune response and is supported by the observation of several astronauts having bacterial or viral infections (see Chap. 11).

40.3.2 Alterations of the Adaptive Immune System

The adaptive immune system is equally affected by spaceflight. Inflight analyses have shown, that μg leads to an inhibition of T-lymphocyte activation and proliferation. Immunosuppression is attributed to the adaptation process of T-lymphocytes to μg, which includes altered phosphorylation of activating signalling pathways, remodelling of the cytoskeleton, and changes at the transcriptional level (see Chap. 17). In general T-lymphocytes fulfil either functions of direct killing of virus-infected cells or tumor cells (cytotoxic T cells) or they have supporting functions by activating other immune cells (T helper($_h$) cells). In this context the T_h1 subset promotes cell-mediated immune responses, whereas the T_h2 subset is involved in the humoral immune response. In space the T_h1:T_h2 cytokine balance is shifted to a decreased IFNγ:IL10 phenotype, which describes a process away from cell-mediated immunity in the direction of immunosuppression and autoimmunity (see Chap. 14). This state further promotes proneness to viral infection, allergies, and autoimmune diseases, some of which become clinically relevant in space and even long after return to Earth. Antibody-producing B-cells show as all immune cells a high susceptibility to changes in gravity. In μg, B-cell lymphopoiesis is severely affected by altered activation and differentiation. Under spaceflight conditions, humoral immune response is disrupted because of lower hypermutation and antibody affinity maturation. Equally affected is the diversity of antibody repertoire because of modified combinations of antibody VH segments (see Chap. 15).

40.4 Monitoring Approaches and Devices

Extensive and multi-disciplinary monitoring of the living environment and well-being of the crew is indispensable for a better understanding of the detrimental effects of spaceflight on physiology, to continuously improve living conditions at space stations and to minimize risks for the crew health and mission failure. In most cases, monitoring procedures have to be performed on-site by the crew. Therefore, recently developed tools accomplish the standard of highest possible sensitivity while they are easy to handle and time-saving. An individual monitoring is of great importance. The following section gives an overview of monitoring tools that have found application on board the ISS and represent promising technologies to improve life on Earth.

40.4.1 Monitoring of the Microbial Burden

On board the ISS, the individual microbiome of each member of the crew is the main source for the variety of microorganisms. Under normal and fully immune-competent conditions, these microorganisms which can be found in the gut or on the skin, etc., are valuable to humans. However, under the conditions of impaired immunity due to stress or other factors, the same microorganisms might have detrimental health effects. Above that, space-induced increased virulence in a majority of microbes represents a further health risk for the crew. To monitor the microbial burden and to prevent distribution, space agencies defined international quality standards for air, surfaces, and water. According to this, air, internal surfaces, drinking water, and food are strictly tested on the ISS and an air-filtration was implemented (see Chap. 25).

40.4.2 Breath Air Analyses

Composition and pollution of ambient air and exhaled breath air give important information about potential health burden and emerging pathophysiological disorders. Gas chromatographic and mass spectrometry-based analytical devices enable monitoring of volatile organic and inorganic compounds (VOICs) in the spacecraft or habitat environment as well as in exhaled breath air of the crew. The E-NOSE for instance is based on a metal oxide sensor technology and has already proven its feasibility in detecting surface contaminations aboard the ISS. Further updating of such devices is needed and on the way to design a monitoring tool to measure quality of ambient air and to define non-invasively crews health by measuring volatile masses in the exhaled breath air. These devices give the opportunity to perform repeated and *point of care* analyses on each individual. By this, health and disease will be assessed as a function of time which is an important step for personalized therapy in space. On Earth, breath gas analysis is already an emerging tool in clinical settings allowing in the future the detection of a large number of disorders like bacterial infection, metabolic disorders, pulmonary disease, and cancer (see Chap. 24).

40.4.3 Monitoring Cardiovascular Deconditioning

Redistribution of body fluids and in particular blood in the upper body may lead to altered autonomic activity and therefore represents another risk factor for astronauts. Assessing heart rate variability (HRV) is a widely used tool to reflect the vegetative state of the body as well as physiological and psychological resilience and adaptability to various settings. HRV can be easily monitored by electrocardiography and blood pressure measurement and gives important information on alterations of autonomic activity and stress status (see Chap. 23).

40.4.4 Monitoring Core Body Temperature

Core body temperature (CBT) is strictly controlled and regulated by endothermic metabolism to ensure a constant body temperature of 36.7°C with slight alterations. During the first 45 days in space, CBT is increased most probably because of a reduced convective heat transfer in μg. During exercise, CBT was shown to rise further. High deviations in CBT have serious effects on physical and cognitive performance as well as on immune functions. Because of this, monitoring of CBT in humans both in illness and in space is indispensable. For a continuous CBT monitoring in all living conditions, measurement tools have to be non-invasive and not disturbing. The *Double Senor* is a combined skin temperature and heat flux sensor, which is placed on the forehead and displays very sensitive and rapid responses. These properties make the *Double Sensor* to a suitable and highly accepted monitoring tool (see Chap. 26).

40.4.5 Monitoring by Advanced Micro-Technologies and Artificial Intelligence

In the last years, technical innovations allowed the development of screening tools to obtain certainty of presence or absence of disease and to judge the demand for therapeutic strategies at the cellular and molecular level (point of care diagnostic). Lab-on-a-chip (micro-) technologies enable to clarify the individual health status by the minimal invasive procedure of blood withdrawal. These monitoring devices will help to evaluate, count, and identify different types of blood cells as well as analyze protein biomarkers in serum and plasma.

The use of artificial intelligence (AI) does not only represent a monitoring tool for psychological well-being but also supports during simple procedures and gives instructions for crew activities. Currently a first prototype of supporting AI robotics called CIMON (crew interactive mobile companion) is tested on the ISS. Besides assistance during crew activities, one of the goals is to develop a companion who assesses unprejudiced stress levels of the crew member and evaluates his potential employment during mission (see Chap. 21).

40.4.6 Monitoring of Immune Cell Performance

A majority of diseases occurring in space and long-term pathophysiology are linked to an inadequate performance of the immune system. For a better understanding of space-related immune alterations in *real-time* and to immediately validate appropriate countermeasures, on-site tools and assays will be implemented on the ISS and will be further extended in future to monitor crew during deep space exploration. These analytic devices will allow together with individualized *in vitro* incubation platforms the evaluation of basic immune phenotypes and functions like distribution shifts of leukocytes, cytokine production, and monitoring virus-specific immunity and overall immune performance (see Chaps. 21 and 27).

40.4.7 Psychological Monitoring

In addition to the enormous work load, the astronauts are exposed to during a space mission, the crew members live together in small groups with only limited privacy and contact to their families. It was shown that the occurrence and intensity of psychological factors correlate with the length of mission. Self-assessment questionnaires of crew members and observations made by ground personal help to evaluate the individual emotional state and need for support (see Chaps. 22 and 39). On ground, an additional quantitative but retrospective analysis of the stress mediators ECs and glucocorticoids in hair samples by high-performance liquid chromatography and mass spectrometry serves as a complementary dataset on acute versus chronic stress and further supports evaluation by covering different time windows (see Chap. 29).

40.4.8 Monitoring Before Deployment

Besides monitoring of hygienic conditions on board spacecrafts and psychological and physiological well-being of the crew, a proper and extensive assessment process of astronauts helps to select the most suitable candidate for space mission and therefore minimizes the risk of detrimental health effects and mission failure. Psychological tests during selection process and preparation phase help to pick candidates who distinguish themselves through a high psychological resilience, professionalism, and reliability (see Chap. 22). Furthermore candidates have to convince by physical fitness and physical capacity. However, susceptibility towards radiation or illness cannot be assessed by these methods. The biological response to space radiation is of critical concern for risk assessment and there is an existing high inter-individual variability in radiosensitivity based on genetic factors, with peripheral blood mononuclear cells (PBMCs) being the most radiosensitive cells among all cell types. Prolonged inadequate immune responses are associated with the emergence of hypersensitivities, autoimmunity, infectious diseases, and malignancies. Therefore ongoing research focusses on the identification of (predictive) biomarkers to determine both the received radiation dose (biodosimetry) as well as the radiosensitivity of individuals (see Chap. 28).

For all introduced strategies it has to be emphasized that the assessment of individual sensitivities and resistances of each person by comprehensive monitoring directs to a specialized medical care and supports the progress of not only *personalized* but also *precision* medicine.

40.5 Using Spaceflight Analogue Environments

Besides of pre-mission screening, on-site monitoring and post-flight assessments and follow-up, fundamental knowledge of stress-induced immune system disturbances under spaceflight conditions was and is still gained in Earth-bound space analogues.

These institutions are located in extreme environments or reflect extreme conditions, thereby allowing to investigate the role of one or more stressors, which ultimately adds to prepare for space missions. In addition, Earth-bound space analogues represent suitable and high fidelity research platforms to test the implementation of new monitoring or read-out parameters and technologies in multiple scientific investigations.

40.5.1 Isolation Studies

The impact of isolation and confinement on emotional well-being and monitoring of neurobiological stress responses and associated alterations of immune functions belong to well-defined experimental conditions. Two large-scaled studies—the Mars500 study and Antarctic overwintering studies—were introduced in this volume. Within the Mars500 study, six healthy male participants lived in a closed habitat simulating a spacecraft for 520 days under standardized nutritional and environmental conditions. In all participants, elevated morning-cortisol levels demonstrated an increased activation on the HPA axis. Furthermore aberrant heightened immune responses indicated a poorly controlled immune system both in and after mission (see Chaps. 36 and 37). Overwintering studies in Antarctica equally revealed an increased neurobiological stress response and altered immune reactions. Main differences between the two studies were duration of isolation (Mars500: 520 days; Antarctica: approx. 9 months) and more standardized conditions in Mars500 whereas more extreme environmental conditions were present during overwintering in Antarctica. At the French–Italian *Concordia* station in inner Antarctica for instance, hypobaric hypoxia prevails due to altitudes of ~3200 m (see Chaps. 36 and 38). Evidence for aberrant immune function at *Concordia* station and comparability to space conditions exists also in matters of reactivation of latent viruses. Reactivation and shedding patterns in saliva samples revealed in both settings an increased shedding during deployment in comparison to before or afterwards, suggesting that overwintering in Antarctica is an excellent analogue to spaceflight (see Chap. 19).

40.5.2 Microgravity-Simulating Studies

40.5.2.1 Simulation of Microgravity and Muscle/Bone Deconditioning

Analysis of effects mediated by *real* μg is realized by Parabolic Flight manoeuvres, a method which is used to test devices determined for implementation in space and to analyse directly effects of μg and hyper-gravity at the cellular level or on the entire organism (see Chap. 36). Simulation of μg for a longer period of time is achieved by dry immersion and bed rest studies, where the effect of longer disuse of the lower extremities and effects on the immune system are investigated. An additional gradient of 6° head tilt down in bed rest studies induces μg-like fluid shifts and allows the analysis of the effect of this condition on cardiovascular functions (see Chaps. 8 and 36).

40.5.2.2 Simulating Microgravity Under Hypoxia

In addition to simulated μg by bed rest, the EU funded "PlanHab"-study included the condition of hypoxia to examine independent or combined effects. The underwater habitat study "NEEMO" implemented in the NASA aqueous habitat served to simulated extravehicular activities under the influence of hyperbaric and hypoxic stress (see Chap. 36).

Analog experiments are of critical importance to investigate effects of space-related environmental factors on the body which are likely to affect immunity. Although a full comparability of the studies cannot be achieved due to different models, study protocols, and methods, these studies revealed that stress due to induction of separate spaceflight conditions leads to various changes in neuroendocrine and immune responsiveness.

40.6 Countermeasures to Prevent and Mitigate Spaceflight Immune Syndromes

Findings obtained from inflight studies on board the ISS and from Earth-bound analogue platforms created a basis for the development of adequate countermeasure options to avoid or mitigate deleterious health effects of spaceflight on humans. For improving and supporting immune performance, two approaches are under consideration. One is based on physical and psychological countermeasures such as exercise and improvement of living quality. Another approach is focussing on metabolism, nutrition or food supplements and pharmaceutical countermeasures (see Chap. 30).

40.6.1 Improving Sleep Quality

On board the ISS the crew has no natural *zeitgeber*. Together with disturbing background noise and occasional uncomfortable sleeping conditions, the circadian rhythm of astronauts is severely misaligned. In addition, sleep pattern is strongly impacted by the chronic stress astronauts are exposed to and sleep time is significantly reduced. Since short sleep and a disrupted circadian rhythm impair general performance and affect negatively metabolism as well as the immune system, countermeasures were designed to improve sleep quality, for circadian entrainment and daytime alertness. For this, lighting conditions at the ISS are adjusted by the use of a solid-state lighting system. Pharmacologically, sleep propensity is tried to be supported by the administration of the hormone melatonin. Furthermore, habitability on the ISS has improved since astronauts are provided with more comfortable sleeping bags and private sleep quarters to minimize sleep interruptions (see Chap. 9). Improved sleep quality is associated with improved mood. However there are still plenty of other factors which endanger psychological well-being.

40.6.2 Psychological and Exercise Countermeasures

Psychological challenges which are associated with spaceflight have been recognized to have decisive effects on the crew and mission success. This is why efforts were invested to make the stay in space more pleasant for the crew, which includes an extensive psychological pre- and post-mission training, a better connection to people on Earth as well as an improved accommodation of working and living. Currently, new technologies like virtual reality and crew supporting artificial intelligence robots are under evaluation (see Chap. 31).

Exercise on board of spacecrafts like in treadmills or centrifuges does not only counteract the physiological degeneration processes of the musculoskeletal and cardiovascular system but also represents a method to facilitate post-flight recovery processes. Above that, exercise significantly improves mood and thus adds to psychological factors (see Chap. 32).

40.6.3 Supplementing the Immune System

Improving the three factors of sleep, emotional well-being and physical exercise has additionally activating properties on the immune system.

However, since cells of the immune system are under the most affected cells types during spaceflight and immune system function is closely linked to presence or absence of infections and malignancies, additional intervention options were developed to counteract immune impairment. The term of *immunonutrition* describes a nutritional state, that enhances immune system function. This kind of supplementation includes antioxidants and enzyme cofactors, amino acids, vitamins, minerals and others (see Chap. 33). These essential compounds are also synthetized by the human microbiome which has also modulating properties on immune system function and *vice versa* (see Chap. 34). Pharmacological countermeasures for the immune system integrate the protection by supplementing with nucleotides or a targeted immune enhancement with AHCC (active hexose-correlated compound) and DHEA (dehydroepiandrosterone) (see Chap. 35).

40.7 How People on Earth Benefit from Space Immune Research: Examples

The different chapters which are collected in this second edition of "Stress Challenges and Immunity in Space" highlight the potential hazard of spaceflight on almost all physiological systems in humans. A special focus was set on the chronic exposure to a variety of stress factors and the effects on the immune system. In the last years, great efforts were invested to understand and to mitigate the phenomenon of immune impairment as an adaptation process to the prevailing conditions in space. This includes fundamental science, large-scaled studies and development of new technologies and covers the field of physics, physiology and psychology.

Findings and innovations of space research and diagnostics on Earth have mutual benefit for people both in health and disease and promote the development of *personalized* and *precision* medicine. Applications of spaceflight technologies and obtained knowledge for people on Earth are summarized in the following section.

The progress in diagnostic and monitoring tools such as breath gas analysis technologies already found its way into medicine applications. So far, the detection of volatile compounds in exhaled breath air facilitates the detection of bacterial infection like with *Helicobacter pylori* and diagnosis of airway diseases like inflammation or asthma. Monitoring of volatile anaesthetic agents like nitrous oxide or propofol gives a better insight into the individual pharmacokinetics of patients and allows an improved dose-adjustment (see Chap. 24).

Both in spaceflight and in overwintering studies, a stress-triggered reactivation of latent viruses was demonstrated. Varicella Zoster Virus (VZV) causes a painful rash and can ultimately result in blindness, paralysis, and stroke. For a quick and sensitive diagnosis of virus reactivation and a proper estimation of adequate countermeasures, NASA developed a polymerase chain reaction (PCR)-based assay to detect the presence of VZV DNA. Based on this assay, NASA designed a kit for physicians on Earth to detect VZV DNA in saliva or urine (see Chap. 19).

Understanding of radiation effects of the whole body like bystander effects and on the immune system in particular already adds to increased treatment success of cancer. Furthermore screening methods to detect individual radiosensitivity of patients might help to adjust radiation therapy to a level of highest efficacy but lowest side effects like radiation toxicity (see Chaps. 20 and 28).

Moreover, space associated research has helped physicians to understand fundamental principles of musculoskeletal and cardiovascular deconditioning and the role of and interconnection to the immune system. This supports the development of countermeasures which are nowadays used in the rehabilitation of patients suffering from the negative effects of immobilization after surgery (see Chap. 32) and that are better understood with their beneficial effects also on improving immunity.

Space-associated investigations on the immune system improve the knowledge of fundamental immune system functions and specific reactions in response to a variety of stress factors in nominal and off-nominal conditions of life. Increased understanding will not only help to identify the disorders which are associated with *hypo* or *hyper*-reactivity of the immune system, it gives also an idea of potential countermeasure options or even preventive measures. Finally, findings from space research on the immune system could open therapeutic ways to treat stress-induced premature and age-related immunosenescence (see Chaps. 11 and 15).

40.8 The Next Steps: Where Are We Going?

The feasibility of human spaceflight to the moon or even one day to Mars attracts the interest from people and space agencies across the world. To realize this endeavour, there is a tight interplay of different disciplines like physics, physiology, and psychology to assess key aspects and to overcome critical hurdles. One big issue is constituted by the energy requirements and consumables which are needed during the journey, but also after reaching the destination. Another point is the duration of mission, where the crew will experience various forms of physical and emotional stress. Considering these main potential showstoppers for long-term manned spaceflight, the induction of astronauts into hibernation represents an *ideal* solution. If it may become possible to induce this state of reduced metabolic rate pharmacologically, the need of consumables and energy consumptions would be strongly reduced and astronauts do not perceive psychological challenges and are less sensitive to radiation (see Chaps. 5 and 39). This field of research is not only of great interest for spaceflight but also in view of potential advantages for medical application.

Another main point is how to handle the effects of adaptation to conditions in space. The principle *one size fits all* is not valid for human physiology, neither in space nor on Earth. The high interindividual variabilities between people prompted a shift away from monitoring whole groups to concentrate on each individual. At the moment we are in a transition period where we start to concentrate on individual analyses and assessments. Further progress in this field will promote the establishment of *personalized* and *precision* medicine (see Fig. 40.2). In space this will mitigate the risk of health difficulties in astronauts and mission failure. On Earth, identified health risk factors may add to avoid the onset of disease or support the development of the optimal treatment strategy in the case of disorders.

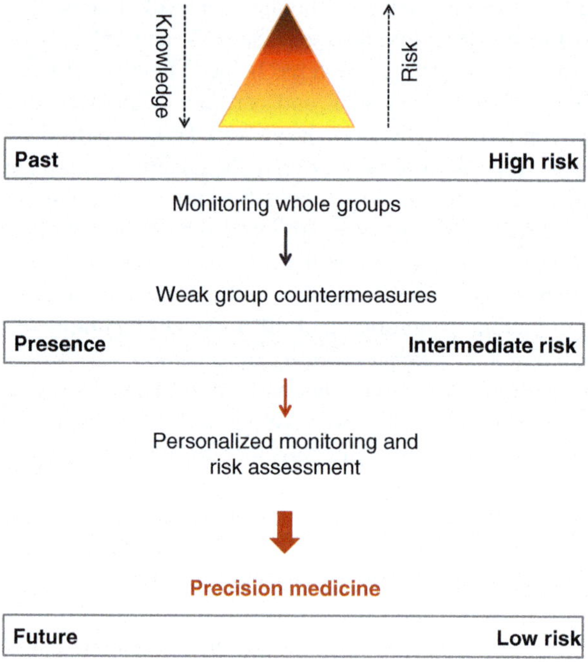

Fig. 40.2 Increase of knowledge reduces risk of failure. In the past, monitoring of only whole groups led to high risk of failure and the development of weak countermeasure options. In the present, knowledge of high variabilities between individuals pushes to personalized monitoring, which supports the development of precision medicine in the future with a continuous risk reduction

40.9 References

This summary is compiling key information from the Chaps. 3–39 and credit is given to all the authors who wrote these chapters. The authors have agreed that this summary can and will contain content extracted from their respective chapter(s). Their referenced chapters in this summary chapter contain all the citations and sources as listed in the respective reference lists accordingly.

Acknowledgements The authors of this summary DM and AC thank all the contributors to this volume and do highly appreciate their intellectual work in each of its chapters and all practical support.

Index

© Springer Nature Switzerland AG 2020
A. Choukér (ed.), *Stress Challenges and Immunity in Space*,
https://doi.org/10.1007/978-3-030-16996-1

B

H

J

K

L

M

Q

R

The manufacturer's authorised representative in the EU is Springer Nature Customer Service Centre GmbH, Europaplatz 3, 69115 Heidelberg, Germany. If you have any concerns regarding our products, please contact ProductSafety@springernature.com

Printed and bound by CPI Group (UK) Ltd, Croydon, CR0 4YY
30/09/2026
02225594-0005